# *Chemistry, Biochemistry, and Biology of (1→3)-β-Glucans and Related Polysaccharides*

# *Chemistry, Biochemistry, and Biology of (1→3)-β-Glucans and Related Polysaccharides*


Edited by

Antony Bacic

Australian Centre for Plant Functional Genomics,
School of Botany, University of Melbourne,
Australia

Geoffrey B. Fincher

Australian Centre for Plant Functional Genomics,
School of Agriculture, Food and Wine, University of Adelaide,
Australia

Bruce A. Stone

Department of Biochemistry, La Trobe University,
Bundoora, Australia



ELSEVIER

AMSTERDAM • BOSTON • HEIDELBERG • LONDON
NEW YORK • OXFORD • PARIS • SAN DIEGO
SAN FRANCISCO • SINGAPORE • SYDNEY • TOKYO

Academic Press is an imprint of Elsevier

Academic Press is an imprint of Elsevier
30 Corporate Drive, Suite 400, Burlington, MA 01803, USA
525 B Street, Suite 1900, San Diego, CA 92101-4495, USA
32 Jamestown Road, London, NW1 7BY, UK
360 Park Avenue South, New York, NY 10010-1710, USA

First edition 2009



**Library of Congress Cataloging-in-Publication Data**
A catalog record for this book is available from the Library of Congress

**British Library Cataloguing in Publication Data**
A catalogue record for this book is available from the British Library

ISBN: 978-0-12-373971-1

For information on all Academic Press publications
visit our website at elsevierdirect.com

Typeset by Macmillan Publishing Solutions
www.macmillansolutions.com

Printed and bound in the USA

09 08 07 06 05 04 03 02 01 10 9 8 7 6 5 4 3 2 1

# Contents

**In Memoriam** ........ ix
**Acknowledgements** ........ xiii
**Contributors** ........ xv

1. ***Introduction and Historical Background*** ........ 1
   *Adrienne E. Clarke*

2. CHEMISTRY AND PHYSICO-CHEMISTRY
   ***2.1 Chemistry of β-Glucans*** ........ 5
   *Bruce A. Stone*
   ***2.2 Physico-chemistry of (1,3)-β-Glucans*** ........ 47
   *Michael J. Gidley and Katsuyoshi Nishinari*

3. BIOCHEMISTRY
   ***3.1 Plant and Microbial Enzymes Involved in the Depolymerization of (1,3)-β-D-Glucans and Related Polysaccharides*** ........ 119
   *Maria Hrmova and Geoffrey B. Fincher*
   ***3.2 Interactions between Proteins and (1,3)-β-Glucans and Related Polysaccharides*** ........ 171
   *D. Wade Abbott and Alisdair B. Boraston*
   ***3.3 Biosynthetic Enzymes***
   ***3.3.1 Enzymology and Molecular Genetics of Biosynthetic Enzymes for (1,3)-β-Glucans: Prokaryotes*** ........ 201
   *Vilma A. Stanisich and Bruce A. Stone*
   ***3.3.2 Biosynthetic Enzymes for (1,3)-β-Glucans and (1,3;1,6)-β-Glucans in Protozoans and Chromistans: Biochemical Characterization and Molecular Biology*** ........ 233
   *Vincent Bulone*
   ***3.3.3 Biosynthetic Enzymes for (1-3)-β-Glucans, (1-3;1-6)-β-Glucans from Yeasts: Biochemical Properties and Molecular Biology*** ........ 259
   *Satoru Nogami and Yoshikazu Ohya*

3.3.4 **Biochemical and Molecular Properties of Biosynthetic Enzymes for (1,3)-β-Glucans in Embryophytes, Chlorophytes and Rhodophytes** ........ 283
*Lynette Brownfield, Monika Doblin, Geoffrey B. Fincher and Antony Bacic*

4. BIOLOGY
4.1 **Functional Roles of (1,3)-β-Glucans and Related Polysaccharides: Prokaryotes** ........ 327
*Vilma A. Stanisich and Bruce A. Stone*
4.2 **Biology of (1,3)-β-Glucans and Related Glucans in Protozoans and Chromistans** ........ 353
*Sverre M. Myklestad and Espen Granum*
4.3 **Organization of Fungal, Oomycete and Lichen (1,3)-β-Glucans** ........ 387
*Cecile Clavaud, Vishukumar Aimanianda and Jean-Paul Latge*
4.4 **Rhodophytes, Chlorophytes and Embryophytes**
4.4.1 **Callose in Cell Division** ........ 425
*Roy C. Brown and Betty E. Lemmon*
4.4.2 **Cytology of the (1-3)-β-Glucan (Callose) in Plasmodesmata and Sieve Plate Pores** ........ 439
*Amit Levy and Bernard L. Epel*
4.4.3 **Callose and its Role in Pollen and Embryo Sac Development in Flowering Plants** ........ 465
*Ed Newbigin, Antony Bacic and Steve Read*
4.4.4 **Callose in Abiotic Stress** ........ 499
*Angelika Stass and Walter J. Horst*
4.4.5 **Callose in Biotic Stress (Pathogenesis) Biology, biochemistry and molecular biology of callose in plant defence: callose deposition and turnover in plant—pathogen interactions** ........ 525
*Christian A. Voigt and Shauna C. Somerville*
4.5 **(1→3)-β-Glucans in Innate Immunity**
4.5.1 **Biological and Immunological Aspects of Innate Defence Mechanisms Activated by (1,3)-β-Glucans and Related Polysaccharides in Invertebrates** ........ 563
*Lage Cerenius, Shun-ichiro Kawabata and Kenneth Söderhäll*
4.5.2 **(1,3)-β-Glucans in Innate Immunity: Mammalian Systems** ........ 579
*Gordon D. Brown and David L. Williams*
4.6 **Distribution, Fine Structure and Function of (1,3;1,4)-β-Glucans in the Grasses and Other Taxa** ........ 621
*Philip J. Harris and Geoffrey B. Fincher*

**4.7 Evolutionary Aspects of (1,3)-β-Glucans and Related Polysaccharides ........ 655**
*Philip J. Harris and Bruce A. Stone*

**Index ........ 663**

# *In Memoriam*

**Bruce Arthur Stone AM FTSE**
*Emeritus Professor*
*4 December 1928–28 June 2008*

We were greatly saddened at the passing of our co-editor Bruce Stone on 28 June 2008, following a two year battle with acute myeloid leukemia. Bruce, with help from his co-author Adrienne Clarke, had almost single-handedly written the forerunner to the current publication. That publication was entitled "Chemistry and Biology of (1→3)-β-Glucans" and was published by La Trobe University Press in 1992. Affectionately known as 'The Book' to Bruce and his colleagues, it represented an encyclopaedic tome of over 800 pages, of which some 280 pages were dedicated to supporting references. 'The Book' quickly found its way to the shelves of offices of carbohydrate chemists, enzymologists and plant and fungal biologists around the world. If one telephoned Bruce to tap into his equally encyclopaedic knowledge of the field and especially to enquire of the early literature, his response was usually 'it's in the Book'. Nevertheless, relevant references and comment usually arrived from Bruce by email within a few hours of the telephone call.

The current publication resulted from Bruce's belief that the field had advanced significantly since 1992, largely through the emergence of new technologies such as molecular biology, functional genomics, and through advances in methods for the chemical, physical and physicochemical analyses of both carbohydrates and the enzymes that synthesise, modify or hydrolyse them. Bruce was realistic enough to realise that the 'second edition' of "Chemistry and Biology of (1→3)-β-Glucans" could not be written by a single person or even by a small group of people. He decided therefore to invite respected experts and colleagues from around the world to write individual review chapters. He also called upon us, as former postgraduate

Excerpts of this obituary are reproduced with permission from the *Journal of Cereal Science*, which published an obituary by GB Fincher in 2008 (*J Cereal Sci* 48, 561–562).

students, to help with the editing process. This we were happy to do, but we need to acknowledge that Bruce was the real driver of this book. Right to the end, it was Bruce who was communicating with the authors and the publishers, it was Bruce who edited all the chapters in detail and chased up late reviews, and it was Bruce who saw this publication as his final contribution in a long and illustrious scientific career.

Bruce's scientific career formally began when he received his Doctor of Philosophy from University College in London in 1954, where he worked on microbial cellulases. He was subsequently appointed to a lectureship at the University of Melbourne in 1958 and quickly rose through the ranks to Reader. In 1972 he was appointed Foundation Professor of Biochemistry in the newly formed department of Biochemistry at La Trobe University in Melbourne, and remained in that position until his retirement in 1993. In 1994, he became Emeritus Professor at La Trobe University and continued his scientific career with unabated energy. Thus, Bruce Stone served international science and training with distinction for more than 50 years and this has been recognised in Australia through his appointment as Fellow of the Australian Academy of Technology, Science and Engineering in 1999, the award of a Centenary Medal for service to Australian society in rural science in 2003 and, most importantly, with the award of the Australia Medal: Member of the Order of Australia in the Queen's Birthday Honours in 2007. The brief citation for the latter was 'for service to science, particularly in the field of biochemistry, as a researcher, academic and administrator'.

During a research career spanning more than 50 years Bruce Stone achieved worldwide recognition for his work in plant cell wall biology. He published over 180 research articles and invited reviews. The extremely high impact of his work was well known and was formally recognised by the international research community through his ISI award in 2001 of an ISI Citation Laureate. This award was attributable to the fact that Bruce's research publications had been cited by national and international scientists close to 4,000 times. In addition, Bruce was awarded the F.B. Guthrie Award of the Royal Australian Chemical Institute's Cereal Chemistry Division in 1985 and the American Association of Cereal Chemists' Thomas Burr Osborne Medal in 2004. These represent the highest awards for longstanding meritorious service and contributions to the Australian and American cereal industries, respectively.

Bruce Stone was an international expert on cereal cell wall chemistry and biochemistry. During his research career, Bruce adopted a multi-disciplinary approach to the definition of cell wall polysaccharide and lignin structure and function in cereals. He was quick to apply emerging technologies and published many seminal papers that have stimulated long-standing and more detailed studies on a broad range of cereals around the world. For example, Professor Stone was the first to develop procedures for the isolation of cell walls from the starchy endosperm and aleurone of wheat, and to provide precise analytical data on their composition. From the isolated walls he was able to extract specific polysaccharides for analysis of fine structure and solution properties, and he initiated programs on the hydrolytic enzymes involved in the depolymerisation of these wall polysaccharides in the germinated grain. His work on (1,3;1,4)-β-glucans and arabinoxylans, the most abundant wall polysaccharides in cereal grains, from wheat and barley set the scene for long term programs in which these properties have been related to industrial applications in oats and other cereals. Bruce was particularly pleased in 2006 when a paper identifying the genes that mediate (1,3;1,4)-β-glucan biosynthesis was published in a top international journal. Bruce was a co-author on that paper. Indeed, (1,3;1,4)-β-glucans always remained one of Bruce's favourite biological molecules.

In other seminal experiments, Bruce's group and colleagues monitored the deposition of cell walls during early grain development. Realizing the value of immunolabelling technology in the definition of grain development, he raised monoclonal antibodies against the most abundant wall polysaccharides. These antibodies are widely used in the international cereals community today and have been applied to specifically describe the spatial and temporal coordination of (1,3;1,4)-β-glucan and arabinoxylan deposition in developing grain and other tissues. Bruce was also extremely interested in the composition of lignins and phenolic acids and the nature of their association with polysaccharides in the walls of grasses. He contributed many novel ideas to this field and these are particular prescient with the renewed interest in ligno-cellulosic grass residues as feedstocks for the biofuels industry.

One of the outstanding features of Bruce Stone's career has been his ability to excite postgraduate students and early career postdoctoral scientists to themselves pursue a career in science and, as indicated above, provide them with the necessary skills to achieve their career goals. Examples of the career achievements of postgraduate students and postdoctoral scientists who were supervised and trained personally by Bruce Stone, and who have since made contributions to science and training in Australia include Professor Marilyn Anderson, Professor of Biochemistry, La Trobe University; Professor Tony Bacic, Director, Plant Cell Biology Research Centre, School of Botany, The University of Melbourne and Director, Bio21 Molecular Science and Biotechnology Institute, The University of Melbourne; Professor Adrienne E. Clarke AC, Laureate Professor, University of Melbourne and Lieutenant Governor, State of Victoria (1997–2001); Professor Geoff Fincher, Professor of Plant Science, University of Adelaide, Director, Waite Agricultural Research Institute and Deputy CEO, Australian Centre for Plant Functional Genomics; Professor Robert J. Henry, Professor of Plant Biotechnology, Southern Cross University, Deputy Director, Cooperative Research Centre for Sustainable Production Forestry and Director, Centre for Plant Conservation Genetics, Southern Cross University and Professor Peter Høj, Vice-Chancellor,

University of South Australia. There are many other graduates of Bruce, too numerous to list here, who have gone on to make equally valuable contributions to scientific knowledge and research management. Indeed, it is unlikely that many, if any, academic staff members from a university in Australia have been able to inspire so many junior scientists to pursue scientific careers and to achieve at the highest level in international science. In this respect Bruce made a special and possibly unprecedented contribution to the Australian community.

Bruce Stone's status as the world authority on cell walls has been recognised through his appointment to editorial boards of key international journals. Bruce had devoted many years of dedicated service to the Journal of Cereal Science, as Regional Editor from 1994–1997, Co-Editor from 1997–1999, and as Editor-in Chief from 1999–2005. In particular, the Journal benefited greatly from his strong but compassionate guidance as Editor-in-Chief, when Bruce's unswerving application of rigorous scientific standards and attention to detail raised both the profile and the impact of the Journal in the field. Bruce also served on numerous national and international committees and review panels, particularly in the Philippines and with the US-Israel Binational Agricultural Research and Development (BARD) Fund. He was involved in international aide programs through his work as the Assistant Director and Director-General of Training for the ATSE-Crawford Fund over the last five years.

At the national level, Bruce had been a member of the Royal Australian Chemical Institute and its Cereal Chemistry Division since 1948, and was Chair of the Cereal Chemistry Division from 1978–1979. His presence and contributions to the Division's annual conferences have been of central importance over many years and invoke fond memories both of his formidable scientific knowledge and his ever-present sense of humour.

In summary, Bruce Stone made an outstanding and long-term contribution to the advancement of our knowledge base in the area of cereal chemistry and biochemistry, both within Australia and internationally. He was a world authority in the field and an outstanding ambassador for cereal chemistry in the international research community. Indeed, it is difficult to identify other individuals who have made such a contribution to the field and a group of Bruce's former students expressed their final appreciation to Bruce as follows: "A pioneering biochemist and teacher. He imbued all his many students with a deep respect for scholarship and truth. He inspired us to choose lives in science. He was a friend, counsel and guide with a quirky and wry sense of humour. His influence on many lives in science globally was profound and will be greatly missed".

Geoffrey B. Fincher and Antony Bacic
6 April 2009

# Acknowledgements

The Editors are enormously grateful to all the Authors for contributing extremely well written chapters and providing them in a timely manner. We also wish to express our gratitude to Ms Joanne Noble, School of Botany, The University of Melbourne, whose considerable organisational skills were critical in guiding us through the administrative logistics of such an enormous undertaking, and also for her expert editorial skills. Professor Stone would also have wanted us to acknowledge the desktop publishing skills of Dr Fung Lay, La Trobe University, for whom any request for yet another figure, no matter its magnitude, was never a problem. We also thank the reviewers of these chapters for their generous time and effort in ensuring high quality contributions by the authors. We are also grateful to our families for their tolerance and understanding in allowing us to indulge in our passion for science.

# Contributors


**D. Wade Abbott** Department of Biochemistry and Microbiology, University of Victoria, Victoria, BC, Canada

**Vishu Kumar Aimanianda** Aspergillus Unit, Institut Pasteur, Paris, France

**Antony Bacic** Australian Centre for Plant Functional Genomics, School of Botany, University of Melbourne, VIC, Australia

**Alisdair B. Boraston** Department of Biochemistry and Microbiology, University of Victoria, Victoria, BC, Canada

**Gordon D. Brown** Institute of Infectious Disease and Molecular Medicine, Division of Immunology, Faculty of Health Sciences, University of Cape Town, Cape Town, South Africa

**Roy C. Brown** Department of Biology, The University of Louisiana at Lafayette, Lafayette, LA, USA

**Lynette Brownfield** Department of Biology, University of Leicester, Leicester, UK

**Vincent Bulone** School of Biotechnology, Royal Institute of Technology (KTH), AlbaNova University Center, Stockholm, Sweden

**Adrienne E. Clarke** School of Botany, University of Melbourne, VIC, Australia

**Lage Cerenius** Department of Physiology and Developmental Biology, Uppsala University, Norbyvagen, Uppsala, Sweden

**Cecile Clavaud** Aspergillus Unit, Institut Pasteur, Paris, France

**Monika Doblin** Plant Cell Biology Research Centre, School of Botany, University of Melbourne, VIC, Australia

**Bernard L. Epel** The Manna Center for Plant Biosciences, Department of Plant Sciences, Tel Aviv University, Tel Aviv, Israel

**Geoffrey B. Fincher** Australian Centre for Plant Functional Genomics, The University of Adelaide, Plant Genomics Centre, Glen Osmond, SA, Australia

**Michael J. Gidley** Centre for Nutrition and Food Sciences, University of Queensland, St Lucia, Brisbane, QLD, Australia

**Espen Granum** Department of Animal and Plant Sciences, University of Sheffield, Sheffield, United Kingdom

**Philip J. Harris** School of Biological Sciences, The University of Auckland, Auckland, New Zealand

**Walter Horst** Institute for Plant Nutrition, Faculty of Natural Sciences, University of Hannover, Hannover, Germany

**Maria Hrmova** Australian Centre for Plant Functional Genomics, The University of Adelaide, Plant Genomics Centre, Glen Osmond, SA, Australia

**Shun-ichiro Kawabata** Department of Biology, Faculty of Sciences, Kyushu University, Fukuoka, Japan

**Jean-Paul Latgé** Aspergillus Unit, Institut Pasteur, Paris, France

**Betty E. Lemmon** Department of Biology, The University of Louisiana at Lafayette, Lafayette, LA, USA

**Amit Levy** The Manna Center for Plant Biosciences, Department of Plant Sciences, Tel Aviv University, Tel Aviv, Israel

**Sverre M. Myklestad** Department of Biotechnology, Norwegian University of Science and Technology (NTNU), Trondheim, Norway

**Ed Newbigin** School of Botany, University of Melbourne, VIC, Australia

**Katsuyoshi Nishinari** Graduate School of Human Life Science, Osaka City University, Sumiyoshi-ku, Osaka, Japan

**Satoru Nogami** Department of Integrated Biosciences, Graduate School of Frontier Sciences, University of Tokyo, Chiba Prefecture, Japan

**Yoshikazu Ohya** Department of Integrated Biosciences, Graduate School of Frontier Sciences, University of Tokyo, Chiba Prefecture Japan

**Steve Read** Forest Research and Development, Forestry Tasmania, Hobart, TAS, Australia

**Kenneth Söderhäll** Department of Physiology and Developmental Biology, Uppsala University, Norbyvagen, Uppsala, Sweden

**Shauna Somerville** Department of Plant Biology, Carnegie Institution of Sciences and, Energy Biosciences Institute, University of California, Berkeley, CA, USA

**Vilma A. Stanisich** Department of Microbiology, La Trobe University, Bundoora, VIC, Australia

**Angelika Stass** Institute of Plant Nutrition, Faculty of Natural Sciences, Leibniz University of Hannover, Hannover, Germany

**Bruce A. Stone**†

**Christian A. Voigt** Department of Plant Biology, Carnegie Institution of Science, Stanford CA, USA and, Energy Biosciences Institute, University of California, Berkeley, CA, USA

**David L. Williams** Departments of Surgery and Pharmacology, James H. Quillen College of Medicine, East Tennessee State University, Johnson City, TN, USA

† Deceased

CHAPTER 1

# Introduction and Historical Background

**Adrienne E. Clarke**
*School of Botany, University of Melbourne, Victoria 3010, Australia*

This book is the final written chapter from Professor Bruce Stone's life work on the (1→3)-β-glucans and related polysaccharides.

It is a journey that started when he took up his first academic appointment at The University of Melbourne in 1958. He embarked on analyses of the cereal glucans and of paramylon from *Euglena gracilis*. This early work, in which I participated as his first PhD student, led to a review of the literature 'Chemistry and biochemistry of β-1,3 glucans' which was published in *Reviews of Pure and Applied Chemistry* in 1963 (Clarke and Stone, 1963). The initial submission for this review was several times the final word count. Bruce felt very strongly that important information would be lost in editing it to an acceptable length. He resolved to write a more extensive work on the subject and to have it published as a book. His initial collaborator on this project was Marilyn Anderson, who was his PhD student at the time. The initial work was interrupted after her graduation when she travelled to the USA for post-doctoral studies. In those days, before email, communication was extremely slow and difficult. After some time, I became involved and took up the challenge of being the co-author with Bruce. This work *Chemistry and Biology of the (1→3)-β-Glucans* by Stone and Clarke was finally published in 1992 (Stone and Clarke, 1992), more than 20 years after its inception.

The volume was, at the time of publication, encyclopaedic. It was characterized by meticulous listing and ordering of information in extensive tables with complete bibliography. These were the hallmarks of Bruce's writing and scholarship. Bruce had the commitment and the drive to track down even the most obscure references. He was not deterred by foreign language references and set about getting translations. In that volume, over 3500 references were listed (in full, at Bruce's insistence). The reference list accounts for 178 pages of a total of 803 pages! His commitment to inclusivity led to a situation which in Australia we refer to as 'painting the Sydney Harbour Bridge'. That is, as soon as application of one coat of paint is complete, the start point looks shabby and the painting starts again at the beginning. And so it was with the book. Given the extensive scope of the book and the fact that more and more research papers were being published in the journals, there were many revisions to include 'the latest'. Finally, a line was drawn and the volume was published with a note in the foreword: 'At the time the final revision was completed, immunological and molecular biological approaches were just being applied to study (1→3)-β-glucan synthesis, the (1→3)-β-glucan hydrolases and their biological functions. The literature in these fields has not been included. It is expanding rapidly and will justify separate reviews in the future.' This volume is such a review.

In the 17 years between the two volumes, the impact of the technologies of molecular genetics on biology in general has been remarkable. For this particular field, application of the technologies has resulted in substantial new knowledge of the enzymes involved in both the biosynthesis and degradation of the (1→3)-β-glucans. The new tools that emerge from this research are making insights into the physiological roles of the (1→3)-β-glucans and the (1→3;1→4)-β-glucans possible. Having these genetic tools has also opened up the way to create plants, particularly cereals, with different content and compositions of β-glucans. Other new techniques, such as atomic force microscopy, have allowed insights into how variation in structure results in variation in solution and gel properties of these β-glucans. Since publication of the first volume, there have been discoveries of specific inhibitors of β-glucan synthesis in fungal cell walls, of how innate immunity in animal systems is modulated, of how the β-glucans complex with other polysaccharides and proteins, and many advances in recording the taxonomic distribution of the (1→3)- and the (1→3;1→4)-β-glucans. All these and other advances are documented in this book. It differs from the format of the earlier work in that it is a collection of 21 chapters each written by experts in the sub-fields. As well as masterminding the whole endeavour, Bruce wrote the chapter on the 'Chemistry of β-Glucans' as sole author, and co-authored two chapters with Vilma Stanisich on the 'Enzymology and Molecular Genetics of Biosynthetic Enzymes for (1,3)-β-Glucans-Prokaryotes' and 'Functional Roles of (1,3)-β-Glucans- and Related Polysaccharides – Prokaryotes', and a chapter on the 'Evolutionary Aspects of (1,3)-β-Glucans and Related Polysaccharides' with Philip Harris. The author list of the remaining chapters reflects the network of personal and professional friendships Bruce made, in many different countries, many of whom he visited in his extensive travels. The other Editors of this volume, Tony Bacic and Geoff Fincher, were his PhD students.

Bruce died in June 2008 after becoming ill with acute myeloid leukaemia in 2006. He had a scientist's insight into his illness, but was buoyed by his work on this volume and the knowledge that it was close to completion. He was the 'wise Elder' of the global β-glucan community to whom all researchers turned, when their work led to questions of β-glucans. He also leaves a 'family' of students and their students, some of whom are active in the β-glucan field and others who have moved to different fields of biology. It surprises many of us who have moved to other fields of biology, how often seemingly unrelated fields suddenly and unexpectedly led back to the ubiquitous (1→3)-β-glucans.

He taught all his students the importance of care and accuracy in everything we wrote, as 'it will be there in print for all time'. This volume reflects this ideal. It will be a personal memorial to Bruce for all the authors, a wonderful resource for many others and a lasting tribute to Professor Bruce Arthur Stone.

## References

Clarke, A. E., & Stone, B. A. (1963). Chemistry and biochemistry of β-1,3-glucans. *Reviews of Pure and Applied Chemistry*, *13*, 134–156.

Stone, B. A., & Clarke, A. E. (1992). *Chemistry and biology of the (1→3)-β-Glucans*. Victoria, Australia: La Trobe University Press, ISBN 1 86324 409 3.

CHAPTER 2.1

# Chemistry of β-Glucans

**Bruce A. Stone**

*Department of Biochemistry, La Trobe University, Bundoora, Victoria, Australia*

The simplest (1,3)-β-glucans are linear, unbranched chains as found in callose, curdlan, paramylon and pachyman. In the side-chain-branched members, exemplified by the chromistan and fungal laminarins and the fungal mucilage glucans, the (1,3)-β-glucosyl chain residues are substituted to varying degrees at C(O)6 by single β-Glc residues or in some instances by short (1,3)-β-oligoglucosyl chains. The cyclic (1,3)-β-glucan from *Bradyrhizobium japonicum* is composed of two blocks of three (1,3)-linked Glc units separated by two blocks of three (1,6)-linked Glc units, and has a single branch (1,6)-linked Glc residue at C(O)6 of one of the cyclic glucoses. Some molecules are substituted by phosphocholine at C(O)6 on one of the cyclic Glc residues. The yeast and fungal cell wall glucans are branch-on-branch molecules in which the linear (1,3)-β-glucosyl chains are joined through (1,6)-linkages. These molecules occur as complexes with other polysaccharides and proteins. The *Streptococcus pneumoniae* S37 polymer has a (1,3)-β-glucan backbone with (1,2)-linked β-Glc side-chain-branches at each main chain glucosyl residue. The (1,3;1,4)-β-glucans from cereals and grasses, other embryophytes, lichens and some other taxa are unsubstituted, linear molecules with sequences mostly of two or three (1,4)-linked β-Glc residues, but with longer sequences of up to 15 β-Glc residues, joined by single (1,3)-linkages. A range of (1,3)-β-glucan derivatives have been prepared by variously esterifying, etherifying or attaching other substituents. Oligosaccharide building units of (1,3;1,6)- and (1,3;1,4)-β-glucans have been synthesized. A number of new β-glucans with (1,3)-linkages have been prepared.

## 1 Chemistry of (1,3)-β-Glucans and Related Polysaccharides[1]

In this chapter the chemical characteristics of (1,3)-β-glucans and related polysaccharides that have been structurally defined are discussed. In addition, records of (1,3)-β-glucans that have been recognized by indirect means are included.

[1] The prefixes D- and L- referring to monosaccharide configurations are omitted throughout, except where ambiguities might arise.

The various structural types found among (1,3)-β-glucans and related polysaccharides and their biological sources are listed in Table 1.

**Table 1: A classification of (1,3)-β-glucans and related polysaccharides based on their origin, linkage types, and organization**

| Structural type | Trivial name | Source | References |
|---|---|---|---|
| **Linear (1,3)-β-glucans** | | | |
| | **Bacteria** | | |
| | curdlan | *Agrobacterium* sp. | Harada et al., 1968; Nakanishi et al., 1976 |
| | | *A. radiobacter* | Nakanishi et al., 1976; Portilho et al., 2006; Saudagar and Singhal, 2004 |
| | | *A. rhizogenes* | Nakanishi et al., 1976 |
| | | *Agrobacterium* sp. | Shivakumar and Vijayendra, 2006 |
| | | *Rhizobium trifolii* | Ghai et al., 1981 |
| | | *Rhizobium* sp. | Footrakul et al., 1981 |
| | | *Cellulomonas flavigena* | Buller and Voepel, 1990; Kenyon and Buller, 2002; Kenyon et al., 2005 |
| | | *C. fimi* | Buller and Voepel, 1990; Kenyon et al., 2005 |
| | | *C. uda* | Buller and Voepel, 1990; Kenyon et al., 2005 |
| | | *Bacillus* sp. | Gummadi and Kumar, 2005 |
| | **Euglenids and haptophytes** | | |
| | paramylon | *Euglena gracilis* | Clarke and Stone, 1960; Kiss et al., 1987, 1988 |
| | | *Pavlova mesolychnon* | Kreger and van der Veer, 1970 |
| | | *Peranema trichophorum* | Cunningham et al., 1962; Archibald et al., 1963 |
| | **Fungi and lichens** | | |
| | pachyman | *Poria cocos* | Warsi and Whelan, 1957; Saito et al., 1968; Wang et al., 2004 |
| | lichen glucans | *Stereocaulon ramulosum* | Baron et al., 1988 |
| | | *Ramalina peruviana* | Cordeiro et al., 2003, 2004 |
| | | *Cladonia* spp. | Carbonero et al., 2001 |
| | | *Ramalina usnea* | Gorin and Iacomini, 1984 |
| | | *Ramalina celastri* | Stuelp et al., 1999 |

**Table 1: (Continued)**

| Structural type | Trivial name | Source | References |
|---|---|---|---|
| | | *Cladina* spp. | Carbonero et al., 2002 |
| | | *Umbilicaria mammulata* | Carbonero et al., 2002 |
| | **Embryophytes** | | |
| | callose (sieve plate) | *Vitis vinifera* | Aspinall and Kessler, 1957 |
| | callose (cotton seed hairs) | *Gossypium arboretum* | Huwyler et al., 1978; Maltby et al., 1979 |
| | callose (pollen tubes) | *Nicotiana alata* | Rae et al., 1985 |
| | laricinan | *Larix laricina* | Hoffmann and Timell, 1970, 1972 |
| **Linear (1,3)-β-glucans with (1,6)-linked-β-glucosyl or β-(1,6)-oligoglucosyl side chains** | | | |
| | **Chromists** | | |
| | **Phaeophytes** | | |
| | laminarin | *Laminaria* spp. | Elyakova et al., 1994; Chizhov et al., 1998 |
| | **Oomycetes** | | |
| | mycolaminarin | *Phytophthora* sp. | Wang and Bartnicki-Garcia, 1973, 1980 |
| | | *Phytophthora parasitica* | Bruneteau et al., 1988 |
| | | *Pythium apinadermatum* | Blaschek et al., 1992 |
| | | *Achyla bisexualis* | Lee et al., 1996 |
| | **Chrysophytes** | | |
| | chrysolaminarin | *Ochromonas malhamensis* | Archibald et al., 1963 |
| | **Diatoms** | | |
| | leucosin | *Phaeodactylum tricornutum* | Beattie et al, 1961; Ford and Percival, 1965 |
| | | *Skeltonema costatum* | Paulsen and Myklestad, 1978 |
| | | *Stauroneis amphioxus* | McConville et al., 1986 |
| | | *Chaetoserus* and other diatoms | Alekseeva et al., 2005 |

(*Continued*)

**Table 1: (Continued)**

| Structural type | Trivial name | Source | References |
|---|---|---|---|
| | | *Phaeodactylum tricornutum, Cylindrotheca fusiformis, Craspedostaurus australas, Thalassiosira pseudonana* | Chiovitti et al., 2004; 2006 |
| | | *Harmomonas dimorpha* | Chiovitti et al., 2006 |
| | | *Coscinodiscus nobilis* | Percival et al., 1980 |
| | | *Thalassiosira weissflogii* | Størseth et al., 2005 |
| | **Fungi** | | |
| | | *Acremonium* spp. | Schmid et al., 2007 |
| | | *Auriciularia judea* | Misaki et al., 1981 |
| | | *Boletus erythropus* | Chauveau et al., 1996 |
| | botryopshaeran | *Botryopshaeria rhodina* | Barbosa et al., 2003; Silva et al., 2008 |
| | cinerean | *Botrytis cinerea* | Stahmann et al., 1995 |
| | | *Claviceps purpurea* | Perlin and Taber, 1963 |
| | coriolan | *Coriolus versicolor* | Miyazaki et al., 1974 |
| | | *Cryponectria parasitica* | Molinaro et al., 2000 |
| | | *Cryptoporus volvatus* | Kitamura et al., 1994 |
| | | *Dreschslera specifera* | Aouadi et al., 1991 |
| | epiglucan | *Epicoccum nigrum* | Schmid et al., 2001 |
| | | *Flammulina velutipes* | Smiderle et al., 2006 |
| | | *Ganoderma applanatum* | Usui et al., 1983 |
| | | *Ganoderma lucidum* | Bao et al., 2002; Chang and Lu, 2004 |
| | | *Ganoderma tsugae* | Wang et al., 1993 |
| | grifolan | *Grifola frondosa* | Ohno et al., 1986; Adachi et al., 1989; Mizuno and Hazama, 1986 |
| | | *Gyrophora esculenta* | Sone et al., 1996 |
| | | *Grifora umbellata* | Miyazaki and Oikawa, 1973 |
| | lentinan | *Lentinus edodes* | Zhang et al., 2004 |
| | | *Nomuraea rileyi* | Latgé et al., 1988 |
| | pestalotan | *Pestalotia* sp. | Misaki et al., 1984 |
| | | *Phytophthora parasitica* | Gandon and Bruneteau, 1998 |
| | | *Phanerochaete chrysosporium* | Ruel and Joseleau, 1991 |

**Table 1: (Continued)**

| Structural type | Trivial name | Source | References |
|---|---|---|---|
| | | *Pleurotus ostreatus* | Yoshioka et al., 1985 |
| | | *Pleurotus eryngii and P. osteatoroseus* | Carbonero et al., 2006 |
| | | *Pleurotus tuber-regium* | Chenghua et al., 2000; Deng et al., 2000 |
| | | *Pleurotus florida* | Rout et al., 2005 |
| | cell walls | *Pneumocystis carnii* | Vassallo et al., 2000 |
| | | *Pythium aphanidermatum* | Blaschek et al., 1992 |
| | | *Poria cocus* | Wang et al., 2004 |
| | pendulan | *Porodisculus pendulus* | Iwamuro et al, 1985 |
| | schizophyllan | *Schizophyllum commune* | Akima et al., 1985 |
| | sclerotan | *Sclerotinia sclerotiorum* | Johnson et al., 1963; Rinaudo and Vincendon, 1982 |
| | | *Sparassis crispus* | Tada et al., 2007 |
| | | *Volvariella volvacea* | Misaki et al., 1986; Kishida et al., 1989, 1992 |
| | **Lichens** | | |
| | | *Teloschistes flavicans* | Reis et al., 2002 |
| | | *Dictyonema glabratum (Cora pavonia)* | Iacomini et al., 1987 |
| | | *Ramalina* spp. | Cordeiro et al., 2003 |
| **Branch-on-branch (1,3;1,6)-β-glucans** | | | |
| | yeast wall glucan | *Saccharomyces cerevisiae* | Kollar et al., 1995; 1997 |
| | | *Candida albicans* | Surarit et al., 1988; Ruiz-Herrera et al., 2006 |
| | fungal wall glucan | *Aspergillus fumigatus* | Hearn and Sietsma, 1994; Fontaine et al., 2000 |
| | oomycete wall glucan | *Pythium aphnidermatum* | Blaschek et al., 1992 |
| **Linear (1,3;1,4)-β-glucans** | | | |
| | cereal glucans | *Hordeum vulgare* | See Table 2 |
| | | *Avena sativa* | See Table 2 |
| | | *Triticum vulgare* | See Table 2 |
| | horsetail glucan | *Equisetum arvense* | Sørensen et al., 2008; Fry et al., 2008 |

(*Continued*)

**Table 1: (Continued)**

| Structural type | Trivial name | Source | References |
|---|---|---|---|
| | liverwort glucan | *Lophocolea bidentata* | Popper and Fry, 2003 |
| | lichenin | *Cetraria islandica* | See Table 2 |
| | | *Parmotrema* spp. | Carbonero et al., 2005 |
| | | *Rimelia* spp. | Carbonero et al., 2005 |
| | fungal cell wall glucan | *Aspergillus fumigatus* | Fontaine et al., 2000 |
| | desmid glucan | *Micrasterias* | Eder et al., 2008 |
| | dinoflagellate glucan | *Peridinium westii* | Nevo and Sharon, 1969 |
| | *Ulva lactuca* glucan | *Ulva lactuca* | Popper and Fry, 2003 |
| | | *Monodus subterraneus* | Ford and Percival, 1985 |
| **Sulfated linear (1,3;1,4)-β-glucans** | | | |
| | | *Kappaphycus alvarezii* | Lechat et al., 2000 |
| **(1,3;1,4)-β-gluco-oligosaccharides** | | | |
| | | *Sarcina ventriculi* | Lee and Hollingsworth, 1997 |
| **Cyclic (1,3;1,6)-β-glucans** | | | |
| | | *Bradyrhizobium japonicum* | Miller et al., 1990; Bhagwat et al., 1999 |
| | | *Rhizobium loti* | Estrella et al., 2000 |
| | | *Azorhizobium caulinodans* | Komaniecka and Choma, 2003 |
| | | *Azospirillum brasiliense* | Altabe et al., 1998 |
| | | *Bradyrhizobium japonicum ndvC* mutant | Bhagwat et al., 1999 |
| | | *Sinorhizobium meliloti ndvC* mutant | Bhagwat et al., 1999 |
| **Branched (1,3;1,2)-β-glucan** | | | |
| | | *Streptococcus pneumoniae* Type 37 | Knecht et al., 1970; Adeyeye et al., 1988 |

### I.A Detection

(1,3)-β-Glucans such as callose, curdlan and related glucans can be specifically detected by staining with the triphenylmethane dye Aniline Blue at pH 8, or by the bright yellow ultraviolet (UV)-induced fluorescence when the Aniline Blue fluorochrome (a benzophenone derivative) is bound to (1,3)-β-glucan and (1,3)-β-xylan chains (Evans et al., 1984; Stone and Clarke, 1992). The fluorochromes Calcofluor White and Congo Red also show UV-induced fluorescence when bound to (1,3)-β-glucans; however, this interaction is not specific for (1,3)-β-glucans as other β-glycans, including cellulose, chitin, (1,3;1,4)-β-glucans and certain bacterial extracellular polysaccharides such as the xanthan and succinoglycan gums, also induce fluorescence with these fluorochromes (Wood and Fulcher, 1984). Other triphenylmethane dyes and the phenoxazone dye Resorcin Blue also appear to be specific for (1,3)-β-glucans (see Stone and Clarke, 1992). Callose is electron-lucent but can be identified in electron micrographs using gold-labelled antibodies specific for (1,3)-β-glucans (Meikle et al., 1991). (1,3;14)-β-Glucans for which no specific staining reaction is available can be identified using gold-labelled antibodies (Meikle et al., 1994). Linear (1,3)-β-glucans do not give the periodate-Schiff reaction because there are no periodate cleavable glycol sites on (1,3)-linked glucose residues in the chain; however, (1,6)-linked glucosyl residues on (1,3;1,6)-β-glucans are periodate reactive.

### I.B Extraction, Purification and Structural Determination

Many (1,3)-β-glucans, in particular the low DP (degree of polymerization) side-chain-branched (1,3;1,6)-β-glucans, are water soluble, but others are only dissolved in aprotic solvents such as dimethyl sulfoxide, formic acid, and aprotic reagents such as N-methylmorpholino-N-oxide and lithium chloride in dimethylacetamide (Yotsuzuka, 2001). Dilute bases (0.25 M NaOH) dissolve linear (1,3)-β-glucans. The ionization of the very weakly acidic hydroxyl groups (pKa 11-12) leads to disruption of the regular organization of the (1,3)-β-glucan chains; however, due to the propensity for (1,3)-β-glucans to undergo quite rapid 'alkaline peeling' (β-elimination reaction) from any unprotected reducing ends (see Stone and Clarke, 1992), inclusion of the reductant sodium borohydride in the alkaline extractant is usually employed to prevent this reaction.

Dissolution of (1,3)-β-glucans is an important step towards purification, which can then be achieved by either fractional precipitation or chromatography on gel permeation matrices. Using MALS (multi-angle laser-light scattering) detection, information about the molecular masses of the components can be obtained. The covalently linked heteropolymer complexes containing the branch-on-branch (1,3;1,6)-β-glucans are recalcitrant to alkaline dissolution unless the covalent interchain linkages are broken, e.g. by acid hydrolysis (Müller et al., 1997) or sodium hypochlorite oxidation (Ohno et al., 1999). Thus, repeated treatment with dilute acetic acid removed the (1,6)-β-glucan from the *S. cerevisiae* heteropolymer complex (Manners et al., 1973). However, acid treatment leads to the loss of fine structure of the branch-on-branch (1,3;1,6)-β-glucan (Ensley et al., 1994).

The structures of (1,3)-β-glucans and their relatives have been determined by conventional methylation techniques and by periodate oxidation procedures (see Stone and Clarke, 1992). The latter have been particularly useful in defining the fine structures of side-chain-branched (1,3)-β-glucans since the interunit residues in (1,3)-β-glucans lacking vicinal hydroxyls are resistant to periodate oxidation. This allows sequential Smith degradation to be used to provide information on branching (see Stone and Clarke, 1992). $^{13}$C-NMR (nuclear magnetic resonance) provides detailed information on anomeric configuration of the Glc units and qualitative and quantitative information on linkage types (e.g. Kim et al., 2000). In certain instances the separation and analysis of products of treatment of the β-glucan with purified β-glucan hydrolases of well defined specificity provides detailed information on fine structure that is not otherwise readily accessible, as shown by treatment of (1,3;1,4)-β-glucans with *Bacillus amyloliquefaciens* (*B. subtilis*) (1,3;1,4)-β-glucan endo-hydrolase (EC 3.2.1.73), a.k.a. 'lichenase' (Woodward et al, 1983; Wood et al., 1994) and *Eisenia bicyclis* (1,3;1,6)-β-glucan with *Sporotrichum dimorphosporum* (1,3)-β-glucan glucohydrolase (Nanjo et al., 1984).

### *1.C Linear (1,3)-β-Glucans*

Linear (1,3)-β-glucans (Fig. 1A) are found in the capsules of some rhizobial species, as intracellular storage polysaccharides in euglenids and some chromistans (see Chapter 4.2), as storage polysaccharide in fungal sclerotia, as wall components of certain zygomycetaceous fungi, as cell wall components in specialized reproductive tissues (see Chapter 4.4.3), and as

Glc*p*β-[(1→3)-Glc*p*β]$_n$-(1→3)-Glc*p*

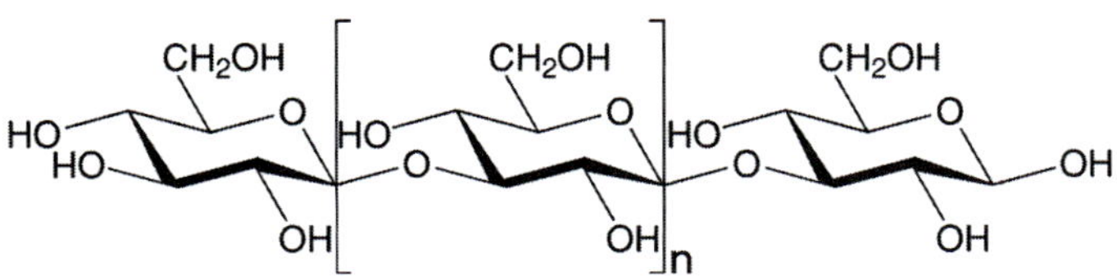


**Fig. 1: Structures of (1,3)-β-glucans and related polysaccharides showing linkage types and their organization. 1A. Linear (1,3)-β-glucans.**

deposits on the plasma membrane in abiotic (see Chapter 4.4.4) and biotic (see Chapter 4.4.5) stress.

### 1.C.1 Curdlan

Curdlan, recognised as a (1,3)-β-glucan by its staining with either the Aniline Blue dye or fluorochrome, is found as a capsular polysaccharide in Gram-negative bacteria belonging to the rhizobiaceae (e.g. *Agrobacterium* and *Rhizobium* spp.) (see Table 1) and the Gram-positive *Cellulomonas falvigena* (Buller and Voepel, 1990; Kenyon and Buller, 2002; Kenyon et al., 2005) and a *Bacillus* sp. (Gummadi and Kumar, 2005) (see Table 1). Curdlan is a linear, unbranched (1,3)-β-glucan (Harada et al., 1968; Nakanishi et al., 1976) (Fig. 1A) which may have as many as 12 000 Glc units (Futatsuyama et al., 1999). Curdlan is insoluble in water, alcohols and most organic solvents but dissolves in dilute bases (0.25 M NaOH) and dimethyl sulfoxide (DMSO).

### 1.C.2 Paramylon

Paramylon is an insoluble, linear (1,3)-β-glucan of high molecular mass occurring naturally in a highly crystalline form (Kiss et al., 1987, 1988) in discrete membrane-bound granules in the cytoplasm of euglenid protozoans (euglenozoans, e.g. *Euglena gracilis*) (Clarke and Stone, 1960) and *Peranema trichophorum* (Cunningham et al., 1962) (see also Chapter 4.2). One chromistan haptophyte *Pavlova mesolychnon* (Kreger and van der Veer, 1970) has cytoplasmic granules that give the same X-ray diffraction pattern as paramylon.

### 1.C.3 Pachyman

The sclerotia of the basidiomycete fungus *Poria cocus* are composed of swollen thin-walled hyphae containing, as the main component, the insoluble linear (1,3)-β-glucan, pachyman (Warsi and Whelan, 1957; Saito et al., 1968; Wang et al., 2004). Among the other polysaccharides that accompany pachyman in the sclerotia are two-side-chain branched (1,3;1,6)-β-glucans (Wang et al., 2004).

The sclerotia of the basidiomycete *Laetiporus sulphureus* contain, together with heteroglycans, a linear (1,3)-β-glucan similar to pachyman (Alquini et al., 2004).

### I.C.4 Conidiobolus and entomophthora (1,3)-β-glucans

Several entomophthoraean genera belonging to the Zygomycete group of fungi have a linear (1,3)-β-glucan in their hyphal walls as shown for *Conidiobolus obscurus* (Latgé et al., 1984). In *Entomophthora aulicae, E. culicis, E. neoaphidis and Zoophthora radicans* the (1,3)-β-glucan is found only in the hyphal walls but not on the protoplast surface (Latgé and Beauvais, 1987; Beauvais et al., 1989). Hyphal walls of *E. aulicae* react with a (1,3)-β-glucan antiserum and with the Aniline Blue fluorochrome (Beauvais et al., 1989).

### 1.C.5 Callose

The (1,3)-β-glucan, callose, occurs widely in embryophyte tissues in specialized walls or wall-associated structures at particular stages of differentiation, and its occurrence as discrete deposits in the wall adjacent to the plasma membrane is characteristically induced by wounding or physiological and pathological stress (see Stone and Clarke, 1992 and Chapters 4.4.4 and 4.4.5). Callose is identified histochemically by its staining properties with either the Aniline Blue dye or fluorochrome or by labelling with the (1,3)-β-glucan specific antibody (Meikle et al., 1991) often combined with its susceptibility to dissolution by specific (1,3)-β-glucan hydrolases. There are few chemical studies on individual callose preparations. Aspinall and Kessler's (1957) examination of the callosic deposits on the sieve plates from *Vitis vinifera* phloem is one of the few definitive structural analyses. Two other callosic structures have been chemically investigated: callose in the innermost wall region bordering the plasma membrane of cotton seed hairs (Huwyler et al., 1978; Maltby et al., 1979) and callose in pollen tube walls of *Nicotiana alata* (Rae et al., 1985) where it occurs with cellulose as the predominant polysaccharide in the inner layer of the pollen tube wall (Meikle et al., 1991; Ferguson et al., 1998). In each case (1,3)-β-Glc linkages were predominant but a small proportion of (1,6)-β-Glc linkages were also found.

Laricinan, a linear (1,3)-β-glucan found in compression wood of *Larix laricina* (Hoffmann and Timell, 1970; 1972), is probably an example of wound-induced callose.

## 1.D Side-Chain-Branched (1,3;1,6)-β-Glucans

Side-chain-branched (1,3;1,6)-β-glucans (Fig. 1B) are found as intracellular storage polysaccharides in the chromistan brown algae (laminarin), oomycetes (mycolaminarin), chrysophytes (chrysolaminarin) and diatoms (leucosin), and occur widely on hyphal surfaces and in sclerotia of ascomycete and basidiomycete fungi.

### 1.D.1 Chromistan side-chain-branched (1,3;1,6)-β-glucans

*1.D.1.a Laminarin* The water-soluble laminarins from species of chromistan brown algae comprise a family of polysaccharides composed of relatively short chains (DP range 31–40) (Chizhov et al., 1998) although in some species the maximum DP is 12 with minor

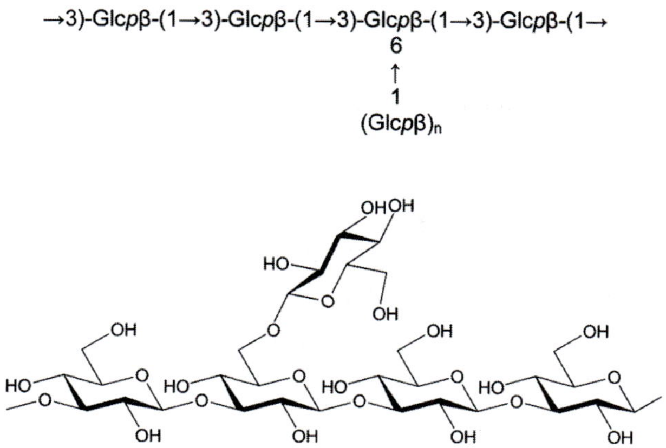


**Fig. 1B: Side-branched (1,3;1,6)-β-glucan.**

components up to DP 38, substituted by occasional (1 in 10) (1,6)-linked β-Glc residues. The content of side-chain-branches is species dependent (Zvyagintseva et al., 2003). Some (1,6)-links may be present in the backbone chain and a proportion of the chains are terminated by mannitol residues (Read et al., 1996; Chizhov et al., 1998) (see also Stone and Clarke, 1992) and in some species by *N*-acetylhexosamine residues (Chizhov et al., 1998).

*1.D.1.b Mycolaminarin* The mycolaminarins are a family of water-soluble side-chain-branched (1,3;1,6)-β-glucans with one, two or three (1,6)-linked Glc units per chain that function as carbohydrate reserves in species of chromistan oomycetes such as *Pythium* and *Achyla* (Table 1). The *Phytophthora parasitica* mycolaminarin also has (1,6)-linked-β-laminaribiose substituents (Bruneteau et al., 1988). Some mycolaminarins are phosphorylated with glucose: phosphate ratios ranging from 18:1 to 30:1 (Wang and Bartnicki-Garcia, 1973, 1980). The *Achyla bisexualis* mycolaminarin is localized in large vesicles in the hyphae and is present in two forms, a small neutral and a large phosphorylated form in which both mono- and di-phosphate esters are present (Lee et al., 1996).

*1.D.1.c Chrysolaminarin* Chrysolaminarins are intracellular carbohydrate reserves in unicellular chrysophycean flagellates (e.g. *Ochromonas malhamensis*). The molecules are similar to laminarin-type laminarins except that no mannitol is present (Archibald et al., 1963). Hot-water extractable polysaccharides from the colonial microalga *Haramonas dimorpha* (Rhaphidophyceae, Ochrophyta) are predominantly (1,3)-β-glucans, with an average DP of 12–16 residues and a relatively low proportion of side-branching with Glc residues (Chiovitti et al., 2006).

*1.D.1.d Leucosin* Leucosin (von Stosch, 1951) is found as a refractile material in the vacuoles of members of the diatom group of unicellular or colonial chromistans, and has been detected by staining with Resorcinol Blue, a (1,3)-β-glucan specific dye (Parker, 1964), or using a (1,3)-β-glucan specific antibody (Chiovitti et al., 2004). Leucosin is a water-soluble side-chain-branched (1,3;1,6)-β-glucan resembling the chrysolaminarins. In addition to the (1,6)-linked side branches, (1,2)- and (1,4)-linked Glc residues were found in some of the four diatoms analysed (Ford and Percival, 1965; Handa and Tominaga, 1969; Percival et al., 1980; Chiovitti et al., 2004). The water-soluble glucans from four diatom species examined by Wustman et al., (1997) consisted predominantly of (1,3)-Glc residues with smaller amounts of (1,6)- and (1,2)-linked Glc residues.

The marine diatom *Chaetoceros mulleri* has a side-chain-branched (1,3;1,6)-β-glucan with a DP of 22–24 and a degree of branching of 0.006–0.009 (Størseth et al., 2005). The glucan from the diatom *Thalassiosira weissflogii* has a DP of 5–13 but is unbranched (Størseth et al., 2005).

*1.D.1.e Ascomycete and Basidiomycete side-chain-branched (1,3;1,6)-β-glucans* Side-chain-branched (1,3;1,6)-β-glucans are found extensively on the surfaces of hyphae and in the scelerotia of ascomycete and basidiomycete fungi. The sources of structurally defined members of this group are listed in Table 1. The degree of substitution of the (1,3)-β-glucan backbone chain with single (1,6)-linked β-Glc residues depends on the source and culture conditions and varies from 1 in 3 (*Schizophyllum* and *Sclerotium*), 2 in 5 (*Lentinus*), 3 in 5 (*Pestalotia*) to 2 in 3 (*Epicocum*). It is often stated that these glucans are composed of repeated (repeating) side-chain-branched units but evidence for this is lacking. It is more likely that the degrees of substitution are average values. In one species, *Botryosphaeria rhodina* (Silva et al., 2008), the appended branches are (1,6)-β-glucosyl units.

## 1.E Branch-on-Branch (1,3;1,6)-β-Glucans

Branch-on-branch (1,3;1,6)-β-glucans are found in the cell walls of fungi, yeasts (see Chapter 4.3), and chromistan oomycetes (see Chapter 4.2).

### 1.E.1 Saccharomyces cerevisiae cell wall glucan

Cell walls of yeasts (hemiascomycetes) have branch-on-branch (1,3;1,6)-β-glucans as major cell wall components. In the yeast *Saccharomyces cerevisiae* a (1,3;1,6)-β-glucan, comprising ~50% of the wall, forms a core (Fig. 1C) whose non-reducing termini are covalently linked

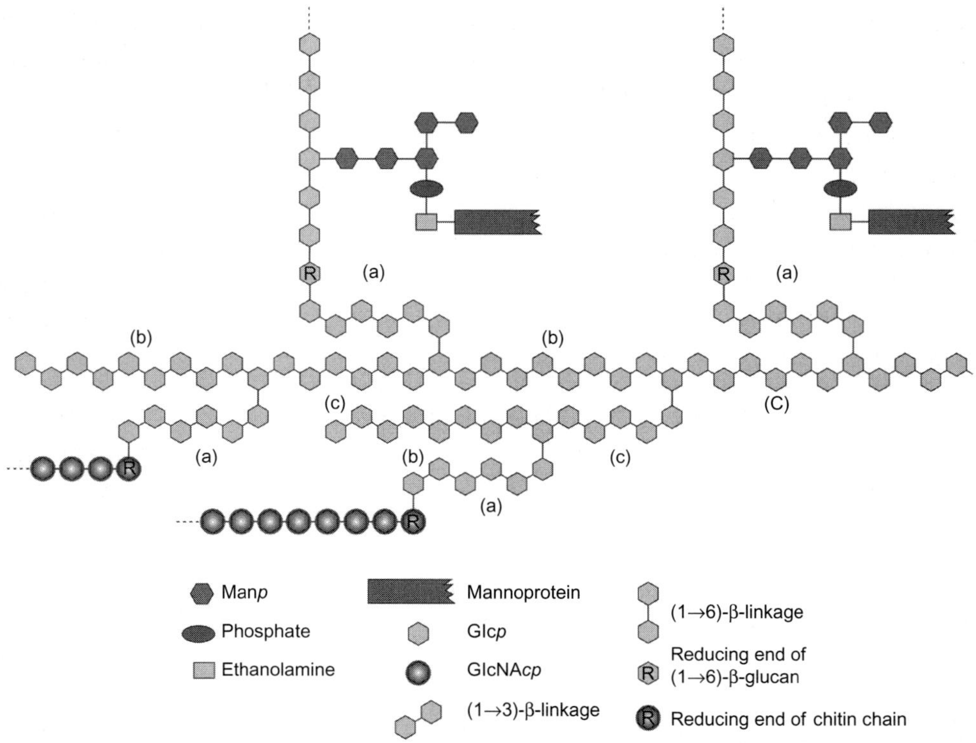


**Fig. 1C: Model of the cell wall of the yeast *Saccharomyces cerevisiae*. The branch-on-branch (1→3;1→6)-β-D-glucan forms the central component of the wall. The proportions of (a), (b) and (c) chains are about equal but their exact lengths are not known. The non-reducing ends of (a) and (b) chains are attachment sites for chitin chains at the plasma membrane surface of the wall and for the reducing ends of (1→6)-β-D-glucan chains at the outer surface of the wall. The (1→6)-β-D-glucan chains are in turn substituted by mannoproteins through the C-terminal amino acid of an amino acid-ethanolamine-phosphodiester-$(Man)_5$ remnant of a GPI anchor. Pir cell wall proteins (not shown) are linked directly to (1→3;1→6)-β-D-glucan. Chitin chains may also be attached to the (1→6)-β-D-glucan chains (not shown). After Manners et al. (1973); de Nobel et al. (2001); Kollar et al. (1997). The colour specifications refer to colours in panels.**

either to chitin, (1,6)-β-glucan or mannoprotein, which together make up ~40% of the wall. The mannoproteins are found mainly at the external surface of the walls linked to (1,6)-β-glucan via remnants of a glycosylphosphatidylinositol anchor. Pir proteins (proteins with internal repeats) are linked directly to the core (1,3;1,6)-β-glucan (Kollar et al., 1995, 1997). The architecture of the heteropolymer complex is discussed in Chapter 4.3. The complex forms a three-dimensional network overlying the protoplast. The *S. cerevisiae* (1,3;1,6)-β-glucan is insoluble in hot alkali (75°C, 0.75 M) due to its covalent association with chitin and other polysaccharides. The fine structure of the core branch-on-branch (1,3;1,6)-β-glucan has been determined by Misaki et al. (1968) and Manners et al. (1973) and is shown in Fig. 1C.

### 1.E.2 *Candida albicans cell wall glucan*

The cell walls of dimorphic yeast (*Candida albicans*) in both the hyphal and yeast forms contain an alkali-insoluble (1,3;1,6)-β-glucan with 30%–39% (1,3)- and 43% (1,6)-linkages, whereas in germ tubes the proportions are reversed: 67% (1,3)- and 14% (1,6)-linkages (Ruiz-Herrera et al., 2006). The (1,3;1,6)-β-glucan is covalently linked to both chitin and (1,6)-β-glucan (Surarit et al., 1988).

### 1.E.3 *Aspergillus fumigatus cell wall glucan*

The alkali-insoluble fraction of the cell wall of *Aspergillus fumigatus* is composed of a heteropolysaccharide complex that, as in *S. cerevisiae*, consists of a core branch-on-branch (1,3;1,6)-β-glucan but lacks the covalently linked (1,6)-β-glucan and protein components (Fontaine et al., 2000). The non-reducing termini are covalently linked either to chitin, a branched galactomannan or a (1,3;1,4)-β-glucan (see Section 1.E.1). The (1,3;1,6)-β-glucan has 4% branch points (Bernard and Latgé, 2001).

### 1.E.4 *Pythium aphanidermatum cell wall glucan*

The wall of the chromistan oomycete *Pythium aphanidermatum* (Blaschek et al., 1992) consists of 18% cellulose and 82% (1,3;1,6)-β-glucan. Of the non-cellulosic glucan 33% is extractable with water at 121°C and is highly branched with 6% (1,6)-linkages. Dilute trifluoroacetic acid treatment of the walls released ~50% of the non-cellulosic glucan which was highly branched, containing 14% (1,6)-linkages and 8% (1,4)-linkages. The extent, if any, of covalent interlinkage between the various glucans in the wall remains to be determined.

## 1.F Cyclic (1,3;1,6)-β-Glucans

Water-soluble cyclic (1,3;1,6)-β-glucans are produced by the legume symbionts *Bradyrhizobium japonicum, Rhizobium loti, Azospirillum brasilense* and *Azorhizobium caulinodans. B. japonicum* strains, growing as free-living cultures or as bacteroids, synthesize a mixture of cyclic (1,3;1,6)-β-glucans that are neutral, unsubstituted and have ring sizes of 10–13 units (Miller et al., 1990; Rolin et al., 1992; Gore and Miller, 1993; Inon de Iannino and Ugalde, 1993). The *B. japonicum* USDA 110 glucan consists of a 12-membered ring composed of two blocks of three (1,3)-β-linked Glc residues each separated by two blocks of three (1,6)-β-linked Glc residues (Fig. 1D) or, less likely, of blocks of two and four or one and five (1,6)-β-linked Glc residues (Rolin et al., 1992). One block of (1,3)-β-linked Glc residues contains a branched Glc at C(O)6 and the other a phosphocholine group at C(O)6 (Fig. 1D).

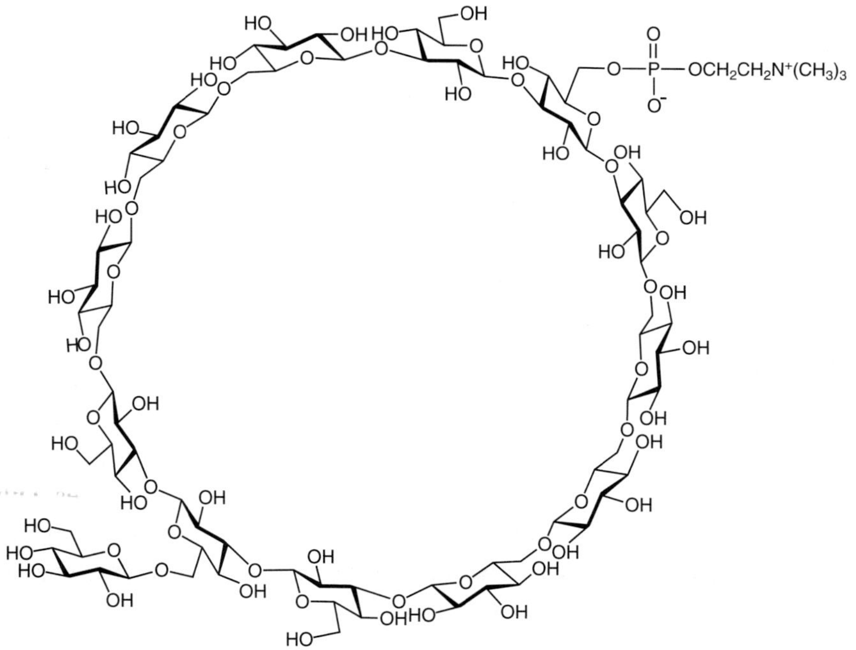

**Fig. 1D: Schematic representation of the structure of cyclic (1,3;1,6)-β-D-glucan, substituted with phosphocholine and Glc at the (1,3)-linked residues. All Glc residues are shown in $^4C_1$ conformation. The molecule was designed by Professor Bruce Stone and drawn by Dr Maria Hrmova (University of Adelaide), using Rolin et al. (1992) as a guide. Dr Spencer Williams (The University of Melbourne) is also acknowledged for his advice in the construction of this model.**

The cyclic glucans produced by *A. caulinodans* are neutral, unbranched and unsubstituted, and like those from *B. japonicum* have ring sizes mainly of 10–13 units, but similar proportions of (1,3)-β- and (1,6)-β-linkages (Komaniecka and Choma, 2003). In contrast, the nine-membered cyclic glucan produced by *R. loti* NZP 2309 differs in the proportion of linkages [three (1,3)-β- and six (1,6)-β-linked Glc units] and has a single (1,6)-β-linked Glc branch (Estrella et al., 2000). *A. brasilense* synthesizes a mixture of cyclic glucans that are all composed of an 11-ring structure containing three (1,3)-β- and eight (1,6)-β-linked residues with a single, (1,4)-β-linked Glc branch. Some molecules have an additional Glc branch [linked (1,3)-β-] that may also carry a 2-*O*-methyl group (Altabe et al., 1998).

Under some circumstances, the production by *B. japonicum* of the native cyclic glucan is replaced by a unique cyclic decaglucan (cyclolaminarinose) composed only of (1,3)-β-linked Glc residues and substituted at a C(O)6 position by a β-laminaribose residue (Pfeffer et al., 1996). This occurs in *B. japonicum* AB-1, a transposon-insertion mutant (*ndvC*::Tn*5*) that lacks the putative (1,6)-β-glucosyltransferase (Bhagwat et al., 1999) and, *in vitro*, when Glc from UDP-[$^{14}$C]Glc is incorporated into inner membranes prepared from *R. loti* (Estrella et al., 2000). Most strikingly, the same cyclic decaglucan is produced by a recombinant strain of *Sinorhizobium meliloti* that cannot produce the 17-25-residue cyclic (1,2)-β-glucans typical of the species because of a defective glucan synthase gene (*ndvB*::Tn*5*), but which has acquired the (1,3;1,6)-β-glucan synthesis locus from *B. japonicum* (Pfeffer et al., 1996).

## 1.G Side-Chain-Branched (1,3;1,2)-β-Glucan

The type 37 capsule of *Streptococcus pneumoniae* (Knecht et al., 1970) is the only homopolysaccharide and one of only two neutral polysaccharides amongst the 90 pneumococcal capsular types (Henrichsen, 1995). The S37 polymer has a (1,3)-β-glucan backbone with (1,2)-linked β-Glc side-branches at each Glc residue giving a crowded, comb-like molecular organization (Fig. 1E). This glucan is soluble in water and DMSO (Adeyeye et al., 1988). Oligosaccharides related to the repeating unit of the type 37 polysaccharide have been chemically synthesized (Larsson et al., 2005).

## 1.H Linear (1,3;1,4)-β-Glucans

(1,3;1,4)-β-Glucans (Fig. 1F) are found in grasses and cereals; liverworts, lichens, fungi and algae; chromalveolates, chromistans and chlorophytes; and in a sulfated form in red algae (see Chapter 4.6).

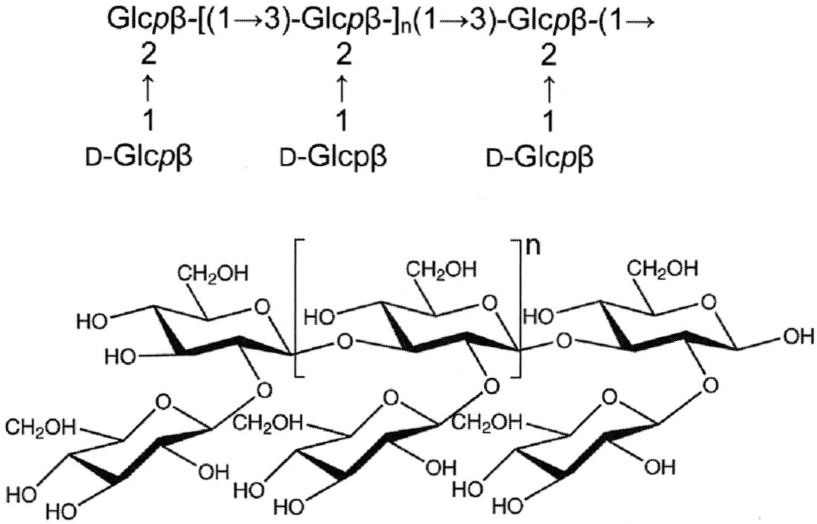


**Fig. 1E: Side-chain-branched (1,3;1,2)-β-glucan.**

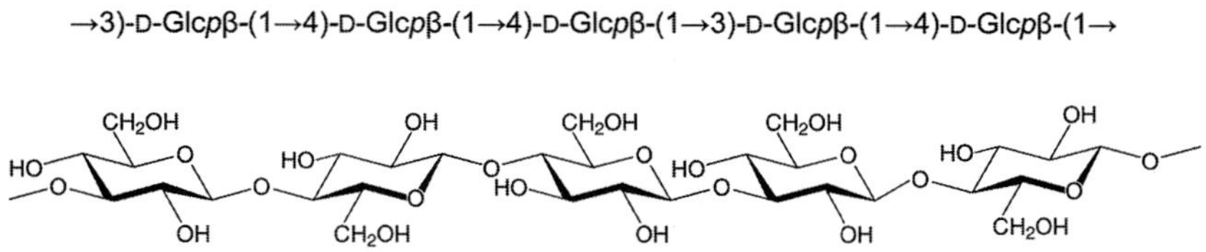


**Fig. 1F: Linear (1,3;1,4)-β-glucan.**

### *1.H.1 Cereal and grass (1,3;1,4)-β-glucans (mixed-linkage glucans)*

(1,3;1,4)-β-Glucans are found characteristically in the cell walls of grasses and cereals (Poaceae) and related Poales families, which form part of the commelinoid monocotyledons (Harris, 2005; Trethewey et al., 2005) (see Chapter 4.6). The Poaceae (1,3;1,4)-β-glucans are linear, unbranched polymers in which the β-Glc residues are joined by both (1,3)- and (1,4)-glucosidic linkages. The sequence of (1,3)- and (1,4)-glucosidic linkages in the chain is not random (Clarke and Stone, 1963). Single (1,3)-linkages separated by two or three (1,4)-linked Glc residues (Fig. 1F) predominate, but longer cello-oligosaccharide units of up to DP 14 may also be present (Table 2). There are few, if any, contiguous (1,3)-linked Glc residues.

**Table 2: Comparative properties of cereal (1,3;1,4)-β-glucans and lichenin**

| Source | Wheat bran[a] (*Triticum vulgare*) | Barley flour[a] (*Hordeum vulgare*) | Oat flour[a] (*Avena sativa*) | Lichenin (*Cetraria islandica*) |
|---|---|---|---|---|
| Trisaccharide/ tetrasaccharide ratio | 4.2–4.5[b], 3.7[c] | 2.7–3.0[d], 2.8[c], 3.0[c] | 2.2–2.4[d], 2.1[c] | 18.6[e], 24.5[c] |
| % trisaccharide + tetrasaccharide | 93.3[b], 91.3[c] | 91.0–92.1[d], 90.9[c], 91.2[c] | 92.4–94.0[d], 90.3[c], 90.5[c] | 77.5[c] |
| % penta-nonasaccharide | 6.7[b] | 7.8[b] | 8.1[b] | Not available |
| % penta-tetradeca-saccharide | 8.7[c] | 9.1[c], 8.8[c] | 9.7[c], 9.5[c] | 22.5[c] |
| Average $M_w \times 10^5$ | 0.49[b], 2.09[c]* | 1.26–2.39[d]*, 2.13[c]*, 1.07[c]* | 0.44–1.10[d]*, 2.03[c]*, 1.05[c]* | 0.55[e], 1.06[c]* |
| Polydispersity ($M_w/M_n$) | 1.65[f] | 1.3–1.9[e] | 1.3–1.5[e] | 1.8[e] |
| Intrinsic viscosity (dL/g) | 4.96[f] | 4.6–6.9[e] | 2.0–9.6[e] | Not available |
| Gelation melting transition (°C) | 72[c] | 65[e], 67.7[d],69.3[d] | 62[e] | 73[e],~89[c] |

*Indicates the peak fraction of the main peak in the HPLC chromatogram.
[a]cereal (1,3;1,4)-β-glucans; [b]Cui et al., 2000; [c]Lazaridou et al., 2004; [d]Papageorgiou et al., 2005; [e]Bohm and Kulicke, 1999; [f]Li et al., 2006.

Among the cereal (1,3;1,4)-β-glucans there are significant differences in the organization of the (1,3)- and (1,4)-glucosidic linkages in the chain, as shown by the differences in the ratio of the 3-*O*-β-cellobiosyl- to 3-*O*-β-cellotriosyl-Glc and the proportion of longer gluco-oligosaccharides released by (1,3;1,4)-β-glucan hydrolase digestion (Table 2). These differences are reflected in their solubility in water; the barley and oat glucans are quite soluble but the wheat glucan is less so. Table 2 lists the molecular sizes and other physical properties of the cereal (1,3;1,4)-β-glucans (see also Chapter 2.2).

### *1.H.2 Equisetum (horsetail) (1,3;1,4)-β-glucan*

Most cell wall types in the horsetail, *Equisetum arvense*, a monilophyte, except those in vascular tissues, contain an abundant (1,3;1,4)-β-glucan. However, there are significant differences in the glucan block structures between Poaceae and *E. arvense* (1,3;1,4)-β-glucans (Sørensen et al., 2008; Fry et al., 2008). In contrast to the Poaceae (1,3;1,4)-β-glucans (see 1.H.1), DP4 residues are the most abundant oligomers released by (1,3;1,4)-β-glucan hydrolase treatment and are 10 or 20 times more abundant than the DP3 units. Furthermore, oligomers with a DP higher than 7 were not detected. Small amounts of a DP2 oligomer that did not co-elute with cellobiose in HPLC were found and proposed to be laminaribiose, suggesting that a few alternating 1,3- and 1,4-linked Glc units are present.

### 1.H.3 Liverwort (1,3;1,4)-β-glucan

Popper and Fry (2003) in a survey of Bryophytes and Charophytes using specific (1,3;1,4)-β-glucan hydrolase digestion (see Chapter 3.1) reported the presence of (1,3;1,4)-β-glucan only in the leafy liverwort *Lophocolea bidentata*. The major oligosaccharides were in the DP 2–6 range and yielded both Glc and Ara on acid hydrolysis.

### 1.H.4 Lichen, fungal and algal (1,3;1,4)-β-glucans

*1.H.4.a Lichen (1,3;1,4)-β-glucans* The (1,3;1,4)-β-glucan lichenin is extractable with hot water from the fronds of Iceland moss (*Cetraria islandica*). The glucan is located in the cell walls of the mycobiont (Honegger and Haisch, 2001). Compared to the cereal counterparts lichenin has a much higher ratio of tri-/tetra-saccharide building units (see Table 2) although the content of these two oligosaccharides is only 75% compared with >90% for the cereal glucans; cello-oligosaccharides DP 5–14 account for 22% of the molecule.

Lichenin-like polysaccharides have been reported from a number of other lichens (Stone and Clarke, 1992; Carbonero et al., 2001, 2002, 2005, 2006).

*1.H.4.b Fungal cell wall (1,3;1,4)-β-glucan* A (1,3;1,4)-β-glucan is a component of the alkali-insoluble hetero-polysaccharide complex of the cell wall of *Aspergillus fumigatus* (Fontaine et al., 2000) (see Chapter 4.3) with a core branch-on-branch (1,3;1,6)-β-glucan (Fig. 1C). The (1,3;1,4)-β-glucan chains represent 10% of the complex but their length has not been determined.

*1.H.4.c Chromalveolate, chromistan and chlorophyte (1,3;1,4)-β-glucans* A putative (1,3;1,4)-β-glucan was reported from the alveolate (dinoflagellate) *Peridinium westii* (Nevo and Sharon, 1969) but has not been further characterized.

The cell walls of the chromistan (xanthophyte) *Monodus subterraneus* contain an alkali-soluble (1,3;1,4)-β-glucan with (1,3)- to (1,4)-linkages in the proportion 15:85 (Ford and Percival, 1985). Both linkages are in the same chain as judged by Smith degradation.

Polysaccharides from the chlorophyte *Ulva lactuca* digested with (1,3;1,4)-β-glucan endohydrolase gave products that differed from the graminoid glucans having higher DPs and containing Xyl in addition to Glc (Popper and Fry, 2003).

The secondary walls and pores of the charophyte (desmid) *Micrasterias* are labelled with a (1,3;1,4)-β-glucan-specific monoclonal antibody (Eder et al, 2008). The glucan is not extracted with water but is successively extracted with 1 M and 4 M KOH, leaving further glucan in the 4 M KOH residue. No detailed structure is available.

*1.H.4.d Rhodophyte sulfated (1,3;1,4)-β-glucans* The matrix of cell walls of the red alga *Kappaphycus alvarezii* (Gigartinales) contain an alkali-soluble (1.5 M NaOH) sulfated (1,3;1,4)-β-glucan, $M_r$ $4.1\times10^4$ Da, composed of ~180 Glc residues of which 92% are (1,4)- and ~8% are (1,3)-linked. The non-sulfated (1,4)-linked Glc residues probably do not occur in long sequences since the polysaccharide is resistant to cellulase treatment. The sulfate esters are located on 64% of the (1,4)-linked Glc residues (Lechat et al., 2000).

The cell walls of several red algae contain hot water or alkali-soluble linear (1,3;1,4)-β-xylans that are homomorphous with linear (1,3;1,4)-β-glucans (see Stone and Clarke, 1992).

*1.H.4.e (1,3;1,4)-β-gluco-oligosaccharides* *Sarcina ventriculi,* a Gram-positive anaerobe, whose cells are surrounded by a cellulosic capsule, when extracted with water yielded two β-oligomers of Glc: a trisaccharide Glc*p*-β-(1,4)-Glc*p*-β-(1,3)-Glc*p* and a dimeric hexasaccharide: Glc*p*-β-(1,4)-Glc*p*-β-(1,3)-Glc*p*-β-(1,4)-Glc*p*-β-(1,4)-Glc*p*-β-(1,3)-Glc*p* (Lee and Hollingsworth, 1997).

## 2 Synthetic β-Gluco-oligosaccharides, Derivatives of (1,3)-β-Glucans and Neo β-Glucans

### 2.A Synthetic β-Gluco-oligosaccharides

A series of linear (1,3)-, (1,3;1,6)- and (1,3;1,4)-β-gluco-oligosaccharides that represent the building units of many naturally occurring β-glucans have been prepared both by chemical synthesis or enzymatically by transglycosylation or using glycosynthases. These are listed in Table 3.

### 2.B Derivatives of (1,3)-β-Glucans

A range of derivatives of (1,3)-β-glucan and related polymers have been prepared by esterification, alkylation, periodate oxidation (and subsequent reduction), glycosylation, tagging with fluorochromes, radioactive isotopes and other compounds. These are listed in Table 4. These derivatives have been variously proposed as anti-viral agents, immunopotentiators, elicitors of plant defence responses, gelling agents, enhancers for complexing with polynucleotides, enzyme inhibitors and as vehicles for drug delivery.

**Table 3: Preparations or syntheses of (1,3)-β-gluco-oligosaccharides**

| Oligosaccharides | Structure | Preparative route | References |
|---|---|---|---|
| **Linear oligosaccharides** | | | |
| | laminaribiose | acetolysis | Fujimoto et al., 1962 |
| | (1,3)-β-gluco-oligosaccharides | synthetic | He et al., 2002 |
| | laminaripentaose | enzymatic | Ebara, 1996 |
| | (1,3)-β-gluco-oligosaccharides | synthetic | Jamois et al., 2005 |
| | (1,3)-β-gluco-oligosaccharides | acid hydrolysis in DMSO | Kamo et al., 1990 |
| | (1,3;1,4)-β-gluco-oligosaccharides | enzymatic | Viladot et al., 1998; Faijes et al., 2001 |
| | *p*-nitrophenyl (1,3)-β-oligoglucosides | enzymatic | Zvyagintseva et al., 1998 |
| | 4-methylumbelliferyl (1,3)-β-oligoglucosides | enzymatic | Zvyagintseva et al., 1998 |
| | sulfated alkyl (1,3)-β-oligoglucosides | synthetic | Katsuraya et al., 1994 |
| | 8-methoxycarbonyloctyl β-glycosides of tri- and tetrasaccharides | synthetic | Takeo and Tei, 1986 |
| | (1,3)-β-megalo-oligosaccharides DP 30–34 | enzymatic | Hrmova et al., 2002 |
| **Branched β-gluco-oligosaccharides** | | | |
| | (1,3;1,6)-β-trisaccharide | synthetic | Heng et al., 2007 |
| | (1,3;1,6)-β-heptasaccharide | synthetic | Zhao et al., 2003 |
| | allyl glycoside of (1,3;1,6)-β-heptasaccharide | synthetic | Zhao et al., 2003 |
| | S-linked (1,3;1,6)-tetrasaccharide | synthetic | ContourGalcera et al., 1996 |
| | three (1,3;1,6)-β-tetrasaccharides related to schizophyllan | synthetic | Takeo and Tei, 1986 |
| | hexasaccharide with a (1,3)-linked α-glucosyl unit | synthetic | Ning et al., 2003 |
| | (1,3)-β-tetra-and penta-saccharides with (1,2)-linked β-glucosyl units | synthetic | Larsson et al., 2005 |
| | methyl-(1,3;1,6)-β-nonasaccharide | synthetic | Du et al., 2004 |
| | five 1,3-dideoxynojirimycin 3-yl glycosides of (1,3)- and (1,6)-β-gluco-oligosaccharides | synthetic | Blattner et al., 2006 |

(*Continued*)

**Table 3: (Continued)**

| Oligosaccharides | Structure | Preparative route | References |
|---|---|---|---|
| **β-Glucans** | | | |
| | (1,3;1,4)-β-glucan with alternating linkages | synthetic | Anderson and Stone, 1975 |
| | (1,3)-β-glucan | synthetic | Okada et al., 1991 |
| | (1,3)-β-glucan (curdlan) with 3-*O*-Me-glucosyl units | *in vivo* | Lee et al., 1997 |
| | I-labelled sclerooglucan | synthetic | Boeykens et al., 2004 |

**Table 4: Derivatives of (1,3)-β-glucans and related polymers**

| | Substituent, reagent or derivative | (1,3)-β-glucan type | References |
|---|---|---|---|
| **Esters** | | | |
| | acetyl | schizophyllan | Albrecht and Rau, 1994 |
| | palmityl | scleroglucan | Carafa et al., 2006 |
| | palmitoylsulfate | curdlan | Lee et al., 2005 |
| | sulfate | pachyman | Chen et al., 2007 |
| | sulfate | curdlan | Gao et al., 1997 |
| | sulfate | laminarin | Hoffman et al., 1995 |
| | sulfate | schizophyllan | Hirata et al., 1994 |
| | sulfate | laminarin | Miao et al., 1995 |
| | sulfate | curdlan | Takano et al., 2000 |
| | sulfate | *Pleurotus* glucan | Zhang et al., 2003 |
| | sulfoalkyl | curdlan | Demleitner et al., 1992 |
| | sulfoalkyl | curdlan | Lee et al., 2001 |
| | cyanoethyl | scleroglucan | Gianni et al., 2002 |
| | sulfoethyl | yeast glucan | Khalikova et al., 2006 |
| | aromatic carbamates | scleroglucan | Vincendon, 1999 |
| **Ethers** | | | |
| | carboxymethyl | pachyman | Stone, 1972 |
| | hydroxyethyl and glyceryl | curdlan | Renn, 1997 |
| | carboxymethyl | scleroglcan | de Nooy et al., 2000 |
| | carboxymethyl | yeast glucan | Soltes et al., 1993 |

**Table 4: (Continued)**

| | Substituent, reagent or derivative | (1,3)-β-glucan type | References |
|---|---|---|---|
| | aminoalkyl | scleroglcan | de Nooy et al., 2000 |
| | formylmethyl | schizophyllan | Usui et al., 1995 |
| | aminoethyl | schizophyllan | Usui et al., 1995 |
| **Oxidative modifications** | | | |
| | periodate | scleroglucan | Alhaique et al., 1986 |
| | periodate | scleroglucan | Christensen et al., 2001 |
| | periodate | scleroglucan | Maeda et al., 2001 |
| | periodate/borohydride polyalcohol | pestalotan | Misaki et al., 1984 |
| | | schizophyllan | Schulz and Rapp, 1991 |
| | TEMPO oxidation of primary alcohols | scleroglcan | de Nooy et al., 2000 |
| **Other substituents** | | | |
| | 6-azido-6-deoxy- | curdlan | Borjhan et al., 2001 |
| | 6-azido-6-deoxy- | curdlan sulphate | Borjihan et al., 2003 |
| | lactosides, ferrocene, pyrene, porphyrin via 6-azido-6-deoxy-curdlan | schizophyllan curdlan | Hasegawa et al., 2005; 2006 |
| | poly(ethylene glycol) | schizophyllan | Karinaga et al., 2005 |
| | glucose 3- and 6-linked via spacers | curdlan | Kiho et al., 1997 |
| | 6-amino groups on side-chain units | schizophyllan | Koumoto et al., 2001 |
| | cholesterol | schizophyllan | Koumoto et al., 2005 |
| | fluorochromes | schizophyllan laminarin *Eisenia bicyclis* laminarin | Kobayashi et al., 1995; Meunier and Wilkinson, 2002; Arnosti, 2003 |
| | polyacrylamide | schizophyllan | Takeo et al., 1993 |
| | phthalic acid | schizophyllan | Muller et al., 1986 |
| | 1,omega-dicarboxylic acid cross-linked glucan | scleroglucan | Casadei et al., 2005 |
| | mitomycin | schizophyllan | Usui et al., 1995 |
| **Complexes** | anti-sense oligonucleotide complexes | schizophyllan | Sakurai and Shinkai, 2000; Sakurai et al., 2001, 2002; Sakurai et al., 2005 |

### 2.C Neo β-Glucans

A new curdlan-based polysaccharide has been produced by direct incorporation *in vivo* of 3-*O*-methyl-D-Glc by *Agrobacterium* ATCC31749 into the curdlan chain to the extent of 8–12 mol% (Lee et al., 1997).

Carboxyl reduction of the Type 3 pneumococcal polysaccharide, which has a repeating →3)GlcAβ(1→4)Glcβ(1→ unit, produced a water-insoluble β-glucan with alternating (1,3)- and (1,4)-glucosidic linkages (Anderson and Stone, 1975).

(1,3)-β-Megalosaccharides (DP 30–45) have been synthesized using (1,3)-β-glucan glucosynthase derived by mutation of a barley (1,3)-β-glucan endohydrolase (Hrmova et al., 2002).

An efficient and stereospecific synthesis of a (1,3)-β-glucan has been achieved by catalytic polymerization of a 1,3-anhydro derivative of Glc (Okada et al., 1991).

## Acknowledgements

I am grateful to Dr Fung Lay, Department of Biochemistry, La Trobe University, Australia and Dr Maria Hrmova, Australian Centre for Plant Functional Genomics, University of Adelaide, Australia for the preparation of figures used in this chapter. I would also like to thank Ms Joanne Noble, Plant Cell Biology Research Centre, School of Botany, University of Melbourne, Australia for her expert editorial assistance.

## References

Adachi, Y., Ohno, N., Ohsawa, M., Sato, K., Oikawa, S., & Yadomae, T. (1989). Physicochemical properties and antitumor activities of chemically modified derivatives of antitumor glucan grifolan Le from *Grifola frondosa*. *Chemical and Pharmaceutical Bulletin*, *37*, 1838–1843.

Adeyeye, A., Jansson, P.-E., Lindberg, B., & Henrichsen, J. (1988). Structural studies of the capsular polysaccharide from *Streptococcus pneumoniae* type 37. *Carbohydrate Research*, *180*, 295–299.

Akima, K., Arika, T., & Amemiya, K. (1985). Purification, physicochemical characterization, and antitumor-activity of a cancer-associated human-serum protein that is increased by treatment with schizophyllan, an antitumor polysaccharide. *Japanese Journal of Cancer Research*, *76*, 541–548.

Albrecht, A., & Rau, U. (1994). Acetylation of a β-1,6-branched β-1,3-glucan, yielding schizophyllan acetate. *Carbohydrate Polymers*, *24*, 193–197.

Alekseeva, S. A., Shevchenko, N. M., Kusaykin, M. I., Ponomorenko, L. P., Isakov, V. V., Zvyagintseva, T. N., & Likhoshvai, E. V. (2005). Polysaccharides of diatoms occurring in Lake Baikal. *Applied Biochemistry and Microbiology*, *41*, 185–191.

Alhaique, F., Riccieri, F. M., Santucci, E., & Crescenzi, V. (1986). Oxidized scleroglucan for the design of a pH-controlled drug delivery system. *Chimica Oggi*, *7-8*, 11–15.

Alquini, G., Carbonero, E. R., Rosado, F. R., Cosentino, C., & Iacomini, M. (2004). Polysaccharides from the fruit bodies of the basidiomycete *Laetiporus sulphureus* (Bull.: Fr.) Murr. *FEMS Microbiology Letters*, *230*, 47–52.

Altabe, S. G., Talaga, P., Wieruszeski, J.-M., Lippens, G., Ugalde, R. A., & Bohin, J.-P. (1998). Periplasmic glucans of *Azospirillum brasilense*. In C. Elmerich, A. Kondorosi, & W. E. Newton (Eds.), *Biological nitrogen fixation for the 21st century*, p. 390. Dordrecht: Kluwer.

Anderson, M. A., & Stone, B. A. (1975). A new substrate for investigating the specificity of β-glucan hydrolases. *FEBS Letters*, *52*, 202–207.

Aouadi, S., Heyraud, A., Seigle-Murandi, F., Steiman, R., & Fournet, B. (1991). Structural analysis and rheological behavior of an extracellular polysaccharide from *Drechslera spicifera*. *Carbohydrate Polymers*, *17*, 177–183.

Archibald, A. R., Cunningham, W. L., Manners, D. J., Stark, J. R., & Ryley, J. F. (1963). Metabolism of the protozoa, X. The molecular structure of the reserve polysaccharides from *Ochromonas malhamensis and Peranema trichophorum*. *Biochemical Journal*, *88*, 444–451.

Arnosti, C. (2003). Fluorescent derivatization of polysaccharides and carbohydrate-containing biopolymers for measurement of enzyme activities in complex media. *Journal of Chromatography, B: Analytical Technologies in the Biomedical and Life Sciences*, *793*, 181–191.

Aspinall, G. O., & Kessler, G. (1957). The structure of callose from the grape vine. *Chemistry and Industry (London)*, 1296.

Bao, X., Wang, X., Dong, Q., Fang, J., & Li, X. (2002). Structural features of immunologically active polysaccharides from *Ganoderma lucidum*. *Phytochemistry*, *59*, 171–181.

Barbosa, A. M., Steluti, R. M., Dekker, R. F. H., Cardoso, M. S., & Corradi da Silva, M. L. (2003). Structural characterization of botryosphaeran: A (1→3;1→6)-β-D-glucan produced by the ascomycetous fungus, *Botryosphaeria* sp. *Carbohydrate Research*, *338*, 1691–1698.

Baron, M., Gorin, P. A. J., & Iacomini, M. (1988). Isolation and identification of a linear (1,3)-linked β-D-glucan and other carbohydrate components of the lichen *Stereocaulon ramulosum* (SW) Rausch. *Carbohydrate Research*, *177*, 235–239.

Beattie, A. B., Hirst, E. L., & Percival, E. (1961). Studies on the metabolism of the Chrysophyceae. Comparative structural investigations on leucosin (chrysolaminarin) separated from diatoms and laminarin from the brown algae. *Biochemical Journal*, *79*, 531–537.

Beauvais, A., Latge, J. P., Vey, A., & Prevost, M-C. (1989). The role of surface components of the entomopathogenic fungus *Entomophaga aulicae* in the cellular immune response of *Galleria mellonella* (Lepidoptera). *Journal of General Microbiology*, *135*, 489–498.

Bernard, M., & Latgé, J. P. (2001). *Aspergillus fumigatus* cell wall: Composition and biosynthesis. *Medical Mycology*, *39*, 9–17.

Bhagwat, A. A., Mithofer, A., Pfeffer, P. E., Kraus, C., Spickers, N., Hotchkiss, A., Ebel, J., & Keister, D. L. (1999). Further studies of the role of cyclic beta-glucans in symbiosis. An ndvC mutant of *Bradyrhizobium japonicum* synthesizes cyclodecakis-(1→3)-β-glucosyl. *Plant Physiology*, *119*, 1057–1064.

Blaschek, W., Kasbauer, J., Kraus, J., & Franz, G. (1992). *Pythium aphanidermatum* – Culture, cell wall composition, and isolation and structure of antitumor storage and solubilized cell wall (1-3), (1-6)-β-D-glucans. *Carbohydrate Research*, *231*, 293–307.

Blattner, R., Furneaux, R. H., & Pakulski, Z. (2006). 1,3-Dideoxynojirimycin-3-yl glycosides of β-(1→3)- and β-(1→6)-linked gluco-oligosaccharides. *Carbohydrate Research*, *341*, 2115–2125.

Boeykens, S. P., Vazquez, C., Temprano, N., & Rosen, M. (2004). Study of a novel labelled scleroglucan macromolecule. *Carbohydrate Polymers*, *55*, 129–137.

Bohm, N., & Kulicke, W. M. (1999). Rheological studies of barley (1→3)(1→4)-β-glucan in concentrated solution: mechanistic and kinetic investigation of the gel formation. *Carbohydrate Research*, *315*, 302–311.

Borjhan, G., Hu, R. C., Liu, G. G., & Zhong, G. Y. (2001). A direct regioselective synthesis of 6-azido-6-deoxy-curdlan. *Chemical Journal of Chinese Universities*, *22*, 706–708.

Borjihan, G., Zhong, G., Baigude, H., Nakashima, H., & Uryu, T. (2003). Synthesis and anti-HIV activity of 6-amino-6-deoxy-(1→3)-β-D-curdlan sulfate. *Polymers for Advanced Technologies*, *14*, 326–329.

Bruneteau, M., Fabre, I., Perret, J., & Michel, G. (1988). Antitumour active β-D-glucans from *Phytophthora parastica*. *Carbohydrate Research*, *175*, 137–143.

Buller, C. S., & Voepel, K. C. (1990). Production and purification of an extracellular polyglucan produced by *Cellulomonas flavigena* strain KU. *Journal of Industrial Microbiology*, *5*, 139–145.

Carafa, M., Marianecci, C., Annibaldi, V., Di Stefano, A., Sozio, P., & Santucci, E. (2006). Novel *O*-palmitoylscleroglucan-coated liposomes as drug carriers: Development, characterization and interaction with leuprolide. *International Journal of Pharmaceutics*, *325*, 155–162.

Carbonero, E., Montai, A., Mellinger, C., Eliasaro, S., Sassaki, G., Gorin, P., & Iacomini, M. (2005). Glucans of lichenized fungi: Significance for taxonomy of the genera *Parmotrema and Rimelia*. *Phytochemistry*, *66*, 929–934.

Carbonero, E. R., Gracher, A. H. P., Smiderle, F. R., Rosado, F. R., Sassaki, G. L., Gorin, P. A. J., & Iacomini, M. (2006). A β-glucan from the fruit bodies of edible mushrooms *Pleurotus eryngii* and *Pleurotus ostreatoroseus*. *Carbohydrate Polymers*, *66*, 252–257.

Carbonero, E. R., Montai, A. V., Woranovicz-Barreira, S. M., Gorin, P. A. J., & Iacomini, M. (2002). Polysaccharides of lichenized fungi of three *Cladina* spp.: significance as chemotypes. *Phytochemistry*, *61*, 681–686.

Carbonero, E. R., Sassaki, G. L., Stuelp, P. M., Gorin, P. A. J., Woranovicz-Barreira, S. M., & Iacomini, M. (2001). Comparative studies of the polysaccharides isolated from lichenized fungi of the genus Cladonia. Significance as chemotypes. *FEMS Microbiology Letters*, *194*, 65–69.

Casadei, M. A., Pitarresi, G., Benvenuti, F., & Giannuzzo, M. (2005). Chemical gels of scleroglucan obtained by cross-linking with 1,omega-dicarboxylic acids: Synthesis and characterization. *Journal of Drug Delivery Science and Technology*, *15*, 145–150.

Chang, Y. W., & Lu, T. J. (2004). Molecular characterization of polysaccharides in hot-water extracts of *Ganoderma lucidum* fruiting bodies. *Journal of Food and Drug Analysis*, *12*, 59–67.

Chauveau, C., Talaga, P., Wieruszeski, J. M., Strecker, G., & Chavant, L. (1996). A water-soluble β-D-glucan from *Boletus erythropus*. *Phytochemistry*, *43*, 413–415.

Chen, Q., Wang, A., Li, C., Li, T., & Jiao, Q. (2007). Spectroscopic characterization of pachyman sulfate and its binding interaction with azur A. *Spectroscopy Letters*, *40*, 83–95.

Chenghua, D., Xiangliang, Y., Xiaoman, G., Yan, W., Jingyan, Z., & Huibi, X. (2000). A β-D-glucan from the sclerotia of *Pleurotus tuber-regium* (Fr.) Sing. *Carbohydrate Research*, *328*, 629–633.

Chiovitti, A., Molino, P., Crawford, S. A., Teng, R., Spurck, T., & Wetherbee, R. (2004). The glucans extracted with warm water from diatoms are mainly derived from intracellular chrysolaminaran and not extracellular polysaccharides. *European Journal of Phycology*, *39*, 117–128.

Chiovitti, A., Ngoh, J. E., & Wetherbee, R. (2006). 1,3-β-D-glucans from *Haramonas dimorpha* (Raphidophyceae). *Botanica Marina*, *49*, 360–362.

Chizhov, A. O., Dell, A., Morris, H. R., Reason, A. J., Haslam, S. M., McDowell, R. A., Chizhov, O. S., & Usov, A. I. (1998). Structural analysis of laminarans by MALDI and FAB mass spectrometry. *Carbohydrate Research*, *310*, 203–210.

Christensen, B. E., Aasprong, E., & Stokke, B. T. (2001). Gelation of periodate oxidised scleroglucan (scleraldehyde). *Carbohydrate Polymers*, *46*, 241–248.

Clarke, A. E., & Stone, B. A. (1960). Structure of the paramylon from *Euglena gracilis*. *Biochimica et Biophysica Acta*, *44*, 161–163.

Clarke, A. E., & Stone, B. A. (1963). Chemistry and biochemistry of β-1,3-glucans. *Reviews of Pure and Applied Chemistry*, *13*, 134–156.

ContourGalcera, M. O., Ding, Y. L., OrtizMellet, C., & Defaye, J. (1996). Stereocontrolled synthesis of sulfur-linked analogues of the branched tetrasaccharide repeating-unit of the immunostimulant polysaccharide schizophyllan and of its β-(1→3)-branched, β-(1→6)-linked isomer. *Carbohydrate Research*, *281*, 119–128.

Cordeiro, L. M. C., Stocker-Woergoetter, E., Gorin, P. A. J., & Iacomini, M. (2004). Elucidation of polysaccharide origin in *Ramalina peruviana* symbiosis. *FEMS Microbiology Letters*, *238*, 79–84.

Cordeiro, L. M. C., Stocker-Worgotter, E., Gorin, P. A. J., & Iacomini, M. (2003). Comparative studies of the polysaccharides from species of the genus Ramalina-lichenized fungi-of three distinct habitats. *Phytochemistry*, *63*, 967–975.

Cui, W., Wood, P. J., Blackwell, B. A., & Nikiforuk, J. (2000). Physicochemical properties and structural characterization by two-dimensional NMR spectroscopy of wheat β-D-glucan – Comparison with other cereal β-D-glucans. *Carbohydrate Polymers*, *41*, 249–258.

Cunningham, W. L., Manners, D. J., & Ryley, J. F. (1962). Structure of a reserve polysaccharide (paramylon) from *Peranema trichophorum*. *Biochemical Journal*, *82*, 12C.

de Nobel, H., Sietsma, J. H., Van Den Ende, H., & Klis, F. M. (2001). Molecular organization and construction of the fungal cell wall. In R. J. Howard & N. A. R. Gow (Eds.), *The Mycota: VIII biology of the fungal cell*, 1st Edition (pp. 181–200). Berlin: Springer.

de Nooy, A. E. J., Rori, V., Masci, G., Dentini, M., & Crescenzi, V. (2000). Synthesis and preliminary characterisation of charged derivatives and hydrogels from scleroglucan. *Carbohydrate Research*, *324*, 116–126.

Demleitner, S., Kraus, J., & Franz, G. (1992). Synthesis and anti-tumour activity of sulfoalkyl derivatives of curdlan and lichenan. *Carbohydrate Research*, *226*, 247–252.

Deng, C. H., Yang, X. L., Gu, X. M., Wang, Y., Zhou, J. Y., & Xu, H. B. (2000). A β-D-glucan from the sclerotia of *Pleurotus tuber-regium* (Fr.) Sing. *Carbohydrate Research*, *328*, 629–633.

Du, Y. G., Gu, G. F., Hua, Y. X., Wei, G. H., Ye, X. S., & Yu, G. L. (2004). Synthesis and antitumor activities of glucan derivatives. *Tetrahedron*, *60*, 6345–6351.

Ebara, T. (1996). *Manufacture of laminaripentaose from β-1,3-glucosyl compounds with glucanase*. Japan: Dainippon Ink & Chemicals, Inc. Application: JP, pp. 6.

Eder, M., Tenhaken, R., Driouich, A., & Lütz-Meindl, U. (2008). Occurrence and characterization of arabinogalactan-like proteins and hemicelluloses in *Micrasterias* (Streptophyta) in press. *Journal of Phycology*.

Elyakova, L. A., Pavlov, G. M., Isakov, V. V., Zaitseva, I., & Stepchenova, T. A. (1994). Molecular characteristics of laminarin subfractions. *Khimiya Prirodnykh Soedinenii*, *2*, 296–298.

Ensley, H. E., Tobias, B., Pretus, H. A., McNamee, R. B., Jones, E. L., Browder, W., & Williams, D. L. (1994). NMR spectral analysis of a water-insoluble (1→3)-β-D-glucan isolated from *Saccharomyces cerevisiae*. *Carbohydrate Research*, *258*, 307–311.

Estrella, M. J., Pfeffer, P. E., Brouillette, J. N., Ugalde, R. A., & De Iannino, N. I. (2000). Biosynthesis and structure of cell associated glucans in the slow growing *Rhizobium loti* strain NZP 2309. *Symbiosis*, *29*, 173–199.

Evans, N. A., Hoyne, P. A., & Stone, B. A. (1984). Characteristics and specificity of the interaction of a fluorochrome from aniline blue (Sirofluor) with polysaccharides. *Carbohydrate Polymers*, *4*, 215–230.

Faijes, M., Fairweather, J. K., Driguez, H., & Planas, A. (2001). Oligosaccharide synthesis by coupled endo-glycosynthases of different specificity: A straightforward preparation of two mixed-linkage hexasaccharide substrates of 1,3/1,4-β-glucanases. *Chemistry*, *7*, 4651–4655.

Ferguson, C., Teeri, T., Siika-aho, M., Read, S., & Bacic, A. (1998). Location of cellulose and callose in pollen tubes and grains of *Nicotiana tabacum*. *Planta*, *206*, 452–460.

Fontaine, T., Simenel, C., Dubreucq, G., Adam, O., Delepierre, M., Lemoine, J., Vorgias, C. E., Diaquin, M., & Latge, J. P. (2000). Molecular organization of the alkali-insoluble fraction of *Aspergillus fumigatus* cell wall. *Journal of Biological Chemistry*, *275*, 27594–27607.

Footrakul, P., Suyanandana, P., Amemura, A., & Harada, T. (1981). Extracellular polysaccharides of *Rhizobium* from the Bangkok MIRCEN collection. *Journal of Fermentation Technology*, *59*, 9–14.

Ford, C. W., & Percival, E. (1965). The carbohydrates of *Phaeodactylum tricornutum*. Preliminary examination of the organism, and characterisation of low molecular weight material and of a glucan. *Journal of the Chemical Society*, 7035–7041.

Ford, C. W., & Percival, E. (1985). Polysaccharides synthesised by *Monodus subterraneus*. Part II. The cell-wall glucan. *Journal of Chemical Society*, 3014–3016.

Fry, S. C., Mohler, K. E., Nesslrod, B. H. W. A., & Franlova, l. (2008). Xyloglucan endotransglucosylase, a novel wall remodelling enzyme from *Equisetum* (horsetails) and charophyte algae. *Plant Journal*, *55*, 240–252.

Fujimoto, K., Matsuda, K., & Aso, K. (1962). Acetolysis of polysaccharides. III. Preparation of laminaribiose from pachyman and laminaran by acetolysis. *Tohoku Journal of Agricultural Research, 13*, 55–59.

Futatsuyama, H., Yui, T., & Ogawa, K. (1999). Viscometry of curdlan, a linear (1→3)-β-D-glucan, in DMSO or alkaline solutions. *Bioscience, Biotechnology and Biochemistry, 63*, 1481–1483.

Gandon, C., & Bruneteau, M. (1998). Structural studies of the extracellular β-D-glucans from *Phytophthora parasitica* Dastur. *Carbohydrate Research, 313*, 259–263.

Gao, Y., Fukuda, A., Katsuraya, K., Kaneko, Y., Mimura, T., Nakashima, H., & Uryu, T. (1997). Synthesis of regioselective substituted curdlan sulfates with medium molecular weights and their specific anti-HIV-1 activities. *Macromolecules, 30*, 3224–3228.

Ghai, S. K., Hisamatsu, M., Amemura, A., & Harada, T. (1981). Production and chemical composition of extracellular polysaccharides of *Rhizobium*. *Journal of General Microbiology, 122*, 33–40.

Gianni, R., Delben, F., Liut, G., Bernstorff, S., Amenitsch, H., & Rizzo, R. (2002). Synthesis and conformational properties of cyanoethyl-scleroglucan. *Carbohydrate Polymers, 47*, 387–391.

Gore, R., & Miller, K. (1993). Cyclic β-1,6 -1,3 glucans are synthesised by *Bradyrhizobium japonicum* bacteroids within soybean (*Glycine max*) root nodules. *Plant Physiology, 102*, 191–194.

Gorin, P. A. J., & Iacomini, M. (1984). Polysaccharides of the lichens *Cetraria islandica* and *Ramalina usnea*. *Carbohydrate Research, 128*, 129–132.

Gummadi, S. N., & Kumar, K. (2005). Production of extracellular water insoluble β-,3-glucan (curdlan) from *Bacillus* sp SNC07. *Biotechnology and Bioprocess Engineering, 10*, 546–551.

Handa, N., & Tominaga, H. (1969). A detailed analysis of carbohydrates in marine particulate matter. *Marine Biology, 2*, 228–235.

Harada, T., Misaki, A., & Saito, H. (1968). Curdlan: a bacterial gel-forming β-D-1,3-glucan. *Archives of Biochemistry and Biophysics, 124*, 292–298.

Harris, P. J. (2005). Diversity in plant cell walls. In R. J. Henry (Ed.), *Plant diversity and evolution: Genotypic and phenotypic variation in higher plants* (pp. 201–227). Wallingford, Oxon, UK: CAB International Publishing.

Hasegawa, T., Umeda, M., Numata, M., Fujisawa, T., Haraguchi, S., Sakurai, K., & Shinkai, S. (2006). Click chemistry on curdlan: A regioselective and quantitative approach to develop artificial β-1,3-glucans with various functional appendages. *Chemistry Letters, 35*, 82–83.

Hasegawa, T., Umeda, M., Numata, M., Li, C., Bae, A.-H., Fujisawa, T., Haraguchi, S., Sakurai, K., & Shinkai, S. (2005). 'Click chemistry' on polysaccharides: A convenient, general, and monitor approach to develop (1→3)-β-D-glucans with various functional appendages. *Carbohydrate Research, 341*, 35–40.

He, H. M., Yang, F., & Du, Y. G. (2002). Synthesis of natural β-D-(1→3)-glucopyranosyl oligosaccharides. *Carbohydrate Research, 337*, 1673–1678.

Hearn, V. M., & Sietsma, J. H. (1994). Chemical and immunological analysis of *Aspergillus fumigatus* cell wall. *Microbiology and Immunology, 140*, 789–795.

Heng, L. S., Zhao, Q. F., Mei, X. D., Lv, H. B., Jiang, X. H., & Ning, J. (2007). Synthesis of alkyl β-D-Glc*p*-(1→3)-[β-D-Glc*p*-(1→6)-β-D-Glc*p*-(1→6)-β-D-glucopyranosides with anti-tumor activity. *Letters in Organic Chemistry, 4*, 142–145.

Henrichsen, J. (1995). Six newly recognized types of *Streptococcus pneumoniae. Journal of Clinical Microbiology, 33*, 2759–2762.

Hirata, A., Itoh, W., Tabata, K., Kojima, T., Itoyama, S., & Sugawara, I. (1994). Anticoagulant activity of sulfated schizophyllan. *Bioscience, Biotechnology, and Biochemistry, 58*, 406–407.

Hoffman, R., Paper, D. H., Donaldson, J., Alban, S., & Franz, G. (1995). Characterization of a laminarin sulfate which inhibits basic fibroblast growth factor binding and endothelial cell proliferation. *Journal of Cell Science, 108*, 3591–3598.

Hoffmann, G. C., & Timell, T. E. (1970). Isolation of a β1,3-glucan (laricinan) from compression wood of *Larix laricina. Wood Science and Technology, 4*, 159–162.

Hoffmann, G. C., & Timell, T. E. (1972). Polysaccharides in compression wood of tamarack (*Larix laricina*). I. Isolation and characterization of laricinan, an acidic glucan. *Svensk Papperstidning, Nordisk Cellulosa, 75*, 135–142.

Honegger, R., & Haisch, A. (2001). Immunocytochemical location of the (1→3) (1→4)-β-glucan lichenin in the lichen-forming ascomycete *Cetraria islandica* (Icelandic moss). *New Phytologist, 150*, 739–746.

Hrmova, M., Imai, T., Rutten, S. J., Fairweather, J. K., Pelosi, L., Bulone, V., Driguez, H., & Fincher, G. B. (2002). Mutated barley (1,3)-β-D-glucan endohydrolases synthesize crystalline (1,3)-β-D-glucans. *Journal of Biological Chemistry, 277*, 30102–30111.

Huwyler, H. R., Franz, C., & Mier, H. (1978). β-(1,3)-glucans in the cell walls of cotton fibres (*Gossypium arboreum*). *Plant Science Letters, 12*, 55–62.

Iacomini, M., Zanin, S. M. W., Fontana, J. D., Hogge, L., & Gorin, P. A. J. (1987). Isolation and characterization of β-D-glucan, heteropolysaccharide, and trehalose components of the basidiomycetous lichen *Cora pavonia. Carbohydrate Research, 168*, 55–65.

Inon de Iannino, N., & Ugalde, R. A. (1993). Biosynthesis of cyclic β-(1-3), β-(1-6) glucan in *Bradyrhizobium* spp. *Archives of Microbiology*, *159*, 30–38.

Iwamuro, Y., Aoki, M., & Mikami, Y. (1985). Purification and some properties of an exo-β-1,3-glucanase from *Porodisculus pendulus*. *Journal of Fermentation Technology*, *63*, 405–409.

Jamois, F., Ferrieres, V., Guegan, J. P., Yvin, J. C., Plusquellec, D., & Vetvicka, V. (2005). Glucan-like synthetic oligosaccharides: Iterative synthesis of linear oligo-β-(1,3)-glucans and immunostimulatory effects. *Glycobiology*, *15*, 393–407.

Johnson, J. J., Kirkwood, S., Misaki, A., Nelson, T. E., Scaletti, J. V., & Smith, F. (1963). Structure of a new glucan. *Chemistry and Industry*, 820–822.

Kamo, Y., Hirayama, M., Sakaeda, T., & Hidaka, H. (1990). *Preparation of oligosaccharides by acid hydrolysis of β-1,3-glucan polysaccharides in dimethyl sulfoxide*. Application: JP, p. 3.

Karinaga, R., Koumoto, K., Mizu, M., Anada, T., Shinkai, S., & Sakurai, K. (2005). PEG-appended β-(1→3)-D-glucan schizophyllan to deliver antisense-oligonucleotides with avoiding lysosomal degradation. *Biomaterials*, *26*, 4866–4873.

Katsuraya, K., Shoji, T., Inazawa, K., Nakashima, H., Yamamoto, N., & Uryu, T. (1994). Synthesis of sulfated alkyl laminara-oligosaccharides having potent anti-HIV activity and the relationship between structure and biological activities. *Macromolecules*, *27*, 6695–6699.

Kenyon, W. J., & Buller, C. S. (2002). Structural analysis of the curdlan-like exopolysaccharide produced by *Cellulomonas flavigena* KU. *Journal of Industrial Microbiology and Biotechnology*, *29*, 200–203.

Kenyon, W. J., Esch, S. W., & Buller, C. S. (2005). The curdlan-type exopolysaccharide produced by *Cellulomonas flavigena* KU forms part of an extracellular glycocalyx involved in cellulose degradation. *Antonie van Leeuwenhoek*, *87*, 143–148.

Khalikova, T. A., Korolenko, T. A., Zhanaeva, S. Y., Kaledin, and V. I., & Kogan, G. (2006). Enhancing effect of new biological response modifier sulfoethylated (1→3)-β-D-glucan on antitumor activity of cyclophosphamide in the treatment of experimental murine leukoses. *Experimental Oncology*, *28*, 308–313.

Kiho, T., Matsushita, M., Usui, S., & Ukai, S. (1997). Preparation of novel (1→3)-β-D-glucans having reducing glucose side chains. *Chemical and Pharmaceutical Bulletin*, *45*, 725–727.

Kim, Y. T., Kim, E. H., Cheong, C., Williams, D. L., Kim, C. W., & Lim, S. T. (2000). Structural characterization of β-D-(1→3,1→6)-linked glucans using NMR spectroscopy. *Carbohydrate Research*, *328*, 331–341.

Kishida, E., Kinoshita, C., Sone, Y., & Misaki, A. (1992). Structures and antitumor activities of polysaccharides isolated from mycelium of *Volvariella volvacea. Bioscience, Biotechnology, and Biochemistry, 56*, 1308–1309.

Kishida, E., Sone, Y., & Misaki, A. (1989). Purification of an antitumor-active, branched (1→3)-β-D-glucan from *Volvariella volvacea*, and elucidation of its fine structure. *Carbohydrate Research, 193*, 227–239.

Kiss, J. Z., Vasconcelos, A. C., & Triemer, R. E. (1987). Structure of the euglenoid storage carbohydrate, paramylon. *American Journal of Botany, 74*, 877–882.

Kiss, J. Z., Vasconcelos, A. C., & Triemer, R. E. (1988). The intramembranous particle profile of the paramylon membrane during paramylon synthesis in *Euglena* (Euglenophyceae). *Journal of Phycology, 24*, 152–157.

Kitamura, S., Hori, T., Kurita, K., Takeo, K., Hara, C., Itoh, W., Tabata, K., Elgsaeter, A., & Stokke, B. T. (1994). An antitumor, branched (1-3)-β-D-glucan from a water extract of fruiting bodies of *Cryptoporus volvatus. Carbohydrate Research, 263*, 111–121.

Knecht, J. C., Schiffman, G., & Austrian, R. (1970). Some biological properties of Pneumococcus type 37 and the chemistry of its capsular polysaccharide. *Journal of Experimental Medicine, 132*, 475–487.

Kobayashi, A., Tai, A., & Kawazu, K. (1995). Structural elucidation of an elicitor-active oligosaccharide, LN-3, prepared from algal laminarin. *Journal of Carbohydrate Chemistry, 14*, 819–832.

Kollar, R., Petrakova, E., Ashwell, G., Robbins, P., & Cabib, E. (1995). Architecture of the yeast cell wall – Linkage between chitin and β-(1-3)-glucan. *Journal of Biological Chemistry, 270*, 1170–11778.

Kollar, R., Reinhold, B., Petrakova, E., Yeh, H., Ashwell, G., Drgonova, J., Kapteyn, J., Klis, F., & Cabib, E. (1997). Architecture of the yeast cell wall – β(1→6)-glucan interconnects mannoprotein, β(1→3)-glucan, and chitin. *Journal of Biological Chemistry, 272*, 17762–17775.

Komaniecka, I., & Choma, A. (2003). Isolation and characterization of periplasmic cyclic β-glucans of *Azorhizobium caulinodans. FEMS Microbiology Letters, 227*, 263–269.

Koumoto, K., Kimura, T., Mizu, M., Sakurai, K., & Shinkai, S. (2001). Chemical modification of schizophyllan by introduction of a cationic charge into the side chain which enhances the thermal stability of schizophyllan-poly(C) complexes. *Chemical Communication*, 1962–1963.

Koumoto, K., Mizu, M., Anada, T., Nagasaki, T., Shinkai, S., & Sakurai, K. (2005). Cholesterol-appended β-(1→3)-D-glucan schizophyllan for antisense oligonucleotides delivery to enhance the cellular uptake. *Bulletin of the Chemical Society of Japan, 78*, 1821–1830.

Kreger, D. R., & van der Veer, J. (1970). Paramylon in a chrysophyte. *Acta Botanica Neerlandica, 19*, 401–402.

Larsson, E. A., Sjöberg, M., & Widmalm, G. (2005). Synthesis of oligosaccharides related to the repeating unit of the capsular polysaccharide from *Streptococcus pneumoniae* type 37. *Carbohydrate Research, 340*, 7–13.

Latgé, J.-P., Boucias, D. G., & Fournet, B. (1988). Structure of the exocellular polysaccharide produced by the fungus, *Nomuraea rileyi*. *Carbohydrate Research, 181*, 282–286.

Latgé, J.-P., Fournet, B., Cole, G., Dubourdieu, D., & Tong, N. (1984). Composition chimique et ultrastructure des parois des corps hyphaux et des azygospores de *Conidiobolus obscurus*. *Canadian Journal of Microbiology, 30*, 1507–1521.

Latgé, J. P., & Beauvais, A. (1987). Wall composition of the protoplastic entomophthorales. *Journal of Invertebrate Pathology, 50*, 53–57.

Lazaridou, A., Biliaderis, C. G., Micha-Screttas, M., & Steele, B. R. (2004). A comparative study on structure–function relations of mixed-linkage (1→3), (1→4) linear β-D-glucans. *Food Hydrocolloids, 18*, 837–855.

Lechat, H., Amat, M., Mazoyer, J., Buleon, A., & Lahaye, M. (2000). Structure and distribution of glucomannan and sulfated glucan in the cell walls of the red alga *Kappaphycus alvarezii* (Gigartinales, Rhodophyta). *Journal of Phycology, 36*, 891–902.

Lee, C. M., Lee, H. C., & Lee, K. Y. (2005). *O*-Palmitoylcurdlan sulfate (OPCurS)-coated liposomes for oral drug delivery. *Journal of Bioscience and Bioengineering, 100*, 255–259.

Lee, J.-H., Mullins, J. T., & Gander, J. E. (1996). Water-soluble reserve polysaccharides from *Achyla* are 1,3-β-glucans. *Mycologia, 88*, 264–270.

Lee, J., & Hollingsworth, R. I. (1997). Oligosaccharide β-glucans with unusual linkages from *Sarcina ventriculi*. *Carbohydrate Research, 304*, 133–141.

Lee, J. W., Yeomans, W. G., Allen, A. L., Gross, R. A., & Kaplan, D. L. (1997). Microbial production of water-soluble noncurdlan type exopolymer-B with controlled composition by *Agrobacterium* sp. *Biotechnology Letters, 19*, 1217–1221.

Lee, K. B., Bae, J. H., Kim, J. S., Yoo, Y. C., Kim, B. S., Kwak, S. T., & Kim, Y. S. (2001). Anticoagulant activity of sulfoalkyl derivatives of curdlan. *Archives of Pharmacal Research, 24*, 109–113.

Li, W., Cui, S. W., & Kakuda, Y. (2006). Extraction, fractionation, structural and physical characterization of wheat ß-D-glucans. *Carbohydrate Polymers, 63*, 408–416.

Maeda, H., Rambone, G., Coviello, T., Yuguchi, Y., Urakawa, H., Alhaique, F., & Kajiwara, K. (2001). Low-degree oxidized scleroglucan and its hydrogel. *International Journal of Biological Macromolecules*, *28*, 351–358.

Maltby, D., Carpita, N. C., Montezinos, D., Kulow, C., & Delmer, D. P. (1979). β-1,3 Glucan in developing cotton fibres. Structure, localization and relationship of synthesis to that of secondary wall formation. *Plant Physiology*, *63*, 1158–1164.

Manners, D. J., Masson, A. J., & Patterson, J. C. (1973). The structure of a β-(13)-D-glucan from yeast cell walls. *Biochemical Journal*, *135*, 19–30.

McConville, M. J., Bacic, A., & Clarke, A. E. (1986). Structural studies of chrysolaminarin from the ice diatom *Stauroneis amphioxus*. *Carbohydrate Research*, *153*, 330–333.

Meikle, P. J., Bonig, I., Hoogenraad, N. J., Clarke, A. E., & Stone, B. A. (1991). The location of (1,3)-β-glucans in the walls of pollen tubes of *Nicotiana alata* using a (1,3)-β-glucan specific monoclonal antibody. *Planta*, *185*, 1–8.

Meikle, P. J., Hoogenraad, N. J., Bonig, I., Clarke, A. E., & Stone, B. A. (1994). A (1→3,1→4)-β-glucan-specific monoclonal antibody and its use in the quantitation and immunocytochemical location of (1→3,1→4)-β-glucans. *Plant Journal*, *5*, 1–9.

Meunier, F., & Wilkinson, K. J. (2002). Nonperturbing fluorescent labeling of polysaccharides. *Biomacromolecules*, *3*, 857–864.

Miao, H. Q., Ishaimichaeli, R., Peretz, T., & Vlodavsky, I. (1995). Laminarin sulfate mimics the effects of heparin on smooth muscle cell proliferation and basic fibroblast growth factor-receptor binding and mitogenic activity. *Journal of Cellular Physiology*, *164*, 482–490.

Miller, K. J., Gore, R. S., Johnson, R., Benesi, A. J., & Reinhold, V. N. (1990). Cell associated oligosaccharides of *Bradyrizobium* spp. *Journal of Bacteriology*, *172*, 136–142.

Misaki, A., Johnson, J. J., Kirkwood, S., Scaletti, J. V., & Smith, F. (1968). Structure of the cell wall glucan of yeast (*Saccharomyces cerevisiae*). *Carbohydrate Research*, *6*, 150–164.

Misaki, A., Kakuta, M., Sasaki, T., Tanaka, M., & Miyaji, H. (1981). Studies on inter-relation of structure and antitumor effects of polysaccharides: Antitumor action of periodate-modified, branched (1-3)-β-D-glucan of *Auricularia auricula judae*, and other polysaccharides containing (1-3)-glycosidic linkages. *Carbohydrate Research*, *92*, 115–129.

Misaki, A., Kawaguchi, K., Miyaji, H., Nagae, H., Hokkoku, S., Kakuta, M., & Sasaki, T. (1984). Structure of pestalotan, a highly branched (1→3)-β-D-glucan elaborated by *Pestalotia* sp. 815. and the enhancement of its antitumor activity by polyol modification of the side chains. *Carbohydrate Research*, *129*, 209–227.

Misaki, A., Nasu, M., Sone, Y., Kishida, E., & Kinoshita, C. (1986). Comparison of structure and antitumor activity of polysaccharides isolated from Fukurotake, the fruiting body of *Volvariella volvacea*. *Agricultural and Biological Chemistry*, *50*, 2171–2183.

Miyazaki, T., & Oikawa, N. (1973). Studies on fungal polysaccharides. XII. Water-soluble polysaccharide of *Grifora umbellata* (Fr.) Pildt. *Chemical and Pharmaceutical Bulletin*, *21*, 2545–2548.

Miyazaki, T., Yadomae, T., Sugiura, M., Ito, H., Fujii, K., Naruse, S., & Kunihisa, M. (1974). Chemical structure of antitumor polysaccharide, coriolan, produced by *Coriolus versicolor*. *Chemical and Pharmaceutical Bulletin*, *22*, 1739–1742.

Mizuno, T., & Hazama, T. (1986). Studies on the host-mediated antitumor polysaccharides. X. Fractionation, formolysis and antitumor activity of fibrous polysaccharides (noncellulose) from Reishi, the fruiting body of *Ganoderma lucidum*. *Shizuoka Daigaku Nogakubu Kenkyu Hokoku*, 77–83.

Molinaro, A., Lanzetta, R., Evidente, A., Rosa, M. D., & Ianaro, A. (2000). Immunostimulant (1-3)-β-glucans from the cell wall of *Cryphonectria parasitica* (Murr.) Barr strain 263. *Carbohydrate Research*, *329*, 441–445.

Müller, A., Ensley, H., Pretus, H. A., McNamee, R. B., Jones, E. L., McLaughlin, E., Chandley, W., Browder, W., Lowman, D., & Williams, D. (1997). The application of various protic acids in the extraction of (1,3)-β-D-glucan from *Saccharomyces cerevisiae*. *Carbohydrate Research*, *299*, 203–208.

Muller, G., Chiron, G., & Levesque, G. (1986). Functional group modification in a non-ionic extracellular schizophyllan polysaccharide. *Polymer Bulletin*, *15*, 1–5.

Nakanishi, L., Kimura, K., Suzuki, T., Ishikawa, M., Banno, L., Sakane, T., & Harada, T. (1976). Demonstration of curdlan-type polysaccharide and some other β-1,3-glucan in microorganisms with aniline blue. *Journal of General and Applied Microbiology*, *22*, 1–11.

Nanjo, F., Usui, T., & Suzuki, T. (1984). Mode of action of an exo-β-(1-3)-D-glucanase on the laminaran from *Eisenia bicyclis*. *Agricultural and Biological Chemistry*, *48*, 1523–1532.

Nevo, Z., & Sharon, N. (1969). The cell wall of *Peridinium westii*, a non cellulosic glucan. *Biochimica et Biophysica Acta*, *173*, 161–175.

Ning, J., Zhang, W. H., Yi, Y. T., Yang, G. B., Wu, Z. K., Yi, J., & Kong, F. Z. (2003). Synthesis of β(1-6)-branched β(1-3) glucohexaose and its analogues containing an α-(1-3) linked bond with antitumor activity. *Bioorganic and Medicinal Chemistry*, *11*, 2193–2203.

Ohno, N., Adachi, Y., Suzuki, I., Sato, K., Oikawa, S., & Yadomae, T. (1986). Characterization of the antitumor glucan obtained from liquid-cultured *Grifola frondosa*. *Chemical and Pharmaceutical Bulletin*, *34*, 1709–1715.

Ohno, N., Uchiyama, M., Tsuzuki, A., Tokunaka, K., Miura, N. N., Adachi, Y., Aizawa, M. W., Tamura, H., Tanaka, S., & Yadomae, T. (1999). Solubilization of yeast cell-wall β-(1→3)-D-glucan by sodium hypochlorite oxidation and dimethyl sulfoxide extraction. *Carbohydrate Research, 316*, 161–172.

Okada, M., Yamakawa, Y., & Sumitomo, H. (1991). Chemical synthesis of (1-3)-β-D-glucopyranan by ring opening polymerization of a 1,3-anhydro sugar derivative. *Macromolecules, 24*, 6797–6799.

Papageorgiou, M., Lakhdara, N., Lazaridou, A., Biliaderis, C. G., & Izydorczyk, M. S. (2005). Water extractable (1-3),(1-4)-β-D-glucans from barley and oats: An intervarietal study on their structural features and rheological behaviour. *Journal of Cereal Science*, *42*, 213–224.

Parker, B. C. (1964). Chemical nature of sieve tube callus in Macrocystis. *Phycologia*, *4*, 27–42.

Paulsen, B. S., & Myklestad, S. (1978). Structural studies of the reserve glucan produced by the marine diatom *Skeletonema costatum* (Grev.) Cleve. *Carbohydrate Research*, *62*, 386–388.

Percival, E., Rahman, M. A., & Weigel, H. (1980). Chemistry of the polysaccharides of the diatom *Coscinodiscus nobilis*. *Phytochemistry*, *19*, 809–811.

Perlin, A. S., & Taber, W. A. (1963). A glucan produced by *Claviceps purpurea*. *Canadian Journal of Chemistry*, *41*, 2278–2282.

Pfeffer, P. E., Osman, S. F., Hotchkiss, A., Bhagwat, A. A., Keister, D. L., & Valentine, K. M. (1996). Cyclolaminarinose. A new biologically active β-(1-3) cyclic glucan. *Carbohydrate Research*, *296*, 23–37.

Popper, Z. A., & Fry, S. C. (2003). Primary cell wall composition of bryophytes and charophytes. *Annales Botanici- London*, *91*, 1–12.

Portilho, M., Matioli, G., Zanin, G. M., Faria de Moraes, F., & Scamparini, A. R. P. (2006). Production of insoluble exopolysaccharide of *Agrobacterium* sp. (ATCC 31749 and IFO 13140). *Applied Biochemistry and Biotechnology*, *131*, 864–869.

Rae, A. L., Harris, P. J., Bacic, A., & Clarke, A. E. (1985). Composition of the cell walls of *Nicotiana alata* Link et Otto pollen tubes. *Planta*, *166*, 128–133.

Read, S. M., Currie, G., & Bacic, A. (1996). Analysis of the structural heterogeneity of laminarin by electrospray ionisation-mass spectrometry. *Carbohydrate Research*, *281*, 187–201.

Reis, R., Tischer, C., Gorin, P., & Iacomini, M. (2002). A new pullulan and a branched (1→3)-, (1→6)-linked beta-glucan from the lichenised ascomycete *Teloschistes flavicans*. *FEMS Microbiology Letters*, *210*, 1–5.

Renn, D. W. (1997). Purified curdlan and its hydroxyalkyl derivatives: Preparation, properties and applications. *Carbohydrate Polymers*, *33*, 219–225.

Rinaudo, M., & Vincendon, M. (1982). Carbon-13 NMR structural investigation of scleroglucan. *Carbohydrate Polymers*, *2*, 135–144.

Rolin, D. B., Pfeffer, P. E., Osman, S. F., Szwergold, B. S., Kappler, F., & Benesi, A. (1992). Structural studies of a phosphocholine substituted β-(1,3);(1,6) macrocyclic glucan from *Bradyrhizobium japonicum* USDA 110. *Biochimica et Biophysica Acta*, *1116*, 215–225.

Rout, D., Mondal, S., Chakraborty, I., Pramanik, M., & Islam, S. S. (2005). Chemical analysis of a new (1→3)-, (1→6)-branched glucan from an edible mushroom, *Pleurotus florida*. *Carbohydrate Research*, *340*, 2533–2539.

Ruel, K., & Joseleau, J-P. (1991). Involvement of extracellular glucan sheath during degradation of *Populus* wood by *Phanerochaete chyrsoporium*. *Applied and Environmental Microbiology*, *57*, 374–384.

Ruiz-Herrera, J., Elorza, M. V., Valentin, E., & Sentandreu, R. (2006). Molecular organization of the cell wall of *Candida albicans* and its relation to pathogenicity. *FEMS Yeast Research*, *6*, 14–29.

Saito, H., Misaki, A., & Harada, T. (1968). A comparison of the structure of curdlan and pachyman. *Agricultural and Biological Chemistry*, *32*, 1261–1269.

Sakurai, K., Iguchi, R., Koumoto, K., Kimura, T., Mizu, M., Hisaeda, Y., & Shinkai, S. (2002). Polysaccharide-polynucleotide complexes VIII. Cation-induced complex formation between poly-uridylic acid and schizophyllan. *Biopolymers*, *65*, 1–9.

Sakurai, K., Mizu, M., & Shinkai, S. (2001). Polysaccharide-polynucleotide complexes. 2. Complementary polynucleotide mimic behavior of the natural polysaccharide schizophyllan in the macromolecular complex with single-stranded RNA and DNA. *Biomacromolecules*, *2*, 641–650.

Sakurai, K., & Shinkai, S. (2000). Molecular recognition of adenine, cytosine, and uracil in a single-stranded RNA by a natural polysaccharide: Schizophyllan. *Journal of the American Chemical Society*, *122*, 4520–4521.

Sakurai, K., Uezu, K., Numata, M., Hasegawa, T., Li, C., Kaneko, K., & Shinkai, S. (2005). β-1,3-Glucan polysaccharides as novel one-dimensional hosts for DNA/RNA, conjugated polymers and nanoparticles. *Chemical Communication*, 4383–4398.

Saudagar, P. S., & Singhal, R. S. (2004). Fermentative production of curdlan. *Applied Biochemistry and Biotechnology*, *118*, 21–31.

Schmid, F., Separovic, F., McDougall, B. M., Stone, B. A., Brownlee, R. T. C., & Seviour, R. J. (2007). Characterisation of the extracellular polysaccharides produced by isolates of the fungus Acremonium. *Carbohydrate Research*, *342*, 2481–2483.

Schmid, F., Stone, B. A., McDougall, B. M., Bacic, A., Martin, K. L., Brownlee, R. T. C., Chai, E., & Seviour, R. J. (2001). Structure of epiglucan, a highly side-chain/branched (1→3;1→6)-β-glucan from the micro fungus *Epicoccum nigrum* Ehrenb. ex Schlecht. *Carbohydrate Research, 331*, 163–171.

Schulz, D., & Rapp, P. (1991). Properties of the polyalcohol prepared from the β-D-glucan schizophyllan by periodate oxidation and borohydride reduction. *Carbohydrate Research, 222*, 223–231.

Shivakumar, S., & Vijayendra, S. V. N. (2006). Production of exopolysaccharides by *Agrobacterium* sp CFR-24 using coconut water – A byproduct of food industry. *Letters in Applied Microbiology, 42*, 477–482.

Silva, I. R., Monterio, N. K., Martinez, P. F., Izeli, N. L., Vasconcelos, A. F. D., Cardoso, M.d. S., de Lourdes, M., da Sillva, C., Barbosa, A.d. M., Dekker, R. F. H., da Silva, G. V. J., & Moraes, L. A. B. (2008). Purificacao e charcaterizacao de uma gentiohexose obtida botriosperana por hyrolise acida parcial (in press). *Quimica Nova.*

Smiderle, F. R., Carbonero, E. R., Mellinger, C. G., Sassaki, G. L., Gorin, P., & Iacomini, M. (2006). Structural characterization of a polysaccharide and a β-glucan isolated from the edible mushroom *Flammulina velutipes. Phytochemistry, 67*, 2189–2196.

Soltes, L., Alfoldi, J., & Sandula, J. (1993). HPLC and C-13-NMR study of carboxymethyl-β-(1-6)-D-Gluco-β-(1-3)-D-glucan derived from *Saccharomyces cerevisiae. Journal of Applied Polymer Science, 48*, 1313–1319.

Sone, Y., IsodaJohmura, M., & Misaki, A. (1996). Isolation and chemical characterization of polysaccharides from Iwatake, *Gyrophara esculenta* Miyoshi. *Bioscience, Biotechnology and Biochemistry, 60*, 213–215.

Sørensen, I., Pettolino, P., Wilson, S. M., Doblin, M., Johansen, B., Bacic, A., & Willats, W. G. T. (2008). Mixed linkage (1→3),(1→4)-β-D-glucan is not unique to the Poales and is an abundant component of *Equisetum arvense* cell walls. *Plant Journal, 54*, 510–521.

Stahmann, K. P., Monschau, N., Sahm, H., Koschel, A., Gawronski, M., Conrad, H., Springer, T., & Kopp, F. (1995). Structural properties of native and sonicated cinerean, a β-(1-3)(1-6)-D-glucan produced by *Botrytis cinerea. Carbohydrate Research, 266*, 115–128.

Stone, B. A. (1972). *O*-Carboxymethylpachyman. In R. V. W. J. BeMiller (Ed.), *Methods carbohydrate chemistry: Vol. VI* (pp. 384–385). Academic.

Stone, B. A., & Clarke, A. E. (1992). *Chemistry and biology of (1→3)-β-Glucans.* Victoria, Australia: La Trobe University Press.

Størseth, T. R., Hansen, T., Reitana, K. I., & Skjermo, J. (2005). Structural characterization of β-D-(1,3)-glucans from different growth phases of the marine diatoms *Chaetoceros mulleri* and *Thalassiosira weissflogii*. *Carbohydrate Research*, *340*, 1159–1164.

Stuelp, P. M., Carneiro Leao, A. M. C., Gorin, P. A. J., & Iacomini, M. (1999). The glucans of *Ramalina celastri*: Relation with chemotypes of other lichens. *Carbohydrate Polymers*, *40*, 101–106.

Surarit, R., Gopal, P. K., & Shepherd, M. G. (1988). Evidence for a glycosidic linkage between chitin and glucan in the cell wall of *Candida albicans*. *Journal of General Microbiology*, *134*, 1723–17830.

Tada, R., Harada, T., Nagi-Miura, N., Adachi, Y., Nakajima, M., Toshiro, Y., & Ohno, N. (2007). NMR characterization of the structure of a β-(1→3)-D-glucan isolate from cultured fruit bodies of *Sparassis crispa*. *Carbohydrate Research*, *342*, 2611–2618.

Takano, R., Nagai, T., Wu, X., Xu, X.-Y., Huy, N. T., Kamei, K., & Hara, S. (2000). Sulfation of polysaccharides using monomethyl sulfate. *Journal of Carbohydrate Chemistry*, *19*, 1185–1190.

Takeo, K., Kawaguchi, M., & Kitamura, S. (1993). Synthesis of polyacrylamide copolymers containing (1,6)-branched (1,3)-β-D-linked trisaccharides and tetrasaccharides related to schizophyllan. *Journal of Carbohydrate Chemistry*, *12*, 1043–1056.

Takeo, K. I., & Tei, S. (1986). Synthesis of the repeating units of schizophyllan. *Carbohydrate Research*, *145*, 293–306.

Trethewey, J. A. K., Campbell, L. M., & Harris, P. J. (2005). (1→3),(1→4)-β-D-Glucans in the cell walls of the Poales (*sensu lato*): An immunogold labeling study using a monoclonal antibody. *American Journal of Botany*, *92*, 1669–1683.

Usui, S., Tomono, Y., Sakai, M., Kiho, T., & Ukai, S. (1995). Preparation and antitumor activities of β-(1→6) branched (1→3)-β-D-glucan derivatives. *Biological and Pharmaceutical Bulletin*, *18*, 1630–1636.

Usui, T., Iwasaki, Y., Mizuno, T., Tanaka, M., Shinkai, K., & Arakawa, M. (1983). Isolation and characterization of antitumor β-D-glucans from the fruit bodies of *Ganoderma applanatum*. *Carbohydrate Research*, *115*, 273–280.

Vassallo, R., Standing, J. E., & Limper, A. H. (2000). Isolated *Pneumocystis carinii* cell wall glucan provokes lower respiratory tract inflammatory responses. *Journal of Immunology*, *164*, 3755–3763.

Viladot, J. L., Stone, B., Driguez, H., & Planas, A. (1998). Expeditious synthesis of a new hexasaccharide using transglycosylation reaction catalyzed by *Bacillus* (1→3),(1→4)-β-D-glucan 4-glucanohydrolase. *Carbohydrate Research*, *311*, 95–99.

Vincendon, M. (1999). Scleroglucan derivatives: Aromatic carbamates. *Journal of Polymer Science Part A – Polymer Chemistry*, *37*, 3187–3192.

von Stosch, H. A. (1951). Uber des Leukosin, den Reservestoff der Chrysophyten. *Naturwissenschaften*, *38*, 192–193.

Wang, G., Zhang, J., Mizuno, T., Zhuang, C., Ito, H., Mayuzumi, H., Okamoto, H., & Li, J. (1993). Antitumor active polysaccharides from the Chinese mushroom Songshan Lingzhi, the fruiting body of *Ganoderma tsugae*. *Bioscience, Biotechnology and Biochemistry*, *57*, 894–900.

Wang, M. C., & Bartnicki-Garcia, S. (1973). Novel phosphoglucans from the cytoplasm of *Phytophthora palmivora* and their selective occurrence in certain life cycle stages. *Journal of Biological Chemistry*, *248*, 4112–4118.

Wang, M. C., & Bartnicki-Garcia, S. (1980). Distribution of mycolaminarans and cell wall β-glucans in the life cycle of Phytophthora. *Experimental Mycology*, *4*, 269–280.

Wang, Y., Zhang, M., Ruan, D., Shashkov, A. S., Kilcoyne, M., Savage, A. V., & Zhang, L. (2004). Chemical components and molecular mass of six polysaccharides isolated from the sclerotium of *Poria cocos*. *Carbohydrate Research*, *339*, 327–334.

Warsi, S. A., & Whelan, W. J. (1957). Structure of pachyman, the polysaccharide component of *Poria cocos*. *Chemistry and Industry*, 1573.

Wood, P. J., & Fulcher, R. G. (1984). Specific interaction of aniline blue with (1-3)-β-D-glucan. *Carbohydrate Polymers*, *4*, 49–72.

Wood, P. J., Weisz, J., & Blackwell, B. A. (1994). Structural studies of (1,3),(1,4)-β-D-glucans by C(13)-nuclear magnetic resonance spectroscopy and by rapid analysis of cellulose-like regions using high-performance anion-exchange chromatography of oligosaccharides released by lichenase. *Cereal Chemistry*, *71*, 301–307.

Woodward, J. R., Fincher, G. B., & Stone, B. A. (1983). Water-soluble (1-3),(10-4)-β-D-glucans from barley (*Hordeum vulgare*) endosperm. II. Fine structure. *Carbohydrate Polymers*, *3*, 207–225.

Wustman, B. A., Gretz, M. R., & Hoagland, K. D. (1997). Extracellular matrix assembly in diatoms (Bacillariophyceae) (I. A model of adhesives based on chemical characterization and localization of polysaccharides from the marine diatom *Achnanthes longipes* and other diatoms). *Plant Physiology*, *113*, 159–1068.

Yoshioka, Y., Tabeta, R., Saitô, H., Uehara, N., & Fukuoka, F. (1985). Antitumor polysaccharides from *P. ostreatus* (Fr.) quél.: Isolation and structure of a β-glucan. *Carbohydrate Research*, *140*, 93–100.

Yotsuzuka, F. (2001). Curdlan. *Food Science and Technology*, *113*, 737–757.

Zhang, M., Zhang, L., Wang, Y., & Cheung, P. C. K. (2003). Chain conformation of sulfated derivatives of β-glucan from sclerotia of *Pleurotus tuber-regium*. *Carbohydrate Research*, *338*, 2863–2870.

Zhang, X. F., Zhang, L. N., & Xu, X. J. (2004). Morphologies and conformation transition of lentinan in aqueous NaOH solution. *Biopolymers*, *75*, 187–195.

Zhao, W., Yang, G. B., & Kong, F. (2003). Synthesis of two heptasaccharide analogues of the lentinan repeating unit. *Carbohydrate Research*, *338*, 2813–2823.

Zvyagintseva, T. N., Makar'eva, T. N., Ermakova, S. P., & Elyakova, L. A. (1998). The production of *p*-nitrophenyl laminarioligosides through the transglycosylation reaction catalyzed by endo-1,3-β-D-glucanase from a marine mollusk. *Bioorganicheskaya Khimiya*, *24*, 219–223.

Zvyagintseva, T. N., Shevchenko, N. M., Chizhov, A. O., Krupnova, T. N., Sundukova, E. V., & Isakov, V. V. (2003). Water-soluble polysaccharides of some far-eastern brown seaweeds, distribution, structure, and their dependence on the developmental conditions. *Journal of Experimental Marine Biology and Ecology*, *294*, 1–13.

CHAPTER 2.2

# Physico-chemistry of (1,3)-β-Glucans

**Michael J. Gidley[1] and Katsuyoshi Nishinari[2]**

[1]*Centre for Nutrition and Food Sciences, The University of Queensland, Australia*

[2]*Graduate School of Human Life Science, Osaka City University, Japan*

The physico-chemical properties exhibited by (1,3)-β-glucans are based primarily on the nature and stability of ordered conformations that are present under hydrated conditions. X-ray fibre diffraction analysis has shown unambiguously that a triple helix can be formed under both anhydrous and hydrated solid state conditions for linear (1,3)-β-glucans; similar diffraction behaviour for side-chain-substituted (1,3)-β-glucans suggests a similar triple helical form with reduced lateral aggregation of glucan backbones due to the presence of side chains. Solid state $^{13}C$ nuclear magnetic resonance (NMR) provides evidence that triple helical conformations are present within gels and some native forms of (1,3)-β-glucans. Light scattering, electron microscopy and atomic force microscopy data show the presence of triple helices for a range of substituted (1,3)-β-glucans in dilute aqueous solution, and are consistent with single-chain forms in solutions of alkali and dimethyl sulfoxide (DMSO). In contrast to (1,3)-β-glucan backbones, there is limited experimental data on ordered conformations in (1,3;1,4)-β-glucans, although analogy with related systems suggests that strict repeating backbone sequences could form the basis for inter-chain interactions.

(1,3)-β-glucans exhibit a range of solution, network and gel properties depending on (1) the chemistry and molecular size of the polysaccharides, (2) their biological origin, and (3) the thermal and solvent histories of extracted polysaccharides. For relatively simple primary structures, the combination of these three variables leads to a diversity of behaviour, and ongoing debates on the chemical and physical bases for observed macroscopic properties in which both thermodynamic and kinetic aspects need to be considered. These properties include not only macroscopic structuring in solutions, gels or cell walls, but also the cellular interactions that underlie some of the promising therapeutic properties of (1,3)-β-glucans.

This chapter will describe methods and results of studies into the conformations adopted by (1,3)-β-glucans in solid, solution and native states (I), and characterization of the physical behaviour of (1,3)-β-glucans in dilute solutions, concentrated solutions and gel networks (II). The intention is to connect features of primary structure (see Chapter 2.1) to the macroscopic rheological and mechanical properties that underlie many of the biological roles and technological applications of (1,3)-β-glucans.

# I Conformations of (1,3)-β-Glucans and Related Polysaccharides

## I.A General Principles of Polysaccharide Conformation

### I.A.1 Chain stiffness

Polysaccharides are composed of cyclic monomers (monosaccharides) typically joined by glycosidic linkages. The cyclic monomer imparts considerable stiffness to polysaccharide chains, not only because of the inherent rigidity of the monomer unit, but also because its bulkiness results in a limited number of possible spatial arrangements of adjacent monomer residues. An exception is provided by (1,6)-linkages that have three-bond glycosidic linkages rather than the usual two; the extra bond allows much greater conformational freedom. The stiffness of polysaccharides has two major consequences.

(i) Once dissolved, stiff chains lead to high viscosities because each polymer sweeps out a large volume of solvent. This ‘hydrodynamic volume’ is typically encapsulated by the measure of intrinsic viscosity (dL/g). Many rheological properties of polysaccharide solutions scale to the intrinsic viscosity (Morris et al., 1981).

(ii) Polysaccharides are typically close to the point of aqueous (in)solubility under ambient conditions. This is because there is little increase in degrees of motional freedom for stiff chains upon dissolution, and hence limited entropic advantage to be gained by dissolving. Where there are enthalpic advantages (e.g. hydrogen bonds) for polysaccharides to be in the solid state, then dissolution will not occur at equilibrium if the overall free energy is not lowered:

$$\Delta G = \Delta H - T\Delta S$$

where G,H,T,S are free energy, enthalpy, temperature and entropy, respectively. The prevalence of the gel state is a manifestation of the borderline solubility of many polysaccharides.

### *I.A.2 Chain shape*

As carbohydrate monomer residues are relatively stiff, the shapes adopted by polymers are determined by geometries around the glycosidic linkage. In particular, the torsion angles $\phi$ and $\psi$ (Fig. 1) define the orientation of adjacent residues, the sum total of which along a chain define the overall polymer shape. In the solid state these shapes can be assumed to be time invariant, whereas in solution local conformations change very rapidly such that it is average chain shapes that are normally investigated.

For each type of linkage between carbohydrate residues there is a geometry imposed by the orientation of the bonds involved in the linkage. For a (1,3)-β-glucosidic linkage a twist in the chain with respect to the monomer residue is introduced because the 3 position is not opposite the 1 position in the ring (Fig. 1). Where there are repetitive 'twist' linkages of the same type, as found for most (1,3)-β-glucans, there is a tendency to form helical chain shapes (Fig. 2).

### *I.A.3 Chain association*

As many polysaccharide chains are on the borderline of solubility in aqueous solvents, association with other chains is a possibility that needs to be considered. From an equilibrium

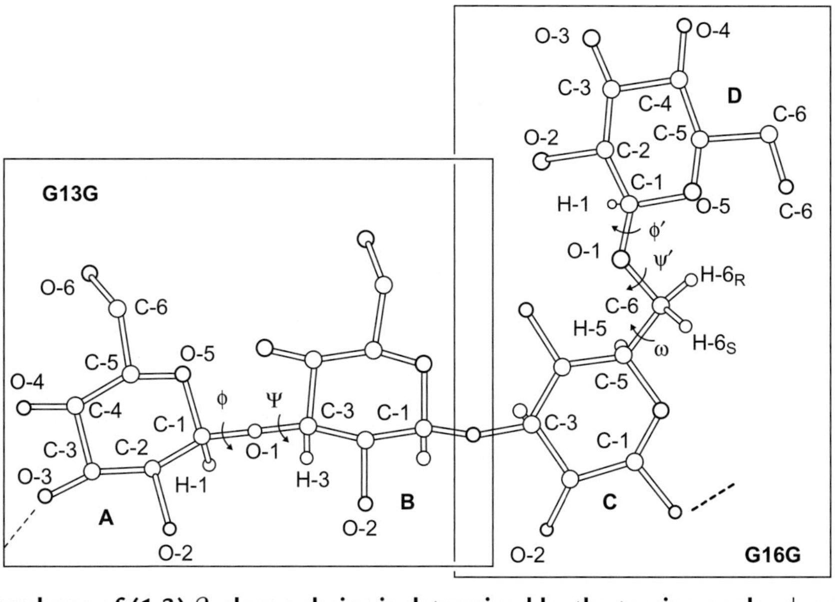


**Fig. 1: The shape of (1,3)-β-glucan chains is determined by the torsion angles $\phi$ and $\psi$, and (1,6) linkages have greater flexibility due to the additional bond linking the monomer units. From Palleschi et al. (2005).**

thermodynamics view point, association/aggregation will occur if the free energy of the associated state is lower than that of the dissolved state. However, the possibility of kinetically trapped structures needs also to be considered. When a stabilized (e.g. associated) state is formed quickly but does not represent the lowest free energy state, it may be unable to access a lower free energy state because of high activation energy, and is described as kinetically trapped. This means that the conformations of (1,3)-β-glucan chains found in their biological source or after extraction may be determined by a combination of time-dependent (kinetic) and energy-dependent (thermodynamic) effects.

The enthalpic stabilization of associations between uncharged polysaccharides is, in essence, a hydrogen bond exchange reaction in which hydrogen bonds between polysaccharides and solvent water are less enthalpically favourable than hydrogen bonds between polysaccharide chains with the concomitant exclusion of water (sometimes referred to as 'hydrophobic bonds'). These interactions are very weak individually and not sufficient to overcome the corresponding loss of entropy that occurs due to a combination of (a) two or more individual chains associating into a single entity and (b) the demixing of solvent water and polysaccharide chains. A repeating primary structure and a well-defined secondary structure (chain shape), however, can facilitate multiple weak chain interactions. Whilst

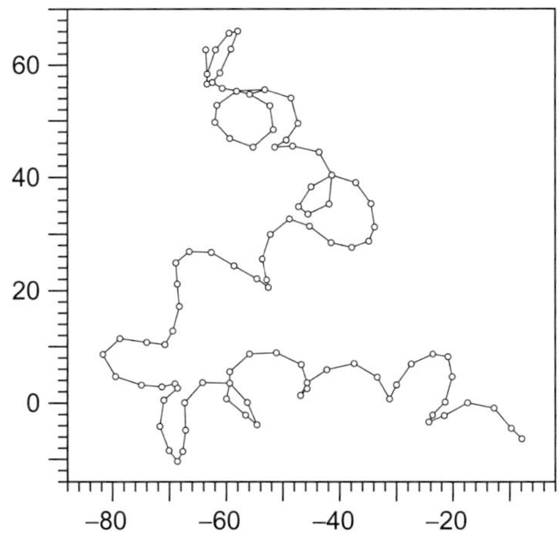


**Fig. 2: Perspective drawing of a segment of a (1,3)-β-glucan chain chosen as representative from a large Monte Carlo sample. Circles represent glycosidic oxygens, and lines are virtual bonds spanning the sugar residues (not shown). View is perpendicular to the x–y plane of an arbitrary coordinate system; the scale is measured in Angstroms. Smaller circles correspond to glycosidic oxygens further away from the viewer. From Burton and Brant (1983).**

much of the loss of entropy on association is due to the conversion of two or more separate chains into a single entity, the enthalpic stabilization of associated states increases with each additional weak interaction. The more repetitive a primary structure is, the more likely it is that sufficient multiple interactions will occur to result in a stable associated state. The length of chains over which this occurs is termed the cooperative length (Fig. 3). In addition, the greater the stiffness of the primary chain, the lower the entropic penalty of chain association, so the lower the number/strength of enthalpic interactions needed for stable association.

For repeating primary structures of polysaccharides (as found in many (1,3)-β-glucans), the tendency for twisted individual chain shapes (Fig. 2) leads to a natural tendency to form associations via helices. In principle, helices can either be of single chains, stabilized by side-to-side associations, or be multi-stranded stabilized by inter-chain intra-helix interactions.

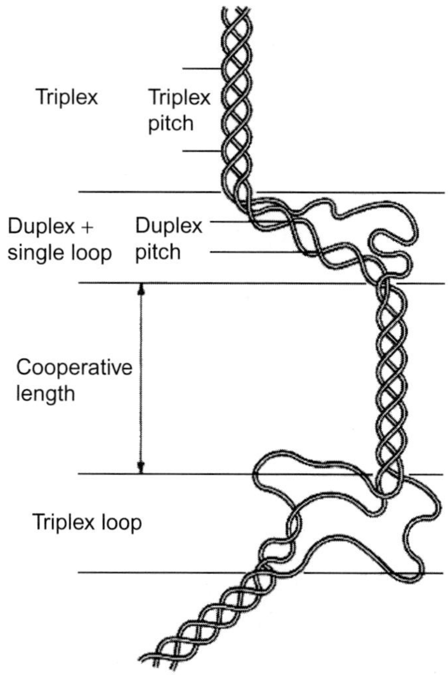


**Fig. 3: Illustration of possible (1,3)-β-glucan conformations, including cooperative length stable triple helices that contribute to enthalpic stabilization, and loop regions of disorder that contribute to entropic stabilization. If individual chains participate in helices with multiple chains, then a three-dimensional network will be formed. From Falch and Stokke (2001).**

Structural irregularities in the form of non-repeating backbone sequences or substituents on the backbone (e.g. (1,6)-side branches, Fig. 1) are expected to destabilize the extensive cooperative individual interactions required for stable associations.

#### *I.A.4 Network formation*

The ability to form stable inter-molecular associations under ambient aqueous conditions is one pre-requisite for the formation of stable three-dimensional networks, most obviously in gels formed under highly hydrated conditions. However, if all of a polysaccharide chain is present in rigid inter-chain associations, then there is an entropic penalty due to the loss of molecular motion. It is therefore common for the lowest free energy to be obtained through an appropriate mix of enthalpically stabilized association ('junction') zones formed from cooperative length interactions and regions of molecular flexibility that contribute to entropy. If multiple inter-chain associations per molecule occur with different molecules, interspersed with entropically stabilized inter-junction segments, then the result will be a network (Fig. 3). Alternatively, inter-molecular associated forms (e.g. triple helices) may associate side by side, leading (under highly hydrated conditions) to liquid crystals or (under low water conditions) to condensed forms such as precipitates.

### *I.B Solid State Conformations in (1,3)-β-Glucans*

As the molecular mobility of polymers in the solid state is limited, it is possible in principle to characterize (the range of) conformations experimentally. This is in contrast to the solution state where molecular mobility is extensive and experimental techniques provide an averaged description of conformation. For solutions and intermediate states such as gels, solid state conformations provide reference points with which experimental observations in more hydrated states can be compared.

#### *I.B.1 Methods*

*I.B.1.a X-ray diffraction* The only current experimental method that can be used to determine detailed polysaccharide chain conformation is X-ray (or electron) fibre diffraction. As polysaccharides do not normally crystallize, the generation of a crystal lattice that can diffract X-rays usually relies on the careful drawing of fibres which are then treated to maximize crystallinity. These treatments are evaluated empirically and are often remote from conditions encountered in biological or technological applications, so methods are needed to check that similar conformations are also present under different environments. Once a diffraction pattern with maximized information has been generated, unit cell indexing and molecular structure determination are carried out using standard methods (Chuah et al., 1983). However, there is normally not enough information in the experimental data to solve the three-dimensional structure explicitly, so molecular modelling is used to limit the number of candidate structures by predicting energetically stable conformations. Through the iterative use of model building and refinement of fits to experimental data, three-dimensional structures can be proposed which minimize residual errors. For some systems, crystallization of oligosaccharide models has also been useful. These lower molecular weight systems have the advantage that they form single crystals more readily than polymers, and so can be analysed to high resolution by diffraction methods. The advent of high power synchrotron sources of X-rays means that micron-sized crystals can now be analysed, removing the previous limiting step of needing to grow millimetre-sized crystals for diffraction by conventional X-ray sources.

As X-ray fibre diffraction of polysaccharides (or X-ray crystallography of oligosaccharides) is a specialized technique available only in a few laboratories worldwide, powder diffraction ‘fingerprints’ are often used to define the presence of a crystalline conformational polymorph in polysaccharide systems. Examples include the A and B polymorphic forms of starch present in native granules and the polymorphs of cellulose encountered naturally (cellulose I) and after technological processing (cellulose II). Powder diffraction can be performed with low power X-rays on standard laboratory equipment, in contrast to the more advanced fibre diffraction and crystallography methods.

*I.B.1.b Solid state NMR* Over the last 30 years, solid state NMR has emerged as a powerful adjunct to X-ray diffraction methods for both characterizing and fingerprinting molecular conformations in polysaccharides. The major technique is $^{13}$C CPMAS NMR. Although $^{13}$C is a less sensitive NMR nucleus than $^{1}$H, it has the major advantage of giving a very wide range of chemical shifts for polysaccharides (50 ppm range cf. 3 ppm range for $^{1}$H). Cross polarization (CP) is a technique for transferring energy from protons to carbons provided that the two atoms involved maintain their relative positions during the transfer process. This has two effects: (i) rigid (‘solid-like’) parts of the system under study are selectively excited, and (ii) sensitivity of carbon spectra from excited sites is increased. Magic angle spinning (MAS) has the effect of averaging the chemical shifts for different molecular orientations within the sample. For solution state NMR, the rapid tumbling of molecules effectively averages any effects of orientation with respect to the magnetic field: MAS achieves nearly the same level of averaging for samples in the solid state.

The solid state NMR technique is based on responses from individual atomic nuclei, and so is a very short range probe of structure (typically sub-nanometre). In contrast, for X-ray diffraction signals to be observed, a number of unit cells need to be present in crystalline register over a minimum of ca. 10 nm. This means that $^{13}C$ CPMAS NMR can be used as a probe of molecular order at the sub-crystalline level. In addition, signals from non-ordered/crystalline solid materials are also observed, so the NMR technique provides an overview of all solid state structures. However, detailed assignment of signals in $^{13}C$ CPMAS NMR currently requires reference to primary conformation information from X-ray fibre diffraction or crystallography. The approach that has been successful in interpreting $^{13}C$ CPMAS NMR (for e.g. cellulose and starch; Atalla and Vanderhart, 1984; Gidley and Bociek, 1985) is to record spectra for samples that have powder diffraction patterns characteristic of a known crystalline polymorph. Once these reference spectra have been reported, they can then be used to quantify the level of molecular (sub-crystalline) order within a polysaccharide sample (Tan et al., 2007). The application of $^{13}C$ CPMAS NMR is not limited to dry solid samples. Wet solids, gels and other soft solid forms can also be examined, but only the 'solid-like' molecular segments within the sample will be detected. Complementary NMR techniques such as $^{13}C$ SPMAS (where SP=single pulse) can detect more mobile molecular segments as solution-like signals, thus giving a description of conformation as a function of polymer flexibility that has been termed mobility-resolved spectroscopy (Foster et al., 1996).

#### *I.B.2 X-ray diffraction of (1,3)-β-glucans*

There have only been a limited number of studies of (1,3)-β-glucans by X-ray diffraction that serve as the reference point for interpretation of conformational features under a range of environmental conditions. Early observations (Bluhm and Sarko, 1977; Marchessault et al., 1977) suggested that a triple helix structure was the most likely, based on density and stereochemical arguments together with limited X-ray diffraction data. Diffraction patterns of greater intensity and resolution were reported by Deslandes et al. (1980) following extrusion of a 10% DMSO solution of curdlan into methanol. The fibres produced were washed with water and annealed under tension at 145°C in the presence of water, and subsequently evacuated to give an anhydrous sample. Despite the artificial nature of the production process, the resulting diffractogram was shown to have similar features to previous lower quality diffraction patterns. X-ray analysis of the annealed anhydrous fibre together with stereochemical model refinement led to the proposal of a detailed molecular structure for a triple helical form of linear (1,3)-β-glucans (Deslandes et al., 1980) which has remained the benchmark. The triple helices were proposed to be composed of parallel strands with extensive hydrogen bonds both within one triple helix and between adjacent helices, thereby stabilizing the

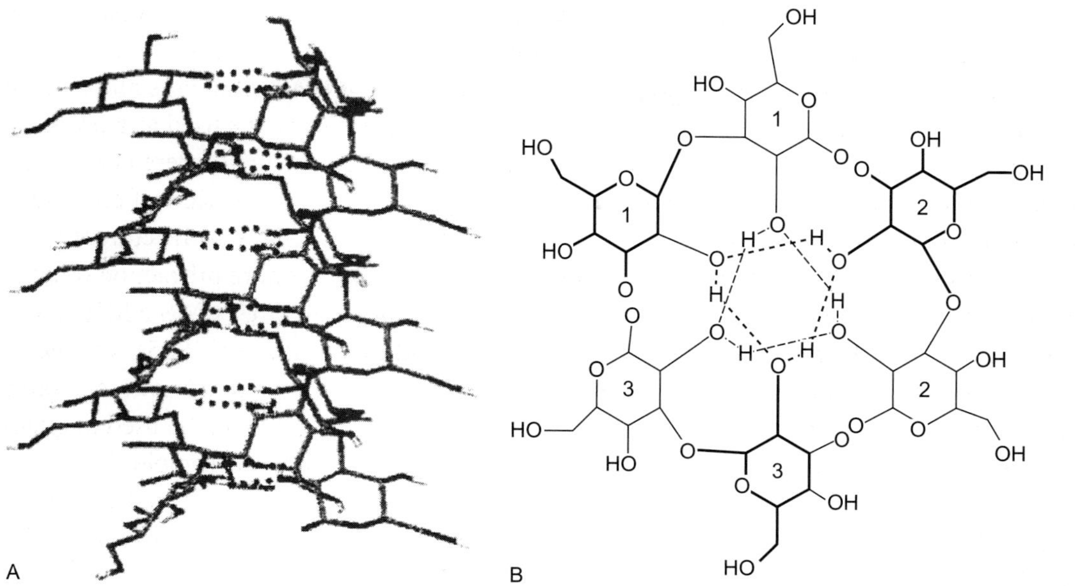


**Fig. 4: (A) Side and (B) top views of the triple helix of (1,3)-β-glucan characterized by Deslandes et al. (1980) and Chuah et al. (1983). The three strands of the triple helix are numbered in (b). The helix is stabilized by multiple inter-strand hydrogen bonds between hydroxyls at C-2, shown as dashed lines. From Miyoshi et al. (2004).**

crystalline structure (Fig. 4). The crystalline unit cell was found to be hexagonal with space group $P6_3$. Subsequently, Chuah et al. (1983) analysed diffraction patterns from the hydrated annealed fibre (i.e. as in Deslandes et al., 1980, without the final evacuation process), together with powder diffraction patterns for paramylon granules (Fig. 5) to propose a model for the hydrated form of triple helical (1,3)-β-glucans. The resulting hexagonal unit cell structure had a high reliability and showed that the individual triple helices were essentially indistinguishable from the geometry proposed for the anhydrous form with the same stabilizing intra-helix hydrogen bonds, although the space group was different (probably $P_1$). The H-bonding between strands shown in Fig. 4 is considered to be the most probable (Chuah et al., 1983) but is not the only possibility (Miyoshi et al., 2004). Further progress in resolving the exact positions of hydrogen atoms may be possible through the use of neutron scattering with selected isotopic replacement of hydrogen with deuterium, as has been demonstrated in the high resolution crystal structures now available for cellulose polymorphs (Nishiyama et al., 2003). The major structural difference from the anhydrous form was in the presence of water molecules between adjacent helices. These water molecules were not part of the crystal structure

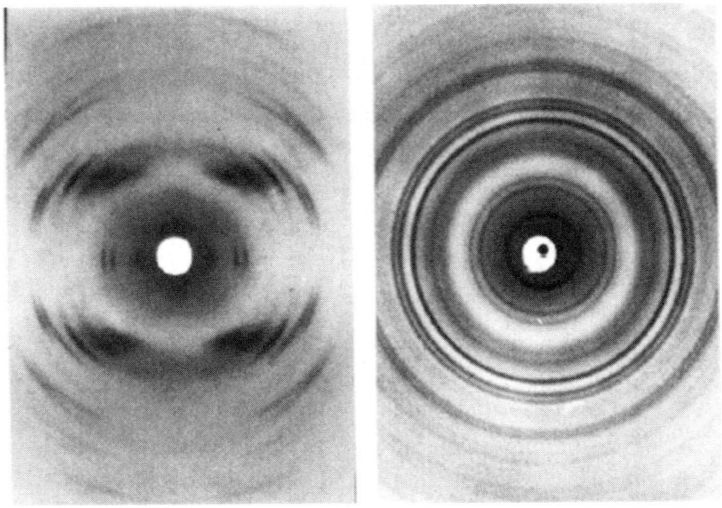

**Fig. 5: X-ray fibre diffraction patterns from hydrated curdlan (left) and paramylon granules (right). The circular form of diffraction for paramylon shows the lack of fibre orientation compared with the drawn fibre of curdlan. From Chuah et al. (1983).**

although were proposed to be clustered around H-bonding oxygen atoms. An important finding was that the exocyclic hydroxymethyl group at C-6 was not constrained to a single conformation in the hydrated form, in contrast to the anhydrous form where a conserved trans-gauche (tg) conformation of the C-6 hydroxyl group about the C-5–C-6 bond was found. The lack of requirement for a specific C-6 conformation to stabilize the triple helix, and the ability of water to occupy spaces between triple helices without destabilizing them, led Chuah et al. (1983) to propose that (1,3)-β-glucans with substituents at the 6-position could adopt similar or identical triple helical structures without an energy penalty. Examples includes the side-chain-branched schizophyllan/scleroglucan/lentinan (with the same average degree of branching but from different sources) which all have similar X-ray diffraction patterns (Bluhm et al., 1982; Chuah et al., 1983) and for which there is additional evidence from light scattering and microscopy for the adoption of a triple helical structure in solution (Kitamura et al., 1996; Sletmoen and Stokke, 2008).

However, there is also evidence that the triple helix may not represent the only ordered conformation found in (1,3)-β-glucans. When alkaline solutions of curdlan are neutralized, gels are formed that can be stretched into (weakly) crystalline structures that differ in diffraction patterns (Fig. 6) from both the well-defined triple helical structures. Two structural interpretations for this highly hydrated form have been proposed. One is based on a triple helical

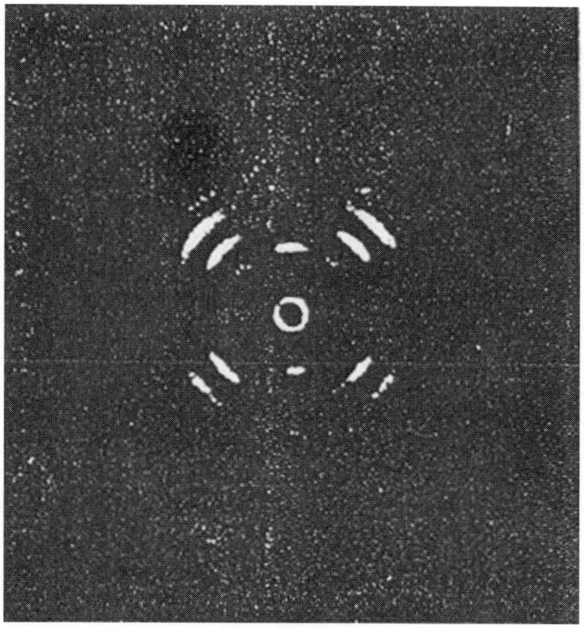

**Fig. 6: X-ray diffraction of curdlan fibres in the 'highly hydrated' form. From Okuyama et al. (1991).**

structure that is less tightly wound than the structures described by Deslandes et al. (1980), Chuah et al. (1983) and Fulton and Atkins (1980), and the other is based on highly hydrated single helices (Okuyama et al., 1991). The diffraction pattern for the highly hydrated form of curdlan (Fig. 6) has a limited information content and was shown to be consistent with many helical arrangements including a number of single, double and triple helices (Okuyama et al., 1991). Subsequent model building data suggested that a six-fold single helix provided the best fit to the experimental data, but other models could not be ruled out. Okuyama et al. (1991) suggested that NMR data for the highly hydrated form was consistent with a single helix as high resolution signals are observed for this form. However, a well-defined (crystalline) single helix would be expected to have sufficient segmental rigidity to be undetectable by high resolution NMR, although a disordered (i.e. non-crystalline) single chain would be expected to give high resolution 'solution state' NMR spectra. As the highly hydrated form can be converted by simple heating to the triple helical forms that are well-established from X-ray diffraction, it seems much more likely that the highly hydrated form is also based on a similar triple helical structure but with accompanying single chain segments (Fig. 3) that give rise to both a modest level of crystallinity and a fraction that is observable by high resolution NMR methods. Subsequent evidence from solid state NMR (Pelosi et al., 2006) and fluorescence spectroscopy (Young et al., 2000) both suggest that a loose and/or highly hydrated triple helix structure is more likely than an ordered single helix for both solid and solution highly hydrated forms of (1,3)-β-glucans. However, as the information content of reported diffractograms is modest, a definitive resolution of the conformational basis for the highly hydrated form of (1,3)-β-glucans is unlikely to be achieved unless more crystalline fibres can be produced. The importance of characterizing the highly hydrated form is that conditions that favour production of this form are associated with a range of biological activities of relevance to potential medical applications of (1,3)-β-glucans (Falch et al., 2000; Saito et al., 1991; Sletmoen and Stokke, 2008; Young et al., 2000; see also Chapter 4.5).

### *I.B.3 Solid state $^{13}C$ NMR of (1,3)-β-glucans*

Because of their multiple different biological origins and their occurrence as crystalline allomorphs, (1,3)-β-glucans were one of the first polysaccharide families to be examined in detail by solid state CPMAS $^{13}C$ NMR. Saito et al. (1981) showed that different spectral features were observed depending on biological origin and interpreted these differences in terms of solid state conformation variations. Fyfe et al. (1984) showed that spectra for the linear (1,3)-β-glucans, curdlan and paramylon, particularly the C-3 resonance, were sensitive to moisture content with narrower lines (indicating a more well-defined conformation) being observed at higher moisture. Consistent with this interpretation, the more crystalline paramylon was found to have narrower NMR lines than curdlan (Fyfe et al., 1984). Saito et al. (1989) extended the study of hydration effects to a range of (1,3)-β-glucans with different molecular weights. The C-3 site was found to be the most variable in chemical shift, with three major forms identified that varied depending on biological origin and sample treatment. These were termed 'annealed', 'hydrate' and 'anhydrous', with characteristic chemical shifts for curdlan. This was confirmed and extended by the studies of Pelosi et al. (2006) on a range of *in vitro* synthesized linear (1,3)-β-glucans. With the exception of the 'annealed' form that seems to have similar spectra for all (1,3)-β-glucans analysed, C-3 chemical shifts for 'hydrate' and 'amorphous' forms differ between, for example, curdlan and laminaran (Saito et al., 1989). 'Amorphous' forms (e.g. anhydrous curdlan) have broad spectral features (Fig. 7), consistent with the lack of a well-defined repeating conformation, and are difficult to interpret in detail without the results of molecular orbital calculations to predict the effect of conformational parameters on $^{13}C$ chemical shifts (Durran et al., 1995; Saito et al., 1987; Swalina et al., 2001). There seems to be agreement that the 'annealed' form corresponds to the triple helical structure characterized by X-ray diffraction (Chuah

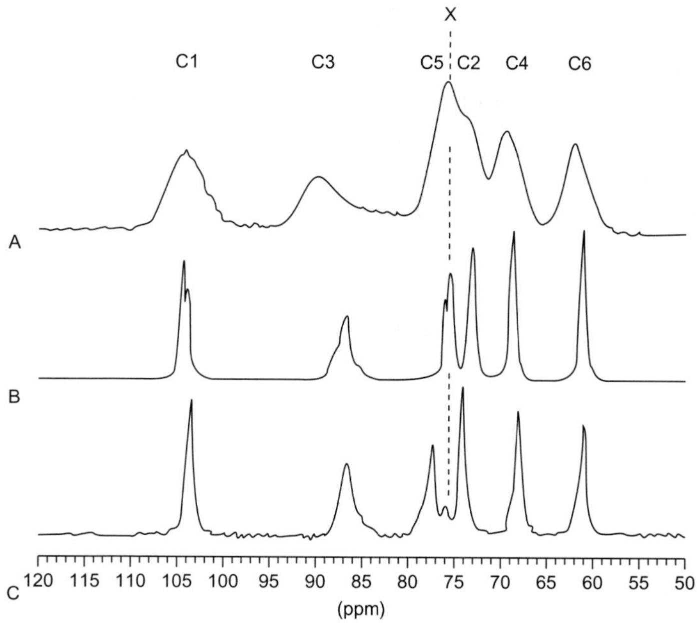


**Fig. 7: Solid state 13C CPMAS NMR spectra for curdlan: (A) anhydrous, (B) hydrated, (C) hydrothermally annealed. The vertical dotted line corresponds to a resonance characteristic of the 'highly hydrated' form whose conformational basis is debated. From Pelosi et al. (2006).**

et al., 1983) as powder diffraction patterns have the same features (Fig. 8A, D). Anhydrous curdlan has limited crystallinity (Fig. 8B) and broad NMR signals (Fig. 7A) but there is continuing debate concerning the conformational basis for the 'hydrate' form that has different diffraction patterns (Fig. 8C) and NMR spectra (Fig. 7B). Saito and co-workers assign this form to a single helix or single chain (Saito et al., 1989; Saito et al., 1990), based on (i) the observation of NMR signals at the characteristic chemical shifts under 'solution state' observation conditions and (ii) the proposal of a single helix form from X-ray diffraction (Okuyama et al., 1991).

However, it is also possible that the 'hydrate' form corresponds to a conformation based on triple helices with either partially open 'frayed' ends (Young et al., 2000) or cyclized forms of triple helices with mismatches between strands producing regions of incomplete helix formation (McIntire and Brant, 1998) (cf. Fig. 3). In both these cases, segmental mobility within the loosely held (1,3)-β-glucan chains can be envisaged to be sufficient for visibility in solution state NMR experiments. This explanation would also be consistent with the enzymic

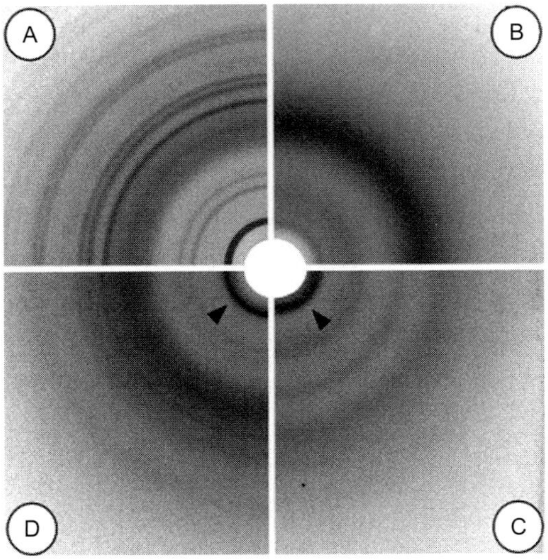


**Fig. 8: X-ray diffraction patterns for (A) paramylon as a triple helical standard, (B) anhydrous curdlan, (C) hydrated curdlan, and (D) hydrothermally annealed curdlan. B,C,D correspond to A,B,C in Fig. 7 respectively. From Pelosi et al. (2006).**

synthesis of (1,3)-β-glucans under similar conditions leading to either the 'annealed' or 'hydrate' forms (Pelosi et al., 2006).

### *I.B.4 Other experimental probes of (1,3)-β-glucan conformation*

To complement X-ray and NMR, a range of microscopy, light-scattering and fluorescence spectroscopy approaches have been utilized. As these studies are either from microscopic samples or in solution state, there is no direct reference that can be made to well-defined solid state structures. Nevertheless, these approaches are very important in providing techniques for studying solutions, gels and other highly hydrated forms.

*I.B.4.a Microscopy* Both electron microscopy and atomic force microscopy (AFM) have provided insights into the behaviour of (1,3)-β-glucans under very dilute conditions where individual species can be observed separately. There has been a particularly intensive study of dilute solutions of schizophyllan, with electron microscopy observations of surface-deposited material showing a range of persistent chains either alone or as loose aggregates (Kitamura et al., 1996; Stokke et al., 1991). The apparent rigidity of the observed molecular species is taken as evidence for the presence of triple helical conformations in conjunction with calorimetry and other supporting information (Kitamura et al., 1996; Stokke et al., 1991). Of particular interest was the observation of cyclized structures, indicative of supercoiling of triple helices and not previously found for any other polysaccharide (Stokke et al., 1991), although known for other stiff multi-stranded helices such as DNA. To observe molecular species by electron microscopy, extensive preparation procedures are needed that might influence the structures finally observed. AFM can, in contrast, be used to directly monitor structures that are, for example, deposited onto a molecularly smooth surface. Using non-contact AFM, both linear and cyclized forms of schizophyllan have been observed (McIntire et al., 1995), allowing detailed study of the inter-conversion between these two states as a function of thermal history (McIntire and Brant, 1998; Sletmoen et al., 2005).

*I.B.4.b Scattering and sedimentation* Information on molecular dimensions under different experimental conditions can be obtained from light scattering, small angle X-ray scattering (SAXS) and sedimentation analysis. Each of these techniques gives data that are consistent with the presence of triple helices as the major conformational form in solutions of (1,3)-β-glucans (Tada et al., 1998; Yanaki et al., 1980) as well as in supercoiled macrocycles (Sletmoen et al., 2005). Whereas very dilute solution conditions are typically used to investigate molecular dimensions from light scattering, SAXS can also be used to provide information on longer range structures such as liquid crystals formed at high concentrations. A study of cinerean, a side-chain-branched (1,3)-β-glucan, with on average, 1 in 3 backbone units substituted (Gawronski et al., 1996) under conditions that lead to liquid crystal formation (high concentrations of rigid rod-like triple helices and a limited molecular weight obtained by sonication), showed not only that the underlying structure was a triple helix of diameter 1.9 nm, but that the phase diagram was quantitatively consistent with classic theories of liquid crystallinity.

*I.B.4.c Fluorescence spectroscopy* One of the most well-established methods for probing (1,3)-β-glucans is the use of dye molecules that show binding specificity (Chapter 2.1). The commercial dye Aniline Blue is used as a histochemical probe for (1,3)-β-glucans (Stone and Clarke, 1992), although the active component is a minor constituent [sodium carbonylbis(4-(phenyleneamino)ben-zenesulphonate) (Fig. 9)] (Evans and Hoyne, 1982; Evans et al., 1984). The fluorescence photophysics of the active component has been characterized (Thistlethwaite et al., 1986), but the exact basis for the specificity of binding to (1,3)-β-glucans remains to be established.

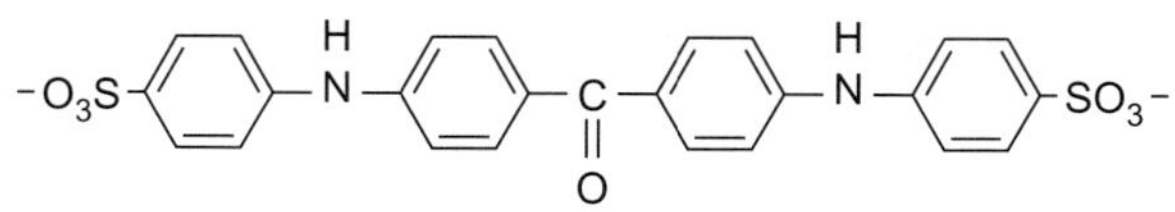


**Fig. 9: The active fluorophore in Aniline Blue.**

An increased fluorescence of dye complexes is observed at pH values that are high enough to disrupt (1,3)-β-glucan triple helices and suggested to correlate with *Limulus* amebocyte activation (Young and Jacobs, 1998). In line with an earlier suggestion (Saito et al., 1991), the relevant conformation was assumed to be a single helix. However, the same authors later modified this conclusion in the light of data from fluorescence resonance energy transfer (FRET) spectroscopy that was not consistent with a single helix conformation, but was consistent with a partially unwound triple helix with single chain reducing ends (Young et al., 2000). These data add to the debate about the conformational form of (1,3)-β-glucans in the highly hydrated form (see above) that is often associated with biological activities. The FRET data are consistent with the model shown in Fig. 3 with single chain (random-coil) forms at the ends as well as in the middle of triplexes. Although the single helix conformation is reported by Young and Jacobs (1998) and Saito et al. (1991), studies on dilute solutions of (1,3)-β-glucans have reported only triple helix and random coil conformations (Sato et al., 1981,1983; Yanaki et al., 1980, 1981,1983, 1985). Despite the uncertainty on the precise molecular details of the interaction with the fluorescent component of Aniline Blue, the reaction has been a mainstay in histological studies of (1,3)-β-glucans, is the basis for a microtitre quantitative assay for (1,3)-β-glucans (Shedletzky et al., 1997), and has been shown to have promise as a method for assessing locations and amounts of (1,3)-β-glucans in foodstuffs (Ko and Lin, 2004).

### *I.B.5 Effects of chemical structure variants on conformation*

There are a number of chemical structure variants of (1,3)-β-glucans (Chapter 2.1), the two most studied of which are the family of side-chain-branched (1,3;1,6)-β-glucans with single (1,6)-linked glucose units attached to the backbone residues, and the (1,3;1,4)-β-glucans in which single (1,3)-β-linkages are interspersed in a mostly (1,4)-β-glucan backbone. Conformational data for these two variants are discussed below.

*I.B.5.a Side-chain-branched (1,3;1,6)-β-glucans* In the solid state, it has been shown (Bluhm et al., 1982; Bluhm and Sarko, 1977; Chuah et al., 1983) that examples of side-chain-branched (1,3)-β-glucans (schizophyllan/scleroglucan/lentinan) adopt triple helical conformations analogous to that characterized for unsubstituted (1,3)-β-glucans, with the single unit side chain adopting a range of local conformations in the hydrated solid state (Chuah et al., 1983). As described above, there is a consistent set of experimental evidence from scattering and sedimentation studies that triple helix structures persist in solution. However, the details of this hydrated solution structure, such as the helix pitch and the presence of an internal cavity within the triple helix, remain to be clarified. One property that side-chain-branched (1,3)-β-glucans have that is not found in non-substituted chains is the ability to complex a range of organic molecules, presumably within the central cavity of the helix (Sletmoen and Stokke, 2008). This implies a widening of the helix for substituted glucans with a concomitant reduction in helical pitch. Molecular dynamics calculations provide support for this, with a decrease in helical pitch predicted as the population of side chains increases (Okobira et al., 2008). Furthermore, substitution at every third backbone glucose (as found in schizophyllan/scleroglucan/lentinan) was found to result in a central cavity of 0.35 nm, whereas no such cavity was predicted for unsubstituted (1,3)-β-glucans (Okobira et al., 2008). Sakurai and Shinkai (Sakurai and Shinkai, 2000, 2001; Sakurai et al., 2001) reported that single-stranded (1,3;1,6)-β-glucans can specifically interact with polynucleotides and form macromolecular complexes at a stoichiometric ratio of two glucan chains to one polynucleotide chain. This is the first clear evidence of the specific interaction between a polysaccharide and a polynucleotide. These (1,3;1,6)-β-glucan/polynucleotide complexes are expected to be applicable to DNA or RNA delivery systems in genetic technologies. A second property that is expected to arise from side-chain-branching is inhibition of the lateral aggregation of triple helices compared with unsubstituted (1,3)-β-glucans. This is because the (1,6)-linked substituents are on the exterior of the triple helix (see Fig. 14) and have a high degree of internal mobility due to the three-bond (1,6)-linkage. Although not normally commented upon, this effect can be inferred from the relative ease with which isolated triple helices are observed in side-chain-branched (1,3)-β-glucans (Kitamura et al., 1996; McIntire and Brant, 1998; Yanaki et al., 1980), and the limited high resolution X-ray diffraction data (indicative of a low tendency for ordered lateral packing of helices). Conversely, unsubstituted (1,3)-β-glucans such as curdlan are normally observed as aggregates of individual triple helices in solution (Tada et al., 1998), and give rise to well-defined X-ray diffraction patterns (e.g. Figs. 5 and 8) demonstrating the regular lateral packing of triple helices into repetitive unit cells.

***I.B.5.b (1,3;1,4)-β-glucans*** (1,3;1,4)-β-Glucans are important components of cell walls of cereal endosperm, and their conformation and molecular organization is of interest since they impact on their food and nutrition properties (see Chapter 4.6). The chemical structures of this family of polysaccharides are described in Chapter 2.1. X-ray diffraction patterns have been reported for (1,3;1,4)-β-glucan (Tvaroska et al., 1983) that are consistent with inter-chain association involving repeating cellotriose units linked by (1,3)-β-linkages. A minimum number of three repeating cellotriosyl units was suggested to be required to form a stable inter-chain association and hence crystallize (Tvaroska et al., 1983). The relatively weak diffraction observed may be due to the imperfect repeat structures in which (1,3)-linkages are inserted in place of usually, but not always, every fourth or fifth (1,4)-linkage (i.e. leading to cellotriosyl or cellotetraosyl units connected by (1,3)-linkages). As might be expected, the predicted solution conformation of (1,3:1,4)-β-glucans (Fig. 10C) is a combination of features due to the pseudo-helical nature of (1,3)-β-glucans (Fig. 10B) and the pseudo-linear nature of (1,4)-β-glucans (Fig. 10A) (Buliga et al., 1986).

As (1,3;1,4)-β-glucans form gels under conditions that would be expected to be due to intermolecular associations, two possible mechanisms can be proposed (see Fig. 22), based on (a) cellulose-like association of extended sequences of (1,4)-β-glucan segments, or (b) multi-stranded helices formed by extended repeat sequences in which a (1,3)-β-linkage occurs

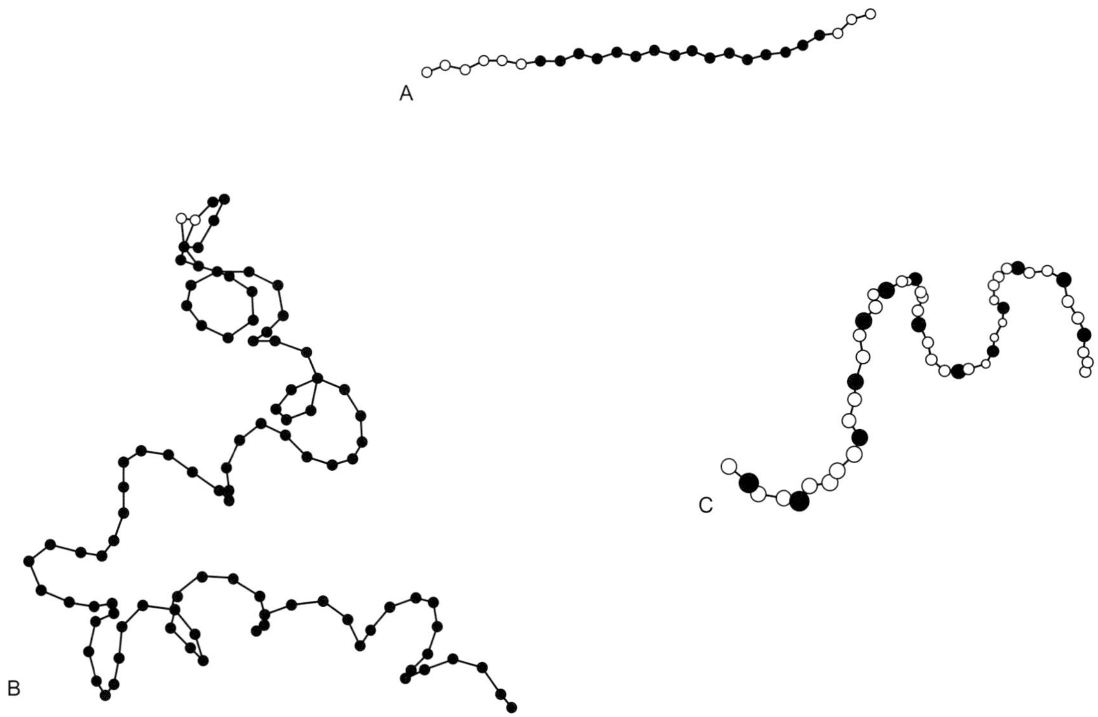


**Fig. 10: Perspective drawing of segments of (A) (1,4)-β-glucan, (B) (1,3)-β-glucan, and (C) (1,3;1,4)-β-glucan chains chosen as representative from a large Monte Carlo sample. Circles represent glycosidic oxygens, and lines are virtual bonds spanning the sugar residues (not shown). Views are perpendicular to the x–y plane of an arbitrary coordinate system. Smaller circles correspond to glycosidic oxygens further away from the viewer. (A) and (B) from Burton and Brant (1983); (C) from Buliga et al. (1986).**

after three (1,4)-β-linkages. Study of gels made from low molecular weight (1,3;1,4)-β-glucan (Morgan et al., 1999) by solid state $^{13}$C NMR (CPMAS) showed the presence of signals not observed for the same polymer in solution. This is good evidence for the presence of a specific conformation that may be related to the inter-chain cross-links that are required for gel formation. The authors speculated that these cross-links may be due to extended runs of consecutive cellotriosyl units (Morgan et al., 1999) as proposed from X-ray diffraction studies (Tvaroska et al., 1983). There was no evidence for the formation of cellulose-like associations between extended sequences of (1,4)-β-oligoglucoside segments, as chemical shifts observed did not match the chemical shifts for cellulose. On the basis of current information, it is therefore plausible that repetitive sequences in (1,3;1,4)-β-glucans can form the basis for inter-chain associations, although details of molecular organization will require greater definition of crystalline structures from X-ray diffraction. One approach to this target is the synthesis of strictly repeating (1,3;1,4)-β-glucan structures. The simplest (1,3;1,4)-β-glucan structure is one in which the two linkages alternate along the chain. Oligomers of this structure containing an average of 12 glucose units have been synthesized using a mutated *Bacillus* glucanase, and shown to form platelet crystals (Faijes et al., 2004). The X-ray diffraction pattern from these crystals suggested a unit cell similar to that of cellulose I, but with the polymer axis dimension being twice as long. This suggests that the alternating (1,3;1,4)-β structure crystallizes with a similar geometry (side-by-side association) to that of cellulose, and unrelated to the geometry of crystalline (1,3)-β-glucans (co-axial triple helix) (Faijes et al., 2004).

### *I.B.6 Conformations of (1,3)-β-glucans in the native state*

The diverse biological origins of (1,3)-β-glucans across eukaryotic and prokaryotic kingdoms (Marchessault and Deslandes, 1979; Stone and Clarke, 1992) encompass examples where the glucan is in a highly crystalline form, e.g. paramylon (Chuah et al., 1983), fibrillar forms with insufficient lateral aggregation to diffract X-rays (e.g. callose, curdlan), non-crystalline (1,3;1,6)-β-glucans such as laminarin (average DP ~30) and the schizophyllan type 1,3;1,6)-β-glucans. However, care must be taken in interpreting reported claims to have studied 'native' forms of (1,3)-β-glucans. This term is frequently used to describe materials that have been obtained after chemical and/or thermal extraction processes from their biological source, so the potential effects of extraction processes need to be taken into account in deciding whether the 'native' state described corresponds with the biological state. For example, commercially available curdlan is often described as 'native'. However, the extraction process involves an alkali treatment to solubilize the polymer followed by drying. Description of 'native' curdlan as granular is misleading as this refers to the extracted and dried form.

(1,3)-β-Glucans from fungal sources are often extracted using a range of solvent conditions (Zhang et al., 2007), reflecting the strength with which the polysaccharide is attached to the cell wall. For *Schizophyllan commune*, some polymers are present as loosely held extracellular materials that can be extracted with hot water, whereas other polymer fractions are held more tightly in the cell wall by ill-defined mechanisms and can be released by hot alkali, e.g. M NaOH, 6h, 80°C (Zhang et al., 2007). A further class of (1,3;1,6)-β-glucans with branch-on-branch structures from yeast and fungal cell walls (see Chapter 2.1 and Chapter 4.3) is covalently linked to chitin and other polysaccharides and proteins and is not released even in concentrated alkali, although treatment with chitinase leads to the release of a soluble polysaccharide. Any description of the 'native' state of (1,3;1,6)-β-glucans therefore needs to take account of these different chemical extraction properties. Solid state $^{13}C$ NMR has been used to examine conformations present in the fruit body of *Grifola frondosa* (Ohno et al., 1986) and a range of edible fungi and yeasts (Ohno et al., 1988) following mild pre-treatment conditions. The results were interpreted to show evidence for the presence of 'curdlan-type' (triple) helices, although further supporting data, from either microscopy or calorimetry for example, would have made the conclusions more certain.

In the plant kingdom, callose is observed microscopically in many tissues, often as a minor cell wall component or one that is produced in response to biological stress. However, there is no detailed data on the molecular conformations present *in vivo*. Similarly, plant (1,3;1,4)-β-glucans have been located microscopically, particularly in the endosperm cell walls of certain cereals, but there is no experimental information available on native conformations.

More detailed information has been obtained from studies of (1,3)-β-glucans synthesized *in vitro* (Pelosi et al., 2003). Synthesis is achieved by isolating the membrane fractions responsible for (1,3)-β-glucan synthesis from sources such as the oomycete *Saprolegnia monoica* or the plant blackberry (*Rubus fruticosa*), extracting the enzyme with detergents, and incubation with UDP-glucose using cellobiose as an activator. Through careful control of isolation, extraction and reaction conditions, sufficient (1,3)-β-glucan can be obtained for spectroscopic, microscopic and diffraction analysis. These studies showed the consistent production of microfibrillar and weakly crystalline forms *in vitro*, although interesting differences of detail in X-ray diffraction diagrams and $^{13}C$ solid state NMR spectra suggested some system and environment sensitivity (Pelosi et al, 2003). In all cases it was proposed that the *in vitro* products were based on triple helices, with subsequent $^{13}C$ solid state NMR analysis (Pelosi et al., 2006) adding further evidence. However, the $^{13}C$ spectra indicate the presence of more than one resonance for C-3 sites, for example, similar to hydrothermally annealed curdlan (Fig. 7), including the resonance assigned to the 'highly hydrated' form of curdlan that is likely to be due to a triple helix but is not yet assigned definitively (see above).

An alternative approach to the *in vitro* synthesis of (1,3)-β-glucans is to use designed mutants of hydrolytic enzymes together with activated (typically fluoridated) substrates to construct the relevant glycosidic bonds, i.e glycosynthases. This approach has been successful in producing crystalline forms of (1,3)-β-glucans (Hrmova et al., 2002) and an alternating (1,3;1,4)-β-glucan (Faijes et al., 2004) using enzymes from barley and *Bacillus,* respectively. The relatively low molecular weight (<5000) of the products from these reactions, together with their precise chemical structure, favours crystallization and provides a useful route to basic information on the molecular basis for polysaccharide interactions.

Before definitive conclusions can be drawn on *in vivo* 'native' conformations, a similar level of detailed analysis is needed for (1,3)-β-glucans in their biological, preferably unextracted, form. This relative lack of direct knowledge on native conformations for (1,3)-β-glucans is in contrast to the detailed and extensive studies that have been performed on solutions and gels of extracted and purified polysaccharides. This is the subject of the next section.

## II Solutions and Gels of (1,3)-β-Glucans

In common with all polymers, (1,3)-β-glucans influence the physical properties of solvated environments. The most obvious effects are on mechanical properties such as viscosity enhancement or gelation. This section will discuss the properties of solutions and gels based on (1,3)-β-glucans, and the proposed molecular basis for the observed effects.

A convenient means of categorizing different forms of solvated polymers such as (1,3)-β-glucans is in terms of dilute solutions, concentrated solutions and gels. As illustrated in Fig. 11, intrinsic (small deformation) mechanical properties are qualitatively and (often) quantitatively different for these three categories.

Characteristic mechanical properties of a gel include an elastic modulus ($G'$) higher than the loss modulus ($G''$) and nearly independent of frequency. Concentrated solutions typically have higher moduli (and viscosity) than dilute solutions and with more elastic character. In structural terms, gels contain a network of polymers connected by specific inter-molecular associations or other long-lasting physical cross-links. Dilute solutions contain polymers with little or no overlap with each other, whereas concentrated solutions contain polymers whose hydrated (hydrodynamic) volumes overlap and therefore entangle with each other resulting in a greatly enhanced level of structuring, but without the long-lasting cross-links required for gel formation. The

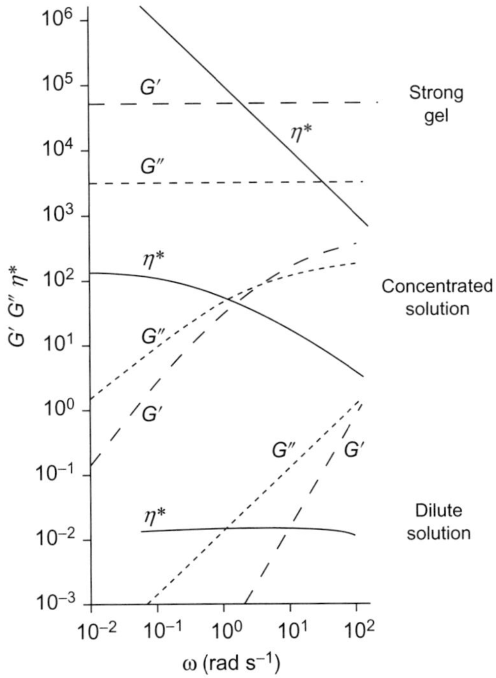


**Fig. 11: Characteristic features of polysaccharide mechanical spectra for a strong gel (2% agar), and concentrated (5% λ carrageenan) and dilute (5% dextran) solutions. Storage modulus $G'$, loss modulus $G''$ (Pa), and dynamic viscosity $\eta^*$ (Pas) are plotted against frequency of small deformation oscillation. Note that 'concentrated' and 'dilute' conditions are defined by reference to intrinsic viscosity (Fig. 12), so 5% λ carrageenan (high intrinsic viscosity) is 'concenrated' whereas 5% dextran (very low intrinsic viscosity) is 'dilute'. From Morris (1984).**

transition between dilute and concentrated solutions is most clearly shown when scaled to the so-called overlap parameter of concentration × intrinsic viscosity, as illustrated in Fig. 12.

## *II.A Dilute Solution Properties*

As discussed in Section I.A.1, the conformation of a polysaccharide chain is dependent on chain stiffness, chain shape and chain association. Dilute solutions of (1,3)-β-glucans have high viscosities due to the stiffness of their chains. The molecular parameters used to understand the functional properties of (1,3)-β-glucans in dilute solution are molar mass (molecular weight), radius of gyration and persistence length. These parameters are determined experimentally by light and SAXS, sedimentation equilibrium, intrinsic viscosity measurements, and AFM (Ross-Murphy, 1994).

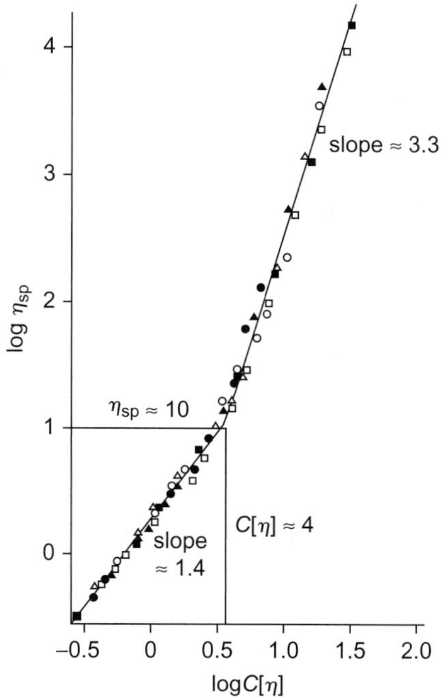


**Fig. 12: Zero shear specific viscosity as a function of the coil overlap parameter, C[η], for a range of random coil polysaccharides. From Morris et al. (1981). The intrinsic viscosity has units of reciprocal concentration, so the product with concentration is dimensionless. At C[η] of ~ 4 the transition from dilute to concentrated solution behaviour is found for a number of random coil polysaccharides (Morris et al., 1981).**

### *II.A.1 Definitions of molecular parameters determined in dilute solution studies*

The mean square end-to-end distance $\langle R^2 \rangle$ for a freely jointed chain is given by $\langle R^2 \rangle = nl^2$, where $n$ is the number of main chain bonds and $l$ is the constant bond length. Flory's characteristic ratio $C_n$ is introduced to take into account the steric hindrance between monomers separated by many bonds $\langle R^2 \rangle = C_n nl^2$ and $\langle R^2 \rangle = C_\infty nl^2$ for long chains ($n \rightarrow \infty$). Flexible polymers have universal properties that are independent of local chemical structure. An equivalent freely jointed chain can be defined so that its mean square end-to-end and the maximum extended length ($R_{max}$ ) are the same with $N_K$ freely jointed effective bonds of length $l_K$. This effective bond length $l_K$ is called the Kuhn's length. The maximum extended length or the length of a polymer chain at its full extension is called contour length ($L$).

Therefore, $L = N_K l_K = nl = R_{max}$. For this equivalent freely jointed chain, the mean square end-to-end distance is given by

$$\langle R^2 \rangle = N_K l_K^{\,2} = C_\infty n l^2$$

Mean square radius of gyration $\langle S^2 \rangle$ is defined by $\langle S^2 \rangle = \sum_0^n s_i^2 / (n+1)$, where $s_i$ is the distance of atom i from the centre of gravity of the chain. $\langle S^2 \rangle$ is used to represent the size of the chain molecule and is related to the square mean end-to-end distance $\langle R^2 \rangle$ for a freely jointed chain $\langle S^2 \rangle = \langle R^2 \rangle / 6 = nl^2/6$. Instead of $\langle S^2 \rangle$, $R_g^2$ is often used; $R_g = \langle S^2 \rangle^{1/2}$ is called root mean square radius of gyration, or simply radius of gyration. Hydrodynamic radius of gyration ($R_h$) is defined by $R_h = kT/6\pi\eta_s D$, where $k$ is Boltzmann constant, $T$ the absolute temperature, $\eta_s$ the viscosity of the solvent, $D$ the diffusion coefficient of the polymer. The ratio $\rho = R_g/R_h$ is often used to discuss the architecture of polymers: it is known that $\rho = 1.50$ for monodisperse random coils in theta condition and 1.78 in good solvent, and that $\rho$ for rigid rod is larger than 2 (Burchard, 1994).

Most (1,3)-β-glucan chains are not as flexible as pullulan, a (1,4;1,6)-α-glucan, which can be well represented by a random coil model, but are rather stiff and therefore described by a worm-like chain (Kratky-Porod chain). The stiffness of chains is usually represented by the persistence length ($q$), which is defined by $q = \kappa l/kT$, where $\kappa$ is the elastic constant of bending. Kuhn's segment length $l_K$ is given by persistence length multiplied by 2; $l_K = 2q$. The contour length $L$ cannot be determined experimentally, while mass per unit length, i.e. the linear mass density $M_L = M/L$, is determined by light-scattering measurements.

In the following sections, solution properties of the family of water-soluble, side-chain-branched (1,3;1,6)-β-glucans, scleroglucan, schizophyllan and lentinan, and their charged derivatives are discussed. The solution properties of the linear (1,3)-β-glucan, curdlan, in DMSO and dilute alkali are also considered.

### *II A.2 Dilute solution properties of linear (1,3)-β-glucans*

Curdlan is a high molecular weight capsular polysaccharide produced by strains of *Agrobacterium* spp., *Bacillus* spp. and *Cellulomonas* spp., and is composed entirely of (1,3)-linked β-glucosyl units (see Chapters 2.1 and 4.1). Curdlan is insoluble in water. This is in contrast to the water solubility of the side-chain-branched schizophyllan, scleroglucan, lentinan, and the *Pleurotus* (1,3;1,6)-β-glucans, which depends on the frequency and distribution of the side-chain branches. Removal of side branches from (1,3;1,6)-β-glucans leads to the precipitation of the linear (1,3)-β-glucan core (Perlin and Taber, 1963).

The insolubility of curdlan is attributed to the existence of extensive intra- and inter-molecular hydrogen-bonded crystalline domains like those found in the (1,4)-β-glucan, cellulose. However, cellulose is not soluble in DMSO or alkaline solvents, whereas curdlan, and other unbranched (1,3)-β-glucans such as paramylon, pachyman and callose are soluble in DMSO and alkaline solvents.

Curdlan fractions with DP $< 25$ are water soluble and take a disordered conformation both in water and alkaline solution, based on the observation of specific rotation, optical rotatory dispersion and complex formation with Congo Red. Solubility is limited for DP $> 25$ fractions but curdlan with DP $> 25$ adopts an ordered structure (Ogawa et al., 1973). The solution conformation of curdlan has been studied using $^{13}$C NMR, where the downfield displacement of $^{13}$C chemical shifts of C-1 and C-3 of curdlan with DP $> 49$ with respect to that with DP $< 14$ and the decrease in peak intensity of C-1,C-3 and C-6 with increasing DP were observed. This indicates that high DP curdlan in solution assumes a different conformation to curdlan with a low DP, and is consistent with results from gels described in a later section (Saito et al., 1978).

Light-scattering and viscosity measurements have been made by Hirano et al. (1979) on nine fractionated samples of curdlan in 1:1 (v/v) water-diluted cadoxen (a cadmium–ammonium complexing reagent) in the range of molecular weight $M_W$ from $6.6\times10^4$ to $6.8\times10^5$ at 25°C. The experimental results of z-average mean-square radius of gyration $< S^2 >_z^{1/2}$ and intrinsic viscosity $[\eta]$ of curdlan were given by the relationships $< S^2 >_z^{1/2} = 3.2\times10^{-2}$ $M_W^{0.53}$ nm, $[\eta]=2.5\times10^{-4} M_W^{0.65}$ cm$^3$g$^{-1}$. Similarly, Nakata et al. (1998) made light-scattering and viscosity measurements on solutions of 18 fractions of curdlan in 0.3 M NaOH. The weight-average molecular weight $M_W$ and $< S^2 >_z^{1/2}$ and $[\eta]$ were given by the following relationships

$$< s^2 >_z^{1/2} = 3.6 \times 10^{-2}\, M_w^{0.53} \text{ nm}$$

$$[\eta] = 7.9 \times 10^{-4}\, M_w^{0.78} \text{ cm}^3\text{g}^{-1}$$

They concluded that the behaviour of curdlan solution cannot be explained by the conventional two-parameter theory of flexible polymer chains by evaluating the Flory universal viscosity constant $\Phi$ $(=[\eta]M/(6<S^2>)^{3/2})$, which is much smaller than the usual value of about $2.5\times10^{21}$ dl/mol • $cm^3$. The values of $< S^2 >_z^{1/2}$ for curdlan are only half of those for cellulose at fixed $M_W$ in 1:1 water-diluted cadoxen, indicating that the curdlan chain is more contracted than cellulose, and that neither curdlan nor cellulose molecules are flexible (Kasai and Harada, 1980).

Ogawa et al. (1972) using optical rotatory dispersion, viscosity and flow birefringence showed that the conformation of curdlan in aqueous NaOH has several transformations interpreted as being from a triple helix to a random coil depending on the concentration of NaOH used, and that the transition from the triple helix occurred at an NaOH concentration between 0.19 and 0.24 M. This transition was confirmed by Saito et al. (1977) by $^{13}C$ NMR analysis.

Steady shear viscosity measurements of curdlan solutions at different NaOH concentrations (Stipanovic and Giammatteo, 1989) showed significant increases in viscosity in the range of 0.05–0.1 M NaOH corresponding to the solvation of triple helices followed by a viscosity reduction above 0.25 M NaOH as the triple molecules are dissociated into single chains of lower molecular weight, which were supported by their $^{13}C$ NMR experiments. Nakata et al. (1998) also found that the intrinsic viscosity of curdlan at 25°C as a function of NaOH concentration decreased sharply at 0.22 M NaOH, and they attributed this to the conformational transition of curdlan chain from helix to a random coil (Fig. 13).

Direct observations of the molecular conformation of curdlan in DMSO and in 5 mM NaOH solution deposited on a mica surface using tapping mode AFM showed that the curdlan is in a single chain in DMSO, but under weakly alkaline conditions formed micelles proposed to be composed of some triple helices (Jin et al., 2006a,b).

Pachyman, essentially a linear (1,3)-β-glucan, is the major constituent of the sclerotia of the fungus *Poria cocos* (see Chapter 2.1). The DP of a native polysaccharide is 255 containing 4 branch points through (1,6)-β-glucosidic linkages (Saito et al., 1963). Three pachyman fractions were examined by static and dynamic light-scattering measurements (Ding et al., 1998). The ratio of the radius of gyration $R_g$ to the hydrodynamic radius $R_h$, $R_g/R_h$, was greater than 2 in DMSO indicating that pachyman chains have an extended conformation in DMSO. Ding et al. (1998) found $R_g/R_h \sim 1.5$ in a 0.05 M NaOH aqueous solution, and concluded that large aggregates with a loose structure are formed in this solvent.

Carboxymethylation of pachyman confers solubility (Wang and Zhang, 2006). Eight fractions of carboxymethylated pachyman were obtained by fractional precipitation, and the relations $[\eta]=1.49\times10^{-2}M_W^{0.75}\,cm^3g^{-1}$, $< S^2 >_z^{1/2}=3.65\times10^{-2}M_W^{0.56}$ nm were obtained (Wang and Zhang, 2006). The data fitted a worm-like chain model, with the molar mass per unit contour length $M_L = 633\,nm^{-1}$, persistence length $q = 5.5\,nm$, and characteristic ratio $C_\infty = 20.2$ in 0.2 M NaCl aqueous solution, respectively. They suggested that the introduction of carboxymethyl groups improved significantly the water solubility and enhanced the stiffness of the chains due to steric hindrance.

### II.A.3 Dilute solution properties of side-chain-branched (1,3;1,6)-β-glucans

The family of side-chain-branched (1,3;1,6)-β-glucans typically produced as extracellular mucilages by fungi vary in the degree of (1,6)-β-linked glucosyl side branching (see Chapter 2.1). In some examples, such as the glucan from *Acremonium*, the ratio of (1,3)- to (1,6)-linkages is as high as 2:3 and the polysaccharide is almost insoluble in water, whereas others, such as schizophyllan, scleroglucan and lentinan, with a ratio of 1:3 are water soluble. Their dilute solution properties are now discussed.

Schizophyllan (SPG) is a side-chain-branched (1,3;1,6)-β-glucan produced by the fungus *Schizophyllan commune* (Kikumoto et al., 1970, 1971) (see Chapter 2.1). The solution conformation of SPG has been studied extensively (Norisuye et al., 1980; Yanaki et al.,1980; Kashiwagi et al., 1981; Sato et al., 1983a, 1983b). SPG dissolves in water as a rod-like triple helix, whereas it disperses in DMSO to a single random coil swollen by the excluded volume effect (the effective raising of local polymer chain concentrations due to the non-available volume occupied by other polymer chains), commonly observed in linear flexible polymers (Yamakawa, 1971). Both the weight-average molar mass ratio $M_w$(water)/$M_w$(DMSO) and the z-average molar mass ratio $M_z$(water)/$M_z$(DMSO) were found to be close to 3 by sedimentation equilibrium and light scattering (Norisuye et al., 1980). In averaging molar mass, z-average molar mass is influenced by high molar mass species even more than the weight-average molar mass. The molar mass per unit cylinder length ($M_L$) and the diameter (*d*) in water were determined as $2150\,nm^{-1}$ and 2.6 nm, respectively, based on the analysis using Yamakawa's theory (Yamakawa, 1971). This $M_L$ gives the pitch of the triple helix per (1,3)-β-glucosyl residue a value of 0.30 nm, in close agreement with the reported pitches in the crystalline region in lentinan (see Section 2.2.1), a similar side-chain-branched (1,3;1,6)-β-glucan (Yanaki et al.,1980). Light-scattering and viscosity measurements of SPG in dilute solution were performed in 0.01 M NaOH which completely dissolves SPG molecules without destroying the triple helical structure and avoiding the problem of removing microgels (Kashiwagi et al., 1981). The persistence length and the pitch per glucose residue in a SPG triple helix were determined as 180 nm and 0.30 nm, respectively, essentially in agreement with the values derived in water.

Using different viscometers appropriate to the molecular weight of the sample, Yanaki et al. (1980) obtained the intrinsic viscosity and the limiting sedimentation coefficients for the schizophyllan samples. The results showed that the SPG triple helix is almost perfectly rigid up to $M_w = 5\times10^5$ g/mol but acquires flexibility at higher molecular weights (Yanaki et al., 1980). In water–DMSO the triple helices of schizophyllan dissociate almost completely into single chains when the DMSO content increases above 87% (Sato et al., 1983a, Kitamura and Kuge, 1989). The conformational transition from triple helices to single chains can also be induced by heating SPG aqueous solutions above 135°C or by the addition of NaOH to >0.2 mol/L at room temperature (Yanaki et al., 1985).

SPG solutions show two highly cooperative conformational transitions on heating, one at ~6°C and the other at ~135°C (Kitamura and Kuge, 1989; Yanaki et al., 1985; Kitamura et al., 1990, 1996; Yoshiba et al., 2002, 2003). The lower temperature order–disorder transition of SPG at 6~7°C in water is shifted to 17~18°C in heavy water ($D_2O$) (Itou et al., 1986; Hayashi et al., 2002). The higher temperature transition is thermo-irreversible, and is attributed to the dissociation of SPG triple helices into single chains (random coils) (Kitamura and Kuge, 1989; Yanaki et al., 1985; Kitamura et al., 1990, 1996). In contrast, the lower temperature transition is thermo-reversible (Itou et al., 1986, 1987), and its origin is still controversial. Whereas some groups (Bot et al., 2001; Tako, 1996) believe that it is associated with the dissociation of aggregates of SPG triple helices, other groups attributed it to an intramolecular transition between two types of SPG triple helices, namely triple helix I and triple helix II (Itou et al., 1986, 1987; Kitamura and Kuge, 1989; Kitamura et al., 1996). The lower temperature version of SPG helices, known as triple helix I, is more ordered than the higher temperature version, triple helix II. Triple helices I and II differ in the manner of the interaction of the β-glucosyl side groups with water molecules. Itou et al. (1987) proposed a molecular model for triple helix I, in which the side groups carried on the same helical strand are connected together, mediated by water molecules, into a helical structure (Fig. 14). Thus, three such side-group–water helices are formed, and play the role of an outer shell, giving rise to the increases in SPG chain stiffness and diameter. If the temperature is raised, the ordered shell structure melts, and triple helix I transforms into triple helix II.

Based on dielectric relaxation and heat capacity measurements, Yoshiba et al. (2003, 2004) found that water molecules in the triple helix I state exist in different structures, classified as bound water, structured water, loosely structured water and free water from the surface of the helix core in increasing order of mobility. Further, they calculated the thickness and the critical weight fraction of these four types of water layers (Yoshiba et al., 2003) (Fig. 15). The helix I–helix II transition exerts a notable influence on the viscoelastic properties of SPG

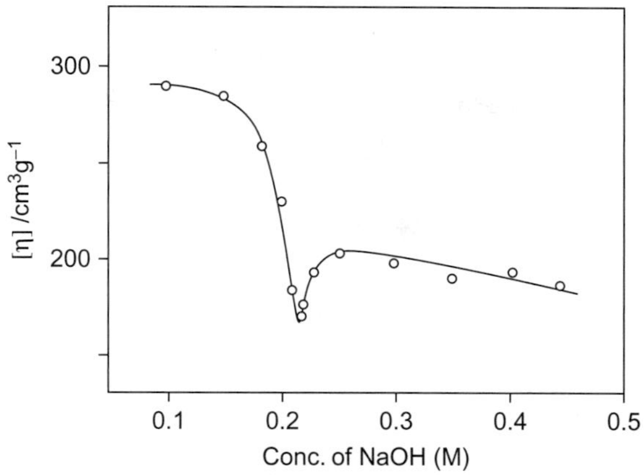


**Fig. 13: Intrinsic viscosity of curdlan ($M_w$=4.5×10$^5$ in DMSO-LiCl solvent) as a function of concentration of NaOH at 25°C (Nakata et al., 1998). At low concentration of NaOH, curdlan is in a triple helix. From Nakata et al. (1998).**

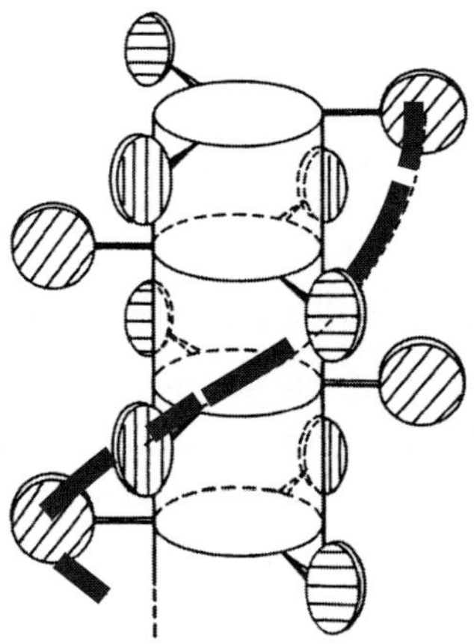

**Fig. 14: Schematic representation of the schizophyllan triple helix. Hatched disks represent side groups of schizophyllan; the central cylinder, the helix core; and the disks along with intervening water molecules form a helical chain as indicated by the dashed line. From Itou et al. (1987).**

solutions (Bot et al., 2001), and in liquid crystalline phase on the phase behaviors of SPG (Yoshiba et al., 2003).

Scleroglucan is a side-chain-branched (1,3;1,6)-β-glucan secreted by the plant fungal pathogen *Sclerotium glucanicum* and *S. rolfsii* with the same degree of side-chain branching as schizophyllan (see Chapter 2.1). In aqueous solution scleroglucan is in a triple helical conformation

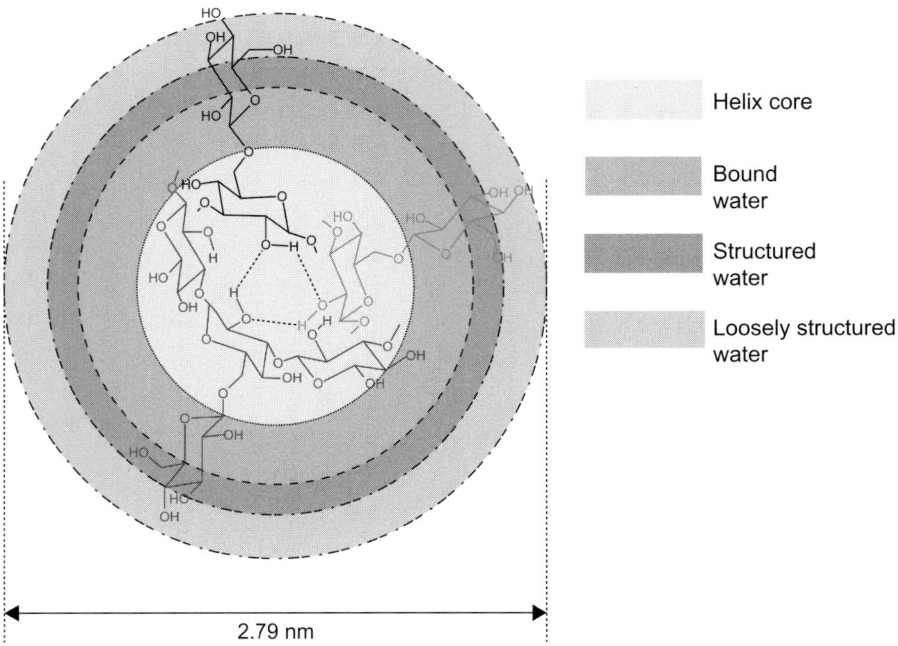


**Fig. 15: Cross-sectional view of the schizophyllan triple helix I in water. The repeat unit of schizophyllan consists of three glucose residues and one side-chain glucose residue (only the red chain residues are shown in full). The circles represent the proposed boundaries of four water structure layers, with approximate diameters and critical weight fractions in parentheses: helix core–bound water (1.68 nm, 1.0), bound water–structured water (2.05 nm, 0.769), structured water–loosely structured water (2.31 nm, 0.644), and loosely structured water–free water (2.79 nm, 0.478). Each cylindrical shell is marked in a specific colour. From Yoshiba et al. (2003). The colour specifications refer to colours in panels.**

(Yanaki et al., 1981, 1983). When dissolved in DMSO or in NaOH (>0.2 M) solution, the conformation changes to a random coil. The intrinsic viscosity of scleroglucan in a water/DMSO mixture increased gradually with decreasing DMSO content up to DMSO content 0.88, and then steeply increased with decreasing DMSO content at 0.85 (Yanaki et al., 1981). The maximum intrinsic viscosity of scleroglucan as a function of DMSO content is around 0.50 DMSO in water/DMSO mixture, but no reasonable interpretation for this maximum has been provided. Yanaki and Norisuye (1983) found that the molar mass ratio $M_w$(0.01 M NaOH)/$M_w$(DMSO) was close to 3 below $M_w(\text{DMSO}) = 2\times10^5$ and an aggregate higher than a trimer for $M_w$(DMSO) above $2\times10^5$, and by light scattering and ultracentrifugation that the number of chains in the aggregate increased with increasing $M_w$(DMSO). They also found that the relationship between the radius of gyration and the molar mass for scleroglucan coincided with that for schizophyllan both in 0.01 M NaOH and in DMSO, and concluded that the scleroglucan trimer approximates to a rigid rod for $M_w$ (0.01 M NaOH) $< 8 \times 10^5$ and semiflexible at $M_w$ (0.01 M NaOH) $> 8 \times 10^5$ (i.e. composed of connected rigid rod segments of $M_w < 8{\times}10^5$). Although scleroglucan was thought to be dissolved as an unperturbed flexible chain in DMSO (Yanaki et al., 1981), the same authors later found that the second virial coefficient of scleroglucan in DMSO is ~10 times larger and concluded that the scleroglucan random coil is perturbed by the excluded volume effect in much the same way as the schizophyllan random coil (Yanaki and Norisuye, 1983). The double logarithmic plot of the intrinsic viscosity versus molar mass, i.e. Mark-Houwink-Sakurada plot, for scleroglucan coincided well with that for schizophyllan (Norisuye et al., 1980; Kashiwagi et al., 1981). The slope of 1.7 was found for $M_w$ (0.01 M NaOH)$<3{\times}10^5$, and in DMSO the slope was 0.69 indicating that both scleroglucan and SPG single chains in this solvent are in the random coil conformation perturbed by the excluded volume effect. The limiting sedimentation coefficient varies linearly with log $M_w$ in 0.01 N NaOH, which also suggests that the scleroglucan trimer is a rigid rod. Bluhm et al. (1982) concluded from X-ray and conformational studies that scleroglucan in the crystalline state has a triple helical structure with a pitch per residue of 0.30 nm, which agrees fairly well with the values obtained in the solution study of Yanaki and Norisuye (1983). On this basis, Yanaki and Norisuye (1983) suggested that the triple helical structure of scleroglucan in the crystalline state is maintained in dilute aqueous NaOH.

Yanaki and Norisuye (1983) further examined the intrinsic viscosity as a function of NaOH concentration, and found a sharp decrease at NaOH concentrations ranging from 0.05 M to 0.1 M, and attributed this to the dissociation of the triple helix into single chains. They found a further decrease in intrinsic viscosity at NaOH concentrations ranging from 0.01 M to 0.05 M for higher molar mass scleroglucan samples ($M_w$(DMSO) $= 5.18 \times 10^5$ and $3.72 \times 10^5$ (Fig. 16), and attributed this to the dissociation of a higher aggregate into trimers. These higher molar mass scleroglucans therefore showed a two-step decrease in intrinsic viscosity as a function of NaOH concentration, dissociation of higher aggregates into trimers and the subsequent sharp decrease caused by the breaking of trimers into single chains. This is in contrast to the conclusion of Bluhm et al. (1982), from a study of a single sample of scleroglucan, that the decrease in the NaOH concentration range from 0.01 M to 0.05 M is due to the breaking of trimers into single chains, rather than the dissociation of higher aggregates into trimers.

Although both scleroglucan and SPG have the same average chemical structure, their solubility behaviour show differences in 0.01 M NaOH when the $M_w$ (DMSO) $> 2 \times 10^5$(Yanaki and Norisuye, 1983). In the higher molar mass region, triple helices of SPG remained intact, while those of scleroglucan tended to associate further with one another to form higher aggregates.

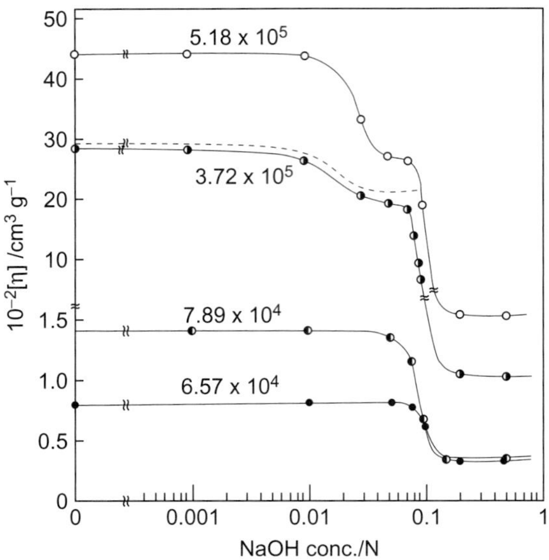


**Fig. 16: Dependence of intrinsic viscosity [$\eta$] on the concentration of NaOH for scleroglucan samples with different molecular weights in aqueous NaOH at 25 °C, by Yanaki and Norisuye (1983). Numbers beside each curve stand for the molecular weight in DMSO. The broken line indicates the data of Bluhm et al. (1982).**

Yanaki and Norisuye (1983) noted the difference in the native states of these glucans: native SPG produced by *S. commune* in a culture medium separates spontaneously from the mycelium and migrates freely into the liquid phase (Kikumoto et al., 1970), whereas scleroglucan produced by *S. rolfsii* adheres to the mycelium as a gel-like aggregate and cannot be dispersed in the liquid phase unless the culture medium is heated and homogenized.

Lentinan isolated from hot water extracts of fruiting bodies of the mushroom *Lentinus edodes* is another side-chain-branched (1,3;1,6)-β-glucan (Saito et al, 1979) with the same degree of side-chain branching as schizophyllan and scleroglucan. The solution properties of lentinan in aqueous NaCl and in water/DMSO have been studied by light scattering, viscosity and NMR using fractionated samples prepared by ultrasonic degradation (Zhang et al., 2001, 2002). It was concluded that lentinan molecules are in a triple helical conformation in aqueous 0.2 M NaCl solution and water/DMSO mixture with over 20% (w/w) water content, whereas they are single flexible chains in DMSO. The C3 signal of C1, C2 and C4 in $^{13}$C NMR spectra for the (1,3)-β-glucosyl backbone in water/DMSO mixtures with different DMSO contents decreased with increasing water content, and disappeared in $D_2O$/DMSO (70/30). The C6 peak profiles in $D_2O$/DMSO mixtures with DMSO contents of 0.85 and 0.80 were asymmetrical and broad, suggesting the coexistence of triple helix and single flexible chains which shifted downfield with increasing water content. The loss of $^{13}$C NMR peak areas and downfield displacement of the carbon atom signals in the backbone with increasing water content indicate the immobilization of the main chain by binding through intra- and inter-molecular hydrogen bonds.

The molar mass ratio $M_w$(0.2 M NaCl)/$M_w$(DMSO) was found to be close to 3 by light scattering, and both the radius of gyration and the intrinsic viscosity were far greater in 0.2 M NaCl than in DMSO (Zhang et al., 2001, 2002). The peak of light-scattering intensity at an angle 90° and differential refractometry at 632.8 nm at 25°C for lentinan in 0.2 M NaCl were unique, indicating that lentinan exists as triple helical chains without obvious lower molar mass single chains. Both these peaks showed a shoulder corresponding to dissociated chains in addition to a main peak of triple helical molecules when lentinan was dissolved in a water/DMSO mixture with a DMSO content of 0.82. These observations support the conclusions from NMR measurements. Further support comes from the intrinsic viscosity which was determined at 25°C in a water/DMSO mixture as a function of DMSO content. The intrinsic viscosity decreases sharply around a DMSO content of 0.87 with increasing DMSO content as was found for SPG (Sato et al., 1983). Zhang et al. (2002) also found the broad maximum of the intrinsic viscosity in a water/DMSO mixture as a function of DMSO content at around DMSO content 0.3, which is similar to the maximum found for scleroglucan (Yanaki et al., 1981). When the DMSO content decreased from 1 to 0.7, the intrinsic viscosity hardly changed, indicating that the shape and size of the single flexible chains remained the same as in pure DMSO. After dissolving lentinan in DMSO and then diluting to the desired DMSO content by adding water, the intrinsic viscosity showed significantly lower values than those observed in the water/DMSO mixture. When the DMSO content decreased below 0.6, the intrinsic viscosity of lentinan increased slowly, indicating renaturation with increasing water content. Thus the helix–coil transition of lentinan in water/DMSO is irreversible, as was found for schizophyllan (Kitamura and Kuge, 1989; Yanaki et al., 1985; Kitamura et al., 1990, 1996).

Surenjav et al. (2006) studied four protein-bound lentinan fractions with protein contents of 5.8%, 5.5%, 4.6%, 15.2% (w/w), and determined the molar mass as $1.48 \times 10^6$, $1.58 \times 10^6$, $1.52 \times 10^6$, $1.68 \times 10^6$ respectively in 0.2 M NaCl aqueous solution, and found that the molar mass decreased to one third in DMSO. By combining these data with the relationship between the molar mass and the radius of gyration, they concluded that these polysaccharides are triple helical chains in 0.2 M NaCl and are single flexible chains in DMSO.

In summary, the solubility of side-chain-branched (1,3;1,6)-β-glucan depends on the frequency and distribution of the side-chain branches. The (1,3;1,6)-β-glucan chains in scleroglucan, schizophyllan and lentinan exist as single random coils in DMSO or in concentrated alkaline solvents, and are in the triple helical conformation in water. Aggregates (higher order structures) of triple helices are found for scleroglucan produced by *S. rolfsii* with high molar mass.

### *II.A.4 Solution properties of charged side-chain-branched (1,3;1,6)-β-glucans*

Whilst scleroglucan, schizophyllan and lentinan are nonionic polysaccharides, they can be converted into polyelectrolytes when carboxylate groups are introduced onto the β-glucosyl side residues by periodate-chlorite oxidation (Crescenzi et al., 1983; Gamini et al., 1984). Thus, the triple-stranded helix of scleroglucan was disentangled to a single chain by the oxidation of β-glucosyl side residues, and became more flexible at high pH (Coviello et al., 1995). It is commonly observed that the intrinsic viscosity of a normal polyelectrolyte decreases monotonically with ionic strength in NaCl aqueous solution, but the intrinsic viscosity of the oxidized scleroglucan (sclerox) showed a minimum in aqueous 0.4 M NaCl and then abruptly increased with a further increase in salt concentration (Coviello et al., 1995). It was concluded that the triple strand helical conformation of sclerox was partly recovered by shielding the charge on the side chains (Coviello et al., 1998).

Controlled oxidation of scleroglucan with sodium periodate gave aldehyde derivatives (scleraldehyde) with a low degree of oxidation (10% and 20%). SAXS studies of scleraldehyde with a high degree of oxidation (>50%) suggested that it disentangled into single chains (Coviello et al., 1998), but retained mainly the conformation of the natural polysaccharide at a low percentage of aldehyde groups (<20%). Thus, the system is proposed to adopt a different conformation where the aldehyde groups are present (Maeda et al., 2001). Subsequent electron microscopy and AFM studies of both aldehyde and carboxylate oxidation products of scleroglucan suggested that even complete periodate modification of scleroglucan side chains is not sufficient to induce dissociation of the triple helix structure, whereas further oxidation of the side chains to carboxylic groups causes dissociation when the degree of substitution is above 0.6. The apparent discrepancy with SAXS data on the periodate-oxidized aldehyde form remains to be resolved.

De Nooy et al. (2000) have modified scleroglucan by selective chemical oxidation of primary (O-6) hydroxyl groups, and found an ordered conformation for modified sclecroglucan with lower DS, whereas the samples with higher DS were found to be in a random coil state, judging from intrinsic viscosity and optical rotation measurements.

The side-chain-branched (1,3;1,6)-β-glucan from sclerotia of *Pleurotus tuber-regium* has a structure similar to schizophyllan, sclerotan and lentinan with, on average, 1 in 3 backbone units substituted (see Chapter 2.1). Zhang et al. (2003a; Zhang et al., 2004, 2006) examined the conformations of the carboxymethyl derivatives of water-soluble fractions by light scattering and viscometry in phosphate buffer solution. The exponent of the Mark-Houwink-Sakurada ( MHS) plot = 0.78, $q = 9.6\,\text{nm}$, $M_L = 790\,\text{nm}^{-1}$, indicating that the chains are more extended than the native polysaccharide due to the enhancement of steric hindrance. The sulfated β-glucan from *P. tuber-regium* is more water soluble and stiffer than the native polysaccharide (the exponent of the MHS plot = 0.70, $q = 8.5\,\text{nm}$, $M_L = 990\,\text{nm}^{-1}$ (Zhang et al., 2003b). Thus, carboxymethylation and sulfation of side-chain-branched (1,3;1,6)-β-glucans induces a conformational transition from a triple helix to a random coil and leads to an increase in water solubility and coil expansion due to steric hindrance.

### *II.A.5 Solution properties of cereal (1,3;1,4)-β-glucans*

Cereal β-glucans extracted from oats, barley and wheat consist of (1,3)-β-linked cellotriosyl and cellotetraosyl units, with lower amounts of consecutive (1,4)-β-linkages up to 14 glucosyl units in length (Cui, 2000; Lazaridou et al., 2003, 2004; Böhm and Kulicke, 1999a) (see Chapter 2.1). Cui et al. (2000) suggested that the predominant molar portion of trisaccharide from wheat (1,3;1,4)-β-glucan leads to a more regular structure, and hence to its greater gelling ability and poorer solubility in water compared with other cereal β-glucans.

The molecular weight of oat (1,3;1,4)-β-glucan ranges from $6.5\times10^4$ to $3\times10^6$ while that of barley is from $1.5\times10^5$ to $2.5\times10^6$ (Lazaridou et al, 2003, 2004). Wang et al. (2003) obtained seven (1,3;1,4)-β-glucan fractions from oat and barley with narrow molecular weight distribution using gradient precipitation with ammonium sulfate, but found no differences in oligosaccharide pattern (DP2–9) derived from each fraction and the parent sample upon hydrolysis with a (1,3;1,4)-β-glucan endohydrolase. In particular, the DP3/DP4 ratio remained constant, indicating no fractionation based on structural differences had taken place.

Grimm et al. (1995) examined the solution properties of (1,3;1,4)-β-glucans isolated from beer by combined static and dynamic light scattering and viscometry. They found that full molecular dissolution was obtained only in cuoxan, a copper-complexing solvent known to dissolve the (1,4)-β-glucan, cellulose, and proposed a fringed micelle model for this cereal glucan in aqueous solvents. They found the minimum aggregation in 5% aqueous maltose solution and attributed it to preferential binding of maltose which partly breaks up the aggregated clusters. Gomez et al. (1997a,b) examined the molecular weight and molecular size, and conformation by the size-exclusion chromatography with multi-angle light scattering/refractive index technique (SEC-MALS/RI), and intrinsic viscosity using barley (1,3;1,4)-β-glucan samples with different molecular weights obtained by enzymatic degradation. They obtained the exponent $n = 0.35$ in $< S^2 >_w^{1/2} \sim M_w^n$ which is much smaller than expected for stiff chains; for example, $n$=0.59 obtained by Varum et al. (1992) for oat (1,3;1,4)-β-glucan with narrow molecular weight distribution; but far larger than $n$=0.22 (Grimm et al., 1995) obtained for barley (1,3;1,4)-β-glucan isolated from beer with a tendency for concentration-dependent aggregation. Gomez et al. (1997a) did not get consistent results for SEC-MALS measurement and batch-mode light scattering in the Zimm plot analysis, and they suggested that this anomaly was due to the formation of labile molecular aggregates previously observed (Varum et al.,1992). Further, in Gomez et al. (1997b), examining the molecular parameters by Stockmayer-Fixman plot on the assumption of a coil conformation, and comparing with previous results from Varum et al. (1992) and Buliga et al. (1986), the characteristic ratio $C_\infty = 7$ was considered too small for stiff (1,3;1,4)-β-glucan chains. Using a worm-like cylinder model and a Bohdanecky plot, they obtained a persistence length $q = 3.47$ nm, cross-section diameter $d = 0.45$ nm and a characteristic ratio $C_\infty = 13.0$.

The Mark-Houwink-Sakurada plot of cereal (1,3;1,4)-β-glucan has been reported by many research groups; exponents reported are shown in Table 1 together with data reported for other (1,3)-β-glucans.

Worm-like chain parameters have also been reported by several research groups for a variety of (1,3)-β-glucan solutions. Results are shown in Table 2.

### II.B Concentrated Solution Behaviour

The concentration dependence of viscosity of flexible polymers is usually represented by the relation between zero shear specific viscosity and the coil-overlap parameter (Fig. 12). Most polymer solutions are non-Newtonian fluids, i.e. the viscosity depends on the shear rate. In most polysaccharide solutions, the viscosity decreases with increasing shear rate, which is called a shear thinning behaviour. At sufficiently low shear rate, the viscosity does not depend on the shear rate and shows a Newtonian plateau, so the viscosity observed at the shear rate extrapolated to zero can be obtained, and it is called simply a zero shear viscosity ($\eta_0$). The double logarithmic plot of the zero shear specific viscosity ($\eta_{sp,0}$) of polymer solutions and polymer concentration shows two straight lines. The slope of these straight lines is smaller at lower concentrations than at higher concentrations, and the cross-over point of these straight lines shifts to lower concentrations with increasing molecular weight of a polymer. Then, instead of the concentration, the coil-overlap parameter, the product ($C[\eta]$) of the concentration (C) with the intrinsic viscosity $[\eta]$, is used as abscissa, as shown in Fig. 12. Irrespective of chain flexibility, the slope at lower concentrations is reported as ca.1 and the

**Table 1: Molecular weight data and Mark – Houwink – Sakurada exponents (α) for (1,3)-β-glucans**

| (1,3)-β-glucan | α value | Mw range (g mol$^{-1}$) | Method* | Solvent | Temp. (°C) | Reference |
|---|---|---|---|---|---|---|
| scleroglucan | 1.7 | $M_w<3\times10^5$ | | 0.01 M NaOH | | Yanaki et al. (1981) |
| scleroglucan | 0.69 | | | DMSO | | Yanaki et al. (1981) |
| lentinan | 1.1 | 97.5–160×10$^4$ | SEC-MALLS | Water | 25 | Y. Zhang et al. (2007) |
| lentinan | 0.72 | 27.1–52×10$^4$ | SEC-MALLS | DMSO | 25 | Y. Zhang et al. (2007) |
| curdlan | 0.78 | 6.6×10$^4$ | | 0.3 M NaOH | | Nakata (1998) |
| curdlan | 0.65 | 6.8×10$^5$ | LS + Vis | 1:1water-cadoxen | 25 | Hirano (1979) |
| oat (1,3:1,4)-β-glucan | 0.75 | 6.3–33×10$^4$ | SEC-MALLS | 1 M LiI | | Varum et al. (1991) |
| beer (1,3;1,4)-β-glucan | 0.72 | 2.8–12×10$^6$ | | maltose solutions | | Grimm et al. (1995) |
| barley (1,3;1,4)-β-glucan | 0.71 | 9.2–573×10$^3$ | LS + Vis | water | | Gomez et al. (1997b) |
| barley (1,3;1,4)-β-glucan | 0.71 | 4.0–37.5×10$^4$ | LS + Vis | water | | Böhm and Kulicke (1999a) |
| barley (1,3;1,4)-β-glucan | 0.65 | | SLS + CV | water | 25 | Wang et al. (2006) |
| oat (1,3;1,4)-β-glucan | 0.62 | 2.55–13.8×10$^5$ | SEC-MALLS + CV | water | 25 | Wang et al. (2006) |

*SEC-MALLS = size exclusion chromatography with multiple angle laser light scattering detection; LS = light scattering; Vis = intrinsic viscosity; SLS = static light scattering; CV = capillary viscometry.

slope at higher concentrations as 3.3 for many polysaccharides, but some other slopes are also reported and tabulated in Lapasin and Pricl (1999). When the zero shear specific viscosity is represented by a power law ($\eta_{sp,0} \sim C^n$), the exponent ($n$) takes different values below and above $C^*$. The exponent $n$ for most polysaccharide solutions ranges from 1.1 to 1.6 below $C^*$, and from 1.9 to 5.6 above $C^*$ (tabulated in Lapasin and Pricl, 1999).

For many flexible polymer solutions, the steady shear viscosity as a function of shear rate and the complex viscosity as a function of frequency are found to coincide; this is called the Cox-Merx rule. Solutions of guar gum and other flexible polysaccharides are shown to obey the Cox-Merz rule, but solutions of stiff chains like xanthan and most (1,3)-β-glucans do not obey the Cox-Merz rule.

**Table 2: Worm-like chain parameters for (1,3)-β-glucans and their derivatives**

| (1,3)-β-glucan | $M_L$ nm$^{-1}$ | q nm | Method* | Solvent# | Temp. | Reference |
|---|---|---|---|---|---|---|
| schizophyllan | 2150 | 180 | LS + Vis | Water | 25 | Norisuye et al. (1980) |
| lentinan | 2180 +/−100 | 120 | SEC-MALLS + Vis | 0.5 M NaCl | 25 | Zhang, Zhang, Zhou et al. (2001) |
| carboxymethylated (1,3;1,6)-β-glucan) | 790 | 9.6 | SEC-MALLS + Vis | PBS | 25 | Zhang, Zhang and Cheung (2003) |
| sulfated (1,3;1,6)-β-glucan | 990 | 8.5 | SEC-MALLS + Vis | Water | 25 | Zhang, Zhang, Wang and Cheung (2003) |
| carboxymethylated (1,3)-β-glucan ) | 633 | 5.5 | SEC-MALLS + Vis | 0.2 M NaCl | 25 | Wang and Zhang (2006) |
| barley (1,3;1,4)-β-glucan | 633 | 3.5–3.8 | SLS + Vis | Water | 25 | Gomez et al. (1997a, b) |
| beer | 4930 | 83 | SLS + DLS + Vis | Water | 20 | Grimm et al. (1995) |
| (1,3;1,4)-β-glucan | 3500 | 72 | | 2 M GHCl | 20 | |

*LS = light scattering; Vis = intrinsic viscosity; SEC-MALLS = size exclusion chromatography with multiple angle laser light scattering detection; SLS = static light scattering; DLS = dynamic light scattering.

#PBS = Phosphate buffer solution; GHC l= Guanidinium hydrochloride.

### *II.B.1 Behaviour of concentrated solutions of curdlan*

When curdlan is dissolved in DMSO, the solution shows a typical mechanical spectrum for a concentrated solution of flexible polymer chains (Fig. 11); $G''$ predominates at lower frequencies because there is enough time for molecular chains to disentangle, whilst $G'$ is dominant at higher frequencies because there is not enough time for chains to disentangle. The storage and loss moduli of aqueous suspensions of curdlan as a function of frequency at 40°C and at 70°C show similar behaviour (Hirashima et al., 1997; Nishinari et al., 1988a); even before gelation at 40°C, $G'$is larger than $G''$ at all the frequencies and hardly dependent on the frequency – typical behaviour of a structured liquid (Fig. 17).

An aqueous curdlan suspension in a test tube flows when the tube is tilted. At 70°C, both moduli increase and show plateau values, and the loss tangent is far smaller than that for the suspension at 40°C. This is a typical behaviour of elastic gels. Tada et al. (1997, 1998, 1999) studied the structure of molecular association of curdlan in the diluted regime in aqueous alkaline solutions by rheological, static light-scattering and SAXS measurements. They found

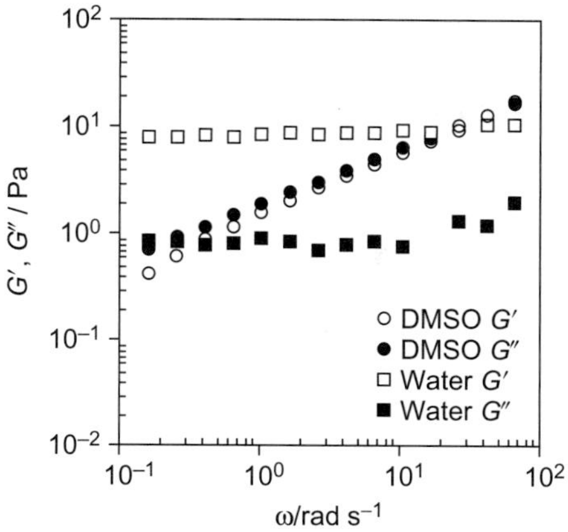


**Fig. 17: Mechanical spectra of 2% (w/w) curdlan in water and in DMSO at 40°C (modified from Hirashima et al., 1997).**

that the degree of association of curdlan molecules increased with decreasing alkaline concentration and the viscoelastic properties depend strongly on the alkaline concentration, i.e., concentrated curdlan solutions show almost a Newtonian flow at high alkali concentrations but a solid-like behaviour at low alkali concentrations. Curdlan in DMSO at above 1% (w/w) behaves like a concentrated polymer solution (Watase and Nishinari, 1994), similar to its behaviour in high concentrations of aqueous NaOH above 0.05 M (Tada et al., 1997).

### *II.B.2 Behaviour of concentrated solutions of side-chain-branched (1,3;1,6)-β-glucans*

*II.B.2.a Concentrated solution properties of side-chain-branched (1,3;1,6)-β-glucans in ordered conformations* The steady shear viscosity of a polymer solution as a function of shear rate shows a Newtonian plateau region at lower shear rates, i.e. the viscosity is independent of the shear rate. However, it begins to decrease at a certain critical shear rate. The phenomenon of the decrease of viscosity with increasing shear rate is called shear thinning, and is more conspicuous with increasing stiffness of the polymer chains.

Generally, the critical shear rate at which the viscosity begins to decrease shifts to low shear rates with increasing concentration. Most side-chain-branched (1,3;1,6)-β-glucans have stiff chains, and therefore the Newtonian region is observed only at very low concentrations and very low shear rates. Enomoto et al. (1985) measured the shear rate dependence of aqueous solutions of schizophyllan from 0.000257 to 0.00411 g/cm$^3$ at the shear rate range $10^{-4}$–$10\,s^{-1}$ using a Zimm-Crothers type rotational viscometer appropriate to the rigid molecular chains such as SPG sample, thereby allowing accurate extrapolation to zero shear not found in many previous measurements of polysaccharide viscosity. They found that the viscosity is almost independent of shear rate over the range studied (Fig. 18), and used the viscosity value as a zero shear viscosity.

The double logarithmic plot of zero shear specific viscosity ($\eta_{sp,0}$) of flexible polymer solutions against polymer concentration (Fig. 12) provides information on the coil-overlap concentration $C^*$ (Morris et al., 1981). When the zero shear specific viscosity is represented by a power law ($\eta_{sp,0} \sim C^n$), the exponent $n$ takes different values below and above $C^*$. The exponent $n$ for most polysaccharide solutions ranges from 1.1 to 1.6 below $C^*$, and from 1.9 to 5.6 above $C^*$ (tabulated in Lapasin and Pricl, 1999). Most random-coil polysaccharides

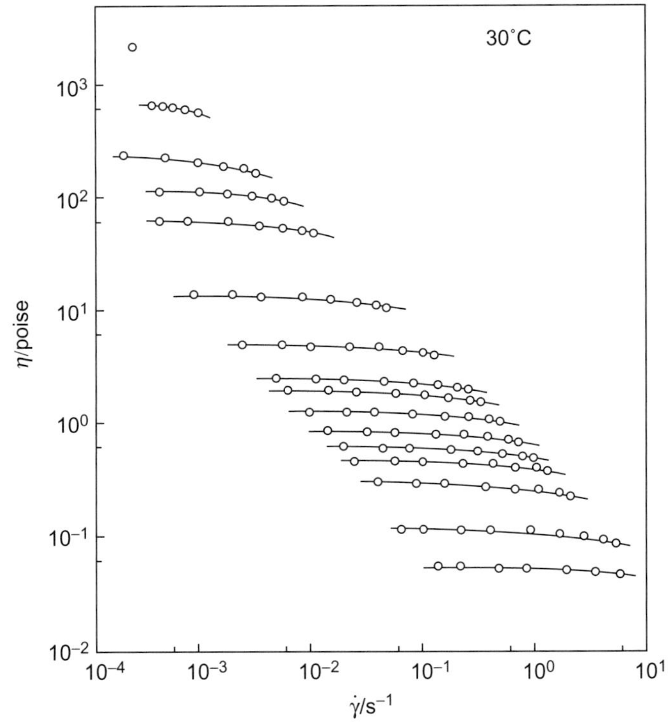


**Fig. 18: Steady shear viscosity ($\eta$/poise) as a function of shear rate ($\dot{\gamma}$/s$^{-1}$) for SPG ($M = 4.3 \times 10^6$) in water at 30°C. The polymer concentration is 0.00411, 0.00313, 0.00284, 0.00246, 0.00216, 0.00159, 0.00125, 0.00104, 0.00102, 0.000894, 0.000796, 0.000716, 0.000652, 0.000551, 0.000378 and 0.000257 g cm$^{-3}$ from top to bottom. From Enomoto et al. (1985).**

have an exponent of 3.3 at high concentrations, but stiff chains show higher values. Although this double logarithmic plot has been used frequently, it should be pointed out that zero shear viscosity as a function of concentration does not change sharply at $C^*$ but rather gradually (Lapasin and Pricl, 1999).

The concentration dependence of the zero shear viscosity of SPG as $\eta_0 \sim C^5$ or $C^8$ has been reported by Enomoto et al. (1985). As for the molecular weight dependence of the zero shear viscosity, the data for SPG showed $\eta_0 \sim M^{6.8}$ or $M^{8.4}$ for the rigid rod regime whereas the exponent was approximately 5 for semiflexible samples, i.e. the molecular weight dependence became weaker with increasing flexibility (Enomoto et al., 1985). However, it should be noted that the molecular weight dependence is much stronger in rod-like polymers such as SPG than in flexible polymers for which $\eta_0 \sim M^{3.4}$ is reported.

It should be noted that stiff chains such as SPG form a liquid crystal above a certain concentration, and the concentration dependence of the viscosity of the anisotropic phase is very different from that in the isotropic phase (Lee and Brant, 2002). Van and coworkers (Van and Teramoto, 1982) observed a cholesteric mesophase in a 14.34% solution SPG ($M_w = 1.7 \times 10^5$), whereas a 2.33% solution remains isotropic at room temperature. They also noticed an abrupt change in optical rotatory dispersion (ORD) when the isotropic solution was cooled to a temperature close to the isotropic–biphasic boundary ca. 5°C, which was ascribed to a pretransition from an isotropic to a cholesteric phase (Van et al., 1981). Fang et al. (2004d) observed a shear thickening phenomenon for 17.22% SPG ($M_w = 4.5 \times 10^5$) solution around a shear rate of $0.1\,s^{-1}$ and observed a steep change in the birefringence indicating an abrupt increase in molecular orientation (Fig. 19).

A phenomenon that the steady shear viscosity at lower shear rate increases and then decreases with increasing concentration of polymers is also observed in xanthan, another polysaccharide with a particularly stiff chain (Lee and Brant, 2002a,b,c; Yin et al., 2007).

Steady shear viscosity as a function of shear rate of aqueous solutions of scleroglucan shows a Newtonian behaviour at very low shear rates, and it changes to shear thinning behaviour (Farina, et al., 2001; Moresi et al., 2001). The shear rate at which the flow behaviour changes from Newtonian to shear thinning shifts to lower shear rates with increasing concentration as is usually observed for other polymers. A similar behaviour is also observed in dilute NaOH (0.01 or 0.1 M) (Bo et al., 1987; Grassi et al., 1996). Normalized relative viscosity, the viscosity at a temperature divided by the viscosity at 15°C, as a function of temperature was shown to increase, and interpreted as the increase in the dimension of the aggregates by intermolecular association or by swelling (Grassi et al., 1996).

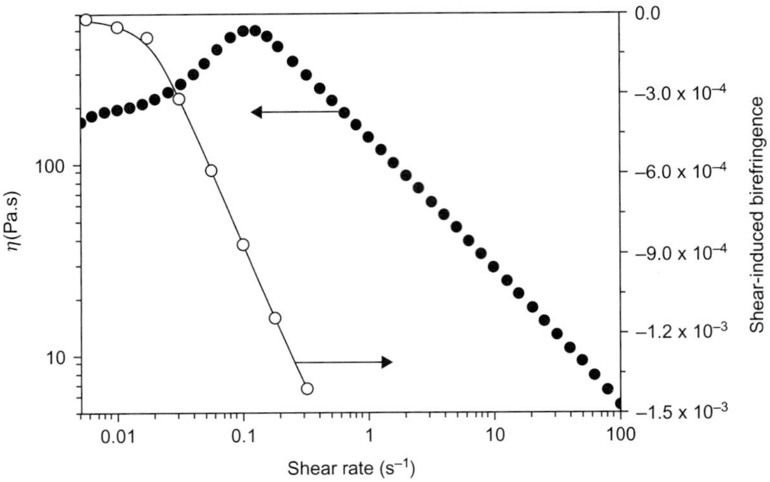


**Fig. 19: Steady shear viscosity (●) and birefringence (○) upon increasing shear rate for 17.22% (w/w) SPG solution. $M_w$=450 000. $T$=25°C. (From Fang et al., 2004d.)**

Xu et al. (2008) examined log$\eta_{sp,0}$ of a lentinan ($M_w = 1.75 \times 10^5$) solution as a function of the logarithm of the coil-overlap parameter ($C[\eta]$), and reported the critical concentration ($C^* = 3.8 \times 10^{-4}$ g/mL), and the slope at lower concentration 1.15 and higher concentration 4.67, which are in general accordance with expected values for these polysaccharides.

*II.B.2.b Concentrated solution properties of denatured side-chain-branched (1,3;1,6)-β-glucans* It is well established that when SPG triple helices are denatured into single (random) coils by heating (Zentz et al., 1992; Yanaki et al., 1985), or by adding strong basic or polar solvents (Sato et al., 1983b; Young and Jacobs, 1998), the SPG chains cannot return to the state of the initial triple helix when their environment is returned to the original state. Many research groups have investigated the chain conformation and morphology of the denatured–renatured SPG (DRSPG) (Vuppu et al., 1997; Kitamura et al., 1996; Stokke et al., 1991, 1993; McIntire et al., 1995, 1998) since the study of the denaturation–renaturation (DR) process has been useful in elucidating some phenomena such as the supercoiling or superhelicity of DNA. DR methods employed for SPG include: i) SPG dissolving in DMSO and then dialysing against $H_2O$; ii) dissolution in NaOH followed by neutralization with HCl; and 3) heating above 135°C and then cooling. Stokke et al. (1991) showed experimentally for the first time that DRSPG could exist in a triple-stranded circular form. The circular triple strand of DRSPG is reminiscent of the supercoiling phenomenon of DNA. Besides the circular form, DRSPG chains have been observed to exist in linear, hairpin and granular forms as well as complex clusters (Stokke et al., 1993; Kitamura et al., 1996; McIntire et al., 1995, 1998; Falch et al., 1999).

The DR process of SPG with a molecular weight of $2.6\times10^6$, designated as SPG-1, has been examined by different rheological methods as well as intrinsic viscosity measurements (Fang et al., 2004c). SPG-1 was first made basic with $[OH^-]=0.2$ mol/L and then neutralized to pH=7, to give DRSPG-1. A sol to weak gel transition occurred in the concentration range 0.65–0.75% (w/w). When the concentration ($C_p$) is above 0.75% (w/w), DRSPG-1 aqueous systems had weak gel (structured liquid) rheological properties. However, for 0.28% (w/w) $\leq C_p \leq$ 0.65% (w/w) and $C_p \leq$ 0.19% (w/w), DRSPG-1 aqueous systems behaved as power law and Newtonian fluids, respectively, which was attributed to the moderate isotropy of DRSPG-1 chains. Furthermore, the critical overlap parameter of $C^*[\eta] = 1.2$ was determined for DRSPG-1 in aqueous solutions, which is close to 1 for intact SPG in water in contrast to the far larger classical value of 4.3 (Fig. 12) for SPG in DMSO. This is considered to be due to the strong interactions or associations of DRSPG-1 chains in water, further confirmed by the intrinsic viscosity measurements in which the DRSPG-1 aqueous solution shows an abnormally large value of the Huggins constant $k'$ (1.57). From multi-run dynamic strain sweep measurements it was suggested that the weak gel structure of DRSPG-1 was due to aggregation by hydrogen-bonding associations of DRSPG-1 chains rather than a permanent three-dimensional network. Although the structure in DRSPG-1 weak gels is not permanent it can be rebuilt very quickly after destruction. One explanation for this behaviour is that during the DR process SPG chains form some imperfect multi-strands that grow to a three-dimensional network by the connection of these multi-strands with coil segments (Stokke et al., 1993, 1996; Kitamura et al., 1996; McIntire and Brant, 1995) (see also Fig. 3). Step-shear rate tests, which were applied to scleroglucan (Grassi et al., 1996), were performed to study the thixotropic behaviour of DRSPG aqueous systems (Fang et al., 2004c). The DRSPG-1 aqueous system was kept flowing continuously at a constant shear rate until a plateau value of stress was obtained. Then the shear rate was changed instantaneously to another value. The information of the transient stress reflects the time dependence of structural breakdown or build-up. Except for $C = 0.37\%$ (w/w), DRSPG-1 aqueous systems exhibited a marked stress overshoot when the shear rate was increased, and a stress undershoot when shear rate was decreased, indicating the thixotropic properties of DRSPG-1 aqueous systems (Grassi et al., 1996). Here the stress overshoot and undershoot correspond to the structural breakdown and build-up of the three-dimensional network, respectively. Moreover, at the same shear rate, the steady stress attained during the increasing shear rate process is somewhat larger than that attained during the decreasing shear rate process, suggesting that the shear flow is slightly affected by the rheological history. It should be noted that the amplitude and kinetics of transient stress depend on the amplitude of shear rate variation and the instantaneous shear rate value. Although the sample with $C$ = 0.55% (w/w) shows sol-like behaviour instead of weak gel-like behaviour, it still exhibits stress overshoot or undershoot when changing shear rate, which is attributed to the higher degree of entanglement of macromolecular chains. The reason why the sample with $C$ = 0.37% (w/w) does not show thixotropic behavior may be attributed to less entanglement. According to a method suggested by Grassi et al. (1996), the transient stress can be described by the following exponent equation:

$$\tau = \left[ \frac{1}{\sqrt{\tau_s}} + \left( \frac{1}{\sqrt{\tau_i}} - \frac{1}{\sqrt{\tau_s}} \right) \exp(-kt) \right]^{-2}$$

where $\tau_i$ and $\tau_s$ represent the initial and stationary values of the stress, respectively, and $k$ is a kinetic constant. Regardless of an increasing or decreasing shear rate process, the extent of transient stress, defined as $\Delta\tau/\tau_s$ ($\Delta\tau=|\tau_i-\tau_s|$), is more pronounced in the low shear rate range, especially for the 0→0.5 $s^{-1}$ process. Relative to the other shear-rate-changing processes, the value of $\Delta\tau/\tau_s$ for the 0→0.5 $s^{-1}$ process is far larger, which means that the shear rate of 0.5 $s^{-1}$ is large enough to destroy the major aggregated structures in DRSPG-1 aqueous systems. This is also revealed by the difference in the parameter of $k$. For $C$=1.59, 1.10 0.75% (w/w), except that the 0→0.5 $s^{-1}$ processes have markedly large values of $k$, $k$ values of the other processes with shear rate larger than 0.5 $s^{-1}$ are approximately in the same order of magnitude. Moreover, these $k$ values are close to that of 0.052 for the 0→0.5 $s^{-1}$ process of the 0.55% (w/w) system which is in the sol state. It tells us indirectly that structural breakdown or build-up taking place above the shear rate of 0.5 $s^{-1}$ mainly comes from the disentanglement or re-entanglement of macromolecular chains, and the weak gel structures have already been destroyed in the 0→0.5 $s^{-1}$ process. It should be stated that disentanglement of macromolecular chains needs more energy than the destruction of macroscopic aggregated structures. Consequently, the 0→0.5 $s^{-1}$ processes show a larger value of $k$ than other shear-rate-changing processes (Fang et al., 2004c).

In summary, at certain critical concentrations the solution behaviour of side-chain-branched (1,3;1,6)-β-glucans changes from Newtonian viscosity to shear-dependent viscosity as is generally observed for many other polysaccharides and also for synthetic polymers (Lapasin and Pricl, 1999). The shear rate dependence is marked for these rod-like polymers which form weak gels, or structured liquids. DR treatment of side-chain-branched (1,3;1,6)-β-glucans also forms weak gels.

### *II.B.3 Solution behaviour of concentrated solutions of cereal (1,3;1,4)-β-glucans*

Böhm and Kulicke (1999a) examined $\log\eta_{sp,0}$ of fresh barley (1,3;1,4)-β-glucan solution as a function of $\log C[\eta]$ and reported the slope at higher concentration to be 5.18, which is much higher than the 3.9 reported for oat (1,3;1,4)-β-glucan (Doublier and Wood, 1995); they attributed the lower value of Doublier and Wood (1995) to the lower range of the maximum of $C[\eta]=7$ in comparison to their maximum of $C[\eta]=50$.

Vaikousi et al. (2004) reported that the double logarithmic plot of $\eta_{sp,0}$ and coil-overlap parameter $C[\eta]$ for solutions of barley β-glucan with different molecular weights was represented by three straight lines at different concentration ranges, and that the slopes 1.07, 1.58 and 4.08 agreed well with previous findings for oat (1,3;1,4)-β-glucan solution, 1.05, 1.60 and 3.90 (Lazaridou et al., 2003), and for solutions of (1,3;1,4)-β-glucan from *Avena sativa* and *A. bysantina,* 1.0, 1.6 and 3.8 (Skendi et al., 2003), and from oats, wheat, barley and lichenin, 1.03, 1.64 and 4.13 (Lazaridou et al., 2004b). Vaikousi et al. (2004) found that the shear thinning behaviour of 10% (w/w) solutions of barley (1,3;1,4)-β-glucan with different molecular weights at 20°C is well described by an equation $\eta = \eta_0 / \left[1 + (\dot{\gamma}/\dot{\gamma}_{1/2})^{0.76}\right]$ proposed by Morris (1984), where $\dot{\gamma}_{1/2}$ represents the shear rate at which the viscosity becomes the half value of the zero shear viscosity $\eta(\dot{\gamma}_{1/2}) = \eta_0/2$.

Vaikousi et al. (2004) examined the effect of storage time on the shear rate dependence of steady shear viscosity of solutions of barley (1,3;1,4)-β-glucan and found that shear thinning behaviour at low shear rates became more pronounced with increasing storage time, and that this effect was stronger for lower molecular weight samples. A similar result was also found for oat (1,3;1,4)-β-glucan (Lazaridou et al., 2003). Vaikousi et al. (2004) also examined the thixotropic loop for solutions of barley (1,3;1,4)-β-glucan and found that hysteresis between the up and down curves became more pronounced with increasing total cycle time, and attributed this to the formation of aggregates during the long period of the test (Fig. 20). A similar phenomenon was also found for oat (1,3;1,4)-β-glucan (Lazaridou et al., 2003). This hysteresis was more pronounced for lower molecular weight samples.

Gomez et al. (1997c) observed a shear thinning behaviour of barley (1,3;1,4)-β-glucan solutions over the shear rate range $10^{-2}-10^{0}\,s^{-1}$ and found a plateau viscosity 100-fold smaller than that at lower shear rate at $10^{0}-10^{2}\,s^{-1}$. The viscosity was higher at 70°C than at 25°C at lower shear rates whereas the situation was reversed at higher shear rates. They concluded

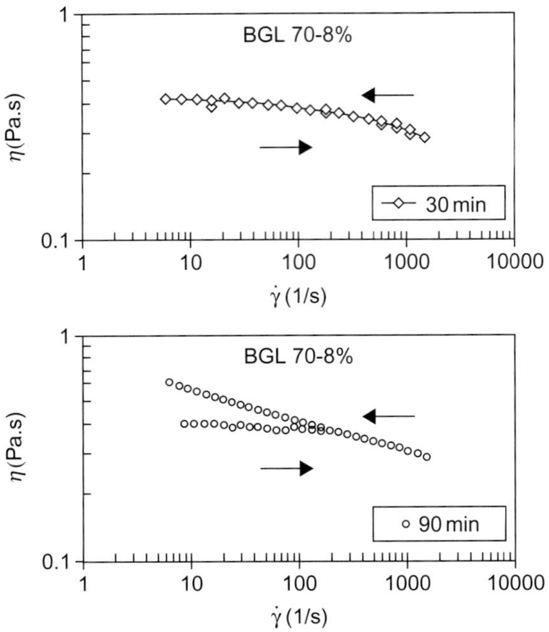


**Fig. 20: Thixotropic loop with different total cycle time for 8% (w/w) barley (1,3;1,4)-$\beta$-glucan at 20°C. Arrows indicate the direction of the applied thixotropic loops. (From Vaikousi et al., 2004.)**

that associations formed at high temperatures are more stress sensitive than those formed at room temperature. The increase in viscosity of liquids with increasing temperature is a rare phenomenon, although it is well known that some cellulose derivatives and curdlan form a gel on heating. Gomez et al. (1997c) found that mechanical spectra, frequency dependence of storage and loss moduli of 1.5% (w/w) barley (1,3;1,4)-β-glucan solutions changed from a concentrated solution behaviour at 25°C to a gel-like behaviour at 70°C. To understand better the mechanism of structure formation, they examined the effect of NaCl on the mechanical spectra of 1.5% (w/w) barley (1,3;1,4)-β-glucan solutions at room temperature and found that NaCl suppressed the structural formation, which was also observed for oat (1,3;1,4)-β-glucan by Wikström et al. (1994). Gomez et al. (1997c) also found that the spectra are very strongly influenced by storage time and temperature, as is always the case for biopolymers; therefore, they could not obtain reproducible results and pointed out the necessity of further study. Johansson et al. (2008) found stronger shear thinning for oats than for barley (1,3;1,4)-β-glucan in water at a concentration of 15 mg/ml, and further examined rheological behaviours in aqueous cuoxam at 15 and 50 mg/ml. They found that viscosity and storage and loss moduli for barley (1,3;1,4)-β-glucan were larger than those for oat (1,3;1,4)-β-glucan, and attributed this to structural differences rather than the size differences, based on previous reports: high molar ratio of DP3 and DP4 increases the possibility of cellotriosyl units to appear regularly (Böhm and Kulicke, 1999; Cui et al., 2000) and hence causes aggregation (Tvaroska et al., 1983).

## II.C Gelation Behaviour of (1,3)-β-Glucans

Gels and gelling processes are still not well defined. It is often easier to recognise a gel than to define a precise measurement-based definition. Since steady shear viscosity measurement destroys the structure being formed during gelation, most researchers use the dynamic viscoelastic measurements in a small deformation range (e.g. Fig. 11) to study the gelling process. Winter and Chambon (1997) proposed that the criteria to determine the critical gelation point should be:

(i) that the storage modulus $G'$ and loss modulus $G''$ show the same frequency dependence $G' \sim G'' \sim \omega^n$, and

(ii) the mechanical loss tangent $\tan\delta = G''(\omega)/G'(\omega) = \tan(n\pi/2)$ is independent of frequency.

(i) and (ii) should be satisfied at the gelation point simultaneously. These criteria have been shown to work well, especially for chemical gelation, and then extended to thermoreversible gelation. However, there have been some arguments that the criteria may not hold for the gelation of some biopolymers, and some papers on the gelation of (1,3)-β-glucans use it whereas others do not. In some colloidal suspensions with a non-zero yield stress, such as polystyrene lattices (Matsumoto and Okubo, 1991), ovalbumin solutions (Matsumoto and Inoue, 1991), curdlan suspensions (Hirashima et al., 1997) and microparticulated cellulose suspensions (Nishinari et al., 1998b; Tatsumi et al., 1999), however, $G' > G''$ and both moduli show plateau values even at low frequencies and well before a gelation point. Both ovalbumin and curdlan dispersions form a gel on heating, but even before the gel formation, $G'$ is already larger than $G''$. These examples together with gelation behaviour of globular proteins (Tobitani and Ross-Murphy, 1997) question the universal validity of the Winter-Chambon criterion.

### II.C.1 Gelation of curdlan

Curdlan can form a gel through thermal processing alone, rather than relying on accompanying conditions such as pH, sugar concentration and the presence of cations. Curdlan aqueous gels can be formed by various methods, such as heating a curdlan aqueous suspension (Konno et al., 1978; Harada et al., 1994), heating a curdlan/DMSO/$H_2O$ suspension (Watase and Nishinari, 1994), neutralizing an alkaline solution of curdlan with carbon dioxide (Kanzawa et al., 1989), or dialysing a solution of curdlan in DMSO (Kanzawa et al., 1987) against water in a quiescent condition at ambient temperature. Furthermore, after heat treatment, an aqueous curdlan suspension is capable of forming two types of gels depending on heating temperature, one of which is a thermo-reversible gel, termed a low-set gel, formed by heating up to about 55°C then cooling, and the other a thermo-irreversible gel, termed a high-set gel, formed by heating at above 80°C. This change is explained by the hypothesis that microfibrils dissociate at 60°C as the hydrogen bonds are broken, but then reassociate at higher temperatures as hydrophobic interactions between the curdlan molecules occur. An additional change to an even more ordered form has been reported as the temperature is raised above 120°C (Harada et al., 1979). Besides temperature, many other factors, such as molecular weight, concentration, heating rate, dispersing method and the addition of inorganic salts, influence the formation and the mechanical properties of curdlan gels (Funami et al., 1999a,b, 2000; Funami and Nishinari, 2007). For example, the transition temperature from a thermo-reversible to an irreversible gel is molecular weight dependent (Zhang et al., 2002) and an increase in concentration also lowers the transition temperature from the previously reported 80°C (Funami et al., 1999a,b).

Low molecular weight curdlan fractions (<DP20) cannot form a helix and thus are not incorporated into a gel network. This is consistent with $^{13}C$ NMR studies (Saito et al., 1978) on curdlan gels prepared by mixing of curdlan and the low molecular weight fractions. They found that a low molecular weight fraction shows sharp $^{13}C$ signals characteristic of the random-coil conformation, suggesting that these molecules are trapped in the interstices of the gel network. Higher molecular weight fractions with helical conformations are believed to be incorporated in the gel network because their $^{13}C$ signals are much broader and the molecular motions are restricted in comparison with those for lower molecular weight fractions.

Heating rate is also an important factor. Thermo-reversible gelation upon cooling is inhibited more on lowering heating rate, and this effect is more pronounced at higher concentrations of curdlan (Funami et al., 2000). Also, the longer the incubation in the swelling temperature range (50–60°C), the smaller is the dynamic storage modulus of gels formed after heating and subsequent cooling processes. These properties can be attributed to the inter-molecular associations created during swelling through hydrophobic interactions, leading to heterogeneous network structures and inhibiting the rearrangement of the molecules upon cooling. An alternative hypothesis suggests that curdlan gels have network structures composed of enthalpically stabilized junction zones with entropic disordered polymer chains, in which particles that are not completely dissolved are embedded, and act as a filler strengthening the gel matrix due to the highly ordered crystalline structure of curdlan, comparable with the molecular organization of starch. Both slower heating rate and longer incubation within the swelling temperature range weaken the crystalline structures, freeing the non-melted crystalline structures composed of triple helices which contribute to the mechanical strength (Zhang et al., 2002).

The addition of inorganic salts decreases the mechanical strength of curdlan gels (Funami et al., 2007). This is attributed to the inhibition of swelling or hydration, and thus the disordering of curdlan molecules through breakage of intra- or inter-molecular hydrogen bonds, resulting in a decreased number of molecules that participate in associations via new or re-natured hydrogen bonds to form junction zones in the network structures of curdlan gels. The gel strength does not change between pH 3 and 10, and can be enhanced greatly by adding borate; but the presence of urea, a reagent which breaks hydrogen bonds, caused a decrease in the elastic modulus (Konno et al., 1979) and in the gelling temperature, and a marked decrease in gel strength as the concentration of urea increased above 2 M (Maeda et al., 1967). However, the reason why the gel strength shows a maximum at certain urea concentrations is not yet understood.

The high-set gel has the properties of being much stronger and more resilient and shows less syneresis than the low-set gel and neutralized gel, and is not broken when frozen and thawed (Kanzawa et al., 1989). The high-set gel was resistant to enzymatic and acidic hydrolysis, but the neutralized gel was not resistant (Kanzawa et al., 1989). Gels set at above 90°C are soluble only in concentrations of NaOH above 1 M, whereas neutralized and 60°C-set gels are soluble in 0.01 M NaOH (Kanzawa et al., 1989). X-ray studies showed a much higher crystalline structure in the resistant part than in the curdlan sample without heat treatment. It has been reported that tannin, sugar and starch effectively reduce the syneresis of curdlan gel (Harada, 1992). From rheology and differential scanning calorimeter (DSC) observations, Zhang et al. (2002) suggested that reducing the molecular weight or holding samples at high temperatures for longer times provides strong evidence that the annealing of triplex structures is the main time-dependent phenomenon occurring (Zhang et al., 2002; Nishinari and Zhang, 2003, 2004). This hypothesis is supported by estimates of renaturing kinetics obtained by low resolution $^{13}$C NMR. Transitions of the un-annealed portion of the sample below 40°C seem to play a crucial role in the kinetic trapping of microstructural states. It is now clear that the high-set and low-set gels described in much of the literature are not discrete and do not form a complete set of possible behaviours, but rather are two possibilities of a whole spectrum of possible gels in which the contribution of annealed triple helical elements to the stress-bearing properties of the generated network can be varied. Holding a sample closer to the transition temperature allows a greater number of imperfect triplexes to be melted out than at correspondingly lower temperatures and, as such, more annealed samples are more resilient and thermo-irreversible.

Interestingly, Dobashi and co-workers (Sato et al., 2005; Dobashi et al., 2004, 2005; Nobe et al., 2005) found that the dialysis of curdlan dissolved in alkaline solution into aqueous solutions of metal salts yielded multifold gel structures. Aqueous calcium salts induced a liquid crystalline gel with a refractive index gradient interpreted as being due to alternating amorphous and gel structures, which is influenced by the molecular weight of curdlan (Nobe et al., 2005). Aqueous salts of trivalent aluminum and ferric cations also induced a rigid liquid crystalline gel, which shrank above a threshold concentration of each salt (Sato et al., 2005). It is possible that these new gel materials could be used as optical components such as polarizers and as indicators (Dobashi et al., 2004).

### II.C.2 Gelation of side-chain-branched (1,3;1,6)-β-glucans

*Scleroglucan* solutions (0.2% w/w) behave as an entangled polymeric solution whereas 1.0% and 2% (w/w) scleroglucan solutions behave as a weak gel (also sometimes referred to as a structured liquid), i.e. the storage shear modulus $G'$ is larger than loss shear modulus $G''$ at the angular frequency range from 0.1 to 100 rad/s (Grassi et al., 1996). The mechanical loss tangent $\delta$ was much smaller in 1.0% and 2% (w/w) scleroglucan solutions than in 0.2% (w/w) scleroglucan solution. The Cox-Merz rule failed for a 2% (w/w) scleroglucan solution, i.e. the complex viscosity as a function of frequency and the steady shear viscosity as a function of shear rate were almost parallel, with the former much larger than the latter. This was interpreted as being due to the presence of a tenuous, transient network, involving non-covalent intermolecular associations, typical of physically weak gel systems.

Hydrogels prepared from scleraldehyde (chemically oxidized scleroglucan) with a low degree of oxidation are brittle and fragmented, in contrast to the elastic/homogeneous hydrogels prepared from scleraldehyde with a high degree of oxidation (Maeda et al., 2001). Based on the SAXS profile analysis, hydrogels prepared from scleraldehyde with a low degree of oxidation possess a network composed of randomly oriented triple helices interlinked at the sites where the aldehyde groups are present (Maeda et al., 2001).

The gelation of scleroglucan–borax complexes has been studied as a potential drug delivery system by Coviello et al. (1999, 2001, 2003a, b, 2006). It is well known that the solutions of poly(hydroxyl) compounds such as poly(vinyl alcohol) (Nijenhuis et al., 1997; Matsuzawa et al., 1987; Shibayama et al., 1992; Koike et al., 1995), and galactomannans (Pezron et al., 1989, 1990) form gels when treated with borate. The gelation results from the formation of a didiol complex between borate and the two pairs of adjacent hydroxyl groups in two different chains (Nijenhuis et al., 1997; Matsuzawa et al., 1987; Pezron et al., 1989, 1990). Scleroglucan–borax gels were prepared in the presence of 0.1 M borax with scleroglucan concentration of 0.7% (w/v) (the lowest polymer content allowing the formation of a self-sustaining gel) and 2.3% (w/v) (the polymer concentration in the completely swollen tablet) (Coviello et al., 1999, 2001, 2003a, b, 2006). Both storage and loss shear moduli were only slightly dependent on the frequency indicating a so-called weak gel formation for scleroglucan with and without borax. Both moduli of these weak gels, or structured liquids, increased more remarkably by the addition of borax at higher concentrations (i.e., 2.3% w/v). The structure of scleroglucan–borax gels was proposed to be due to a combination of chemical and physical cross-linking: parallel arrangements of triple helices are held together by covalent linkage and partially by physical interactions, with borate ions (Shibayama et al., 1988). The diol groups of the side chains of the polymer can react with borate ions to give a di-ligand complex in which the borate ions act as a bridge between the polysaccharide chains.

*Lentinan*: The dynamic viscoelastic behaviour of lentinan in water has been studied (Zhang et al., 2008) and found to be similar to that reported for the schizophyllan–sorbitol system (Fang and Nishinari, 2004a). The applicability of Winter-Chambon's criterion (Winter and Mours, 1997) was examined to determine the critical gelation point, and it was found that the exponent $n'$ and $n''$ in $G' \sim \omega^{n'}$ and $G'' \sim \omega^{n''}$ tended to coincide below ca. 20°C, and that $\tan\delta$ measured at various frequencies as a function of temperature passed through the common point at a certain temperature, which was defined as the gel point or critical gelation temperature. Thus the Winter-Chambon's criterion is applicable to the lentinan–water system. The gelation mechanism proposed for lentinan in water by Zhang et al. (2007) envisages that the extremely entangled lentinan chains make a continuous network, conferring the gel-like properties of the system; however, this lentinan–water gel does not contain junction zones or pronounced aggregates as inferred from dynamic strain sweep measurement as employed for the schizophyllan–sorbitol system (Fang and Nishinari, 2004a).

*Schizophyllan* (SPG) is only able to form a weak gel when aqueous solutions are cooled below 6°C (Bot et al., 2001). The addition of some small molecules, e.g. borax or sorbitol, can lead to a relatively strong SPG gel (Fuchs et al., 1997; Maeda et al., 1999; Grisel and Muller, 1996, 1997, 1998; Fang et al, 2004a, b). In the case of borax this is due to the complexation reaction between the hydroxyl groups on the SPG side chain and borate ions as discussed for scleroglucan. The effects of SPG concentration, borax concentration, pH and temperature on the gelation behaviours of the SPG–borax system have been qualitatively investigated (Grisel et al., 1997, 1998). The cross-links formed in SPG gels in the presence of borax are non-permanent and somewhat dynamic in nature because the cross-linking reaction is governed by complexation equilibrium. The gelation processes can be monitored by dynamic viscoelastic measurements to examine the effects of borax content, SPG concentration, temperature, salt concentration, salt type and strain. The first order kinetic model containing three parameters, $t_0$ (gelation time), $1/\tau_c$ (gelation rate), and $G'_{sat}$ (saturated storage modulus), has been successfully applied to describe gelation of the SPG–borax system. The gelation occurs faster at higher borax content, higher SPG concentration, higher salt concentration or lower temperature. Moreover, the gelation is cation-type specific. The storage modulus is a linear function of both borax content and SPG concentration. The former relationship can be explained by the ideal rubber elasticity theory in which the elasticity is proportional to the concentration of cross-links. On the other hand, the latter could result from the fact that the number of interchain contacts and hence the formation of cross-links may scale linearly with SPG concentration for rigid chains. The apparent activation energy and cross-linking enthalpy are calculated to be $-74.5$ and $-32.4$ kJ/mol. The strain sweep measurement showed that the elasticity behaviour of this gel starts to deviate from the Gaussian chain network at a small strain of 10% (Fang et al., 2004b).

In contrast to the SPG–borate system, the gelation mechanism of SPG in the presence of sorbitol is still being debated. Fuchs et al. (1995, 1997) proposed at first that the presence of sorbitol reduces the mobility of water molecules and drives SPG molecules together to aggregate, leading to a three-dimensional gel network (Fig. 21).

Later Maeda et al. (1999) proposed an alternative gelation mechanism in which sorbitol partially breaks the SPG triple helix, and the free chains reassociate to form junction zones by hydrogen bonds. The gelation behaviour of aqueous SPG–sorbitol solutions upon cooling (Fang and Nishinari, 2004a) has been further investigated by rheology, DSC and ORD. The ORD profiles and DSC curves indicate that the gelation of a SPG–sorbitol aqueous solution is accompanied by a highly cooperative conformational transition of SPG triple helix from a disordered state to an ordered state. The Winter-Chambon criterion was found to be valid for determining the gelation point of a SPG–sorbitol aqueous solution although the system already behaved as a weak gel before gelation. A small exponent $n$ of 0.15 in the Winter-Chambon criterion was found, which may be attributed to the higher molecular weight used and the presence of entanglements of SPG triple helices. At constant sorbitol content, the gelation temperature of the SPG–sorbitol aqueous solution is independent of SPG concentration, and slightly decreases with lowering SPG molecular weight. With increasing sorbitol content, the gelation temperature increases remarkably, whereas the enthalpy change accompanying the gelation is independent of sorbitol content.

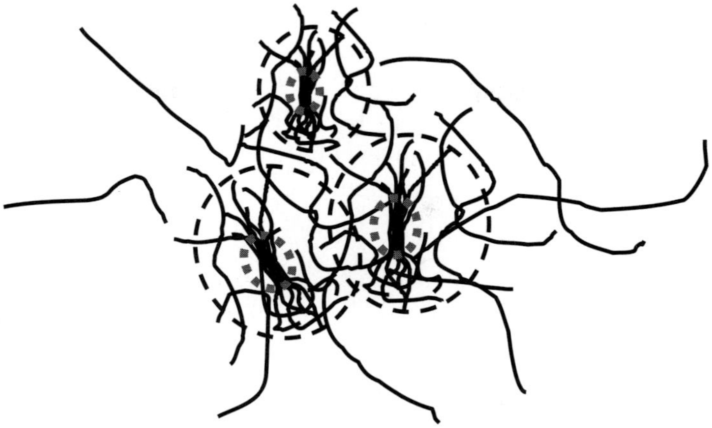

**Fig. 21: Junction zone formation in SPG–sorbitol gelation. Clusters (small broken line circles) grow until inter-penetration is prevented by the impenetrable parts of the heterogeneous clusters (large broken line circles). (From Fuchs et al., 1997.)**

Based on these experimental findings, Fang and Nishinari (2004a) proposed a new gelation mechanism. (i) Cooling induces an intramolecular conformational transition of SPG chains from triple helix II to triple helix I (Itou et al., 1987; Hirao et al.,1990; Hayashi et al., 2002; Yoshiba et al., 2002, 2003). (ii) This transition increases the diameter and stiffness of the triple helix (Kitamura et al., 1996; Itou et al., 1987), and decreases the mobility of SPG chains and the surrounding water molecules (Yoshiba et al., 2002, 2003), which greatly enhances the entanglement of SPG chains. (iii) The extremely entangled SPG chains make a three-dimensional network which confers gel-like properties on the SPG–sorbitol system. This is analogous with the concentration-independent gelation mechanism proposed by Gidley et al. (1987) for a neutral *Rhizobium* capsular polysaccharide, which is induced by a cooperative disordered to ordered conformation transition involving side chains. Furthermore, the structure of SPG–sorbitol gel does not contain junction zones or aggregations of the SPG triple helix I form. The aqueous SPG–sorbitol gel is in fact like a very concentrated solution which is unable to flow within a timescale of usual observations; therefore, it should be called a structured liquid rather than a weak gel.

Bot et al. (2001) also studied the melting behaviour of SPG gels formed in the presence of glucose. The melting temperature shifted from 5°C to 20°C with increasing concentration of glucose (0%–50% w/w). Transmission electron microscopic observations suggest that lateral aggregates of SPG are present below the melting temperature.

### *II.C.3 Gelation of cereal (1,3;1,4)-β-glucans*

The molecular basis for the gelation of cereal (1,3;1,4)-β-glucans has been the subject of debate. Woodward et al. (1983) and Doublier and Wood (1995) proposed that the infrequent long sequences of (1,4)-β-linked glucopyranosyl units would associate by hydrogen bonds to form junction zones (Fig. 22A), and that gelation would occur in a sufficiently concentrated solution. However, Doublier and Wood (1995) did not detect the difference in the amount of cellulose-like regions between a partially hydrolysed gelling (1,3;1,4)-β-glucan and a high molecular weight non-gelling sample. Böhm and Kulicke (1999a) showed by dynamic viscoelastic measurements that a freshly prepared 6% (w/w) solution of barley (1,3;1,4)-β-glucan behaves as a viscoelastic liquid, i.e. frequency-independent complex viscosity, storage modulus with a slope of 2 and loss modulus with a slope of 1, and that the solution changes into a gel with time. They observed shear-induced gelation for concentrated (1,3;1,4)-β-glucan solutions, and attributed gelation to molecular orientation by shear forces and the consequent

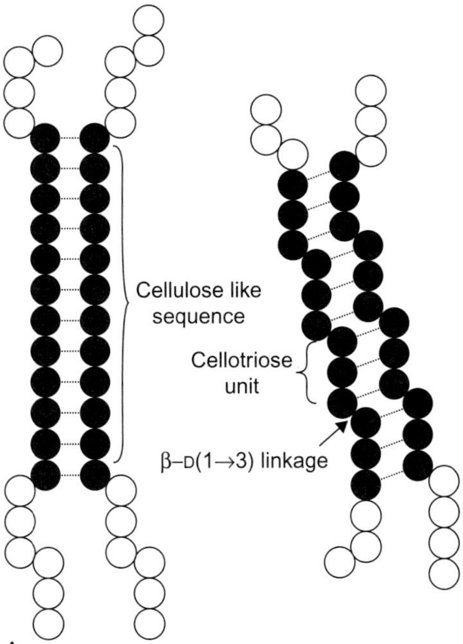


A B

**Fig. 22: Chain interactions in a (1,3;1,4)-β-glucan leading to gelation. (A) Sequences of consecutive (1,4)-β-linkages associate. (B) Association of consecutive cellotriose units joined by (1,3)-β-linkages, which forms a helical structure. From Böhm and Kulicke (1999).**

lateral association via hydrogen bonds during rest periods. Böhm and Kulicke (1999b) also found that the gelation rate decreases with decreasing concentration and increasing molar mass of (1,3;1,4)-β-glucan. The effect of concentration was ascribed to the increase of the contact between coil chains and that of molar mass was attributed to the higher mobility of shorter chains, as suggested by Doublier and Wood (1995). They compared the gelation rate of 10% (w/w) solutions of the (1,3;1,4)-β-glucans from barley, oat, and lichenin, a similar (1,3;1,4)-β-glucan (Table 3) with the same molar mass ($M_n$=30000 g/mol), and found that the gelation rate was the fastest with lichenin and slowest with oat (1,3;1,4)-β-glucan. By enzymatic degradation with a (1,3;1,4)-β-glucanase, lichenin was found to have the most regular chain structure and oat (1,3;1,4)-β-glucan the least. On the basis of these data, they proposed that the gelation is due to formation of stable junction zones through hydrogen bonds formed between consecutive cellotriosyl units (Fig. 22B). This is also consistent with the results of solid state $^{13}$C NMR data (Morgan et al., 1999) which showed the presence of chemical shifts from gels of a low molecular weight (1,3;1,4)-β-glucans different from both solution and gel forms of either (1,3)-β-glucan or (1,4)-β-glucan (Section I.B.5.b).

Vaikousi et al. (2004) studied the gelation of 8% (w/w) solutions of barley (1,3;1,4)-β-glucans with different molecular weights by observing the evolution of storage and loss shear moduli at 25°C, and found that the gelation rate and the plateau modulus after a sufficient time decreased with increasing molecular weight. Oat (1,3;1,4)-β-glucans with different molecular weights behaved in a similar way (Lazaridou et al., 2003). The gelation rate increased with increasing concentration both in barley and oat (1,3;1,4)-β-glucans (Vaikousi et al., 2004; Lazaridou et al., 2003). The temperature dependence of the storage modulus of 8% gels of these barley and oat (1,3;1,4)-β-glucans formed at 25°C was examined by heating at 3°C/min to 92°C. The storage modulus stayed almost constant up to 60°C and then showed a melting

**Table 3: Compression modulus (E), true stress ($\tau_{TR}$) and true strain ($\varepsilon_{TR}$) at failure for (1,3;1,4)-β-glucan gels at a concentration of 8% (w/v)**

| Sample | DP3/DP4 | E (kPa) | $\tau_{TR}$ (kPa) | $\varepsilon_{TR}$ |
|---|---|---|---|---|
| Oat200 | 2.1 | 36.6 | 4.5 | 0.22 |
| Barley200 | 3.0 | 36.7 | 6.3 | 0.38 |
| Wheat200 | 3.1 | 44.6 | 18.5 | 0.5 |
| Oat100 | 2.1 | 39.7 | 2.5 | 0.08 |
| Barley100 | 2.8 | 55.7 | 4.6 | 0.16 |
| Lichenin100 | 24.5 | 91.1 | 24.3 | 0.27 |

200 and 100 represent the approximate molar mass×$10^{-3}$. Gel curing and measurement temperature 25°C. From Lazaridou et al. (2004).

behaviour in the range from 60°C to 70°C. Whereas the storage modulus at lower temperatures was highest for the lowest molecular weight samples, the storage modulus at higher temperatures was highest for the highest molecular weight samples. Melting temperature was found to shift to higher temperatures with increasing concentration, especially for the lowest molecular weight sample, but the concentration dependence was weaker in the higher molecular weight sample both for oat and barley (1,3;1,4)-β-glucans (Vaikousi et al., 2004; Lazaridou et al., 2003). Lazaridou et al. (2003) examined the large deformation behaviour of 8% gels of oat (1,3;1,4)-β-glucans with different molecular weights, and found that compression (Young's) modulus increased up to a certain molecular weight and then decreased, whilst both true stress and strain at failure increased monotonically with increasing molecular weight. Vaikousi et al. (2004) found that for 8% gels of barley (1,3;1,4)-β-glucans with different molecular weights, compression (Young's) modulus decreased whilst both true stress and strain at failure increased monotonically with increasing molecular weight. Tosh et al. (2004b) confirmed the previous assertion that lower molecular weight (1,3;1,4)-β-glucans form gels faster than higher molecular weight samples irrespective of whether the depolymerization is by acid, cellulase or (1,3;1,4)-β-glucanase. They observed storage modulus as a function of aging time up to 7 days. After aging at 5°C for 7 days, the mechanical spectrum of 6% solutions of hydrolysed (1,3;1,4)-β-glucan with $M_w$ $3{\sim}4\times10^4$ showed typical elastic gel behaviour, while that with $M_w$ $1.7{\sim}2.3\times10^5$ showed the behaviour of a concentrated solution, i.e. strongly frequency-dependent storage and loss moduli (Fig. 11). They recognised that partial hydrolysates produced by cellulase, which cleaves the cellulose-like blocks and leaves more of the DP3 units intact, showed a stronger gelling ability than those produced by a (1,3;1,4)-β-glucanase which specifically cleaves (1,4)-β-glucosidic linkages adjacent to 3-*O*-substituted glucosyl units. This supports the gelation model in which the junction zones in gels are formed by (1,3)-β-linked cellotriose units (Böhm and Kulicke,1999b; Cui and Wood, 2000). It was clearly shown that the storage modulus of 6% glucan gels aged 7 days increased with increasing mol % cellotriosyl units (DP3) from the observation of increasing order of DP3 oat < barley < rye < wheat < lichenin, but no correlation between $G'$ and the mol % (1,4)-β-oligoglucoside DP6–9 was found. DSC also showed that the onset and peak temperature shifted to higher temperatures with increasing DP3 content (Tosh et al., 2004a). This is in accordance with the results reported by Lazaridou et al. (2004), who used the samples with different molar ratios of tri- to tetrasaccharides (DP3/DP4), lichenin (24.5) > wheat (3.7) > barley (2.8–3.0) > oat (2.1). Compressive modulus and true stress and strain at failure were shown to increase with increasing DP3/DP4. They also showed that both storage modulus and melting enthalpy from DSC increased with decreasing molecular size and with increasing DP3/DP4.

Cryogelation of some water-soluble polymers such as poly(vinyl alcohol) (Watase and Nishinari, 1988; Lozinsky and Damshkaln, 2001) and locust bean gum (Tanaka et al., 1998) have been reported, and Lazaridou and Biliaderis (2004a) reported for the first time the cryogelation of cereal (1,3;1,4)-β-glucans. They examined the effects of initial solution concentration, number of freeze–thaw cycles, and molecular size on the rheological and thermal properties of cryogels of barley and wheat (1,3;1,4)-β-glucans. Storage modulus increased and tanδ decreased with decreasing polysaccharide molecular size, and with increasing initial solution concentration, number of freeze–thaw cycles and number of trisaccharide segments in the polymeric chains. The DSC endothermic enthalpy of these cryogels of (1,3;1,4)-β-glucans increased with decreasing polysaccharide molecular size, and with increasing amount of cellotriose units, but was independent of the number of freeze–thaw cycles. The DSC melting temperature of the gel was found to increase with the molecular size and amount of DP3 units of (1,3;1,4)-β-glucan. Mechanical tests revealed an increase in compressive Young's modulus and true stress at 40% deformation with increasing molecular size and decreasing trisaccharide units, which agrees with the previous results from the same group (Lazaridou et al., 2003) and also in some points with Böhm and Kulicke (1999a,b). However, this contrasts with results from rheological and thermal observations that in poly(vinyl alcohol) cryogels and non-cryogels of most polymers the gel network tends to be strengthened with increasing molecular weight (Clark and Ross-Murphy, 1987; Lapasin and Pricl, 1999; Nijenhuis, 1997; Nishinari, 1997, 2000).

## III Conclusions

(1,3)-β-Glucans exhibit a complexity of physico-chemical and rheological properties that belies their apparently simple molecular structures. Three inter-related features of (1,3)-β-glucans are largely responsible for the material properties that are characteristic of many biological roles and technological applications. These are: (i) borderline solubility in aqueous systems leading to a propensity to form molecular associations, (ii) relatively stiff polymer chains in solution resulting in non-Newtonian viscoelastic behaviour, (iii) a stable triple-stranded helical conformation that further increases chain stiffness and acts as a cross-link in gels and other solidified forms. Chemical structure features are linked to physical properties through their effects on solubility, chain stiffness and triple helix formation/aggregation. Linear (1,3)-β-glucans are the least soluble, and form triple helices that can aggregate to form extended solid structures as found in gels and granules. (1,3)-β-Glucans with carbohydrate and/or synthetic side chains tend to be more soluble. They can still adopt triple helical conformations but these do not aggregate due to the steric hindrance of side chains on the outside of the helix. In contrast, backbone modifications as in (1,3;1,4)-β-glucans disrupt polymer chain conformational regularity and prevent the formation of triple helices. This results in the need to invoke an alternative mechanism for cross-link formation, the most likely hypothesis for which is association of regular trisaccharide repeat sequences.

Although some general principles of physico-chemical properties can be related to chemical structure features, the phase behaviour of (1,3)-β-glucans is often complex and incompletely described. In particular, the interplay between kinetic features (e.g. time/temperature/solvent history) and macroscopic phase behaviour (gelation, liquid crystalline behaviour, short-lived chain associations, etc.) is often not well understood. As the biological functions of (1,3)-β-glucans in, for example, cell wall architecture, extracellular structuring, molecular recognition and energy storage are closely linked to polymer conformation and higher level structures, it is not surprising that Nature has evolved to make use of this richness of physico-chemical behaviour.

## References

Atalla, R. H., & Vanderhart, D. L. (1984). Native cellulose–a composite of 2 distinct crystalline forms. *Science*, *223*, 283–285.

Bluhm, T. L., Deslandes, Y., Marchessault, R. H., Perez, S., & Rinaudo, M. (1982). Solid-state and solution conformation of scleroglucan. *Carbohydate Research*, *100*, 117–130.

Bluhm, T. L., & Sarko, A. (1977). Triple helical structure of lentinan, a linear beta (1→3)-D-glucan. *Canadian Journal of Chemistry-Revue Canadienne De Chimie*, *55*, 293–299.

Bo, S., Milas, M., & Rinaudo, M. (1987). Behaviour of scleroglucan in aqueous solution containing sodium hydroxide. *International Journal of Biological Macromolecules*, *9*, 153–157.

Böhm, N., & Kulicke, W-M. (1999a). Rheological studies of barley (1→3)(1→4)-β-glucan in concentrated solution: investigation of the viscoelastic flow behaviour in the sol state. *Carbohydate Research*, *315*, 293–301.

Böhm, N., & Kulicke, W-M. (1999b). Rheological studies of barley (1→3)(1→4)-β-glucan in concentrated solution: mechanistic and kinetic investigation of the gel formation. *Carbohydate Research*, *315*, 302–311.

Bot, A., Smorenburg, H. E., Vreeker, R., Paques, M., & Clark, A. H. (2001). Melting behaviour of schizophyllan extracellular polysaccharide gels in the temperature range between 5 and 20 degree. *Carbohydrate Polymers*, *45*, 363–372.

Buliga, G. S., Brant, D. A., & Fincher, G. B. (1986). The sequence statistics and solution conformation of a barley (1→3,1→4)-β-D-glucan. *Carbohydrate Research*, *157*, 139–156.

Burchard, W. (1994). Light scattering. In S. B. Ross-Murphy (Ed.), *Physical techniques for the study of food biopolymers* (pp. 151–213). London: Blackie Academic & Professional.

Burton, B. A., & Brant, D. A. (1983). Comparative flexibility, extension, and conformation of some simple polysaccharide chains. *Biopolymers*, *22*, 1769–1792.

Chuah, C. T., Sarko, A., Deslandes, Y., & Marchessault, R. H. (1983). Packing analysis of carbohydrates and polysaccharides. Part 14. Triple-helical crystalline structure of curdlan and paramylon hydrates. *Macromolecules*, *16*, 1375–1382.

Clark, A. H., & Ross-Murphy, S. B. (1987). Structural and mechanical properties of biopolymer gels. *Advances in Polymer Science*, *130*, 51–57.

Coviello, T., Coluzzi, G., Palleschi, A., Grassi, M., Santucci, E., & Alhaique, F. (2003a). Structural and rheological characterization of scleroglucan/borax hydrogel for drug delivery. *International Journal of Biological Macromolecules*, *32*, 83–92.

Coviello, T., Palleschi, A., Grassi, M., Matricardi, P., Bocchinfuso, G., & Alhaique, F. (2006). Scleroglucan: A versatile polysaccharide for modified drug delivery. *Molecules*, *10*, 6–33.

Coviello, T., Maeda, H., Yuguchi, Y., Urakawa, H., Kajiwara, K., Dentini, M., & Crescenzi, V. (1998). Conformational characteristics of oxidized scleroglucan. *Macromolecules*, *31*, 1602–1607.

Coviello, T., Dentini, M., Crescenzi, V., & Vincent, A. (1995). Ionic strength and temperature dependence of oxidized scleroglucan solution properties: optical activity and viscosity data. *Carbohydrate Polymers*, *26*, 5–10.

Coviello, T., Grassi, M., Lapasin, R., Marino, A., & Alhaique, F. (2003b). Scleroglucan/borax: characterization of a novel hydrogel system suitable for drug delivery. *Biomaterials*, *24*, 2789–2798.

Coviello, T., Grassi, M., Rambone, G., & Alhaique, F. (2001). A crosslinked system from scleroglucan derivative: preparation and characterization. *Biomaterials*, *22*, 1899–1909.

Cui, W., & Wood, P. J. (2000). Relationships between structural features, molecular weight and rheological properties of cereal β-D-glucans. In K. Nishinari (Ed.), *Hydrocolloids, physical chemistry and industrial applications of gels, polysaccharides and proteins: Vol. 1* (pp. 159–168). Amsterdam: Elsevier.

Cui, W., Wood, P. J., Blackwell, B., & Nikiforuk, J. (2000). Physicochemical properties and structural characterization by two dimensional NMR spectroscopy of wheat β-D-glucan–comparison with other cereal β-D-glucans. *Carbohydrate Polymers*, *41*, 249–258.

De Nooy, A. E. J., Rori, V., Masci, G., Dentini, M., & Crescenzi, V. (2000). Synthesis and preliminary characterization of charged derivatives and hydrogels from scleroglucan. *Carbohydrate Research*, *324*, 116–126.

Deslandes, Y., Marchessault, R. H., & Sarko, A. (1980). Packing analysis of carbohydrates and polysaccharides triple-helical structure of (1-3)-beta-D-glucan. *Macromolecules*, *13*, 1466–1471.

Ding, Q., Jiang, S., Zhang, L., & Wu, S. (1998). Laser light scattering studies of pachyman. *Carbohydrate Research*, *308*, 339–343.

Dobashi, T., Nobe, M., Yoshihara, H., Yamamato, T., & Konno, A. (2004). Liquid crystalline gel with refractive index gradient of curdlan. *Langmuir*, *20*, 6530–6534.

Dobashi, T., Yoshihara, H., Nobe, M., Koike, M., Yamamoto, T., & Konno, A. (2005). Liquid crystalline gel beads of curdlan. *Langmuir*, *21*, 2–4.

Durran, D. M., Howlin, B. J., Webb, G. A., & Gidley, M. J. (1995). Ab-initio nuclear shielding calculations of a model alpha–(1-4)-glucan. *Carbohydrate Research*, *271*, C1–C5.

Evans, N. A., & Hoyne, P. A. (1982). A fluorochrome from aniline blue–structure, synthesis and fluorescence properties. *Australian Journal of Chemistry*, *35*, 2572–2575.

Evans, N. A., Hoyne, P. A., & Stone, B. A. (1984). Characteristics and specificity of the interaction of a fluorochrome from aniline blue (sirofluor) with polysaccharides. *Carbohydrate Polymers*, *4*, 215–230.

Faijes, M., Ima, T., Bulone, V., & Planas, A. (2004). In vitro synthesis of a crystalline (1 -> 3,1 -> 4)-beta-D-glucan by a mutated (1 -> 3,1 -> 4)-beta-D-glucanase from Bacillus. *Biochemical Journal*, *380*, 635–641.

Falch, B. H., & Stokke, B. T. (2001). Structural stability of (1 -> 3)-beta-D-glucan macrocycles. *Carbohydrate Polymers*, *44*, 113–121.

Falch, B. H., Elgsaeter, A., & Stokke, B. T. (1999). Exploring the (1→3)-β-D-glucan conformational phase diagrams to optimize the linear to macrocycle conversion of the triple-helical polysaccharide scleroglucan. *Biopolymers*, *50*, 496–512.

Fang, Y., & Nishinari, K. (2004a). Gelation behaviors of schizophyllan-sorbitol aqueous solutions. *Biopolymers*, *73*, 44–60.

Fang, Y., Takahashi, R., & Nishinari, K. (2004b). A gel network constituted by rigid schizophyllan chains and nonpermanent cross-links. *Biomacromolecules*, *5*, 126–136.

Fang, Y., Takahashi, R., & Nishinari, K. (2004c). Rheological characterization of Schizophyllan aqueous solutions after denaturation–renaturation treatment. *Biopolymers*, *74*, 302–315.

Fang, Y., Takemasa, M., Katsuta, K., & Nishinari, K. (2004d). Rheology of schizophyllan solutions in isotropic and anisotropic phase regions. *Journal of Rheology*, *48*, 1147–1166.

Farina, J. I., Sineriz, F., Molina, O. E., & Perotti, N. I. (2001). Isolation and physicochemical characterization of soluble scleroglucan from Sclerotium rolfsii. Rheological properties, molecular weight and conformational characteristics. *Carbohydrate Polymers*, *44*, 41–50.

Foster, T. J., Ablett, S., McCann, M. C., & Gidley, M. J. (1996). Mobility-resolved C-13-NMR spectroscopy of primary plant cell walls. *Biopolymers*, *39*, 51–66.

Fuchs, T., Richtering, W., & Burchard, W. (1995). Thermoreversible gelation of a polysaccharide with immunological activity: rheology and dynamic light scattering. *Macromolecular Symposia*, *99*, 227–238.

Fuchs, T., Richtering, W., Burchard, W., Kajiwara, K., & Kitamura, S. (1997). Gel point in physical gels: rheology and light scattering from thermoreversibly gelling schizophyllan. *Polymer Gels and Networks*, *5*, 541–559.

Funami, T., Funami, M., Yada, H., & Nakao, Y. (1999a). Gelation mechanism of curdlan by dynamic viscoelasticity measurements. *Journal of Food Science*, *64*, 129–132.

Funami, T., Funami, M., Yada, H., & Nakao, Y. (1999b). Rheological and thermal studies on gelling characteristics of curdlan. *Food Hydrocolloids*, *13*, 317–324.

Funami, T., Funami, M., Yada, H., & Nakao, Y. (2000). A rheological study on the effects of heating rate and dispersing method on the gelling characteristics of curdlan aqueous dispersions. *Food Hydrocolloids*, *14*, 509–518.

Funami, T., & Nishinari, K. (2007). Gelling characteristics of curdlan aqueous dispersions in the presence of salts. *Food Hydrocolloids*, *21*, 59–65.

Fyfe, C. A., Stephenson, P. J., Taylor, M. G., Bluhm, T. L., Deslandes, Y., & Marchessault, R. H. (1984). Hydration effects in the C-13 CP/MAS NMR-spectra of solid (1-3)-beta-D-glucans. *Macromolecules*, *17*, 501–502.

Gawronski, M., Aguirre, G., Conrad, H., Springer, T., & Stahmann, K. P. (1996). Molecular structure and precipitates of a rodlike polysaccharide in aqueous solution by SAXS experiments. *Macromolecules*, *29*, 1516–1520.

Gidley, M. J., & Bociek, S. M. (1985). Molecular organization in starches–a C-13 CP MAS NMR-study. *Journal of the American Chemical Society*, *107*, 7040–7044.

Gidley, M. J., Dea, I. C. M., Eggleston, G., & Morris, E. R. (1987). Structure and gelation of Rhizobium capsular polysaccharide. *Carbohydrate Research*, *160*, 381–396.

Gómez, C., Navarro, A., Manzanares, P., Horta, A., & Carbonell, J. V. (1997a). Physical and structural properties of barley (1→3), (1→4)-β-D-glucan. I. Determination of molecular weight and macromolecular radius by light scattering. *Carbohydrate Polymers*, *32*, 7–15.

Gómez, C., Navarro, A., Manzanares, P., Horta, A., & Carbonell, J. V. (1997b). Physical and structural properties of barley (1→3), (1→4)-β-D-glucan. II. Viscosity, chain stiffness and macromolecular dimensions. *Carbohydrate Polymers*, *32*, 17–22.

Grassi, M., Lapasin, R., Pricl, S., & Colombo, I. (1996). Apparent non-Fickian release from a scleroglucan gel matrix. *Chemical Engineering Communications*, *155*, 89–112.

Grassi, M., Lapasin, R., & Pricl, S. (1996). A study of the rheological behavior of scleroglucan weak gel systems. *Carbohydrate Polymers*, *29*, 169–181.

Grimm, A., Krüger, E., & Burchard, W. (1995). Solution properties of β-D-(1, 3)(1, 4)-glucan isolated from beer. *Carbohydrate Polymers*, *27*, 205–214.

Grisel, M., & Muller, G. (1996). Rheological properties of schizophyllan in presence of borate ions. *Progress in Colloid and Polymer Science*, *102*, 32–37.

Grisel, M., & Muller, G. (1997). The salt effect over the physical interactions occurring for schizophyllan in the presence of borate ions. *Macromolecular Symposia*, *114*, 127–132.

Grisel, M., & Muller, G. (1998). Rheological properties of the schizophyllan-borax system. *Macromolecules*, *31*, 4277–4281.

Harada, T., Koreeda, A., Sato, S., & Kasai, N. (1979). Electron microscopic study on the ultrastructure of curldan gel: Assembly and dissociation of fibrils by heating. *Journal of Electron Microscopy*, *28*, 147–153.

Harada, T. (1992). The story of research into curdlan and the bacteria producing it. *Trends in Glycoscience and Glycotechnology*, *4*, 309–317.

Harada, T., Okuyama, K., Konno, A., Koreeda, A., & Harada, A. (1994). Effect of heating on formation of curdlan gels. *Carbohydrate Polymers*, *24*, 101–106.

Hayashi, Y., Shinyashiki, N., Yagihara, S., Yoshiba, K., Teramoto, A., Nakamura, N., Miyazaki, Y., Sorai, M., & Wang, Q. (2002). Ordering in aqueous polysaccharide solutions. I. dielectric relaxation in aqueous solutions of a triple-helical polysaccharide schizophyllan. *Biopolymers*, *63*, 21–31.

Hirano, I., Einaga, Y., & Fujita, H. (1979). Curdlan (bacterial β-1,3-glucan) in a cadoxen-water mixture. *Polymer Journal*, *11*, 901–904.

Hirao, T., Sato, T., Teramoto, A., Matsuo, T., & Suga, H. (1990). Solvent effects on the cooperative order-disorder transition of aqueous solutions of schizophyllan, a triple helical polysaccharide. *Biopolymers*, *29*, 1867–1876.

Hirashima, M., Takaya, T., & Nishinari, K. (1997). DSC and rheological studies on aqueous dispersions of curdlan. *Thermochimica Acta*, *306*, 109–114.

Hrmova, M., Imai, T., Rutten, S. J., Fairweather, J. K., Pelosi, L., Bulone, V., Driguez, H., & Fincher, G. B. (2002). Mutated barley (1,3)-beta-D-glucan endohydrolases synthesize crystalline (1,3)-beta-D-glucans. *Journal of Biological Chemistry*, *277*, 30102–30111.

Itou, T., Teramoto, A., Matsuo, T., & Suga, H. (1986). Ordered structure in aqueous polysaccharide .5. cooperative order-disorder transition in aqueous schizophyllan. *Macromolecules*, *19*, 1234–1240.

Itou, T., Teramoto, A., Matsuo, T., & Suga, H. (1987). Isotope effect on the order-disorder transition in aqueous schizophyllan. *Carbohydrate Research*, *160*, 243–257.

Jin, Y., Zhang, H., Yin, Y., & Nishinari, K. (2006a). Comparison of curdlan and its carboxymethylated derivative by means of Rheology, DSC, and AFM. *Carbohydrate Research*, *341*, 90–99.

Jin, Y., Zhang, H., Yin, Y., & Nishinari, K. (2006b). Conformation of curdlan as observed by tapping mode atomic force microscopy. *Colloid and Polymer Science*, *284*, 1371–1377.

Kanzawa, Y., Harada, T., Koreeda, A., & Harada, A. (1987). Curdlan gel formed by neutralizing its alkaline solution. *Agricultural and Biological Chemistry*, *51*, 1839–1843.

Kanzawa, Y., Harada, T., Koreeda, A., Harada, A., & Okuyama, K. (1989). Difference of molecular association in two types of curdlan gel. *Carbohydrate Polymers*, *10*, 299–313.

Kasai, N., & Harada, T. (1980). Ultra structure of curdlan. ACS symposium Series, 141. In A. D. French & K. H. Gardner (Eds.), *Fiber diffraction methods* (pp. 363–383). Washington, DC: The American Chemical Society.

Kashiwagi, Y., Norisuye, T., & Fujita, H. (1981). Triple helix of a Schizophyllum Commune polysaccharide in aqueous solution. 4. Light scattering and viscosity in dilute aqueous sodium hydroxide. *Macromolecules*, *14*, 1220–1225.

Kikumoto, S., Miyajima, T., Yoshizumi, S., Fujimoto, S., & Kimura, K. (1970). Polysaccharide produced by *Schizophyllum Commune*. 1. Formation and some properties of an extracellular polysaccharide. *Journal of the Agricultural Chemical Society of Japan*, *44*, 337–342.

Kitamura, S., & Kuge, T. (1989). A differential scanning calorimetric study of the conformational traqnsition of Schizophyllan in mixtures of water and dimethylsulfoxide. *Biopolymers*, *28*, 639–654.

Kitamura, S., Ozasa, M., Tokioka, H., Hara, C., Ukai, S., & Kuge, T. (1990). A differential scanning calorimetric study of the conformational transitions of several kinds of (1-6) branched (1-3)-beta-D-glucans in a mixture of water and dimethylsulfoxide. *Thermochimica Acta*, *163*, 89–96.

Kitamura, S., Hirano, T., Takeo, K., Fukada, H., Takahashi, K., Falch, B. H., & Stokke, B. T. (1996). Conformational transitions of schizophyllan in aqueous alkaline solution. *Biopolymers*, *39*, 407–416.

Ko, Y. T., & Lin, Y. L. (2004). 1,3-beta-Glucan quantification by a fluorescence microassay and analysis of its distribution in foods. *Journal of Agricultural and Food Chemistry*, *52*, 3313–3318.

Koike, A., Nemoto, N., Inoue, T., & Osaki, K. (1995). Dynamic light scattering and dynamic viscoelasticity of poly(vinyo alcohol) in aqueous borax solutions. 1.Concentration effects. *Macromolecules*, *28*, 2339–2344.

Konno, A., Azechi, Y., & Kimura, H. (1979). Properties of curdlan gel. *Agricultural and Biological Chemistry*, *43*, 101–104.

Konno, A., & Harada, T. (1991). Thermal properties of curdlan in aqueous suspension and curdlan gel. *Food Hydrocolloids*, *5*, 427–434.

Lapasin, R., & Pricl, S. (1999). *Rheology of industrial polysaccharides. Theory and applications*. USA: An Aspen Publ.

Lazaridou, A., & Biliaderis, C. G. (2004a). Cryogelation of cereal beta-glucans: structure and molecular size effects. *Food Hydrocolloids*, *18*, 933–947.

Lazaridou, A., Biliaderis, C. G., & Izydorczyk, M. S. (2003). Molecular size effects on rheological properties of oat beta-glucans in solution and gels. *Food Hydrocolloids*, *17*, 693–712.

Lazaridou, A., Biliaderis, C. G., Micha-Screttas, M., & Steele, B. R. (2004b). A comparative study on structure-function relations of mixed-linkage (1→3), (1→4) linear beta-d-glucans. *Food Hydrocolloids*, *18*, 837–855.

Lee, H. C., & Brant, D. A. (2002). Rheology of concentrated isotropic and anisotropic xanthan solutions. 1. A rodlike low molecular weight sample. *Macromolecules*, *35*, 2212–2222.

Lee, H. C., & Brant, D. A. (2002). Rheology of concentrated isotropic and anisotropic xanthan solutions. 2. a semiflexible wormlike intermediate molecular weight sample. *Macromolecules*, *35*, 2223–2234.

Lee, H. C., & Brant, D. A. (2002). Rheology of concentrated isotropic and anisotropic xanthan solutions: 3. temperature dependence. *Biomacromolecules*, *3*, 742–753.

Maeda, H., Rambone, G., Coviello, T., Yuguchi, Y., Urakawa, H., Alhaique, F., & Kajiwara, K. (2001). Low-degree oxidized scleroglucan and its hydrogel. *International Journal of Biological Macromolecules*, *28*, 351–358.

Maeda, H., Yuguchi, Y., Kitamura, S., Urakawa, H., Kajiwara, K., Richtering, W., Fuchs, T., & Burchard, W. (1999). Structural aspect of gelation in schizophyllan/sorbitol aqueous solution. *Polymer Journal*, *31*, 530–534.

Maeda, I., Saito, H., Masada, M., Misaki, A., & Harada, T. (1967). Properties of gels formed by heat treatment of curdlan, a bacterial β-1,3 glucan. *Agricultural and Biological Chemistry*, *31*, 1184–1188.

Marchessault, R. H., Deslandes, Y., Ogawa, K., & Sundararajan, P. R. (1977). X-ray diffraction data for beta–(1-3)–D-glucan. *Canadian Journal of Chemistry-Revue Canadienne De Chimie*, *55*, 300–303.

Matsumoto, T., & Okubo, T. (1991). Viscoelastic investigation of crystal-liquid transition in concentrated monodisperse latices. *Journal of Rheology*, *35*, 135–148.

Matsumoto, T., & Inoue, H. (1991). Effect of heat denaturation on the structure and rheological properties of ovalbumin aqueous colloids. *Journal of the Chemical Society, Faraday Transactions*, *87*, 3385–3388.

Matsuzawa, S., Yamaura, K., Tanigami, T., Nomura, T., & Nakata, M. (1987). Reaction of borate ions with isotactic polyvinyl alcohol. *Polymer Communications*, *28*, 105–106.

McIntire, T. M., Penner, R. M., & Brant, D. A. (1995). Observations of a circular, triple-helical polysaccharide using noncontact atomic force microscopy. *Macromolecules*, *28*, 6375–6377.

McIntire, T. M., & Brant, D. A. (1998). Observations of the (1→3)-β-D-glucan linear triple helix to macrocycle interconversion using noncontact atomic force microscopy. *Journal of American Chemical Society*, *120*, 6909–6919.

Miyoshi, K., Uezu, K., Sakurai, K., & Shinkai, S. (2004). Proposal of a new hydrogen-bonding form to maintain curdlan triple helix. *Chemistry & Biodiversity*, *1*, 916–924.

Moresi, M., Lo Presti, S., & Mancini, M. (2001). Rheology of scleroglucan dispersions. *Journal of Food Engineering*, *50*, 235–245.

Morgan, K. R., Roberts, C. J., Tendler, S. J. B., Davies, M. C., & Williams, P. F. (1999). A C-13 CP MAS NMR spectroscopy and AFM study of the structure of Glucagel (TM), a gelling beta-glucan from barley. *Carbohydrate Research*, *315*, 169–179.

Morris, E. R., Cutler, A. N., Ross-Murphy, S. B., Rees, D. A., & Price, J. (1981). Concentration and shear rate dependence of viscosity in random coil polysaccharide solutions. *Carbohydrate Polymers*, *1*, 5–21.

Morris, E. R. (1984). Applications of hydrocolloids. In G. O. Phillips, D. J. Wedlock, & P. A. Williams (Eds.), *Gums and stabilisers for the food industry 2* (pp. 57–78). Oxford: Pergamon Press.

Nakata, M., Kawaguchi, T., Kodama, Y., & Konno, A. (1998). Characterization of curdlan in aqueous sodium hydroxide. *Polymer*, *39*, 1475–1481.

Nijenhuis, K.te. (1997). Thermoreversible Networks. *Advances in Polymer Science*, *130*, 51–57.

Nishinari, K. (1997). Rheological and DSC study of sol-gel transition in aqueous dispersions of industrially important polymers and colloids. *Colloid and Polymer Science*, *275*, 1093–1107.

Nishinari, K. (2000). Rheology of physical gels and gelling processes. *Reports on Progress in Polymer Physics in Japan*, *43*, 163–192.

Nishinari, K., Hirashima, M., Miyoshi, E., & Takaya, T. (1998a). Rheological and DSC studies of aqueous dispersions and gels of curdlan. In P. A. Williams & G. O. Phillips (Eds.), *Gum and stabilizers for the food industry, 9*, pp. 26–33.

Nishinari, K., Miyoshi, E., & Takaya, T. (1998b). Rheological studies of aqueous dispersions of microparticulated cellulose. In P. A. Williams & G. O. Phillips (Eds.), *Gum and stabilizers for the food industry, 9,* pp. 16–25.

Nishinari, K., & Zhang, H. (2000). Curdlan. In G. O. Phillips & P. A. Williams (Eds.), *Handbook of hydrocolloids* (pp. 270–286). Cambridge England: Woodhead Publishing Limited.

Nishinari, K., & Zhang, H. (2004). Recent advances in the understanding of heat set gelling polysaccharides. *Trends in Food Science and Technology*, *15*, 305–312.

Nishiyama, Y., Sugiyama, J., Chanzy, H., & Langan, P. (2003). Crystal structure and hydrogen bonding system in cellulose I$\alpha$, from synchrotron X-ray and neutron fiber diffraction. *Journal of America Chemical Society*, *125*, 14300–14306.

Nobe, M., Kuroda, N., Dobashi, T., Yamamoto, T., Konno, A., & Nakata, M. (2005). Molecular weight effect on liquid crystalline gel formation of curdlan. *Biomacromolecules*, *6*, 3373–3379.

Norisuye, T., Yanaki, T., & Fujita, H. (1980). Triple helix of a Schizophyllum Commune polysaccharide in aqueous solution. *Journal of Polymer Science, Polymer Physics Edition*, *18*, 547–558.

Ogawa, K., Tsurugi, J., Watanabe, T., & Ono, S. (1972). Conformational behavior of a gel-forming $(1\rightarrow3)$-$\beta$-D-glucan in alakaline solution. *Carbohydrate Research*, *23*, 399–405.

Ogawa, K., Tsurugi, J., & Watanabe, T. (1973). The dependence of the conformation of a $(1\rightarrow3)$-$\beta$-D-glucan on chain-length in alkaline solution. *Carbohydrate Research*, *29*, 397–403.

Ohno, N., Adachi, Y., Suzuki, I., Oikawa, S., Sato, K., Suzuki, Y., Ohsawa, M., & Yadomae, T. (1986). 2 different conformations of antitumor glucans obtained from *Grifola-frondosa*. *Chemical & Pharmaceutical Bulletin*, *34*, 2555–2560.

Ohno, N., Adachi, Y., & Yadomae, T. (1988). Conformations of fungal beta-D-glucans in the fruit body of edible fungi assessed by CPMAS C-13 nuclear magnetic resonance spectroscopy. *Chemical & Pharmaceutical Bulletin*, *36*, 1198–1204.

Okobira, T., Miyoshi, K., Uezu, K., Sakurai, K., & Shinkai, S. (2008). Molecular dynamics studies of side chain effect on the beta-1,3-D-glucan triple helix in aqueous solution. *Biomacromolecules*, *9*, 783–788.

Okuyama, K., Otsubo, A., Fukuzawa, Y., Ozawa, M., Harada, T., & Kasai, N. (1991). Single-helical structure of native curdlan and its aggregation state. *Journal of Carbohydrate Chemistry*, *10*, 645–656.

Palleschi, A., Bocchinfuso, G., Coviello, T., & Alhaique, F. (2005). Molecular dynamics investigations of the polysaccharide scleroglucan: first study on the triple helix structure. *Carbohydrate Research, 340*, 2154–2162.

Pelosi, L., Bulone, V., & Heux, L. (2006). Polymorphism of curdlan and (1 -> 3)-beta-D-glucans synthesized in vitro: A C-13 CP-MAS and X-ray diffraction analysis. *Carbohydrate Polymers, 66*, 199–207.

Perlin, A. S., & Taber, W. A. (1963). A glucan produced by *Claviceps purpurea*. *Canadian Journal of Chemistry, 41*, 2278–2282.

Pezron, E., Leibler, L., & Lafuma, F. (1989). Complex formation in polymer-ion solutions. 2. Polyelectrolyte effects. *Macromolecules, 22*, 2656–2662.

Pezron, E., Richard, A., & Leibler, L. (1990). Rheology of galactomannan-borax gels. *Journal of Polymer Scince, Polymer Physics Edition, 28*, 2445–2461.

Ross-Murphy, S. B. (Ed.), (1994). *Physical techniques for the study of food biopolymers*. London: Blackie Academic & Professional.

Saito, H., Miyata, E., & Sasaki, T. (1978). A $^{13}$C nuclear magnetic resonance study of gel-forming (1→3)-β-D-Glucans: Molecular-weight dependence of helical conformation and of the presence of junction zones for association of primary molecules. *Macromolecules, 11*, 1244–1251.

Saito, H., Ohki, T., & Sasaki, T. (1977). A $^{13}$C nuclear magnetic resonance study of gel-forming (1-3)-β-D-glucans. Evidence of the presence of single-helical conformation in a resilient gel of a curdlan-type polysaccharide 13140 from *Alcaligenes faecalis var. myxogenes*. *Biochemistry, 16*, 908–914.

Saito, H., Ohki, T., & Sasaki, T. (1979). A $^{13}$C nuclear magnetic resonance study of polysaccharide gels. Molecular architecture in the gels consisting of fungal, branched (1→3)-β-D-glucans (lentinan and schizophyllan) as manifested by conformational changes induced by sodium hydroxide. *Carbohydrate Research, 74*, 227–240.

Saito, H., Yoshioka, Y., Uehara, N., Aketagawa, J., Tanaka, S., & Shibata, Y. (1991). Relationship between conformation and biological response for (1,3)-β-glucans in the activation of coagulation Factor G from limulus amebocyte lysate and host-mediated antitumor activity. Demonstration of single-helix conformation as a stimulant. *Carbohydrate Research, 217*, 181–190.

Saito, H., Tabeta, R., Yokomi, M., & Erata, T. (1987). A high-resolution solid state C-13 NMR-study of the secondary structure of linear (1-3)-beta-D-glucans–a condformational elucidation of non-crystalline and crystalline forms by means of conformation-dependent C-13 chemical shifts. *Bulletin of the Chemical Society of Japan, 60*, 4259–4266.

Saito, H., Yokoi, M., & Yoshioka, Y. (1989). Effect of hydration on conformational change or stabilisation of (1-3)-beta-D-glucans of various chain lengths in the solid state as studied by high-resolution solid-state C-13 NMR-spectroscopy. *Macromolecules*, *22*, 3892–3898.

Saito, H., Yoshioka, Y., Yokoi, M., & Yamada, J. (1990). Distinct gelation mechanism between linear and branched (1-3)-beta-D-glucans as revealed by high resolution solid-state C-13 NMR. *Biopolymers*, *29*, 1689–1698.

Sakurai, K., & Shinkai, S. (2000). Molecular recognition of adenine, cytosine, and uracil in a single-stranded RNA by a natural polysaccharide: Schizophyllan. *Journal of American Chemical Society*, *122*, 4520–4521.

Sakurai, K., Mizu, M., & Shinkai, S. (2001). Polysaccharide-polynucleotide complexes. 2. complementary polynucleotide mimic behavior of the natural polysaccharide schizophyllan in the macromolecular complex with single-stranded RNA and DNA. *Biomacromolecules*, *2*, 641–650.

Sato, M., Nobe, M., Dobashi, T., Yamamoto, T., & Konno, A. (2005). Multifold curdlan gel formation by dialysis into aqueous solutions of metal salts. *Colloid and Polymer Science*, *284*, 293–300.

Sato, T., Norisuye, T., & Fujita, H. (1983). Triple helix of a Schizophyllum Commune polysaccharide in aqueous solution. 5. Light scattering and refractometry in mixtures of water and dimethyl sulfoxide. *Macromolecules*, *16*, 185–189.

Sato, T., Norisuye, T., & Fujita, H. (1981). Melting behavior of schizophyllum Commune polysaccharide in mixtures of water and dimethyl sulfoxide. *Carbohydrate Research*, *95*, 195–204.

Sato, T., Sakurai, K., Norisuye, T., & Fujita, H. (1983). Collapse of randomly coiled Schizophyllan in mixtures of water and dimethylsulfoxide. *Polymer Journal*, *15*, 87–96.

Shedletzky, E., Unger, C., & Delmer, D. P. (1997). A microtiter-based fluorescence assay for (1,3)-beta-glucan synthases. *Analytical Biochemistry*, *249*, 88–93.

Shibayama, M., Kurokawa, H., Nomura, S., Muthukumar, M., Stein, R. S., & Roy, S. (1992). Small-angle neutron scattering from poly(vinyl alcohol)-borate gels. *Polymer*, *33*, 2883–2890.

Skendi, A., Biliaderis, C. G., Lazaridou, A., & Izydorczyk, M. S. (2003). Structure and rheological properties of water soluble β-glucans from oat cultivars of *Avena sativa* and *Avena bysantina*. *Journal of Cereal Science*, *38*, 15–31.

Sletmoen, M., Christensen, B. E., & Stokke, B. T. (2005). Probing macromolecular architectures of nanosized cyclic structures of (1 -> 3)-beta-D-glucans by AFM and SEC-MALLS. *Carbohydrate Research*, *340*, 971–979.

Sletmoen, M., & Stokke, B. T. (2008). Review: Higher order structure of (1,3)-beta-D-glucans and its influence on their biological activities and complexation abilities. *Biopolymers*, *89*, 310–321.

Stipanovic, A. J., & Giammatteo, P. J. (1989). In J. Edward Glass (Ed.), Polymers in aqueous media. pp. 73–87.

Stokke, B. T., Elgsaeter, A., Brant, D. A., & Kitamura, S. (1991). Supercoiling in circular triple-helical polysaccharides. *Macromolecules*, *24*, 6349–6351.

Stokke, B. T., Elgsaeter, A., Brant, D. A., Kuge, T., & Kitamura, S. (1993). Macromolecular cyclization of (1-6)-branched-(1-3)-beta-D-glucans observed after denaturation renaturation of the triple-helical structure. *Biopolymers*, *33*, 193–198.

Surenjav, U., Zhang, L., Xu, X., Zhang, X., & Zeng, F. (2006). Effects of molecular structure on antitumor activities of (1-3)-b-D-glucan from different *Lentinus Edodes*. *Carbohydrate Polymers*, *63*, 97–104.

Swalina, C. W., Zauhar, R. J, DeGrazia, M. J., & Moyna, G. (2001). Derivation of C-13 chemical shift surfaces for the anomeric carbons of oligosaccharides and glycopeptides using ab initio methodology. *Journal of Biomolecular NMR*, *21*, 49–61.

Tada, T., Matsumoto, T., & Masuda, T. (1997). Network structure of curdlan in DMSO and mixture of DMSO and water. *Biopolymers*, *42*, 479–487.

Tada, T., Matsumoto, T., & Masuda, T. (1998). Structure of molecular association of curdlan at dilute regime in alkaline aqueous systems. *Chemical Physics*, *228*, 157–166.

Tada, T., Matsumoto, T., & Masuda, T. (1999). Dynamic viscoelastocity and small-angle X-ray scattering studies on the gelation mechanism and network structure of curdlan gels. *Carbohydrate Polymers*, *39*, 53–59.

Tako, M. (1996). Molecular origin for the thermal stability of schizophyllan. *Polymer Gels and Networks*, *4*, 303–313.

Tan, I., Flanagan, B. M., Halley, P. J., Whittaker, A. K., & Gidley, M. J. (2007). A method for estimating the nature and relative proportions of amorphous, single, and double-helical components in starch granules by C-13 CP/MAS NMR. *Biomacromolecules*, *8*, 885–891.

Tatsumi, D., Ishioka, S., & Matsumoto, T. (1999). Effect of particle and salt concentrations on the rheological properties of cellulose fibrous suspensions. *Nihon Reoroji Gakkaishi*, *27*, 243–248.

Thistlethwaite, P., Porter, I., & Evans, N. (1986). Photophysics of the aniline blue fluorophore–a fluorescent probe showing specificity toward (1-3)–beta-D-glucans. *Journal of Physical Chemistry*, *90*, 5058–5063.

Tobitani, A., & Ross-Murphy, S. B. (1997). Heat-induced gelation of globular proteins. 2. Effect of environmental factors on single-component and mixed-protein gels. *Macromolecules*, *30*, 4855–4862.

Tosh, S. M., Brummer, Y., Wood, P. J., Wang, Q., & Weisz, J. (2004a). Evaluation of structure in the formation of gels by structurally diverse (1→3) (1→4)-beta-d-glucans from four cereal and one lichen species. *Carbohydate Polymers*, *55*, 249–259.

Tosh, S. M., Wood, P. J., Wang, Q., & Weisz, J. S. M. (2004b). Structural characteristics and rheological properties of partially hydrolyzed oat beta-glucan: effects of molecular weight and hydrolysis method. *Carbohydate Polymers*, *55*, 425–436.

Tvaroska, I., Ogawa, K., Deslandes, Y., & Marchessault, R. H. (1983). Crystalline conformation and structure of lichenan and barley β-D-glucan. *Canadian Journal of Chemistry*, *61*, 1608–1616.

Vaikousi, H., Biliaderis, C. G., & Izydorczyk, M. S. (2004). Solution flow behaviour and gelling properties of water-soluble barley (1→3) (1→4)-beta-glucans varying in molecular size. *Journal of Cereal Science*, *39*, 119–137.

Van, K., Norisuye, T., & Teramoto, A. (1981). Liquid crystal formation in aqueous solutions of a polysaccharide schizophyllan. *Molecular Crystals & Liquid Crystals*, *78*, 123–134.

Van, K., & Teramoto, A. (1982). Isotropic-liquid crystal phase equilibrium in aqueous solutions of a triple-helical polysaccharide schizophyllan. *Polymer Journal*, *14*, 999–1005.

Vårum, K. M., Martinsen, A., & Smidsrod, O. (1991). Fractionation and viscometric characterization of a (1→3), (1→4)-β-D-glucan from oat, and universal calibration of a high-performance size-exclusion chromatographic system by the use of fractionated β-D-glucans, alginates and pullulans. *Food Hydrocolloids*, *5*, 363–374.

Vårum, K. M., Smidsrod, O., & Brant, D. A. (1992). Light scattering reveals micelle-like aggregation in the (1→3), (1→4)-β-D-glucans from oat aleurone. *Food Hydrocolloids*, *5*, 497–511.

Vuppu, A. K., Garcia, A. A., & Vernia, C. (1997). Tapping mode atomic force microscopy of scleroglucan networks. *Biopolymers*, *42*, 89–100.

Wang, Q., Wood, P. J., Huang, X., & Cui, S. W. (2003). Preparation and characterisation of molecular weight standards of low polydispersity from oat and barley (1→3) (1→4)-beta-d-glucan. *Food Hydrocolloids*, *17*, 845–853.

Wang, Y., & Zhang, L. (2006). Chain conformation of carboxymethylated derivatives of (1→3)-β-D-glucan from Poria cocos sclerotium. *Carbohydrate Polymers*, *65*, 504–509.

Watase, M., & Nishinari, K. (1988). Thermal and rheological properties of poly(vinyl alcohol) hydrogels prepared by repeated cycles of freezing and thawing. *Makromolekulare Chemie*, *189*, 871–880.

Watase, M., & Nishinari, K. (1994). Rheology and DSC of curdlan-DMSO-water systems. In K. Nishinari & E. Doi (Eds.), *Food hydrocolloids: Structures, properties and functions* (pp. 125–129). New York: Plenum Press.

Wikström, K., Lindahl, L., Andersson, R., & Westerlund, E. (1994). Rheological studies of water-soluble (1–3), (1-4)-β-d-glucans from milling fractions of oat. *Journal of Food Science*, *59*, 1077–1080.

Winter, H. H., & Mours, M. (1997). Rheology of polymers near liquid-solid transitions. *Advances in Polymer Science*, *134*, 165–234.

Woodward, J. R., Fincher, G. B., & Stone, B. A. (1983). Water soluble (1→3) (1→4)-β-d-glucan from barley (*Hordeum vulgare*) endosperm. II. Fine structure. *Carbohydrate Polymers*, *3*, 207–225.

Xu, X., Xu, J., Zhang, Y., & Zhang, L. (2008). Rheology of triple helical Lentinan in solution: Steady shar viscosity and dynamic oscillatory behavikor. *Food Hydrocolloids*, *22*, 735–741.

Yamakawa, H. (1971). *Modern theory of polymer solutions*. New York: Harper & Row.

Yanaki, T., Kojima, T., & Norisuye, T. (1981). Triple helix of scleroglucan in dilute aqueous sodiumhydroxide. *Polymer Journal*, *153*, 1135–1143.

Yanaki, T., & Norisuye, T. (1983). Triple helix and random coil of scleroglucan in dilute solution. *Polymer Journal*, *15*, 187–396.

Yanaki, T., Norisuye, T., & Fujita, H. (1980). Triple helix of a Schizophyllum Commune polysaccharide in aqueous solution. 3. Hydrodynamic properties in water. *Macromolecules*, *13*, 1462–1466.

Yanaki, T., Tabata, K., & Kojima, T. (1985). Melting behaviour of a triple helical polysaccharide Schizophyllan in aqueous solution. *Carbohydrate Polymers*, *5*, 275–283.

Yanaki, T., Ito, W., Tabata, K., Kojima, T., Norisuye, T., Takano, N., & Fujita, H. (1983). Correlation between the anti-tumor activity of a polysaccharide schizophyllan and its triple helical conformation in dilute aqueous solution. *Biophysical Chemistry*, *17*, 337–342.

Yin, Y., Zhang, H., & Nishinari, K. (2007). Voltametric characterization of the hydrophobic interaction in polysaccharide hydrogels. *Journal of Physical Chemistry B*, *111*, 1590–1596.

Young, S. H., & Jacobs, R. R. (1998). Sodium hydroxide-induced conformational change in schizophyllan detected by the fluorescence dye, aniline blue. *Carbohydrate Research*, *310*, 91–99.

Young, S. H., Dong, W. J., & Jacobs, R. R. (2000). Observation of a partially opened triple-helix conformation in 1 -> 3-beta-glucan by fluorescence resonance energy transfer spectroscopy. *Journal of Biological Chemistry*, *275*, 11874–11879.

Yoshiba, K., Ishino, T., Teramoto, A., Nakamura, N., Miyazaki, Y., Sorai, M., Wang, Q., Hayashi, Y., Shinyashiki, N., & Yagihara, S. (2002). Ordering in aqueous polysaccharide solutions II. optical rotation and heat capacity of aqueous solutions of a triple-helical polysaccharide schizophyllan. *Biopolymers*, *63*, 370–381.

Yoshiba, K., Teramoto, A., & Nakamura, N. (2003). Static water structure detected by heat capacity measurements on aqueous solutions of a triple-helical polysaccharide schizophyllan. *Biomacromolecules*, *4*, 1348–1356.

Zentz, F., Verchere, J. F., & Muller, G. (1992). Thermal-denaturation and degradation of schizophyllan. *Carbohydrate Polymers*, *17*, 289–297.

Zhang, H., Nishinari, K., Williams, M. A. K., Foster, T. J., & Norton, I. T. (2002). A molecular description of the gelation mechanism of curdlan. *International Journal of Biological Macromolecules*, *30*, 7–16.

Zhang, Y., Xu, X., & Zhang, L. (2008). Dynamic viscoelastic behavior of triple helical Lentinan in water: Effect of temperature. *Carbohydrate Polymers*, *73*, 26–34.

Zhang, Y., Xu, X., Xu, J., & Zhang, L. (2007). Dynamic viscoelastic behavior of triple helical Lentinan in water: Effects of concentration and molecular weight. *Polymer*, *48*, 6681–6690.

Zhang, L., Zhang, X., Zhou, Q., Zhang, P., Zhang, M., & Li, X. (2001). Triple helix of b-D-glucan from *Lentinus Edodes* in 0.5 M NaCl aqueous solution characterized by light scattering. *Polymer Journal*, *33*, 317–321.

Zhang, L., Li, X., Zhou, Q., Zhang, X., & Chen, R. (2002). Transition from triple helix to coil of lentinan in solution measured by SEC, viscometry, and $^{13}C$ NMR. *Polymer Journal*, *34*, 443–449.

Zhang, L., Zhang, M., Dong, J., Guo, J., Song, Y., & Cheung, P. C. K. (2001). Chemical structure and chain conformation of the water-insoluble glucan isolated from Pleurotus tuber-regium. *Biopolymers*, *59*, 457–464.

Zhang, M., Zhang, L., & Cheung, P. C. K. (2003). Molecular mass and chain conformation of carboxymethylated derivatives of β-glucan from sclerotia of *Pleurotus tuber-regium*. *Biopolymers*, *68*, 150–159.

Zhang, M., Zhang, L., Wang, Y., & Cheung, P. C. K. (2003). Chain conformation of sulfated derivatives of β-glucan from sclerotia of *Pleurotus tuber-regium*. *Carbohydrate Research*, *338*, 2683–2870.

Zhang, M., Cheung, P. C. K., Zhang, L., Chiu, C., & Ooi, V. E. C. (2004). Carboxymethylated β-glucans from mushroom sclerotium of *Pleurotus tuber-regium* as novel water-soluble anti-tumor agent. *Carbohydrate Polymers*, *57*, 319–325.

Zhang, M., Cui, S. W., Cheung, P. C. K., & Wang, Q. (2006). Antitumor polysaccharides from mushrooms: a review on their isolation process, structural characteristics and antitumor activity. *Trends in Food Science & Technology*, *18*, 4–19.

CHAPTER 3.1

# *Plant and Microbial Enzymes Involved in the Depolymerization of (1,3)-β-D-Glucans and Related Polysaccharides*

**Maria Hrmova and Geoffrey B. Fincher**
*Australian Centre for Plant Functional Genomics, School of Agriculture, Food and Wine, University of Adelaide, Waite Campus, Glen Osmond, SA 5064, Australia*

## 3.1.1 Introduction

(1,3)-β-D-Glucans are widely distributed in plants, algae, fungi, euglenoid protozoans and bacteria, where they are involved in cell wall structure and in a range of other biological functions. In higher plants, (1,3)-β-D-glucans are commonly referred to as callose and are detectable under ultraviolet light following staining with the Aniline Blue fluorochrome (Stone et al., 1985). During normal growth and development of plants, callose is found in the cell plate in dividing cells. It is also a major component of pollen mother cell walls and pollen tubes, and is found as a structural component of plasmodesmatal canals. Callose is deposited in abscission zones and on sieve plates in dormant phloem (Stone and Clarke, 1992). In the developing endosperm of cereal grains, callosic material is deposited 3–6 days after pollination, when the syncytium is compartmentalized by the centripetal synthesis of cell walls around individual nuclei (Wilson et al., 2006). In this process, which results in cellularization of the endosperm, (1,3)-β-D-glucans appear as one of the first components of the growing cell walls that appear between 3 and 4 days after pollination (Wilson et al., 2006). Apart from its role in normal growth and development, callose is deposited between the plasma membrane and the cell wall after plants are exposed to abiotic and biotic stresses such as wounding, desiccation, metal toxicity, and microbial attack (Stone and Clarke, 1992). There has been considerable interest in the role of callose in plant–microbe interactions. Following microbial attack, one common response of plant host cells is to rapidly synthesize and deposit callose in close proximity to the invading pathogen (Ryals et al., 1996; Donofrio and Delaney, 2001; Jacobs et al., 2003).

The papillary callose so deposited is thought to contain (1,3)-β-D-glucans, other polysaccharides, phenolic compounds, reactive oxygen intermediates, and some proteins (Smart et al., 1986; Bolwell, 1993; Bestwick et al., 1997; Thordal-Christensen et al., 1997; Heath et al., 2002).

In fungi, (1,3)-β-D-glucans occur as major components of cell walls, in combination with a range of other polysaccharides, including chitin, cellulose and mannans. In addition to their role as structural components of the wall, (1,3)-β-D-glucans may function as intra- and extracellular storage polysaccharides, and extracellular (1,3)-β-D-glucans and/or oligosaccharides derived from them may cause wilting and other effects during pathogenic attack on higher plants (Stone and Clarke, 1992). Curdlan is a bacterial (1,3)-β-D-glucan found as a protective capsule around *Agrobacterium* and related rhizobia, and the Gram-positive *Cellulomonas* spp.

The wide distribution and diverse functional roles of (1,3)-β-D-glucans and related polysaccharides in biological systems are well documented. In many systems, the (1,3)-β-D-glucans may be transitory in nature. For example, in higher plants wound callose, sieve plate callose, plasmodesmatal callose, cell plate callose and the (1,3)-β-D-glucans that are deposited during cellularization of the endosperm can disappear or may be substantially reduced in amount following their initial deposition. This indicates that there are (1,3)-β-D-glucan hydrolases operating as integral components of these systems. Here, we will summarize the major classes of enzymes that are responsible for the depolymerization of (1,3)-β-D-glucans and related polysaccharides, with an emphasis on the enzymes found in higher plants. Although (1,3)-β-D-glucans are generally linear homopolymers of (1,3)-linked β-D-glucopyranosyl residues, structural variants of linear (1,3)-β-D-glucans include cyclic (1,3;1,6)-β-D-glucans, branch-on-branch (1,3;1,6)-β-D-glucans, side-chain-branched (1,3;1,6)-β-D-glucans, (1,3;1,6)-β-D-glucans with both (1,3)- and (1,6)-linkages in the main chain, and side-chain-branched (1,3;1,2)-β-D-glucans (Stone and Clarke, 1992; Stone et al., 2009). In addition, (1,3;1,4)-β-D-glucans are found as major cell wall components in members of the monocotyledon family Poaceae, to which the cereals and grasses belong, and in related families of the order Poales (Trethewey et al., 2005). There have been recent reports of the presence of a (1,3;1,4)-β-D-glucan in walls of Equisetum (Fry et al., 2008a; Sørenson et al., 2008), which is a primitive plant in the horsetails group. The (1,3;1,4)-β-D-glucans of the Poaceae are linear, unbranched polysaccharides containing β-D-glucopyranosyl monomers polymerized through both (1,4)- and (1,3)-linkages (Fincher and Stone, 2004). A structurally related (1,3;1,4)-β-D-glucan, lichenin, is present in the walls of the fungal component of the lichen, Iceland moss (*Cetraria islandica*) (Honneger and O'Haisch, 2001), and is covalently attached to the non-reducing ends of the core of (1,3;1,6)-β-D-glucans of the cell wall polysaccharide complex in the yeast *Saccharomyces cerevisiae* (Kapteyn et al., 1997) and the fungus *Aspergillus fumigatus* (Bernard and Latgé, 2001).

In discussing enzymes that depolymerize (1,3)-β-D-glucans and related polysaccharides it is therefore necessary to consider enzymes that hydrolyse a range of linkage types between β-D-glucopyranosyl residues. Furthermore, enzymes can catalyse the hydrolysis of the glycosidic linkages through an endo- or exo-action pattern and, in the case of (1,3;1,4)-β-D-oligoglucosides, enzymes more traditionally classified as β-D-glucosidases might be involved. To achieve the complete hydrolysis of these polysaccharides several types of enzyme are usually required. In some cases, cleavage of linkages between β-D-glucopyranosyl residues in (1,3)-β-D-glucans and related polysaccharides is achieved through phosphorylase activity. Each of these major classes of enzymes, and in some cases closely related enzymes, are discussed below.

### 3.1.2 (1,3)-β-D-Glucan endohydrolases and related enzymes

Enzymes that catalyse the hydrolysis of (1,3)-β-D-glucans with an endo-action pattern have traditionally been classified in the EC 3.2.1.39 group, although representatives of this group have been classified according to their structural features in families GH16, GH17, GH55, GH64 and GH81 of the CAZY database (Coutinho and Henrissat, 1999; http://www.cazy.org/). Most information is available on members of the GH16 and GH17 families, in which enzymes from the Archae, bacteria, Eukaryota and viruses are found. Most attention here will be focused on the (1,3)-β-D-glucan endohydrolases and related enzymes from the GH17 family of glycoside hydrolases.

The GH17 enzymes are classified in clan GH-A, a superfamily of enzymes with common $(\beta/\alpha)_8$ folds, catalytic machinery and with net retention of configuration, but they exhibit a divergence of substrate specificities (Henrissat and Davies, 1997; Ryttersgaard et al., 2002; Taylor et al., 2005). The location of the catalytic apparatus is conserved in these enzymes with the catalytic acid/base and nucleophile glutamates positioned on strands β-4 and β-7, respectively (Jenkins et al., 1995; Henrissat et al., 1995; Taylor, 2005). It is therefore intriguing how these enzymes differentiate between substrates (Taylor et al., 2005). Clan GH-A represents the largest of the clans classified in the CAZY database (Henrissat and Davies, 1997), where as many as 18 GH family members are listed, among them GH1, GH5, GH17 and GH26 that will be the subject of this review.

### 3.1.2.1 Family GH17 (1,3)-β-D-glucan endohydrolases

*Classification and occurrence*

The (1,3)-β-D-glucan endohydrolases are widely distributed in plants, where they participate in the turnover of (1,3)-β-D-glucans during normal growth and development or are members of the PR2 group of pathogenesis-related proteins that are believed to protect the plant against invading fungi by hydrolysing the (1,3)-β-D-glucans and related polysaccharides in fungal walls (Boller, 1987; Meins et al., 1992). The genome sequences of rice and Arabidopsis have shown that the (1,3)-β-D-glucan endohydrolase gene families of higher plants are very large, with up to 70 members (Coutinho and Henrissat, 1999). The large gene families presumably allow the plant to exert independent control of expression of individual (1,3)-β-D-glucan endohydrolases in different tissues, during different stages of growth and development, and in response to different abiotic or biotic stresses.

*Properties*

Three (1,3)-β-D-glucan glucanohydrolase (EC 3.2.1.39) isoenzymes GI, GII and GIII have been purified from young leaves of barley (*Hordeum vulgare*) using $(NH_4)_2SO_4$ fractional precipitation, ion-exchange chromatography, chromatofocusing and gel-filtration chromatography, and have been characterized in detail (Hrmova and Fincher, 1993) (Table 1). The three (1,3)-β-D-glucan endohydrolases are monomeric proteins of apparent $M_r$ 32 000 with pI values in the range 8.6–9.8. Amino acid sequence analyses confirmed that the three isoenzymes represent the products of separate genes. Isoenzymes GI and GII are less stable at elevated temperatures and are active over a narrower pH range than is isoenzyme GIII, which is a glycoprotein containing 20–30 mol of hexose equivalents/mol of enzyme. Amino acid sequences indicate that isoenzyme GII has no potential *N*-glycosylation site, but that isoenzymes GI and GIII have one and five potential *N*-glycosylation sites, respectively (Hrmova and Fincher, 1993).

*Substrate specificities and action patterns*

The preferred substrate for the three characterized barley enzymes is laminarin from the brown alga *Laminaria digitata*, an essentially linear (1,3)-β-D-glucan with a low degree of glucosyl substitution and a degree of polymerization (DP) of approx. 25. The three enzymes are classified as endohydrolases, because they yield (1,3)-β-D-oligoglucosides with DP 3–8 in the initial stages of hydrolysis of laminarin (Hrmova and Fincher, 1993). These oligosaccharides can be completely hydrolysed to glucose by various β-D-glucan exohydrolases and β-D-glucosidases, as described in the following sections.

**Table 1: Properties of barley family GH17 (1,3)-β-D-glucan endohydrolases**

| Property | Isoenzyme GI | Isoenzyme GII | Isoenzyme GIII |
|---|---|---|---|
| Apparent molecular mass | 33 000 | 32 300 | 32 400 |
| Amino acid residues | 310 | 306 | 305 |
| Isoelectric point | 8.6 | 9.5 | 9.8 |
| *N*-glycosylation sites | 1 | 0 | 5 |
| Substrate specificity | (1,3)- and (1,3;1,6)-β-D-glucans | (1,3)- and (1,3;1,6)-β-D-glucans | (1,3)- and (1,3;1,6)-β-D-glucans |
| Anomeric configuration | Retained during hydrolysis | Retained during hydrolysis | Retained during hydrolysis |
| Glycosyl hydrolase classification | Family GH17 | Family GH17 | Family GH17 |
| Protein fold | $(\alpha/\beta)_8$ barrel | $(\alpha/\beta)_8$ barrel | $(\alpha/\beta)_8$ barrel |
| Catalytic acid | Glu96 or Glu291 | Glu94 or Glu288 | Glu92 or Glu287 |
| Catalytic nucleophile | Glu234 | Glu231 | Glu230 |
| Substrate binding sites | 7 | 7 | 6 |
| Expression sites | Root, leaves | Aleurone | Root, leaves |

Source: Høj et al., 1988; Høj et al., 1989; Xu et al., 1992; Hrmova and Fincher, 1993; Wang et al., 1992; Chen et al., 1993b; Varghese et al., 1994; Chen et al., 1995b; Hrmova et al., 1995.

Kinetic analyses of the three barley (1,3)-β-D-glucan endohydrolases indicated apparent $K_m$ values in the range 200 μM, $k_{cat}$ constants of 40–160 $s^{-1}$ and pH optimum of 4.8 (Hrmova and Fincher, 1993). The three isoenzymes hydrolyse substituted (1,3)-β-D-glucans with DP 25–31 and various high-$M_r$ substituted and side-branched fungal (1,3;1,6)-β-D-glucans. However, the isoenzymes differ in their rates of hydrolysis of a (1,3;1,6)-β-D-glucan from baker's yeast and their specific activities against laminarin vary significantly. The enzymes do not hydrolyse (1,3;1,4)-β-D-glucans, (1,6)-β-D-glucans, CM-cellulose, insoluble (1,3;1,6)-β-D-glucans or aryl β-D-glycosides (Hrmova and Fincher, 1993). The properties of the three barley (1,3;1,4)-β-D-glucan endohydrolases are summarized in Table 1.

### *Structure*

The three-dimensional (3D) structure of barley (1,3)-β-D-glucan endohydrolase isoenzyme GII was the first solved for a (1,3)-β-D-glucan endohydrolase (Varghese et al., 1994). Crystals were prepared by hanging drop vapour diffusion at 4°C in the presence of $(NH_4)_2SO_4$ or polyethylene glycol (Chen et al., 1993a). The 3D structure of the enzyme was solved to 2.3 Å resolution and showed that the enzyme assumed a $(\beta/\alpha)_8$ barrel conformation (Varghese et al., 1994). Eight parallel β-strands constitute the core of the enzyme and are connected via short loops to α-helices that form the outer surfaces of the enzyme (Varghese et al., 1994). A deep substrate-binding cleft traverses the surface of the enzyme, perpendicular to the barrel axis (Fig. 1) and is consistent with its endo-action pattern, whereby the enzyme can bind at most positions along the polysaccharide substrate and hydrolyse internal glycoside linkages. The dimensions of the cleft suggest that it could accommodate eight or nine residues of a (1,3)-β-D-glucan chain (Varghese et al., 1994), consistent with subsite mapping analyses (Hrmova et al., 1995) (Fig. 2). In the latter work, oligo-(1,3)-β-D-glucosides with DP 2–9 were labelled at their reducing terminal residues by catalytic tritiation and were used in kinetic and thermodynamic analyses to examine substrate binding in (1,3)-β-D-glucan glucanohydrolase isoenzymes GI, GII, and GIII from young seedlings of barley. Bond-cleavage frequencies and the kinetic parameter $k_{(cat)}/K_m$ were calculated as a function of substrate chain length to define the number of subsites that accommodate individual β-D-glucosyl residues and to estimate binding energies at each subsite. Each isoenzyme was shown to have eight β-D-glucosyl-binding subsites (Hrmova et al., 1995) (Figure 2). The catalytic amino acids, which were identified on the basis of the 3D structure and site-directed mutagenesis (Chen et al., 1993b; Chen et al., 1995a), were shown to be located between the third and fourth subsite from the non-reducing terminus of the substrate. Negative binding energies in subsites adjacent to the hydrolysed glycosidic linkage suggested that substrate distortion may occur in this region during binding, and that the resultant strain induced in the substrate might facilitate hydrolytic cleavage (Hrmova et al., 1995). It was concluded, on the basis of the 3D conformation of the enzyme and molecular modelling of substituted or branched (1,3)-β-D-glucan substrates into the structure, that the (1,3)-β-D-glucan endohydrolases require relatively extended regions of unsubstituted or unbranched (1,3)-β-D-glucan backbone for activity. This has implications with respect to their ability to hydrolyse the branched or substituted (1,3)-β-D-glucans or (1,3;1,6)-β-D-glucans of fungal walls (Hrmova et al., 1995).

### *Catalytic mechanism*

The availability of the purified (1,3)-β-D-glucan endohydrolases from barley, together with their 3D structures, enabled their catalytic mechanisms to be investigated. The family GH17 enzymes are believed to retain anomeric configuration during hydrolysis (Coutinho and Henrissat, 1999; http://www.cazy.org/), and this was confirmed in the case of the barley (1,3)-β-D-glucan endohydrolases by NMR analyses of products released during the reaction (Chen et al., 1995b).

All of the barley β-D-glucan endo- and exohydrolases described in this chapter are retaining enzymes (Chen et al., 1995b; Hrmova et al., 1996) and the following description of the catalytic

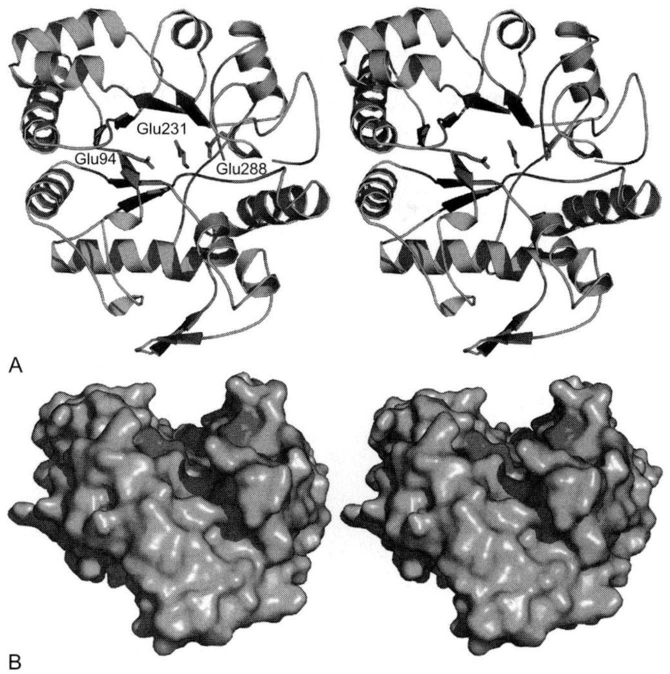


**Fig. 1: Stereo representation of the 3D structure of a barley family GH17 (1,3)-β-D-glucanase isoenzyme GII (Protein Data Bank accession code 1GHS). (A) Ribbon diagram of the enzyme with three catalytic amino acids Glu94, Glu231 and Glu288 shown in sticks and coloured in red. (B) Molecular surface drawing of the enzyme with three catalytic amino acids Glu94, Glu231 and Glu288 (in red) positioned at the bottom of substrate-binding cleft. Adapted with permission from the National Academy of Sciences of USA (Varghese et al., 1994). The colour specifications refer to colours in panels.**

mechanism of retaining glycoside hydrolases will therefore be used by most of these enzymes. When a substrate is bound to a retaining glycoside hydrolase, initiation of hydrolysis occurs through the protonation of the glycosidic oxygen atom by an appropriately positioned amino acid, known as the catalytic acid/base (White and Rose, 1997; Zechel and Withers, 1999; Zechel and Withers, 2000) (Fig. 3). The proton donor is generally the unionized carboxylic acid group of an Asp or Glu residue (Legler and Herrchen, 1981). Protonation of the glycosidic oxygen results in the cleavage of the C1-O bond of the glycosidic linkage, whereupon the aglycone portion of the substrate diffuses away from the catalytic site (Heightman and Vasella, 2000). At the same time a positively charged oxocarbenium ion-like transition state is generated and this forms a covalent glycosyl–enzyme intermediate with an inverted configuration at the anomeric centre. The covalent bond of the intermediate links the glycone portion of the substrate

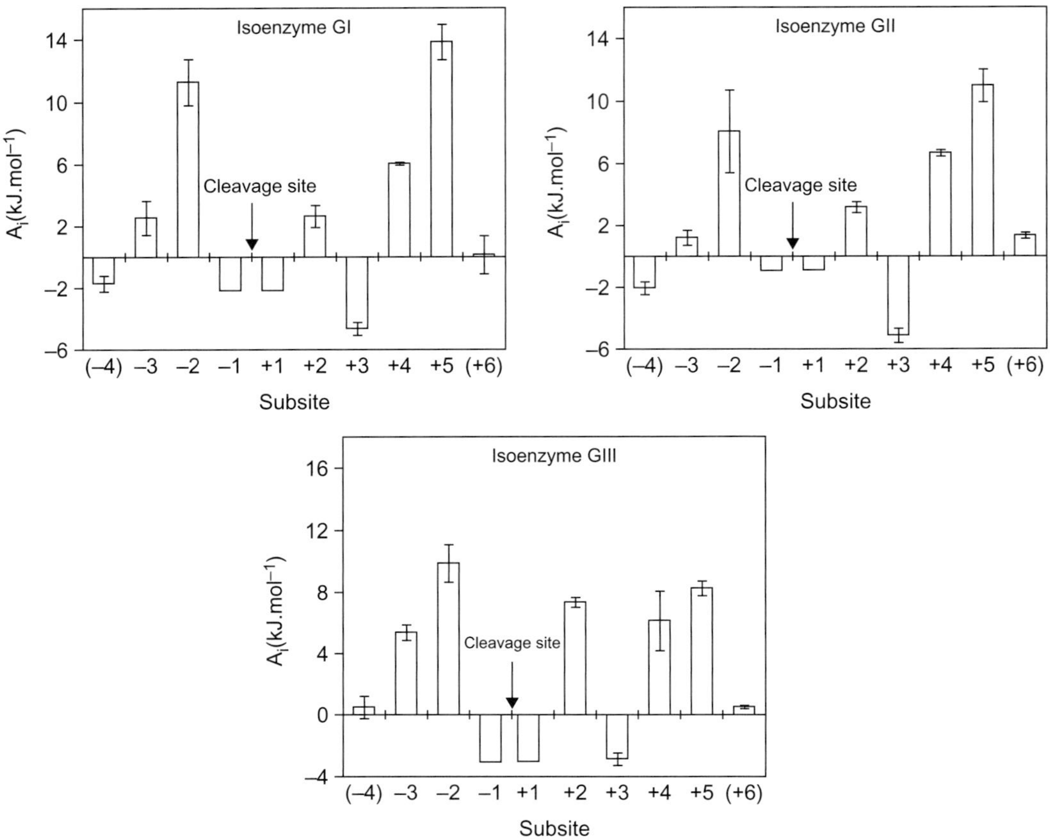


**Fig. 2: Subsite maps of the barley (1,3)-β-D-glucanase isoenzymes GI, GII and GIII. The number of subsites and the positions of catalytic amino acid residues were determined as described (Hrmova et al., 1995). The arrow shows the position of the catalytic amino acid residues. The error bars indicate standard deviations that were calculated for pairs of positional isomers; in some cases only one pair of positional isomers will bind at a particular subsite; thus in these cases standard deviations are not given. Adapted with permission from the American Chemical Society (Hrmova et al., 1995).**

to a different, nucleophilic amino acid, but again this is usually an Asp or Glu residue (Street et al., 1992; McCarter and Withers, 1994; Svensson and Stone, 2001). Finally, hydrolysis of the covalent glycosyl–enzyme linkage through a water molecule liberates the glycone portion of the hydrolysed substrate. The catalytic acid on the enzyme is simultaneously re-protonated. A diagrammatical representation of the likely mechanism of substrate hydrolysis by retaining glycoside hydrolases, occurring by a double displacement mechanism, is shown in Fig. 3.

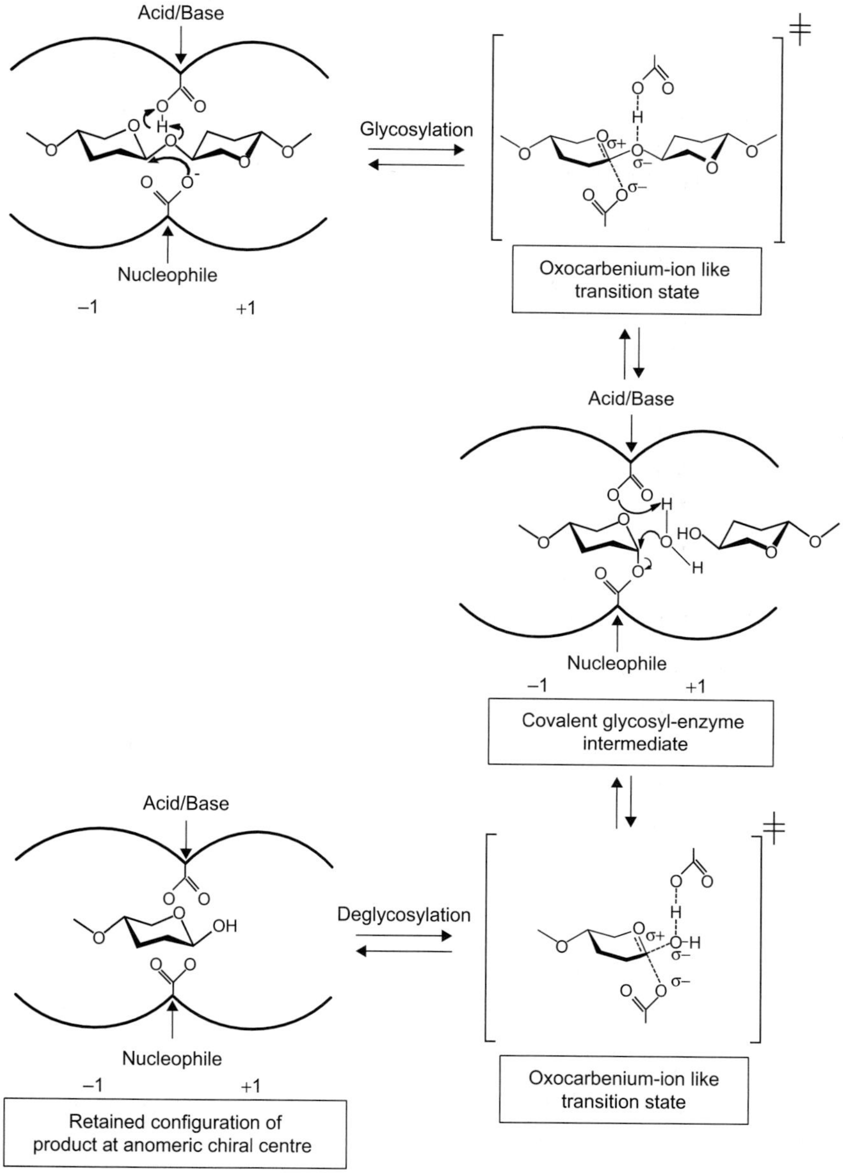


**Fig. 3: Mechanism of catalysis of retaining plant β-D-glucan endo- and exohydrolases. The double-displacement reaction (Koshland, 1953) at the anomeric chiral centre proceeds through the protonation of the glucosidic oxygen, the formation of an oxocarbenium-ion like transition state, a covalent β-glucosyl–enzyme intermediate, a second oxocarbenium-ion like transition state, and finally the regeneration of the two catalytic amino acid residues. The anomeric configuration of the released product is retained. Substrate-binding subsites −1 and +1 are indicated. Adapted with permission from Kluwer Academic Publishers (Hrmova and Fincher, 2001).**

The catalytic nucleophile of the barley (1,3)-β-D-glucan endohydrolases isoenzyme GII is almost certainly Glu231, based on observations that this residue is tagged with specific epoxyalkyl -β-D-oligoglucoside inhibitors (Chen et al., 1993b) and is highly conserved in family 17 glycoside hydrolases. It is positioned about two thirds of the way along the bottom of the substrate-binding cleft (Varghese et al., 1994). Although the catalytic acid/base was initially identified as Glu288 by carbodiimide-mediated labelling procedures (Chen et al., 1993b), it was subsequently suggested that the catalytic acid/base was more likely to be Glu93 (Jenkins et al., 1995; Henrissat et al., 1995). Both residues are highly conserved in family 17 glycoside hydrolases (Høj and Fincher, 1995). The Glu288 residue is located about 8 Å from the catalytic nucleophile Glu231 (Varghese et al., 1994), but this is considered to be too far for retaining glycoside hydrolases. The distance between Glu232 and Glu93 is 5–6 Å and this is certainly more typical of retaining glycoside hydrolases (Jenkins et al., 1995; Henrissat et al., 1995). The spatial dispositions of these residues in the catalytic site region are shown in Fig. 1. As pointed out by Hrmova and Fincher (2001), it is not yet clear whether Glu93 or Glu288, or possibly both, contribute to protonation of the glycosidic oxygen during (1→3)-β-D-glucan hydrolysis by this enzyme, because there are often several highly conserved acidic amino acids in the catalytic region of glycoside hydrolases, together with conserved basic amino acids (Chen et al., 1995a) and coordinated water molecules. Hrmova and Fincher (2001) suggested that the proton that hydrolyses the glycosidic linkage of the bound substrate might be relatively mobile in this conserved region of acidic and basic amino acid residues.

#### *Transglycosylation reactions*

It was demonstrated during the characterization of the barley (1,3)-β-D-glucan endohydrolases that they would catalyse transglycosylation reactions in the presence of high concentrations of substrate; the major products of these reactions appeared to be gentiobiose and higher (1,6)-β-D-oligoglucosides (Hrmova and Fincher, 1993). Although similar reactions have been detected with other (1,3)-β-D-glucan endohydrolases, it is not clear whether these have any functional relevance or are simply the mechanistic consequences of the catalytic mechanism, through which sugars released in earlier catalytic events can substitute for water molecules as acceptors of the glycone product of the hydrolysis.

### 3.1.2.2 Family GH17 (1,3;1,4)-β-D-glucan endohydrolases

#### *Classification and occurrence*

The family GH17 (1,3;1,4)-β-D-glucan endohydrolases are classified as EC 3.2.1.73 enzymes and are largely restricted to higher plants. As noted earlier, (1,3;1,4)-β-D-glucans are found mainly in the Poaceae, and it would appear that they have evolved relatively recently given the position of the Poaceae in phylogenic trees (Høj and Fincher, 1995; Buckeridge et al., 2001). Nevertheless, there are isolated instances where they occur in other plants, as in the primitive land plant Equisetum (Fry et al., 2008a; Sørenson et al., 2008). The evolution of (1,3;1,4)-β-D-glucans as major constituents of cell walls in the Poaceae presumably resulted in a concomitant need for enzymes that could specifically hydrolyse the polysaccharide during normal development and degradation of walls. For example, (1,3;1,4)-β-D-glucans account for up to 70% of the walls in the starchy endosperm of barley (Fincher, 1975) and require the action of specific (1,3;1,4)-β-D-glucan endohydrolases for their removal in germinated grain (Woodward and Fincher, 1982a).

There is a body of circumstantial evidence to suggest that members of the Poaceae recruited pre-existing (1,3)-β-D-glucan endohydrolases, which are widely distributed in plants, for the evolution of (1,3;1,4)-β-D-glucan endohydrolases (Høj and Fincher, 1995). Not only do the two enzymes have closely related substrate specificities and amino acid sequences, but the (1,3)-β-D-glucan endohydrolases and (1,3;1,4)-β-D-glucan endohydrolases also have almost identical 3D structures, as described in more detail below.

In contrast to the (1,3)-β-D-glucan endohydrolases, which are usually encoded by large gene families in higher plants, the gene families encoding the (1,3;1,4)-β-D-glucan endohydrolases appear to be relatively small. In barley there are only two genes encoding the (1,3;1,4)-β-D-glucan endohydrolases, and the high degree of amino acid sequence identity between the two extant barley enzymes (approximately 92%; Slakeski et al., 1990) further suggests that the two corresponding genes evolved via gene duplication relatively recently. The two genes in barley (*HvGlb1* and *HvGlb2*) are located on chromosome 1H and 7H, respectively (Burton et al., 2008). The gene encoding isoenzyme EI is transcribed in various tissues, including the scutellum and aleurone of germinated grain, and in young vegetative tissues (Slakeski and Fincher, 1992). Transcription of the gene encoding (1,3;1,4)-β-D-glucan endohydrolase isoenzyme EII appears to be restricted to the aleurone layer of germinated grain (Slakeski and Fincher, 1992).

In rice, there is at least one gene encoding a (1,3;1,4)-β-D-glucan endohydrolase, but the similarity of sequences between the (1,3)-β-D-glucan and (1,3;1,4)-β-D-glucan endohydrolases, and the absence of supporting biochemical evidence makes it difficult to determine whether or not there is more than one (1,3;1,4)-β-D-glucan endohydrolase in rice.

### *Properties*

The properties of the two barley (1,3;1,4)-β-D-glucan endohydrolases are summarized in Table 2. Both enzymes are basic proteins with a pI value of 8.5 and 10.6 for isoenzyme

**Table 2: Properties of barley family GH17 (1,3;1,4)-β-D-glucan endohydrolases**

| Property | Isoenzyme EI | Isoenzyme EII |
|---|---|---|
| Apparent molecular mass | 30 000 | 32 000 |
| Amino acid residues | 306 | 306 |
| Isoelectric point | 8.5 | 10.6 |
| Carbohydrate | 0 | 4% by weight |
| Substrate specificity | Absolute for (1,3;1,4)-β-D-glucans | Absolute for (1,3;1,4)-β-D-glucans |
| Anomeric configuration | Retained during hydrolysis | Retained during hydrolysis |
| Glycosyl hydrolase classification | Family GH17 | Family GH17 |
| Protein fold | $(\alpha/\beta)_8$ barrel | $(\alpha/\beta)_8$ barrel |
| Catalytic acid | Glu93 or Glu288 | Glu93 or Glu288 |
| Catalytic nucleophile | Glu232 | Glu232 |
| Substrate binding sites | 4-6 | 4-6 |
| Expression sites | Scutellum, young vegetative tissue, aleurone | Aleurone |

Source: Woodward and Fincher, 1982a; Woodward and Fincher, 1982b; Stuart et al., 1988; Slakeski and Fincher, 1992; Chen et al., 1993b; Varghese et al., 1994; Chen et al., 1995b.

EI and EII, respectively. They have 306 amino acid residues and molecular masses of 30 000–32 000 (Woodward and Fincher, 1982a). Isoenzyme EII carries about 4% by weight carbohydrate and is somewhat more stable at higher temperatures than isoenzyme EI (Doan and Fincher, 1992).

*Substrate specificity and action pattern*

The EC 3.2.1.73 barley (1,3;1,4)-β-D-glucan endohydrolases are absolutely specific for (1,3;1,4)-β-D-glucans of the type found in walls of the Poaceae. They exhibit an endo-action pattern and hydrolyse internal (1,4)-β-glucosidic linkages where these linkages are adjacent to a (1,3)-β-D-glucosyl residue, as follows:

```
              ↓           ↓                   ↓
G 4 G 4 G 3 G 4 G 4 G 3 G 4 G 4 G 4 G 4 G 3 G 4 G 4 … red
```

where G represents a β-D-glucosyl residue, 3 and 4 are (1,3)- and (1,4)-linkage, respectively, and red indicates the reducing terminus (Parrish et al., 1960; Anderson and Stone, 1975; Woodward and Fincher, 1982b). Thus, the EC 3.2.1.73 enzymes require adjacent (1,3)- and (1,4)-β-D-glucosyl residues and therefore release (1,3;1,4)-β-D-tri- and tetrasaccharides ($G4G3G_{red}$ and $G4G4G3G_{red}$) as major hydrolysis products. However, they also release higher oligosaccharides of up to 10 or more (1,4)-β-D-glucosyl residues with a single reducing

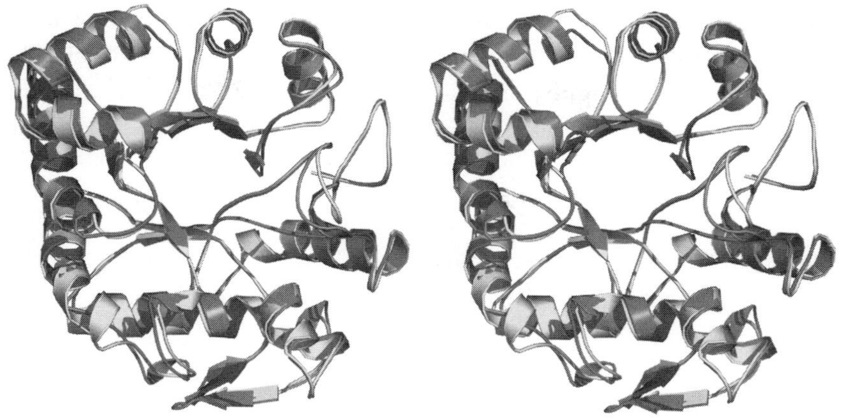

**Fig. 4: Stereo representation of 3D structures of a barley (1,3)-β-D-glucanase isoenzyme GII (green) superposed on (1,3;1,4)-β-D-glucanase isoenzyme EII (yellow). The rmsd deviation in Cα positions between the two enzymes is 0.65 A for 278 residues. Adapted with permission from the National Academy of Sciences of USA (Varghese et al., 1994). The colour specifications refer to colours in panels.**

terminal (1,3)-β-D-glucosyl residue (e.g. $G4G4G4G4G4G4G3G_{red}$) from the longer regions of adjacent (1,4)-linkages mentioned earlier (Woodward et al., 1983b; Wood et al., 1994).

In addition to the well-characterized (1,3;1,4)-β-D-glucan endohydrolases of the EC 3.2.1.73 class, an unusual (1,3;1,4)-β-D-glucan endohydrolase, referred to as 'β-D-glucan solubilase' in barley, but also reported in maize coleoptiles, is believed to release larger (1,3;1,4)-β-D-glucan molecules from walls (Bamforth and Martin, 1981; Inouhe et al., 1999). In maize the enzyme releases (1,3;1,4)-β-D-glucans with DP 60–100 from isolated polysaccharide substrates (Thomas et al., 2000). However, neither enzyme has been purified and the complete amino acid sequence of the maize coleoptile enzyme bears no similarity to amino acid sequences of (1,3;1,4)-β-D-glucan endohydrolases, cellulases or other plant glycoside hydrolases (Thomas et al., 1998; 2000).

### *Three-dimensional structure and evolution*

The 3D structure of barley (1,3;1,4)-β-D-glucan endohydrolase isoenzyme EII has been defined by X-ray crystallography to 2.2–2.3 Å resolution (Varghese et al., 1994). The enzyme adopts a $(\beta/\alpha)_8$ barrel fold (Fig. 4). As with the (1,3)-β-D-glucan endohydrolases, the substrate-binding region consists of a deep cleft about 40 Å long that extends across the surface of the enzyme and is long enough to accommodate 6–8 glucosyl-binding subsites (Fig. 4).

Although overall amino acid sequence identities between the barley (1,3)-β-D-glucan endohydrolases and the (1,3;1,4)-β-D-glucan endohydrolases are only about 50% (Høj and Fincher, 1995), the crystal structures of the enzymes show that their Cα polypeptide chains are almost perfectly superimposable (Figure 4; Varghese et al., 1994). This can be taken as strong evidence that the (1,3)- and (1,3;1,4)-β-D-glucan endohydrolases from barley originated from a common ancestral enzyme (Høj and Fincher, 1995).

Attempts to diffuse polysaccharide and oligosaccharide substrates into crystals have not been successful, so there is no diffraction data for the enzyme–substrate complex (M. Hrmova, Chen, L., J. N. Varghese, G. B. Fincher, unpublished data) and the nature of chemical interactions between amino acid residues and reactive groups on the substrate cannot be defined. If crystal structures of enzyme–substrate complexes could be obtained, then the details of substrate binding that explain the respective specificities for (1,3)-β-D-glucans or (1,3;1,4)-β-D-glucans should become apparent. Site-directed mutagenesis has been suggested as a means of interchanging the specificities of the two classes of enzymes (Høj and Fincher, 1995), but so far we have been unable to alter the specificities or to define residues that are important in defining the quite distinct specificities of the two barley enzymes.

#### *Catalytic mechanism*

Anomeric configuration is retained during hydrolysis of (1,4)-β-glucosyl linkages in (1,3;1,4)-β-D-glucans by EC 3.2.1.73 (1,3;1,4)-β-D-glucan endohydrolases (Chen et al., 1995b). The catalytic nucleophile of the enzyme is probably Glu232, which is highly conserved in family 17 glycoside hydrolases (Chen et al., 1995a), and the catalytic acid/base is likely to be either Glu288 or Glu93 (Chen et al., 1993b; Jenkins et al., 1995; Henrissat et al., 1995). As with the (1,3)-β-D-glucan endohydrolases isoenzyme GII, the proton that eventually hydrolyses the glycosidic linkage of the bound substrate might be relatively mobile in the conserved region of acidic and basic amino acid residues.

### *3.1.2.3 Family GH26 (1,3;1,4)-β-D-glucan endohydrolases*

The GH26 family of enzymes of the CAZy classification (Coutinho and Henrissat, 1999) contains predominantly bacterial and eukaryotic (1,4)-β-D-mannan endohydrolases and bacterial (1,3)-β-D-xylan endohydrolases, in addition to bacterial (1,3;1,4)-β-D-glucan endohydrolases. The GH26 family (1,3;1,4)-β-D-glucan endohydrolase from *Clostridium thermocellum* forms a polypeptide chain of 900 amino acid residues that folds into at least three independent modules (Carvalho et al., 2004; Taylor et al., 2005). Carvalho et al. (2004) characterized the binding properties and architecture of a COOH-terminal family 11 carbohydrate binding module β-sandwich that bound preferentially (1,3;1,4)-β-D-glucans and (1,4)-β-D-oligoglucosides, but no (1,3)-β-D-glucans. Taylor et al. (2005) determined the 3D structure of a GH26 catalytic module that possessed (1,3;1,4)-β-D-glucan endohydrolase activity, but no (1,4)-β-D-mannan endohydrolase activity or activity against the insoluble cellulosic substrate Avicel. In contrast, the GH5 module was active on (1,3;1,4)-β-D-glucan and Avicel. Investigations of hydrolytic properties of a two-domain GH26–GH5 bifunctional enzyme revealed its activity against (1,3;1,4)-β-D-glucan and Avicel; the latter activity was significantly higher than that of the GH5 module alone (Taylor et al., 2005). The latter authors showed that interactions at subsite −2 play a critical role in stabilizing the transition state. The consequences of these interactions during (1,3;1,4)-β-D-glucan binding thereby dictate why the *Clostridium* enzyme specifically binds and hydrolyses (1,3;1,4)-β-D-glucan-like substrates (Money et al., 2008). Finally, the GH26 family contains bacterial (1,3)-β-D-xylan endohydrolases that specifically hydrolysed (1,3)-β-D-xylan, but not other polysaccharides such as (1,4)-β-D-xylan, carboxymethylcellulose, curdlan, glucomannan and (1,4)-β-D-mannan (Araki, 2000; Okazaki et al., 2002). The 3D structure of a *Vibrio* sp. (1,3)-β-D-xylan endohydrolase catalytic module has recently been reported (Protein Data Bank accession number 2ddx; Sakaguchi, K., Kawamura, T., Watanabe, N., Kiyohara, M., Yamaguchi, K., Ito, M., Tanaka, I., unpublished data).

#### 3.1.2.4 Family GH16 (1,3)- and related β-D-glucan endohydrolases

Various microbial (1,3)- and (1,3;1,4)-β-D-glucan endohydrolases have been classified by Coutinho and Henrissat (1999) into the GH16 family of glycosyl hydrolases that are members of the GH-B clan in the CAZy classification and fold into β-jelly roll architectures (Henrissat and Davies, 1997). Members of this large and complex GH16 family can be divided into several subgroups on the basis of differences in substrate specificities (Strohmeier et al., 2004; Figure 5). The 'true' (1,3)-β-D-glucan endohydrolases are represented in the laminarinase and laminarinase-like subgroups (Fig. 5). A subgroup that is able to hydrolyse both (1,3)-β-D-glucans and (1,3;1,4)-β-D-glucans is labelled 'non-specific (1,3/1,3;1,4)-β-D-glucanase', while the 'true' (1,3;1,4)-β-D-glucan endohydrolases are labelled 'lichenases' in Fig. 5. The only subgroup that contains enzymes from higher eukaryotes is the xyloglucan endotransglycosylases/hydrolase (XTH) group from plants (Fig. 5). Each of these subgroups will be described further below.

Other subgroups in the GH16 family include bacterial β-agarases and κ-carrageenases (Fig. 5), which are (1,3)-β-D-galactan endohydrolases (EC 3.2.1.81) that hydrolyse (1,3)-β-D-galactosyl linkages in complex polysaccharides such as agarose (Coutinho and Henrissat, 1999; http://www.cazy.org/), and enzymes (EC 3.2.1.83) that are specific for (1,4)-β-D-galactosyl linkages in other complex polysaccharides that contain sulfated and anhydro-galactosyl residues, including κ-carrageenan and keratin sulfate (Kloareg and Quatrano 1988; Potin et al., 1991). In addition, subgroups in the GH16 family include fungal GPI-glucanosyl transferases (Bruneau et al., 2001) that might be important in covalently cross-linking different

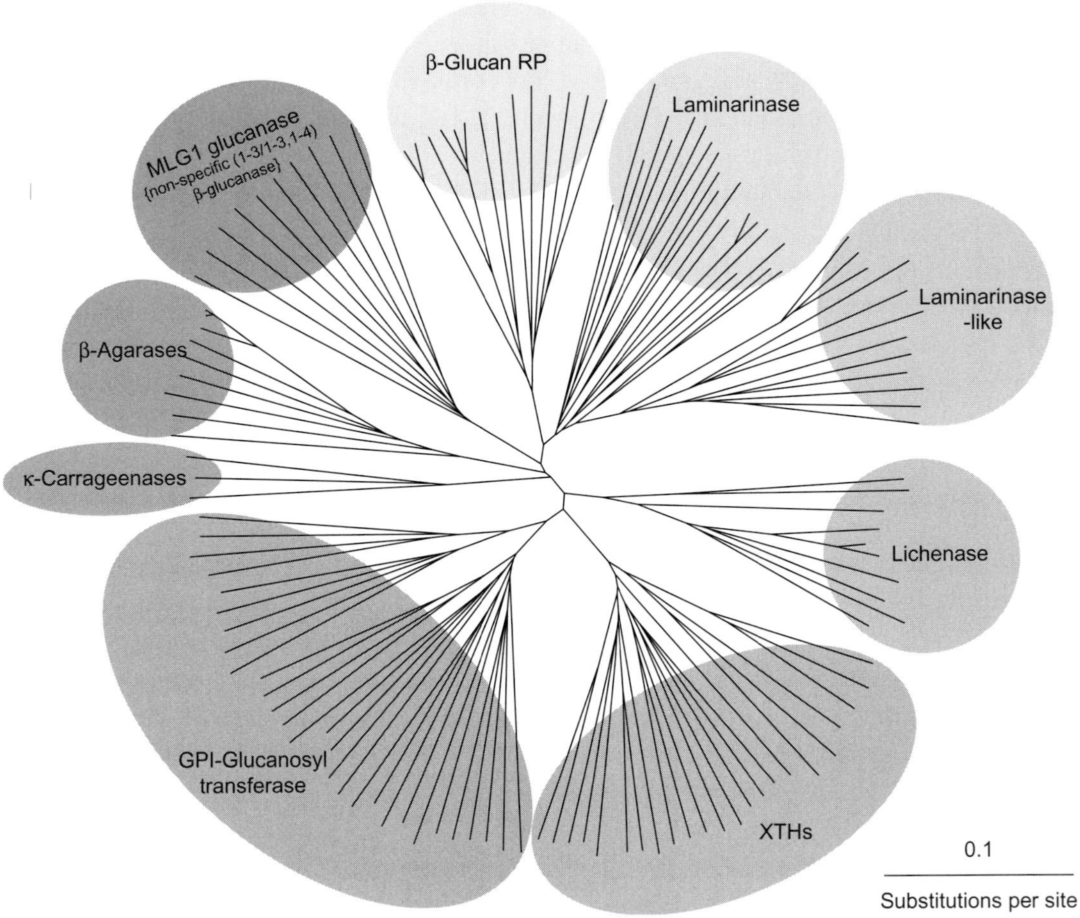


**Fig. 5: Unrooted radial phylogenetic tree of selected family GH16 members. Amino acid sequences were aligned with ClustalX and branch lengths are drawn to scale. Nine subgroups are clustered into non-catalytic and catalytic proteins, the latter according to their substrate specificity. The colour specifications refer to colours in panels.**

polysaccharides such as β-D-glucans and chitin in fungal walls during spore formation and under conditions of stress (Purugganan et al., 1997). A relatively small group of β-D-glucan recognition proteins is also found in the GH16 family (Brown and Gordon, 2005). The β-agarase, κ-carrageenase, GPI-glucanosyl transferases and β-D-glucan recognition protein subgroups will not be considered further here.

### *(1,3)-β-D-glucan endohydrolases (EC 3.2.1.39)*

The family GH16 (1,3)-β-D-glucan endohydrolases are found predominantly in microorganisms (Fig. 5) and have essentially the same substrate specificity as described for the plant family GH17 (1,3)-β-D-glucan endohydrolases in Section 3.1.2.1. They are retaining enzymes that use Glu residues for their catalytic nucleophile and catalytic acid/base, but they differ from the GH17 family insofar as they adopt the β-jelly roll conformation (Fibriansah et al., 2007) that is characteristic of family GH16 enzymes (http://www.cazy.org/). Similarly, the solved 3D structure of the family GH16 β-carrageenase from *Pseudoalteromonas carragenovora* (Michel et al., 2001) has been used successfully as a template for molecular modelling of the catalytically active regions of (1,3)-β-D-glucanases from *Bacillus circulans* (Yahata et al., 1990) and the sea urchin *Strongylocentrotus purpuratus* (Bachman and McClay, 1996). Both models showed the typical family GH16 β-jelly roll fold. The substrate-binding clefts of both enzymes stretch across the surface of the protein, and in the case of the *Strongylocentrotus purpuratus* (1,3)-β-D-glucanase model, the cleft appears to be closed at the top (Strohmeier et al., 2004). The 3D arrangements of the putative catalytic region are conserved in both molecular models (Strohmeier et al., 2004).

The presence of (1,3)-β-D-glucan endohydrolases with both β-jelly roll (family GH16) and $(\beta/\alpha)_8$ folds (family GH17) indicates that this specificity has been developed in a wide variety of biological systems through convergent evolution.

### *(1,3/1,3;1,4)-β-D-Glucan endohydrolases (EC 3.2.1.6)*

The subgroup of GH16 glycoside hydrolases that includes the 'non-specific' (1,3/1,3;1,4)-β-D-glucan endohydrolases (EC 3.2.1.6) appears to contain enzymes that have an 'intermediate' or relatively 'loose' specificity. The enzymes in this subgroup have been termed 'non-specific' because they can hydrolyse both (1,3)- and (1,4)-linkages in β-D-glucans, provided there is an adjacent (1,3)-β-D-glucosyl residue on the non-reducing terminal side of the linkage hydrolysed (Anderson and Stone, 1975; Høj and Fincher, 1995). The so-called non-specific (1,3/1,3;1,4)-β-D-glucan endohydrolase (EC 3.2.1.6) from *Rhizopus arrhizus* has been purified and characterized in detail (Parrish et al., 1960; Anderson and Stone, 1975). The enzyme can hydrolyse both

(1,3)-β-D-glucans and (1,3;1,4)-β-D-glucans, via an endo-action pattern. Products released from (1,3)-β-D-glucans include (1,3)-β-D-oligoglucosides of DP 3 and 4, while the (1,3;1,4)-β-D-oligoglucosides $G4G3G_{red}$ and $G4G4G3G_{red}$ are the major products released from (1,3;1,4)-β-D-glucans. The latter products are the same as those released from (1,3;1,4)-β-D-glucans by the family GH17 (1,3;1,4)-β-D-glucan endohydrolases, as described in Section 3.1.2.2. Thus, the *Rhizopus arrhizus* (1,3/1,3;1,4)-β-D-glucan endohydrolase (EC 3.2.1.6) will hydrolyse either a (1,3)-β-D-glucosyl linkage or a (1,4)-β-D-glucosyl linkage, provided that linkage has an adjacent (1,3)-β-D-glucosyl residue on the non-reducing terminal side of the linkage that is hydrolysed (Anderson and Stone, 1975). The family GH16 (1,3/1,3;1,4)-β-D-glucan endohydrolase from the basidiomycete *Phanerochaete chrysosporium* permits 6-O-glucosyl substitution at subsite +1 during the hydrolysis of (1,3;1,6)-β-D-glucans (Kawai et al., 2006).

### *(1,3;1,4)-β-D-Glucan endohydrolases (EC 3.2.1.73)*

Another subgroup of GH16 glycoside hydrolases comprises 'true' (1,3;1,4)-β-D-glucan endohydrolases (EC 3.2.1.73) that are structurally distinct from the GH17 (1,3;1,4)-β-D-glucan endohydrolases. These GH16 enzymes are designated 'lichenases' in Fig. 5. The family GH16 (1,3;1,4)-β-D-glucan endohydrolases are also absolutely specific for the hydrolysis of a (1,4)-β-D-glucosyl linkage, but only if there is an adjacent (1,3)-β-D-glucosyl residue towards the non-reducing end of the substrate (Planas et al., 1992), as described for the family GH17 (1,3;1,4)-β-D-glucan endohydrolases from barley in Section 3.1.2.2. These enzymes can therefore hydrolyse only (1,3;1,4)-β-D-glucans such as those found in the cell walls of the Poaceae, and the artificial RSIII polysaccharide that has alternating (1,3)- and (1,4)-β-D-glucosyl linkages (Parrish et al., 1960; Anderson and Stone, 1975; Høj and Fincher, 1995). They have no activity on (1,3)-β-D-glucans.

The 3D structure of the (1,3;1,4)-β-D-glucan endohydrolases from *Bacillus subtilis* has been solved (Keitel et al., 1993). In contrast to the barley (1,3;1,4)-β-D-glucan endohydrolases, which adopt a $(\beta/\alpha)_8$ 'TIM barrel' fold (Varghese et al., 1994), the family GH16 *Bacillus subtilis* (1,3;1,4)-β-D-glucan endohydrolases have a quite different 'jelly-roll' β-barrel fold (Keitel et al., 1993). The enzyme *from Bacillus subtilis* also has a much deeper substrate-binding cleft than the family GH17 barley (1,3;1,4)-β-D-glucan endohydrolase. These data indicate that the apparently identical substrate specificities of the barley and *Bacillus* (1,3;1,4)-β-D-glucan endohydrolases have also arisen by convergent evolution (Høj and Fincher, 1995). The *Bacillus* (1,3;1,4)-β-D-glucan endohydrolase follows a double-displacement reaction mechanism, by which the configuration of the anomeric C1 of the glucosyl unit in subsite −1 is retained (Mallet et al., 1993). In a mutated E105Q/E109Q *Bacillus* enzyme that was modified in the two catalytic residues, Gaiser et al. (2006) could identify at least 10 active site residues that formed a network of hydrogen bonds and hydrophobic stacking interactions with the 3-O-β-cellotriosyl-β-D-glucopyranose substrate positioned at subsites −4 to −1.

*Xyloglucan endotransglycosylases/hydrolases (XTHs; EC 2.4.1.207)*

The final subgroup of the GH16 family to be considered here contains xyloglucan-modifying enzymes that are collectively referred to as xyloglucan endotransglycosylases/hydrolases (XTHs; EC 2.4.1.207). The XTH enzymes include both xyloglucan endotransglycosylases (XETs) and xyloglucan endohydrolases (XEHs) (Farkas et al., 1992; Fanutti et al., 1993; Rose et al., 2002b). These enzymes hydrolyse (1,4)-β-D-glucosyl linkages specifically in xyloglucans, but those with XET activity can also catalyse transglycosylation reactions, in which the non-reducing terminal product of the hydrolysis reaction can subsequently be transferred onto another xyloglucan molecule (Rose et al., 2002b). Thus, enzymes within this group can have XET activity or both XET and XEH activities (Thomson and Fry, 2000; Hahn et al., 1995). Quantitation of ratios of XET and xyloglucosides under pseudo-equilibrium conditions indicated that the free energy of the (1,4)-β-glucosidic bond in xyloglucans is preserved in the glycosyl–enzyme intermediate, and that this energy is harnessed for subsequent re-ligation of the xyloglucan polysaccharide (Piens et al., 2008).

Although the xyloglucans and (1,3;1,4)-β-D-glucans of plant cell walls are chemically quite distinct, the molecular backbones of both polysaccharides have contiguous (1,4)-β-D-glucosyl residues. Xyloglucans consist of a backbone of (1,4)-β-D-glucan substituted with xylosyl, galactosyl and fucosyl residues (Fry, 1989). In attempting to reconcile the low levels of xyloglucans in cell walls of most barley tissues with the large XTH gene family and the high expression levels of these genes in barley, Strohmeier et al. (2004) suggested that some of the XTHs might be active on the more abundant arabinoxylans and the (1,3;1,4)-β-D-glucans found in barley walls. Initial molecular modelling of the family GH16 enzymes revealed evolutionary links between higher plant XTHs and microbial (1,3;1,4)-β-D-glucan endohydrolases (Strohmeier et al., 2004) that were later supported by the 3D structure of the *Populus tremula x tremuloides* XET (Johansson et al., 2004) and the *Tropaeolum majus* XET/XEH enzyme (Baumann et al., 2007). The enzyme adopts a curved β-sandwich, or β-jelly roll fold that is similar to other family GH16 glycoside hydrolases, including the *Bacillus* (1,3;1,4)-β-D-glucan endohydrolase. The XET enzymes have COOH-terminal extensions consisting of a β-strand and a short α-helix. However, regions of its substrate-binding cleft are similar to the more distantly related family GH7 enzymes (Johansson et al., 2004). Saura-Valls et al. (2008) have undertaken a detailed analysis of substrate specificity of a poplar XET using a xyloglucoside library. Their findings indicated that the active site of the enzyme is composed of four negative and three positive subsites (Saura-Valls et al., 2008). The existence of an extended acceptor-binding site in barley XET enzymes was also confirmed by Hrmova et al. (2009).

In attempts to determine whether barley XET enzymes could catalyse transfer of xyloglucan onto donors other than xyloglucans, as originally suggested by Strohmeier et al. (2004), Hrmova et al. (2007; 2009) purified two XET isoenzymes from extracts of barley seedlings to a near monodisperse form and showed that the barley HvXET5 and HvXET6 enzymes catalyse the *in vitro* formation of covalent linkages between xyloglucan oligosaccharide acceptors and celluloses, and between xyloglucan oligosaccharide acceptors and (1,3;1,4)-β-D-glucans, albeit at different rates. The polysaccharides are linked from reducing to non-reducing ends of donor and acceptor substrates, probably through (1,4)-β-linkages, but it has not yet been demonstrated that XETs covalently link different polysaccharides *in muro* (Hrmova et al., 2007; 2008). A precedent for such a role might be provided by studies on the re-modelling of fungal cell walls during spore formation, which suggest that GPI-anchored glucanosyl transferase and other transglycosylase enzymes, some of which are members of family GH16 (Fig. 5), might be involved in linking different polysaccharides such as β-D-glucans and chitin in fungal cell walls (Purugganan et al., 1997). The recent findings of Cabib et al. (2008) have demonstrated that the putative Crh transglycosylases, which are abundant in bud scars of yeast, indeed transfer chitin chains to (1,6)-β-D-glucans at linear rates. Fry et al. (2008b) have reported enzymes in unpurified extracts from *Equisetum* spp. and charophytic algae that have '(1,3;1,4)-β-D-glucan:xyloglucan endotransglucosylase' activity.

### *Evolution of enzymes in the GH16 family*

While the (1,3)-β-D-glucan endohydrolases and the (1,3;1,4)-β-D-glucan endohydrolases of the GH17 family of enzymes appear to have evolved in higher plants to mediate biological processes involving callose metabolism and cell wall metabolism in the Poaceae, those in the GH16 family are found mainly in microorganisms (Fig. 5). The latter are presumably saprophytic fungi and bacteria that degrade plant residues containing (1,3)-β-D-glucans and (1,3;1,4)-β-D-glucans, or plant pathogenic microorganisms that use these enzymes to penetrate the physical defences of plants, which might include cell walls and papillary callose. Setting aside these different functions, the GH16 family of enzymes has subgroups that, in evolutionary terms, appear to represent a progression of substrate specificity. Thus, the specificities move from simple (1,3)-β-D-glucan-binding proteins with no hydrolytic activity (the β-glucan recognition proteins), to enzymes with strict (1,3)-β-D-glucan endohydrolase activity (the laminarinase subgroups), through enzymes with 'loosened' specificity that allows them to hydrolyse both (1,3)- and (1,3;1,4)-β-D-glucans [the non-specific (1,3/1,3;1,4)-β-D-glucan endohydrolases subgroup], to strict (1,3;1,4)-β-D-glucan endohydrolases (the lichenase group) (Fig. 5). In addition, the XETs have specificities that can be related to the 'lichenase' subgroup (Strohmeier et al., 2004) and to the GPI-glucanosyl transferase group (Hrmova et al., 2007).

#### 3.1.2.5 (1,6)-β-D-Glucan endohydrolases (EC 3.2.1.75)

Relatively few (1,6)-β-D-glucan endohydrolases have been purified and characterized in detail. One well-characterized (1,6)-β-D-glucan endohydrolase, which was purified from culture filtrates of the filamentous fungus *Acremonium persicinum*, had an apparent molecular mass of 42.7 kDa, a pI of 4.9 and a pH optimum of 5.0 (Pitson et al., 1996). The enzyme hydrolysed the (1,6)-β-D-glucans pustulan and lutean in an endo-hydrolytic manner. The final hydrolysis products from these substrates were β-gentiobiose and β-gentiotriose, which suggested the enzymic mechanism proceeds with retention of anomeric configuration (Pitson et al., 1996). The purified enzyme also hydrolysed the (1,3;1,6)-β-D-glucan, laminarin, from *Eisenia bicyclis*, liberating glucose, gentiobiose, and a range of larger oligoglucosides, through the apparent hydrolysis of (1,6)- and some (1,3)-β-linkages. The authors concluded that the *Acremonium persicinum* (1,6)-β-D-glucan endohydrolase should, more correctly, be described as a (1,3;1,6)-β-D-glucan glucanohydrolase (Pitson et al., 1996). The *Acremonium persicinum* (1,6)-β-D-glucan endohydrolase is probably a member of the GH5 family of glycosyl hydrolases.

### 3.1.3 (1,3)-β-D-Glucan exohydrolases and related enzymes

Two major groups of exo-hydrolytic enzymes will be discussed here and include enzymes belonging to the GH1 and GH3 families of glycoside hydrolases, based on their classification in the Carbohydrate-Active enZymes (CAZy) database (Coutinho and Henrissat, 1999). While various members of the GH1 family of glycoside hydrolases will be discussed, most attention will be focused on a barley β-D-glucan glucohydrolase that can be considered a paradigm for other members of the GH3 group of enzymes. The properties of representatives of the GH1 and GH3 families from barley are summarized in Tables 3 and 4.

#### 3.1.3.1 β-D-Glucosidases of Family GH1

##### *Classification and occurrence*

The GH1 group of hydrolytic enzymes includes more than 2200 individual entries in the CAZy database (Coutinho and Henrissat, 1999). These entries are represented mainly in eubacteria and are less abundant in Archaea, fungi, plants and animals. It has been concluded that in this group of enzymes, compensated frame-shift mutation events such as base-pair substitution and/or insertions and deletions of nucleotides have been maintained throughout evolution and significantly contributed to the evolution of extant enzymes (Rojas et al., 2003).

### *Properties*

The broad distribution of family GH1 members in various kingdoms suggests that β-D-glucosidases play key roles in numerous fundamental biological processes. For example, plant β-D-glucosidases from barley, rice, oat and maize have been shown to hydrolyse oligosaccharides derived from cell wall (1,3;1,4)-β-D-glucans and (1,3)-β-D-glucans (Hrmova et al., 1998; Opassiri et al., 2003; Opassiri et al., 2004; Marana, 2006; Opassiri et al., 2006). Possible functions of the β-D-glucosidases in developing and germinated barley grain have been suggested by Leah et al. (1995), and some of these are related to cell wall metabolism. The genes encoding the barley enzymes are expressed only in the maturing endosperm of the grain at a time when (1,3;1,4)-β-D-glucan is being deposited in the walls of starchy endosperm cells (Leah et al., 1995), and thus the enzymes might be involved in trimming or turnover of wall (1,3;1,4)-β-D-glucans during their synthesis. Plant β-D-glucosidases have also been implicated in activation of defence compounds (Poulton, 1990; Nielsen et al., 2006), production of phytohormones (Falk and Rask, 1995), and lignin precursor formation (Dharmawardhana et al., 1995). Most eubacterial β-D-glucosidases, which are secreted enzymes, are critical for releasing free glucose for growth of microorganisms in diversified environments (Christakopoulos et al., 1994, Li and Lee, 1999, Brunner et al., 2002, Park et al., 2005; Tsukada et al., 2006). The properties of the two barley β-D-glucosidases are summarized in Table 3.

### *Substrate specificity*

The generic term 'β-D-glucosidases' covers a diverse group of enzymes with at least 18 known substrate specificities (Coutinho and Henrissat, 1999), including for example β-D-glucosidase (EC 3.2.1.21), 6-phospho-β-D-glucosidase (EC 3.2.1.86), β-D-galactosidase (EC 3.2.1.23), β-D-mannosidase (EC 3.2.1.25), β-D-glucuronidase (EC 3.2.1.31) and β-D-fucosidase (EC 3.2.1.38) (Coutinho and Henrissat, 1999). A detailed examination of substrate specificity of the barley β-D-glucosidase isoenzyme βII revealed that the enzyme exhibits a marked preference for (1,4)-β-D-oligoglucosides (cello-oligosaccharides) and that the rate of hydrolysis increases with the DP of the (1,4)-β-D-oligoglucosides (Hrmova et al., 1998). The barley enzyme also hydrolyses (1,3)-β-D-oligoglucosides of DP 2–4 with decreasing efficiencies, while laminaripentaose and higher oligosaccharides are not hydrolysed (Hrmova et al., 1998). Similarly, the enzyme does not hydrolyse (1,3;1,4)-, (1,3)- and (1,4)-β-D-glucans at significant rates. Thus,

**Table 3: Properties of barley family GH1 β-D-glucosidases**

| Property | Isoenzyme βI | Isoenzyme βII |
|---|---|---|
| Apparent molecular mass | 62 000 | 62 000 |
| Amino acid residues | 471 | 471 |
| Isoelectric point | 8.9 | 9.0 |
| Carbohydrate | Not known | Not known |
| Substrate specificity: | | |
| 4NPGlc | Active | Active |
| laminarin | Not active | Not active |
| (1,3;1,4)-β-D-glucans | Not active | Not active |
| (1,3;1,4)-β-D-oligosaccharides | Active | Active |
| Anomeric configuration | Retained during hydrolysis | Retained during hydrolysis |
| Glucosyl hydrolase classification | Family GH1 | Family GH1 |
| Protein fold | $(\beta/\alpha)_8$ barrel | $(\beta/\alpha)_8$ barrel |
| Catalytic acid | Glu181 | Glu181 |
| Catalytic nucleophile | Glu391 | Glu391 |
| Subsite binding sites | 6 | 6 |
| Expression sites | Developing endosperm | Developing endosperm |

Source: Leah et al., 1995; Hrmova et al., 1996; Hrmova et al., 1998.

the substrate specificity and action patterns of the barley β-D-glucosidase are characteristic of polysaccharide exohydrolases of the (1,4)-β-D-glucan glucohydrolase group (EC 3.2.1.74), rather than of an enzyme with a preference for low molecular mass (1,4)-β-D-oligoglucosides. The preference of the barley (1,4)-β-D-glucan glucohydrolases for longer chain (1,4)-β-D-oligoglucosides (Table 3) is consistent with subsite mapping data (Hrmova et al., 1998), which indicate that the enzymes have five to six glucosyl-binding subsites (Fig. 6). The term 'subsite' represents an arrangement of amino acid residues that binds a single glycosyl residue of the polymeric substrate (Suganuma et al., 1978). Similar conclusions were drawn for a rice β-D-glucosidase, although the individual binding energies at each subsite were somewhat different (Opassiri et al., 2004). The extended series of subsites in plant β-D-glucosidases indicate that the biological functions of the enzymes are in the hydrolysis of longer oligosaccharides, possibly derived from cell wall (1,3;1,4)-β-D-glucans and (1,3)-β-D-glucans (Hrmova et al., 1998; Opassiri et al., 2003; Marana, 2006). Subsite mapping of a family GH1 β-D-glucosidase from *Aspergillus niger* showed a much shorter substrate-binding region, consisting of only three subsites (Yazaki et al., 1997). It remains to be confirmed if a lower number of subsites in microbial β-D-glucosidases represents a general trend with microbial enzymes, although this is often the case (Yazaki et al., 1997).

Much attention has been devoted to dissecting the 3D structural features that are responsible for aglycon substrate specificity in this large group of enzymes (Day and Withers, 1986; Sinnot, 1990; McCarter and Withers, 1994; Withers, 1995; Czjzek et al., 2000; Czjzek et al., 2001; Vallmitjana et al., 2001; Verdoucq et al., 2004; Marana, 2006; Isorna et al., 2007). However, it is not known what precise conformations gluco- and manno-configured substrates adopt when bound to the active sites of the GH1 enzymes, and what transition states develop during catalytic cycles of these enzymes. It has been suggested that the most realistic information on these events is obtained with inhibitors resembling putative transition states, e.g. with non-hydrolysable substrate analogues using Saturation Transfer Difference NMR spectroscopy techniques (Martin-Pastor et al., 2006).

### *Action pattern*

The retention or inversion of anomeric configuration during hydrolysis of a glycosidic linkage represents an important property in the classification of glycoside hydrolases (Henrissat and Davies, 1997). In the case of the family GH1 β-D-glycosidases, single glucose molecules are released from the non-reducing termini of substrates, with retention of anomeric configuration (e→e). This has been shown for the sweet-almond β-D-glucosidase (Eveleigh and Perlin, 1969), the *Agrobacterium* sp. β-D-glucosidase/β-D-galactosidase (Day and Withers, 1986), the barley

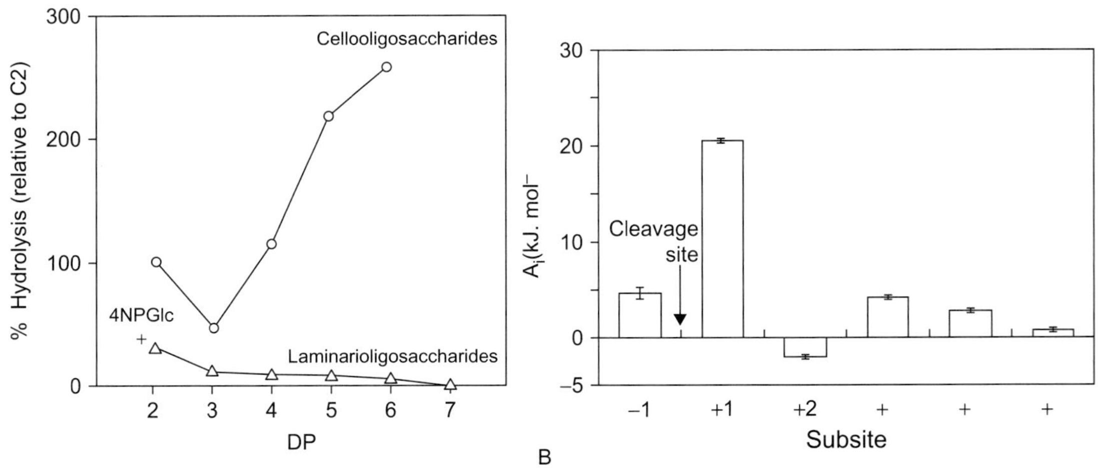


**Fig. 6: Relative rates of hydrolysis of (1,3)- and (1,4)-β-D-oligoglucosides by the barley β-D-glucosidase (A), and subsite map of β-D-glucosidase isoenzyme βII, evaluated for the hydrolysis of (1,4)-β-D-gluco-oligosaccharides (B). Adapted with permission from the American Chemical Society (Hrmova et al., 1998).**

β-D-glucosidase isoenzyme βII (Hrmova et al., 1996) (Table 3), and other β-D-glucosidases that belong to this class of enzymes (Coutinho and Henrissat, 1999). As typical retaining hydrolases (Sinnot, 1990), some GH1 family enzymes exhibit transglycosylating activities, for example the barley β-D-glucosidase at high 4-nitrophenyl β-D-glucoside (4-NPGlc) concentrations is known to synthesize 4-NP-β-laminaribioside, 4-NP-β-cellobioside and 4-NP-β-gentiobioside (Hrmova et al., 2008), and Thai rosewood and cassava β-D-glucosidases show transglycosylation activities with alcohols (Hommalai et al., 2004).

### *Catalytic mechanism*

Glycoside hydrolysis with retention of anomeric configuration proceeds in a two-step double-displacement catalytic mechanism (Fig. 3). This catalytic mechanism is a characteristic feature for the entire GH1 group of hydrolases. The catalytic event advances with participation of two key amino acid residues, a catalytic acid/base and a catalytic nucleophile, which represent the most highly conserved amino acid residues in family GH1 glycoside hydrolases. The exceptions to this rule are plant myrosinases, for example from *Sinapsis alba* (Burmeister et al., 1997) or *Brevicoryne brassicae* (Husebye et al., 2005), which do not possess catalytic proton donors. Much of the information about the catalytic mechanism has been derived from chemical modification studies (McCarter and Withers, 1994; Withers, 1995; Vallmitjana et al., 2001), mutagenesis (Ly and Withers, 1999), around 100 structural studies of GH1 enzymes (e.g. Barrett et al., 1995; Coutinho and Henrissat, 1999), or from molecular models based on 3D structures of homologous enzymes (Hrmova et al., 1998; Cicek et al., 2000, Berrin et al., 2003). The catalytic nucleophile of the β-D-glucosidase from *Agrobacterium* sp. has been determined using 2',4'-dinitrophenyl-2-deoxy-2-fluoro-β-D-glucoside (Street et al., 1992), and the mechanism-based inhibitor conduritol B epoxide, which binds covalently to Glu391 in the barley enzyme (Hrmova et al., 1998). The two catalytic amino acid residues Glu181 and Glu391 in the barley enzyme (Table 3) are 5–6 Å apart and are positioned near the bottom of the substrate-binding pocket. However, it is not clear, at least in the barley enzyme, how the non-reducing end of the substrate is selected in preference to the reducing end for correct orientation of the substrate in the active site funnel. It is also not obvious how the substrate dissociates, at least partly, from the enzyme surface after each hydrolytic cycle, to provide enough space at the bottom of the pocket for the glucose product to diffuse out after each hydrolytic cycle (Hrmova et al., 1998).

### *Three-dimensional structures*

Approximately 100 3D structures have been so far determined for members of this class of glycoside hydrolase, and in all instances the GH1 β-D-glucosidase enzymes fold into $(\beta/\alpha)_8$

barrel projections (Fig. 7). According to the SCOP structural protein classification (Murzin et al., 1995), the proteins belong to alpha and beta (class), TIM β/α-barrel (fold), transglycosylases (superfamily), and family 1 glycoside hydrolase proteins. It has also been postulated that some members of the GH1 group of enzymes form tetrameric to hexameric quaternary assemblies that presumably reduce overall surface areas of the proteins and result in higher enzyme thermostabilities (Chi et al., 1999). Alternatively, the aggregation might regulate substrate binding and/or catalytic efficiency (Kim et al., 2005; Sue et al., 2006). Some of the β-D-glucosidases contain large hydrophobic patches on their surfaces that most likely help these enzymes associate with membrane surfaces (Akiba et al., 2004). In contrast to the open-cleft structures of endo-hydrolases that facilitate the hydrolysis of internal linkages of polysaccharide substrates, exo-hydrolases such as the β-D-glucosidases from clover, maize, oat and barley align their substrate in dead-end funnels into which several glycosyl residues of the substrates are bound (Aguilar et al., 1997; Hrmova et al., 1998). The non-reducing

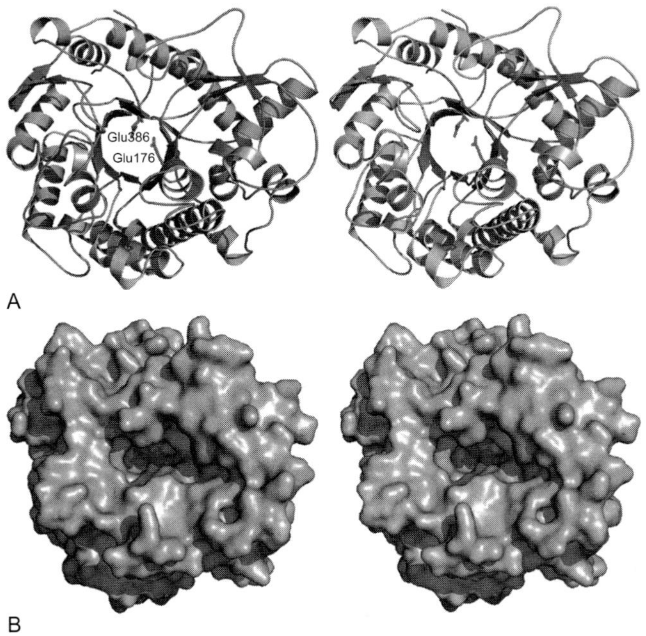


**Fig. 7: Stereo representation of 3D structure of a rice β-D-glucosidase (Protein Data Bank accession code 2RGL). (A) Ribbon diagram of the enzyme with two catalytic amino acids Glu176 and Glu386 shown in sticks and coloured in red. (B) Molecular surface drawing of the enzyme with two catalytic amino acids Glu176 and Glu386 coloured in red. Adapted with permission from Elsevier Science (Chuenchor et al., 2008). The colour specifications refer to colours in panels.**

terminal linkages of the substrates are thereby brought into juxtaposition with catalytic amino acid residues (Barrett et al., 1995; Hrmova et al., 1998) located at the bottom of the funnel, close to the glycosidic linkage of the non-reducing terminal residue (Barrett et al., 1995; Czjzek et al., 2000; Verdoucq et al., 2004; Marana, 2006; Isorna et al., 2007).

### 3.1.3.2 β-D-Glycoside Exo-Hydrolases of Family GH3

*Classification and occurrence*

The GH3 group of hydrolytic enzymes includes more than 1700 individual entries in the CAZy database (Coutinho and Henrissat, 1999). The entries mainly represent nucleotide sequences from genome-sequencing programs, and therefore substrate specificities and biological functions of these entries have rarely been tested (Harvey et al., 2000; Hrmova and Fincher, 2007). The GH3 family members are distributed predominantly in eubacteria, fungi, and plants, and are scarce or absent in Archea and animals (Coutinho and Henrissat, 1999; Harvey et al., 2000; Cournoyer and Faure, 2003). Phylogenetic analyses of GH families indicate that the GH3 group of enzymes is one of the three groups, together with the GH13 and GH23 families, that are the most highly represented in bacterial genomes (Cournoyer and Faure, 2003).

*Properties*

The broad distribution of family GH3 members in various kingdoms suggests that they play key roles in fundamental biological processes. These functions include for example the microbial degradation of plant residues, the modification of structures of glycosides, bacterial antibiotics and plant-derived antifungal molecules, the turnover, recycling and re-modelling of cellular components in bacteria, fungi and plants, and the modification of host–pathogen interactions during microbial infection of plants (Cournoyer and Faure, 2003; Hrmova and Fincher, 2001). A great deal of attention has been focused on the characterization of the barley enzymes (Table 4). Two isoenzymes were isolated from germinated barley seedlings (Hrmova et al., 1996), and additional isoforms have been detected in barley by Labrador and Nevins (1989) and Kotake et al. (1997). These enzymes are also abundant in maize coleoptiles (Kim et al., 2000) and dicotyledonous plants (Cline and Albersheim, 1981; Crombie et al., 1998). The genes encoding the family GH3 β-D-glucan glucohydrolases are transcribed in the scutellum of germinated grain, but their mRNAs are most abundant in elongating coleoptiles (Harvey et al., 2001). The latter observation has led to the suggestion that β-D-glucan glucohydrolases function in auxin-mediated cell elongation in growing coleoptiles (Kotake et al., 1997; Hoson and Nevins, 1989; Harvey et al., 2001), where the amount of (1,3;1,4)-β-D-glucan in walls decreases markedly during coleoptile growth and wall loosening

(Sakurai and Masuda, 1978); the latter process is believed to be necessary for cell elongation (Labrador and Nevins, 1989). Another possible function for the β-D-glucan glucohydrolases could be in defence strategies developed by plants to counter pathogen attack (Wessels, 1993; Hrmova and Fincher, 1998). In other plants, such as lily and maize, the β-D-glucan glucohydrolases play roles in pollen development (Takeda et al., 2004), while in barley seedlings the orthologous enzymes are suggested to participate in hydrolysis of cell wall poly- and oligosaccharides (Hrmova and Fincher, 2007) (Table 4).

### *Substrate specificity*

More than seven known substrate specificities and around 200 enzymes are classified in the GH3 family. The enzymes are variously annotated as β-D-glucosidases, (1,3;1,4)- and (1,3)-β-D-glucan exohydrolases, β-D-xylosidases, α-L-arabinofuranosidases and *N*-acetyl β-D-glucosaminidases (Coutinho and Henrissat, 1999). Because the substrate specificity

**Table 4: Properties of barley family GH3 β-D-glucan glucohydrolases**

| Property | Isoenzyme ExoI | Isoenzyme ExoII |
|---|---|---|
| Apparent molecular mass | 69 000 | 71 000 |
| Amino acid residues | 605 | 602 |
| Isoelectric point | 7.8 | 8.0 |
| Carbohydrate | 4.7% by weight at 3 *N*-glycosylation sites | Not known |
| Substrate specificity: | | |
| 4NPGlc | Active | Active |
| (1,3)-, (1,3;1,4)-, (1,3;1,6)-β-D-glucans | Active | Active |
| (1,2)-, (1,3)-, (1,4)- and (1,3;1,4)-β-D-oligosaccharides | Active | Active |
| Anomeric configuration | Retained during hydrolysis | Retained during hydrolysis |
| Glucosyl hydrolase classification | Family GH3 | Family GH3 |
| Protein fold | $(\alpha/\beta)_8$ barrel and $(\alpha/\beta)_6$ sandwich | $(\alpha/\beta)_8$ barrel and $(\alpha/\beta)_6$ sandwich |
| Catalytic acid | Glu491 | Glu491 |
| Catalytic nucleophile | Asp285 | Asp284 |
| Subsite binding sites | 2 | 2 |
| Expression sites | Mainly in scutellum and coleoptiles, also in young leaves and roots | Mainly in scutellum and coleoptiles, also in young roots and leaves |

Source: Hrmova et al., 1996; Kotake et al., 1997; Hrmova and Fincher, 1998; Varghese et al., 1999; Hrmova et al., 2001; Hrmova et al., 2002.

assignments of the GH3 family members have mainly been based on similarities between nucleotide sequences of the genes, there are only a small number of enzymes with experimentally defined substrate specificities and the substrate specificity annotations of the majority of GH3 hydrolases are probably unreliable. The substrate specificities of the two barley β-D-glucan glucohydrolase isoforms, designated isoenzymes ExoI and ExoII, are broad (Hrmova and Fincher, 1996; Harvey et al., 2001) (Table 4). The barley enzymes hydrolyse unbranched and unsubstituted β-D-glucans such as (1,3)-β-D-glucans and (1,3;1,4)-β-D-glucans. The latter substrate represents an abundant component of cell walls in the Poaceae family of monocotyledonous plants, while the former is not only found in many plant tissues, but is also an abundant component in fungal cell walls (Hrmova and Fincher, 1998). The two barley β-D-glucan glucohydrolases also hydrolyse β-D-gluco-oligosaccharides with (1,2)-, (1,3)-, (1,4)- or (1,6)-linked sugar moieties, as well as aryl β-D-glucosides such as 4-nitrophenyl β-D-glucoside (Hrmova and Fincher, 1998). The barley β-D-glucan glucohydrolases therefore exhibit broad substrate specificity and are difficult to classify in existing Enzyme Commission classes (Hrmova and Fincher, 2007). Nevertheless, these enzymes are fundamentally different from the β-D-glucosidases of the GH1 group of enzymes.

### *Action pattern*

The GH3 family enzymes remove single glucose units from the non-reducing termini of polymeric and oligomeric substrates, with retention of anomeric (e→e) configuration (Koshland, 1953; Hrmova et al., 1996). The catalytic regions of the barley β-D-glucan glucohydrolases consist of 2–3 subsites (Hrmova et al., 1996, 2002) (Fig. 8). As typical retaining hydrolases (Sinnot, 1990), the GH3 family enzymes exhibit transglycosylation activity (Hrmova and Fincher, 1998; Kawai et al., 2004; Seidle and Huber, 2005) at higher than 3 mM substrate concentrations. Predominantly (1,6)-linked, but also (1,3)- and (1,4)-linked products are formed during these transglycosylation reactions (Hrmova et al., 2002). The enzyme binds various substrates at their non-reducing termini, and the restricted depth of the dead-end pocket, coupled with the spatial disposition of catalytic amino acid residues, ensures that only the glycosidic linkage at the non-reducing end of the substrate can be hydrolysed. The shape of the active site in the barley β-D-glucan glucohydrolases, which has been likened to a short 'coin slot' sandwiched between the two domains of the enzyme (Varghese et al., 1999), can again be contrasted with those of endo-hydrolases, which usually have open-cleft- or tunnel-like topologies that allow random binding of the enzyme to internal regions of polysaccharide substrates.

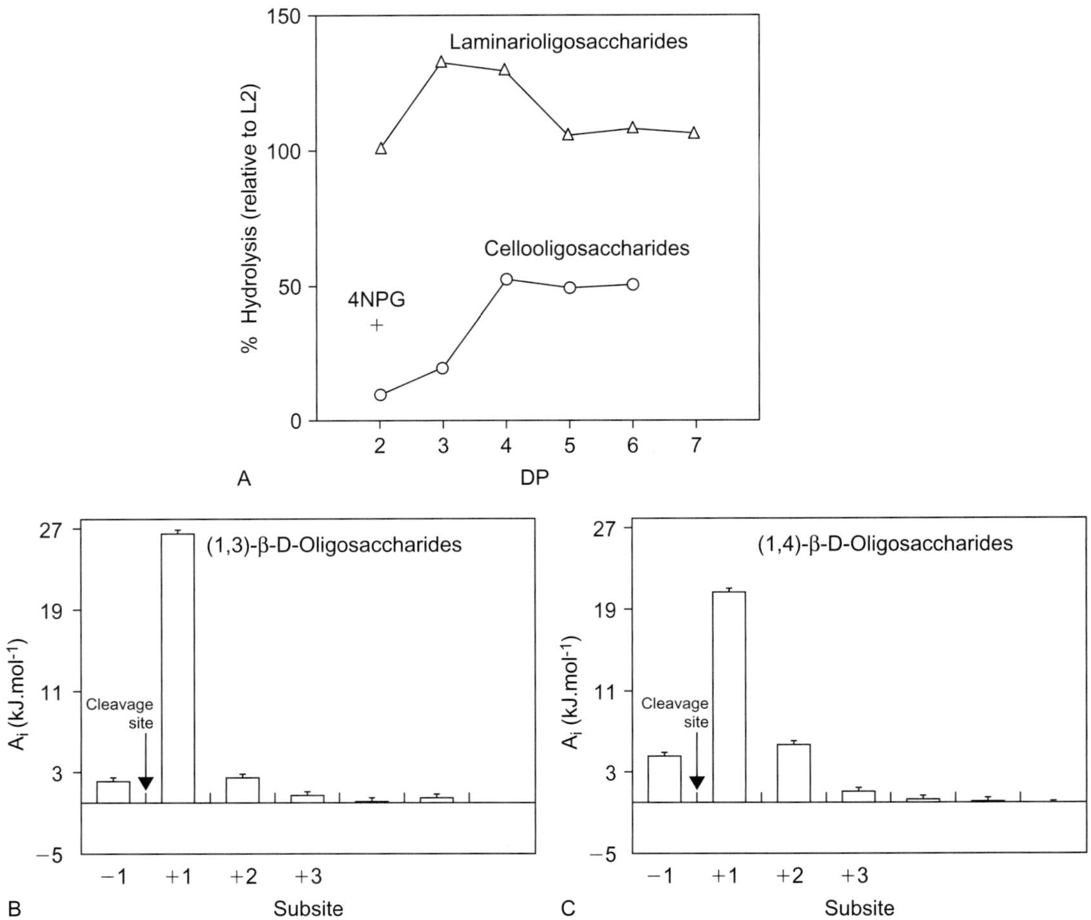


**Fig. 8: Relative rates of hydrolysis of (1,3)- and (1,4)-β-D-gluco-oligosaccharides by the β-D-glucan glucohydrolase (A), and subsite maps of β-D-glucan glucohydrolase evaluated for the hydrolysis of (1,3)- (B), and (1,4)-β-D-gluco-oligosaccharides (C). Adapted with permission from the American Chemical Society (Hrmova et al., 1996), and the American Society of Plant Biologists (Hrmova et al., 2002).**

### *Catalytic mechanism*

The family GH3 enzymes contain an aspartic acid as the catalytic nucleophile amino acid residue in a highly conserved sequence motif (Harvey et al., 2000). Molecular modelling and hydrophobic cluster analysis indicate that the catalytic nucleophile in the plant GH3 family enzymes is always found near the COOH-terminus of β-strand g in the first domain (Coutinho and Henrissat, 1999), and that this β-strand is always positioned in the shallow active site pocket that is located at the interface of the two domains (Varghese et al., 1999). Labelling the β-D-glucan glucohydrolase with conduritol B epoxide or 2,4-dinitrophenyl 2-deoxy-2-fluoro-β-D-glucopyranoside, coupled with comparative tryptic peptide mapping and other structural analyses, show that Asp285 is the catalytic nucleophile in the barley β-D-glucan glucohydrolase isoenzyme ExoI (Hrmova et al., 2001) (Table 4, Fig. 9). Earlier, Asp12 in a 13.2-kDa β-D-glucosidase fragment isolated from *Aspergillus wentii* was identified as a catalytic nucleophile in the pioneering work of Legler (1980), and the catalytic nucleophile has been identified as Asp242 in the *N*-acetyl-β-D-glucosaminidase from *Vibrio furnisii* (Vocadlo et al., 2000), Asp261 in a *Aspergillus niger* β-D-glucosidase (Dan et al., 2000), Asp247 in the β-D-glucosidase from *Flavobacterium meningosepticum* (Chir et al., 2002), Asp223 in the glucosylceraminidase from *Paenibacillus* sp. TS12 (Paal et al., 2004), and Asp283 in the bufunctional β-*N*-acetyl-D-glucosaminidase/β-D-glucosidase from *Cellulomonas fimi* (Meyer et al., 2006). These amino acid residues are all equivalent to Asp285 in the barley enzyme, and have been identified experimentally as catalytic nucleophiles. From the 3D structure of the barley β-D-glucan glucohydrolases, it became clear that the most likely candidate for the catalytic acid/base was Glu491 (Varghese et al., 1999). This was confirmed experimentally from the crystal structure of the enzyme bound to the non-hydrolysable S-glycosyl substrate analogue $4^{I}$, $4^{III}$, $4^{V}$-S-trithiocellohexaose (Hrmova et al., 2001). The likely catalytic acid/base Glu491 was deduced from the distances between the S-atom of the S-glycosidic linkage and conserved amino acid residues at the catalytic site. The catalytic acid/base in the barley enzyme, Glu491, was present only in closely related members of the GH3 family. It logically followed that if Glu491 were variable in the family GH3 glycoside hydrolases, then the role of the catalytic acid/base could be adopted by equivalently positioned amino acid residues in more distant members of the GH3 family (Coutinho and Henrissat, 1999).

The structural analysis of the crystallized barley enzyme revealed that a glucose molecule is bound in the active site pocket (Figure 9), where about 18 amino acid residues contribute to binding of this single glucose molecule (Varghese et al, 1999; Hrmova et al, 2001). For this reason the catalytic mechanism of a family GH3 retaining barley β-D-glucan glucohydrolase is somewhat different from a typical retaining glycoside hydrolase. In the case of the barley enzyme, the final hydrolysis product glucose, in which anomeric configuration is retained, remains bound in the active site and represents the enzyme–product complex E~Glc.

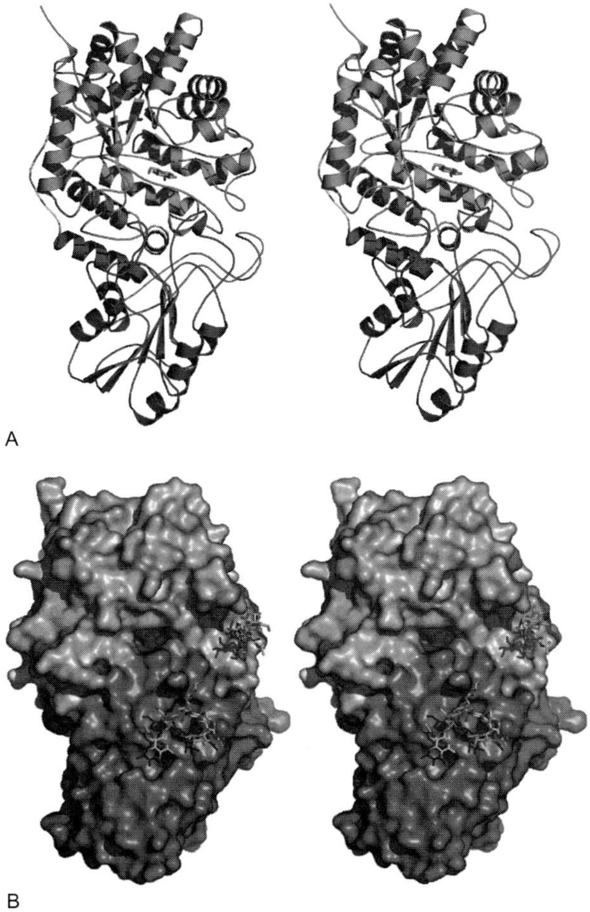

Fig. 9: Stereo representation of 3D structure of a barley β-D-glucan glucohydrolase (Protein Data Bank accession code 1EX1 and 1IEQ). (A) Ribbon diagram of the enzyme with bound glucose (in cpk colours), where domain 1, linker, domain 2, and the COOH-terminal antiparallel loop are in cyan, yellow, magenta and green, respectively. (B) Molecular surface drawing of the enzyme (colours as specified in panel A) with two occupied Asn221 and Asn498-linked glycosylation sites (cpk colours). Adapted with permission from Elsevier Science (Varghese et al., 1999; Hrmova et al., 2001). The colour specifications refer to colours in panels.

*Three-dimensional structures*

Almost all members of the GH3 family possess multiple, individually folded domains (Harvey et al., 2000; Rojas et al., 2005; Cournoyer and Faure, 2003), although the sequential arrangement of the domains can differ. The only complete 3D crystal structure of the GH3 family enzymes in the public databases is that of the barley β-D-glucan glucohydrolase isoenzyme ExoI (Varghese et al., 1999; Hrmova et al., 2001, 2002, 2005, 2006) (Figs. 9 and 11), although the structure of the $(\alpha/\beta)_8$ TIM barrel domain 1 of an *N*-acetyl-β-glucosaminidase from *Vibrio cholerae* has recently been reported in its free form and in complex with *N*-acetyl-β-D-glucosamine (J. Gorman and L. Shapiro, unpublished).

The barley β-D-glucan glucohydrolase enzyme adopts a globular, two-domain modular structure. The first domain folds into a $(\alpha/\beta)_8$ TIM-barrel conformation, while the second domain forms a $(\alpha/\beta)_6$ sandwich. A pocket about 13 Å deep at the interface of the two domains has been identified as the active site of the enzyme. The dimensions of the pocket indicate that it could accommodate two glucosyl residues (Hrmova et al., 2001; Hrmova et al., 2002) and these data are consistent with subsite mapping results (Fig. 8).

### *Structural determinants of substrate specificity*

The 3D structures of the barley β-D-glucan glucohydrolase isoenzyme ExoI in complex with $4^{I}$, $4^{III}$, $4^{V}$-S-trithiocellohexaose (Hrmova et al., 2001) and 4'-nitrophenyl

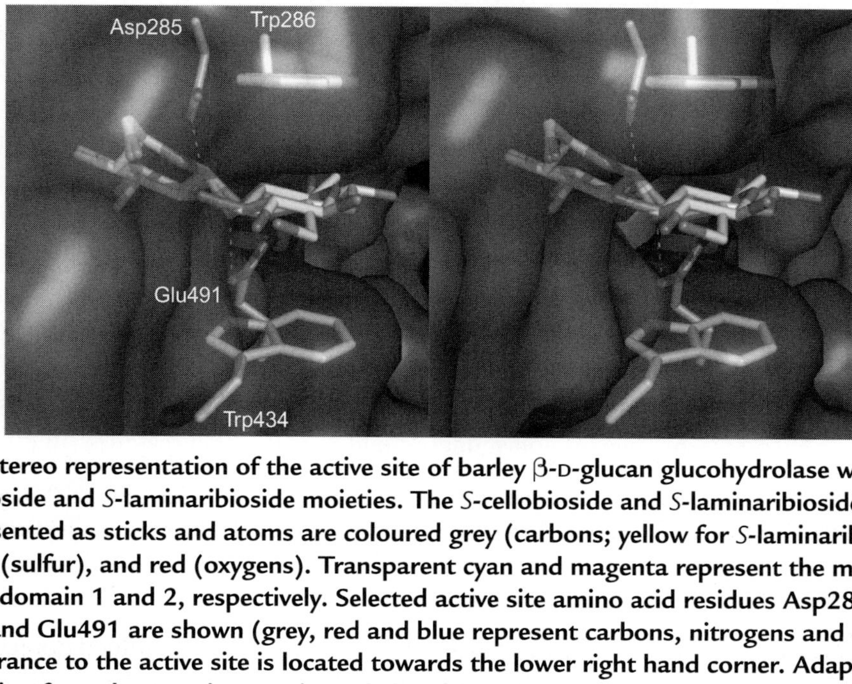


**Fig. 10: Stereo representation of the active site of barley β-D-glucan glucohydrolase with bound *S*-cellobioside and *S*-laminaribioside moieties. The *S*-cellobioside and *S*-laminaribioside moieties are presented as sticks and atoms are coloured grey (carbons; yellow for *S*-laminaribioside), orange (sulfur), and red (oxygens). Transparent cyan and magenta represent the molecular surface of domain 1 and 2, respectively. Selected active site amino acid residues Asp285, Trp286, Trp434 and Glu491 are shown (grey, red and blue represent carbons, nitrogens and oxygens). The entrance to the active site is located towards the lower right hand corner. Adapted with permission from the American Society of Plant Biologists (Hrmova et al., 2002). The colour specifications refer to colours in panels.**

*S*-(β-D-glucopyranosyl)-(1,3)-(3-thio-β-D-glucopyranosyl)-(1,3)-β-D-glucopyranoside (Hrmova et al., 2002) demonstrated that both ligands displace the glucose molecule that is bound in the active site pocket, and that the two non-reducing end residues of the inhibitors are positioned in the pocket at the −1 and +1 subsites (Fig. 10). While the glucosyl residue at subsite −1 is tightly constrained through extensive hydrogen bonding with multiple amino acid residues, the glucosyl residue at subsite +1 is sandwiched between two tryptophan residues, Trp286 and Trp434. The hydrophobic interactions between the glucosyl residue at subsite +1 and the tryptophan residues are not as precise as the multiple hydrogen bonding interactions at subsite −1 and, as a result, differences in the spatial positions of non-reducing and penultimate glucosyl residues in say (1,3)- and (1,4)-β-D-linked glucoside substrates can be accommodated through the flexibility of binding at the +1 subsite.

From the superpositions of the *S*-laminaribioside– and the *S*-cellobioside–enzyme complexes, and from the sophorose– and gentiobiose–enzyme models, it is possible to draw a structural rationale for a broad substrate specificity of the enzyme (Fig. 10) (Hrmova et al., 2002). For all ligand–enzyme structures, the glucopyranosyl residues at the −1 subsite are bound in almost identical positions. In contrast, the glucopyranosyl residues of the *S*-laminaribioside and gentiobiose and *S*-cellobioside and sophorose moieties occupy subsite +1 that is located between the Trp286 and Trp434 residues. In the case of the *S*-laminaribioside and gentiobiose moieties, the apolar face of the glucopyranosyl residue at subsite +1 is geometrically complementary with the pyrrole ring of Trp286, while the polar face of the glucopyranosyl residue positions itself over the phenyl ring of Trp434. In the *S*-cellobioside and sophorose moieties, the polar and apolar faces of the glucopyranosyl residue at subsite +1 are in contact with the pyrrole ring of Trp286 and the phenyl ring of Trp434, respectively. That is, the four positional sugar isomers can adopt two different orientations with respect to the phenyl/pyrrole rings of Trp286 or Trp434. It therefore follows that if a substrate binds to the broad-specificity barley β-D-glucan glucohydrolase, binding will be largely independent of polysaccharide conformation and the glycosidic linkage positions between adjacent non-reducing-end β-D-glucosyl residues. In other words, the relative flexibility of binding at subsite +1 by the broad-specificity barley β-D-glucan glucohydrolase, coupled with the projection of the remainder of bound substrate away from the enzyme's surface, means that the overall active site can accommodate a range of substrates with variable spatial dispositions of adjacent β-D-glucosyl residues (Hrmova et al., 2002).

The flexibility in substrate positioning allowed by the two relatively wide Trp residues at subsite +1, and hence the broad specificity of these family GH3 enzymes for β-D-glucosides, can be contrasted to the situation in a family GH5 (1,3)-β-D-glucan glucohydrolase from

*Candida albicans* (Chambers et al., 1993). The latter enzyme has an active site that accommodates two β-D-glucosyl residues, but in this case the penultimate β-D-glucosyl residue at subsite +1 is sandwiched between two Phe residues at the entrance to the pocket (Cutfield et al., 1999). One might suggest, based on the central role of the Trp286/Trp434 clamp in allowing binding of substrates with different relative orientations of adjacent glucopyranosyl residues in the barley GH3 β-D-glucan glucohydrolase, that the narrower Phe144/Phe258 clamp of the *Candida* enzyme would significantly tighten its substrate specificity. This is indeed the case, and the relative catalytic efficiencies $k_{cat}/K_m$ for G3OG, G4OG and G6OG by the *Candida* enzyme are 100%, 0.06% and 0.14%, respectively (Stubbs et al., 1999), whereas the corresponding values for the barley enzyme are 100%, 3% and 19%, respectively (Hrmova et al., 2002).

In summary, the contrasting broad substrate specificity of the barley enzyme from family GH3 and the relatively tight substrate specificity of GH1 and GH5 enzymes can be explained through 3D structural information available for these β-D-glucan exohydrolases.

### 3.1.4 (1,3)-β-D-Glucan phosphorylases and related enzymes

The (1,3)-β-D-glucan phosphorylases and related enzymes include those with laminaribiose phosphorylase (2.4.1.31), (1,3)-β-D-glucan phosphorylase (2.4.1.97) and (1,3)-β-D-oligoglucan phosphorylase (2.4.1.30) activities. The enzymes are phosphorolytic insofar as an exo-phosphorolytic mechanism is used to cleave (1,3)-β-D-glucosidic linkages in (1,3)-β-D-glucans and (1,3)-β-D-oligosaccharides that serve as glucosyl donors. While the acceptor in most glycoside hydrolases is a water molecule, the acceptor in phosphorylase reactions is an inorganic phosphate molecule to which is transferred a glucosyl residue, to produce α-glucose-1-phosphate (Kitaoka and Hayashi, 2002). The phosphorylase enzymes have an inverting mechanism of action with no apparent hydrolytic activities. The reactions they catalyse are reversible, because they can operate in both the phosphorolytic and the synthetic direction (Hidaka et al., 2006). The main roles of these enzymes are in the utilization of paramylon, which is a storage polysaccharide in algae and euglenoid cells (Manners and Taylor, 1967; Marechal, 1967). Cell-free extracts of *Euglena gracilis* contain at least two phosphorylase enzyme activities, namely laminaribiose phosphorylase and (1,3)-β-D-oligoglucan phosphorylase (Kitaoka et al., 1991). The laminaribiose phosphorylase from *Astasia ocellata* is capable of utilizing laminarin as an acceptor substrate in the synthetic direction (Manners and Taylor, 1967). This observation indicates that the active site of laminaribiose phosphorylase possesses multiple subsites. Using cell-free extracts of *E. gracilis* and a α-glucose-1-phosphate:glucose ratio of 20, formation of laminari-oligosaccharides of DP 2–13 were detected with laminaribiose phosphorylase; that is, oligosaccharides larger than 10 were readily obtained. Laminaribiose phosphorylase from *E. gracillis* occurs in three isoforms and these isoforms catalyse the reversible phosphorolysis of laminaribiose and other laminari-oligosaccharides (Kitaoka et al., 1993). Under native conditions, the enzyme occurs in a dimeric form with a molecular mass around 160 kDa, as determined by SDS–PAGE and size-exclusion chromatography (Kitaoka et al., 1991; 1993). The crystal structure of laminaribiose phosphorylase has recently been determined to 2.0 Å (Hidaka et al., 2007). The solution of the structure revealed an $(\alpha/\alpha)_6$-β-sheet fold (http://www.nfri.affrc.go.jp/research/overseas/pdf/CBM-hidaka.pdf) that is typical for a family GH94 glycoside hydrolase (Coutinho and Henrissat, 1999). A comparison of the active site structures of laminaribiose phosphorylase and cellobiose phosphorylase, which are both classified within the GH94 group of enzymes, revealed that the amino acids binding sugar donors of both substrates are conserved. The substrate specificities of laminaribiose phosphorylase and cellobiose phosphorylase are controlled by the spatial positioning of glucosyl moieties at the sugar-acceptor binding sites, where they can adopt two, opposite orientations (Hidaka et al., 2007).

## 3.1.5 Proteinaceous inhibitors of plant (1,3)-β-D-glucanases produced by fungal pathogens

Plant cells are surrounded by complex cell walls composed of polysaccharides and proteins that provide structural support for the cells (Farrokhi et al., 2006) and protect them against fungal pathogens (Juge, 2006). Plant pathogens secrete an array of degradative enzymes that have the capacity to disassemble the cell walls, colonize the plant host and use plant cell walls and other cell components as a source of nutrients. Plants have developed a variety of defence mechanisms to protect themselves against a wide variety of fungal pathogens. One such mechanism is the production of hydrolytic enzymes, such as (1,3)-β-D-glucanases and chitinases, that degrade core components of fungal cell walls. These plant glycanases are known as pathogenesis-related proteins (Kauffmann et al., 1987). It has been shown that (1,3)-β-D-glucanases and chitinases, working in concert, inhibit growth of fungal cultures *in vitro* (Roberts and Selitrennikoff, 1998). However, it has also been demonstrated (Ludwig and Boller, 1990) that certain fungi that are normally sensitive to (1,3)-β-D-glucanases and/or chitinases, could become resistant after several hours of exposure to these enzymes. On the basis of these experiments it has been anticipated that certain fungi could secrete proteinaceous molecules that are responsible for acquiring this resistance. For example, Albersheim and Valent (1974) reported that the fungus *Colletotrichum lindemutianum* secretes a proteinaceous inhibitor that inhibits the activity of a (1,3)-β-D-glucanase isolated from French beans, although the authors have not identified or characterized the proteinaceous inhibitor. A similar phenomenon was observed by Ham et al. (1997) with a soybean fungal pathogen *Phytopthora sojae*. Here, the authors identified a fungal protein GIP-1 of 34 kDa that was able to inhibit one of the two soybean (1,3)-β-D-glucanases. These observations indicated that the GIP-1 protein could be highly selective as to the isoforms of the soybean (1,3)-β-D-glucanases it inhibits. The primary amino acid sequences of the soybean GIP family proteinaceous inhibitors were further identified by Rose et al. (2002a). The GIP proteins were homologous with a trypsin class of serine proteases, but were not functioning proteolytically. The GIP-1 protein inhibited (1,3)-β-D-glucanase-mediated release of elicitor-active gluco-oligosaccharides from *Phytopthora sojae* cell walls *in vitro* (Rose et al., 2002a). This observation suggested that these classes of fungal proteinaceous inhibitors operate at a molecular level in order to suppress a certain type of a plant defense response.

## 3.1.6 Concluding remarks

Given that (1,3)-β-D-glucans are widely distributed in nature it is not surprising that a diversity of (1,3)-β-D-glucan exo- and endohydrolases have evolved for their depolymerization. These mainly include enzymes from higher plants and fungi, although in many cases the precise substrate specificities or biological functions of the enzymes have not been described. The barley (1,3)-β-D-glucan endohydrolases have been characterized in detail and their action patterns, substrate specificities and kinetics have been explained at the molecular level through the availability of 3D crystal structures. However, the key biological functions of individual enzymes are not always known. This is the case with the barley (1,3)-β-D-glucan exoglucanases, which are surmised to be involved in the conversion of various oligosaccharides to glucose. Some attempts have been made to define the functions of (1,3)-β-D-glucan endohydrolases, but there is still a good deal of work that needs to be done to define the precise functions of some of the (1,3)-β-D-glucan exohydrolases, particularly those from higher plants. With the recent interest in conversion of lignocellulosic material to fermentable sugars for bioethanol production, these enzymes could find a role in future biofuels industries.

## Acknowledgement

The work described here has been supported over many years by grants from the Australian Research Council.

## References

Aguilar, C. F., Sanderson, I., Moracci, M., Ciaramella, M., Nucci, R., Rossi, M., & Pearl, L. H. (1997). Crystal structure of the β-glycosidase from the hyperthermophilic archeon *Sulfolobus solfataricus*: resilience as a key factor in thermostability. *Journal of Molecular Biology*, *271*, 789–802.

Akiba, T., Nishio, M., Matsui, I., & Harata, K. (2004). X-ray structure of a membrane-bound β-glycosidase from the hyperthermophilic archaeon *Pyrococcus horikoshii*. *Proteins*, *57*, 422–431.

Albersheim, P., & Valent, B. S. (1974). Host-pathogen interactions VII. Plant pathogens secrete proteins which inhibit enzymes of the host capable of attacking the pathogen. *Plant Physiology*, *53*, 684–687.

Anderson, M. A., & Stone, B. A. (1975). A new substrate for investigation the specificity of β-D-glucan hydrolases. *FEBS Letters*, *52*, 202–207.

Bachman, E. S., & McClay, D. R. (1996). Molecular cloning of the first metazoan β-1,3 glucanase from eggs of the sea urchin *Strongylocentrotus purpuratus*. *Proceedings of the National academy of Sciences of the United States of America*, *93*, 6808–6813.

Barrett, T., Suresh, C. G., Tolley, S. P., Dodson, E. J., & Hughes, M. A. (1995). The crystal structure of a cyanogenic β-glucosidase from white clover, a family 1 glycosyl hydrolase. *Structure*, *3*, 951–960.

Bamforth, C. W., & Martin, H. L. (1981). The development of β-D-glucan solubilase during barley germination. *Journal of the Institute of Brewing*, *87*, 81–84.

Berrin, J. G., Czjzek, M., Kroon, P. A., McLauchlan, W. R., Puigserver, A., Williamson, G., & Juge, N. (2003). Substrate (aglycone) specificity of human cytosolic β-glucosidase. *Biochemical Journal*, *373*, 41–48.

Bernard, M., & Latgé, J.-P. (2001). *Aspergillus fumigatus* cell wall: composition and biosynthesis. *Medical Mycology*, *39*, 9–17.

Bestwick, C. S., Brown, I. R., Bennett, M. H. R., & Mansfield, J. W. (1997). Localization of hydrogen peroxide accumulation during the hypersensitive reaction of lettuce cells to *Pseudomonas syringae pv phaseolicola*. *Plant Cell*, *9*, 209–221.

Boller, T. (1987). Hydrolytic enzymes in plant disease resistance. In T. Kosuge & E. W. Nester (Eds.), *Plant-Microbe Interactions* (pp. 385–413). New York: Macmillan.

Bolwell, G. P. (1993). Dynamic aspects of the plant extracellular matrix. *International Review of Cytology*, *146*, 261–323.

Brown, G. D., & Gordon, S. (2005). Immune recognition of fungal β-glucans. *Cellular Microbiology*, *7*, 471–479.

Bruneau, J. M., Magnin, T., Tagat, E., Legrand, R., Bernard, M., Diaquin, M., Fudali, C., & Latgé, J. P. (2001). Proteome analysis of *Aspergillus fumigatus* identifies glycosylphosphatidylinositol-anchored proteins associated to the cell wall biosynthesis. *Electrophoresis*, *22*, 2812–2823.

Brunner, F., Wirtz, W., Rose, J. K. C., Darvill, A. G., Govers, F., Scheel, D., & Nurnberger, T. (2002). A β-glucosidase/xylosidase from the phytopathogenic oomycete, *Phytophthora infestans*. *Phytochemistry*, *59*, 689–696.

Buckeridge, M. S., Vergara, C. E., & Carpita, N. C. (2001). Insight into multi-site mechanisms of glycosyl transfer in (1→4)-β-D-glycans provided by the cereal mixed-linkage (1→3),(1→4)-β-D-glucan synthase. *Phytochemistry*, *57*, 1045–1053.

Burmeister, W. P., Cottaz, S., Driguez, H., Iori, R., Palmieri, S., & Henrissat, B. (1997). The crystal structures of *Sinapis alba* myrosinase and a covalent glycosyl-enzyme intermediate provide insights into the substrate recognition and active-site machinery of an S-glycosidase. *Structure*, *5*, 663–675.

Burton, R. A., Jobling, S. A., Harvey, A. J., Shirley, N. J., Mather, D. E., Bacic, A., & Fincher, G. B. (2008). The genetics and transcriptional profiles of the cellulose synthase-like HvCslF gene family in barley. *Plant Physiology*, *146*, 1821–1833.

Cabib, E., Farkas, V., Kosik, O., Blanco, O., Arroyo, J., & McPhie, P. (2008). Assembly of the yeast cell wall Crh1p and Crh2p act as transglycosylases in vivo and in vitro. *Journal of Biological Chemistry*, *283*, 29859–29872.

Carvalho, A. L., Goyal, A., Prates, J. A. M., Bolam, D. N., Gilbert, H. J., Pires, V. M. R., Ferreira, L. M. A., Planas, A., Romao, M. J., & Fontes, C. M. G. A. (2004). The family 11 carbohydrate-binding module of *Clostridium thermocellum* Lic26A-Cel5E accommodates β-1,4- and β-1,3-1,4-mixed linked glucans at a single binding site. *Journal of Biological Chemistry*, *279*, 34785–34793.

Chambers, R. S., Broughton, M. J., Cannon, R. D., Carne, A., Emerson, G. W., & Sullivan, P. A. (1993). An exo-β-(1,3)-glucanase of *Candida albicans*: purification of the enzyme and molecular cloning of the gene. *Journal General Microbiology*, *139*, 325–334.

Chen, L., Garrett, T. J. P., Varghese, J. N., Fincher, G. B., & Høj, P. B. (1993a). Crystallization and preliminary X-ray analysis of (1,3)- and (1,3;1,4)-β-D-glucanases from germinating barley. *Journal of Molecular Biology*, *234*, 888–889.

Chen, L., Fincher, G. B., & Høj, P. B. (1993b). Evolution of polysaccharide hydrolase substrate specificity: catalytic amino acids are conserved in barley 1,3-1,4- and 1,3-β-glucanases. *Journal of Biological Chemistry*, *268*, 13318–13326.

Chen, L., Garrett, T. P. J., Fincher, G. B., & Høj, P. B. (1995a). A tetrad of ionizable amino acids is important for catalysis in barley β-glucanases. *Journal of Biological Chemistry*, *270*, 8093–8101.

Chen, L., Sadek, M., Stone, B. A., Brownlee, R. T. C., Fincher, G. B., & Høj, P. B. (1995b). Stereochemical course of glucan hydrolysis by barley 1,3- and 1,3;1,4-β-glucan endohydrolases. *Biochimica et Biophysica Acta, 1253*, 112–116.

Chi, Y.-I., Martinez-Cruz, L. A., Jancarik, J., Swanson, R. V., Robertson, D. E., & Kim, S.-H. (1999). Crystal structure of the β-glycosidase from the hyperthermophilic archeon *Thermosphaera aggregans*: insights into its activity and thermostability. *FEBS Letters, 445*, 375–383.

Chir, J., Withers, S., Wan, C.-F., & Li, Y.-K. (2002). Identification of the two essential groups in the family 3 β-glucosidase from *Flavobacterium meningosepticum* by labelling and tandem mass spectrometric analysis. *Biochemical Journal, 365*, 857–863.

Christakopoulos, P., Goodenough, P. W., Kekos, D., Macris, B. J., Claeyssens, M., & Bhat, M. K. (1994). Purification and characterisation of an extracellular β-glucosidase with transglycosylation and exo-glucosidase activities from *Fusarium oxysporum. European Journal of Biochemistry, 224*, 379–385.

Cicek, M., Blanchard, D., Bevan, D. R., & Esen, A. (2000). The aglycone specificity-determining sites are different in 2,4-dihydroxy-7-methoxy-1,4-benzoxazin-3-one (DIMBOA)-glucosidase (maize β-glucosidase) and dhurrinase (sorghum β-glucosidase). *Journal of Biological Chemistry, 275*, 20002–20011.

Cline, K., & Albersheim, P. (1981). Host-pathogen interactions. XVI. Purification and characterization of a glucosyl hydrolase/transferase present in the walls of soybean cells. *Plant Physiology, 68*, 207–220.

Cournoyer, B., & Faure, D. (2003). Radiation and functional specialization of the family 3 glycoside hydrolases. *Journal of Molecular Microbiology and Biotechnology, 5*, 190–198.

Coutinho, P., & Henrissat, B. (1999). Carbohydrate-active enzymes: an integrated database approach. In H. Gilbert, G. Davies, B. Henrissat, & B. Svensson (Eds.), *Recent Advances in Carbohydrate Bioengineering* (pp. 3–12). Cambridge: The Royal Society of Chemistry.

Crombie, H., Chengappa, S., Hellyer, A., & Reid, J. S. G. (1998). A xyloglucan oligosaccharide-active, transglycosylating β-D-glucosidase from the cotyledons of nasturtium (*Tropaeolum majus* L) seedlings - purification, properties and characterization of a cDNA clone. *Plant Journal, 15*, 27–38.

Czjzek, M., Cicek, M., Zamboni, V., Bevan, D. R., Henrissat, B., & Esen, A. (2000). The mechanism of substrate (aglycone) specificity in β-glucosidases is revealed by crystal structures of mutant maize β-glucosidase-DIMBOA, -DIMBOAGlc, and -dhurrin complexes. *Proceedings of the National academy of Sciences of the United States of America, 97*, 13555–13560.

Czjzek, M., Cicek, M., Zamboni, V., Burmeister, W. P., Bevan, D. R., Henrissat, B., & Esen, A. (2001). Crystal structure of a monocotyledon (maize ZMGlu1) β-glucosidase and a model of its complex with p-nitrophenyl β-D-thioglucoside. *Biochemistry Journal, 354*, 37–46.

Cutfield, S. M., Davies, G. J., Marshudov, G., Anderson, B. F., Moody, P. C. E., Sullivan, P. A., & Cutfield, J. F. (1999). The structure of the exo-β-(1,3)-glucanase from *Candida albicans* in native and bound forms: relationship between a pocket and groove in family 5 glycosyl hydrolases. *Journal of Molecular Biology*, *294*, 771–783.

Dan, S., Marton, I., Dekel, M., Bravdo, B. A., He, S., Withers, S. G., & Shoseyov, O. (2000). Cloning, expression, characterization, and nucleophile identification of family 3, *Aspergillus niger* β-glucosidase. *Journal of Biological Chemistry*, *275*, 4973–4980.

Day, A. G., & Withers, S. G. (1986). The purification and characterisation of a β-D-glucosidase from *Alcaligenes faecalis*. *Canadian Journal of Biochemistry*, *64*, 914–922.

Dharmawardhana, D. P., Ellis, B. E., & Carlson, J. E. (1995). A β-glucosidase from lodgepole pine xylem specific for the lignin precursor coniferin. *Plant Physiology*, *107*, 331–339.

Donofrio, N. M., & Delaney, T. P. (2001). Abnormal callose response phenotype and hypersusceptibility to *Peronospora parasitica* in defense-compromised Arabidopsis *nim1-1* and salicylate hydroxylase plants. *Molecular Plant-Microbe Interactions*, *14*, 439–450.

Eveleigh, D. E., & Perlin, A. S. (1969). A proton magnetic resonance study of the anomeric species produced by β-D-glucosidases. *Carbohydrate Research*, *10*, 87–95.

Falk, A., & Rask, L. (1995). Expression of a zeatin-*O*-glucoside-degrading β-glucohydrolase in *Brassica napus*. *Plant Physiology*, *108*, 1369–1377.

Fanutti, C., Gidley, M. J., & Reid, J. S. G. (1993). Action of a pure xyloglucan endo-transglycosylase (formerly called xyloglucan-specific endo-(1-4)-β-D-glucanase) from the cotyledons of germinated nasturtium seeds. *Plant Journal*, *3*, 691–700.

Farkas, V., Sulova, Z., Stratilova, E., Hanna, R., & Maclachlan, G. (1992). Cleavage of xyloglucan by nasturtium seed xyloglucanase and transglycosylation to xyloglucan subunit oligosaccharides. *Archives of Biochemistry and Biophysics*, *298*, 365–370.

Farrokhi, N., Burton, R. A., Brownfield, L., Hrmova, M., Wilson, S. M., Bacic, A., & Fincher, G. B. (2006). Plant cell wall biosynthesis: genetic, biochemical and functional genomics approaches to the identification of key genes. *Plant Biotechnology Journal*, *4*, 145–167.

Fibriansah, G., Masuda, S., Koizumi, N., Nakamura, S., & Kumasaka, T. (2007). The 1.3 Å crystal structure of a novel endo-β-1,3-glucanase of glycoside hydrolase family 16 from alkaliphilic *Nocardiopsis sp.* strain F96. *Proteins*, *15*, 683–690.

Fincher, G. B. (1975). Morphology and chemical composition of barley endosperm cell walls. *Journal of the Institute of Brewing*, *81*, 116–122.

Fincher, G. B., & Stone, B. A. (2004). Chemistry of non-starch polysaccharides from cereal grains. In C. Wrigley, H. Corke, & C. E. Walker, (Eds.) *Encyclopedia of Grain Science: Vol 1* (pp. 206–223). Elsevier Academic Press.

Fry, S. C. (1989). Cellulases, hemicellulases and auxin-stimulated growth: A possible relationship. *Physiologia Plantarum*, *72*, 532–536.

Fry, S. C., Nesselrode, B. H., Miller, J. G., & Mewburn, B. R. (2008a). Mixed-linkage (1→3,1→4)-β-D-glucan is a major hemicellulose of Equisetum (horsetail) cell walls. *New Phytologist*, *179*, 104–115.

Fry, S. C., Mohler, K. E., Nesselrode, B. H., & Franková, L. (2008b). Mixed-linkage β-glucan:xyloglucan endotransglucosylase, a novel wall-remodelling enzyme from Equisetum (horsetails) and charophytic algae. *Plant Journal*, *55*, 240–252.

Gaiser, O. J., Piotukh, K., Ponnuswamy, M. N., Planas, A., Borriss, R., & Heinemann, U. (2006). Structural basis for the substrate specificity of a *Bacillus,* 1,3-1,4-β-glucanase. *Journal of Molecular Biology*, *357*, 1211–1225.

Ham, K.-S., Wu, S.-C., Darvill, A. G., & Albersheim, P. (1997). Fungal pathogens secrete an inhibitor that distinguishes isoforms of plant pathogenesis-related endo-(1,3)-β-D-glucanases. *Plant Journal*, *11*, 169–179.

Harvey, A. J., Hrmova, M., DeGori, R., Varghese, J. N., & Fincher, G. B. (2000). Comparative modeling of the three-dimensional structures of family 3 glycoside hydrolases. *Proteins: Structure Functions Genetics*, *41*, 257–269.

Harvey, A. J., Hrmova, M., & Fincher, G. B. (2001). Regulation of genes encoding β-D-glucan glucohydrolases in barley. *Physiologia Plantarum*, *113*, 108–120.

Heath, L. S., Ramakrishnan, N., Sederoff, R. R., Whetten, R. W., Chevone, B. I., Struble, C. A., Jouenne, V. Y., Chen, D. W., van Zyl, L., & Grene, R. (2002). Studying the functional genomics of stress responses in loblolly pine with the Expresso microarray experiment management system. *Comparisons and Functional Genomics*, *3*, 226–243.

Heightman, T. D., & Vasella, A. T. (2000). Recent insight into inhibition, structure and mechanism of configuration of retaining glycosidases. *Angew. Rev. Int. Ed.*, *38*, 750–770.

Henrissat, B., Callebaut, I., Fabrega, S., Lehn, P., Mornon, J. P., & Davies, G. J. (1995). Conserved catalytic machinery and the prediction of a common fold for several families of glycosyl hydrolases. *Proceedings of the National academy of Sciences of the United States of America*, *92*, 7090–7094.

Henrissat, B., & Davies, G. (1997). Structural and sequence-based classification of glycoside hydrolases. *Current Opinion in Structural Biology*, *7*, 637–644.

Hidaka, M., Kitaoka, M., Hayashi, K., Wagaki, T., Shoun, H., & Fushinobu, S. (2006). Structural dissection of the reaction mechanism of cellobiose phosphorylase. *Biochemistry Journal*, *398*, 37–43.

Hidaka, M., Fushinobu, S., Nishimoto, M., and Kitaoka, M. (2007). Structural basis for a substrate specificity of a GH-94 laminaribiose phosphorylase. 7th Carbohydrate Bioengineering Meeting, Braunschweig, Germany, 22-25 April.

Høj, P. B., Slade, A. M., Wettenhall, R. E. H., & Fincher, G. B. (1988). Isolation and characterization of a (1→3)-β-glucan endohydrolase from germinating barley (*Hordeum vulgare*): Amino acid sequence similarity with barley (1→3,1→4)-β-glucanases. *FEBS Letters*, *230*, 67–71.

Høj, P. B., Hartman, D. J., Morrice, N. A., Doan, D. N. P., & Fincher, G. B. (1989). Purification of (1→3)-β-glucan endohydrolase isoenzyme II from germinated barley and determination of its primary structure from a cDNA clone. *Plant Molecular Biology*, *13*, 31–42.

Høj, P. B., & Fincher, G. B. (1995). Molecular evolution of plant β-glucan endohydrolases. *Plant Journal*, *7*, 367–379.

Hommalai, G., Chaiyen, P., & Svasti, J. (2004). Studies on the transglucosylation reactions of cassava and Thai rosewood β-glucosidases using 2-deoxy-2-fluoro-glycosyl-enzyme intermediates. *Archives of Biochemistry and Biophysics*, *442*, 11–20.

Hoson, T., & Nevins, D. J. (1989). β-D-Glucan antibodies inhibit auxin-induced cell ngation and changes in cell wall of *Zea* coleoptile segments. *Plant Physiology*, *90*, 1353–1358.

Hrmova, M., & Fincher, G. B. (1993). Purification and properties of three (1,3)-β-D-glucanase isoenzymes from young leaves of barley (*Hordeum vulgare*). *Biochemistry Journal*, *289*, 453–461.

Hrmova, M., Garrett, T. P. J., & Fincher, G. B. (1995). Subsite affinities and disposition of catalytic amino acids in the substrate-binding region of barley 1,3-β-glucanases: Implications in plant-pathogen interactions. *Journal of Biological Chemistry*, *270*, 14556–14563.

Hrmova, M., & Fincher, G. B. (1998). Barley β-D-glucan exohydrolases. Substrate specificity and kinetic properties. *Carbohydrate Research*, *305*, 209–221.

Hrmova, M., & Fincher, G. B. (2001). Structure-function relationships of β-D-glucan endo- and exohydrolases from higher plants. *Plant Molecular Biology*, *47*, 73–91.

Hrmova, M., & Fincher, G. B. (2007). Dissecting the catalytic mechanism of a plant β-D-glucan glucohydrolase through structural biology using inhibitors and substrate analogues. *Carbohydrate Research*, *342*, 1613–1623.

Hrmova, M., DeGori, R., Smith, B. J., Driguez, H., Varghese, J. N., & Fincher, G. B. (2002). Structural basis for broad substrate specificity in higher plant β-D-glucan glucohydrolases. *Plant Cell*, *14*, 1033–1052.

Hrmova, M., Harvey, A. J., Wang, J., Shirley, N. J., Jones, G. P., Høj, P. B., & Fincher, G. B. (1996). Barley β-D-glucan exohydrolases with β-D-glucosidase activity. Purification and determination of primary structure from a cDNA clone. *Journal of Biological Chemistry*, *271*, 5277–5286.

Hrmova, M., MacGregor, E. A., Biely, P., Stewart, R. S., & Fincher, G. B. (1998). Substrate binding and catalytic mechanism of a barley β-D-glucosidase/(1,4)-β-D-glucan exohydrolase. *Journal of Biological Chemistry*, *273*, 11134–11143.

Hrmova, M., Varghese, J. N., DeGori, R., Smith, B., Driguez, H., & Fincher, G. B. (2001). Catalytic mechanisms and reaction intermediates along the hydrolytic pathway of plant β-D-glucan glucohydrolase. *Structure with Folding & Design*, *9*, 1005–1016.

Hrmova, M., Farkas, V., Lahnstein, J., & Fincher, G. B. (2007). A barley xyloglucan xyloglucosyl transferase covalently links xyloglucan, cellulose and (1,3;1,4)-β-D-glucans. *Journal of Biological Chemistry*, *282*, 12951–12962.

Hrmova, M., Farkas, V., Harvey, A. J., Lahnstein, J., Wischmann, B., Kaewthai, N., Ezcurra, I., Teeri, T. T., & Fincher, G. B. (2009). Substrate specificity and catalytic mechanism of a xyloglucan xyloglucosyl transferase HvXET6 from barley (*Hordeum vulgare* L.) in the press. *FEBS Journal*.

Husebye, H., Arzt, S., Burmeister, W. P., Hartel, F. V., Brandt, A., Rossiter, J. T., & Bones, A. M. (2005). Crystal structure at 1.1 Angstroms resolution of an insect myrosinase from *Brevicoryne brassicae* shows its close relationship to β-glucosidases. *Insect Biochemistry and Molecular Biology*, *35*, 1311–1320.

Inouhe, M., Hayashi, K., & Nevins, D. J. (1999). Polypeptide characteristics and immunological properties of exo- and endoglucanases purified from maize coleoptile cell walls. *Journal of Plant Physiology*, *154*, 334–340.

Isorna, P., Polaina, J., Latorre-Garcia, L., Canada, F. J., Gonzalez, B., & Sanz-Aparicio, J. (2007). Crystal structures of *Paenibacillus polymyxa* β-glucosidase B complexes reveal the molecular basis of substrate specificity and give new insights into the catalytic machinery of family I glycosidases. *Journal of Molecular Biology*, *371*, 1204–1218.

Jacobs, A. K., Lipka, V., Burton, R. A., Panstruga, R., Strizhov, N., Schulze-Lefert, P., & Fincher, G. B. (2003). An *Arabidopsis thaliana* callose synthase, GSL5, is required for wound and papillary callose formation. *Plant Cell*, *15*, 2503–2513.

Jenkins, J., Lo Leggio, L., Harris, G., & Pickersgill, R. (1995). β-Glucosidase, β-galactosidase, family A cellulases, family F xylanases and two barley glycanases for a superfamily of enzymes with 8-fold β/α architecture and with two conserved glutamates near the carboxy-terminal ends of β-strands four and seven. *FEBS Letters*, *362*, 281–285.

Johansson, P., Brumer, H., III, Baumann, M. J., Kallas, Å. M., Henriksson, H., Denman, S. E., Teeri, T. T., & Jones, T. A. (2004). Crystal structures of a poplar xyloglucan endotransglycosylase reveal details of transglycosylation acceptor binding. *Plant Cell*, *16*, 874–886.

Juge, N. (2006). Plant protein inhibitors of cell wall degrading enzymes. *Trends in Plant Science*, *11*, 359–367.

Kapteyn, J. C., Ram, A. F. J., Groos, E. M., Kollar, R., Montijn, R. C., VandenEnde, H., Llobell, A., Cabib, E., & Klis, F. M. (1997). Altered extent of cross-linking of β-1,6-glucosylated mannoproteins to chitin in *Saccharomyces cerevisiae* mutants with reduced cell wall β-1,3-glucan content. *Journal of Bacteriology*, *179*, 6279–6284.

Kauffmann, S., Legrand, M., Geoffroy, P., & Fritig, B. (1987). Biological functions of 'pathogenesis-related' proteins: four PR proteins of tobacco have (1,3)-β-D-glucanase activity. *EMBO Journal*, *6*, 3209–3213.

Kawai, R., Igarashi, K., Kitaoka, M., Ishii, T., & Samejima, M. (2004). Kinetics of substrate transglycosylation by glycoside hydrolase family 3 glucan (1,3)-β-glucosidase from the white-rot fungus *Phanerochaete chrysosporium*. *Carbohydrate Research*, *339*, 2851–2857.

Kawai, R., Igarashi, K., Yoshida, M., Kitaoka, M., & Samejima, M. (2006). Hydrolysis of β-1,3/1,6-glucan by glycoside hydrolase family 16 endo-1,3(4)-β-glucanase from the basidiomycete *Phanerochaete chrysosporium*. *Applied Microbiology and Biotechnology*, *71*, 898–906.

Keitel, T., Simon, O., Borriss, R., & Heinemann, U. (1993). Molecular and active-site structure of a *Bacillus* 1,3-1,4-β-D-glucanase. *Proceedings of the National academy of Sciences of the United States of America*, *90*, 5287–5291.

Kim, J. B., Olek, A. T., & Carpita, N. C. (2000). Cell wall and membrane-associated exo-β-D-glucanases from developing maize seedlings. *Plant Physiology*, *123*, 471–485.

Kim, S. Y., Kim, Y. W., Hegerl, R., Cyrklaff, M., & Kim, I. S. (2005). Novel type of enzyme multimerization enhances substrate affinity of oat β-glucosidase. *Journal of Structural Biology*, *150*, 1–10.

Kitaoka, M., & Hayashi, K. (2002). Carbohydrate-processing phosphorolytic enzymes. *Trends in Glycoscience Glycotechnology*, *14*, 35–50.

Kitaoka, M., Sasaki, T., & Taniguchi, H. (1991). Synthesis of laminarioligosaccharides using crude extract of *Euglena gracilis* Z cells. *Agricultural and Biological Chemistry*, *55*, 1431–1432.

Kitaoka, M., Sasaki, T., & Taniguchi, H. (1993). Purification and properties of laminaribiose phosphorylase (EC 2.4.1.31) from *Euglena gracilis* Z. *Archives of Biochemistry and Biophysics*, *304*, 508–514.

Kloareg, B., & Quatrano, R. S. (1988). Structure of the cell walls of marine algea and ecophysical functions of the matrix polysaccharides. *Oceanography and Marine Biology - An Annual Review*, *26*, 259–315.

Koshland, D. E., Jr. (1953) Stereochemistry and the mechanism of enzymatic reations. *Biological Reviews*, *28*, 416–436.

Kotake, T., Nagawa, N., Takeda, K., & Sakurai, N. (1997). Purification and characterization of wall-bound exo-1,3-β-D-glucanase from barley (*Hordeum vulgare* L) seedlings. *Plant and Cell Physiology, 38*, 194–200.

Labrador, E., & Nevins, D. J. (1989). An exo-β-D-glucanase derived from Zea coleoptile walls with a capacity to elicit cell elongation. *Physiologia Plantarum, 77*, 479–486.

Leah, R., Kigel, J., Svendsen, I., & Mundy, J. (1995). Biochemical and molecular characterization of a barley seed β-glucosidase. *Journal of Biological Chemistry, 270*, 15789–15797.

Legler, G., & Herrchen, M. (1981). Active site directed inhibition of galactosidase by conduritol C epoxides (1,2-anhydro-epi- and neo-inositol). *FEBS Letters, 135*, 139–144.

Li, Y.-K., & Lee, J.-A. (1999). Cloning and expression of β-glucosidase from *Flavobacterium meningosepticum*: A new member of family B β-glucosidase. *Enzyme and Microbial Technology, 24*, 144–150.

Ludwig, A., & Boller, T. (1990). A method for the study of fungal growth inhibition by plant proteins. *FEMS Microbiology Letters, 69*, 61–66.

Ly, H. D., & Withers, S. G. (1999). Mutagenesis of glycosidases. *Annual Review of Biochemistry, 68*, 487–522.

Mallet, C., Jiménez-Barbero, J., Bernabé, M., Brosa, C., & Planas, A. (1993). Stereochemical course and structure of the products of the enzymic action of endo-1,3-1,4- β-D-glucan 4-glucanohydrolase from *Bacillus licheniformis. Biochemistry Journal, 296*, 7563–7758.

Manners, D. J., & Taylor, D. C. (1967). Studies on carbohydrate metabolizing enzymes. XVI. Specificity of laminaribiose phosphorylase from *Astasia ocellata. Archives of Biochemistry and Biophysics, 121*, 443–451.

Marana, S. R. (2006). Molecular basis of substrate specificity in family 1 glycoside hydrolases. *IUBMB Life, 58*, 6–73.

Marechal, L. R. (1967). β-1,3-Oligoglucan: orthophosphate glucosyltransferases from *Euglena gracilis*. II. Comparative studies between laminaribiose- and β-1,3-oligoglucan phosphorylase. *Biochimica et Biophysica Acta, 146*, 431–442.

Martin-Pastor, M., Vega-Vázquez, M., Capua, A., Canales, A., André, S., Gabius, H.-J., & Jiménez-Barbero, J. (2006). Enhanced signal dispersion in saturation transfer difference experiments by conversion to a 1D-STD-homodecoupled spectrum. *Biomolecular NMR, 36*, 103–109.

McCarter, J., & Withers, S. G. (1994). Mechanisms of enzymatic glycoside hydrolysis. *Current Opinion in Structural Biology, 4*, 885–892.

Meins, F., Jr., Neuhaus, J.-M., Sperisen, C., & Ryals, J. (1992). The primary structure of plant pathogenesis-related glucanohydrolases and their genes. In T. Boller & F. Meins (Eds.), Jr., *Genes Involved in Plant Defense* (pp. 245–282) Vienna: Springer.

Michel, G., Chantalat, L., Duee, E., Barbeyron, T., Henrissat, B., Kloareg, B., & Dideberg, O. (2001). The kappa-carrageenase of *P. carrageenovora* features a tunnel-shaped active site: a novel insight in the evolution of Clan-B glycoside hydrolases. *Structure*, *9*, 513–525.

Money, V. A., Cartmell, A., Guerreiro, C. I. P. D., Ducros, V. M. A., Fontes, C. M. G. A., Gilbert, H. J., & Davies, G. J. (2008). Probing the β-1,3:1,4 glucanase, CtLic26A, with a thio-oligosaccharide and enzyme variants. *Organic & Biomolecular Chemistry*, *6*, 851–853.

Murzin, A. G., Brenner, S. E., Hubbard, T., & Chothia, C. (1995). SCOP: a structural classification of proteins database for the investigation of sequences and structures. *Journal of Molecular Biology*, *247*, 536–540.

Nielsen, K. A., Hrmova, M., Nielsen, J. N., Forslund, K., Ebert, S., Olsen, C. E., Fincher, G. B., & Møller, B. L. (2006). Reconstitution of cyanogenesis in barley and its implications for resistance against the barley powdery mildew fungus. *Planta*, *223*, 1010–1023.

Okazaki, F., Tamaru, Y., Hashikawa, S., Li, Y.-T., & Araki, T. (2002). Novel carbohydrate-binding module of β-1,3-xylanase from a marine bacterium, *Alcaligenes* sp. strain XY-234. *Journal of Bacteriology*, *184*, 2399–2403.

Opassiri, R., Hua, Y., Wara-Aswapati, O., Akiyama, T., Svasti, J., Esen, A., & Ketudat Cairns, J. R. (2004). β-Glucosidase, exo-β-glucanase and pyridoxine transglucosylase activities of rice BGlu1. *Biochemistry Journal*, *379*, 125–131.

Opassiri, R., Ketudat Cairns, J. R., Akiyama, T., Wara-Aswapati, O., Svasti, J., & Esen, A. (2003). Characterization of a rice β-glucosidase highly expressed in flower and germinating shoot. *Plant Science*, *165*, 627–638.

Opassiri, R., Pomthong, B., Onkoksoong, T., Akiyama, T., Esen, A., & Ketudat Cairns, J. R. (2006). Analysis of rice glycosyl hydrolase family 1 and expression of Os4bglu12 β-glucosidase. *BMC Plant Biology*, *6*, 33–51.

Paal, K., Ito, M., & Withers, S. G. (2004). *Paenibacillus* sp. TS12 glucosylceramidase - Kinetic studies of a novel sub-family of family 3 glycosidases and identification of the catalytic residues. *Biochemistry Journal*, *378*, 141–149.

Park, T.-H., Choi, K.-W., Park, C.-S., Lee, S.-B., Kang, H.-Y., Shon, K.-J., Park, J.-S., & Cha, J. (2005). Substrate specificity and transglycosylation catalyzed by a thermostable β-glucosidase from marine hyperthermophile *Thermotoga neapolitana*. *Applied Microbiology and Biotechnology*, *69*, 411–422.

Parrish, F. W., Perlin, A. S., & Reese, E. T. (1960). Selective enzymolysis of poly-β-D-glucans, and the structure of the polymers. *Canadian Journal of Chemistry*, *38*, 2094–2104.

Piens, K., Faure, R., Sundqvist, G., Baumann, B. J., Saura-Valls, M., Teeri, T. T., Cottaz, S., Planas, A., Driguez, H., & Brumer, H. (2008). Mechanism-based labeling defines the free energy change for formation of the covalent glycosyl-enzyme intermediate in a xyloglucan endo-transglycosylase. *Journal of Biological Chemistry*, *283*, 21864–21872.

Pitson, S. M., Seviour, R. J., Mcdougall, M. C., Stone, B. A., & Sadek, M. (1996). Purification and characterization of an extracellular (1→6)-β-glucanase from the filamentous fungus *Acremonium persicinum*. *Biochemistry Journal*, *316*, 841–846.

Planas, A., Juncosa, M., Lloberas, J., & Querol, E. (1992). Essential catalytic role of $Glu^{134}$ in endo β-1,3-1,4-D-glucan 4-glucanohydrolase from *B. licheniformis* as determined by site-directed mutagenesis. *FEBS Letters*, *308*, 141–145.

Potin, P., Sanseau, A., Le Gall, Y., Rochas, C., & Kloareg, B. (1991). Purification and characterization of a new kappa-carrageenase from a marine Cytophaga-like bacterium. *European Journal of Biochemistry*, *201*, 241–247.

Poulton, J. E. (1990). Cyanogenesis in plants. *Plant Physiology*, *94*, 401–405.

Purugganan, M. M., Braam, J., & Fry, S. C. (1997). The Arabidopsis TCH4 xyloglucan endotransglycosylase - substrate specificity, pH optimum and cold tolerance. *Plant Physiology*, *115*, 181–190.

Roberts, W. K., & Selitrennikoff, C. P. (1998). Plant and bacterial chitinases differ in antifungal activity. *Journal of General Microbiology*, *134*, 169–176.

Rojas, A. L., Fischer, H., Eneiskaya, E. V., Kulminskaya, A. A., Shabalin, K. A., Neustroev, K. N., Craievich, A. F., Golubev, A. M., & Polikarpov, I. (2005). Structural insights into the β-xylosidase from *Trichoderma reesei* obtained by synchrotron small-angle X-ray scattering and circular dichroism spectroscopy. *Biochemistry*, *44*, 15578–15584.

Rojas, A., Garcia-Vallve, S., Montero, M. A., Arola, L., & Romeu, A. (2003). Frameshift mutation events in β-glucosidases. *Genetics*, *314*, 191–1999.

Rose, J. C. K., Ham, K.-S., Wu, S.-C., Darvill, A. G., & Albersheim, P. (2002a). Molecular cloning and characterization of glucanase inhibitor proteins: coevolution of a counterdefense mechanism by plant pathogens. *Plant Cell*, *14*, 1329–1345.

Rose, J. K. C., Braam, J., Fry, S. C., & Nishitani, K. (2002b). The XTH family of enzymes involved in xyloglucan endotransglucosylation and endohydrolysis: current perspectives and a new unifying nomenclature. *Plant and Cell Physiology*, *43*, 1421–1435.

Ryals, J. A., Neuenschwander, U. H., Willits, M. G., Molina, A., Steiner, H. Y., & Hunt, M. D. (1996). Systemic acquired resistance. *Plant Cell*, *8*, 1809–1819.

Ryttersgaard, C., Lo Leggio, L., Coutinho, P., Henrissat, B., & Larsen, S. (2002). *Aspergillus aculeatus* β-1,4-galactanase: substrate recognition and relations to other glycoside hydrolases in clan GH-A. *Biochemistry*, *41*, 15135–15143.

Sakurai, N., & Masuda, Y. (1978). Auxin-induced changes in barley coleoptile cell wall position. *Plant and Cell Physiology*, *19*, 1217–1223.

Saura-Valls, M., Faure, R., Brumer, H., Teeri, T. T., Cottaz, S., & Driguez, H. (2008). Active-site mapping of a Populus xyloglucan endo-transglycosylase with a library of xylogluco-oligosaccharides. *Journal of Biological Chemistry*, *283*, 21853–21863.

Seidle, H. F., & Huber, R. E. (2005). Transglucosidic reactions of the *Aspergillus niger* family 3 β-glucosidase: qualitative and quantitative analyses and evidence that the transglucosidic rate is independent of pH. *Archives of Biochemistry and Biophysics*, *436*, 254–264.

Seidle, H. F., Marten, I., Shoseyov, O., & Huber, R. E. (2004). Physical and kinetic properties of the family 3 β-glucosidase from *Aspergillus niger* which is important for cellulose breakdown. *Protein Journal*, *23*, 11–23.

Slakeski, N., Baulcombe, D. C., Devos, K. M., Ahluwalia, B., Doan, D. N. P., & Fincher, G. B. (1990). Structure and tissue-specific regulation of genes encoding barley (1→3,1→4)-β-glucan endohydrolases. *Molecular and General Genetics*, *224*, 437–449.

Slakeski, N., & Fincher, G. B. (1992). Developmental regulation of (1→3,1→4)-β-glucanase gene expression in barley. Tissue-specific expression of individual isoenzymes. *Plant Physiology*, *99*, 1226–1231.

Smart, M. G., Aist, J. R., & Israel, H. W. (1986). Structure and function of wall appositions. 1. General histochemistry of papillae in barley (*Hordeum vulgare*) coleoptiles attacked by *Erysiphe graminis f. sp. hordei*. *Canadian Journal of Botany*, *64*, 793–801.

Stone, B. A., Evans, N. A., Bonig, I., & Clarke, A. E. (1985). The application of Sirofluor, a chemically defined fluorochrome from aniline blue for the histochemical detection of callose. *Protoplasma*, *122*, 191–195.

Stone, B. A., & Clarke, A. E. (1992). *Chemistry and biology of (1-3)-β-glucans* (808 pp.) Bundoora, Australia: La Trobe University Press.

Stone, B. A., Clarke, A. E., Fincher, G. B., & Bacic, A. (2009). *Chemistry and biology of (1-3)-β-glucans*. Bundoora, Australia: La Trobe University Press.

Street, I. P., Kempton, J. B., & Withers, S. G. (1992). Inactivation of a β-glucosidase through the accumulation of a stable 2-deoxy-2-fluoro-β-D-glucopyranosyl-enzyme intermediate: a detailed investigation. *Biochemistry*, *31*, 9970–9978.

Stuart, I. M., Loi, L., & Fincher, G. B. (1988). Varietal and environmental variations in (1→3,1→4)-β-glucan levels and (1→3,1→4)-β-glucanase potential in barley: Relationships to malting quality. *Journal of Cereal Science*, *7*, 61–71.

Stubbs, H. J., Brasch, D. J., Emerson, G. W., & Sullivan, P. A. (1999). Hydrolase and transferase activities of the β-1,3-exoglucanase of *Candida albicans*. *European Journal of Biochemistry*, *263*, 889–895.

Strohmeier, M., Hrmova, M., Fischer, M., Harvey, A. J., Fincher, G. B., & Pleiss, J. (2004). Molecular modelling of family GH16 glycoside hydrolases: Potential roles for xyloglucan endotransglucosylases/hydrolases in cell wall modification in the Poaceae. *Protein Science*, *13*, 3200–3213.

Sue, M., Yamazaki, K., Yajima, S., Nomura, T., Matsukawa, T., Iwamura, H., & Miyamoto, T. (2006). Molecular and structural characterization of hexameric β-D-glucosidases in wheat and rye. *Plant Physiology*, *141*, 1237–1247.

Suganuma, T., Matsuno, R., Ohnishi, M., & Hiromi, K. (1978). A study of the mechanism of action of Taka-amylase A on linear oligosaccharides by product analysis and computer simulation. *Journal of Biochemistry*, *84*, 293–316.

Takeda, H., Yoshikawa, T., Liu, X.-Z., Nakagawa, N., Li, Y.-Q., & Sakurai, N. (2004). Molecular Cloning of two exo-β-glucanases and their *in vivo* substrates in the cell walls of lily pollen tubes. *Plant and Cell Physiology*, *45*, 436–444.

Taylor, E. J., Goyal, A., Guerreiro, C. I. P. D., Prates, J. A. M., Money, V. A., Ferry, N., Morland, C., Planas, A., Macdonald, J. A., Stick, R. V., Gilbert, H. J., Fontes, C. M. G. A., & Davies, G. J. (2005). How family 26 glycoside hydrolases orchestrate catalysis on different polysaccharides - Structure and activity of a *Clostridium hermocellum* lichenase, CtLic26A. *Journal of Biological Chemistry*, *280*, 32761–32767.

Thomas, B. R., Simmons, C. R., Inouhe, M., & Nevins, D. J. (1998). Maize coleoptile endoglucanase is encoded by a novel gene family (Accession No. AF072326) (PGR98-143). *Plant Physiology*, *117*, 1525.

Thomas, B. R., Inouhe, M., Simmons, C. R., & Nevins, D. J. (2000). Endo-1,3;1,4-β-D-glucanase from coleoptiles of rice and maize: role in the regulation of plant growth. *International Journal of Biological Macromolecules*, *27*, 145–149.

Thordal-Christensen, H., Zhang, Z., Wei, Y., & Collinge, D. B. (1997). Subcellular localization of $H_2O_2$ in plants: $H_2O_2$ accumulation in papillae and hypersensitive response during the barley-powdery mildew interaction. *Plant Journal*, *11*, 1187–1194.

Trethewey, J. A. K., Campbell, L. M., & Harris, P. J. (2005). (1→3),(1→4)-β-D-Glucans in the cell walls of the Poales (sensu lato): An immunogold labeling study using a monoclonal antibody. *American Journal of Botany*, *92*, 1660–1674.

Tsukada, T., Igarashi, K., Yoshida, M., & Samejima, M. (2006). Molecular cloning and characterization of two intracellular. β-glucosidases belonging to glycoside hydrolase family 1 from the basidiomycete *Phanerochaete chrysosporium*. *Applied Microbiology and Biotechnology*, *73*, 807–814.

Vallmitjana, M., Ferrer-Navarro, M., Planell, R., Abel, M., Querol, E., Planas, A., & Pérez-Pons, J.-A. (2001). Mechanism of the family β-glucosidase from *Streptomyces* sp: catalytic residues and kinetic studies. *Biochemistry*, *40*, 5975–5982.

Varghese, J. N., Garrett, T. P. J., Colman, P. M., Chen, L., Høj, P. B., & Fincher, G. B. (1994). The three-dimensional structures of two plant β-glucan endohydrolases with distinct substrate specificities. *Proceedings of the National academy of Sciences of the United States of America*, *91*, 2785–2789.

Varghese, J. N., Hrmova, M., & Fincher, G. B. (1999). Three-dimensional structure of a barley β-D-glucan exohydrolase, a family 3 glycosyl hydrolase. *Structure with Folding & Design*, *7*, 179–190.

Verdoucq, L., Moriniere, J., Bevan, D. R., Esen, A., Vasella, A., Henrissat, B., & Czjzek, M. (2004). Structural determinants of substrate specificity in family 1 β-glucosidases: novel insights from the crystal structure of sorghum dhurrinase-1, a plant β-glucosidase with strict specificity, in complex with its natural substrate. *Journal of Biological Chemistry*, *279*, 31796–31803.

Vocadlo, D. J., Mayer, C., He, S., & Withers, S. G. (2000). Mechanism of action and identification of Asp242 as the catalytic nucleophile of *Vibrio furnisii* N-acetyl-β-D-glucosaminidase using 2-acetamido-2-deoxy-5-fluoro-a-L-idopyranosyl fluoride. *Biochemistry*, *39*, 117–126.

Wang, J., Xu, P., & Fincher, G. B. (1992). Purification, characterization and gene structure of (1→3)-β-glucanase isoenzyme GIII from barley (*Hordeum vulgare*). *European Journal of Biochemistry*, *209*, 103–109.

Wessels, J. G. F. (1993). Wall growth, protein excretion and morphogenesis in fungi. *New Phytologist*, *123*, 397–413.

White, A., & Rose, D. R. (1997). Mechanism of catalysis by retaining β-glycosyl hydrolases. *Current Opinion in Structural Biology*, *7*, 645–651.

Wilson, S. M., Burton, R. A., Doblin, M. S., Stone, B. A., Newbigin, E. J., Fincher, G. B., & Bacic, A. (2006). Temporal and spatial appearance of wall polysaccharides during cellularization of barley (*Hordeum vulgare*) endosperm. *Planta*, *224*, 655–667.

Withers, S. G. (1995). Probing of glycosidase active sites through labeling, mutagenesis and kinetic studies. Proceedings of the Carbohydrate Bioengineering Meeting, Elsinore, Denmark.

Wood, P. J., Weisz, J., & Blackwell, B. A. (1994). Structural studies of (1→3,1→4)-β-D-glucans by $^{13}$C-nuclear magnetic resonance spectroscopy and by rapid analysis of cellulose-like regions. *Cereal Chemistry*, *71*, 301–307.

Woodward, J. R., & Fincher, G. B. (1982a). Purification and chemical properties of two 1,3:1,4-β-glucan endohydrolases from germinating barley. *European Journal of Biochemistry*, *121*, 663–669.

Woodward, J. R., & Fincher, G. B. (1982b). Substrate specificities and kinetic properties of two (1→3), (1→4)-β-D-glucan endo-hydrolases from germinating barley (*Hordeum vulgare*). *Carbohydrate Research*, *106*, 111–122.

Xu, P., Wang, J., & Fincher, G. B. (1992). Evolution and differential expression of the (1→3)-β-glucan endohydrolase gene family in barley. *Genetics*, *120*, 157–165.

Yahata, N., Watanabe, T., Nakamura, Y., Yamamoto, Y., Kamimiya, S., & Tanaka, H. (1990). Structure of the gene encoding β-1,3-glucanase A1 of *Bacillus circulans* WL-12. *Genetics*, *86*, 113–117.

Yazaki, T., Ohnishi, M., Rokushika, S., & Okada, G. (1997). Subsite structure of the β-glucosidase from *Aspergillus niger*, evaluated by steady-state kinetics with cello-oligosaccharides as substrates. *Carbohydrate Research*, *298*, 51–57.

Zechel, D. L., & Withers, S. G. (1999). Glycosyl transferase mechanisms. In K. Barton, K. Nakanishi, & C. D. Poulter, (Eds.) *Comprehensive Natural Products Chemistry: vol 5* (pp. 279–314). New York: Elsevier.

Zechel, D. L., & Withers, S. G. (2000). Glycosidase mechanisms: anatomy of a finely tuned catalyst. *Accounts of Chemical Research*, *33*, 11–18.

CHAPTER 3.2

# Interactions between Proteins and (1,3)-ß-Glucans and Related Polysaccharides

**D. Wade Abbott[1] and Alisdair B. Boraston[2]**

*[1,2]Biochemistry & Microbiology, University of Victoria, STN CSC, Victoria, BC, Canada*

## I.A. Introduction

(1,3)-β-Glucans are linear glucose polymers linked through (1,3)-β-glycosidic linkages. Depending upon their biological source these polysaccharides can display heterogeneity within the polymer backbone as in the (1,3;1,4)-β−glucans and along it as (1,6)-β-linked branches. (1,3)-β-Glucans are found in bacteria, plants and fungi, where they are involved in specialized functions. For example, (1,3)-β-glucans are components of the extracellular polysaccharide and biofilm matrixes in *Agrobacterium* sp., contribute to the heteropolysaccharide matrix of the cell wall in yeasts and fungi, and operate as energy stores in certain types of brown algae and euglenoid protozoa (McIntosh et al., 2005). Recently, this class of polysaccharide has come under intense study as they have been determined to be potent activators of the innate immune system in invertebrates and vertebrates, and are currently being investigated as promising pharmacological agents for fungal infections and the treatment of many debilitating pathologies such as diabetes and cancer (Berdal et al., 2007; Chen and Seviour, 2007; Muta, 2006; Zekovic et al., 2005). See also chapters on innate immunity in invertebrates and vertebrates (Chapters 4.5.1 and 4.5.2).

(1,3)-β-Glucans and related polysaccharides are recognised by a wide range of proteins, including carbohydrate-binding modules (CBMs), non-catalytic proteins from glycoside hydrolase family 16 (GH16s), and members of the NK-cell-receptor-like C-type lectin family (i.e. dectin-1). These different proteins vary significantly in biological activities, ranging from biomass recycling to host–pathogen immunity, and overall structure and binding site architecture. For the purposes of this review we will limit our discussion to what is currently known about the biology of these different protein classes, with a concerted focus on the molecular determinants of ligand recognition.

## I.B. General Structural Properties of (1,3)-β-Glucan-Binding Proteins

Most proteins known to bind (1,3)-β-glucans are CBMs (see Table 1). CBMs are loosely defined as contiguous amino acid sequences that form a non-catalytic entity with an independent fold dedicated to the recognition and binding of carbohydrates. Most commonly,

**Table 1: Ligand binding profiles for (1,3)-β-glucan-binding CBMs**

| NAME | LIGANDS |
|---|---|
| **Family 4** | |
| *Tm*CBM4-2 | **(1,3)-β-oligoglucosides, laminarin, oat β-glucan, pustulan, curdlan,** hydroxyethyl cellulose (Boraston et al., 2002c; Zverlov et al., 2001) |
| **Family 6** | |
| *Bh*CBM6 | **(1,3)-β-oligoglucosides, laminarin**, xylose, xylo-oligosaccharides, glucose, sophorose, wheat arabino-xylan, birchwood glucurono-xylan, pectic galactan (van Bueren et al., 2005) |
| *Cm*CBM6-2 | **Barley (1,3;1,4)-β-glucan, lichenin, laminarin, (1,3)-β-oligoglucosides,** (1,4)-β-oligoglucosides, cellulose, xylo-oligosaccharides (Henshaw et al., 2004; Pires et al., 2004) |
| **Family 11** | |
| *Ct*CBM11 | **Oat (1,3;1,4)-β-glucan, lichenin**, hydroxyethyl cellulose, glucomannan, oat spelt xylan (Malburg et al., 1997) |
| **Family 13** | |
| *Bc*CBM13 | **Pachyman, lichenin,** *Aspergillus oryzae* cell walls (Asano et al., 2002) |
| **Family 17** | |
| *Cc*CBM17 | **Oat β-(1,3;1,4)-β-glucan,** cello-oligosaccharides, amorphous cellulose, microcrystalline cellulose, plant cell wall sections (Blake et al., 2006; Boraston et al., 2000; Notenboom et al., 2001) |
| **Family 22** | |
| *Ct*CBM22-2 | Oat spelt xylan, wheat arabinoxylan, rye arabinoxylan, **β-glucan,** and hydroxyethyl, (1,4)-β-oligoglucosides (Dias et al., 2004) |
| **Family 28** | |
| *Bsp*CBM28 | **Barley (1,3;1,4)-β-glucan**, regenerated cellulose, (1,4)-β-oligoglucosides, plant cell wall sections (Blake et al., 2006; Boraston et al., 2002a; Jamal et al., 2004) |
| **Family 39** | |
| *Pi*CBM39 | **Curdlan,** lipoteichoic acid (Fabrick et al., 2004) |
| **Family 43** | |
| *Oe*CBM43B | **Laminarin** (Barral et al., 2005) |

Carbohydrates containing (1,3)- or (1,3;1,4)-β-linkages are shown in bold.

CBMs are found appended to a catalytic module; however, there are several well characterized examples that operate independently (Abbott et al., 2008; Abbott et al., 2007; Flint et al., 2005; Flint et al., 2004; Vaaje-Kolstad et al., 2005a; Vaaje-Kolstad et al., 2005b). CBMs can be classified functionally based upon the structural properties of the target ligand (Boraston, 2007) (Fig. 1): (1) Type A CBMs adsorb to the surface of flat crystalline ligands, (2) Type B CBMs bind soluble carbohydrate chains, and (3) Type C CBMs interact with short oligosaccharides (mono-, di- and trisaccharides). Not surprisingly, each of these different CBM

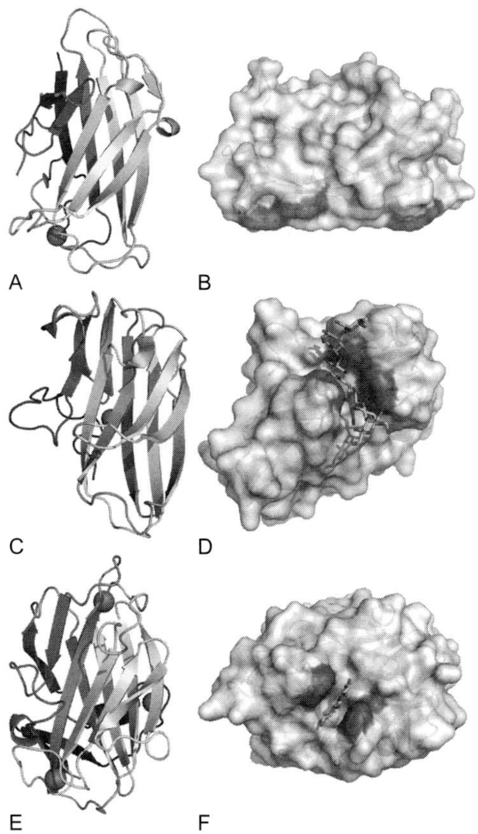


**Fig. 1: Three-dimensional structure and surface topographies of Type A, B and C CBMs. (A, B) Type A CBM3 from a *Clostridium thermocellum* scafoldin protein (PDB ID: 1NBC). (C, D) Type B CBM4-2 from a predicted *Thermatoga maritima* (1,3)-β-glucanase in complex with laminarihexose (PDB ID: 1GUI). (E, F) Type C CBM9 from a *T. maritima* xylanase in complex with cellobiose (PDB ID: 1I82). Calcium atoms are displayed as spheres (A, C, and E) and aromatic residues involved in ligand recognition are shaded (B, D, and F).**

classes has a structurally distinct binding site architecture that is tailored to recognize the signature configurations and substituent stereochemistries of each ligand type. In this regard, all of the characterized (1,3)-β-glucan-binding CBMs to date have all turned out to be Type B, generally displaying a prominent binding cleft lined with aromatic residues to complement the three-dimensional (3D) surface of the carbohydrate (Fig. 1D).

CBMs are most commonly grouped into families on the basis of amino acid similarity. At the time of writing, this ever-expanding list has grown to include over 50 different families that have been biochemically characterized (Coutinho and Henrissat, 1999). The majority of these families adopt a 'β-sandwich fold', which consists of two β-sheets stacked upon each other (Boraston et al., 2004). The β-sandwich has proven to be an extremely plastic scaffold for carbohydrate recognition as it is found in diverse CBM families (Blake et al., 2006), appended to unrelated enzyme classes (Boraston, 2007; Boraston et al., 2004), to interact with a myriad of different carbohydrate ligands (Abbott et al., 2008; Blake et al., 2006), and to contain binding sites in unique and sometimes multiple locations (Czjzek et al., 2001; Henshaw et al., 2004).

There are several other fold classifications that have been described for CBMs (Boraston et al., 2004), including the β-trefoil from family 13 (Ferrer, 2006) and an α/β fold from family 43 (Trevino et al., 2007). In addition, the structures of non-CBM proteins GH16s and dectin-1 display a β-jelly roll and a structurally unrelated α/β fold. These structures will be considered in further detail below within the context of their biological and biochemical properties.

## I.C. (1,3)-β-Glucan Structure

(1,3)-β-Glucans have a unique binding surface that differs substantially in 3D space from that of the (1,4)-linked β-glucans (Fig. 2). The $\phi$ and $\psi$ glycosidic bond angles of (1,3)-β-oligoglucosides in solution have been determined to be $-72°$ and $108°$ respectively by molecular mechanics and molecular dynamic simulations (Frecer et al., 2000; Buliga et al., 1986). This stereochemistry induces the β-glucan backbone to adopt a signature helical conformation (Fig. 2B and C). Individual helixes then intertwine with each other to form a triple-helix superstructure (Marchessault et al., 1980) which creates a unique surface topography for recognition by β-glucan binding proteins and degrading enzymes.

The family of (1,3)-β-glucans have various structural and functional roles in nature, and show distinct differences in solubility. For example, the high DP (degree of polymerization) (1,3)-β-homopolymers (e.g. pachyman from *Poria cocos* and curdlan from *Agrobacterium* sp.)

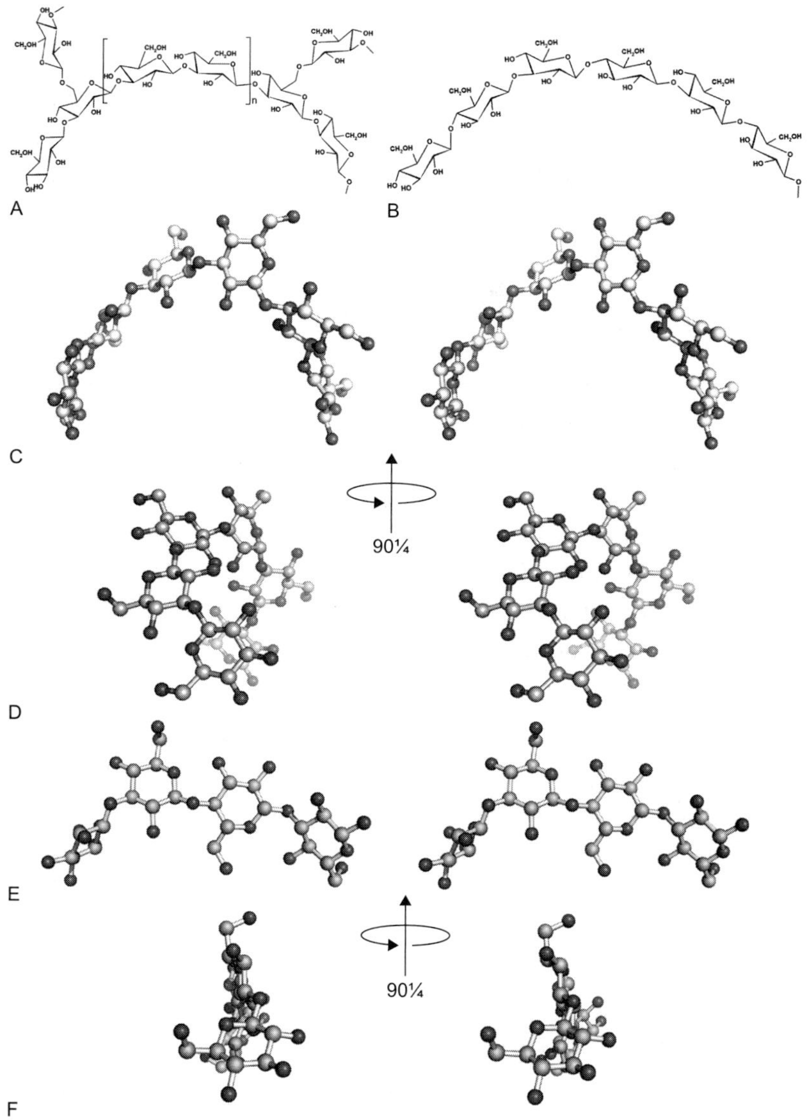


**Fig. 2: Structures of (1,3)-β-glucans. Schematic representation of a (1,3)-β-glucan with (1,6)-β-linked glucosyl side branches (A) and a (1,3;1,4)-β-glucan (B). The degree of polymerization between the brackets and the presence of (1,6)-β-linkages in (A) varies with carbohydrate source. Three-dimensional representation of laminarihexaose extracted from the *Bh*CBM6-3 complex coordinates (PDB ID: 1W9W) displayed in wall-eyed stereo viewed from the side (C) and non-reducing end (D) (van Bueren et al., 2005). Three-dimensional representation of a mixed (1,3)-β-(1,4)-β-(1,3)-β-linked glucosyl tetraose ligand extracted from the *Cm*CBM6-2 complex coordinates (PDB ID: 1UY0) displayed in wall-eyed stereo viewed from the side (E) and non-reducing end (F) (Pires et al., 2004).**

are insoluble in water; whereas, the low DP (33) laminarin from *Laminaria saccharina*, a (1,3)-β-linked polysaccharide which contains modest β-glucosyl substitutions at the C6 position (approximately 1 glucose/10 backbone subunits), is soluble in aqueous solutions.

Related (1,3;1,4)-β-glucans are found in the cell walls of grasses and cereals and cell walls of fungi and lichens (Carpita, 1996, see Chapter 1 Chemistry). These unbranched molecules consist mainly of cellotriose and cellotetraose subunits joined through (1,3)-β-glucosidic linkages, although longer runs up to DP 14 of (1,4)-linked residues are encountered. The frequency of the cellotriosyl and cellotetraosyl subunits depends upon the source. For example, oat (1,3),(1,4)-β-glucan has a ratio of ~ 2 cellotriosyl for every 2.4 cellotetraosyl subunits; whereas, barley is ~ 2.7–3.0 (Papageorgiou, 2005). In lichenin the ratio of cellotriosyl and cellotetraosyl subunits is much higher at 24.5, and 22.5% of the molecule is composed of (1,4)-linked residues DP 5–14 (Lazaridou, 2004). The insertion of (1,3)-linkages in an otherwise cellulosic chain results in an extended twisted/reptate ribbon-like polysaccharide conformation, which presents unique binding surfaces for proteins (Fig. 2E and F).

In the following sections we will discuss the molecular determinants of the recognition of (1,3)-β glucans and related polysaccharides. Firstly, we will consider the available structural data for CBM families 4, 6, 11, 17, 22, 28 and 43 (see Table 2 for a complete list). This will be followed by an analysis of structurally uncharacterized β-glucan-binding CBMs from families 13 and 39, and the non-CBM proteins, GH16s and dectin-1. In conclusion, we will compare and contrast what is currently known about the binding site architecture of (1,3)-β-glucan-binding CBMs.

## I.D. Structure-Function Relationships of β-Glucan Binding CBMs

### *I.D.a CBM4*

Family 4 CBMs are found in enzymes with specificities for various cell wall polysaccharides and have been shown to bind diverse ligands, including xylan, amorphous cellulose, and (1,3)-, (1,3;1,4)- and (1,6)-β-glucans (Abou Hachem et al., 2000; Boraston et al., 2002c; Boraston et al., 2001; Simpson et al., 2002; Zverlov et al., 2001) (Table 1). This family has proven to be a versatile scaffold for the display of ligand binding diversity in nature, and biotechnological efforts have been successful in engineering a CBM4 from *Rhodothermus marinus* (*Rm*CBM4-2) to interact specifically with selected carbohydrates (Cicortas Gunnarsson et al., 2007; Cicortas Gunnarsson et al., 2004; Gunnarsson et al., 2006b) and even the protein component of a glycoprotein (Gunnarsson et al., 2006a).

**Table 2: Three-dimensional structures for (1,3)-β–glucan-binding CBMs**

| Family | Name | Fold | Complex | Reference |
|---|---|---|---|---|
| **4** | *Tm*CBM4-2 | β-sandwich | 1CX1 (unliganded)<br>1GU3 (cellopentaose)<br>**1GUI** (laminarihexaose) | (Brun et al., 2000)<br>(Boraston et al., 2002c)<br>(Boraston et al., 2002c) |
| **6** | *Bh*CBM6 | β-sandwich | 1W9S (unliganded)<br>**1W9W** (laminarihexaose)<br>1W9T (xylobiose) | (van Bueren et al., 2005)<br>(van Bueren et al., 2005)<br>(van Bueren et al., 2005) |
| | *Cm*CBM6-2 | β-sandwich | 1UXZ (unliganded)<br>**1UY0** (G3G4G3G)<br>**1UZ0** (G4G3G4G)<br>1UYX (cellobiose)<br>1UYY (cellotriose)<br>1UYZ (xylotetraose) | (Pires et al., 2004)<br>(Pires et al., 2004)<br>(Pires et al., 2004)<br>(Pires et al., 2004)<br>(Pires et al., 2004)<br>(Pires et al., 2004) |
| | *Cs*CBM6-3 | β-sandwich | 1O8P (unliganded)<br>**1OD3** (laminaribiose)<br>1O8S (cellobiose)<br>1NAE (xylotriose) | (Boraston et al., 2003b)<br>(Boraston et al., 2003b)<br>(Boraston et al., 2003b)<br>(Boraston et al., 2003b) |
| **11** | *Ct*CBM11 | β-sandwich | 1V0A (unliganded) | (Carvalho et al., 2004) |
| **17** | *Cc*CBM17 | β-sandwich | 1J83 (unliganded)<br>1J84 (cellotetraose) | (Notenboom et al., 2001)<br>(Notenboom et al., 2001) |
| **22** | *Ct*CBM22-2 | β-sandwich | 1H6X (unliganded)<br>1H6Y (unliganded) | (Xie et al., 2001)<br>(Xie et al., 2001) |
| **28** | *Bsp*CBM28 | β-sandwich | 1UWW (unliganded) | (Jamal et al., 2004) |
| **43** | *Oe*CBM43A | α/β | 2JON (unliganded) | (Trevino et al., 2007) |

Co-crystal structures containing a ligand with (1,3)-β-linkage are shown in bold.

The first two CBM4 structures determined were the tandem N-terminal cellulose binding domains from endoglucanase C from *Cellulomonas fimi* (Brun et al., 2000; Johnson et al., 1996). The fold of these CBMs is a conventional β-sandwich consisting of 10 β-strands organized into two anti-parallel sheets with a jellyroll topology (Fig. 3A). The binding sites are clefts that traverse the face of the concave β-sheet perpendicular to the directionality of the β-strands (Fig. 3B and C).

Comparative structure–function studies on *Cf*CBM4-1 and *Tm*CBM4-2, in complex with cellopentaose and laminarihexaose, respectively, have illuminated the molecular determinants of β-glucan selectivity in family 4 CBMs (Boraston et al., 2002c) (Fig. 3B and C). *Tm*CBM4-2

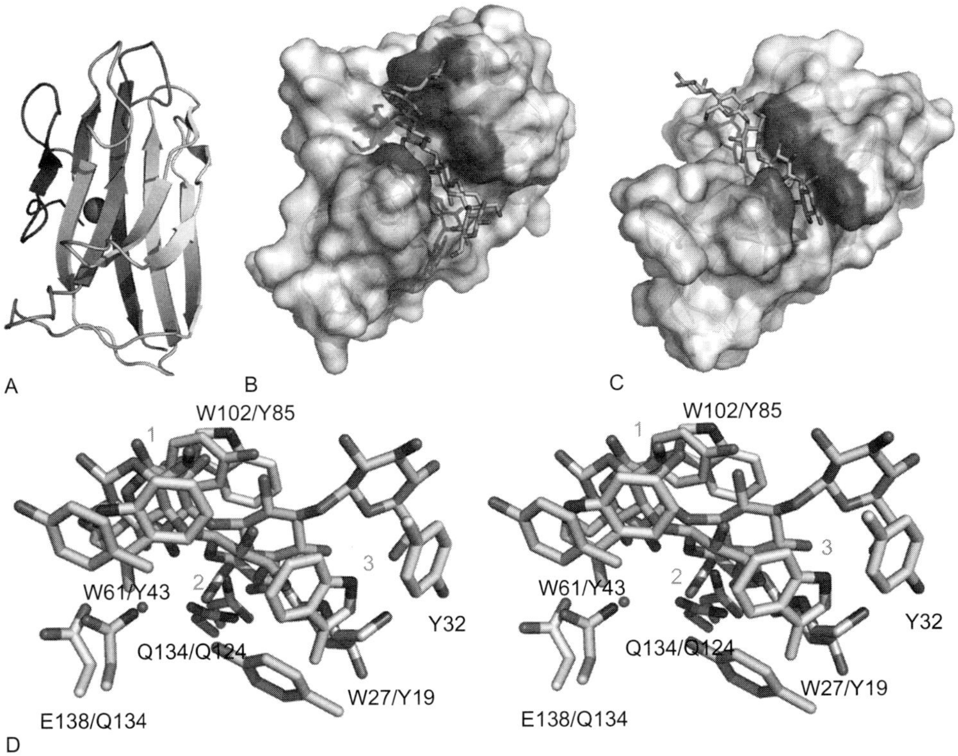


**Fig. 3: Differential β-glucan recognition by family 4 CBMs. (A) *Tm*CBM4-2 (PDB ID: 1GUI) displayed in cartoon format. The structural calcium is shown as a sphere. (B) Solvent-accessible surface representation of *Tm*CBM4-2, a (1,3)-β-glucan binding CBM in complex with laminarihexaose (PDB ID: 1GUI). (C) *Cf*CBM4-1, a (1,4)-β-glucan binding CBM in complex with cellopentaose (PDB ID: 1GU3). Aromatic residues involved in ligand binding in (B) and (C) are shown. (D) Wall-eyed stereo view of superimposed complexes from (B) and (C). Amino acids involved in ligand binding from *Tm*CBM4-2 (Trp27, Trp61, Trp102, Gln134 and Glu138) and *Cf* CBM4-1 are shown (Tyr19, Tyr43, Tyr85, Gln124 and Gln134). Ligands are trimmed down to the three residues that occupy subsites 1 to 3 for clarity. Laminarihexaose and cellopentaose is rendered as sticks with the subsites numbered. The colour specifications refer to colours in panels.**

is 'well-designed' to interact with the signature configuration of (1,3)-β-glucans. The binding cleft resembles a U-shaped depression with high-sided walls formed by connecting loops between β-strands. One end of the cleft is closed off by an extended loop comprised of residues 18 to 25. These two macromolecular formations restrict the binding of carbohydrate ligands with an extended linear conformation such as (1,4)-β-glucans.

Within the core of the binding cleft is an 'aromatic cradle' that consists of Trp27, Trp61 and Trp102, which interact with the central residues of laminarihexaose (Fig. 3D). Interestingly, this cluster of aromatic residues is conserved within the shallow binding cleft of *Cf*CBM4-1 (Tyr19, Tyr43, and Tyr85); notably however in *Tm*CBM4-2, Trp27 is rotated ~60° to accommodate the curvature of a (1,3)-β-glucan. In addition *Tm*CBM4-2 contains an extra aromatic residue, Tyr32, which is strategically disposed to provide a hydrophobic interaction with the ligand as it exits the binding cleft. Previously, site-directed mutagenesis and biochemical analysis demonstrated that each of these aromatic residues makes significant contributions to complex formation ($\sim$8–13 kJ $\text{moL}^{-1}$) (Boraston et al., 2001; Kormos et al., 2000). In the *Tm*CBM4-2 complex sugars in positions 3, 4 and 5 and bound in subsites 1, 2 and 3 (labelled from the reducing end) are in co-planar alignment, which produces a flat binding surface that stacks with the aromatic triad at the core of the binding cleft (Fig. 3B and D). This co-planarity indicates that (1,3)-β-oligoglucosides undergo a small conformational change upon binding.

Closer analysis of the binding site reveals that in addition to the aromatic cradle there are also two polar amino acids positioned at equivalent locations within the binding site that selectively interact with their target ligands (Fig. 3D). Gln124 and Gln134 form hydrogen bonds with the O6 of sugar 2 and the O6 of sugar 4 (subsite 2) in *Cf*CBM4-1 and *Tm*CBM4-2, respectively; whereas, Gln128 and Glu138 hydrogen bond to the O3 of the reducing-end sugar and make a water-mediated interaction with the O4 of sugar 3 (subsite 1) in *Cf*CBM4-1 and *Tm*CBM4-2, respectively. Several other polar amino acids exclusive to the function of *Tm*CBM4-2, including Asn64, Asn66, Lys94 and Asn104, also contribute hydrogen bonds to the protein–carbohydrate complex. These amino acids are selective for sugar subunits at positions 3 and 4 of the (1,3)-β-glucan within the core of the binding site.

### *I.D.b. CBM6*

Enzymes containing appended CBM6s have predicted activities against cellulose, β-glucans, xylan and agarose (Coutinho and Henrissat, 1999). Accordingly, there have been several studies reporting diverse ligand-binding profiles ranging from cello- and xylo-configured (1,4)-glycosyl linked carbohydrates (Czjzek et al., 2001; Henshaw et al., 2004; Pires et al., 2004) to (1,3;1,4)-β-glucans (Henshaw et al., 2004; Hong et al., 2002) and (1,3)-β-oligoglucosides (van Bueren et al., 2005). CBM6 is the most intensively studied β-glucan-binding CBM family. Currently there are over 20 crystal structures, including complexes with cello-, xylo-, laminari- and neoagaro-configured oligosaccharides, within the database (Coutinho and Henrissat, 1999).

In 2001, the first 3D structure of a CBM6 from a *Clostridium thermocellum* xylanase (*Ct*CBM6) was reported (Czjzek et al., 2001) (Fig. 4A). Significantly, this study documented the presence of multiple binding sites within the same CBM and established the presence of a CBM superfamily comprised of families 4, 6, and 22 – all of which have members involved in β-glucan recognition. This discovery was corroborated by a thorough two-part structure–function analysis of the CBM6-2 from a *Cellvibrio mixtus* endoglucanse 5 A (*Cm*CBM6-2) (Henshaw et al., 2004; Pires et al., 2004) that described the ligand specificity of each binding site designated 'Cleft A' and 'Cleft B' in this organism (Fig. 4B and C).

Cleft A is located within the loop regions of the module and contains key solvent-exposed aromatic residues, Tyr33 and Trp92 (Fig. 4B). This binding site was previously shown to interact preferentially with (1,4)-β-linked xylo-configured ligands in *Ct*CBM6 (Boraston et al., 2003b; Czjzek et al., 2001) and has demonstrated structural similarities with several lectins (Boraston et al., 2003b). Cleft B contains only one solvent-exposed aromatic, Trp39 (Fig. 4C). Site-directed mutagenesis and biochemical analysis demonstrated that this residue interacts with the (1,3;1,4)-glucans, lichenin and barley β-glucan, but is not essential for binding laminarin (Henshaw et al., 2004). Similar results were observed with mutations to Glu73, a key polar residue for ligand binding in Cleft B, suggesting that this binding site is specific for the (1,3;1,4)-β-glucans but not (1,3)-β-glucans. Analysis of several oligosaccharide–*Cm*CBM6-2 complexes reveals subtle ligand specificity within this binding site (Fig. 4D) (Pires et al., 2004). Cleft B contains four subsites, with subsite 2 and 3 providing significant binding energy. Discrimination between xylo- and gluco-configured ligands occurs at subsite 2, as Glu73 and the backbone of Glu74 form hydrogen bonds with the O6 of glucose – both interactions that would be absent in xylo-oligosaccharides. Discrimination between (1,3)- and (1,4)-β-glucans occurs at subsite 4, which is in a conformation that accommodates a (1,3)-β-glucosyl moiety.

The first structure of CBM6 in complex with a (1,3)-β-oligoglucoside was reported in 2003 (Boraston et al., 2003b). The CBM was the third CBM6 from a *Clostridium stercorarium* xylanase (*Cs*CBM6-3). *Cs*CBM6-3 displays a promiscuous ligand binding profile, interacting with cellulose, (1,4)-β-xylan and barley (1,3;1,4)-β-glucan (Boraston et al., 2002b; Boraston et al., 2003b). Structural superimposition of *Cs*CBM6-3 in complex with cellobiose (PDB ID: 1O8S) and laminaribiose (PDB ID: 1OD3) revealed that both sugars were orientated in the same direction with the reducing-end monosaccharide in subsite 1 slightly disordered (Fig. 4E). Within subsite 2 the residue was sandwiched by two aromatic residues (Tyr56 and Phe112) and O3 was hydrogen bonded to Asn140. The occupation of O3 precludes the extension of another (1,3)-β-glucose beyond subsite 2 (into subsite 3). When compared to the cellobiose complex this subsite could potentially accommodate another (1,4)-linked β-glucosyl

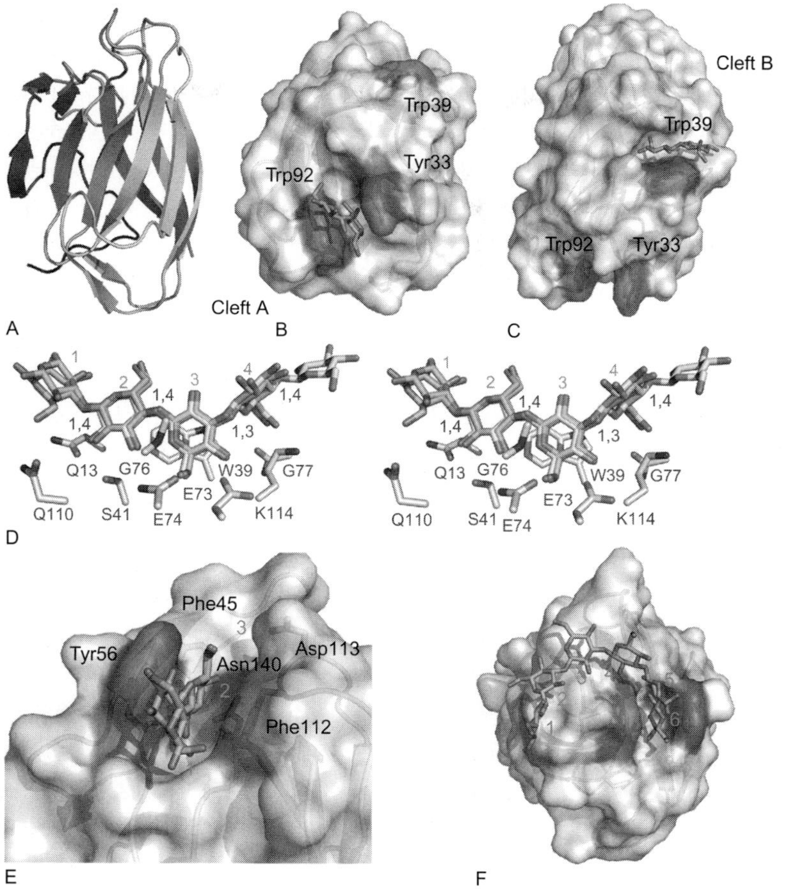


Fig. 4: (A) Secondary structure representation of unliganded *Cm*CBM6-2 in cartoon format (PDB ID: 1UXZ). (B) Cleft 'A' in *Cm*CBM6-2 in complex with cellobiose from chain B (PDB ID: 1UYX). The three surface aromatic residues Tyr33, Trp39 and Trp92 are shown. (C) Cleft 'B' is displayed in complex with cellobiose from chain A (PDB ID: 1UYX). (D) Superimposition of ligands bound in *Cm*CBM6-2: cellotriose (PDB ID: 1UYY), G3G4G3 (PDB ID: 1UY0), and G4G3G4 (PDB ID: 1UZ0). Only the amino acids from the cellotriose complex that define the subsite architecture (numbered 1 to 4) are shown. The linkages are labelled. (E) Solvent surface representation of CsCBM6-3 of superimposed complexes with cellobiose and laminaribiose. The aromatic residues Tyr56 and Phe112 that define subsite 1 are shown. The oxygen and nitrogen of the Asn140 functional group which interacts with O3 and O4 of both ligands are displayed. Phe45 and Asp113, which form subsite 3 in the xylo-oligosaccharide complexes, are indicated. (F) Solvent-accessible surface model of BHCBM6 in complex with laminarihexaose. The colour specifications refer to colours in panels.

residue, as the O4 non-reducing sugar is accessible, an observation which is in agreement with the ability of the CBM to bind cellulose (Boraston et al., 2002b; Boraston et al., 2003b). At the other end of the disaccharide both ligands are correctly poised to accommodate further polymerizations. These binding patterns suggest that although xylan is its preferential ligand, *Cs*CBM6-3 can bind to the non-reducing end of (1,3)-β-glucans or the internal (1,4)-β-linked glucosyl residues in (1,3;1,4)-β-glucans.

A recent study has put a new twist on our understanding of β-glucan recognition by family 6 CBMs (van Bueren et al., 2005). *Bh*CBM6 is a C-terminal CBM found within the family GH 81 (1,3)-β-glucanase from *Bacillus halodurans*. The affinity of this CBM for (1,3)-β-oligoglucosides was determined to be directly related to the degree of ligand polymerization (determined using ligands ranging in size from laminaribiose to laminarihexaose), while binding to the laminarin polymer (DP ~25) in a ratio of approximately ~0.8:1 (CBM: polysaccharide chain). This suggested that *Bh*CBM6 may bind to the end of the polysaccharide, a result that was supported by direct structural evidence. Co-crystallization of the CBM in complex with laminarihexaose revealed the presence of six subsites that are oriented along the surface of the CBM in a pattern consistent with the natural curvature of the (1,3)-β-glucan chain (Fig. 4F). This topology would restrict the binding of linear and extended ligands such as (1,4)-β-glucans and (1,3;1,4)-β-glucans, respectively. The non-reducing end of the oligosaccharide is accommodated in subsite 6 (the subsites are numbered reducing → non-reducing) and is sandwiched by the two surface aromatics (Trp42 and Trp99) found within Cleft A. The cleft is closed off at one end by Asn132 which makes two hydrogen bond contacts with the O3 and O4 of the non-reducing sugar, supporting the observation that *Bh*CBM6 binds to the end of laminarin chains. The subsites curve outward across the protein surface and are exclusively composed of polar amino acids that make direct and water-mediated contacts with the ligand. This study reveals that, in addition to the two previously described binding clefts in CBM6s, there is also evidence for the utilization of novel surface features of the protein evolved to interact specifically with the helical conformation of the target ligand.

### *I.D.c CBM11*

CBM11 is a small family with only eight members reported – all of which are modules in endo-1,4-β-glucanases (Coutinho and Henrissat, 1999). There is currently one structure available for a CBM11 from the *C. thermocellum* Lic26A-CelE (*Ct*CBM11, PDB ID: 1V0A) (Malburg et al., 1997). This enzyme contains both a family 5 and 26 catalytic domain that has differential activities on (1,3)-β-glucan and (1,3;1,4)-β-glucans, respectively. Whether the solitary CBM in this context displays preferential binding for either one of these predicted substrates or interacts promiscuously with both is an important biological question for our understanding of CBM phylogeny and β-glucan recognition.

To investigate the binding profile of *Ct*CBM11 the module was tested against a library of different complex polysaccharides (Carvalho et al., 2004). Using affinity gel electrophoresis, *Ct*CBM11 was shown to bind the (1,3;1,4)-β-glucans, oat β-glucan and lichenin with the highest affinity; demonstrated weaker binding to hydroxylethyl cellulose, glucomannan and oat spelt xylan; and did not interact with pure (1,3)-β-glucans. This observation was supported using isothermal titration calorimetry, which determined that (1,3;1,4)-β-glucans, including lichenin, oat β-glucan and (1,3;1,4)-β-glucotetrasaccharides, bound with ~5-fold greater affinity than (1,4)-β-oligoglucosides (Carvalho et al., 2004). Preferential interactions with (1,3;1,4)-β-glucans and (1,4)-β-glucans parallel the activities of the appended family GH 5 and 26 catalytic modules and indicate that *Ct*CBM11 is tailored to recruit both glycosidases with differing specificities to amenable target substrates. Whether downstream enzymatic activities are concomitant (GH5A and GH26 active at the same time) or successive (one enzyme lags behind the other dependent upon substrate concentration) remains to be established.

The structure of *Ct*CBM11 is a β-sandwich with jelly-roll topology containing two structural calcium atoms (Fig. 5A) (Malburg et al., 1997). The first calcium lies within the convex β-sheet and the second is coordinated within the loop regions near the N- and C-termini, which provides stability to this region that lacks a significant secondary structure. The binding site was mapped to the concave β-sheet by site-directed mutagenesis, which demonstrated the importance of three aromatic residues in complex formation: Y22, Y53 and Y129. Subsite dissection and elucidation of the molecular determinants of β-glucan recognition await the report of protein complexes containing (1,3;1,4)- and (1,4)-β-oligoglucosides.

### *I.D.d CBM17 and CBM22*

Family 17 CBMs are found exclusively in endo-(1,4)-ß-glucanases from *Bacillus* sp. and *Clostridium* sp. (Coutinho and Henrissat, 1999), and have been shown to bind different forms of (1,4)-β-gluco-configured carbohydrates, including (1,4)-β-oligoglucosides, amorphous cellulose, microcrystalline cellulose, and sections of plant cell walls (Blake et al., 2006; Boraston et al., 2000; Notenboom et al., 2001). The ability of a CBM17 from a *Clostridium cellulovorans* cellulase (*Cc*CBM17) to bind oat β-glucan was demonstrated by qualitative and competitive affinity gel electrophoresis, ultraviolet difference and fluorescence spectroscopy

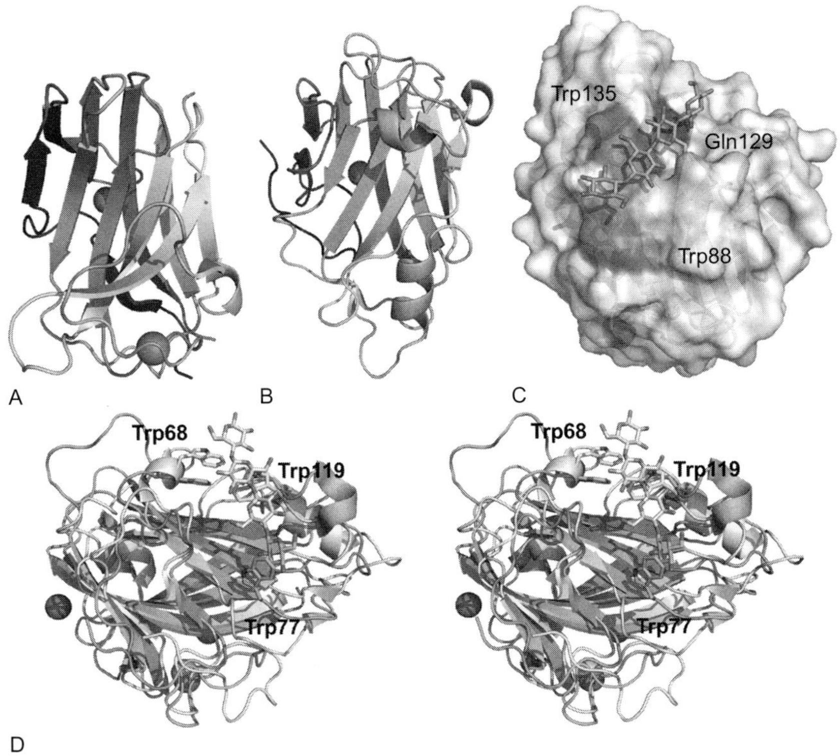


**Fig. 5: (A) Cartoon representation of *Ct*CBM11 in complex with two calcium atoms shown as spheres (PDB ID: 1V0A). (B) Cartoon representation of *Cc*CBM17 in complex with a calcium atom shown as a sphere (PDB ID: 1J83). (C) Solvent accessible surface model of *Cc*CBM17 in complex with cellotetraose (PDB ID: 1J84). The shallow binding cleft is lined with two surface aromatics residues, Trp88 and Trp135, and Gln129. (D) Wall-eyed stereo view of *Bsp* CBM28 (PDB ID: 1UWW) overlaid with *Ct* CBM17 (yellow, PDB ID: 1V0A). The aromatic residues implicated in binding for *Bsp* CBM28 (Trp68, Trp77, and Trp119) are displayed. The cellotetraose ligand from *Ct*CBM17 is shown. The colour specifications refer to colours in panels.**

(Boraston et al., 2000). In these cases the affinity of *Cc*CBM17 for glucans is in the range of both (1,4)-β-oligoglucosides and (1,4)-β-linked polysaccharides.

The structure of *Cc*CBM17 in complex with cellotetraose was determined in 2001 (Notenboom et al., 2001). The CBM adopts the conventional β-sandwich fold with two antiparallel sheets comprised of four and five strands each (Fig. 5B). There is one structural calcium near the N-terminus that is coordinated within the convex β-sheet. In addition, there is a short α-helix connecting β-strands 7 and 8 that encroaches upon the binding cleft within the concave β-sheet. In contrast to CBMs described above, the binding site of *Cc*CBM17 has been described as a 'wide shallow groove'. The solvent accessible basin contains two aromatic residues (Trp88 and Trp135) that stack with the two glucose moieties at the reducing end (subsites 1 and 2), causing the ligand to sit flat on the surface of the protein. In total, there were four identifiable subsites that were comprised mainly of polar amino acids that made stereo-specific hydrogen bonds with the gluco-configured ligand. Significantly, Gln129 forms a direct interaction with the glycosidic oxygen between the non-reducing sugar (subsite 4) and its neighbour (subsite 3), which precludes the binding of a (1,3)-β-glucosyl residue at this position (Fig. 5C).

Currently, there is no structural information available to illuminate the mechanism behind *Cc*CBM17 binding of (1,3;1,4)-β-glucans. However, based upon the cellotetraose complex one can provide a hypothesis. The binding cleft of *Cc*CBM17 is relatively unobstructed at either end, which suggests that there may be some allowable conformational freedom in the ligand residues that exit in either. Thus, the (1,3)-β-linkages flanking a cellotetraosyl subunit in (1,3;1,4)-β-glucans may be accommodated by extending into solvent. Co-crystallization of *Cc*CBM17 in complex with (1,3;1,4)-β-oligoglucosides and (1,3)-β-oligoglucosides would be helpful to elucidate this possibility.

CBM28s are almost always found in tandem with CBM17s (Boraston et al., 2002a). Although these two CBM families have similar primary structures and are in fact distantly related, there is noticeable disparity in the amino acids involved in binding (Boraston et al., 2003a). Using localization experiments, the binding profiles of CBM17 (*Bsp*CBM17) and CBM28 (*Bsp*CBM28) from a *Bacillus* sp. 1139 cellulase were determined to interact with differential cellulose substructure and have a cooperative effect in amorphous cellulose recognition (Boraston et al., 2003a). CBM28 was also shown to bind barley β-glucans with an affinity similar to that of cello-oligosaccharides and in addition there was a synergism observed for the *Bsp*CBM17/*Bsp*CBM28 dimodule interactions with barley β-glucan (Boraston et al., 2002a).

*Bsp*CBM28 is similar in three-dimensional structure to *Cc*CBM17 (Jamal et al., 2004). A structural superimposition with the *Cc*CBM17 complex (overlapping with an r.m.s. of 2.70 $Å^2$ for 179 aligned $C^a$) identifies three key solvent-accessible tryptophan residues (Trp67, Trp77 and Trp119) (Fig. 5D). The surface topography of this putative binding site resembles the shallow groove observed in *Cc*BM17. The molecular determinants of recognition for (1,3;1,4)-β-glucans and differential substructures within non-crystalline cellulose await further investigation.

### I.D.e CBM22

CBM22s are found exclusively in xylanases (Coutinho and Henrissat, 1999), which is consistent with the ability of CBM22 to bind xylo-oligosaccharides with high affinity (Charnock et al., 2000; Xie et al., 2001). (1,3;1,4)-β-Glucan binding has been demonstrated for *Ct*CBM22-2 (Dias et al., 2004). Using affinity gel electrophoresis, the module demonstrated optimal binding to oat spelt, wheat and rye arabinoxylan, had high affinity for oat β-glucan, and weak interactions were observed for hydroxyethyl cellulose. There was no observable binding with various galacto- and gluco-configured heterogeneous polysaccharides (Dias et al., 2004). Contradictory results were obtained using calorimetry; however, as the affinity of *Ct*CBM22-2 for barley β-glucan was determined to be very low and in the range of cellohexaose (Dias et al., 2004). Further experiments are required to establish if *Ct*CBM22-2 is a true β-glucan binding module.

The structure of *Ct*CBM22-2 was determined in 2000 (Dias et al., 2004) and used to map the amino acids involved in xylo-oligosaccharide recognition (Xie et al., 2001). The overall fold is a β-sandwich jelly roll formed by two opposing sheets of four antiparallel β-strands each (Fig. 6A). There is a structural calcium present near the N- and C- termini on the convex face of the sandwich. The binding site is positioned with a shallow groove formed by three aromatic residues: Trp53, Tyr103 and Tyr134 (Dias et al., 2004) (Fig. 6B). Two key polar amino acids, Arg25 and Glu138, are also important for ligand recognition as determined by mutagenesis and xylo-oligosaccharide and xylan binding assays (Xie et al., 2001). A co-crystal structure would be helpful to understand the properties of *Ct*CBM22-2 with regard to xylo-configured and (1,3;1,4)-β-glucan-configured ligand binding.

### I.D.f. CBM43

Family 43 CBMs are found at the C-terminus of GH17 and GH72 glycoside hydrolases or independent of a catalytic module. These enzyme families are responsible for the hydrolysis and transglycosylation of (1,3)-β-glucans. CBM43s are restricted to certain species of fungi and plants, and are quite commonly present in many paralogous copies. As an extreme example, *Arabidopsis thaliana* contains 58 different genes containing a CBM43 (Coutinho and Henrissat, 1999).

Our understanding of the structure–function relationship of this CBM family has centred on two *Olea europaea* (olive) pollen allergens: Ole e 9 (*Oe*CBM43A) and Ole e 10 (*Oe*CBM43B) (Barral et al., 2005; Huecas et al., 2001; Trevino et al., 2007). *Oe*CBM43A

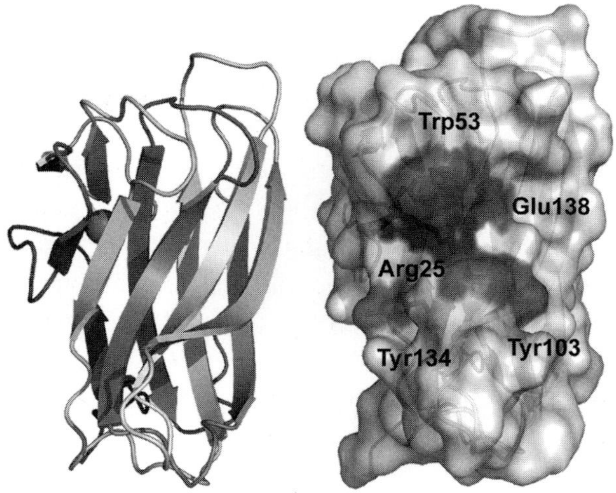


**Fig. 6: (A) Cartoon representation of *Ct*CBM22-2 (PDB ID: 1H6Y – the mutated Glu138 was reintroduced for reference). Solvent-accessible surface model of *Cc*CBM22-2 in complex with cellotetraose (PDB ID: 1J84). (B) The residues implicated in ligand binding are shown.**

is a C-terminal module within a protein that was recently shown to be a functional GH17 (1,3)-β-endo-glucanase active on laminarin (Huecas et al., 2001). Comparison of the two CBM43 primary structures reveals that there is a high level of amino acid identity (53%) between the C-terminus of *Oe*CBM43A and the full-length *Oe*CBM43B (Barral et al., 2004) (Fig. 7A). Using affinity gel electrophoresis, analytical ultracentrifugation and UV fluorescence, *Oe*CBM43B was shown to interact specifically with laminarin (Barral et al., 2005). Binding induced a conformational change and increased the thermal stability of the protein.

Recently, the solution structure of *Oe*CBM43A was determined (PDB ID: 2JON, Trevino et al., 2007) (Fig. 7B). The fold consists of two parallel α-helices (residues: 22–32 and 56–69) juxtaposed at an angle of 55° and a small two-stranded antiparallel β-sheet. Interestingly, epitope mapping of the allergen revealed that most of the IgG and IgE immunogenic structures are encoded within the loop regions. The surface of the CBM contains two hydrophobic patches containing aromatic residues: (1) Phe49, Tyr60, Try91 and Phe96; (2) Tyr31, Try65 and Phe78 (Fig. 7C and D). All of these amino acids are functionally conserved in *Oe*CBM43A except for Tyr91, which is replaced with an asparagine. A role of these residues in ligand recognition has not been determined.

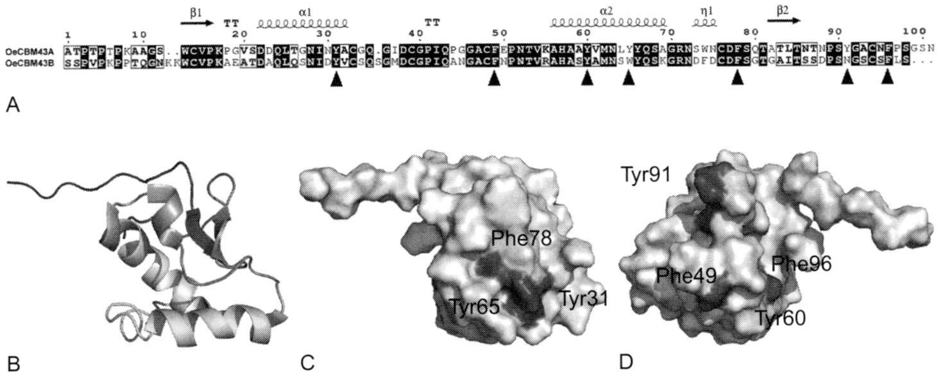


**Fig. 7: (A) Structural alignment of *Oe*CBM43A and *Oe*CBM43B using the secondary structures from *Oe*CBM43A (PDB ID: 1OJN). The sequences include residues 360–461 from *Oe*CBM43A and 23–123 from *Oe*CBM43B. The aromatic residues suggested to be involved in binding are indicated with black triangles. (B) Cartoon representation of *Oe*CBM43A. (C–D) Solvent-accessible surface models of OeCBM43A showing the two exposed hydrophobic patches described within the text. Aromatic residues are shaded.**

## I.E. Biochemical Analysis of Other β-Glucan Binding Proteins

### *I.E.a CBM13*

Family 13 CBMs are most commonly associated with plant lectins such as ricin chain B and agglutinin; however, in nature they actually represent a functionally diverse family found in many different enzyme classes, including xylanases, arabinofuranosidases, β-agarases, pectate lyases and glycosyltransferases (Coutinho and Henrissat, 1999). In this regard, the potential of CBM13s to interact with many unique carbohydrate ligands is vast; however, there are only a handful of characterized enzymes containing CBM13s with predicted or demonstrated activity on (1,3)- or (1,3;1,4)-β-glucans. These include the secreted βgl II and βgl proteins from *Cellulosimicrobium cellulans* (Ferrer, 2006) and BglM from *Bacillus circulans* (Asano et al., 2002), enzymes that are all involved in the induction of fungal cell lysis.

The modular architectures of βgl II, βgl, and BglM are quite similar. Each enzyme contains an N-terminal catalytic domain and a C-terminal CBM13 with three internal repeats. These repeats create the three-fold symmetry observed in other CBM13s, such as ricin chain B (Rutenber and Robertus, 1991). The CBM13 from BglM (*Bc*CBM13) was shown to bind pachyman, lichenin and *Aspergillus oryzae* cell walls (Asano et al., 2002), consistent with its role in fungal cell wall lysis. Indeed, the lytic activity of a truncated from of βgl II (GH16) was compromised and displayed a higher $K_m$ for insoluble yeast glucan in the absence of its CBM13 (Ferrer, 2006). A detailed thermodynamic analysis of β-glucan-binding CBM13 and structural investigation into the architecture of its binding site(s) is required to further our understanding of the molecular mechanisms in these processes.

### I.E.b. Immunostimulatory Proteins: CBM39, GH16s and Dectin-1

As mentioned above, (1,3)-β-glucans comprise a significant portion of fungal cell walls. These carbohydrates have turned out to be an immunostimulatory biomarker to the innate immune system in invertebrates and vertebrates (Cerenius and Soderhall, 2004; Chen and Seviour, 2007; Descroix et al., 2006; and discussed in detail in Chapters 4.5.1 and 4.5.2). Such microbial structures that are essential for the survival of the microorganism, and thus are immutable, are referred to as Pathogen Associated Molecular Patterns (PAMPs). Recognition of PAMPs by pattern-recognition receptors (PRRs) triggers an immune response, culminating in microbial cell phagocytosis and destruction (Brown, 2006; Willment and Brown, 2008). PRRs and their implications for biomedical application has become an intensively studied area of health and pharmacological research (Chen and Seviour, 2007; Descroix et al., 2006). Here, we will limit our discussion to family 39 CBMs, non-catalytic GH16s and the vertebrate β-glucan binding protein, dectin-1.

Although β-glucan recognition proteins (BGRPs) have been identified in several different species, the CAZy classification of the family 39 (CBM39) and family 16 (GH16) BGRPs is not currently observed within the literature. For the purposes of consistency within this review we would like to introduce this nomenclature. CBM39s and GH16s are PRRs that have an integrated relationship in nature as they are commonly found within the same polypeptide (Fabrick et al., 2004; Lee et al., 2000; Ma and Kanost, 2000; Ochiai and Ashida, 2000; Zhang et al., 2003). There are some CBM39s, however, that are independent CBMs and lack a catalytic module entirely (Coutinho and Henrissat, 1999). BGRPs are components of the innate immune system in invertebrates that bind to (1,3)-β-glucans to activate the prophenoloxidase pathway (Fabrick et al., 2004; Jiang et al., 2004; Ma and Kanost, 2000; Ochiai and Ashida, 1988; Ochiai and Ashida, 2000; Soderhall et al., 1994; Wang et al., 2006; Zhang et al., 2003) (see Chapter 4.5.1). Comparative analysis of both GH16 and CBM39 from the Indian meal moth, *Plodia interpunctella*, demonstrated that the N-terminal CBM39 (*Pi*CBM39) binds the (1,3)-β-glucan (curdlan) and lipoteichoic acid, and activates the prophenoloxidase pathway, whereas the C-terminal GH16 domain demonstrated specificity for laminarin (Fabrick et al., 2004).

Structural determination of the overall fold and binding site of CBM39s and a non-catalytic GH16 is currently lacking; although there are several models available for catalytic GH16s, which adopt a β-jelly roll fold (Coutinho and Henrissat, 1999). These examples include enzymes active on (1,3;1,4)-β-glucans (Tsai et al., 2005) and (1,3)-β-glucans (Fibriansah et al., 2007), and xyloglucan (Brown et al., 2007). Interestingly, circular dichroism suggested that the *Pi*CBM39 domain is rich in α-helical character and contains a minor β-structure (Fabrick et al., 2004). These preliminary findings suggest that the protein fold found in this family substantially deviates from the common β-sandwich scaffold.

The PRR protein dectin-1 is expressed by many vertebrate immune cell types, including dendritic cells, macrophages, monocytes, neutrophils and some T cells (Ariizumi et al., 2000; Taylor et al., 2002). Consistent with its role in immune surveillance, it is highly produced on the surfaces of cells at sites of entry of microbial pathogens, such as the mucosal layers in the lungs and intestinal tract (Reid et al., 2004; Taylor et al., 2002). Dectin-1 binds soluble (1,3)-β-glucans, (1,6)-β-glucans, and (1,3;1,6)-β-glucans. This makes it the first protein known to bind mixed (1,3;1,6)-β-glucans and the first member of the NK-cell-receptor like C-type lectin (CTL) family shown to specifically bind a carbohydrate despite the complete lack of any amino acid conservation known to be involved in carbohydrate binding by other lectins (Weis et al., 1998).

Recently, the structure of dectin-1 was determined (Brown et al., 2007). The protein fold is similar to other CTL domains (Zelensky and Gready, 2005), containing two antiparallel β-sheets and two α-helices (Fig. 8A). Previously, mutagenesis experiments had demonstrated that Trp221 and His223 were important for dectin-1 binding to various forms of

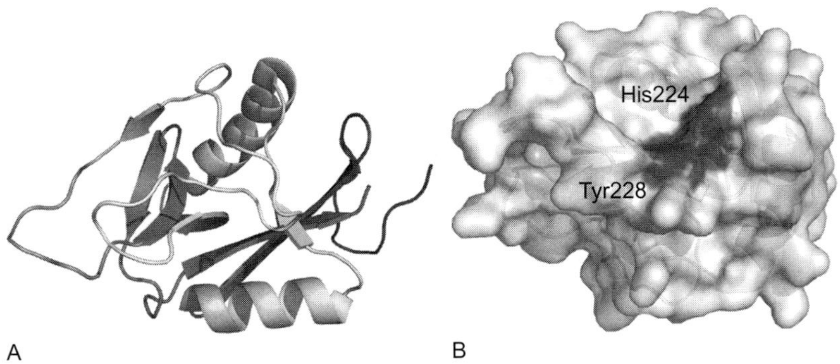


**Fig. 8: (A) Cartoon representation of dectin-1 (PDB ID: 2BPD). (B) Solvent-accessible surface model of dectin-1 displaying the putative β-glucan binding site.**

(1,3)-β-glucans (Adachi et al., 2004) and are conserved in all homologs that have been sequenced. These amino acids map to the surface of dectin-1 and are connected through a shallow apolar groove (Brown et al., 2007) (Fig. 8B). Although a protein–carbohydrate complex is currently lacking to help explain the molecular determinants of ligand recognition, the process may involve oligomerization of dectin-1 (Brown et al., 2007).

### I.E.c. Structural Comparison of Sequence-Unrelated CBM β-Glucan Binding Sites

Although there is still a paucity of (1,3)-β-oligoglucoside–CBM complexes available within the database, we have performed a comparative analysis on the binding site architectures of known structures to try and uncover any potential conserved or convergent features. For CBMs there are only three known co-structures that contain a (1,3)-β-glucan: *Tm*CBM4-2, *Bh*CBM6 and *Cs*CBM6-3. *Tm*CBM4-2 and *Bh*CBM6 were solved in complex with laminarihexaose (these structures superimposed with an r.m.s. of 3.05 $Å^2$ for 109 aligned $C^a$), whereas *Cs*CBM6-3 is bound to laminaribiose (which aligned with an r.m.s. of 1.05 $Å^2$ for 125 aligned $C^a$ with *Bh*CBM6 and 3.08 $Å^2$ for 106 aligned $C^a$ with *Tm*CBM4-2).

Direct comparison of the architecture within the CBM4 and CBM6 binding sites is complicated by the fact that they are located at distinctly different positions on the β-sandwich scaffold, and furthermore there are translational and rotational differences observed between the two CBM6 ligands even though they bind at similar locations. *Bh*CBM6 interacts with the non-reducing end of the (1,3)-β-glucan through a series of six subsites that initiate within Cleft A and extend outward along the surface of the CBM (Fig. 4F). *Cs*CBM6-3 also binds the reducing end of laminaribiose in Cleft A in two subsites (1 and 2); a third subsite that accommodates xylo- and cello-configured carbohydrates remains unoccupied. In contrast to the family 6 CBMs, *Tm*CBM4-2 binds the internal residues of laminarihexaose at a position that is equivalent to Cleft B in family 6 CBMs (Fig. 3A). Despite these significant differences, however, structural alignments can be performed using the ligand as a reference. This analysis revealed that there is some structural conservation and an argument for functional convergence in two key aromatic residues that sandwich a glucose residue (Fig. 9). These aromatics constitute subsite 1 in *Tm*CBM4-2, subsite 6 in *Bh*CBM6, and subsite 2 in *Cs*CBM6-3. In this orientation the Asn132 from *Bh*CBM6 that seals the end of binding site overlaps with Asn140 from CsCBM6-3. Both residues are functionally conserved to form hydrogen bonds with the O3 and O4 of glucose (Fig. 9). In particular the O3 interaction is critical for binding the non-reducing ends of (1,3)-β-gluco-configured carbohydrates. The absence of a structurally conserved asparagine in *Tm*CBM4-2 enables it to interact with internal residues of the (1,3)-β-glucan.

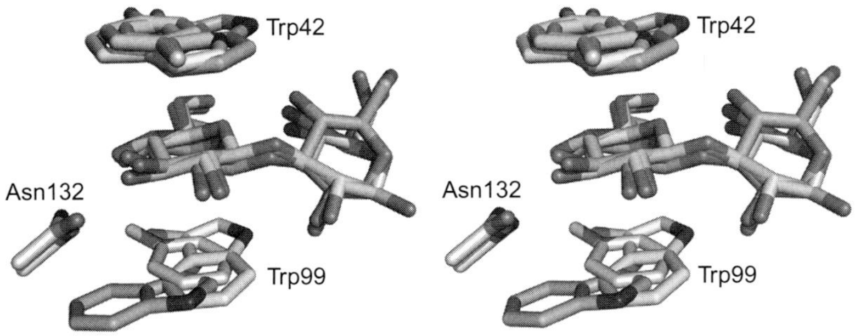


**Fig. 9: Wall-eyed stereo view of structural alignment of the binding sites from (1,3)-β-glucanbinding CBMs. The ligands from *Bh*CBM6 (PDB ID: 1W9W), *Cs*CBM6-3 (PDB ID: 1OD3) and *Tm*CBM4-2 (PDB ID: 1GUI) have been superimposed and displayed along with key amino acids within the binding sites. Only the residues from *Bh*CBM6 are labelled. The two asparagines from the family 6 CBMs, which make critical contacts with the non-reducing O3 and O4, are shown. The colour specifications refer to colours in panels.**

Although these observations provide some preliminary insight into the molecular determinants of (1,3)-β-glucan substructure recognition, further co-crystal structures will be required to establish if these examples reflect a general mechanism.

## I.F. Overview

The recognition of (1,3)-β-glucans and related polysaccharides by CBMs and other proteins is emerging as an important factor in biological processes such as nutrient recycling and immune surveillance. To date, the structural evidence has painted a somewhat cloudy picture about the mechanisms behind the molecular recognition of this unique class of polysaccharides by proteins; however, some conclusions can be drawn.

(1) Most CBMs that bind β-glucans appear to be selective for (1,3;1,4)-β-glucan ligands. Dedicated (1,3)-β-glucan binding families such as CBM39 seem to be the exception.

(2) Within the protein-(1,3)-β-glucan complexes that have been determined (*Tm*CBM4-2, *Bh*CBM6, and *Cs*CBM6-3) there are noticeable mechanistic distinctions in ligand recognition. For example, when the two CBM families are compared, their binding sites are located at different locations within the β-sandwich, there is diversity in subsite composition and trajectory, and they interact with different substructures in the laminarin ligand. Despite these differences, however, there does seem to some convergence in binding through an aromatic sandwich.

(3) There are several different protein folds that have been identified in (1,3)-β-glucan-binding proteins. Although the vast majority of these are the β-sandwich, consistent with CBMs in general, the β-trefoil from CBM13s, an α/β fold from CBM43s, and the C-type lectin-like domain of dectin-1 have also been described. It is quite possible that in the future even more fold diversity will emerge for this functionally related group of proteins.

## References

Abbott, D. W., Eirin-Lopez, J. M., & Boraston, A. B. (2008). Insight into ligand diversity and novel biological roles for family 32 carbohydrate-binding modules. *Molecular Biology and Evolution*, *25*, 155–167.

Abbott, D. W., Hrynuik, S., & Boraston, A. B. (2007). Identification and characterization of a novel periplasmic polygalacturonic acid binding protein from *Yersinia enterolitica*. *Journal of Molecular Biology*, *367*, 1023–1033.

Abou Hachem, M., Nordberg Karlsson, E., Bartonek-Roxa, E., Raghothama, S., Simpson, P. J., Gilbert, H. J., Williamson, M. P., & Holst, O. (2000). Carbohydrate-binding modules from a thermostable *Rhodothermus marinus* xylanase: cloning, expression and binding studies. *Biochemical Journal*, *345*, 53–60.

Adachi, Y., Ishii, T., Ikeda, Y., Hoshino, A., Tamura, H., Aketagawa, J., Tanaka, S., & Ohno, N. (2004). Characterization of beta-glucan recognition site on C-type lectin, dectin 1. *Infection and Immunity*, *72*, 4159–4171.

Ariizumi, K., Shen, G. L., Shikano, S., Xu, S., Ritter, R., 3rd, Kumamoto, T., Edelbaum, D., Morita, A., Bergstresser, P. R., & Takashima, A. (2000). Identification of a novel, dendritic cell-associated molecule, dectin-1, by subtractive cDNA cloning. *Journal of Biological Chemistry*, *275*, 20157–20167.

Asano, T., Taki, J., Yamamoto, M., & Aono, R. (2002). Cloning and structural analysis of bglM gene coding for the fungal cell wall-lytic beta-1,3-glucan-hydrolase BglM of Bacillus circulans IAM1165. *Bioscience Biotechnology and Biochemistry*, *66*, 1246–1255.

Barral, P., Batanero, E., Palomares, O., Quiralte, J., Villalba, M., & Rodriguez, R. (2004). A major allergen from pollen defines a novel family of plant proteins and shows intra- and interspecies [correction of interspecie] cross-reactivity. *Journal of Immunology*, *172*, 3644–3651.

Barral, P., Suarez, C., Batanero, E., Alfonso, C., Alche Jde, D., Rodriguez-Garcia, M. I., Villalba, M., Rivas, G., & Rodriguez, R. (2005). An olive pollen protein with allergenic activity, Ole e 10, defines a novel family of carbohydrate-binding modules and is potentially implicated in pollen germination. *Biochemical Journal*, *390*, 77–84.

Berdal, M., Appelbom, H. I., Eikrem, J. H., Lund, A., Zykova, S., Busund, L. T., Seljelid, R., & Jenssen, T. (2007). Aminated beta-1,3-D-glucan improves wound healing in diabetic db/db mice. *Wound Repair and Regeneration*, *15*, 825–832.

Blake, A. W., McCartney, L., Flint, J. E., Bolam, D. N., Boraston, A. B., Gilbert, H. J., & Knox, J. P. (2006). Understanding the biological rationale for the diversity of cellulose-directed carbohydrate-binding modules in prokaryotic enzymes. *Journal of Biological Chemistry*, *281*, 29321–29329.

Boraston, A., Lammerts van Bueren, A., Ficko-Blean, E., & Abbott, D. W. (2007). Carbohydrate-protein interactions: Carbohydrate-binding modules *Carbohydrate Glycoscience* (In press).

Boraston, A. B., Bolam, D. N., Gilbert, H. J., & Davies, G. J. (2004). Carbohydrate-binding modules: fine-tuning polysaccharide recognition. *Biochemical Journal*, *382*, 769–781.

Boraston, A. B., Chiu, P., Warren, R. A., & Kilburn, D. G. (2000). Specificity and affinity of substrate binding by a family 17 carbohydrate-binding module from Clostridium cellulovorans cellulase 5A. *Biochemistry*, *39*, 11129–11136.

Boraston, A. B., Ghaffari, M., Warren, R. A., & Kilburn, D. G. (2002a). Identification and glucan-binding properties of a new carbohydrate-binding module family. *Biochemical Journal*, *361*, 35–40.

Boraston, A. B., Kwan, E., Chiu, P., Warren, R. A., & Kilburn, D. G. (2003a). Recognition and hydrolysis of noncrystalline cellulose. *Journal of Biological Chemistry*, *278*, 6120–6127.

Boraston, A. B., McLean, B. W., Chen, G., Li, A., Warren, R. A., & Kilburn, D. G. (2002b). Co-operative binding of triplicate carbohydrate-binding modules from a thermophilic xylanase. *Molecular Microbiology*, *43*, 187–194.

Boraston, A. B., Notenboom, V., Warren, R. A., Kilburn, D. G., Rose, D. R., & Davies, G. (2003b). Structure and ligand binding of carbohydrate-binding module CsCBM6-3 reveals similarities with fucose-specific lectins and "galactose-binding" domains. *Journal of Molecular Biology*, *327*, 659–669.

Boraston, A. B., Nurizzo, D., Notenboom, V., Ducros, V., Rose, D. R., Kilburn, D. G., & Davies, G. J. (2002c). Differential oligosaccharide recognition by evolutionarily-related beta-1,4 and beta-1,3 glucan-binding modules. *Journal of Molecular Biology*, *319*, 1143–1156.

Boraston, A. B., Warren, R. A., & Kilburn, D. G. (2001). Beta-1,3-Glucan binding by a thermostable carbohydrate-binding module from *Thermotoga maritima*. *Biochemistry*, *40*, 14679–14685.

Brown, G. D. (2006). Dectin-1: a signalling non-TLR pattern-recognition receptor. *Nature Reviews in Immunology*, *6*, 33–43.

Brown, J., O'Callaghan, C. A., Marshall, A. S., Gilbert, R. J., Siebold, C., Gordon, S., Brown, G. D., & Jones, E. Y. (2007). Structure of the fungal beta-glucan-binding immune receptor dectin-1: implications for function. *Protein Science*, *16*, 1042–1052.

Brun, E., Johnson, P. E., Creagh, A. L., Tomme, P., Webster, P., Haynes, C. A., & McIntosh, L. P. (2000). Structure and binding specificity of the second N-terminal cellulose-binding domain from Cellulomonas fimi endoglucanase C. *Biochemistry*, *39*, 2445–2458.

Buliga, G. S., Brant, D. A., & Fincher, G. B. (1986). The sequence statistics and solution conformation of a barley (1,3),(1,4)-β-glucan. *Carbohydrate Research*, *157*, 139–156.

Carpita, N. C. (1996). Structure and biogenesis of the cell walls of grasses. *Annual Reviews of Plant Physiology and Plant Molecular Biology*, *47*, 445–476.

Carvalho, A. L., Goyal, A., Prates, J. A., Bolam, D. N., Gilbert, H. J., Pires, V. M., Ferreira, L. M., Planas, A., Romao, M. J., & Fontes, C. M. (2004). The family 11 carbohydrate-binding module of *Clostridium thermocellum* Lic26A-Cel5E accommodates beta-1,4- and beta-1,3-1,4-mixed linked glucans at a single binding site. *Journal of Biological Chemistry*, *279*, 34785–34793.

Cerenius, L., & Soderhall, K. (2004). The prophenoloxidase-activating system in invertebrates. *Immunological Reviews*, *198*, 116–126.

Charnock, S. J., Bolam, D. N., Turkenburg, J. P., Gilbert, H. J., Ferreira, L. M., Davies, G. J., & Fontes, C. M. (2000). The X6 "thermostabilizing" domains of xylanases are carbohydrate-binding modules: structure and biochemistry of the *Clostridium thermocellum* X6b domain. *Biochemistry*, *39*, 5013–5021.

Chen, J., & Seviour, R. (2007). Medicinal importance of fungal beta-(1-->3), (1-->6)-glucans. *Mycological Research*, *111*, 635–652.

Cicortas Gunnarsson, L., Montanier, C., Tunnicliffe, R. B., Williamson, M. P., Gilbert, H. J., Nordberg Karlsson, E., & Ohlin, M. (2007). Novel xylan-binding properties of an engineered family 4 carbohydrate-binding module. *Biochemical Journal*, *406*, 209–214.

Cicortas Gunnarsson, L., Nordberg Karlsson, E., Albrekt, A. S., Andersson, M., Holst, O., & Ohlin, M. (2004). A carbohydrate binding module as a diversity-carrying scaffold. Protein Engineering. *Design and Selection*, *17*, 213–221.

Coutinho, P. M., & Henrissat, B. (1999). Carbohydrate-active enzymes: an integrated database approach. In G. Davies, H. J. Gilbert, B. Henrissat, & B. Svensson (Eds.), *Recent Advances in Carbohydrate Bioengineering* (pp. 3–12). Cambridge: The Royal Society of Chemistry.

Czjzek, M., Bolam, D. N., Mosbah, A., Allouch, J., Fontes, C. M., Ferreira, L. M., Bornet, O., Zamboni, V., Darbon, H., Smith, N. L., Black, G. W., Henrissat, B., & Gilbert, H. J. (2001). The location of the ligand-binding site of carbohydrate-binding modules that have evolved from a common sequence is not conserved. *Journal of Biological Chemistry*, *276*, 48580–48587.

Descroix, K., Ferrieres, V., Jamois, F., Yvin, J. C., & Plusquellec, D. (2006). Recent progress in the field of beta-(1,3)-glucans and new applications. *Mini-Reviews in Medicinal Chemistry*, *6*, 1341–1349.

Dias, F. M., Goyal, A., Gilbert, H. J., Jose, A. M. P., Ferreira, L. M., & Fontes, C. M (2004). The N-terminal family 22 carbohydrate-binding module of xylanase 10B of *Clostridium themocellum* is not a thermostabilizing domain. *FEMS Microbiological Letters*, *238*, 71–78.

Fabrick, J. A., Baker, J. E., & Kanost, M. R. (2004). Innate immunity in a pyralid moth: functional evaluation of domains from a beta-1,3-glucan recognition protein. *Journal of Biological Chemistry*, *279*, 26605–26611.

Ferrer, P. (2006). Revisiting the *Cellulosimicrobium cellulans* yeast-lytic beta-1,3-glucanases toolbox: a review. *Microbial Cell Factories*, *5*, 10.

Fibriansah, G., Masuda, S., Koizumi, N., Nakamura, S., & Kumasaka, T. (2007). The 1.3 A crystal structure of a novel endo-beta-1,3-glucanase of glycoside hydrolase family 16 from alkaliphilic *Nocardiopsis sp*. strain F96. *Proteins*, *69*, 683–690.

Flint, J., Bolam, D. N., Nurizzo, D., Taylor, E. J., Williamson, M. P., Walters, C., Davies, G. J., & Gilbert, H. J. (2005). Probing the mechanism of ligand recognition in family 29 carbohydrate-binding modules. *Journal of Biological Chemistry*, *280*, 23718–23726.

Flint, J., Nurizzo, D., Harding, S. E., Longman, E., Davies, G. J., Gilbert, H. J., & Bolam, D. N. (2004). Ligand-mediated dimerization of a carbohydrate-binding molecule reveals a novel mechanism for protein-carbohydrate recognition. *Journal of Molecular Biology*, *337*, 417–426.

Frecer, V., Rizzo, R., & Miertus, S. (2000). Molecular dynamics study on the conformational stability of laminaran oligomers in various solvents. *Biomacromolecules*, *1*, 91–99.

Gunnarsson, L. C., Dexlin, L., Karlsson, E. N., Holst, O., & Ohlin, M. (2006a). Evolution of a carbohydrate binding module into a protein-specific binder. *Biomolecular Engineering*, *23*, 111–117.

Gunnarsson, L. C., Zhou, Q., Montanier, C., Karlsson, E. N., Brumer, H., 3rd, & Ohlin, M. (2006b). Engineered xyloglucan specificity in a carbohydrate-binding module. *Glycobiology*, *16*, 1171–1180.

Henshaw, J. L., Bolam, D. N., Pires, V. M., Czjzek, M., Henrissat, B., Ferreira, L. M., Fontes, C. M., & Gilbert, H. J. (2004). The family 6 carbohydrate binding module CmCBM6-2 contains two ligand-binding sites with distinct specificities. *Journal of Biological Chemistry*, *279*, 21552–21559.

Hong, T. Y., Cheng, C. W., Huang, J. W., & Meng, M. (2002). Isolation and biochemical characterization of an endo-1,3-beta-glucanase from *Streptomyces sioyaensis* containing a C-terminal family 6 carbohydrate-binding module that binds to 1,3-beta-glucan. *Microbiology*, *148*, 1151–1159.

Huecas, S., Villalba, M., & Rodriguez, R. (2001). Ole e 9, a major olive pollen allergen is a 1,3-beta-glucanase. Isolation, characterization, amino acid sequence, and tissue specificity. *Journal of Biological Chemistry*, *276*, 27959–27966.

Jamal, S., Nurizzo, D., Boraston, A. B., & Davies, G. J. (2004). X-ray crystal structure of a non-crystalline cellulose-specific carbohydrate-binding module: CBM28. *Journal of Molecular Biology, 339*, 253–258.

Jiang, H., Ma, C., Lu, Z. Q., & Kanost, M. R. (2004). Beta-1,3-glucan recognition protein-2 (betaGRP-2)from *Manduca sexta*; an acute-phase protein that binds beta-1,3-glucan and lipoteichoic acid to aggregate fungi and bacteria and stimulate prophenoloxidase activation. *Insect Biochemistry and MolecularBiolog*, *34*, 89–100.

Johnson, P. E., Joshi, M. D., Tomme, P., Kilburn, D. G., & McIntosh, L. P. (1996). Structure of the N-terminal cellulose-binding domain of *Cellulomonas fimi* CenC determined by nuclear magnetic resonance spectroscopy. *Biochemistry, 35*, 14381–14394.

Kormos, J., Johnson, P. E., Brun, E., Tomme, P., McIntosh, L. P., Haynes, C. A., & Kilburn, D. G. (2000). Binding site analysis of cellulose binding domain CBD(N1) from endoglucanse C of *Cellulomonas fimi* by site-directed mutagenesis. *Biochemistry, 39*, 8844–8852.

Lazaridou, A., Biliaderis, C. G., Micha-Screttas, M., & Steele, B. R. (2004). A comparative study on structure–function relations of mixed-linkage (1→3), (1→4) linear β-D-glucans. *Food Hydrocolloids, 18*, 837–855.

Lee, S. Y., Wang, R., & Soderhall, K. (2000). A lipopolysaccharide- and beta-1,3-glucan-binding protein from hemocytes of the freshwater crayfish *Pacifastacus leniusculus*. Purification, characterization, and cDNA cloning. *Journal of Biological Chemistry, 275*, 1337–1343.

Ma, C., & Kanost, M. R. (2000). A beta1,3-glucan recognition protein from an insect, *Manduca sexta*, agglutinates microorganisms and activates the phenoloxidase cascade. *Journal of Biological Chemistry, 275*, 7505–7514.

Marchessault, R. H., & Deslandes, Y. (1979). Fine structure of (1-3)-β-D-glucans: curdlan and paramylon. *Carbohydrate Research, 75*, 231–242.

Malburg, S. R., Malburg, L. M., Jr., Liu, T., Iyo, A. H., & Forsberg, C. W. (1997) Catalytic properties of the cellulose-binding endoglucanase F from *Fibrobacter succinogenes* S85. *Applied and Environmental Microbiology, 63*, 2449–2453.

Muta, T. (2006). Molecular basis for invertebrate innate immune recognition of (1-->3)-beta-D-glucan as a pathogen-associated molecular pattern. *Current Pharmaceutical Design, 12*, 4155–4161.

Notenboom, V., Boraston, A. B., Chiu, P., Freelove, A. C., Kilburn, D. G., & Rose, D. R. (2001). Recognition of cello-oligosaccharides by a family 17 carbohydrate-binding module: an X-ray crystallographic, thermodynamic and mutagenic study. *Journal of Molecular Biology*, 797–806.

Ochiai, M., & Ashida, M. (1988). Purification of a beta-1,3-glucan recognition protein in the prophenoloxidase activating system from hemolymph of the silkworm, *Bombyx mori*. *Journal of Biological Chemistry, 263*, 12056–12062.

Ochiai, M., & Ashida, M. (2000). A pattern-recognition protein for beta-1,3-glucan. The binding domain and the cDNA cloning of beta-1,3-glucan recognition protein from the silkworm, *Bombyx mori*. *Journal of Biological Chemistry*, *275*, 4995–5002.

Papageorgiou, M., Lakhdara, N., Lazaridou, A., Biliaderis, C. C., & Izydorczyk, M. S. (2005). Water extractable (1,3),(1,4)-ß-D-glucans from barley and oats: An intervarietal study on their structural features and rheological behaviour. *Journal of Cereal Science*, *42*, 213–224.

Pires, V. M., Henshaw, J. L., Prates, J. A., Bolam, D. N., Ferreira, L. M., Fontes, C. M., Henrissat, B., Planas, A., Gilbert, H. J., & Czjzek, M. (2004). The crystal structure of the family 6 carbohydrate binding module from *Cellvibrio mixtus* endoglucanase 5a in complex with oligosaccharides reveals two distinct binding sites with different ligand specificities. *Journal of Biological Chemistry*, *279*, 21560–21568.

Reid, D. M., Montoya, M., Taylor, P. R., Borrow, P., Gordon, S., Brown, G. D., & Wong, S. Y. (2004). Expression of the beta-glucan receptor, Dectin-1, on murine leukocytes in situ correlates with its function in pathogen recognition and reveals potential roles in leukocyte interactions. *Journal of Leukocyte Biology*, *76*, 86–94.

Rutenber, E., & Robertus, J. D. (1991). Structure of ricin B-chain at 2.5 A resolution. *Proteins*, *10*, 260–269.

Simpson, P. J., Jamieson, S. J., Abou-Hachem, M., Karlsson, E. N., Gilbert, H. J., Holst, O., & Williamson, M. P. (2002). The solution structure of the CBM4-2 carbohydrate binding module from a thermostable *Rhodothermus marinus* xylanase. *Biochemistry*, *41*, 5712–5719.

Soderhall, K., Cerenius, L., & Johansson, M. W. (1994). The prophenoloxidase activating system and its role in invertebrate defence. *Annals of the New York Academy of Sciences*, *712*, 155–161.

Taylor, P. R., Brown, G. D., Reid, D. M., Willment, J. A., Martinez-Pomares, L., Gordon, S., & Wong, S. Y. (2002). The beta-glucan receptor, dectin-1, is predominantly expressed on the surface of cells of the monocyte/macrophage and neutrophil lineages. *Journal of Immunology*, *169*, 3876–3882.

Trevino, M. A., Palomares, O., Castrillo, I., Villalba, M., Rodriguez, R., Rico, M., Santoro, J., & Bruix, M. (2008). Solution structure of the C-terminal domain of Ole e 9, a major allergen of olive pollen. *Protein Science*, *17*, 371–376.

Tsai, L. C., Shyur, L. F., Cheng, Y. S., & Lee, S. H. (2005). Crystal structure of truncated *Fibrobacter succinogenes* 1,3-1,4-beta-D-glucanase in complex with beta-1,3-1,4-cellotriose. *Journal of Molecular Biology*, *354*, 642–651.

Tvaroska, I., Ogawa, K., Deslandes, Y., & Marchessault, R. H. (1983). Crystalline conformation and structure of lichenan and barley beta-glucan. *Canadian Jornal of Chemistry*, *61*, 1608–1616.

Vaaje-Kolstad, G., Horn, S. J., van Aalten, D. M., Synstad, B., & Eijsink, V. G. (2005a). The non-catalytic chitin-binding protein CBP21 from *Serratia marcescens* is essential for chitin degradation. *Journal of Biological Chemistry*, *280*, 28492–28497.

Vaaje-Kolstad, G., Houston, D. R., Riemen, A. H., Eijsink, V. G., & van Aalten, D. M. (2005b). Crystal structure and binding properties of the *Serratia marcescens* chitin-binding protein CBP21. *Journal of Biological Chemistry*, *280*, 11313–11319.

van Bueren, A. L., Morland, C., Gilbert, H. J., & Boraston, A. B. (2005). Family 6 carbohydrate binding modules recognize the non-reducing end of beta-1,3-linked glucans by presenting a unique ligand binding surface. *Journal of Biological Chemistry*, *280*, 530–537.

Wang, X., Rocheleau, T. A., Fuchs, J. F., & Christensen, B. M. (2006). Beta 1, 3-glucan recognition protein from the mosquito, *Armigeres subalbatus*, is involved in the recognition of distinct types of bacteria in innate immune responses. *Cell Microbiology*, *8*, 1581–1590.

Weis, W. I., Taylor, M. E., & Drickamer, K. (1998). The C-type lectin superfamily in the immune system. *Immunological Reviews*, *163*, 19–34.

Willment, J. A., & Brown, G. D. (2008). C-type lectin receptors in antifungal immunity. *Trends in Microbiology*, *16*, 27–32.

Xie, H., Gilbert, H. J., Charnock, S. J., Davies, G. J., Williamson, M. P., Simpson, P. J., Raghothama, S., Fontes, C. M., Dias, F. M., Ferreira, L. M., & Bolam, D. N. (2001). *Clostridium thermocellum* Xyn10B carbohydrate-binding module 22-2: the role of conserved amino acids in ligand binding. *Biochemistry*, *40*, 9167–9176.

Zekovic, D. B., Kwiatkowski, S., Vrvic, M. M., Jakovljevic, D., & Moran, C. A. (2005). Natural and modified (1-->3)-beta-D-glucans in health promotion and disease alleviation. *Critical Reviews in Biotechnology*, *25*, 205–230.

Zelensky, A. N., & Gready, J. E. (2005). The C-type lectin-like domain superfamily. *FEBS Journal*, *272*, 6179–6217.

Zhang, R., Cho, H. Y., Kim, H. S., Ma, Y. G., Osaki, T., Kawabata, S., Soderhall, K., & Lee, B. L. (2003). Characterization and properties of a 1,3-beta-D-glucan pattern recognition protein of *Tenebrio molitor* larvae that is specifically degraded by serine protease during prophenoloxidase activation. *Journal of Biological Chemistry*, *278*, 42072–42079.

Zverlov, V. V., Volkov, I. Y., Velikodvorskaya, G. A., & Schwarz, W. H. (2001). The binding pattern of two carbohydrate-binding modules of laminarinase Lam16A from *Thermotoga neapolitana*: differences in beta-glucan binding within family CBM4. *Microbiology*, *147*, 621–629.

CHAPTER 3.3.1

# Enzymology and Molecular Genetics of Biosynthetic Enzymes for (1,3)-β-Glucans: Prokaryotes

**Vilma A. Stanisich[1] and Bruce A. Stone[2]**
[1]*Department of Microbiology, La Trobe University, Melbourne, Australia*
[2]*Department of Biochemistry, La Trobe University, Melbourne, Australia*

The enzymes involved in the synthesis of prokaryote (1,3)-β-glucans - curdlan, the cyclic (1,3;1,6)-β-glucans and (1,3;1,2)-β-glucan - are all members of the inverting GT family 2. The curdlan [(1,3)-β-glucan] synthase catalytic subunit (72.9 kDa) is the product of the *Agrobacterium crdS* gene and was identified on the basis of its sequence and structural similarity with various β-glycan synthases. These include bacterial and embryophyte cellulose synthases, and the chito-oligosaccharide and hyaluronan synthases. The *crdS* gene encodes a polytopic, inner membrane protein with a cytoplasmic region carrying the catalytic and UDPGlc substrate-binding motifs. This gene and the two others in the *crdASC* cluster are essential for curdlan production. The cyclic (1,3;1,6)-β-glucans of *Bradyrhizobium japonicum* are products of three, clustered, monocistronic genes, *ndvB-ndvC-ndvD*, and an unlinked gene, *ndvA*. The *ndvB* gene encodes a (1,3)-β-glucan synthase (98.5 kDa) and *ndv*C a GH family 17 (1,6)-glucosyl transferase (61.6 kDa). The (1,3;1,2)-β-glucan synthase of Type 37 *Streptococcus pneumoniae*, the product of the *tts* gene, is a dual specificity integral membrane protein (58.9 kDa) capable of synthesizing both (1,3)- and (1,2)-glucosidic linkages in the side-chain-branched glucan.

## A. Introduction

(1,3)-β-Glucans and related glucans produced by prokaryotes include the linear (1,3)-β-glucan, curdlan, found in *Agrobacterium, Cellulomonas and Bacillus* spp., the rhizobial cyclic (1,3;1,6)-β-glucans found in *Bradyrhizobium, Azospirillum* and *Azorhizobium* spp., and the *Streptococcus pneumoniae* side-chain-branched (1,3;1,2)-β-glucan. The enzymes involved in their synthesis all belong to family GT2 glycosyl transferases (http://afmb.cnrs-mrs.fr/CAZY/), which also includes synthases for bacterial and embryophyte cellulose, invertebrate chitin and chito-oligosaccharides, and bacterial and vertebrate hyaluronan. In addition, a GH17 (1,6)-β-glucosyl transglycosylase is involved in the synthesis of the cyclic (1,3;1,6)-β-glucan. In this chapter the enzymology and molecular genetics of the biosynthesis of the three bacterial β-glucans is discussed. An account of their biology and functional roles is given in Chapter 4.1.

# B. Curdlan

## B.1. The curdlan biosynthesis locus

The production of curdlan was first detected in isolates of *Agrobacterium*, and is a property that may be limited to only a few bacterial species (see Chapter 4.1). The molecular genetics of curdlan biosynthesis has been studied in *Agrobacterium* LTU50, a spontaneous chloramphenicol-resistant mutant of the curdlan hyper-producing biovar 1 strain, *Agrobacterium* sp. ATCC 31749. These studies were facilitated by the ability of curdlan to be stained specifically by the triphenylmethane dye Aniline Blue (Nakanishi et al., 1976) and by the Aniline Blue fluorochrome (Evans et al., 1984). Thus, a qualitative assessment of curdlan production (or the lack of production) could be made from the staining response of colonies grown on Aniline Blue-supplemented agar medium (dark-blue-staining vs non-staining colonies, respectively), and by fluorescence microscopy of bacteria grown in liquid culture. Furthermore, initial studies on genes cloned from LTU50 revealed high sequence similarity (>98%) to genes of *Agrobacterium tumefaciens* strain C58, a fully sequenced member of the *Agrobacterium* genus (Goodner et al., 2001; Wood et al., 2001).

The close relatedness of *Agrobacterium* sp. LTU50 and *A. tumefaciens* C58 enabled curdlan-production genes to be identified not only by random transposon (Tn)-mediated mutagenesis and linkage analysis, but also by directed (targeted) disruption of selected LTU50 genes based on the sequences of the gene homologues in C58. The precise location of newly identified genes could also be determined by reference to the genome composition of C58. This strain, and others of biovar 1, has a multi-replicon genome (Jumas-Bilak et al., 1998; Urbanczyk et al., 2003) which, in C58, consists of two chromosomes (one circular of 2.84 Mb and one linear of 2.07 Mb) and two mega-plasmids (pAtC58 and pTiC58). All the genes currently known to affect curdlan production in LTU50 are chromosomally located, with no evidence of plasmid involvement (Table 1). The plasmid content of LTU50 is not known; however, it either lacks, or has a non-functional, pTi plasmid, since it is non-pathogenic when tested on tomato stems, leaves of *Bryophyllum daigremontiana* and carrot root discs (A. Matthysse, personal communication), and lacks T-DNA-encoded virulence determinants (*ipt* and *virB9*) based on PCR analysis (A. Matthysse personal communication; F. Taner and V.A. Stanisich, unpublished data).

**Table 1: Genes involved in the production of bacterial (1,3)-β-glucans**

| Gene name (ID)[a] | Location[b] | Size (bp) | Protein (amino acids)/Mass | Role |
|---|---|---|---|---|
| ***Agrobacterium* sp. LTU50 – production of linear (1,3)-β-glucan** | | | | |
| *crdA* (Atu3057) | L | 1458 | 485 (53.3 kDa) | putative membrane protein (possible transport protein) |
| *crdS* (Atu3056) | L | 1965 | 654 (72.9 kDa) | (1,3)-β-glucosyl transferase (GT2) |
| *crdC* (Atu3055) | L | 1266 | 421 (45.5 kDa) | putative periplasmic protein (possible transport protein) |
| *crdR* (Atu0361) | C | 420 | 139 (15.1 kDa) | helix-turn-helix transcriptional regulator protein |
| *pssA* (Atu1062) | C | 825 | 288 (31.5 kDa) | membrane-associated phosphatidylserine synthase |
| *ntrB* (Atu1445) | C | 1149 | 382 (41.3 kDa) | two-component sensor kinase |
| *ntrC* (Atu1446) | C | 1452 | 483 (53.8 kDa) | two-component response regulator |
| *ntrY* (Atu1447) | C | 2262 | 753 (82.5 kDa) | membrane-associated two-component sensor kinase |
| *ntrX* (Atu1448) | C | 1365 | 454 (50.3 kDa) | two-component response regulator |
| *relA* (Atu1030) | C | 2235 | 744 (83.7 kDa) | GTP pyrophosphorylase/synthetase |
| ***Bradyrhizobium japonicum* USDA 110 – production of cyclic (1,3;1,6)-β-glucans** | | | | |
| *ndvC* (bll4612) | C | 1665 | 554 (61.6 kDa) | putative (1,6)-β-glucan synthase (GH17) |
| *ndvD* (blr4613) | C | 597 | 198 (20.6 kDa) | putative membrane protein (possible synthesis/regulatory protein) |
| *ndvB* (blr4614) | C | 2670 | 889 (98.5 kDa) | (1,3)-β-glucosyl transferase (GT2) |
| *ndvA* (bll1321) | C | 1809 | 602 (65.9 kDa) | putative cyclic β-glucan ABC-transporter protein |
| ***Streptococcus pneumoniae* 1235/89 (Type 37) – production of side-chain-branched (1,3;1,2)-β-glucan** | | | | |
| *tts* | C | 1530 | 509 (58.9 kDa) | (1,3;1,2)-β-glucosyl transferase (GT2) |

[a]Gene identification (ID), and details on gene and product size are taken from the annotated sequences of *Agrobacterium tumefaciens* C58 (Wood et al., 2001; NC_003062 and NC_003063) and *Bradyrhizobium japonicum* USDA 110 (Kaneko et al., 2002; NC_004463). Information on *S. pneumoniae tts* is from AJ131985.

[b]L= linear chromosome; C = circular chromosome.

### *B.1.1. The crdASC biosynthesis genes*

The curdlan biosynthesis locus consists of the *crdASC* gene cluster which was recognised as such based on the presence of the curdlan synthase gene, *crdS*, whose deduced product (654 amino acids) contains characteristic conserved motifs typical of GT2 glycosyl transferases with repetitive action patterns (Stasinopoulos et al., 1999) (see Section B.4). The genes are contiguous in a 4863-bp region located close to one end of the linear chromosome and were first isolated on an 8.8-kb *Eco*RI genomic clone (Stasinopoulos et al., 1999). They are transcribed in the same direction and have an operon-like organization with sequences of 132 bp and 45 bp separating *crdA* from *crdS* and *crdS* from *crdC*, respectively (Fig. 1). The 423-bp intergenic region upstream of *crdA* is A-T-rich (56%) compared to the *crdASC*-coding region (43%), indicative of a role in the transcriptional regulation of the *crd* gene cluster. The genes that flank *crdASC* are transcribed in the opposite orientation and hence are not part of the putative operon. The only other complete open reading frame (ORF) that occurs in the 8.8-kb *Eco*RI fragment, *orf4* (= Atu3054), is located immediately downstream of *crdC*. This hypothetical gene is not required for curdlan production since an *orf4* disruptant of LTU50 was unaffected in polymer production.

Evidence that the *crdASC* genes are all required for curdlan production came from several sources. Tn-insertion mutants that produced non-staining colonies on Aniline Blue agar usually had mutations in *crdA* or *crdS* that could be complemented for curdlan production by the 8.8-kb *Eco*RI fragment. The few mutations that were not complemented occurred in an unlinked locus, *crdR* (Stasinopoulos et al., 1999) (see Section B.2). Complementation of *crdA* or *crdS* point mutations with the respective individually cloned genes confirmed that each gene was essential for curdlan production (Anguillesi, 2003; Karnezis et al., 2003). Mutants of *crdC* were exceptional in that they produced pale-blue-staining colonies. These mutations also could be complemented by the 8.8-kb fragment or by a cloned *crdC* gene, resulting in the formation of the dark-blue-staining colonies typical of the wild type (M. Hermans and V.A. Stanisich, unpublished data).

In gravimetric assays in which the bacteria were grown in an N-limited, glucose-rich salts medium (Stasinopoulos et al., 1999), none of the *crdA*, *crdS* or *crdC* mutants produced recoverable curdlan, whereas this occurred if the strain had the complementing gene fragment. Under microscopic examination, *crdC* mutants, but not *crdA* or *crdS* mutants, produced curdlan-like material (perhaps oligosaccharides) detectable by the sensitive Aniline Blue fluorochrome (McIntosh, 2004). These data suggested that all three genes have essential roles in curdlan production.

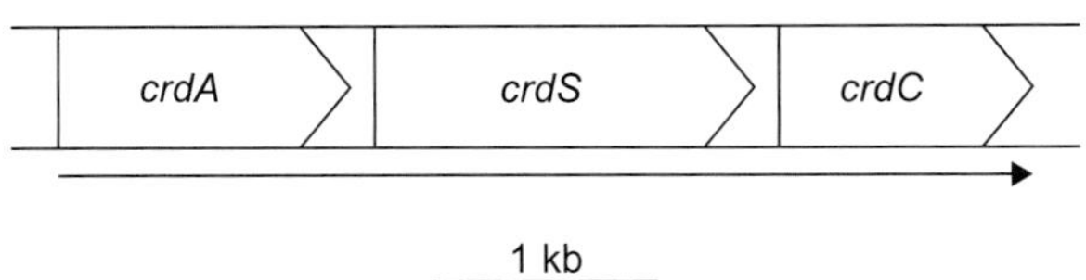


**Fig. 1: Organization of the (1,3)-β-glucan (curdlan) synthesis gene cluster in *Agrobacterium* sp. LTU50 and *A. tumefaciens* C58.**

Analysis of the nucleotide sequence of the *crdASC* region revealed that the deduced CrdA (485 amino acids) and CrdC proteins (421 amino acids) contain no conserved domains suggestive of the roles of these proteins. This is in contrast to the CrdS protein which shares similarity with numerous GT2 glycosyl transferases. Nonetheless, database entries include sequences of putative proteins that share similarity with CrdA or CrdC (27–45% identity; 46–57% similarity). These proteins are encoded by members of a small number of bacterial genera (e.g. species of *Methylobacterium*, *Bradyrhizobium*, *Burkholderia* and *Ralstonia*) all of which are α-or β-proteobacteria. When expressed in *Escherichia coli*, the cloned *crdA* and *crdC* genes yielded proteins of $M_r \sim 55$ kDa and $\sim 42$ kDa, respectively, which are close to those predicted (Table 1). In the case of CrdS, the experimentally determined mass, $M_r$ 60 kDa (Karnezis et al., 2003), was significantly lower than that predicted (72.9 kDa), probably because of the overall hydrophobicity of the protein, a feature that can increase protein mobility in a gel matrix.

## B.2. Other Curdlan-Related Loci

### B.2.1. The *pssA* gene encoding a phospholipid synthase

A number of genes at loci that are distant from the *crdASC* cluster have been found to variously affect the production of curdlan. One such gene, *pssA* (825 bp) in the circular chromosome, is required for maximal yields of the polymer. The gene encodes a membrane-associated phospholipid serine synthase that could be expressed as an enzymatically active protein in *E. coli* (Karnezis et al., 2002). Mutants of *pssA* produced a significantly reduced yield of curdlan (20% of wild type) that was reflected in the incomplete (mottled) Aniline Blue-staining response on agar medium, and the absence of the dry curdlan pellicle that is typical of the wild type on prolonged incubation (see Chapter 4.1, Fig. 1). The reduced production of curdlan correlated with the absence of the major membrane phospholipid, phosphatidyl ethanolamine, which is synthesized from the short-lived phosphatidyl serine precursor. Correspondingly, both curdlan yield and phospholipid profile were restored in the mutant by complementation with a cloned *pssA* gene. It is possible that the reduced yield of curdlan caused by the lack of phosphatidyl ethanolamine is due to direct or indirect effects on the structure or activity of the CrdS synthase (and/or the other Crd proteins), as has been observed in other polysaccharide production systems (see Section B.7).

### B.2.2. The *crdR* regulatory gene

Several regulators of curdlan production have been identified. The first gene recognized, *crdR* (420 bp) is located in the circular chromosome and yielded a cytoplasmic protein of $M_r \sim 15$ kDa when expressed in *E. coli* (Anguillesi, 2003; Stasinopoulos et al., 1999). This cellular location is consistent with sequence predictions, which also identified the presence of a putative helix-turn-helix DNA-binding motif in the $NH_2$-terminal portion of the protein (residues 29–59). Mutants of *crdR* were phenotypically indistinguishable from *crdA* and *crdS* mutants and displayed restored curdlan production on reintroduction of a plasmid-borne *crdR* gene. However, the *crdR* complemented strain, unlike strains complemented with the *crdA* or *crdS* genes, displayed early onset of curdlan production on Aniline Blue agar medium and elevated curdlan yield (150%) from liquid culture (Aracic, 2009), both features presumably reflecting the increased *crdR* gene dosage. These findings, and the curdlan-deficient phenotype of *crdR* mutants, suggested that the CrdR protein is a regulatory activator of curdlan production, although whether it acts directly on the *crdASC* locus is not known.

Genes related to *crdR* occur in other species of the α-proteobacteria, including rhizobia (e.g. *Sinorhizobium meliloti* and *Mesorhizobium loti*), *Brucella melitensis* and *Caulobacter cresentus*. In all of these bacteria, as in *Agrobacterium*, the *crdR* homologue is part of a conserved gene arrangement with *lnt* (encoding apolipoprotein-N-acyltransferase) on the upstream flank and *sam* (encoding S-adenosylmethionine synthase) located downstream. None of these bacteria have a recognisable *crdASC* gene cluster suggesting that the *crdR* homologues have roles other than those related to curdlan production. Only the *crdR* homologue from *S. meliloti*, denoted *phrR* (90% identity at the protein level), has been studied experimentally. In the acid-tolerant strain, *S. meliloti* WSM419, low-level, constitutive expression of *phrR* was enhanced 5-fold at reduced pH ($< 6.2$) and to lesser extents by other stresses (e.g. $Cu^{2+}$, 100 mM; $Zn^{2+}$, 40 mM; $H_2O_2$, 400 mM) (Reeve et al., 1998). Despite pH responsiveness, *phrR* is not involved in the acid-tolerance of WSM419 (Reeve et al., 1998) which is controlled by several genes including the adjacent *lnt* (= *actA*) gene (Tiwari et al., 1996). Transcriptome analysis of *S. meliloti* 1021 also detected constitutive expression of *phrR* (relative value = 1) in liquid cultures under aerobic and micro-oxic conditions; however, the gene was down-regulated in bacteroids (relative values of 0.3–0.4) where N is limiting (Ampe et al., 2003). No roles have been ascribed to *phrR* in *S. meliloti,* and inactivation of the gene does not affect nodule formation on alfalfa (Dilworth et al., 2002).

### B.2.3. The *ntrBC* and *ntrYX* nitrogen regulatory genes

Other regulatory genes that affect curdlan production have well-characterized homologues in other species of bacteria, where they serve as global gene regulators. One of these is the gene pair, *ntrBC*, which encodes the terminal components of an extensively studied signal transduction pathway that is responsive to nitrogen starvation (Magasanik, 1996). During N-limited growth, NtrB, a cytosolic protein kinase/phosphoprotein phosphatase, undergoes autophosphorylation at a histidine residue; the phosphate is then transferred to an aspartate residue in the amino-terminal domain of NtrC, a DNA-binding response regulator protein. The presence of phospho-NtrC leads to an adaptive response that typically enhances the binding of the $\sigma^{54}$ ($\sigma^{N}$) RNA polymerase holoenzyme to cognate promoters, amongst which is the $\sigma^{54}$-promoter that controls the nitrate assimilation genes (Lin and Stewart, 1998). Components of the N-starvation response would be expected to affect curdlan production, since the onset of its production coincides with depletion of the N-source and the cessation of cell growth (Lee, 2002).

Random transposon mutagenesis of *Agrobacterium* LTU50 detected a class of curdlan-deficient mutants that failed to stain with the Aniline Blue dye and were further distinguished by their inability to grow with nitrate as the sole N source (Aracic et al., 2008). The insertion mutations were found to be distributed in the genes of the *orf1-ntrBC* operon (Fig. 2) which is located in the circular chromosome and is transcribed from a putative $\sigma^{54}$ promoter (Dombrecht et al., 2002). Complementation of the *ntrC* insertion mutations with the cloned *ntrC* gene restored the two phenotypic traits. This is consistent with the known role of the *ntrC* regulatory gene in nitrate assimilation (Lin and Stewart, 1998) and also implicated *ntrC* in the regulation of curdlan production.

The possibility that curdlan production is activated by NtrC rather than by phospho-NtrC was raised by the behaviour of the *ntrB* insertion mutants. These produce neither NtrB (the cognate phosphoryl donor of NtrC) nor NtrC. Such mutants were unexpectedly complemented for curdlan production both by an intact *ntrC* gene and also by an *ntrC* fragment that no longer encoded the $NH_2$-terminal phosphorylation domain. This observation implied that unphosphorylated NtrC can serve as a transcriptional activator in the curdlan production pathway. A comparable situation has been described in the production of the capsular polysaccharide, alginate, by *Pseudomonas aeruginosa*. In this system, two response regulators, AlgB (belonging to the NtrC subfamily of proteins) and AlgR (belonging to the LysR subfamily) can promote polymer production in the absence of their cognate sensor histidine kinases (Leech, et al., 2008; Ma et al., 1998).

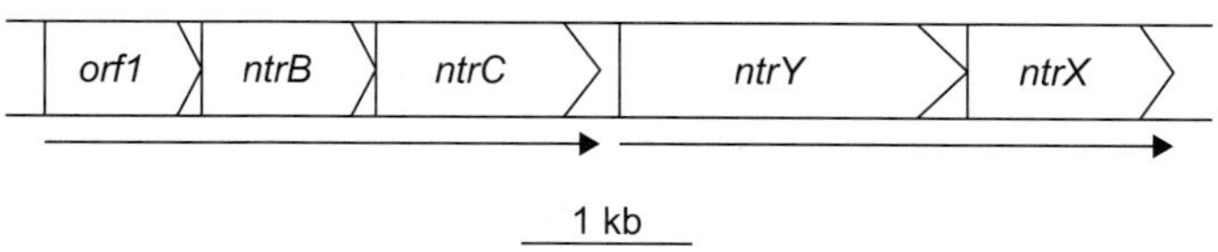


**Fig. 2: Organization of two-component regulatory genes involved in nitrogen metabolism in *Agrobacterium* sp. LTU50 and *A. tumefaciens* C58.**

In contrast to the findings regarding curdlan production, complementation of the *ntrB* insertion mutants for nitrate assimilation displayed a different outcome. In this case, nitrate assimilation was restored by the intact *ntrC* gene but not by the *ntrC* fragment. This implied that activation of the nitrate assimilation genes was strictly dependent on phospho-NtrC, which is in accord with numerous other studies. Since the complemented strain lacked NtrB, an alternative phosphoryl donor such as acetyl phosphate (Klein et al., 2007; McCleary and Stock, 1994) must have served as a substitute to activate NtrC. The significant distinction raised by these studies is that the curdlan production pathway can be activated by NtrC (and presumably also by phospho-NtrC) whereas activation of the nitrate assimilation pathway is restricted to phospho-NtrC.

Whether the first gene of the operon, *orf1* (=Atu1444) (also denoted *nifR3*; Morett and Bork, 1998; Patriarca et al., 2002) has a specific role in curdlan production (or nitrate assimilation) is not known and the phenotypic effects of *orf1* inactivation may merely have been the result of polarity of the transposon insertion on *ntrBC*. It is noteworthy that the *orf1-ntrBC* gene organization that occurs in *Agrobacterium* is conserved in a variety of other plant-associated bacteria [e.g. *Rhizobium etli* (Patriarca et al., 1993); *Mesorhizobium loti* (Kaneko et al., 2002); *Azospirillum brasilense* (Machado et al., 1995); *Azorhizobium caulinodans* (Pawlowski et al., 1991)]. No role has been ascribed to *orf1* in any of these species although in sequence annotations the homologues are predicted to encode TIM-barrel enzymes that are possibly dehydrogenases involved in nitrogen metabolism. However, the *yhdG* gene of *E. coli* that is highly similar to the *R. etli* and related NifR3 proteins (Morett and Bork, 1998) has been shown to encode a tRNA-dihydrouridine synthase (Bishop et al., 2002). This enzyme acts on tRNA transcripts and reduces specific uridines in the D-loop to the modified base 5,6-dihyrdouridine, thereby affecting conformational flexibility of tRNA (Dalluge et al., 1996). This functional role of YhdG raises the possibility that NifR-family proteins may serve as post-transcriptional regulators.

The global N-metabolism network in *Agrobacterium* includes a second two-component system (NtrYX) encoded by conserved genes that lie immediate downstream of *orf1-ntrBC* (Fig 2). NtrY (an NtrB-related protein) is predicted to encode a membrane-associated sensor kinase that may detect extracellular N and modulate the activity of NtrX (an NtrC-like protein). It has not been possible to disrupt the *Agrobacterium ntrX* gene (suggesting that loss of NtrX function is lethal) although disruption can occur for *ntrY* provided that transcriptional read-through of *ntrX* continues (Aracic et al., 2008). The resulting mutants are defective both for curdlan production and nitrate assimilation, although the phenotypes are unstable. These findings provide preliminary evidence that *ntrYX* has a role in curdlan production. They also support findings first made in *A. caulinodans* (Pawlowski et al., 1991) and also in *A. brasilense* (Ishida et al., 2002; Vitorino et al., 2001) that NtrYX are components of the bacterial global N-metabolism system.

### B.2.4. The *relA* gene encoding the (p)ppGpp alarmone

Another well-known global regulatory system is the stringent response which enables microorganisms to adapt to suboptimal growth conditions (Chatterji and Ojha, 2001). The response is mediated through the synthesis of guanosine pentaphosphate and guanosine tetraphosphate, collectively named (p)ppGpp, which function as alarmone signals to trigger the stringent stress response and co-ordinate entry of the bacterial cells into stationary phase. In *Agrobacterium,* the production of (p)ppGpp involves a bifunctional (p)ppGpp synthetase and hydrolase enzyme that has been studied experimentally in *A. tumefaciens* A6 where, as expected, the encoding gene (*relA*) is transcribed only in the stationary growth phase (Zhang et al., 2004). The corresponding gene in LTU50, which shares 90% identity with $relA_{A6}$ and 96% identity with $relA_{C58}$, was detected by random Tn mutagenesis, since the defect resulted in curdlan deficiency (Taner et al., 2008). The colonial morphology of the mutant on Aniline Blue agar was distinctive in that, on subculture, the bacteria segregated as a white colony lineage that was stable, and a pale-blue colony lineage that was unstable and perpetuated both white and pale-blue forms. In accordance with the staining response, both lineages produced little recoverable curdlan (6% vs 2% from the pale-blue and white lineages, respectively).

In complementation studies, curdlan production by the mutant was restored on re-introduction of a cloned *relA* gene, thereby confirming that the defect in *relA* was responsible for the curdlan-deficient phenotype and not polarity on the downstream genes which are transcribed in the same direction as *relA*. The involvement of *relA* in polymer production is consistent with curdlan's formation as a secondary metabolite after the cessation of cell growth. The use of constitutively expressed *relA* in the complementation study further revealed that low-level expression of the gene resulted in colonies that were all uniformly stained with Aniline Blue, whereas augmented expression from the $P_{lac}$ promoter of the vector plasmid restored curdlan production in some colonies (~50% stained dark blue) but not others (which stained pale blue). This heterogeneity presumably reflects a significant re-adjustment of the cell's complex and finely tuned physiology to the unnaturally elevated levels of a key global regulator. Not surprisingly, the multiple roles performed by RelA in various bacteria, and which include the control of biofilm formation, cell division and long-term viability (Braeken et al., 2006), were evident in pleiotropic consequences of the *relA* mutation in LTU50. In addition to the down-regulation of curdlan production, the mutant exhibited a reduced growth rate in minimal salts medium, formed unusually large flocs of cells and produced a brown pigmented metabolite on prolonged incubation. Also, the cells of the *relA* mutant were elongated compared to those of LTU50 ($\sim$3.6 μm vs 2.5 μm). All of these properties were restored to wild type by complementation with the *relA* gene, except for flocculation which persisted to a reduced extent in the complemented strain (Taner et al., 2008).

In summary, a relatively large number of genes have been shown to affect curdlan production and, undoubtedly, others are yet to be discovered. Studies on the optimization of curdlan yield in batch- and commercial-scale culture have shown that curdlan production occurs in the post-stationary growth phase and is influenced by numerous factors that include C and N sources, phosphate, sulphate and cation composition of the culture medium, and pH and degree of aeration of the medium (Lee, 2002; McIntosh et al., 2005). It is likely that the RelA-mediated stress response initiates the cascade of events leading to curdlan production. The initiation event and entry into stationary phase are, however, insufficient to elicit curdlan production, since this occurs only in a chemically defined medium when the N source is depleted, and not at all in a nutrient-rich medium. The involvement of NtrBC and NtrYX as activators in the cascade accords with the observed relationship between curdlan production and N supply and with the roles of the Ntr proteins as sensors of N starvation. CrdR is likely to act at a still later stage of the cascade, since *ntrC* mutants that have a plasmid-borne *crdR* gene can produce some curdlan on Aniline Blue agar medium (Aracic, 2009). The target gene(s) of each of the activators is not known, and it seems unlikely that NtrC directly activates either *crdR* or *crdASC* since the regions upstream of these loci contain no obvious $\sigma^{54}$-like promoter consensus sequence (Dombrecht et al., 2002).

Curdlan production is also a strain-specific attribute and the requisite genes appear to be permanently repressed in many wild-type *Agrobacterium* strains. Such strains typically produce succinoglycan, an acidic heteropolysaccharide, and either no curdlan or very little (Nakanishi et al., 1976). *A. tumefaciens* strain C58 is a case in point: it is a succinoglycan producer (Changelosi et al., 1987) and although it has homologues of the *crdASC* genes (and others described above), it produces little curdlan. Conversely, spontaneous curdlan-producing mutants that arise from wild-type strains (for example, the LTU50 lineage) lose the ability to produce succinoglycan (Hisamatsu et al., 1977) suggesting a negative correlation in production of these polysaccharides. Succinoglycan production and regulation have been little studied in *Agrobacterium* (Aird et al., 1991; Changelosi et al., 1987; Tiburtius et al., 1996), but are well documented in *S. meliloti* where the succinoglycan and galactoglucan production pathways are negatively co-regulated (Becker and Pühler, 1998). The superficially similar regulatory interaction between the succinoglycan and curdlan production pathways, resulting in repression of the latter, perhaps explains why curdlan production has been overlooked as a feature of agrobacterial biology and poses questions as to its roles (see Chapter 4.1).

## B.3. Enzymology

The CrdS enzyme (72.9 kDa) shares sequence and structural homology with other (1,3)-β-glycan synthases, including synthases for bacterial and plant cellulose, chitin and chito-oligosaccharide and hyaluronan, all members of glycosyl transferase family GT2 (Coutinho and Henrissat, 1999). However, CrdS has no homology with the plant cellulose synthase-like proteins that direct the synthesis of plant (1,3)-β-glucans (Li et al., 2003) or the (1,3)-β-glucan synthase-related FSK1 and FSK2 proteins from yeasts and fungi (e.g. *Saccharomyces, Candida, Aspergillus*) which are both classified as GT48 glycosyltransferases (see Chapters 3.3.3 and 3.3.4, respectively).

Although various members of the GT2 family have been shown to produce their respective polysaccharides in cell-free systems in protozoa and chromistans (see Chapter 3.3.2), and in embryophytes (see Chapter 3.3.4), by contrast, membrane preparations from *Agrobacterium* LTU50 incorporate glucose from UDP[$^{14}$C]-α-glucose into ethanol-insoluble (1,3)-β-glucan with low efficiency (McIntosh, 2004).

## B.4. Topology of the Membrane-Bound Curdlan Synthase and Associated Proteins

*In silico* analysis predicts that CrdS is a polytopic membrane protein. This was confirmed experimentally (Karnezis et al., 2003) using C-terminal deletion fusions of CrdS to the LacZ (β-galactosidase) and PhoA (alkaline phosphatase) reporter enzymes. CrdS was shown to comprise six transmembrane helices, one non-membrane-spanning amphipathic helix, a 42-amino acid C-terminus that lies in the cytoplasm and an $N_{out}$–$C_{in}$ disposition (Karnezis et al., 2000, 2003) (Fig. 3).

A large central and relatively hydrophilic cytoplasmic region of ~ 300 residues situated between transmembrane helices 3 and 4 (Fig. 3) carries various conserved motifs including the UDPGlc substrate-binding $Yx_{17}Dx$ Px DD(G/X) motif found in all cellulose synthases, the *S. meliloti* NodC synthase and the *S. pyogenes* hylauronan synthases, and the catalytic D,D,$D_{35}QxxRW$ motif (Karnezis et al., 2000) (Table 2). This region shares highest homology (42% similarity) with bacterial cellulose [(1,4)-β-glucan] synthases such as those from *A. tumefaciens* (CelA) and *Gluconacetobacter xylinus* (AcsA) (Stasinopoulos et al., 1999). Two other motifs $Yx_2Rx_6KAG$ and $QTPx_6D$ are also shared with the cellulose synthases but not other GT2 enzymes (Table 2).

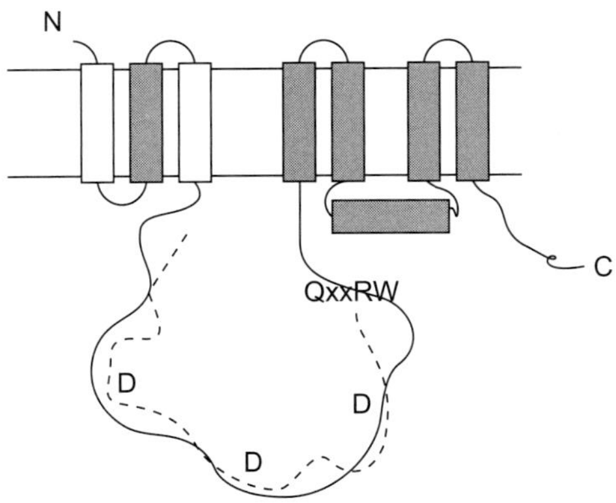


**Fig. 3: Experimentally determined membrane topology of the *Agrobacterium* sp. LTU50 CrdS protein showing a large cytoplasmic region, seven transmembrane segments and a membrane-associated region. The dotted line shows the location of the catalytic D,D,D$_{35}$QxxRW motif in the cytoplasmic region. (From Karnezis et al., 2003.)**

Although CrdA and CrdC have not been subjected to experimental topological analysis, when expressed in *E. coli* the CrdA ($M_r$ 48 kDa) and CrdC ($M_r$ 42 kDa) radiolabelled proteins were detected only in the membrane and periplasmic fractions of the cells, respectively (Anguillesi, 2003; B. Russ and V.A. Stanisich, unpublished data). For CrdA this conforms to predictions that it is a membrane-bound protein that contains a single putative $NH_2$-terminal transmembrane helix from residues 23–41 and has a predicted $N_{in}$–$C_{out}$ topology. In contrast, the CrdC protein occurs only in the periplasmic cell fraction, consistent with the presence of a putative cleavable signal peptide (residues 1–21) in the deduced protein sequence.

## B.5. Structure of the Active Site and Mechanism of Formation and Specification of Glycosidic Linkage

The 654-residue amino acid sequence of the bacterial (1,3)-β-glucan synthase (CrdS) (Stasinopoulos et al., 1999) which contains the extended $D^1D^2xDD^3{}_{35}$QxxRW motif has been threaded onto the structure of the family GT2 SpsA from *Bacillus subtilis* (Charnock and Davies, 1999; Tabouriech et al., 2001) and shows a similar folding pattern in the UDPGlc (TDPGlc) binding site of the catalytic region of SpsA (Karnezis et al., 2003) (Fig. 4). In the 256-amino-acid SpsA there are two domains, a nucleotide-binding domain and an acceptor-binding domain, and it features a disordered loop spanning the active site. Both $UDPMn^{2+}$ and $UDPMg^{2+}$ bind in a deep cleft situated in the $NH_2$-terminal domain. The hydrogen bonding and ionic interactions

**Table 2: Motifs common to putative catalytic site sequences of (1,3)-β-glycan synthases**

| | | |
|---|---|---|
| CEL_AG | --DYRPTVDVFVPS**Y**NEDAELLANTLAAAKNM**D**Y**P**ADRFTVWLL**DD**$^1$GGSVQKRNAANIVE | 314 |
| CRD_AG | YLAGNPHVDVFICT**Y**NEPLNVLEKSIIAAQAM**D**Y**P**--RLRVFVC**DD**------------- | 156 |
| BRJ_3 | -------VSIHIPA**Y**FEPVEMLKQTLDALSRLNY**P**NYECVVIINNTP**D**P----------- | 460 |
| HASA_6 | --PHDYKVAAVIPS**Y**NEDAESLLETLKSVLAQTY**P**--LSEIYIV**DD**GSS---------- | 105 |
| TTS_SP | KSSSISEAKKVILL**Y**CTANDFVPECLVESMQQDYAN--FETVIL**DD**SKS---------- | 136 |
| SPS_A | ----MPKVSVIMTS**Y**NK-SDYVAKSISSILSQTFS--DFELFIM**DD**NSN---------- | 42 |
| CEL_AG | AQAAQRRHEELKKLCEDLDVR--**Y**LT**R**ERNVHA**KAG**NLNNGLAHS-----TGELVTVF**D**$^2$A | 366 |
| CRD_AG | -----TRRGEVRTYCEAAGVN--**Y**VT**R**PDNKHA**KAG**NLNNALLHTNALEEVSDFIMVL**D**A | 209 |
| BRJ_3 | -----AFWQPIQDHCRALGERFKFINAEKVQGF**KAG**ALRIAMDRT---AVDAEIIGIL**D**A | 512 |
| HASA_6 | ---NTDAIQLIEEYVNREVDICRNVIVHRSLVN**K**GKRHAQAWAFE---RSDADVFLTV**D**S | 159 |
| TTS_SP | ----EVYKQQVDEFAKKYNVS--VIRRDDRNGF**KAG**NINNYLKNK----NDYDYFVLL**D**S | 186 |
| SPS_A | ----EETLNVIRPFLNDNRVRFYQSDISGVKERTEKTRYAALINQAIEMAEGEYITYAT**D** | 98 |
| CEL_AG | DHAPARDFLLETVGYFDEDPRLFLV**QTP**HFFVNP**D**PIERNLRTFETMPSENEMFYGIIQR | 427 |
| CRD_AG | DFAPQANFLRRVTGLFS-DPKVAVV**QTP**QFYFNS**D**PIQHNLGIDKSFVDDQRVFFDDFQP | 265 |
| BRJ_3 | DYVVDPDWLKDLVPAFA-DPRVGLV**Q**A**P**QEHRDG**D**LSIMHYIMNGEYAG----FFDIGMV | 567 |
| HASA_6 | DTYIYPNALEELLKSFN-DETVYAATGHLNARNRQTNLLTRLTDIRYDN----AFGVERA | 214 |
| TTS_SP | DEIIPSNFIKKSLAYFEKNRNLGIL**Q**ATHVASRNRNLFMDTLAIGVDSH-----WPVYQK | 242 |
| SPS_A | DNIYMPDRLLKMVRELDTHPEKAVIYSASKTYHLNENRDIVKETVRPAA-----QVTWNA | 153 |
| CEL_AG | GLDKWNGA**FFCG**-**S**AAVLRREALQ-DSDGFS-------GVSITE**D**$^3$CETALA-LHSRGWNS | 477 |
| CRD_AG | AKDAVGCA**F**RV**G**-TSFVVRRAAVN-GIGGFP-------TDALTE**D**MLLTYR-LMERGYVT | 318 |
| BRJ_3 | QRNEANAIIVHG-TMCLIRRAAMD-MAGGWS-------SDTICE**D**SDLGLA-IQELGWVT | 617 |
| HASA_6 | AQSLTGNILVCSGPLSIYRREVIIPNLERYKNQTFLGLPVSIGD**D**RCLTNY-AIDLG-RT | 272 |
| TTS_SP | VKHYYGFLSLLG-HGAMISKDCYQ-AAGGFP--------HVVAE**D**LCFSIESRIKGDYHV | 291 |
| SPS_A | PCAIDHCSVMHRYSVLEKVKEKFGSYWDESP------AFYRIG**D**ARFFWRVNHFYPFYPL | 207 |
| CEL_AG | VYVDKPLIAGLQPATFASFIG**Q**RS**RW**AQGMMQILIFRQ-PLFKRGLS-- | 523 |
| CRD_AG | RWLNEKWSVGLSAEGVPEYIT**Q**RT**RW**CLGTIQIGLLRTGPLWRGNF--- | 364 |
| BRJ_3 | HYTNHRYGQGLLPDTYEAFKK**Q**RH**RW**AYGGLQIVKKHWRHFLPGRS--- | 663 |
| HASA_6 | VYQSTARCDTDVPFQLKSYLK**Q**QN**RW**NKSFFRESIISVKKILSNPI--- | 318 |
| TTS_SP | GFADDIVCQEEYPVDYLAFKKRHSK**W**TQGNMEFIKRYTPIILKSKLKWQ | 340 |
| SPS_A | DEELDLNYITDQSIHFQLFELEKNEFVRNLPPQRNCRELRESLKKLGMG | 256 |

Computer-generated alignment (PILEUP) of the amino acid regions indicated for the following polypeptides: CelA cellulose synthase from *Agrobacterium tumefaciens* C58, (CEL_AG, Q44418); CrdS curdlan synthase from *Agrobacterium* sp. ATCC 31749, (CRD_AG, AAD20440); NdvB (1,3)-ß-glucan synthase from *Bradyrhizobium japonicum* USDA 110, (BRJ_3, NP_771254.1); HasA hyaluronan synthase from *Streptococcus pyogenes* serotype M6, (HASA_6, Q5X9A9); Tts (1,3;1,2)-ß-glucan synthase from *Streptococcus pneumoniae* serotype 37, (TTS_SP, Q9X9S1); and SpsA glycosyl transferase from *Bacillus subtilis*, (SPS_A, P39621). Components of various motifs are in boldface: namely, Y..D.. P..DD$^1$ (the UDPGlc-binding domain), Y..R..KAG and QTP..D (Stasinopoulos et al., 1999); FFCGS (Römling, 2002) and the extended $D^1$,$D^2$,$D^3{}_{35}$QxxRW motif (Saxena and Brown, 1997).

of the enzyme with the UDPMn$^{2+}$ at the SpsA binding site are shown in Fig. 4; the equivalent amino acids in CrdS involved with substrate-binding are numbered. The analysis revealed that D$^2$(Asp$^{208}$) sits adjacent to the distal phosphate of UDP in the binding site and coordinates with the leaving group Mn$^{2+}$ of the conjectured aspartate involved in metal binding.

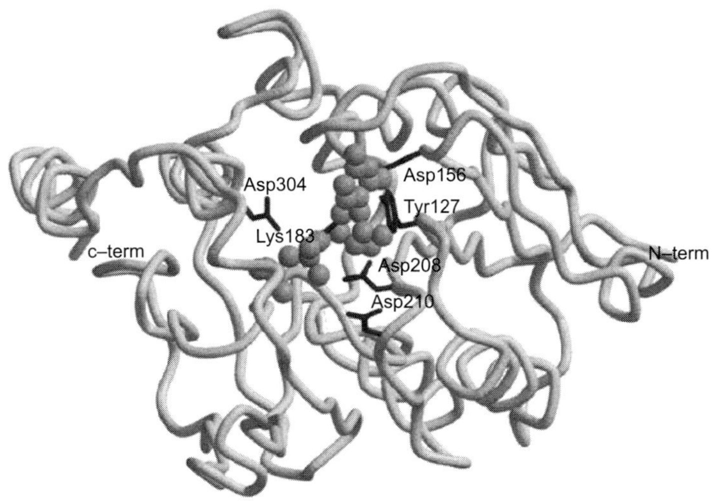


**Fig. 4: Three-dimensional modelling of the UDP-$Mn^{2+}$ binding to the CrdS protein obtained by threading the CrdS protein onto the structure of the GT2 glycosyl transferase SpsA. The conserved aspartic acid residues in CrdS equivalent to those in SpsA are indicated. (From Karnezis et al., 2003.)**

## *B.6. Initiation, Direction of Chain Growth and Specification of Linkage Position*

The initiator of curdlan chain growth is not known nor has the direction of chain growth been established. Many β-glycans are believed to elongate from the non-reducing end of the growing chain, but others have been shown to grow from the reducing end (Karnezis et al., 2000). The latter is almost invariably the case when the glycosyl lipid donor is involved, as appears to be so for cellulose synthesis in *A. tumefaciens* C58 strain A1045 (Matthysse et al., 1995). No glycosyl–lipid intermediates have been implicated in curdlan biosynthesis.

The features of the CrdS protein in its folded state that specify the insertion of (1,3)- rather than (1,4)-β-glucosidic linkages into the growing chain remain to be determined. The subtle features of the active site that allow the presentation of the appropriate hydroxyl on the growing polysaccharide chain to orient in the correct position to accept the transferred glycosyl residue presumably determines linkage type. The motifs (FFCGS and $Rx_2FLx_2PL$) found in known or putative bacterial cellulose synthases and proposed to have a role in determining (1,4)-β-linkage specificity (Römling, 2002) are barely recognisable in CrdS [$Fx_2Gx$ (Table 2) and $x_4Lx_2Px$ (residues 385–393) (Karnezis et al., 2003)]. The determinants of (1,3)-specificity await elucidation of the CrdS structure.

### B.7. Deposition of Capsular Curdlan

The process of polymerization of the curdlan chain mediated by CrdS is initiated on the cytoplasmic face of the inner membrane (Karnezis et al., 2003) where the UDPGlc substrate is available, and so is unlikely to involve the periplasmic CrdC and CrdA directly. The growing chain must traverse the inner membrane, the periplasmic space and the outer membrane before being deposited in the capsule. CrdA might assist translocation of the nascent polymer across the cytoplasmic membrane and CrdC its passage across the periplasm, consistent with their experimentally determined locations. This scenario is supported by the distinctive ability of *crdC* mutants, as opposed to *crdA* mutants, to produce some curdlan detectable by the sensitive Aniline Blue fluorochrome (McIntosh, 2004). Both classes of mutants can synthesize curdlan (they have *crdS*), but only *crdC* mutants can externalize the polymer if, as proposed, CrdA enables transport across the inner membrane. The possibility that CrdASC forms a membrane-associated, oligomeric, biosynthetic complex is suggested by the occurrence of native protein aggregates of $M_r$ ~420 and ~500 that contain CrdS (Karnezis et al., 2003).

The importance of membrane phospholipid composition in curdlan synthesis is highlighted by the reduced production of curdlan when the phosphatidyl serine synthase (*pssA*) gene responsible for the synthesis of the progenitor of membrane phosphatidyl ethanolamine is disabled (Karnezis et al., 2002). It is conjectured that the presence of phosphatidyl ethanolamine in the membrane is required for the correct folding of the nascent CrdS protein, as has been found for the *E. coli* lactose permease (Bogdanov and Dowhan, 1998), or is required for the stabilization of the membrane-associated CrdS. In the case of the membrane-bound hyaluronan synthase from *Streptococcus pyogenes* (Tlapak-Simmons et al., 1998; 1999), 16 cardiolipin molecules are needed to give a functional synthase.

Although the mechanism of externalization of curdlan is not yet clear, it is in stark contrast to the elaborate machinery that operates in the externalization of the lipid–polysaccharide outer membrane complexes of Gram-negative bacteria (Whitfield, 2006). On the other hand, it is more complex than the systems proposed for the production of hyaluronan by Gram-positive *S. pyogenes* (Heldermon et al., 2001) (see above) and the S37 polysaccharide from *Streptococcus pneumoniae* (see Section D), where the passage of their polymeric products to the bacterial surface does not appear to require ancillary proteins. The difference may lie in insolubility of curdlan chains requiring the presence of a molecule, such as CrdC, to chaperone them across the periplasm and outer membrane.

The form of the nascent curdlan chains as they emerge from the outer membrane is of interest in relation to the events following chain synthesis. Examination of 'native' (never-heated) curdlan by NMR and X-ray diffraction shows that it is composed of polymer chains that have: a high level of conformational definition at the individual chain level (narrow NMR lines), significant stabilizing features at the chain aggregate level (insolubility in alkali), and weak organization over longer distance scales (10's nm: X-ray diffraction) (Stanisich et al., 2006). These results are compatible with a mixture of single (1,3)-β-glucan chains and multiple helices, and suggest that curdlan is produced first as a single chain and then associates to form helices in the capsule.

## C. Cyclic (1,3;1,6)-β-Glucans

The cyclic (1,3;1,6)-β-glucan of *Bradyrhizobium japonicum* (de Iannino et al., 1993) is composed of a 12-membered ring composed of two blocks of three (1,3)-β-linked glucose units each separated by a block of three (1,6)-β-linked glucose units. One block of (1,3)-β-linked units contains a branched glucose at C(O)6 and the other a phosphocholine group linked at C(O)6 (see Chapter 2.1 and Figure 2.1.D). These periplasmic, cyclic (1,3;1,6)-β-glucans produced by *Bradyrhizobium japonicum*, *Azospirillum brasilense* and *Azorhizobium caulinodans* (see Chapter 2.1) have roles in adaptation of the bacteria to hypo-osmotic stress and interactions with host plants (see Chapter 4.1).

### *C.1. The* ndv *Biosynthesis Genes*

Information on the molecular genetics of production of the cyclic (1,3;1,6)-β-glucans comes from studies conducted on *B. japonicum* strain USDA 110, where a cluster of three monocistronic genes *ndvC*, *ndvD* and *ndvB* (Fig. 5) has been shown to affect cyclic glucan production (Chen et al., 2002) and a fourth, unlinked gene, *ndvA*, encodes the probable cyclic glucan transporter protein (Wiedemann and Müller, 2004). The complete genome sequence of USDA 110 has been determined (Kaneko et al., 2002) and the genes are all located on the single, large chromosome present in this strain.

The initial identification of *B. japonicum* genes required for (1,3;1,6)-β-glucan synthesis was based on the suspected common role in osmoregulation of the bradyrhizobial glucans and

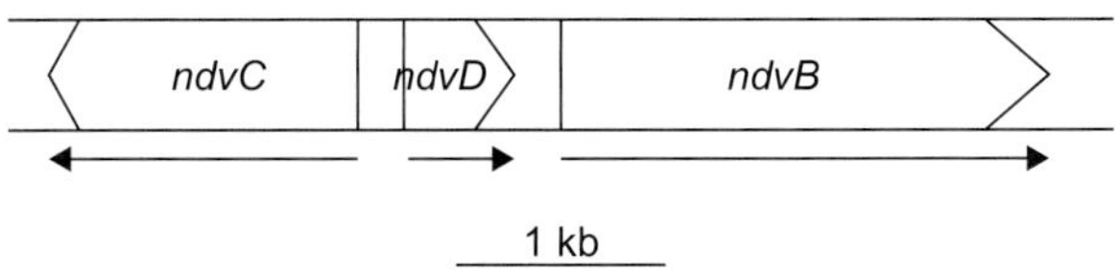


**Fig. 5: Organization of the cyclic (1,3;1,6)-β-glucan synthesis gene cluster in *Bradyrhizobium japonicum* USDA 110.**

the cyclic, periplasmic (1,2)-β-glucans produced by strains of *Sinorhizobium* (and *Agrobacterium*) and for which the (1,2)-β-glucan synthase gene, *ndvB* (and *chvB* in *Agrobacterium*), was known (Ielpi et al., 1990; Zorreguieta and Ugalde, 1986). This information led to the isolation of a 13-kb *Eco*RI-*Hin*dIII fragment from *B. japonicum* that complemented an *ndvB* mutation in *S. meliloti*, resulting in the restoration of swarming growth on low-osmolarity agar medium and production of cyclic glucan (Bhagwat et al., 1993; Bhagwat and Keister, 1995). Random transposon mutagenesis of the fragment, combined with testing for loss of complementation of the *S. meliloti ndvB* mutation, resulted in the detection of two *ndv* genes in *B. japonicum* that were transcribed in opposite directions: *ndvB*, located in a 5.2-kb *Eco*RI fragment (Bhagwat and Keister, 1995) and the upstream *ndvC*, located in a 2.0-kb *Eco*RI fragment (Bhagwat et al., 1996). Subsequent sequence analysis of the region revealed the presence of *ndv*D, which is located between *ndvC* and *ndvB* and is transcribed in the same direction as *ndvB* (Chen et al., 2002) (Fig. 5). Complementation analysis showed that the genes are each functionally independent, since clones of the individual genes complemented only the respective transposon (Tn*5*) insertion mutations.

Genetic analysis of the *B. japonicum ndv* genes involved replacement of the wild-type genes with the respective mutated gene. In the case of the *ndvB*-mutated gene (i.e. *ndvB*::Tn*5*), the resulting pleiotropic changes in the mutant (strain AB-14) were similar to those of *S. meliloti ndvB* mutants. That is, AB-14 was non-motile, displayed impaired growth in low-osmolarity medium, produced no glucans (either *in vivo* or *in vitro*) and formed ineffective nodules on soybean (Bhagwat and Keister, 1995; Dunlap et al., 1996). Despite the functional similarity between the *B. japonicum* and *S. meliloti ndvB* genes, they are not related at the sequence level, as demonstrated by the lack of DNA:DNA hybridization (Bhagwat et al., 1992; Dylan et al., 1986), and the respective gene products share homology with glycosyl transferases of family GT2 (*B. japonicum* NdvB) and family GT36 (*S. meliloti* NdvB). Significantly, complementation of the *S. meliloti ndvB* defect by the bradyrhizobial *ndvB* gene did not restore production of the cyclic (1,2)-β-glucan typical of *S. meliloti* but rather a functionally analogous cyclic (1,3)-β-linked glucan was produced (Pfeffer et al., 1996), suggesting that the bradyrhizobial *ndvB* gene encodes a (1,3)-β-glucosyl transferase.

In contrast to the pleiotropic phenotype of the *B. japonicum ndvB*::Tn*5* mutant, the *ndvC*::Tn*5* replacement mutant (strain AB-1) did not differ markedly from the wild type in terms of motility, osmoadaptation, and levels of glucan production (both intra- and extra-cellularly). Despite these features, AB-1 was symbiotically ineffective and the small, white, pseudonodules that occurred on soybean were much delayed in formation and contained no viable bacteroids (Bhagwat et al., 1996; Dunlap et al., 1996). Analysis of the glucan showed that it was structurally altered and primarily (>95%) (1,3)-β-linked, like that formed by the complemented *S. meliloti ndvB* strain (Bhagwat et al., 1996; Pfeffer et al., 1996). These findings indicated that the NdvC protein was involved in synthesis of the (1,6)-β-linkages. The 'new' glucan (termed cyclodecakis-(1,3)-β-glucosyl) was clearly not effective in supporting nodule development by *B. japonicum* on soybean, but was able to provide osmoprotection.

The *B. japonicum ndvD*::Tn*5* replacement mutant (strain RC-2), like AB-14 (*ndvB*::Tn*5*) was defective in cell motility, osmoadaptation and glucan synthesis *in vivo*, and formed small ineffective nodules (Chen et al., 2002). The bacteroid content of RC-2-nodules, although less than for the wild type (100-fold), was higher (10-fold) than for AB-14, and in other comparisons (antibiotic resistance and conjugal efficiency) RC-2 also differed phenotypically from the other two strains. The most striking feature, however, was that RC-2 continued to synthesize glucan *in vitro* at levels equivalent to the wild type and in contrast to AB-14, which produced no glucan either *in vivo* or *in vitro*.

The *ndvA* gene of *B. japonicum* that most probably encodes a specific transporter of cyclic (1,3;1,6)-β-glucans was identified during saturation transposon (Tn*KPK2*) mutagenesis of a randomly cloned 10-kb *Pst*I genomic fragment of strain 110*spc4* (Wiedemann and Müller, 2004). The gene comprised one end of the fragment and, on translation, its sequence resembled the NdvA/ChvA ATP-binding transport proteins that are required for production of the respective cyclic (1,2)-β-glucans of *S. meliloti* and *A. tumefaciens* (Breedveld and Miller, 1998). All three transporters are characterized by an $NH_2$-terminal region containing six transmembrane helices and a C'-terminal functional domain associated with ATPase activity. Unlike the situation in *S. meliloti* and *A. tumefaciens* where the transporter (*ndvA*) and synthesis (*ndvB*) genes form an operon, *B. japonicum ndvA* is separated from the *ndvC-D-B* cluster by 3.67 Mb; *ndvA* is apparently monocistronic and is flanked upstream by an unknown gene and downstream by one encoding a probable penicillin-binding protein. Replacement of the *B. japonicum ndvA* gene with a mutant allele (*ndvA*::Tn*KPK2*#17) resulted in phenotypic changes resembling those described for the *ndvB*::Tn*5* mutant, AB-14 (Bhagwat et al., 1993; Dunlap et al., 1966). That is, the *ndvA*::Tn*KPK2*#17 mutant exhibited reduced tolerance to osmotic stress and formed small ineffective nodules on soybean that had abnormal nodule ultrastructure. Although no studies were conducted that specifically demonstrated that the *ndvA*::Tn*KPK2*#17 mutant failed to produce periplasmic cyclic β-glucans, the characteristics of the mutant and features of the deduced NdvA product make this likely.

Finally, no *B. japonicum* gene(s) has been reported to be responsible for transfer of the phosphocholine substituents to the bradyrhizobial glucans. In the case of *S. meliloti*, mutation of the *cgm* (cyclic glucan modification) gene prevents addition of sn-1-phosphoglycerol substituents to the cyclic (1,2)-β-glucan backbone, a reaction that occurs in the periplasm (Breedveld and Miller, 1994; Wang et al., 1999).

A study on *Azospirillum brasilense* Sp7 that may be relevant to cyclic (1,3;1,6)-β-glucan production in this species concerns the gene, *chvB* (chromosomal virulence). This gene was identified in *A. brasilense* genomic fragments that were able to complement the defect in an *A. tumefaciens chvB* (1,2)-β-glucan synthase mutant with respect to tumour formation in a leaf-disc assay (Raina et al., 1995). One of the *A. brasilense* fragments contained the 5' region of *chvB*, and its encoded product (480 amino acid residues) shared significant similarity (60% identity) with the $NH_2$-terminal portion of the *S. meliloti* NdvB protein. These observations implicated the *chvB* locus in adsorption of *A. brasilense* to plant roots, given the established roles of the rhizobial *ndvB* and *chvB* genes in plant nodule and tumor formation, respectively (Breedveld and Miller, 1998). Whether the *A. brasilense chvB* gene is involved in (1,3;1,6)-β-glucan production was not examined, but this may be the case since *A. brasilense* strains do not synthesize (1,2)-β-glucans (Altabe et al., 1990).

## C.2. Membrane Topology and Enzymology

### C.2.1. The NdvB protein

*In silico* topological modelling (Fig. 6) of the NdvB protein (889 amino acids) shows a polytopic protein with seven to eight possible transmembrane segments and two large cytoplasmic loops, one (202 amino acids) situated in the $NH_2$-terminal portion and the other (299 amino acids) located in the central portion. The first hydrophilic loop shares features of glycosyl transferases belonging to the GH17 family, which includes a number of (1,3)-β-glucan endohydrolases (http://afmb.cnrs-mrs.fr/CAZY/), whereas the second loop shares features of glycosyl transferases belonging to the GT2 family. Thus, structurally, NdvB (= blr4614) is a two-domain enzyme with a 'polymerizing' or 'repetitive' GT2 module appended to a GH17 module. If NdvB is part of the 90-kDa gluco-protein intermediate detected by de Iannino and Ugalde (1993) (see below), then it appears that the cyclic (1,3;1,6)-β-glucan product of NdvB is synthesized from UDPGlc by the concerted action of the GT2-like domain with the GH17 domain. Endo-glucanases are known to have transglycosylating ability under some circumstances (see Chapter 3.1). Both Miller and Gore (1992) and Rolin et al. (1992) found that cyclic (1,3;1,6)-β-glucans synthesized *in vitro* have more (1,3)-linkages than those *in vivo* so it is possible to modulate this feature. The NdvB-mediated polymerization events are, however, restricted to the formation of a (1,3)-β-linked product, since *Bradyrhizobium* mutants

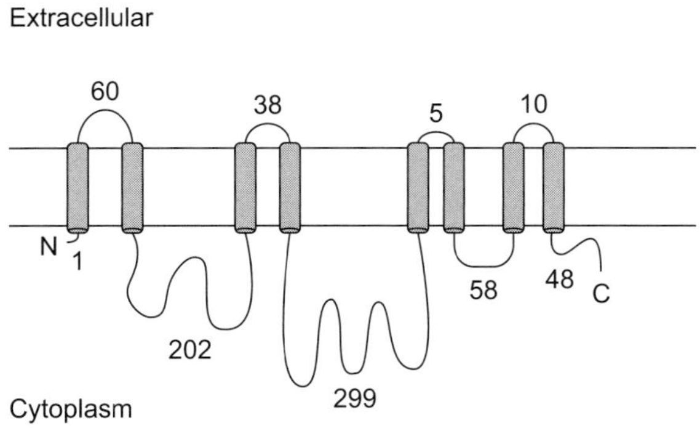


**Fig. 6: Predicted membrane topology of the *Bradyrhizobium japonicum* USDA 110 NdvB protein showing two hydrophilic regions separated by several transmembrane segments. The first hydrophilic region is predicted to contain a family GH17 glycosyl transferase domain and the second a GT2-like domain.**

(*ndvC*::Tn*5*) that have NdvB but lack NdvC produce an all (1,3)-β-cyclic glucan (Pfeffer et al., 1996).

The mechanism of cyclization is not known, but for analogous cyclic (1,2)-β-glucans from rhizobia it has been proposed that non-repetitive glucosylation of a high molecular mass membrane protein produces an oligoglucan whose terminal non-reducing residue is able to accept the 'reducing' glucose of the growing chain on the enzyme to release the cyclic molecule (Williamson et al., 1992). However, the NdvB (and ChvB) proteins responsible both for synthesis and cyclization of the cyclic (1,2)-β-glucans (Breedveld and Miller, 1998) are larger (~319 kDa) and quite different proteins from the NdvB and NdvC enzymes involved in the synthesis of the cyclic (1,3;1,6)-β-glucans.

Biochemical studies in *B. japonicum* (Bhagwat and Keister, 1992; Cohen and Miller, 1991; de Iannino and Ugalde, 1993) showed that incorporation of glucose from UDP[$^{14}$C]Glc into cyclic (1,3;1,6)-β-glucans was supported by inner membranes but not outer membranes from strains USDA 110, BR8404 and BR4406. $Mg^{2+}$ or $Mn^{2+}$ (10 mM) stimulated *in vitro* synthesis but could be substituted less efficiently by $Co^{2+}$ (31%) or $Ca^{2+}$ (16%); EDTA (10 mM) abolished incorporation. The addition of GTP (10 mM) had no effect on glucan production. The labelled product from USDA 110 was associated with a compound that migrated at ~90 kDa during gel electrophoresis and on treatment with laminarinase released the label, indicating the presence of a laminarin-type glucan. When inner membranes were treated with trypsin, a labelled 80-kDa fragment was released indicating that the 90-kDa compound was a glucoprotein. Inner membranes from strains BR8404 and BR4406 formed a 100-kDa protein.

Altabe et al. (1994) studied *in vitro* synthesis of cyclic (1,3;1,6)-β-glucans by *Azospirillum brasilense* CdRif. Inner membranes incorporated 10-fold more labelled glucose from UDPGlc than did outer membranes or permeabilized cells, and no incorporation occurred from ADPGlc as substrate donor. Divalent cations were required for incorporation which was abolished in the presence of EDTA (10 mM) and restored by addition of $Mg^{2+}$, $Mn^{2+}$ or $Ca^{2+}$. Highest activity was obtained with (40 mM) $MgCl_2$. β-Glucose disaccharides at 20 mM stimulated incorporation: gentiobiose, 40%; laminaribiose, 27%; sophorose, 18%. Nigerose was ineffective. The optimal temperature for formation of soluble glucan was 30°C. Interestingly, glucan synthesis was subject to osmoregulation since membrane preparations from cells grown in the presence of 250 mM NaCl or 500 mM mannitol showed 85% and 50% reduction, respectively, in synthesis compared to those from Luria Broth-grown cells.

### *C.2.2. The NdvC and NdvD proteins*

Sequence analysis of the deduced NdvC protein (61.6 kDa) revealed a polytopic protein with eight potential transmembrane segments. The first and second segments (residues 22–39 and 339–357, respectively) are separated by a large cytoplasmic loop (298 amino acids); the second and remaining segments are evenly distributed in the C'-terminal half of the protein and are connected by only very short loops on the cytoplasmic or periplasmic face of the membrane. The large $NH_2$-terminal cytoplasmic loop shares features of GH17-family enzymes, which include glucanosyl transferases from *Candida albicans* and (1,3)-β-endo-glucanases from *Saccharomyces cerevisiae*. This is consistent with a role for NdvC in β-glucan metabolism and with experimental findings (Bhagwat et al., 1996; Pfeffer et al., 1996) that suggest that *ndvC* encodes a (1,6)-β-glucosyl transferase.

The deduced NdvD protein (20.6 kDa) has a single $NH_2$-terminal transmembrane domain (residues 16–36) and a probable $N_{in}$–$C_{out}$ disposition. Similar proteins of unknown function occur in several *Bradyrhizobium* spp. and also in species of two other α-proteobacteria, *Nitrobacter* and *Rhodopseudomonas* (61–68% identity; 66–75% similarity). NdvD is essential for glucan synthesis *in vivo* but not *in vitro*, suggesting that it may modify or direct the synthesis of the glucans by NdvB and/or NdvC, or have a role in glucan transport to the periplasm (Chen et al., 2002). If NdvD is involved in transport, the process in *Bradyrhizobium* and its relationship to synthesis is different from that of the NdvA transport protein in *S. meliloti* since *ndvA* mutants accumulate (1,2)-β-glucan intracellularly (Breedveld and Miller, 1994). This is not the case in *B. japonicum* where cyclic glucans were not detected *in vivo* in an *ndvD*::Tn*5* mutant (strain RC-2). Chen et al. (2002) suggested that NdvD might modify or direct *in vivo* synthesis of the bradyrhizobial glucan by serving as a template or effector for the NdvB and/or NdvC proteins during β-glucan synthesis. At present there is no explanation for the observation that the *ndvD* mutant synthesized glucan *in vitro* but not *in vivo*.

## D. Pneumococcal Type 37 (1,3;1,2)-β-Glucan

### D.1. Molecular Biology

Serotype 37 strains of *Streptococcus pneumoniae* uniquely produce a capsular polysaccharide composed exclusively of glucose in which every (1,3)-linked glucosyl unit in the backbone is substituted by a (1,2)-β-linked unit (see Chapter 2.1 and Figure 2.1.D). The capsule is characteristically very thick and is encoded by a single gene, *tts*, separated from the typical *cap* locus responsible for capsular formation in all of the other numerous pneumococcal serotypes. The chromosomally located *cap* gene cluster is bounded by *dexB* (encoding a dextran synthase) and *aliA* (encoding an oligopeptide permease) (Yother, 1999), and in type 37 strains it is almost identical to that of serotype 33F except that it is non-functional (cryptic) due to many deletions and point mutations (Llull et al., 1999).

The DNA region containing *tts* was initially isolated as a 7.3-kb *Pst*I genomic fragment capable of converting a non-capsulated strain of *S. pneumoniae* to the S37 capsular type (Llull et al., 1999). Analysis of the sequence of the *Pst*I fragment revealed *tts* which had no significant similarity to other genes in the *S. pneumoniae* database. The *tts* gene is situated downstream of *gmpA* (encoding a putative phosphoglyceromutase) and adjacent to a defective transposase gene ($trp_{1167}$). Insertional inactivation of *tts* resulted in loss of capsule formation, as deduced by the failure of type 37 antiserum to agglutinate the mutant; reintroduction of a cloned *tts* gene restored the S37 serotype. These findings indicated that the encoded protein is the type 37-specific polysaccharide synthase, a conclusion that was supported by analysis of the deduced Tts protein sequence (509 amino acids; 58.9 kDa) which exhibited similarities with plant and bacterial cellulose synthases and other GT2 glycosyl transferases (see Table 2).

The Tts synthase also has dual specificity, synthesizing both (1,3)- and (1,2)-linkages in the side-chain-branched polymer. This feature was demonstrated when *tts* was introduced on a plasmid into heterologous Gram-positive bacteria (i.e. *Streptococcus oralis*, *Streptococcus gordonii* and *Bacillus subtilis*) or when a single copy of *tts* was introduced into the chromosome of *S. oralis*. In each instance the bacterial cells became agglutinable in the presence of type 37-specific antiserum (Llull et al., 2001). Such dual specificity is a feature shared with the GT2 synthase producing the type 3 pneumococcal Cap3B polysaccharide (Arrecubieta et al., 1996), hyaluronan synthase (HasA) of *S. pyogenes* (De Angelis and Weigel, 1994;

Tlapak-Simmons et al., 1998, 1999), and the synthases from *P. multocida* for chondroitin (De Angelis and Padgett-McCue, 2000) and heparosan (De Angelis and White, 2002). A further example is the bifunctional GT2 transferase for the *E. coli* K5 capsular KfiC polysaccharide, identical to *N*-acetylheparosan, that catalyses the sequential addition of (1,4)-β-glucuronyl and (1,4 )-α-*N*-acetylglucosaminyl units to the non-reducing end of the growing chain (Griffiths et al., 1998). However, for this synthesis it appears that two enzymes, KfiA and KfiC, act in tandem to form the disaccharide repeat. Another deduced protein in the operon (KfiB) is suggested to stabilize the enzymic complex during elongation *in vivo* but is not catalytic (Hodson et al., 2000).

### D.2. Membrane Topology

Tts is predicted to be an integral membrane protein (509 amino acids; 58.9 kDa) with six transmembrane segments, the first of which may be a cleavable signal sequence (Fig. 7). Consistent with the predicted membrane location was the incorporation of label from UDP[$^{14}$C]Glc into a macromolecular product by the membrane fraction, but not by the soluble fraction, of an *S. pneumoniae* strain (M24) with a plasmid-borne *tts* gene (Llull et al., 2001). Moreover, gel electrophoresis of the membrane fraction revealed the presence of a protein of $M_r$ ~50 kDa which was not present in strains lacking the *tts* plasmid. The large loop (286 amino acids) separating transmembrane segments 2 and 3 is relatively hydrophilic and is predicted to reside in the cytoplasm. It contains conserved sequences and structural features

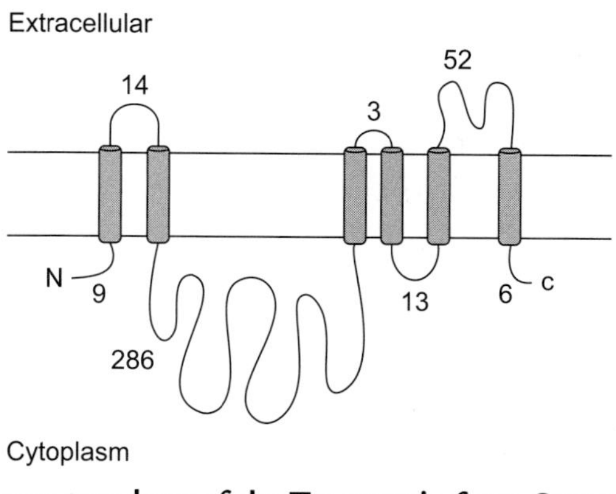


**Fig. 7: Predicted membrane topology of the Tts protein from *Streptococcus pneumoniae* Type 37 showing six transmembrane segments and a large hydrophilic region containing a GT2-like catalytic domain(s).**

of repetitive GT2 enzymes: the D,D,$D_{35}$QxxRW motif is recognisable (Llull et al., 1999) (see Table 2); however, in Tts the characteristic QxxRW motif found in repetitive, inverting GT2 enzymes is replaced by an RxxKW sequence. Other motifs, $Yx_2Rx_6$KAG and $QTPx_6D$, found in cellulose synthases and CrdS (Stasinopoulos et al., 1999) are not strongly conserved in Tts.

The observation that *tts* alone, when introduced into different bacteria, is sufficient to produce the glucan makes Tts the first described inverting GT2 enzyme able to synthesize a side-chain-branched polysaccharide. Future mutational studies may determine whether synthesis of the (1,3)-β-glucan backbone precedes the addition of the (1,2)-linked glucosyl residues or whether formation of (1,3)- and (1,2)-linkages takes place simultaneously with polysaccharide chain extension. Production of the glucan in different bacteria also suggests that the nascent glucan does not use specific transporters to cross the membrane. Llull et al. (2001) speculate that unspecific transporters are used, that several Tts monomers make a pore or there is an interaction of membrane phospholipids with the synthase as described for the *S. pyogenes* hyaluronan synthase (Tlapak-Simmons et al., 1998, 1999).

### D.3. Enzymology

Cell-free membrane preparations support the synthesis of the type 37 polysaccharide from UDP[$^{14}$C]-Glc (Llull et al., 2001). Tts activity is stimulated in the presence of 10 mM $MgCl_2$ (or $MnCl_2$) but only slightly by $Ca^{2+}$ at low concentration in the absence of $Mg^{2+}$; 10 mM EDTA completely inhibited the reaction. Tts activity is optimal between pH 6.8 and 7.5 and is maximal at 30°C in the presence of substrate UDPGlc. When membranes were preincubated in the presence of UTP, UDP, UMP and TMP there was strong inhibition. Uronic acids did not affect incorporation of Glc. The nucleotide sugars UDPGal, UDPXyl and UDPMan gave complete inhibition whereas CDPGlc, GDPGlc, GMP, CMP were only moderately inhibitory. Both ionic and anionic detergents are powerful inhibitors of Tts, suggesting a close association between Tts and the cell membrane. Bacitracin, an inhibitor of undecaprenyl pyrophosphate-based polysaccharide synthesis (Siewert and Strominger, 1967), is ineffective, suggesting that such lipid intermediates are not involved in the biosynthesis of the type 37 glucan.

## Conclusion

The production of (1,3)-β-glucans by eubacteria is restricted. Only three types are known, exemplified by the linear, agrobacterial (1,3)-β-glucan, curdlan; the bradyrhizobial cyclic (1,3;1,6)-β-glucans; and the streptococcal side-chain-branched (1,3;1,2)-β-glucan. Notwithstanding, there is significant information available about the enzymology and molecular biology of their formation. In each case, the biosynthetic genes involved are related to one another and to a large group of genes involved in the biosynthesis of cellulose, chitin and hyaluronan. In the case of the synthesis of the agrobacterial and bradyrhizobial products, genes additional to those encoding glucosyl transferases are involved. The functional roles of curdlan and the S37 polysaccharide in capsules, and the cyclic (1,3;1,6)-β-glucans as osmoregulators are discussed in Chapter 4.1.

## Acknowledgements

We thank Sanja Aracic, Ferdiye Taner and Danielle Tromp for critically reading the manuscript and Dr Fung Lay and Sanja Aracic for preparation of Figures. Work in our laboratories was supported, in part, by Australian Research Council Grants (AO9925079, LX 0211339).

## References

Aird, E. L. H., Brightwell, G., Jones, M. A., & Johnston, A. W. B. (1991). Identification of the *exo* loci required for exopolysaccharide synthesis in *Agrobacterium radiobacter* NCIB 11883. *Journal of General Microbiology*, *137*, 2287–2297.

Altabe, S., de Iannino, N., de Mendoza, D., & Ugalde, R. A. (1990). Expression of the *Agrobacterium tumefaciens chvB* virulence region in *Azospirillum* spp. *Journal of Bacteriology*, *172*, 2563–2567.

Altabe, S., de Iannino, N., de Mendoza, D., & Ugalde, R. A. (1994). New osmoregulated β (1–3), β (1–6) glucosyltransferase(s) in *Azospirillum brasilense*. *Journal of Bacteriology*, *196*, 4890–4898.

Ampe, F., Kiss, E., Sadourdy, F., & Batut, J. (2003). Transcriptome analysis of *Sinorhizobium meliloti* during symbiosis. *Genome Biology*, *4*, R15 (Open Access).

Anguillesi, A. (2003). *Molecular biology of curdlan biosynthesis by Agrobacterium sp. ATCC31749 with an emphasis on crdR*. MSc thesis. Australia: La Trobe University.

Aracic, S. (2010). *Studies on the molecular genetics of curdlan [(1,3)-β-glucan] production in Agrobacterium sp. strain LTU50 and the role of this extracellular polysaccharide*. PhD thesis. Australia: La Trobe University.

Aracic, A., Matthysse, A. G., Stone, B. A., & Stanisich, V. A. (2008). The biological role of agrobacterial (1,3)-β-glucan and regulation of its production. *Annual scientific meeting of the Australian Society for Microbiology*, Melbourne, 6–10th July, Abstract P04.05.

Arrecubieta, C., López, R., & García, E. (1996). Type 3-specific synthase of *Streptococcus pneumoniae* (Cap3B) directs type 3 polysaccharide biosynthesis in *Escherichia coli* and in pneumococcal strains of different serotypes. *Journal of Experimental Medicine*, *184*, 449–455.

Becker, A., & Pühler, A. (1998). Production of exopolysaccharides. In H. P. Spaink, A. Kondorosi, & P. J. J. Hooykaas (Eds.), *The rhizobiaceae: Molecular biology of model plant-associated bacteria* (pp. 97–118). Kluwer Academic Publishers.

Bhagwat, A. A., & Keister, D. L. (1992). Synthesis of β-glucans by *Bradyrhizobium japonicum* and *Rhizobium fredii*. *Canadian Journal of Microbiology*, *38*, 510–514.

Bhagwat, A. A., Tully, R. E., & Keister, D. L. (1992). Isolation and characterization of an *ndvB* locus from *Rhizobium fredii*. *Molecular Microbiology*, *6*, 2159–2165.

Bhagwat, A. A., Tully, R. E., & Keister, D. L. (1993). Identification and cloning of a cyclic β-(1–3), β-(1–6)-D-glucan synthesis locus from *Bradyrhizobium japonicum*. *FEMS Microbiology Letters*, *114*, 139–144.

Bhagwat, A. A., & Keister, D. L. (1995). Site-directed mutagenesis of the β-(1–3), β-(1–6)-D-glucan synthesis locus of *Bradyrhizobium japonicum*. *Molecular Plant-Microbe Interactions*, *3*, 366–370.

Bhagwat, A. A., Gross, K. C., Tully, R. E., & Keister, D. L. (1996). β-glucan synthesis in *Bradyrhizobium japonicum*: Characterization of a new locus (*ndvC*) influencing β-(1–6) linkages. *Journal of Bacteriology*, *178*, 4635–4642.

Bishop, A. C., Xu, J., Johnson, R. C., Schimmel, P., & de Crécy-Lagard, V. (2002). Identification of the tRNA-dihydrouridine synthase family. *Journal of Biological Chemistry*, *277*, 25090–25095.

Bogdanov, M., & Dowhan, W. (1998). Phospholipid-assisted protein folding: Phosphatidylethanolamine is required at a late step of the conformational maturation of the polytopic membrane protein lactose permease. *EMBO Journal*, *17*, 5255–5264.

Braeken, Y. R., Moris, M., Daniels, R., Vanderleyden, J., & Michels, J. (2006). New horizons for (p)ppGpp in bacterial and plant physiology. *Trends in Microbiology*, *14*, 45–54.

Breedveld, M. W., & Miller, K. J. (1994). Cyclic β-glucans of members of the family Rhizobiaceae. *Microbiology Reviews*, *58*, 145–161.

Breedveld, M. W., & Miller, K. J. (1998). Cell-surface β-glucans. In H. P. Spaink, A. Kondorosi, & P. J. J. Hooykaas (Eds.), *The rhizobiaceae: Molecular biology of model plant-associated bacteria* (pp. 81–96). Kluwer Academic Publishers.

Changelosi, G. A., Hung, L., Puvanesarajah, M. V., Stacey, G., Ozga, D. A., Leigh, J. A., & Nester, E. W. (1987). Common loci for *Agrobacterium tumefaciens* and their roles in plant interactions. *Journal of Bacteriology*, *169*, 2086–2091.

Charnock, S. J., & Davies, G. J. (1999). Structure of the nucleotide-diphospho-sugar transferase, SpsA from *Bacillus subtilis*, in native and nucleotide-complexed forms. *Biochemistry*, *38*, 6380–6385.

Chatterji, D., & Ojha, A. K. (2001). Revisiting the stringent response, ppGpp and starvation signalling. *Current Opinion in Microbiology*, *4*, 160–165.

Chen, R., Bhagwat, A. A., Yaklich, R., & Keister, D. L. (2002). Characterisation of *ndvD*, the third gene involved in the synthesis of cyclic-β-(1–3),(1–6)-D-glucans in *Bradyrhizobium japonicum*. *Canadian Journal of Microbiology*, *48*, 1008–1016.

Cohen, J. L., & Miller, K. J. (1991). A novel membrane-bound glucosyltransferase from *Bradyrhizobium japonicum*. *Journal of Bacteriology*, *173*, 4271–4276.

Coutinho, P. M., & Henrissat, B. (1999). Carbohydrate-active enzymes: An integrated database approach. In H. J. Gilbert, G. Davies, B. Henrissat, & B. Svensson (Eds.), *Recent advances in carbohydrate bioengineering* (pp. 3–12). Cambridge: The Royal Society of Chemistry.

Dalluge, J. J., Hashizume, T., Sopchik, A. E., McClosky, J. A., & Davis, D. R. (1996). Conformational flexibility in RNA: The role of dihydrouridine. *Nucleic Acids Research*, *24*, 1073–1079.

De Angelis, P. L., & Weigel, P. H. (1994). Immunochemical confirmation of the primary structure of streptococcal hyaluronan synthase and synthesis of high molecular weight product by the recombinant enzyme. *Biochemistry*, *33*, 9033–9039.

De Angelis, P. L., & Padgett-McCue, A. J. (2000). Identification and molecular cloning of a chondroitin synthase from *Pasteurella multocida* Type F. *Journal of Biological Chemistry*, *275*, 24124–24129.

De Angelis, P. L., & White, C. L. (2002). Identification and molecular cloning of a heparosan synthase from *Pasteurella multocida* type D. *Journal of Biological Chemistry*, *277*, 7213–7290.

de Iannino, N. I., & Ugalde, R. A. (1993). Biosynthesis of cyclic β-(1–3),β-(1–6) glucan in *Bradyrhizobium* spp. *Archives of Microbiology*, *159*, 30–38.

Dilworth, M. J., Reeve, W. G., Tiwari, R. P., Vivas-Marfisi, A. I., Fenner, B. J., & Glenn, A. R. (2002). Understanding acid tolerance in root nodule bacteria. In T. L. Finan, M. R. O'Brian, D. B. Layzell, J. K. Vessey, & W. Newton (Eds.), *Nitrogen fixation: Global perspectives* (pp. 204–208). Proceedings of the 13th international congress on nitrogen fixation. Hamilton, Canada: CABI Publishing.

Dombrecht, B., Marchal, K., Vanderleyden, J., & Michelis, J. (2002). Prediction and overview of the RpoN-regulon in closely related species of the Rhizobiales. *Genome Biology*, *3*, 1–11.

Dunlap, J., Minami, E., Bhagwat, A. A., Keister, D. L., & Stacey, G. (1966). Nodule development induced by mutants of *Bradyrhizobium japonicum* defective in cyclic β-glucan synthesis. *Molecular Plant-Microbe Interactions*, *9*, 546–555.

Dylan, T., Ielpi, L., Stanfield, S., Kashyap, L., Douglas, C., Yanofsky, M., Nester, E. W., Helinski, D. R., & Ditta, G. (1986). *Rhizobium meliloti* genes required for nodule development are related to chromosomal virulence genes in *Agrobacterium tumefaciens*. *Proceedings of the National Academy of Sciences of the United States of America*, *83*, 4403–4407.

Evans, N. A., Hoyne, P. A., & Stone, B. A. (1984). Characteristics and specificity of the interaction of a fluorochrome from aniline blue (Sirofluor) with polysaccharides. *Carbohydrate Polymers*, *4*, 215–230.

Goodner, B., Hinkle, G., Gattung, S., Miller, N., Blanchard, M., Qurollo, B., Goldman, B. S., Cao, Y., Askenazi, M., Halling, C., Mullin, L., Houmiel, K., Gordon, J., Vaudin, M., Iartchouk, O., Epp, A., Liu, F., Wollam, C., Allinger, M., Doughty, D., Scott, C., Lappas, C., Markelz, B., Flanagan, C., Crowell, C., Gurson, J., Lomo, C., Sear, C., Strub, G., Cielo, C., & Slater, S. (2001). Genome sequence of the plant pathogen and biotechnology agent *Agrobacterium tumefaciens* C58. *Science*, *294*, 2323–2328.

Griffiths, G., Cook, N. J., Gottfridson, E., Lind, T., Lidholt, K., & Roberts, I. S. (1998). Characterisation of the glycosyltransferase enzyme from the *Escherichia coli* K5 capsule gene cluster and identification and characterization of the glucuronyl active site. *Journal of Biological Chemistry*, *273*, 11752–11757.

Heldermon, C., De Angelis, P. L., & Weigel, P. H. (2001). Topological organization of the hyaluronan synthase from *Streptococcus pyogenes*. *Journal of Biological Chemistry*, *276*, 2037–2046.

Hisamatsu, M., Ott, I., Amemura, A., Harada, T., Nakanishi, I., & Kimura, K. (1977). Change in ability of *Agrobacterium* to produce water-soluble and water-insoluble β-glucans. *Journal of General Microbiology*, *103*, 375–379.

Hodson, N., Griffiths, G., Cook, N., Pourhossein, M., Gottfridson, E., Lind, T., Lidholt, K., & Roberts, I. S. (2000). Identification that KfiA, a protein essential for the biosynthesis of the *Escherichia coli* K5 capsular polysaccharide, is an α-UDP-GlcNAc glycosyltransferase. The formation of a membrane-associated K5 biosynthetic complex requires KfiA, KfiB and KfiC. *Journal of Biological Chemistry*, *275*, 27311–27315.

Ielpi, L., Dylan, T., Ditta, G. S., Helinski, D. S., & Stanfield, S. W. (1990). The *ndvB* locus of *Rhizobium meliloti* encodes a 319-kDa protein involved in the production of β-(1–2)-glucan. *Journal of Biological Chemistry*, *265*, 2843–2851.

Ishida, M. L., Assumpcao, M. C., Machado, H. B., Benelli, E. M., Souza, E. M., & Pedrosa, F. O. (2002). Identification and characterisation of the two-component NtrY/NtrX regulatory system in *Azospirillum brasilense*. *Brazilian Journal of Medical and Biological Research*, *3*, 651–661.

Jumas-Bilak, E., Michaux-Charachon, S., Bourg, G., Ramuz, M., & Allardet-Servent, A. (1998). Unconventional genomic organisation in the alpha subgroup of the Proteobacteria. *Journal of Bacteriology*, *180*, 2749–2755.

Kaneko, T., Nakamura, Y., Sato, S., Minamisawa, K., Uchiumi, T., Sasmoto, S., Watanabe, A., Idesawa, K., Iriguchi, M., Kawashima, K., Kohara, M., Matsumoto, M., Shimpo, S., Tsuruoka, H., Wada, T., Yamada, M., & Tabata, S. (2002). Complete genomic sequence of nitrogen-fixing symbiotic bacterium *Bradyrhizobium japonicum* USDA 110. *DNA Research*, *9*, 189–197.

Karnezis, T., McIntosh, M., Wardak, A. Z., Stanisich, V. A., & Stone, B. A. (2000). The biosynthesis of β-glycans. *Trends in Glycoscience Glycotechnology, 12*, 211–227.

Karnezis, T., Fisher, H. C., Neumann, G. M., Stone, B. A., & Stanisich, V. A. (2002). Cloning and characterization of the phosphatidylserine synthase gene of *Agrobacterium* sp. strain ATCC 31749 and effect of its inactivation on the production of high-molecular-mass (1-3)-β-D-glucan (curdlan). *Journal of Bacteriology, 184*, 4114–4123.

Karnezis, T., Epa, V. C., Stone, B. A., & Stanisich, V. A. (2003). Topological characterisation of an inner membrane (1–3)-β-D-glucan (curdlan) synthase from *Agrobacterium* sp. strain ATCC31749. *Glycobiology, 13*, 693–706.

Klein, A. H., Shulla, A., Reimann, S. A., Keating, D. H., & Wolfe, A. J. (2007). The intracellular concentration of acetyl phosphate in *Escherichia coli* is sufficient for direct phosphorylation of two-component response regulators. *Journal of Bacteriology, 189*, 5574–5581.

Lee, I.-Y. (2002). Curdlan. In E. J. Vandamme, S. De Baets, & A. Steinbuchel (Eds.), *Polysaccharides I: Polysaccharides from prokaryotes* Vol. 5. (pp. 135–158). Biopolymers. Wiley.

Leech, A. J., Sprinkle, A., Wood, L., Wozniak, D. J., & Ohman, D. E. (2008). The NtrC family regulator AlgB, which controls alginate biosynthesis in mucoid *Pseudomonas aeruginosa*, binds directly to the *algD* promoter. *Journal of Bacteriology, 190*, 581–589.

Li, J., Burton, R. A., Harvey, A. J., Hrmova, M., Wardak, A. Z., Stone, B. A., & Fincher, G. B. (2003). Biochemical evidence linking a putative callose synthase gene with (1–3)-β-D-glucan biosynthesis in barley. *Plant Molecular Biology, 53*, 213–225.

Lin, J. T., & Stewart, V. (1998). Nitrate assimilation by bacteria. *Advances in Microbial Physiology, 38*, 1–30.

Llull, D., Muñoz, R., López, R., & García, E. (1999). A single gene (*tts*) located outside the *cap* locus directs the formation of *Streptococcus pneumoniae* Type 37 capsular polysaccharide: Type 37 pneumococci are natural, genetically binary strains. *Journal of Experimental Medicine, 190*, 241–251.

Llull, D., García, E., & López, R. (2001). Tts, a processive β-glucosyltransferase of *Streptococcus pneumoniae*, directs the synthesis of the branched type 37 capsular polysaccharide in pneumococcus and other Gram-positive species. *Journal of Biological Chemistry, 276*, 21053–21061.

Ma, S., Selvaraj, U., Ohman, D. E., Quarless, R., Hassett, D. J., & Wozniak, D. J. (1998). Phosphorylation-independent activity of the response regulators AlgB and AlgR in promoting alginate biosynthesis in mucoid *Pseudomonas aeruginosa. Journal of Bacteriology, 180*, 956–968.

Machado, H. B., Yates, M. G., Funayama, S., Rigo, L. U., Steffens, M. B., Souza, E. M., & Pedrosa, F. O. (1995). The *ntrBC* genes of *Azospirillum brasilense* are part of a *nifR3*-like-*ntrB*-*ntrC* operon and are negatively regulated. *Canadian Journal of Microbiology, 41*, 674–684.

Magasanik, B. (1996). Regulation of nitrogen utilization. In F. C. Neidhardt, R. Curtiss III, J. L. Ingraham, E. C. C. Lin, K. B. Low, B. Magasanik, W. S. Reznikoff, M. Riley, M. Schaechter, & H. E. Umbarger (Eds.), *Escherichia coli* and *Salmonella: Cellular and molecular biology* (2nd ed., pp.1344–1356). Washington, DC: ASM Press.

Matthysse, A. G., Thomas, D. L., & White, A. R. (1995). Mechanism of cellulose synthesis in *Agrobacterium tumefaciens*. *Journal of Bacteriology*, *177*, 1076–1081.

McCleary, W. R., & Stock, J. B. (1994). Acetyl phosphate and the activation of two-component response regulators. *Journal of Biological Chemistry*, *269*, 31567–31572.

McIntosh, M. (2004). *An investigation of the production of curdlan, a (1–3)-β-glucan, by an Agrobacterium sp.* PhD thesis. Australia: La Trobe University.

McIntosh, M., Stone, B. A., & Stanisich, V. A. (2005). Curdlan and other bacterial (1-3)-β-D-glucans. *Applied Microbiology and Biotechnology*, *68*, 163–173.

Miller, J. K., & Gore, R. S. (1992). Cyclic beta-1,6–1,3 glucans of *Bradyrhizobium*: Functional analogs of the cyclic beta-1,2-glucans of *Rhizobium*? *Current Microbiology*, *24*, 101–104.

Morett, E., & Bork, P. (1998). Evolution of new protein function: Recombination enhancer Fis originated by horizontal gene transfer from the transcriptional regulator NtrC. *FEBS Letters*, *433*, 108–112.

Nakanishi, I., Kimura, K., Suzuki, T., Ishikawa, M., Banno, I., Sakane, T., & Harada, T. (1976). Demonstration of curdlan-type polysaccharide and some other β-1,3-glucans in microorganisms with aniline blue. *The Journal of General and Applied Microbiology*, *22*, 1–11.

Patriarca, E. J., Riccio, A., Tatè, R., Colonna-Romano, S., Isccarino, M., & Defez, R. (1993). The *ntrBC* genes of *Rhizobium leguminosarum* are part of a complex operon subject to negative regulation. *Molecular Microbiology*, *9*, 569–577.

Patriarca, E. J., Tatè, R., & Iaccarino, M. (2002). Key role of bacterial $NH_4^+$ metabolism in *Rhizobium*-plant symbiosis. *Microbiology and Molecular Biology Reviews*, *66*, 203–222.

Pawlowski, K., Klosse, U., & de Bruijn, F. J. (1991). Characterisation of a novel *Azorhizobium caulinodans* ORS571 two-component regulatory system, NtrY/NtrX, involved in nitrogen fixation and metabolism. *Molecular and General Genetics*, *231*, 124–138.

Pfeffer, P. E., Osman, S. F., Hotchkiss, A., Bhagwat, A. A., Keister, D. L., & Valentine, K. M. (1996). Cyclolaminarinose. A new biologically active β-(1-3) cyclic glucan. *Carbohydrate Research*, *296*, 23–37.

Raina, S., Raina, R., Venkatesh, T. V., & Das, H. K. (1995). Isolation and characterization of a locus from *Azospirillum brasilense* SP7 that complements the tumorigenic defect of *Agrobacterium tumefaciens chvB* mutant. *Molecular Plant-Microbe Interactions*, *8*, 322–326.

Reeve, W. G., Tiwari, R. P., Wong, C. M., Dilworth, M. J., & Glenn, A. R. (1998). The transcriptional regulator gene *phrR* in *Sinorhizobium meliloti* WSM419 is regulated by low pH and other stresses. *Microbiology, 144*, 3335–3342.

Rolin, D. B., Pfeffer, P. E., Osman, S. F., Szwergold, B. S., Kappler, F., & Benesi, A. (1992). Structural studies of a phosphocholine substituted β-(1-3);(1-6) macrocyclic glucan from *Bradyrhizobium japonicum* USDA 110. *Biochimica et Biophysica Acta, 1116*, 215–225.

Römling, U. (2002). Molecular biology of cellulose production in bacteria. *Research in Microbiology, 153*, 205–212.

Saxena, I. M., & Brown, R. M. (1997). Identification of cellulose synthase(s) in higher plants: Sequence analysis of processive β-glycosyl transferases with the common motif "D,D,D35Q(RQ)XRW". *Cellulose, 4*, 33–49.

Siewert, G., & Strominger, J. L. (1967). Bacitracin: An inhibitor of the dephosphorylation of lipid pyrophosphate, an intermediate in biosynthesis of the peptidoglycan of bacterial cell walls. *Proceedings of the National Academy of Sciences of the United States of America, 57*, 767–773.

Stanisich, V. A., Stone, B. A., Brown, R. M., McIntosh, M., Flanagan, B. M., & Gidley, M. (2006). *The curdlan capsule of Agrobacterium sp. strain ATCC31749: its organization and the molecular form of the curdlan chains*. XXIII international carbohydrate symposium. Whistler Canada, July 23–28, Abstract #332.

Stasinopoulos, S., Fisher, P. R., Stone, B. A., & Stanisich, V. A. (1999). Detection of two loci involved in (1-3)-β-glucan (curdlan) biosynthesis by *Agrobacterium* sp. ATCC31749, and comparative sequence analysis of the putative curdlan synthase gene. *Glycobiology, 9*, 31–41.

Taboureich, N., Charnock, S. J., & Davies, G. J. (2001). Three dimensional structure of the Mn and Mg dTDP complexes of the family GT-2 glycosyltransferase SpsA: A comparison with the related NDP-sugar glycosyltransferases. *Journal of Molecular Biology, 314*, 655–661.

Taner, F., Glanfield, S., Aracic, S., Stone, B., & Stanisich, V. (2008). *The stringent response is required for the production of capsular (1,3)-β-glucan (curdlan) by Agrobacterium sp. LTU50.* Annual scientific meeting of the Australian Society for Microbiology. Melbourne 6–10th July, Abstract P04.36.

Tiburtius, A., de Luca, N. G., Hussain, H., & Johnston, A. W. B. (1996). Expression of the *exoY* gene, required for exopolysaccharide synthesis in *Agrobacterium*, is activated by the regulatory *ros* gene. *Microbiology, 142*, 2621–2629.

Tiwari, R. P., Reeve, W. G., Dilworth, M. J., & Glenn, A. R. (1996). An essential role for *actA* in acid tolerance of *Rhizobium meliloti*. *Microbiology, 142*, 601–610.

Tlapak-Simmons, V. L., Kempner, E. S., Baggenstoss, B. A., & Weigel, P. H. (1998). The active streptococcal hyaluronan synthases (HASs) contain a single HAS monomer and multiple cardiolipin molecules. *Journal of Biological Chemistry, 273*, 26100–26109.

Tlapak-Simmons, V. L., Baggenstoss, B. A., Clyne, T., & Weigel, P. H. (1999). Purification and lipid dependence of the recombinant hyaluronan synthases from *Streptococcus pyogenes* and *Streptococcus equisimilis*. *Journal of Biological Chemistry*, *274*, 4239–4245.

Urbanczyk, H., Suzuki, K., Yoshida, K., & Kondo, K. (2003). Physical and gene maps of *Agrobacterium* biovar 2 strains and their relationship to biovar 1 chromosomes. *Microbiology*, *149*, 3035–3042.

Vitorino, J. C., Steffens, M. B. R., Machado, H. B., Yates, M. G., Souza, E. M., & Pedrosa, F. O. (2001). Potential roles for the *glnB* and *ntrYX* genes in *Azospirillum brasilense*. *FEMS Microbiology Letters*, *201*, 199–204.

Wang, P., Ingram-Smith, C., Hadley, J. A., & Miller, K. J. (1999). Cloning, sequencing and characterization of the *cgmB* gene of *Sinorhizobium meliloti* involved in cyclic β-glucan biosynthesis. *Journal of Bacteriology*, *181*, 4576–4583.

Whitfield, C. (2006). Biosynthesis and assembly of capsular polysaccharides in *Escherichia coli*. *Annual Review of Biochemistry*, *75*, 39–68.

Wiedemann, G., & Müller, P. (2004). Use of Tn*KPK2* for sequencing a 10.6-kb *Pst*I DNA fragment of *Bradyrhizobium japonicum* and for the construction of *aspA* and *ndvA* mutants. *Archives of Microbiology*, *181*, 418–427.

Williamson, G., Damani, K., Devenney, P., Faulds, C. B., Morris, V. J., & Stevens, B. J. (1992). Mechanism of action of cyclic β-1,2-glucan synthetase from *Agrobacterium tumefaciens*: Competition between cyclization and elongation reactions. *Journal of Bacteriology*, *174*, 7941–7947.

Wood, D. W., Setubal, J. C., Kaul, R., Monks, D. E., Kitajima, J. P., Okura, V. K., Zhou, Y., Chen, L., Wood, G. E., Almeida, N. F., Jr., Woo, L., Chen, Y., Paulsen, I. T., Eisen, J. A., Karp, P. D., Bovee, D., Sr., Chapman, P., Clendenning, J., Deatherage, G., Gillet, W., Grant, C., Kutyavin, T., Levy, R., Li, M. J., McClelland, E., Palmieri, A., Raymond, C., Rouse, G., Saenphimmachak, C., Wu, Z., Romero, P., Gordon, D., Zhang, S., Yoo, H., Tao, Y., Biddle, P., Jung, M., Krespan, W., Perry, M., Gordon-Kamm, B., Liao, L., Kim, S., Hendrick, C., Zhao, Z. Y., Dolan, M., Chumley, F., Tingey, S. V., Tomb, J. F., Gordon, M. P., Olson, M. V., & Nester, E. W. (2001). The genome of the natural genetic engineer *Agrobacterium tumefaciens* C58. *Science*, *294*, 2317–2323.

Yother, J. (1999). Common themes in the genetics of streptococcal capsular polysaccharides. In J. B. Goldberg (Ed.), *Genetics of bacterial polysaccharides* (pp. 161–184). CRC Press.

Zhang, H. B., Wang, C., & Zhang, L. H. (2004). The quormone degradation system of *Agrobacterium tumefaciens* is regulated by starvation signal and stress alarmone (p)ppGpp. *Molecular Microbiology*, *52*, 1389–1401.

Zorreguieta, A., & Ugalde, R. A. (1986). Formation in *Rhizobium* and *Agrobacterium* spp. of a 235-kilodalton protein intermediate in β-D-(1-2)-glucan synthesis. *Journal of Bacteriology*, *167*, 947–951.

CHAPTER 3.3.2

# Biosynthetic Enzymes for (1,3)-β-Glucans and (1,3;1,6)-β-Glucans in Protozoans and Chromistans: Biochemical Characterization and Molecular Biology

**Vincent Bulone**

*School of Biotechnology, Royal Institute of Technology (KTH), AlbaNova University Center, Stockholm, Sweden*

## I. Introduction

The biosynthesis of linear β-glucans is generally considered to involve several steps (Fig. 1). The process starts with the synthesis of an initiator of polymerization or primer, followed by the transfer of the primer to a membrane-bound synthase. The latter catalyses the repetitive transfer of glucosyl units from an activated sugar donor to the acceptor – i.e. the primer or the elongating chain – until the polymerization stops (Fig. 1). Most molecular events occurring during the biosynthesis of (1,3)-β-glucans and (1,3;1,6)-β-glucans in protozoans and chromistans are not well understood (Fig. 1). The process for which most biochemical data have been accumulated is the polymerization of linear (1,3)-β-glucan chains. A lot less is known about the mode of synthesis of primers, and in fact whether such primers are absolutely required for β-glucan biosynthesis can be questioned (Fig. 1). In addition, the linear (1,3)-β-glucans synthesized by cell-free extracts usually reach a certain degree of polymerization (Pelosi et al., 2003), but how this is controlled and why the polymerization reaction stops is unknown (Fig. 1).

The ability to synthesize β-glucans *in vitro* represents a very useful tool for assaying glucan synthases during enzyme purification. This is also useful in identifying the type of effectors that might influence enzyme activity *in vivo* and thus regulate the biosynthesis of the polysaccharides. Such biochemical approaches have been successfully applied to a number of (1,3)-β-glucan synthases from protozoans and chromistans, as detailed in the next sections. They typically involve the isolation of particulate fractions which correspond to suspensions of plasma

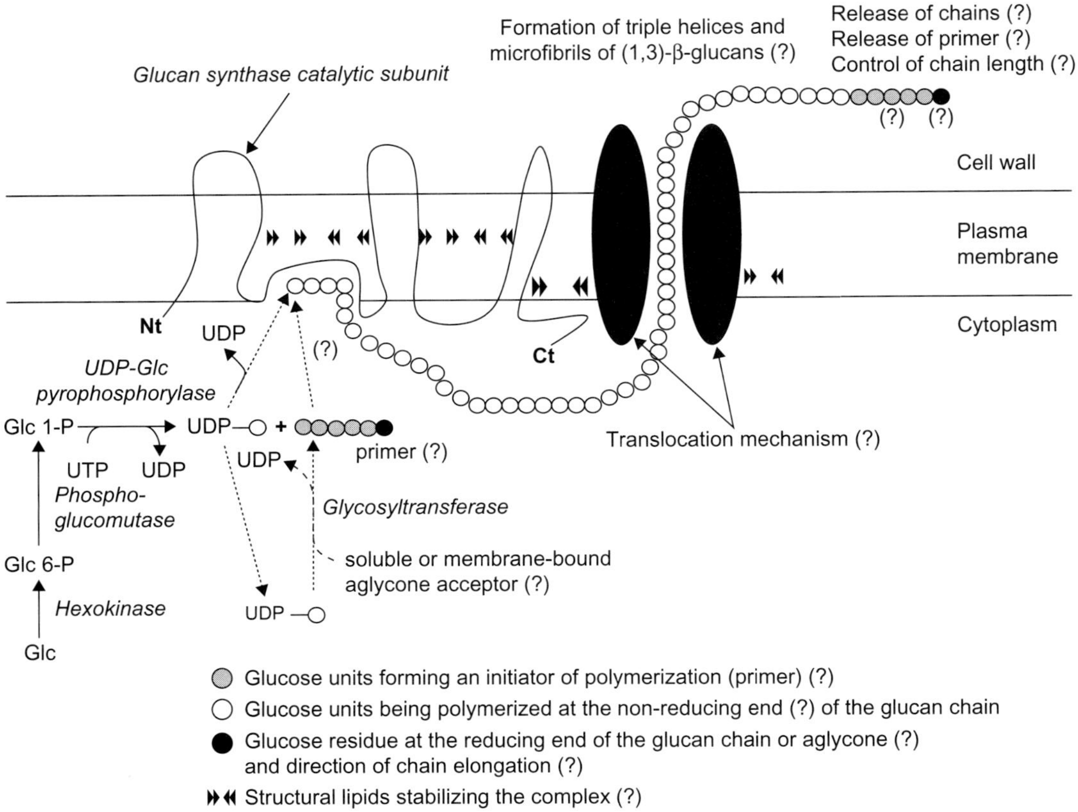


**Fig. 1: Hypothetical model for the biosynthesis of linear β-glucans. (?) refers to aspects that remain to be clarified including: (a) involvement of a primer to initiate polymerization (broken arrows); (b) orientation of the glucan chain being extruded; (c) mechanism of translocation of the glucan chains across the plasma membrane; (d) mode of formation of triple helices and microfibrils in the case of (1,3)-β-glucans; (e) control of chain length; (f) mode of release of the primer; (g) mode of release of the chains in the cell wall; (h) involvement of structural lipids stabilizing the complex. Other potential protein components of the complex, such as regulation subunits, are not represented. Their stoichiometry in the complex is not known.**

membranes, usually as a mixture with intracellular membrane compartments, or the preparation of so-called solubilized enzymatic fractions, which are detergent extracts of particulate fractions. In addition to linear (1,3)-β-glucans, numerous protozoans and chromistans synthesize side-chain-branched β-glucans that usually consist of a main chain of (1,3)-β-glucan in which some glucosyl units are substituted by a single glucose or a (1,3)-β-oligoglucosyl chain through a (1,6)-β-linkage (see Chapter 2.1). In some examples their structure is quite heterogeneous and the main (1,3)-β-glucan chain is sometimes interrupted in several places by non-contiguous (1,6)-β-linked glucosyl residues (Handa and Nisizawa, 1961; Maeda and Nisizawa, 1968; Yamamoto and Nevins, 1983). The modes of insertion of the (1,6)-β-linkages, either as part of the main (1,3)-β-glucan chain or as branching points, is unknown. As opposed to the situation in fungi (Kollar et al., 1997) (see Chapter 4.3), the occurrence of linear (1,6)-β-glucans that link other wall polymers has not been demonstrated in the walls of protozoans and chromistans.

## II. Biochemistry of β-Glucan Biosynthesis in Protozoans and Chromistans

### II.A. Model Organisms

Even though β-glucan synthesis occurs in a large number of protozoans and chromistans, the corresponding enzymes have been studied only in a limited number of species. Pioneering work was conducted on the protozoan *Euglena gracilis* (Goldemberg and Marechal, 1963; Marechal and Goldemberg, 1964) and the oomycete *Phytophthora cinnamomi* (Wang and Bartnicki-Garcia, 1966). *E. gracilis* produces a linear (1,3)-β-glucan, paramylon, as an intracellular storage carbohydrate (Clarke and Stone, 1960). This polymer was first identified as a polysaccharide that is isomeric to starch but that cannot be stained with iodine (Gottlieb, 1850). It was shown to be first synthesized as elementary microfibrils of a lateral size of 3–4 nm that accumulate in immature granules in a poorly crystalline form (Kiss et al., 1987; 1988a). During maturation of the granules, the paramylon microfibrils are thought to line up to form highly ordered discrete arrays that are characterized by a high crystallinity (Booy et al., 1981; Kiss et al., 1988a; Marchessault and Deslandes, 1979). The granules are surrounded by a membrane that has an organization similar to the plasma membrane (Kiss et al., 1988b). Paramylon granules occur in all euglenoids but their size, shape and number vary widely depending on the species considered. The latter characteristics, together with pyrenoid and chloroplast morphology and structure, have been used as criteria for the classification of Euglenophyceae (Brown et al., 2003). Since paramylon can be accumulated inside *E. gracilis* cells in high amounts (up to 90% dry weight) by optimizing the growth conditions and selecting spontaneous non-photosynthetic mutant strains (Barsanti et al., 2001), this species represents an optimal model for the study of (1,3)-β-glucan synthesis.

Oomycetes, especially species belonging to the *Phytophthora* and *Saprolegnia* genera, are amongst the most studied organisms from the chromistan class. Some of these microorganisms are plant or animal pathogens responsible for severe environmental damage and economic loss (Margulis and Schwartz, 2000) (Chapter 4.2). They synthesize (1,3)-β-glucans as wall polysaccharides and (1,3;1,6)-β-glucans as intracellular storage carbohydrates called mycolaminarins because of their structural similarity with the laminarins found in most brown algae (Bartnicki-Garcia, 1968; Blaschek et al., 1992; Wang and Bartnicki-Garcia, 1974; 1980; Zevenhuizen and Bartnicki-Garcia, 1969). Mycolaminarins occur as neutral β-glucans in the mycelium of oomycetes, but both neutral and phosphorylated forms are present at other developmental stages, e.g. in sporangia, zoospores, cysts, chlamydospores and oospores (Wang and Bartnicki-Garcia, 1973; 1980). The enzymes involved in the phosphorylation of mycolaminarins are unknown. Most of the data available on β-glucan synthesis in oomycetes were obtained on the cell wall (1,3)-β-glucan synthases. In addition to their importance in fundamental processes such as the growth and morphogenesis of mycelia, the latter enzymes represent potential targets of specific growth inhibitors, which is currently of great relevance for the agriculture and aquaculture industries.

An important collection of data is available on the structure of (1,3)-β-glucans and (1,3;1,6)-β-glucans produced by different species of brown algae and marine diatoms (Bacillariophyceae) as intracellular storage carbohydrates (see Chapters 2.1 and 4.2). These β-glucans exhibit low molecular weights (degrees of polymerization (DPs) in the range 5–50) compared to cell wall (1,3)-β-glucans and (1,3;1,6)-β-glucans and, in addition to their variable molecular size, they differ from each other by their degree of branching. The trivial names for the intracellular β-glucans from brown algae and diatoms are laminarins and chrysolaminarins, respectively. The former may contain mannitol, mannuronic or guluronic acid as end groups (Chizhov et al., 1998; Elyakova and Zvyagintseva, 1974; Størseth et al., 2006). None of the enzymes involved in the addition of these unusual terminal groups are known. In addition, so far none of the laminarin synthases have been studied *in vitro* and one report only is available on chrysolaminarin biosynthesis in diatoms (Roessler, 1987). In this case, cell-free extracts of *Cyclotella cryptica* were successfully used to incorporate glucose from UDP-glucose into a (1,3)-β-glucan with a DP of nearly 30 (Roessler, 1987).

## II.B. Subcellular Localization of β-Glucan Synthases and Preparation of Cell-Free Extracts

All known β-glucan synthases are membrane-bound enzymes that are typically assayed using particulate fractions or detergent-extracted enzymes. The cells used as a source of membranes are commonly disrupted using a French press [e.g. *E. gracilis* (Bäumer et al., 2001; Goldemberg and Marechal, 1963; Marechal and Goldemberg, 1964) and *C. cryptica* (Roessler, 1987)], a glass-bead homogenizer [mycelium, zoospores and cysts of oomycetes (Wang and Bartnicki-Garcia, 1976; 1982)], a Waring or Virtis type of blender [mycelial forms of oomycetes (Antelo et al., 1998; Fèvre and Rougier, 1981; Pelosi et al., 2003)] or, in fewer cases, a mortar or an ultrasonicator (Hill and Mullins, 1982). The last is however quite severe and may lead to a loss of activity, especially in the case of some rather unstable (1,3)-β-glucan synthase complexes. Particulate fractions are pelleted at 50 000–100 000 *g* following the removal of cell debris and total cell walls by a preliminary low-speed centrifugation (2000–5000 *g*) of the homogenized cells. This type of preparation was used almost systematically as a source of enzyme in early work on protozoans and chromistans (Fèvre and Rougier, 1981; Goldemberg and Marechal, 1963; Marechal and Goldemberg, 1964; Wang and Bartnicki-Garcia, 1966; 1976; 1982). The localization of cell-wall-synthesizing (1,3)-β-glucan synthases in the plasma membrane has been demonstrated in oomycetes (Girard and Fèvre, 1984). However, a lower activity has also been reported in the endoplasmic reticulum and Golgi membranes enriched by ultracentrifugation of total cellular membranes from *Saprolegnia monoica* on linear density gradients (Girard and Fèvre, 1984). The latter intracellular activities likely correspond to immature forms of the plasma-membrane-bound enzymes, although contamination of the intracellular membrane fractions by plasma membranes cannot be ruled out. It is however noteworthy that no membrane compartment has been identified as responsible for the biosynthesis of mycolaminarins in oomycetes. It is thus possible that the intracellular (1,3)-β-glucan synthase activity detected in the enriched endoplasmic reticulum and Golgi fractions from *S. monoica* may in fact correspond to the enzyme attached to the membrane structures responsible for the synthesis of mycolaminarins. Interestingly, zoospores of *Phytophthora palmivora* have been shown to be devoid of a cell wall (Sing and Bartnicki-Garcia, 1975), while the corresponding membrane fractions are able to synthesize *in vitro* soluble β-glucans that exhibit the same chemical properties as neutral and phosphorylated mycolaminarins (Wang and Bartnicki-Garcia, 1982). In addition, the encystment of zoospores is dependent on mycolaminarins and accompanied by the formation of walls similar to those observed at other developmental stages of *Phytophthora* species (Tokunaga and Bartnicki-Garcia, 1971a; 1971b; Zevenhuizen and Bartnicki-Garcia, 1969). Thus, in addition to serving as a reserve of energy, mycolaminarins may also contribute to the biosynthesis of wall β-glucans, possibly by being used as primers. In this hypothesis, the intracellular biosynthesis of mycolaminarins would be followed by their secretion to the cell wall, for instance when the concentration of the glucan synthase substrate (UDP-glucose) is insufficient to account by itself for the synthesis of wall β-glucans. Such a mechanism has been proposed to explain the formation of a wall during zoospore encystment in *P. palmivora* (Wang and Bartnicki-Garcia, 1982).

The (1,3)-β-glucan synthases of oomycetes can be extracted in an active form from the plasma membrane using various detergents. For instance, the enzyme from *S. monoica* can be solubilized using digitonin (Fèvre, 1979), the zwitterionic detergent 3-[(3-cholamidopropyl)-dimethylammonio]-1-propane sulfonate (CHAPS) (Bulone et al., 1990) or a mixture of CHAPS and octyl-β-D-glucopyranoside (Girard et al., 1992). More recently similar solubilization procedures were repeated using *Phytophthora sojae*, and CHAPS was confirmed to be an efficient detergent for solubilizing (1,3)-β-glucan synthases from oomycetes (Antelo et al., 1998). Other detergents such as CHAPSO (3-[(3-cholamidopropyl)-dimethylammonio]-2-hydroxy-1-propane sulfonate), which is a hydroxylated form of CHAPS, nonanoyl-*N*-methylglucamide (MEGA 9), octyl-β-D-glucopyranoside, octyl-α-D-glucopyranoside and octyl-β-D-galactopyranoside solubilized the (1,3)-β-glucan synthase from *S. monoica* with levels of activity higher than those observed with CHAPS, whereas other detergents including the sulfobetain Zwittergent 3-12 and detergents from the Brij family (polyoxyethylene ethers) were inefficient (Pelosi and Bulone, unpublished data). Similar data were obtained with Zwittergent 3-12 for the *P. sojae* enzyme, which, as opposed to the *S. monoica* glucan synthase, was shown to be inhibited by octyl-β-D-glucopyranoside (Antelo et al., 1998). It is noteworthy that detergents that allow solubilization of (1,3)-β-glucan synthases usually provoke a concomitant stimulation of the activity. This has been reported consistently not only in the case of the oomycete enzymes (Antelo et al., 1998; Girard et al., 1992) but also for the comparable (1,3)-β-glucan (callose) synthases from higher plants (Lai Kee Him et al., 2001; Li et al., 1997). Thus, the true solubilization efficiency of a given detergent is usually difficult to determine accurately as the activity measured is the combined result of both solubilization and stimulation, the extent of the latter being quite variable from one detergent to another.

Early work on paramylon synthase showed the possibility of synthesizing *in vitro* (1,3)-β-glucan from cell-free extracts of *E. gracilis* (Goldemberg and Marechal, 1963; Marechal and Goldemberg, 1964). Even though it was concluded that the use of deoxycholate allowed a true solubilization of the paramylon synthase, it seems that the preparation was in fact particulate as all the activity pelleted when the fraction recovered after deoxycholate treatment was centrifuged at 100 000 *g* (Marechal and Goldemberg, 1964). The authors suggested that the enzyme is not bound to the polysaccharide fraction, but that it is associated to unidentified particles, since most of the activity was recovered in a fraction apparently devoid of paramylon granules (Marechal and Goldemberg, 1964). It is only more recently that paramylon synthase was shown to be bound to the membrane that surrounds the granules (Bäumer et al., 2001). In the latter report, the authors purified paramylon granules from which they solubilized paramylon synthase in an active form using CHAPS. Thus, the membrane fractions used by Marechal and Goldemberg (1964) most likely contained membranes arising from paramylon granules.

## *II.C. Assay Conditions and Kinetic Parameters*

β-Glucan synthase activities in particulate fractions or detergent extracts of protozoan and chromistan membranes are typically assayed between 20°C and 30°C, and the pH is maintained in the range 7–8. The assays are based on the measurement of the incorporation of radioactive glucose from UDP-α-D-[$^{14}$C]glucose or UDP-α-D-[$^{3}$H]glucose into ethanol-insoluble polysaccharides. The total concentration of the UDP-α-D-glucose substrate is in the range 5 μM–10 mM (most often 1 mM or a few mM). The stoichiometry of the reaction, which usually proceeds linearly for up to 1 h depending on the source of enzyme and composition of the reaction mixture, has been particularly studied in the case of the paramylon synthase from *E. gracilis*, for which a good agreement between UDP-glucose consumption and UDP formation was observed (Marechal and Goldemberg, 1964). This study was completed by the demonstration that the amount of glucose released from the *in vitro* synthesized (1,3)-β-glucan upon digestion with a specific (1,3)-β-glucanase was similar to that of UDP released by the synthase. Other nucleotide-sugars such as ADP-glucose, TDP-glucose, GDP-glucose, dADP-glucose and UDP-β-D-glucose were tested on enzymes from different species, but either very low or no significant levels of incorporation were detected (Fèvre and Rougier, 1981; Marechal and Goldemberg, 1964; Wang and Bartnicki-Garcia, 1966). It is generally accepted that UDP-α-D-glucose is the natural substrate of β-glucan synthases, as it is for similar enzymes from higher plants such as the callose and cellulose synthases (Delmer, 1999). Since UDP-α-D-glucose is a cytosoluble substrate, the active sites of β-glucan synthases are most likely located on the cytoplasmic side of the plasma membrane (Fig. 1). In the case of plants, there is some evidence that sucrose synthase may contribute to the synthesis of UDP-glucose as a substrate for β-glucan synthases, in addition to the cytosoluble UDP-glucose pyrophosphorylase (Haigler et al., 2001). However, this has not been shown to occur in protozoans and chromistans. Since the highest *in vitro* glucan synthase activity is systematically obtained using UDP-α-D-glucose as a sugar donor and the final product is a glucan that contains exclusively β-linkages, the synthases are referred to as *inverting* glycosyltransferases. Depending on the source of β-glucan synthase, the reaction mixtures may be supplemented by one or several bivalent cations such as $Mg^{2+}$, $Ca^{2+}$ or $Mn^{2+}$ and/or a β-glucose disaccharide that activates the enzyme, such as cellobiose, laminaribiose or gentiobiose (see for instance Billon-Grand et al., 1997; Marechal and Goldemberg, 1964; Wang and Bartnicki-Garcia, 1982).

The Michaelis–Menten constant ($K_m$) of protozoan and chromistan (1,3)-β-glucan synthases for UDP-glucose has generally been determined in typical assay reaction mixtures that contain all the minimum required factors for an optimal activity. The published $K_m$ values are generally in the mM range (0.5–7 mM depending on the source of enzyme), but it must be kept in mind that they have often been determined using impure enzyme preparations (particulate fractions or detergent extracts; Billon-Grand et al., 1997; Cerenius and Söderhäll, 1984; Marechal and Goldemberg, 1964; Wang and Bartnicki-Garcia, 1982). Such extracts may contain several isoforms of the synthases that exhibit different kinetic properties, and/or other enzymes that may use UDP-glucose as a substrate. Therefore, the measured $K_m$ values merely represent an estimation of the true intrinsic $K_m$ of the considered enzyme. In addition, the data are systematically analysed as if the enzyme was fully soluble and following a classical one-substrate Michaelis–Menten type of kinetics. This may not be correct in many cases, particularly when using particulate preparations for which the type of kinetics should rather be considered as occurring in a heterogeneous phase. Furthermore, (1,3)-β-glucan synthases are most likely organized as complexes that possibly consist of several catalytic subunits, which may be regulated in a cooperative manner by the substrate itself. In this type of mechanism, enzyme kinetics are expected to be of the allosteric type, as in the case of some plant (1,3)-β-glucan synthases (Lai Kee Him et al., 2001). Measurements of $V_{max}$ give an indication of the level of activity of a given preparation and represent a good description of the extent of incorporation of glucose into (1,3)-β-glucan chains per unit of volume and time. However, the (1,3)-β-glucan synthase preparations used for kinetics are most often crude or enriched fractions in which the enzyme concentration is not known. Since $V_{max}$ depends directly on this parameter, comparisons of values obtained using preparations from various origins are not that useful.

## II.D. Modulation of (1,3)-β-Glucan Synthase Activities In vitro

A number of effectors have been shown to influence (1,3)-β-glucan synthase activities *in vitro*. The compound to be tested is typically added to the reaction mixture at different concentrations, either alone or in combination with other potential effectors, thus allowing the detection of possible synergistic or antagonistic effects.

### II.D.1. Divalent cations

A simple view of the effect of divalent metal ions on (1,3)-β-glucan synthase activities from protozoans and chromistans consists of dividing the enzymes into two groups, one being stimulated by cations like $Mg^{2+}$, $Ca^{2+}$ and/or $Mn^{2+}$ while the second group is either insensitive or inhibited by these ions. The situation is in fact more ambiguous. For instance, the (1,3)-β-glucan synthase activities of the oomycetes *P. sojae* (Antelo et al., 1998) and *Aphanomyces astaci* (Cerenius and Söderhäll, 1984) are either not stimulated or inhibited by

$Mg^{2+}$ or $Ca^{2+}$, whereas the enzymes from the closely related species *P. cinnamomi* (Wang and Bartnicki-Garcia, 1966; 1976) and *P. palmivora* (Wang and Bartnicki-Garcia, 1982) are stimulated by $Mg^{2+}$, $Ca^{2+}$ and $Mn^{2+}$. A possible explanation for these controversial observations is that the crude extracts used for the assays may contain different forms of (1,3)-β-glucan synthases which exhibit different sensitivities to divalent metal ions. This hypothesis is supported by the occurrence in *S. monoica* of two (1,3)-β-glucan synthase activities, which can each be specifically assayed in the presence of the other: the first activity is optimal at pH 6 and inhibited in vitro by $Ca^{2+}$, $Mn^{2+}$ and $Mg^{2+}$, whereas the second activity is maximal at pH 9 and stimulated by these divalent cations (Billon-Grand et al., 1997). Since the $Ca^{2+}$ concentration is higher at the elongating apex than in the non-growing parts of the hyphae (Jackson and Heath, 1993; Levina et al., 1995), it has been proposed that the alkaline enzyme would be more specifically located at the apex, whereas the second isoform would have a sub-apical localization (Billon-Grand et al., 1997). This hypothesis is in keeping with the observation that (1→3)-β-glucans are present in all regions of both elongating and non-elongating hyphae in *Achlya bisexualis* (Shapiro and Mullins, 2002).

The assay mixture described in the pioneering work performed on paramylon synthase contained $Mg^{2+}$ (Marechal and Goldemberg, 1964). However, a more recent report showed that $Mg^{2+}$ has no effect on enzyme activity, while $Ca^{2+}$ slightly stimulates paramylon synthase *in vitro* (Bäumer et al., 2001). In the early work of Marechal and Goldemberg (1964), the incorporation of glucose into (1,3)-β-glucan was linear over a longer time in the presence of $Mg^{2+}$. Since UDP, the second product of the reaction, is a competitive inhibitor of UDP-glucose for the synthase, it was suggested that the cation may act as an activator of a UDP-degrading enzyme (Marechal and Goldemberg, 1964).

### *II.D.2. Nucleotides*

As opposed to fungal (1,3)-β-glucan synthases, which are generally stimulated by nucleotides (Szaniszlo et al., 1985), the enzymes from the most studied protozoans and chromistans are insensitive or inhibited by this type of compound (Antelo et al., 1998; Billon-Grand et al., 1997; Fèvre, 1983a; 1984; Girard et al., 1991; Marechal and Goldemberg, 1964). Uracil-based nucleotides strongly inhibit the (1,3)-β-glucan synthase from *S. monoica*, the strongest effect being observed with UTP and UDP (Fèvre, 1983a). The latter is a competitive inhibitor of the reaction (Fèvre, 1983a), as in the case of the paramylon synthase from *E. gracilis* (Marechal and Goldemberg, 1964). ATP and GTP have no effect on the (1,3)-β-glucan synthases from *S. monoica* at concentrations lower than 0.1 mM, while higher concentrations are inhibitory (Billon-Grand et al., 1997; Fèvre, 1983a). This contrasts with the (1,4)-β-glucan synthase from the same organism, which is activated by ATP and GTP in the concentration range 0.01–0.1 mM (Fèvre, 1984). Interestingly, unlike (1,3)-β-glucan synthase, the latter enzyme is also stimulated by cyclic diguanylic acid (Girard et al., 1991), which is the natural activator of the cellulose synthase from the bacterium *Gluconacetobacter xylinus* (Ross et al., 1987; 1991). With the exception of the (1,3)-β-glucan synthase from *A. astaci*, which is stimulated by ATP and GTP but to a much lower extent than the fungal enzymes (Cerenius and Söderhäll, 1984), ATP and GTP have no effect on the synthases from other oomycete species (Szaniszlo et al., 1985). Again, this contrasts with the situation in fungi where (1,3)-β-glucan synthase activities are regulated by a GTP-dependent mechanism involving GTP-binding proteins belonging to the Rho family (Mazur and Baginsky, 1996; Qadota et al., 1996) (see also Chapter 3.3.3). Thus, unlike their fungal counterparts and other β-glucan synthases such as cellulose synthases, oomycete (1,3)-β-glucan synthases do not seem to be regulated by a nucleotide-dependent mechanism.

### *II.D.3. Activation by free and bound saccharides and requirement of a primer*

Monosaccharides, α-methyl-D-glucoside, disaccharides and some trisaccharides have been shown to activate (1,3)-β-glucan synthases from *E. gracilis* (Marechal and Goldemberg, 1964), *P. palmivora* (Wang and Bartnicki-Garcia, 1982), *S. monoica* (Fèvre and Dumas, 1977) and *A. astaci* (Cerenius and Söderhäll, 1984). The highest stimulatory effect is systematically obtained with homodisaccharides consisting of β-linked glucoses, such as cellobiose, laminaribiose or gentiobiose. The mode of action of these free disaccharides has not been elucidated. It seems unlikely that they act as initiators of polymerization since glucan synthases catalyse the polymerization of β-glucan chains *in vitro* even in the absence of added disaccharides in the reaction mixture (Antelo et al., 1998; Blanton and Northcote, 1990; Fèvre and Rougier, 1981; Marechal and Goldemberg, 1964; Pelosi et al., 2003; Wang and Bartnicki-Garcia, 1976; 1982). Similarly, even though paramylon has been shown to activate the paramylon synthase from *E. gracilis in vitro* and is referred to as a primer (Bäumer et al., 2001), the polymer is apparently not an initiator of the polymerization process, as evidenced by the following observations: paramylon is not required for enzyme activity; laminaribiose and laminaritriose are not incorporated by paramylon synthase into (1,3)-β-glucans of a higher DP; and a preliminary depletion of (1,3)-β-glucans from the synthase preparation by action of a (1,3)-β-glucanase does not affect the activity of paramylon synthase (Marechal and Goldemberg, 1964). Thus, oligosaccharides or polymers such as paramylon are most likely true activators of (1,3)-β-glucan synthases rather than primers. It is possible however that the enzymatic extracts contain, at least in some cases, endogenous initiator(s) of polymerization of another type. Interestingly, the *in vitro* reaction mixtures recovered after incubation of paramylon synthase with UDP-glucose were shown to contain (1,3)-β-linked gluco-oligosaccharides that were covalently bound to an aglycone, possibly through a pyrophosphate linkage (Tomos and Northcote, 1978). The aglycone was proposed to be a protein because it could be precipitated by trichloroacetic acid, it exhibited a high molecular weight (estimated to be of at least 180 kDa) and was mobile in SDS–PAGE gels (Tomos and Northcote, 1978). However, it was not possible to stain it directly with Coomassie Blue and protease treatment did not release any of the bound (1,3)-β-glucans to lower molecular weight fractions (Tomos and Northcote, 1978). It is noteworthy that the involvement of a protein as a primer for polysaccharide biosynthesis has been demonstrated in the case of glycogen (reviewed by Smythe and Cohen, 1991). Furthermore, a heat-stable and non-dialysable membrane-bound activator of the (1,3)-β-glucan synthase from *S. monoica* was shown to affect the rate of the synthetic reaction *in vitro*, without modifying the apparent $K_m$ of the enzyme for UDP-glucose (Girard and Fèvre, 1991). The molecular weight of the activator was decreased by proteolysis but the digested activator remained active and non-dialysable (Girard and Fèvre, 1991). It was concluded that it probably corresponds to the glycan chain of a glycoprotein which might act as a primer or as an allosteric activator (Girard and Fèvre, 1991). However, its precise structure and mode of action remain to be determined.

#### *II.D.4. Proteases*

The (1,3)-β-glucan synthases from *S. monoica* (Fèvre, 1979) and *P. cinnamomi* (Wang and Bartnicki-Garcia, 1976) are stimulated *in vitro* in the presence of trypsin, suggesting that these synthases occur as inactive zymogens that are activated by partial proteolysis. In addition, since the cellulose synthase from *S. monoica* is not stimulated by trypsin under the same conditions, it has been proposed that proteolysis may be involved in the regulation of wall biosynthesis *in vivo* (Fèvre, 1979). Interestingly, the (1,3)-β-glucan synthase from another oomycete, *A. astaci*, does not seem to exist as a zymogen that can be activated by partial proteolysis (Cerenius and Söderhäll, 1984). This contrasts not only with the *S. monoica* and *P. cinnamomi* enzymes but also with other polysaccharide synthases from higher plants and yeasts, such as the developmentally regulated (1,3)-β-glucan synthase from pollen tubes of *Nicotiana* (Schlüpmann et al., 1993), the soyabean (1,3)-β-glucan synthase (Kauss et al., 1983) and chitin synthases from *Saccharomyces cerevisiae* (Bulawa, 1993; Cabib et al., 2001).

#### *II.D.5. Other compounds affecting (1,3)-β-glucan synthase activities in vitro*

The (1,3)-β-glucan synthase from *S. monoica* was recently shown to be activated by a 36-kDa annexin that systematically co-purifies with enzyme activity (Bouzenzana et al., 2006; Bulone and Fèvre, 1996; Bulone et al., 1990; Girard et al., 1992). Unlike the membrane-bound activator (glycoprotein?) characterized earlier by Girard and Fèvre (1991), the stimulating activity of the annexin was lost upon denaturation by heat, and the sequence deduced from the corresponding gene strongly suggested that the protein does not contain any glycan chain (Bouzenzana et al., 2006). Thus, several proteins seem to be involved in the regulation of this oomycete (1,3)-β-glucan synthase.

An inhibitor of paramylon synthase was shown to be present in cell-free extracts of *E. gracilis* (Marechal and Goldemberg, 1964), but its identity is still unknown. Numerous natural inhibitors produced by fungi have been shown to affect (1,3)-β-glucan synthase activities in yeast (Baguley et al., 1979; Duran et al., 1984; Varona et al., 1983). However, apart from papulacandin B, which does not inhibit the enzyme from *S. monoica* (Fèvre, 1983b), none of these compounds seem to have been tested on enzymes from protozoans and chromistans.

A number of chemicals have been shown to affect (1,3)-β-glucan synthase activities *in vitro*. Thiol reagents are of particular importance as they generally provoke a strong inhibition of (1,3)-β-glucan synthases. For instance, the paramylon synthase from *E. gracilis* is inactivated by 1 mM *p*-hydroxymercuribenzoate (Marechal and Goldemberg, 1964). Thus, thiol groups seem to play an important role in maintaining (1,3)-β-glucan synthases in an active form. Accordingly, dithiothreitol is often added as a protective reagent in the buffers used for enzyme extraction (Bäumer et al., 2001; Cerenius and Söderhäll, 1984; Fèvre and Rougier, 1981; Roessler, 1987) and/or in the synthase reaction mixtures (e.g. Bulone et al., 1990; Fèvre and Rougier, 1981). The diazo dye Congo Red inhibits *in vitro* the (1,3)-β-glucan and cellulose synthases from *S. monoica* in a non-competitive manner (Nodet et al., 1990). Since the dye can form complexes with (1,3)-β-glucans and cellulose (Wood, 1980), it was proposed that it prevents initiation of β-glucan synthesis and the attachment of additional glucose units by coating the growing chains (Nodet et al., 1990).

## II.E. Purification of (1,3)-β-Glucan Synthases

(1,3)-β-Glucan synthases are generally considered to occur as multiprotein complexes, essentially because enriched preparations systematically contain several proteins of different apparent molecular weights. However, there is no evidence that all the proteins in a given preparation are actually physically associated as a complex. In addition, it is not known whether the catalytic subunits are active by themselves or if any of the putative associated partners are absolutely required for activity. As for most membrane-bound complexes, the purification of (1,3)-β-glucan synthases is complicated by the fact that these enzymes are unstable upon detergent extraction (Antelo et al., 1998; Bulone et al., 1990). Thus, the purification protocol should be as rapid as possible and involve as few steps as possible to minimize enzyme inactivation (Antelo et al., 1998; Bulone et al., 1990). Enriched preparations have been obtained in a few instances. This is the case of the paramylon synthase from *E. gracilis* that was enriched from the membrane associated to paramylon granules after extraction with CHAPS, ultracentrifugation on a discontinuous sucrose gradient and anion-exchange chromatography (Bäumer et al., 2001). The specific activity was increased 177-fold compared to the cell-free homogenate and the native complex exhibited an estimated molecular weight of 670 kDa (Bäumer et al., 2001). SDS-PAGE analysis and Coomassie Blue staining of the gels revealed the presence of multiple proteins with apparent molecular weights between 16 and 80 kDa (Bäumer et al., 2001). Photoaffinity labeling with α-[$^{32}$P]UDP-glucose showed that two of these proteins, which exhibited an apparent molecular weight of 37 and 54 kDa, were able to bind the paramylon synthase substrate (Bäumer et al., 2001). However, none of the proteins present in the enriched fraction have been sequenced to date. Gradient centrifugation coupled to product entrapment was used to purify the (1,3)-β-glucan synthase from *S. monoica* (Bulone et al., 1990). Product entrapment is a kind of affinity purification technique in which the matrix is generated by the glucan synthase itself. It consists of incubating the detergent-extracted enzyme with UDP-glucose to form an insoluble (1,3)-β-glucan that can be sedimented by low-speed centrifugation together with the associated enzyme complex. This method was first successfully applied to the purification of the yeast chitin synthase (Kang et al., 1984). In the case of *S. monoica*, three major protein bands of 34, 48 and 55 kDa were shown to co-purify with enzyme activity by using product entrapment (Bulone et al., 1990). Other approaches based on the lectin-binding activity of the glucan synthase complex from *S. monoica* (Girard et al., 1992) or on immunochemical techniques involving antibodies directed against the proteins in the 34-, 48- and 55-kDa bands (Bulone and Fèvre, 1996) further supported the involvement of the latter proteins in (1,3)-β-glucan synthase activity. Recent work has shown that one of the proteins in the 34-kDa band is an annexin that activates (1,3)-β-glucan synthase, whereas major proteins in the 48- and 55-kDa bands correspond to ATP synthase subunits which most likely arise from contaminations by mitochondria (Bouzenzana et al., 2006). 2D-PAGE analysis has shown that the 34- and 55-kDa bands contain numerous spots in addition to the already identified annexin and ATP synthase subunits, some of which might correspond to the actual components of the glucan synthase complex (Bouzenzana et al., 2006). However, their identity remains to be determined. Interestingly, the apparent molecular weights of the 34- and 55-kDa proteins from *S. monoica* (Bouzenzana et al., 2006; Bulone and Fèvre, 1996; Bulone et al., 1990; Girard et al., 1992) are similar to those of the UDP-glucose-binding proteins identified in the paramylon synthase complex (Bäumer et al., 2001). The procedure originally devised for the purification of the *S. monoica* enzyme, and based on CHAPS solubilization and product entrapment (Bulone et al., 1990), was used more recently to obtain a fraction enriched in (1,3)-β-glucan synthase from *P. sojae* (Antelo et al., 1998). A protein band of 108 kDa was enriched, but the fraction contained numerous other bands of up to 100 kDa (Antelo et al., 1998). None of these proteins have been identified so far.

### II.F. Product Characterization

The type of linkage formed *in vitro* by the synthase is typically demonstrated by determining the sensitivity of the polysaccharides synthesized in the presence of radioactive substrate (UDP-α-D-[$^{14}$C]glucose or UDP-α-D-[$^{3}$H]glucose) to the action of specific β-glucanases. In addition, this approach demonstrates that the polysaccharides analysed were newly synthesized and do not arise from the original cell walls. The products released upon enzymatic hydrolysis are typically identified by a technique such as gel-filtration, paper or thin-layer chromatography. This biochemical approach is simple to use and convenient since specific hydrolytic enzymes are readily available. For this reason it has been widely applied to the identification of the polysaccharides synthesized *in vitro* by β-glucan synthases from a number of protozoan and chromistan species (Antelo et al., 1998; Billon-Grand et al., 1997; Bulone et al., 1990; Bäumer et al., 2001; Fèvre and Rougier, 1981; Marechal and Goldemberg, 1964; Roessler, 1987; Wang and Bartnicki-Garcia, 1966; 1976; 1982). A more complete structural characterization of the β-glucan synthase products can be achieved by using physical and chemical techniques, whereas transmission electron microscopy is the tool of choice to obtain morphological information (reviewed by Bulone, 2007; Colombani et al., 2004). The latter methods have been used much less frequently than the biochemical approach.

Cell-free extracts from *E. gracilis* (Bäumer et al., 2001; Marechal and Goldemberg, 1964), *C. cryptica* (Roessler, 1987), *S. monoica* ( Billon-Grand et al., 1997; Bulone et al., 1990; Fèvre and Rougier, 1981; Pelosi et al., 2003), *A. astaci* (Cerenius and Söderhäll, 1984), *P. sojae* (Antelo et al., 1998), *P. cinnamomi* (Wang and Bartnicki-Garcia, 1966; 1976) and *P. palmivora* (Wang and Bartnicki-Garcia, 1982) have all been shown to synthesize *in vitro* linear (1,3)-β-glucans when incubated in the presence of UDP-glucose. The particulate enzyme preparations from the latter two species were also able to synthesize (1,3;1,6)-β-glucans (Wang and Bartnicki-Garcia, 1966; 1982), but the mode of formation of the (1,6)-β-linkages has yet to be determined, as briefly discussed in Section II.G. Interestingly, the slime mold *Dictyostelium discoideum* does not seem to synthesize any (1,3)-β-glucan *in vivo* (Blanton et al., 2000; Freeze and Loomis, 1978), but membrane extracts from this species were shown to synthesize *in vitro* β-glucans consisting of 12% (1,3)-linkages and 78% (1,4)-linkages (Blanton and Northcote, 1990). It was not determined whether both linkages occurred in the same or different polymers. By analogy with an early hypothesis that the same enzyme in higher plants may be able to form (1,4)-β and (1,3)-β-linkages (Delmer, 1999), it was proposed that the *in vitro* formation of both types of linkages is catalysed by a single glucan synthase in the extracts from *D. discoideum* (Blanton and Northcote, 1990). This was further supported by the observation that membrane extracts from a *D. discoideum* strain in which a cellulose synthase gene has been disrupted do not synthesize any β-glucan at all (Blanton et al., 2000). Interestingly, the situation seems to be different in higher plants. Indeed, in the latter case a series of data strongly suggest that the catalytic subunits of the (1,4)-β-glucan and (1,3)-β-glucan synthases correspond to different proteins coded by genes that do not share any significant sequence similarity (Brownfield et al., 2007; Cui et al., 2001; Doblin et al., 2001; Hong et al., 2001; Li et al., 2003; Pear et al., 1996) and belong to glycosyltransferase families 2 and 48, respectively (http://www.cazy.org) (see Chapter 3.3.4).

The (1,3)-β-glucans synthesized *in vitro* by particulate fractions from *P. cinnamomi* (Wang and Bartnicki-Garcia, 1976) and *S. monoica* (Fèvre and Rougier, 1981; Pelosi et al., 2003) exhibited a microfibrillar morphology. More recent work revealed the existence of two (1→3)-β-glucan synthases in *S. monoica* (Billon-Grand et al., 1997). The product synthesized *in vitro* by the enzyme that has an optimal pH of 6 exhibited a DP of more than 20 000 and formed endless ribbon-like microfibrils (Pelosi et al., 2003). When the synthesis was performed by the synthase that has an optimal pH of 9, much shorter microfibrils (200–300 nm) corresponding to an average DP of 5000 and exhibiting a lower degree of crystallinity were synthesized (Pelosi et al., 2003). All microfibrillar (1,3)-β-glucans consisted of triple helices associated in a hexagonal packing, but the chains synthesized by the alkaline enzyme were less tightly bound and more susceptible to intrachain hydration and enzymatic hydrolysis (Pelosi et al., 2003; 2006). These observations are consistent with the lack of well-organized microfibrillar polymers at the tip of the hyphae and with the hypothesis that the alkaline enzyme is more specifically located at the apical part of the cells (Billon-Grand et al., 1997; Pelosi et al., 2003). Conversely, a more crystalline and tightly packed polysaccharide, like the one synthesized at pH 6 by the other (1,3)-β-glucan synthase, would be more characteristic of the non-growing mature or sub-apical regions (Pelosi et al., 2003). However, care must be taken when making such hypotheses since the structure of the β-glucans synthesized *in vitro* do not necessarily reflect the structure of the polymers produced in the cell walls. The mode of formation of triple helices of (1,3)-β-glucans is probably spontaneous since these structures are self-stabilizing (Bluhm et al., 1982). In addition, it may be proposed that the corresponding molecular events involve the coordinated action of three catalytic subunits (or groups of three catalytic subunits) associated in a single complex, as previously suggested for plant callose synthases (Lai Kee Him et al., 2001).

### II.G. Biosynthesis of Side-Chain-Branched (1,3;1,6)-β-Glucans

There is virtually no information available on the mode of formation of (1,6)-β-linkages in side-chain-branched (1,3;1,6)-β-glucans from protozoan and chromistan species. *In vitro* synthesis of this class of polysaccharides has been achieved using particulate fractions from two *Phytophthora* species only (Wang and Bartnicki-Garcia, 1966; 1982), and none of the corresponding synthesizing enzymes have been characterized to date. The types of mechanisms that have been proposed for the biosynthesis of fungal (1,3;1,6)-β-glucans (Schmid et al., 2006) may also apply to protozoan and chromistan species. They involve the action of glycosyltransferases and transglycosylases, and assume that chain elongation occurs from the non-reducing end, by analogy with the elongation processes of extracellular polysaccharides such as fungal (1,3;1,6)-β-glucans (Batra et al., 1969), bacterial cellulose (Koyama et al., 1997) and chitin from the vestimentiferan *Lamellibrachia satsuma* and the diatom *Thalassiosira weissflogii* (Imai et al., 2003). Interestingly, analyses of the recently sequenced genomes of the diatoms *Phaeodactylum tricornutum* and *Thalassiosira pseudonana* (Kroth et al., 2008) and of oomycetes from the *Phytophthora* genus (Meijer et al., 2006) indicate the occurrence of genes that code for proteins similar to KRE6, which is required for (1,6)-β-glucan biosynthesis in yeast (Lesage and Bussey, 2006). Based on its sequence and structure similarities with enzymes in glycoside hydrolase family 16, KRE6 is most likely a glycoside hydrolase or a transglycosylase, which suggests an involvement in a glucan-processing step or in the formation of branching points in β-glucans by transglycosylation (Montijn et al., 1999). The characterization of the diatom and oomycete homologues of KRE6 represents the next important step to determine whether these proteins play a similar role to the yeast enzyme in (1,6)-β-glucan biosynthesis.

## III. Molecular Biology of β-Glucan Biosynthesis in Protozoans and Chromistans

The full-length sequence of the cDNA encoding the annexin activator of the (1,3)-β-glucan synthase from *S. monoica* was determined recently (Bouzenzana et al., 2006). This represents the only report to date describing the use of molecular biology for unraveling the mechanisms of β-glucan biosynthesis in protozoans and chromistans. For a long time, major limitations have been the absence of sequenced genomes from these organisms and the difficulties in preparing stable transformants and generating gene knockouts. However, the situation is changing, as illustrated by the development of an increasing number of EST databases (e.g. Gajendran et al., 2006) and by the recent publication of the genomes of three oomycetes (*P. sojae*, *P. ramorum* and *P. infestans*; Tyler et al., 2006), of the slime mold *D. discoideum* (Eichinger et al., 2005) and of the diatoms *Thalassiosira pseudonana* (Armbrust et al., 2004) and *Phaeodactylum tricornutum* (Kroth et al., 2008) (an updated list of sequenced genomes and sequencing projects in progress is available at http://www.ncbi.nlm.nih.gov/genomes/leuks.cgi). In addition, molecular tools including transformation procedures for diatoms (Dunahay et al., 1995; Poulsen et al., 2006; Roessler, 2000; Walker et al., 2005) and transient gene silencing in oomycetes using dsRNA (Whisson et al., 2005) are becoming available, allowing the functional characterization of specific genes. There is no doubt that these molecular biological approaches will help to shed light on the molecular mechanisms of β-glucan biosynthesis in a number of model organisms from the protozoan and chromistan families. This is illustrated by the recent functional characterization of a novel family of cellulose synthase genes in the oomycete *P. infestans* using gene silencing (Grenville-Briggs et al., 2008). In addition, searches in the genomes of *Phaeodactylum tricornutum* and *Thalassiosira pseudonana* (Kroth et al., 2008), and *P. ramorum*, *P. sojae* and *P. infestans* (Fugelstad and Bulone, unpublished data; Meijer et al., 2006) have allowed the identification of several genes that are similar to genes coding for putative catalytic subunits of plant and fungal (1,3)-β-glucan synthases. However, the latter proteins exhibit a much higher molecular weight (190–220 kDa) than the proteins that co-purify with the (1,3)-β-glucan synthases from *S. monoica* (Bouzenzana et al., 2006; Bulone and Fèvre, 1996; Bulone et al., 1990; Girard et al., 1992) and *P. sojae* (Antelo et al., 1998) (see Section II.E). Functional analyses of the genes identified through searches in the genomes of *P. sojae* and *P. infestans* will demonstrate whether the corresponding proteins are responsible for the synthesis of (1,3)-β-glucans. In addition, the availability of the *Phytophthora* genomes (Tyler et al., 2006) will greatly facilitate the identification of the proteins that co-purify with the (1,3)-β-glucan synthases from *P. sojae* and closely related species such as *S. monoica*. This will help in determining which of these proteins are truly involved in (1,3)-β-glucan synthesis.

## Acknowledgement

This work and related research in the author's laboratory were supported by the Swedish Centre for Biomimetic Fiber Engineering (Biomime).

## References


Antelo, L., Cosio, E. G., Hertkorn, N., & Ebel, J. (1998). Partial purification of a GTP-insensitive (1,3)-β-glucan synthase from *Phytophthora sojae*. *FEBS Letters*, *433*, 191–195.

Armbrust, E. V., Berges, J. A., Bowler, C., Green, B. R., Martinez, D., Putnam, N. H., Zhou, S. G., Allen, A. E., Apt, K. E., Bechner, M., Brzezinski, M. A., Chaal, B. K., Chiovitti, A., Davis, A. K., Demarest, M. S., Detter, J. C., Glavina, T., Goodstein, D., Hadi, M. Z., Hellsten, U., Hildebrand, M., Jenkins, B. D., Jurka, J., Kapitonov, V. V., Kroger, N., Lau, W. W. Y., Lane, T. W., Larimer, F. W., Lippmeier, J. C., Lucas, S., Medina, M., Montsant, A., Obornik, M., Parker, M. S., Palenik, B., Pazour, G. J., Richardson, P. M., Rynearson, T. A., Saito, M. A., Schwartz, D. C., Thamatrakoln, K., Valentin, K., Vardi, A., Wilkerson, F. P., & Rokhsar, D. S. (2004). The genome of the diatom *Thalassiosira pseudonana*: Ecology, evolution, and metabolism. *Science*, *306*, 79–86.

Baguley, B. C., Rommele, G., Gruner, J., & Wehrli, W. (1979). Papulacandin B – Inhibitor of glucan synthesis in yeast spheroplasts. *European Journal of Biochemistry*, *97*, 345–351.

Barsanti, L., Vismara, R., Passarelli, V., & Gualtieri, P. (2001). Paramylon ((1,3)-β-glucan) content in wild type and WZSL mutant of *Euglena gracilis*. Effects of growth conditions. *Journal of Applied Phycology*, *13*, 59–65.

Bartnicki-Garcia, S. (1968). Cell wall chemistry, morphogenesis and taxonomy of fungi. *Annual Reviews of Microbiology*, *22*, 87–108.

Batra, K. K., Nordin, J. H., & Kirkwood, S. (1969). Biosynthesis of the β-D-glucan of *Sclerotium rolfsii* Sacc. Direction of chain propagation and the insertion of the branch residues. *Carbohydrate Research*, *9*, 221–229.

Billon-Grand, C., Marais, M. F., Joseleau, J. P., Girard, V., Gay, L., & Fèvre, M. (1997). A novel (1,3)-β-glucan synthase from the Oomycete *Saprolegnia monoica*. *Microbiology*, *143*, 3175–3183.

Blanton, R. L., Fuller, D., Iranfar, N., Grimson, M. J., & Loomis, W. F. (2000). The cellulose synthase gene of *Dictyostelium*. *Proceedings of the National Academy of Sciences of the United States of America*, *97*, 2391–2396.

Blanton, R. L., & Northcote, D. H. (1990). A (1,4)-β-D-glucan synthase system from *Dictyostelium discoideum*. *Planta*, *180*, 324–332.

Blaschek, W., Käsbauer, J., Kraus, J., & Franz, G. (1992). *Pythium aphanidermatum*: Culture, cell wall composition, and isolation and structure of antitumour storage and solubilised cell wall (1,3;1,6)-β-D-glucans. *Carbohydrate Research*, *231*, 293–307.

Bluhm, T. L., Deslandes, Y., Marchessault, R. H., Perez, S., & Rinaudo, M. (1982). Solid-state and solution conformation of scleroglucan. *Carbohydrate Research*, *100*, 117–130.

Booy, F. P., Chanzy, H., & Boudet, A. (1981). An electron diffraction study of paramylon storage granules from *Euglena gracilis*. *Journal of Microscopy*, *121*, 133–140.

Bouzenzana, J., Pelosi, L., Briolay, A., Briolay, J., & Bulone, V. (2006). Identification of the first Oomycete annexin as a (1,3)-β-D-glucan synthase activator. *Molecular Microbiology*, *62*, 552–565.

Brown, P. J. P., Zakrys, B., & Farmer, M. A. (2003). Plastid morphology, ultrastructure, and development in *Colacium* and the Loricate Euglenophytes (Euglenophyceae). *Journal of Phycology*, *39*, 115–121.

Brownfield, L., Ford, K., Doblin, M. S., Newbigin, E., Read, S., & Bacic, A. (2007). Proteomic and biochemical evidence links the callose synthase in *Nicotiana alata* pollen tubes to the product of the NaGSL1 gene. *Plant Journal*, *52*, 147–156.

Bulawa, C. E. (1993). Genetics and molecular biology of chitin synthesis in fungi. *Annual Review of Microbiology*, *47*, 505–534.

Bulone, V. (2007). *In vitro* synthesis and analysis of plant (1,3)-β-D-glucans and cellulose: A key step towards the characterization of glucan synthases. In R. M. Brown, Jr. & I. M. Saxena (Eds.), *Cellulose: Molecular and structural biology* (pp. 123–145) Dordrecht: Springer.

Bulone, V., & Fèvre, M. (1996). A 34-kDa polypeptide is associated with (1,3)-β-glucan synthase activity from the fungus *Saprolegnia monoica*. *FEMS Microbiology Letters*, *140*, 145–150.

Bulone, V., Girard, V., & Fèvre, M. (1990). Separation and partial purification of (1,3)-β-glucan and (1,4)-β-glucan synthases from *Saprolegnia*. *Plant Physiology*, *94*, 1748–1755.

Bäumer, D., Preisfeld, A., & Ruppel, H. G. (2001). Isolation and characterization of paramylon synthase from *Euglena gracilis* (Euglenophyceae). *Journal of Phycology*, *37*, 38–46.

Cabib, E., Roh, D. H., Schmidt, M., Crotti, L. B., & Varma, A. (2001). The yeast cell wall and septum as paradigms of cell growth and morphogenesis. *Journal of Biological Chemistry*, *276*, 19679–19682.

Cerenius, L., & Söderhäll, K. (1984). Isolation and properties of β-glucan synthetase from the aquatic fungus, *Aphanomyces astaci*. *Physiologia Plantarum*, *60*, 247–252.

Chizhov, A. O., Dell, A., Morris, H. R., Reason, A. J., Haslam, S. M., McDowell, R. A., Chizhov, O. S., & Usov, A. I. (1998). Structural analysis of laminarins by MALDI and FAB mass spectrometry. *Carbohydrate Research*, *310*, 203–210.

Clarke, A. E., & Stone, B. A. (1960). Structure of the paramylon from *Euglena gracilis*. *Biochimica et Biophysica Acta*, *44*, 161–163.

Colombani, A., Djerbi, S., Bessueille, L., Blomqvist, K., Ohlsson, A., Berglund, T., Teeri, T. T., & Bulone, V. (2004). *In vitro* synthesis of (1,3)-β-D-glucan (callose) and cellulose by detergent extracts of membranes from cell suspension cultures of hybrid aspen. *Cellulose*, *11*, 313–327.

Cui, X. J., Shin, H. S., Song, C., Laosinchai, W., Amano, Y., & Brown, R. M. (2001). A putative plant homolog of the yeast (1,3)-β-glucan synthase subunit FKS1 from cotton (*Gossypium hirsutum L.*) fibers. *Planta*, *213*, 223–230.

Delmer, D. P. (1999). Cellulose biosynthesis: Exciting times for a difficult field of study. *Annual Review of Plant Physiology and Plant Molecular Biology*, *50*, 245–276.

Doblin, M. S., De Melis, L., Newbigin, E., Bacic, A., & Read, S. M. (2001). Pollen tubes of *Nicotiana alata* express two genes from different β-glucan synthase families. *Plant Physiology*, *125*, 2040–2052.

Dunahay, T. G., Jarvis, E. E., & Roessler, P. G. (1995). Genetic transformation of the diatoms *Cyclotella cryptica* and *Navicula saprophila*. *Journal of Phycology*, *31*, 1004–1012.

Duran, A., Varona, R., Perez, P., & Garcia-Acha, I. (1984). Biosynthesis of fungal wall β-glucan and its inhibition by antifungal agents. In C. Nombela (Ed.), *Microbial cell wall synthesis and autolysis* (pp. 297–305). Amsterdam: Elsevier Science.

Eichinger, L., Pachebat, J. A., Glockner, G., Rajandream, M. A., Sucgang, R., Berriman, M., Song, J., Olsen, R., Szafranski, K., Xu, Q., Tunggal, B., Kummerfeld, S., Madera, M., Konfortov, B. A., Rivero, F., Bankier, A. T., Lehmann, R., Hamlin, N., Davies, R., Gaudet, P., Fey, P., Pilcher, K., Chen, G., Saunders, D., Sodergren, E., Davis, P., Kerhornou, A., Nie, X., Hall, N., Anjard, C., Hemphill, L., Bason, N., Farbrother, P., Desany, B., Just, E., Morio, T., Rost, R., Churcher, C., Cooper, J., Haydock, S., van Driessche, N., Cronin, A., Goodhead, I., Muzny, D., Mourier, T., Pain, A., Lu, M., Harper, D., Lindsay, R., Hauser, H., James, K., Quiles, M., Babu, M. M., Saito, T., Buchrieser, C., Wardroper, A., Felder, M., Thangavelu, M., Johnson, D., Knights, A., Loulseged, H., Mungall, K., Oliver, K., Price, C., Quail, M. A., Urushihara, H., Hernandez, J., Rabbinowitsch, E., Steffen, D., Sanders, M., Ma, J., Kohara, Y., Sharp, S., Simmonds, M., Spiegler, S., Tivey, A., Sugano, S., White, B., Walker, D., Woodward, J., Winckler, T., Tanaka, Y., Shaulsky, G., Schleicher, M., Weinstock, G., Rosenthal, A., Cox, E. C., Chisholm, R. L., Gibbs, R., Loomis, W. F., Platzer, M., Kay, R. R., Williams, J., Dear, P. H., Noegel, A. A., Barrell, B., & Kuspa, A. (2005). The genome of the social amoeba *Dictyostelium discoideum*. *Nature*, *435*, 43–57.

Elyakova, L. A., & Zvyagintseva, T. N. (1974). A study of the laminarins of some far-eastern, brown seaweeds. *Carbohydrate Research*, *34*, 241–248.

Fèvre, M. (1979). Digitonin solubilization and protease stimulation of β-glucan synthetases of *Saprolegnia*. *Zeitschrift Für Pflanzenphysiologie*, *95*, 129–140.

Fèvre, M. (1983a). Nucleotide effects on glucan synthesis activities of particulate enzymes from *Saprolegnia*. *Planta*, *159*, 130–135.

Fèvre, M. (1983b). Inhibitors of synthesis of lipid-linked saccharides also inhibit β-glucan synthesis by cell-free extracts of the fungus *Saprolegnia monoica*. *Journal of General Microbiology*, *129*, 3007–3013.

Fèvre, M. (1984). ATP and GTP stimulate membrane-bound but not digitonin-solubilized β-glucan synthases from *Saprolegnia monoica*. *Journal of General Microbiology*, *130*, 3279–3284.

Fèvre, M., & Dumas, C. (1977). β-Glucan synthetases from *Saprolegnia monoica*. *Journal of General Microbiology*, *103*, 297–306.

Fèvre, M., & Rougier, M. (1981). (1,3)-β-Glucan and (1,4)-β-glucan synthesis by membrane fractions from the fungus *Saprolegnia*. *Planta*, *151*, 232–241.

Freeze, H., & Loomis, W. F. (1978). Chemical analysis of stalk components of *Dictyostelium discoideum*. *Biochimica et Biophysica Acta*, *539*, 529–537.

Gajendran, K., Gonzales, M. D., Farmer, A., Archuleta, E., Win, J., Waugh, M. E., & Kamoun, S. (2006). *Phytophthora* functional genomics database (PFGD): Functional genomics of *Phytophthora*-plant interactions. *Nucleic Acids Research*, *34*, D465–D470.

Girard, V., Bulone, V., & Fèvre, M. (1992). Separation and partial peptide characterization of (1,3)-β-glucan synthase from *Saprolegnia*. *Plant Science*, *82*, 145–153.

Girard, V., & Fèvre, M. (1984). (1,4)-β-Glucan and (1,3)-β-glucan synthases are associated with the plasma membrane of the fungus *Saprolegnia*. *Planta*, *160*, 400–406.

Girard, V., & Fèvre, M. (1991). Solubilization of a membrane-bound stimulator of (1,3)-β-glucan synthase from *Saprolegnia*. *Plant Science*, *76*, 193–200.

Girard, V., Fèvre, M., Mayer, R., & Benziman, M. (1991). Cyclic diguanylic acid stimulates (1,4)-β-glucan synthase from *Saprolegnia monoica*. *FEMS Microbiology Letters*, *82*, 293–296.

Goldemberg, S. H., & Marechal, L. R. (1963). Biosynthesis of paramylon in *Euglena gracilis*. *Biochimica et Biophysica Acta*, *71*, 743–744.

Gottlieb, J. (1850). Ueber eine neue, mit stärkmehl isomere substanz. *Annalen der Chemie und Pharmacie*, *75*, 51–61.

Grenville-Briggs, L. J., Anderson, V. L., Fugelstad, J., Avrova, A. O., Bouzenzana, J., Williams, A., Wawra, S., Whisson, S. C., Birch, P. R., Bulone, V., & van West, P. (2008). Cellulose synthesis in *Phytophthora infestans* is required for normal appressorium formation and successful infection of potato. *Plant Cell*, *20*, 720–738.

Haigler, C. H., Ivanova-Datcheva, M., Hogan, P. S., Salnikov, V. V., Hwang, S., Martin, K., & Delmer, D. P. (2001). Carbon partitioning to cellulose synthesis. *Plant Molecular Biology*, *47*, 29–51.

Handa, N., & Nisizawa, K. (1961). Structural investigation of a laminarin isolated from *Eisenia bicyclis*. *Nature*, *192*, 1078–1080.

Hill, T. W., & Mullins, J. T. (1982). Association of UDPG transferase activity with cell walls of *Achlya ambisexualis*. *Mycologia*, *74*, 851–854.

Hong, Z. L., Delauney, A. J., & Verma, D. P. S. (2001). A cell plate specific callose synthase and its interaction with phragmoplastin. *Plant Cell*, *13*, 755–768.

Imai, T., Watanabe, T., Yui, T., & Sugiyama, J. (2003). The directionality of chitin biosynthesis: A revisit. *Biochemical Journal*, *374*, 755–760.

Jackson, S. L., & Heath, I. B. (1993). Roles of calcium ions in hyphal tip growth. *Microbiological Reviews*, *57*, 367–382.

Kang, M. S., Elango, N., Mattia, E., Auyoung, J., Robbins, P. W., & Cabib, E. (1984). Isolation of chitin synthetase from *Saccharomyces cerevisiae* – Purification of an enzyme by entrapment in the reaction product. *Journal of Biological Chemistry*, *259*, 4966–4972.

Kauss, H., Kohle, H., & Jeblick, W. (1983). Proteolytic activation and stimulation by $Ca^{2+}$ of glucan synthase from soybean cells. *FEBS Letters*, *158*, 84–88.

Kiss, J. Z., Roberts, E. M., Brown, R. M., & Triemer, R. E. (1988a). X-Ray and dissolution studies of paramylon storage granules from *Euglena*. *Protoplasma*, *146*, 150–156.

Kiss, J. Z., Vasconcelos, A. C., & Triemer, R. E. (1987). Structure of the euglenoid storage carbohydrate, paramylon. *American Journal of Botany*, *74*, 877–882.

Kiss, J. Z., Vasconcelos, A. C., & Triemer, R. E. (1988b). The intramembranous particle profile of the paramylon membrane during paramylon synthesis in *Euglena* (Euglenophyceae). *Journal of Phycology*, *24*, 152–157.

Kollar, R., Reinhold, B. B., Petrakova, E., Yeh, H. J. C., Ashwell, G., Drgonova, J., Kapteyn, J. C., Klis, F. M., & Cabib, E. (1997). Architecture of the yeast cell wall – (1,6)-β-Glucan interconnects mannoprotein, (1,3)-β-glucan, and chitin. *Journal of Biological Chemistry*, *272*, 17762–17775.

Koyama, M., Helbert, W., Imai, T., Sugiyama, J., & Henrissat, B. (1997). Parallel-up structure evidences the molecular directionality during biosynthesis of bacterial cellulose. *Proceedings of the National Academy of Sciences of the United States of America*, *94*, 9091–9095.

Kroth, P. G., Chiovitti, A., Gruber, A., Martin-Jezequel, V., Mock, T., Schnitzler Parker, M., Stanley, M. S., Kaplan, A., Caron, L., Weber, T., Maheswari, U., Armbrust, E. V., & Bowler, C. (2008). A model for carbohydrate metabolism in the diatom *Phaeodactylum tricornutum* deduced from comparative whole genome analysis. *PLoS ONE*, *3*, e1426.

Lai Kee Him, J., Pelosi, L., Chanzy, H., Putaux, J. L., & Bulone, V. (2001). Biosynthesis of (1,3)-β-D-glucan (callose) by detergent extracts of a microsomal fraction from *Arabidopsis thaliana*. *European Journal of Biochemistry*, *268*, 4628–4638.

Lesage, G., & Bussey, H. (2006). Cell wall assembly in *Saccharomyces cerevisiae*. *Microbiology and Molecular Biology Reviews*, *70*, 317–343.

Levina, N. N., Lew, R. R., Hyde, G. J., & Heath, I. B. (1995). The roles of $Ca^{2+}$ and plasma-membrane ion channels in hyphal tip growth of *Neurospora crassa*. *Journal of Cell Science*, *108*, 3405–3417.

Li, H. J., Bacic, A., & Read, S. M. (1997). Activation of pollen tube callose synthase by detergents – Evidence for different mechanisms of action. *Plant Physiology*, *114*, 1255–1265.

Li, J., Burton, R. A., Harvey, A. J., Hrmova, M., Wardak, A. Z., Stone, B. A., & Fincher, G. B. (2003). Biochemical evidence linking a putative callose synthase gene with (1,3)-β-D-glucan biosynthesis in barley. *Plant Molecular Biology*, *53*, 213–225.

Maeda, M., & Nisizawa, K. (1968). Fine structure of laminarin of *Eisenia bicyclis*. *Journal of Biochemistry*, *63*, 199–206.

Marchessault, R. H., & Deslandes, Y. (1979). Fine structure of (1,3)-β-D-glucans – Curdlan and paramylon. *Carbohydrate Research*, *75*, 231–242.

Marechal, L. R., & Goldemberg, S. H. (1964). Uridine diphosphate glucose-β-(1,3)-glucan β-3-glucosyltransferase from *Euglena gracilis*. *Journal of Biological Chemistry*, *239*, 3163–3167.

Margulis, L., & Schwartz, K. V. (2000). Five kingdoms: An illustrated guide to the phyla of life on earth. New York: Freeman and Co.

Mazur, P., & Baginsky, W. (1996). *In vitro* activity of (1,3)-β-D-glucan synthase requires the GTP-binding protein Rho1. *Journal of Biological Chemistry*, *271*, 14604–14609.

Meijer, H. J. G., van de Vondervoort, P. J. I., Yin, Q. Y., de Koster, C. G., Klis, F. M., Govers, F., & de Groot, P. W. J. (2006). Identification of cell wall-associated proteins from *Phytophthora ramorum*. *Molecular Plant Microbe Interactions*, *19*, 1348–1358.

Montijn, R. C., Vink, E., Müller, W. H., Verkleij, H., van den Ende, H., Henrissat, B., & Klis, F. M. (1999). Localization of synthesis of β-(1,6)-glucan in *Saccharomyces cerevisiae*. *Journal of Bacteriology*, *181*, 7414–7420.

Nodet, P., Girard, V., & Fèvre, M. (1990). Congo Red inhibits *in vitro* β-glucan synthases of *Saprolegnia*. *FEMS Microbiology Letters*, *69*, 225–228.

Pear, J. R., Kawagoe, Y., Schreckengost, W. E., Delmer, D. P., & Stalker, D. M. (1996). Higher plants contain homologs of the bacterial *celA* genes encoding the catalytic subunit of cellulose synthase. *Proceedings of the National Academy of Sciences of the United States of America*, *93*, 12637–12642.

Pelosi, L., Bulone, V., & Heux, L. (2006). Polymorphism of curdlan and (1,3)-β-D-glucans synthesized *in vitro*: A C-13 CP-MAS and X-ray diffraction analysis. *Carbohydrate Polymers*, *66*, 199–207.

Pelosi, L., Imai, T., Chanzy, H., Heux, L., Buhler, E., & Bulone, V. (2003). Structural and morphological diversity of (1,3)-β-D-glucans synthesized *in vitro* by enzymes from *Saprolegnia monoica*. Comparison with a corresponding *in vitro* product from blackberry (*Rubus fruticosus*). *Biochemistry*, *42*, 6264–6274.

Poulsen, N., Chesley, P. M., & Kroger, N. (2006). Molecular genetic manipulation of the diatom *Thalassiosira pseudonana* (Bacillariophyceae). *Journal of Phycology*, *42*, 1059–1065.

Qadota, H., Python, C. P., Inoue, S. B., Arisawa, M., Anraku, Y., Zheng, Y., Watanabe, T., Levin, D. E., & Ohya, Y. (1996). Identification of yeast Rho1p GTPase as a regulatory subunit of (1,3)-β-glucan synthase. *Science*, *272*, 279–281.

Roessler, P. G. (1987). UDP-glucose pyrophosphorylase activity in the diatom *Cyclotella cryptica* – Pathway of chrysolaminarin biosynthesis. *Journal of Phycology*, *23*, 494–498.

Roessler, P. G. (2000). More tools for diatom molecular biology research. *Journal of Phycology*, *36*, 259–260.

Ross, P., Mayer, R., & Benziman, M. (1991). Cellulose biosynthesis and function in bacteria. *Microbiological Reviews*, *55*, 35–58.

Ross, P., Weinhouse, H., Aloni, Y., Michaeli, D., Weinberger-Ohana, P., Mayer, R., Braun, S., Devroom, E., Vandermarel, G. A., Vanboom, J. H., & Benziman, M. (1987). Regulation of cellulose synthesis in *Acetobacter xylinum* by cyclic diguanylic acid. *Nature*, *325*, 279–281.

Schlüpmann, H., Bacic, A., & Read, S. M. (1993). A novel callose synthase from pollen tubes of *Nicotiana*. *Planta*, *191*, 470–481.

Schmid, F., Stone, B. A., Brownlee, R. T. C., McDougall, B. M., & Seviour, R. J. (2006). Structure and assembly of epiglucan, the extracellular (1,3;1,6)-β-glucan produced by the fungus *Epicoccum nigrum* strain F19. *Carbohydrate Research*, *341*, 365–373.

Shapiro, A., & Mullins, J. T. (2002). Hyphal tip growth in *Achlya bisexualis*. I. Distribution of (1,3)-β-glucans in elongating and non-elongating regions of the wall. *Mycologia*, *94*, 267–272.

Sing, V. O., & Bartnicki-Garcia, S. (1975). Adhesion of *Phytophthora palmivora* zoospores: Electron microscopy of cell attachment and cyst wall fibril formation. *Journal of Cell Science*, *18*, 123–132.

Smythe, C., & Cohen, P. (1991). The discovery of glycogenin and the priming mechanism for glycogen biogenesis. *European Journal of Biochemistry*, *200*, 625–631.

Størseth, T. R., Kirkvold, S., Skjermo, J., & Reitan, K. I. (2006). A branched (1,3;1,6)-β-D-glucan from the marine diatom *Chaetoceros debilis* (Bacillariophyceae) characterized by NMR. *Carbohydrate Research*, *341*, 2108–2114.

Szaniszlo, P. J., Kang, M. S., & Cabib, E. (1985). Stimulation of (1,3)-β-glucan synthetase of various fungi by nucleoside triphosphates – Generalized regulatory mechanism for cell wall biosynthesis. *Journal of Bacteriology*, *161*, 1188–1194.

Tokunaga, J., & Bartnicki-Garcia, S. (1971a). Cyst wall formation and endogenous carbohydrate utilization during synchronous encystment of *Phytophthora palmivora* zoospores. *Archiv für Mikrobiologie*, *79*, 283–292.

Tokunaga, J., & Bartnicki-Garcia, S. (1971b). Structure and differentiation of the cell wall of *Phytophthora palmivora*: Cysts, hyphae and sporangia. *Archiv für Mikrobiologie*, *79*, 293–310.

Tomos, A. D., & Northcote, D. H. (1978). Protein-glucan intermediate during paramylon synthesis. *Biochemical Journal*, *174*, 283–290.

Tyler, B. M., Tripathy, S., Zhang, X. M., Dehal, P., Jiang, R. H. Y., Aerts, A., Arredondo, F. D., Baxter, L., Bensasson, D., Beynon, J. L., Chapman, J., Damasceno, C. M. B., Dorrance, A. E., Dou, D. L., Dickerman, A. W., Dubchak, I. L., Garbelotto, M., Gijzen, M., Gordon, S. G., Govers, F., Grunwald, N. J., Huang, W., Ivors, K. L., Jones, R. W., Kamoun, S., Krampis, K., Lamour, K. H., Lee, M. K., McDonald, W. H., Medina, M., Meijer, H. J. G., Nordberg, E. K., Maclean, D. J., Ospina-Giraldo, M. D., Morris, P. F., Phuntumart, V., Putnam, N. H., Rash, S., Rose, J. K. C., Sakihama, Y., Salamov, A. A., Savidor, A., Scheuring, C. F., Smith, B. M., Sobral, B. W. S., Terry, A., Torto-Alalibo, T. A., Win, J., Xu, Z. Y., Zhang, H. B., Grigoriev, I. V., Rokhsar, D. S., & Boore, J. L. (2006). *Phytophthora* genome sequences uncover evolutionary origins and mechanisms of pathogenesis. *Science*, *313*, 1261–1266.

Varona, R., Perez, P., & Duran, A. (1983). Effect of papulacandin B on β-glucan synthesis in *Schizosaccharomyces pombe*. *FEMS Microbiology Letters*, *20*, 243–247.

Walker, T. L., Collet, C., & Purton, S. (2005). Algal transgenics in the genomic era. *Journal of Phycology*, *41*, 1077–1093.

Wang, M. C., & Bartnicki-Garcia, S. (1980). Distribution of mycolaminarins and cell wall β-glucans in the life cycle of *Phytophthora*. *Experimental Mycology*, *4*, 269–280.

Wang, M. C., & Bartnicki-Garcia, S. (1966). Biosynthesis of (1,3)-β- and (1,6)-β-linked glucan by *Phytophthora cinnamomi* hyphal walls. *Biochemical and Biophysical Research Communications*, *24*, 832–837.

Wang, M. C., & Bartnicki-Garcia, S. (1973). Novel phosphoglucans from the cytoplasm of *Phytophthora palmivora* and their selective occurrence in certain life cycle stages. *Journal of Biological Chemistry*, *248*, 4112–4118.

Wang, M. C., & Bartnicki-Garcia, S. (1974). Mycolaminarins: Storage (1,3)-β-glucans from the cytoplasm of the fungus *Phytophthora palmivora*. *Carbohydrate Research*, *37*, 331–338.

Wang, M. C., & Bartnicki-Garcia, S. (1976). Synthesis of (1,3)-β-glucan microfibrils by a cell-free extract from *Phytophthora cinnamomi*. *Archives of Biochemistry and Biophysics*, *175*, 351–354.

Wang, M. C., & Bartnicki-Garcia, S. (1982). Synthesis of noncellulosic cell-wall β-glucan by cell-free extracts from zoospores and cysts of *Phytophthora palmivora*. *Experimental Mycology*, *6*, 125–135.

Whisson, S. C., Avrova, A. O., van West, P., & Jones, J. T. (2005). A method for double-stranded RNA-mediated transient gene silencing in *Phytophthora infestans*. *Molecular Plant Pathology*, *6*, 153–163.

Wood, P. J. (1980). Specificity in the interaction of direct dyes with polysaccharides. *Carbohydrate Research*, *85*, 271–287.

Yamamoto, R., & Nevins, D. J. (1983). Degradation of a glucan containing (1,3)-β and (1,6)-β linkages by exo-(1,3)-β-D-glucanase. *Carbohydrate Research*, *122*, 217–226.

Zevenhuizen, L. P., & Bartnicki-Garcia, S. (1969). Chemical structure of the insoluble hyphal wall glucan of *Phytophthora cinnamomi*. *Biochemistry*, *8*, 1496–1502.

CHAPTER 3.3.3

# Biosynthetic Enzymes for (1-3)-β-Glucans, (1-3;1-6)-β-Glucans from Yeasts: Biochemical Properties and Molecular Biology

**Satoru Nogami and Yoshikazu Ohya**

*Department of Integrated Biosciences, Graduate School of Frontier Sciences, University of Tokyo, 5-1-5 Kashiwanoha, Kashiwa, Chiba Prefecture 277-8562 Japan*

## 1. Introduction

In yeast cells, the plasma membrane is surrounded by a layered structure called the cell wall, which is mainly composed of polysaccharides and proteins (Fig. 1). The cell wall is integral in retaining cell shape and protecting organisms from environmental changes such as osmotic shock, physical stress and exposure to chemicals (see also Chapter 4.3). In addition, the cell wall functions as a scaffold for proteins that function outside of the cell. The cell wall can be compared to a ferroconcrete structure, as both consist of a fiber-like network and a glue-like structure that confer tensile strength. However, unlike static ferroconcrete structures, the yeast cell wall is flexible and can dynamically remodel its structure, enabling yeast cells to rapidly change shape.

Synthesis of the cell wall is tightly regulated through a process that is intimately associated with cell growth and morphogenesis. Cell wall precursors are polymerized at the cell surface and then modified and connected to each other. Other materials are also made in the endomembrane system. Glucans are the major filamentous components of the cell wall structure that account for more than half of the dry weight of a budding yeast cell wall, and are required for correct functioning of the cell wall. Therefore, regulation of the synthesis of glucans is an essential step for the construction of a functional cell wall. (1,3)-β-Glucan synthase

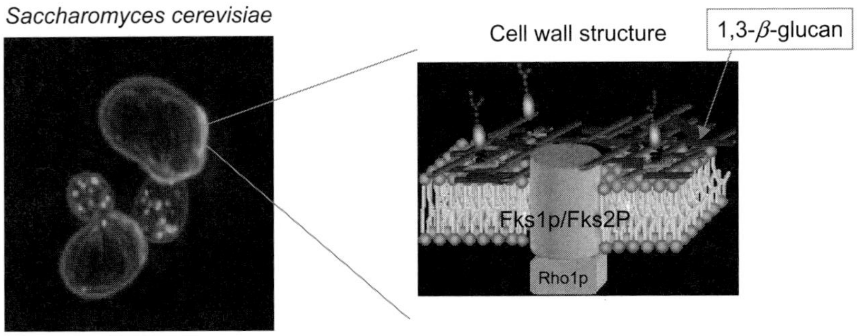


**Fig. 1: Yeast cell wall structure and (1,3)-β-glucan. In the left panel, yeast cells are stained with FITC-conjugated concanavalin A and Rhodamin phalloidin to visualize cell wall and actin filament. In the right panel, glucans and chitin are depicted in brown and yellow fibers, respectively, and mannoproteins are depicted white balls with a red tail. The colour specifications refer to colours in panels.**

activity is modulated by either directly controlling the quantity or membrane localization of regulatory subunits, or indirectly through transcriptional regulation.

In this chapter, we review the biosynthesis of β-glucans: the polymerizing enzymes and regulatory components required for β-glucan synthesis and their regulation.

## 2. Cell Walls and β-glucans in Yeast

The cell wall of yeast is not static, but rather a flexible structure that dynamically remodels in response to various exogenous and endogenous stimuli. For example, yeast cells shrink when exposed to high osmolarity and this response is reversible (Morris et al., 1986). The elasticity of the cell wall enables this shrinkage. As another example, yeast cell shape can change within 1 h, from ellipsoidal to pear-shaped, when exposed to mating pheromones (Mackay and Manney, 1974). The morphology of some yeast cells with special functions also dramatically changes to a pseudohyphal shape when nutrient availability is limited (Gimeno et al., 1992). Thus, cell wall defects, induced by either exogenous stimuli or endogenous mutations, lead to morphological abnormality and weaken cell responses to various stresses, resulting in a crisis for cell survival.

β-Glucan is a major component of yeast glucans required for correct functioning of the cell wall. β-Glucan contains (1,3) and (1,6) linkages, the former being mainly responsible for rigidity of the yeast cell wall (Fig. 2). (1,3)-β-Glucan is a polymer of (1,3)-linked β-glucan with approximately 1500 glucose residues (see also Chapter 2.1). (1,3)-β-Glucan chains are branched with (1,6)-β linkages and are connected to each other to form a fibrillar 1,3-β-glucan.

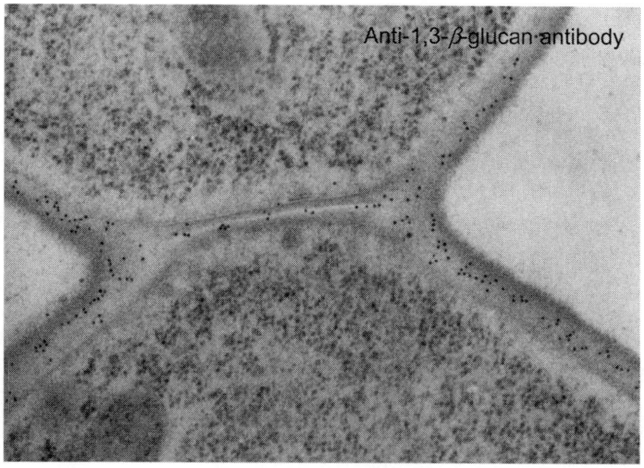


**Fig. 2: Localization of (1,3)-β-glucan in yeast cell wall. (1,3)-β-Glucan in the cell wall is visualised with antibody conjugated with gold particles.**

(1,3)-β-Glucan is also covalently linked to the other wall components. Chitin, a N-acetylglucosamine (GlcNAc) polymer, links to glucose residues of (1,3)-β-glucan through (1,4)-β-bonds. Mannoproteins can be linked to (1,6)-β-glucose chains through a processed glycosyl-phosphatidylinositol (GPI) anchor or to (1,3)-β-glucan through an alkali-sensitive bond.

## 3. Biosynthetic Enzymes for (1-3)-β-Glucans

Early biochemical studies by Enrico Cabib and colleagues in the 1980s revealed that (1,3)-linked β-glucan polymers formed from UDP-glucose in the presence of GTP and that the key enzyme, (1,3)-β-glucan synthase, is located on the plasma membrane (Shematek et al., 1980; Shematek and Cabib, 1980). Enzymatic analysis using a yeast cell membrane fraction has shown that (1,3)-β-glucan synthase activity can be separated into two fractions by treating the membrane preparation with detergent and salt, suggesting that the enzyme consists of at least two subunits: a membrane-bound catalytic subunit and a GTP-binding regulatory subunit (Kang and Cabib, 1986; Mol et al., 1994; Szaniszlo et al., 1985). However, purification and identification of the enzymatic subunits was very difficult.

### *3.1. Catalytic Subunit*

In 1995, glucan synthase was purified from a plasma membrane fraction of *Saccharomyces cerevisiae* using a special procedure called "product entrapment" (Inoue et al., 1995).

The method, a form of affinity purification based on the affinity of the enzyme for its own product, was first applied to the purification of chitin synthase (Kang et al., 1984), and later to the purification of (1,3)-β-glucan synthase from various species including *Neurospora crassa*, *Aspergillus nidulans*, *A. fumigatus*, *Candida albicans* and *S. cerevisiae* (Awald et al., 1994; Beauvais et al., 2001; Inoue et al., 1995; Kelly et al., 1996; Mio et al., 1997). After the addition of UDP-glucose as substrate for (1,3)-β-glucan synthase, the solubilized membrane preparation supplemented with GTPγS becomes turbid in a time-dependent manner, and a (1,3)-β-glucan aggregate is formed. The collected (1,3)-β-glucan aggregate contains (1,3)-β-glucan synthase that is easily released from the aggregate in the absence of UDP-glucose. A 700-fold purification can be achieved through two cycles of this simple purification procedure. Monoclonal antibodies prepared from the purified fraction identified a large 200-kDa protein that is highly enriched in the purified fraction. Determination of the partial amino acid sequence followed by cloning identified two highly homologous genes, *GSC1* and *GSC2*, encoding 1876 and 1895 amino acid proteins, respectively (Inoue et al., 1995). *GSC1* and *GSC2* are identical to *FKS1* and *FKS2*, respectively, cloned during a study of the immunosuppressant FK506 sensitive mutant (Douglas et al., 1994; Eng et al., 1994; Garrett-Engele et al., 1995; Parent et al., 1993). Hereafter, we use the names *FKS1* and *FKS2* for these genes. *FKS1* and *FKS2* encode a pair of integral membrane proteins with 16 transmembrane domains that share 88% identity.

The gene encoding the catalytic subunit of glucan synthase was identified using several different methods. Cloning of the *FKS1* gene was first reported from an FK506 hypersensitive mutant study (Parent et al., 1993). *ETG1*, *PBR1* and *CWH53* were obtained from an echinocandin B derivative (L-733, 560)-resistant mutant, a papulacandin B-resistant mutant, and a calcofluor white hypersensitive mutant, respectively, and were found to be identical to the *FKS1* gene (Castro et al., 1995; Douglas et al., 1994; Ram et al., 1994). This reflects the functional importance of this gene.

Several lines of evidence, not all direct, indicate that Fks1p and Fks2p are catalytic subunits of (1,3)-β-glucan synthase. First, *in vitro* glucan synthase activity is inhibited with the addition of an anti-200-kDa monoclonal antibody, suggesting that these proteins are required for glucan synthase activity (Mazur et al., 1995). Second, gene disruption of the *FKS1* gene leads to reduced *in vitro* (1,3)-β-glucan synthase activity and a decrease in the level of glucan in the cell (Douglas et al., 1994; Inoue et al., 1995). Deletion of the *FKS2* gene has little effect on (1,3)-β-glucan synthase activity or (1,3)-β-glucan content during vegetative growth on glucose (Inoue et al., 1995; Mazur et al., 1995). Although deletion of one of these genes is not lethal, double-deletion of the two genes is lethal, suggesting that they have overlapping and essential functions for yeast growth (Inoue et al., 1995; Mazur et al., 1995). Third, a protein homologous to Fks1p from several fungi co-purifies with enzyme activity during solubilization and product entrapment (Beauvais et al., 2001; Inoue et al., 1995; Kelly et al., 1996; Mio et al., 1997). Finally, in a product entrapment fraction, *N. crassa* Fks1p protein with high similarity to yeast Fks1p and Fks2p cross-linked to a photo-activated UDP-glucose analogue (Schimoler-O'Rourke et al., 2003).

Fks1p and its homologues in other fungi lack the UDP-glucose binding site, R/K-X-G-G, commonly found in other glycosyltransferases that use UDP-glucose as a substrate (Farkas et al., 1990). Analysis of temperature-sensitive mutants of Fks1p identified functional domains in the catalytic subunit of 1,3-β-glucan synthase (Dijkgraaf et al., 2002). Fks1p has a four-domain structure: an N-terminal ~400 amino acid cytosolic tail involved in (1,3)-β-glucan synthase activity, a six-transmembrane domain (TMD) containing an ~300 amino acid region required for correct localization to polarized growth sites, a cytosolic ~600 amino acid putative catalytic region, and a C-terminal ~600 amino acid ten-TMD-containing domain involved in localization to the cell surface (Dijkgraaf et al., 2002). Based on the cell wall β-glucan content, mutants can be placed into three groups. Group A consists of mutants with reduced β-glucan content and all carry mutations within the central cytoplasmic loop of Fks1p. Group B consists of mutants without net alterations in their cell wall β-glucan content and mutation points are distributed on the C-terminal three domains of Fks1p. Group C consists of one mutant with a higher content of both (1,3)-β- and (1,6)-β-glucan, suggesting hyperactivity for the synthesis of both polymers (Dijkgraaf et al., 2002).

### 3.2. Regulatory Subunit

Biochemical analysis in early studies suggested the involvement of GTPase in (1,3)-β-glucan synthesis (Kang and Cabib, 1986; Mol et al., 1994). The GTP-binding regulatory subunit was partially purified and its molecular weight was estimated to be 20 kDa by photoaffinity labeling (Mol et al., 1994), but impurity of the sample prevented further detailed analysis of this regulatory component. However, genetic analysis in *Schizosaccharomyces pombe* suggested that geranylgeranyltransferase I (GGTase I) plays a role in the regulation of glucan synthase (Diaz et al., 1993; Ribas et al., 1991). One of the most important substrates of GGTase I is a Rho-type GTPase, acting as a molecular switch that monitors and receives upstream signals for cell morphogenesis (Inoue et al., 1996; Takai et al., 2001). These preceding studies suggested that the Rho type GTPase is the regulatory subunit of glucan synthase. Our group and that of Cabib's revealed that Rho1p, one of the Rho-type GTPases in yeast, regulates glucan synthase activity (Drgonova et al., 1996; Qadota et al., 1996). Rho1 was detected in the purified glucan synthase fraction. Some alleles of the *rho1* conditional mutant result in reduced glucan synthase activity and low glucan content (Qadota et al., 1996). The low glucan synthase activity can be restored by adding recombinant Rho1 to the reaction mixture (Mazur and Baginsky, 1996; Qadota et al., 1996). Low-glucan synthesis activity of a temperature-sensitive *FKS1* mutant was restored by overexpressing the constitutive active allele of *RHO1* (Sekiya-Kawasaki et al., 2002).

## 3.3. Other Proteins Associated with Glucan Synthase Activity

Besides Fks1p/Fks2p and Rho1p, several other proteins were found to locate near the (1,3)-β-glucan synthase complex. Photo-crosslinking experiments using a UDP-glucose analog revealed that Pma1p appears in close association with the glucan synthase complex in *N. crassa* (Schimoler-O'Rourke et al., 2003). Pma1p, a proton pump on the plasma membrane, maintains transmembrane electrochemical proton gradients. It may provide the driving force for either translocating products or maintaining an acidic cell wall environment near the membrane, both of which are important for glucan synthesis. Two proteins of 40 and 18 kDa in the membrane fraction in yeast were also identified as photo-activatable cross-linking echinocandin LY303366 (Radding et al., 1998). The 40-kDa protein could be a homolog of Pil1p and Lsp1p, which are sphingolipid-dependent regulators of cell wall integrity signaling, although their role in (1,3)-β-glucan synthase activity is unknown (Edlind and Katiyar, 2004). Attempts to reconstitute (1,3)-β-glucan synthase activity from independently purified components has not yet succeeded because of incomplete homogeneous purification.

## 3.4. Regulation of (1,3)-β-Glucan Synthesis in Yeast

Evidence is accumulating that (1,3)-β-glucan synthesis is precisely regulated in response to temporally preceding upstream signals by multiple mechanisms.

### 3.4.1. Factors involved in the regulation of glucan synthase activity via Rho1p

Like other small G proteins, Rho1 cycles between its active GTP-bound state and its inactive GDP-bound state. Although the glucan synthase from wild-type cells requires GTPγS for maximum activity, the enzyme from dominant active *RHO1* mutant cells has full activity even in the absence of GTPγS, indicating that activation of Rho1p is important in glucan synthesis (Qadota et al., 1996). The GTP-fixed forms of Rho1 (Q68L and G19V) suppress the glucan synthesis defect of the temperature-sensitive *FKS1* mutant (Sekiya-Kawasaki et al., 2002). Seven multicopy suppressors of the temperature-sensitive *FKS1* mutant, Wsc1p, Wsc3p, Mtl1p, Rom2p, Lre1p, Zds1p, and Msb1p, were isolated (Sekiya-Kawasaki et al., 2002).

A defect in (1,3)-β-glucan staining with Aniline Blue in the bud, reduction of glucan content, and reduced (1,3)-β-glucan incorporation in temperature-sensitive *FKS1* mutant cells were restored by overexpression of these proteins, indicating that (1,3)-β-glucan synthesis was restored. Suppression of (1,3)-β-glucan incorporation by active Rho1p was not enhanced by additional overexpression of *ROM2*, *WSC3*, *MTL1*, *LRE1*, and *ZDS1*, suggesting that these suppressors act on (1,3)-β-glucan synthase through Rho1p activation (Sekiya-Kawasaki et al., 2002). The results are consistent with literature reports showing that the GDP/GTP exchange factor of Rho1p, Rom2p, is a putative surface sensor on the cell wall; Wsc family proteins (Wsc1p and Wsc3p) and Mtl1p are upstream of the Rho1p signal pathway (Ketela et al., 1999; Ozaki et al., 1996; Rajavel et al., 1999; Verna et al., 1997). In contrast, *MSB1* showed a significant increase in *in vitro* (1,3)-β-glucan synthesis even in the presence of GTPγS, suggesting that Msb1p does not act on the GDP/GTP exchange on Rho1p (Sekiya-Kawasaki et al., 2002).

### *3.4.2. Post-translational Modification of Rho1p*

A post-translational modification of Rho1p is required for (1,3)-β-glucan synthesis. The C-terminus of the Rho-type small GTPase is modified with a geranylgeranyl group by type I geranylgeranyl transferase (GGTase I) (Ohya et al., 1993; Qadota et al., 1992). In *cal1-1* mutant cells carrying a mutation in the b subunit of GGTase I (Ohya et al., 1991), unmodified Rho1p accumulates and the proportion of Rho1p in the soluble fraction increases, indicating geranylgeranylation of Rho1p is required for Rho1p localization to the membrane (Ohya et al., 1996). In *cal1-1* mutant cells, (1,3)-β-glucan synthase activity is reduced (Inoue et al., 1999). A geranylgeranylation-deficient recombinant mutant Rho1p cannot interact with Fks1p or rescue the low *in vitro* (1,3)-β-glucan synthase activity of membranes prepared from *rho1-3* mutant cells (Inoue et al., 1999). These results indicate Rho1 modification by the geranylgeranyl group is required for both binding with and activation of Fks1p. The amount of Fks1p in *cal1-1* cells is low and not restored by overexpressing constitutive active Rho1, suggesting that Fks1p is easily degraded when not bound to geranylgeranylated Rho1p (Inoue et al., 1999).

### *3.4.3. Transcriptional Regulation of the FKS Genes*

Although *FKS1* is constitutively expressed in the yeast cell cycle, levels of the *FKS1* transcript peak at the late G1/S phase of the cell cycle in a Swi4-dependent manner, a component of the Swi4-Swi6 cell cycle box-binding factor (SBF) transcription factor (Igual et al., 1996; Mazur et al., 1995; Ram et al., 1998). The Chromatin immunoprecipitation (ChIP)-on-chip assay revealed that SBF binds to the *FKS1* promoter, indicating that *FKS1* expression is regulated by SBF in growing cells during the cycle (Iyer et al., 2001).

*FKS2* expression is induced by signals for cell integrity impairment, calcineurin, and carbon source depletion (Zhao et al., 1998). Crz1p is a transcription factor that activates transcription in a calcineurin-dependent manner (Matheos et al., 1997; Stathopoulos and Cyert, 1997) and the upstream sequence of the *FKS2* gene has a Crz1p-binding sequence.

Regulation of the *FKS2* expression by calcineurin explains why a deletion of the *FKS1* gene encoding a catalytic subunit of (1,3)-β-glucan synthase results in hypersensitivity against FK506 and shows synthetic lethality with the calcineurin defect (Douglas et al., 1994; Eng et al., 1994; Garrett-Engele et al., 1995; Parent et al., 1993). In mammalian cells, FK506 is an immunosuppressant that binds to an immunophilin, FKBP12, and represses phosphatase activity of calcineurin, a protein phosphatase 2B that plays a central role in T-cell activation (Clipstone and Crabtree, 1992; Fruman et al., 1992; Liu et al., 1991; O'Keefe et al., 1992). FK506 also inhibits calcineurin activity in yeast, resulting in transcriptional repression of various genes, including *FKS2* (Foor et al., 1992; Mazur et al., 1995). Since *FKS2* is the only gene for the catalytic subunit of (1,3)-β-glucan synthase in *FKS1* gene-deleted cells required for vegetative growth, transcriptional down-regulation of the *FKS2* gene by FK506 causes depression of (1,3)-β-glucan synthase, resulting in hypersensitivity to FK506.

### *3.4.4. Location of (1,3)-β-Glucan Synthase*

The catalytic and regulatory subunits of (1,3)-β-glucan synthase are co-localized at the growing bud site and neck of the large bud (Qadota et al., 1996), positions where glucan synthesis is required. An immunofluorescence study with an antibody that specifically reacts with the active form of Rho1p revealed that Rho1p is activated at the bud tip, suggesting that not all Rho1p bound to Fks1 at the bud is activated (Abe et al., 2003). Since both Fks1p and Fks2p have transmembrane domains, newly translated Fks1p and Fks2p are thought to be transported to the plasma membrane via secretory vesicles. Indeed, Fks1p and Fks2p accumulate in secretory vesicles when vesicular transport is blocked by *sec* mutations (Abe et al., 2003). Fks1p and Fks2p co-fractionate with Rho1p, but not with the GTP-bound form of Rho1p, suggesting that (1,3)-β-glucan synthase is inactivated (Abe et al., 2003). Overexpression of Rom2p, a GDP/GTP exchange factor of Rho1p, induces unusual activation of Rho1p in secretory vesicles, resulting in ectopic synthesis of (1,3)-β-glucan in the vesicles (Abe et al., 2003). These results indicate that the GTP-bound active form of Rho1p is required for the active form of (1,3)-β-glucan synthase.

### *3.4.5. Movement of Fks1p in the Plasma Membrane*

Besides the growing bud site and the neck of the large bud, Fks1p is located at the presumed bud site of the mother cell (Utsugi et al., 2002). These locations overlap with cortical actin patches, which are localized in the cell cortex around areas of cell wall remodeling and associated with the plasma membrane (Pruyne and Bretscher, 2000a; Pruyne and Bretscher, 2000b).

Similar to the movement of actin patches (Doyle and Botstein, 1996; Waddle et al., 1996), Fks1p-GFP (-green fluorescent protein) spots move randomly in the plasma membrane, frequently changing direction (Fig. 3) (Utsugi et al., 2002). The randomness of (1,3)-β-glucan synthase movements explains a previous description of (1,3)-β-glucan microfibrils: yeast (1,3)-β-glucan microfibrils assembled in the cell wall are intertwined in all directions (Kopecká et al., 1974; Kreger and Kopecká, 1976). The movement of actin and Fks1p-GFP is abolished in *las*17 and *arc*18 mutants, whose product activates or constitutes a component of the Arp2/3 complex that is required for actin patch movement (Utsugi et al., 2002). This indicates the movement of Fks1p is actin patch-dependent. Cell staining with Aniline Blue, a specific fluorescent dye for (1,3)-β-glucan, revealed that *las17* and *arc18* mutant cells have an uneven thickness of β-glucan (Utsugi et al., 2002). Rough filamentous structures were observed in electron micrographs of *las17* mutant cells, suggesting that not only the thickness but also the structure of the cell wall is altered in *las17* cells. *las17* and *arc18* mutant cells have a significantly higher tendency for cell lysis under normal growth conditions, suggesting that the mutant cells lyse at a region of the cell wall with lowered (1,3)-β-glucan synthetic activity (Utsugi et al., 2002). The results indicate that movement of Fks1p in the plasma membrane is required for correct cell wall remodeling.

### 3.5. Glucan Synthesis in Spore Formation

Diploid yeast cells enter the meiotic cell cycle and form an ascus containing four haploid spores when nutrients become limited (Kupiec et al., 1997; Neiman, 2005). Each spore is surrounded by a spore wall, which plays a central role in protecting the cell from environmental damage. The spore wall consists of four distinct layers: the inner two layers consists of β-glucan and mannan, components that are similar to those found in the vegetative cell wall; the outer layer consists of

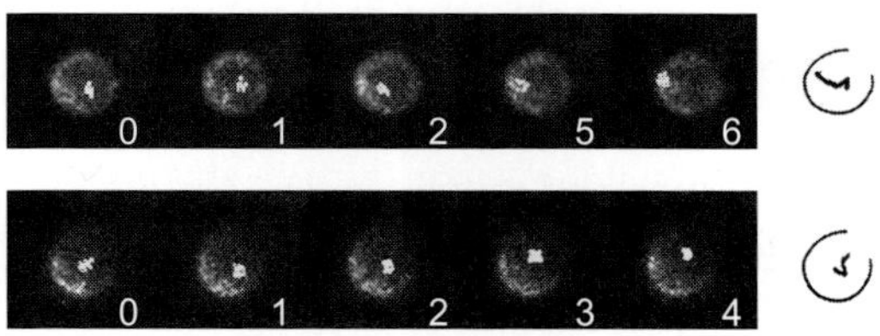


**Fig. 3: Movement of (1,3)-β-glucan synthase on the cell surface. Fluorescent spots that are coloured in sequential frames represent images of Fks1p-GFP mobility. The images were taken at 1-s intervals except that the 4th image in the upper column was taken 3-s after the 3rd image. The path of the coloured dot is shown on the right.**

chitosan, a polymer of (1,4)-β-linked glucosamine; and the outermost layer consists of dityrosine polymers. Both outer layers are specific to the spore wall (Briza et al., 1988; Briza et al., 1986).

A genome-wide analysis revealed the reduced expression of *FKS1* and elevated expression of *FKS2* and *FKS3* during meiosis (Chu et al., 1998) and a sporulation defect of the *fks2* or *fks3* deletion diploid (Deutschbauer et al., 2002), suggesting a role of (1,3)-β-glucan synthase in meiosis. Fks3p, a third Fks1p homolog protein, was found by homology searching and shares 56% identity with Fks1p and Fks2p (Mazur et al., 1995) although (1,3)-β-glucan synthase activity of Fks3p has not been reported. Deletion of the *FKS3* gene does not cause obvious cell wall defects during vegetative growth and no genetic interactions occur between *FKS3* and *FKS1* or *FKS2* (Dijkgraaf et al., 2002; Lesage et al., 2004).

A biochemical and cell biological analysis of Fks2p and Fks3p revealed a function for these proteins in sporulation (Ishihara et al., 2007). Deletion of *FKS2* or *FKS3* results in apparently normal meiosis, forming four viable spores in one ascus; however, the spore wall of these mutants is aberrant. *fks2* spores have abnormal tube-like structures between the plasma membrane and the outmost layers in the spore wall. Using immuno-electron microscopy, β-glucan was detected only in the innermost layers in *fks2* mutant spores, in contrast to the ubiquitous distribution in the spore wall layers in wild-type spores. In contrast, the spore wall of *fks3* spores has an uneven thickness, and in some cases, includes cytoplasm between the inner and outer layers. β-Glucan was found to be located between the inner and outer layers of the spore walls of *fks3* mutants (Ishihara et al., 2007). Consistent with the microscopic observations, the reduced hexose concentrations of the alkali-insoluble fraction of *fks2* and *fks2 fks3* spores, mainly composed of β-glucan and chitosan, and the abnormal release of (1,3)-β-glucan from the alkali-insoluble fraction of *fks3* and *fks2 fks3* spores by recombinant (1,3)-β-glucanase suggest a less organized β-glucan structure (Ishihara et al., 2007). These results suggest that Fks2p and Fks3, but not Fks1p, have distinct roles in spore wall formation.

An abnormality in morphology and organization of β-glucan polymers in spore walls relates to stress resistance of the spores. Spores of *fks2, fks3*, and *fks2 fks3* double mutants are sensitive to diethyl ether, high temperature (55°C), and ethanol, as compared to the wild type and *fks1* mutant, although all wild-type and mutant spores have high germination ability under normal conditions (Ishihara et al., 2007). A multi-copy of the *FKS3* gene does not suppress stress sensitivity of *fks2* mutant spores, and vice versa, suggesting that the two genes have distinct functions in spore wall formation (Ishihara et al., 2007).

The role of Fks2p in meiosis is thought to be as a catalytic subunit of (1,3)-β-glucan synthase, since a multi-copy of both *FKS2* and *FKS2* promoter-driven *FKS1* suppresses stress sensitivity of *fks2* mutant spores (Ishihara et al., 2007). In contrast, the role of Fks3p is unclear. The expression of *FKS1* promoter-driven Fks3p during vegetative growth does not affect (1,3)-β-glucan synthase activity *in vitro* but effectively suppresses the growth defect of the temperature-sensitive *fks1* mutant by stabilizing Rho1p, suggesting that *FKS3* is required for normal spore wall formation because it affects the upstream regulation of (1,3)-β-glucan synthase (Ishihara et al., 2007). One interesting speculation is that Fks3p interacts with accessory components of (1,3)-β-glucan synthase that modify (1,3)-β-glucan synthase activity and increase (1,3)-β-glucan synthase activity during β-glucan layer assembly.

### 3.6. Cell Wall Integrity Checkpoint in Yeast

High β-glucan synthesis is required during bud growth, since the cell wall must be newly synthesized in the early stages of the cell cycle. A mutant carrying a temperature-sensitive allele of the *FKS1* gene, *fks1-1154*, accumulates cells with tiny buds at restrictive temperatures (Suzuki et al., 2004). Since the DNA content doubled but nuclear division did not take place in these cells, this suggests that the cell cycle is arrested at the G2/M boundary. The spindle pole body (SPB), a microtubule organization center in yeast, duplicated but did not separate (Fig. 4). These results indicate that a defect in (1,3)-β-glucan synthesis results in cell cycle arrest at G2/M, and suggest the presence of checkpoint mechanisms. A cell cycle checkpoint either transiently arrests or delays at a specific period of the cell cycle until events required for correct cell cycle function are in order (Lew et al., 1997). The expression of a mitotic cell-cycle engine, *CLB2*, was repressed in *fks1-1154* cells at restrictive temperatures, suggesting the checkpoint mechanism regulates the cell cycle via *CLB2* at the expression level, unlike other cell cycle checkpoints in yeast (Suzuki et al., 2004). To identify factors that are involved in this checkpoint system, mutants with lowered viability in defective (1,3)-β-glucan synthesis were screened. *wac1* is a mutant allele of the *ARP1* gene (Suzuki et al., 2004). Arp1p is a component of the dynactin complex. Dynactin is an activator of cytoplasmic dynein and is required for nuclear migration and correct orientation of spindles in mitosis (Muhua et al., 1994). In *wac1* mutant cells, nuclear migration was not affected, suggesting that this special allele of *ARP1* is defective only in the cell wall integrity checkpoint. The dynactin complex consists of Arp1p, Jnm1p, and Nip100p in yeast (Hildebrandt and Hoyt, 2000). Moreover, the regions of Arp1 that are responsible for checkpoint activity differ from those for positioning the nucleus (Igarashi et al., 2005). Overexpression of Fkh2p, a component of the transcription factor complex of Mcm1p–Fkh2p that regulates transcription of *CLB1* and *CLB2* in G2/M, overrides the cell cycle arrest caused by the (1,3)-β-glucan synthesis defect, indicating that G2 arrest is associated with *FKH2* regulation (Suzuki et al., 2004).

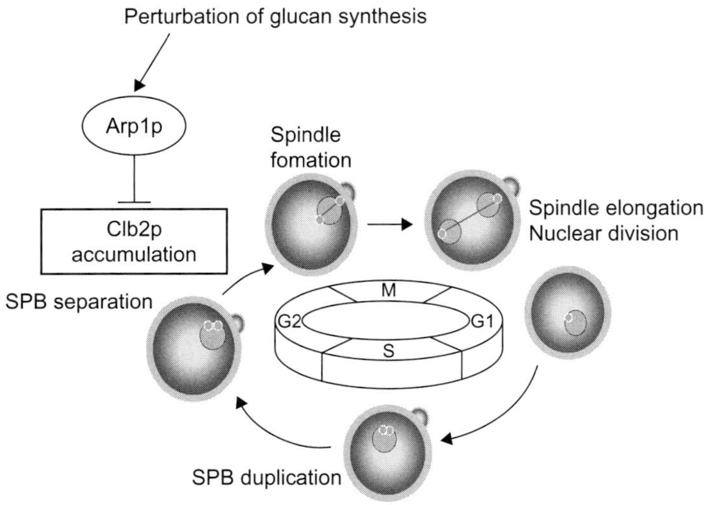


**Fig. 4: Cell wall integrity checkpoint as a novel checkpoint control. If cell wall synthesis is perturbed, Arp1p represses CLB2 mRNA and inhibits accumulation of Clb2p. This results in cell cycle arrested at G2/M boundary, with a tiny bud (because of defect in bud growth) and duplicated SPBs. The colour specifications refer to colours in panels.**

### *3.7. Cross Talk in the Regulation of Other Cell Wall Components*

Disruption of the *FKS1* gene leads to a decrease in the level of β-glucan and an increase in chitin and mannoprotein levels in cells (Dallies et al., 1998; Dijkgraaf et al., 2002; Ram et al., 1998). *FKS2* mRNA is elevated by various environmental stimuli, including heat shock, high calcium concentrations, addition of a mating pheromone or the reducing agent dithiothreitol, and exposure to cell wall- and cell surface-damaging agents such as caspofungin and amphotericin B (Agarwal et al., 2003; de Nobel et al., 2000; Lagorce et al., 2003; Mazur et al., 1995; Ram et al., 1994; Travers et al., 2000; Zhao et al., 1998). Deletions of genes required for cell wall assembly, such as *fks1, gas1, kre6, mnn9* and *smi1*, also results in high levels of *FKS2* mRNA (Lagorce et al., 2003; Terashima et al., 2000). These observations suggest roles for Fks2p in response to environmental stresses.

Collection of deletion mutants of all 6000 yeast genes has enabled large-scale double-mutant construction and screening for interacting double-mutant combinations. This synthetic genetic interaction analysis was applied to *FKS1*, *CHS1*, *CHS3*, *GAS1*, *FKS2* and *SMI1*, and networks of genetic interactions have been analysed (Lesage et al., 2004; Lesage et al., 2005). Large-scale phenotypic analyses, surveying genes required for growth in the presence of cell wall-damaging reagents, including caspofungin, Congo red, K1 killer toxin, and Calcofluor white, have also been conducted (Lesage et al., 2004; Lussier et al., 1997; Markovich et al., 2004; Page et al., 2003; Sanchez-Perez et al., 2004). Genes found in these studies are cell wall genes, genes encoding cell wall stress sensors and components of the PKC cell integrity pathway, and genes required for either cell polarization or with roles in vesicular transport or endocytosis. These interactions reflect the underlying compensatory processes between wall polymers.

### 3.8. Glucan Synthase Inhibitors

Whereas (1,3)-β-glucan is essential and widely distributed in the fungal kingdom, it is absent in mammals; therefore, glucan synthase is thought to be a favored antifungal target. Echinocandin drugs, such as caspofungin, were the first class of antifungal compounds found to target (1,3)-β-glucan synthase. They are cyclic hexapeptides, N-linked to a fatty acyl side chain (Nyfeler and Keller-Schierlein, 1974). Three echinocandins, caspofungin, anidulafungin, and micafungin, are FDA-approved in the USA and other countries for treatment of fungal infections. Their broad-spectrum antifungal activity against *Candida* and *Aspergillus* without cross-resistance to existing antifungal agents (including amphotericin and nystatin) is effective against azole-resistant yeasts and molds. Papulacandins (glycolipids) are a second class of (1,3)-β-glucan synthase inhibitors consisting of modified disaccharides linked to two fatty acyl chains (Traxler et al., 1977). Despite efforts in medicinal chemistry to improve the efficacies of these compounds, the papulacandins have not been developed because they have limited potency in animal models. Acidic terpenoids are a third class of (1,3)-β-glucan synthase inhibitors (Onishi et al., 2000) and include enfumafungin, ascosteroside, arundifungin, and ergokonin A.

Recent large-scale surveillance studies revealed the outstanding potency of echinocandin drugs against clinical isolates of *Candida* species (Pfaller et al., 2008). As the use of echinocandin drugs broadens, a reduction in susceptibility to echinocandins is expected to become a serious problem. As described in a previous section, a library of yeast knockout mutants was analyzed for either hypersensitivity or reduced susceptibility to caspofungin, and revealed that drug susceptibility is influenced by complex cellular pathways (Lesage et al., 2005; Markovich et al., 2004). Although these pathways have the potential to contribute to clinical resistance, they are themselves unlikely to be the cause of treatment failure because of their modest effect on susceptibility to drugs (Lesage et al., 2005; Markovich et al., 2004).

While other cell response/adaptive mechanisms can result in elevated minimum inhibitory concentration values, only the *FKS1* mechanism is firmly associated with clinical resistance. Early genetic studies with caspofungin in *S. cerevisiae* and *C. albicans* and recent studies in clinical isolates of *Candida* species indicate that mutations that confer reduced echinocandin susceptibility map to *FKS1* (reviewed in Perlin, 2007). Most amino acid substitutions mapped to a highly conserved region in the *FKS* family (Gardiner et al., 2005; Kahn et al., 2007; Laverdière et al., 2006; Miller et al., 2006). Fks1 hot-spot mutations that confer resistance to echinocandins in *C. albicans* and other *Candida* spp. alter the kinetic inhibition properties of glucan synthase (Garcia-Effron et al., 2008; Gardiner et al., 2005). The interaction site of Fks1p/Fks2p with echinocandins has not yet been determined. Elucidating the mode of action of echinocandins on (1,3)-β-glucan synthase will improve the effectiveness of the above drugs.

## 4. Biosynthetic Enzymes for (1-6)-β-Glucans

(1,6)-β-Glucan is a second β-linked glucan of the cell wall in *S. cerevisiae*. In vegetative cells, the (1,6)-β-polymer comprises ~12% of the cell wall polysaccharides. Whereas (1,3)-β-glucan forms a microfibrillar structure, (1,6)-β-glucan forms a branched amorphous structure. The polymer has an average chain length of ~350 glucose residues, with the (1,6)-β-backbone branched with (1,6)-β-side chains via (3,6)-substituted glucose residues on ~15% of the residues (Kollár et al., 1997; Magnelli et al., 2002; Manners et al., 1973) (see also Chapter 2.1). The (1,6)-β-glucan in yeast is thought to act as a molecular glue that connects components of the cell wall.

Despite the identification of genes that affect (1,6)-β-glucan synthesis using genetic approaches, the catalytic enzyme for (1,6)-β-glucan has not been identified. Synthesis of (1,6)-β-glucan seems to take place at the plasma membrane (Montijn et al., 1999), but several endoplasmic reticulam and Golgi proteins are also in some unknown way involved in (1,6)-β-glucan synthesis. Many genes have been identified that affect the amounts of (1,6)-β-glucan in *S. cerevisiae* throughout the secretory pathway to the cell wall (Page et al., 2003; Shahinian and Bussey, 2000). Of these genes, *KRE5* encodes a putative ER protein with significant similarity to UDP-glucose: glycoprotein glucosyltransferase (UGGT) enzymes, and loss of Kre5p results in no detectable (1,6)-β-glucan (Azuma et al., 2002; Meaden et al., 1990). Although Kre5p is important for (1,6)-β-glucan synthesis in *S. cerevisiae*, the cell wall from a deletion mutant of Kre5p in *C. albicans* contains low but residual (1,6)-β-glucan, suggesting that Kre5p is not a (1,6)-β-glucan synthase (Herrero et al., 2004). *KRE6* and *SKN1* encode a pair of homologous Golgi proteins with significant similarity to family 16 glycoside hydrolases, and simultaneous loss of both Kre6p and Skn1p also results in a severe (1,6)-β-glucan defect (Montijn et al., 1999).

One reason for the difficulty in the identification of the (1,6)-β-glucan synthase has been the absence of an assay system. Recently, an assay system based on a specific antibody against

(1,6)-β-glucan was developed (Vink et al., 2004). This *in vitro* assay requires a crude cell-free membrane preparation, UDP-glucose, and GTP, suggesting the involvement of a GTP-binding protein in regulating (1,6)-β-glucan synthesis. Consistently, overexpression of Rho1p enhances *in vitro* activity, suggesting the involvement of Rho1p in (1,6)-β-glucan synthesis, as well as in (1,3)-β-glucan synthesis. Hopefully, novel insights into (1,6)-β-glucan synthesis, including the catalytic enzyme, will be revealed using this assay.

## 5. Conclusions

β-Glucans, major polymers of the yeast cell wall structure, have important roles in the function of the cell wall. (1,3)-β-glucan is the main β-glucan synthesized by a (1,3)-β-glucan synthase complex, consisting of a catalytic subunit encoded by the *FKS1* and *FKS2* genes, and a regulatory subunit encoded by the *RHO1* gene. (1,3)-β-Glucan synthesis is spatio-temporally modulated by directly controlling enzyme activity or indirectly regulating expressions. These regulatory mechanisms enable organisms to respond to exogenous and endogenous stimuli to protect against environmental changes.

## References

Abe, M., Qadota, H., Hirata, A., & Ohya, Y. (2003). Lack of GTP-bound Rho1p in secretory vesicles of *Saccharomyces cerevisiae*. *Journal of Cell Biology*, *162*, 85–97.

Agarwal, A. K., Rogers, P. D., Baerson, S. R., Jacob, M. R., Barker, K. S., Cleary, J. D., Walker, L. A., Nagle, D. G., & Clark, A. M. (2003). Genome-wide expression profiling of the response to polyene, pyrimidine, azole, and echinocandin antifungal agents in *Saccharomyces cerevisiae*. *Journal of Biological Chemistry*, *278*, 34998–35015.

Awald, P., Frost, D., Drake, R., & Selitrennikoff, C. (1994). (1,3) β-Glucan synthase activity of *Neurospora crassa*: identification of a substrate-binding protein. *Biochimica et Biophysica Acta*, *1201*, 312–320.

Azuma, M., Levinson, J. N., Page, N., & Bussey, H. (2002). *Saccharomyces cerevisiae* Big1p, a putative endoplasmic reticulum membrane protein required for normal levels of cell wall β-1,6-glucan. *Yeast*, *19*, 783–793.

Beauvais, A., Bruneau, J., Mol, P., Buitrago, M., Legrand, R., & Latgé, J. (2001). Glucan synthase complex of *Aspergillus fumigatus*. *Journal of Bacteriology*, *183*, 2273–2279.

Briza, P., Ellinger, A., Winkler, G., & Breitenbach, M. (1988). Chemical composition of the yeast ascospore wall. The second outer layer consists of chitosan. *Journal of Biological Chemistry*, *263*, 11569–11574.

Briza, P., Winkler, G., Kalchhauser, H., & Breitenbach, M. (1986). Dityrosine is a prominent component of the yeast ascospore wall. A proof of its structure. *Journal of Biological Chemistry*, *261*, 4288–4294.

Castro, C., Ribas, J. C., Valdivieso, M. H., Varona, R., del Rey, F., & Duran, A. (1995). Papulacandin B resistance in budding and fission yeasts: isolation and characterization of a gene involved in (1,3)β-D-glucan synthesis in *Saccharomyces cerevisiae*. *Journal of Bacteriology*, *177*, 5732–5739.

Chu, S., DeRisi, J., Eisen, M., Mulholland, J., Botstein, D., Brown, P. O., & Herskowitz, I. (1998). The transcriptional program of sporulation in budding yeast. *Science*, *282*, 699–705.

Clipstone, N., & Crabtree, G. (1992). Identification of calcineurin as a key signalling enzyme in T-lymphocyte activation. *Nature*, *357*, 695–697.

Dallies, N., François, J., & Paquet, V. (1998). A new method for quantitative determination of polysaccharides in the yeast cell wall. Application to the cell wall defective mutants of *Saccharomyces cerevisiae*. *Yeast*, *14*, 1297–1306.

de Nobel, H., Ruiz, C., Martin, H., Morris, W., Brul, S., Molina, M., & Klis, F. (2000). Cell wall perturbation in yeast results in dual phosphorylation of the Slt2/Mpk1 MAP kinase and in an Slt2-mediated increase in *FKS2*-lacZ expression, glucanase resistance and thermotolerance. *Microbiology*, *146*(Pt 9), 2121–2132.

Deutschbauer, A., Williams, R., Chu, A., & Davis, R. (2002). Parallel phenotypic analysis of sporulation and postgermination growth in *Saccharomyces cerevisiae*. *Proceedings of the National Academy of Sciences of the United States of America* , *99*, 15530–15535.

Diaz, M., Sanchez, Y., Bennett, T., Sun, C. R., Godoy, C., Tamanoi, F., Duran, A., & Perez, P. (1993). The *Schizosaccharomyces pombe* $cwg2^+$ gene codes for the β subunit of a geranylgeranyltransferase type I required for β-glucan synthesis. *EMBO Journal*, *12*, 5245–5254.

Dijkgraaf, G. J., Abe, M., Ohya, Y., & Bussey, H. (2002). Mutations in Fks1p affect the cell wall content of β-1,3- and β-1,6-glucan in *Saccharomyces cerevisiae*. *Yeast*, *19*, 671–690.

Douglas, C. M., Foor, F., Marrinan, J. A., Morin, N., Nielsen, J. B., Dahl, A. M., Mazur, P., Baginsky, W., Li, W., el-Sherbeini, M., et al. (1994). The *Saccharomyces cerevisiae FKS1 (ETG1)* gene encodes an integral membrane protein which is a subunit of 1,3-β-D-glucan synthase. *Proceedings of the National Academy of Sciences of the United States of America*, *91*, 12907–12911.

Doyle, T., & Botstein, D. (1996). Movement of yeast cortical actin cytoskeleton visualized *in vivo*. *Proceedings of the National Academy of Sciences of the United States of America*, *93*, 3886–3891.

Drgonova, J., Drgon, T., Tanaka, K., Kollar, R., Chen, G. C., Ford, R. A., Chan, C. S., Takai, Y., & Cabib, E. (1996). Rho1p, a yeast protein at the interface between cell polarization and morphogenesis. *Science*, *272*, 277–279.

Edlind, T. D., & Katiyar, S. K. (2004). The echinocandin "target" identified by cross-linking is a homolog of Pil1 and Lsp1, sphingolipid-dependent regulators of cell wall integrity signaling. *Antimicrobial Agents and Chemotherapy*, *48*, 4491.

Eng, W. K., Faucette, L., McLaughlin, M. M., Cafferkey, R., Koltin, Y., Morris, R. A., Young, P. R., Johnson, R. K., & Livi, G. P. (1994). The yeast *FKS1* gene encodes a novel membrane protein, mutations in which confer FK506 and cyclosporin A hypersensitivity and calcineurin-dependent growth. *Gene*, *151*, 61–71.

Farkas, I., Hardy, T. A., DePaoli-Roach, A. A., & Roach, P. J. (1990). Isolation of the *GSY1* gene encoding yeast glycogen synthase and evidence for the existence of a second gene. *Journal of Biological Chemistry*, *265*, 20879–20886.

Foor, F., Parent, S. A., Morin, N., Dahl, A. M., Ramadan, N., Chrebet, G., Bostian, K. A., & Nielsen, J. B. (1992). Calcineurin mediates inhibition by FK506 and cyclosporin of recovery from alpha-factor arrest in yeast. *Nature*, *360*, 682–684.

Fruman, D., Klee, C., Bierer, B., & Burakoff, S. (1992). Calcineurin phosphatase activity in T lymphocytes is inhibited by FK 506 and cyclosporin A. *Proceedings of the National Academy of Sciences of the United States of America*, *89*, 3686–3690.

Garcia-Effron, G., Katiyar, S. K., Park, S., Edlind, T. D., & Perlin, D. S. (2008). A naturally occurring proline-to-alanine amino acid change in *Candida parapsilosis*, *Candida orthopsilosis* and *Candida metapsilosis* accounts for reduced echinocandin susceptibility. *Antimicrobial Agents and Chemotherapy*, *52*, 2305–2312.

Gardiner, R. E., Souteropoulos, P., Park, S., & Perlin, D. S. (2005). Characterization of *Aspergillus fumigatus* mutants with reduced susceptibility to caspofungin. *Medical Mycology*, *43*(Suppl 1), S299–S305.

Garrett-Engele, P., Moilanen, B., & Cyert, M. S. (1995). Calcineurin, the $Ca^{2+}$/calmodulin-dependent protein phosphatase, is essential in yeast mutants with cell integrity defects and in mutants that lack a functional vacuolar $H^+$-ATPase. *Molecular and Cellular Biology*, *15*, 4103–4114.

Gimeno, C. J., Ljungdahl, P. O., Styles, C. A., & Fink, G. R. (1992). Unipolar cell divisions in the yeast S. cerevisiae lead to filamentous growth: regulation by starvation and *RAS*. *Cell*, *8*, 1077–1090.

Herrero, A. B., Magnelli, P., Mansour, M. K., Levitz, S. M., Bussey, H., & Abeijon, C. (2004). *KRE5* gene null mutant strains of *Candida albicans* are avirulent and have altered cell wall composition and hypha formation properties. *Eukaryotic Cell*, *3*, 1423–1432.

Hildebrandt, E., & Hoyt, M. (2000). Mitotic motors in *Saccharomyces cerevisiae*. *Biochimica et Biophysica Acta*, *1496*, 99–116.

Igarashi, R., Suzuki, M., Nogami, S., & Ohya, Y. (2005). Molecular dissection of *ARP1* regions required for nuclear migration and cell wall integrity checkpoint functions in *Saccharomyces cerevisiae*. *Cell Structure and Function*, *30*, 57–67.

Igual, J., Johnson, A., & Johnston, L. (1996). Coordinated regulation of gene expression by the cell cycle transcription factor Swi4 and the protein kinase C MAP kinase pathway for yeast cell integrity. *EMBO Journal*, *15*, 5001–5013.

Inoue, S., Qadota, H., Arisawa, M., Watanabe, T., & Ohya, Y. (1999). Prenylation of Rho1p is required for activation of yeast 1, 3-β-glucan synthase. *Journal of Biological Chemistry*, *274*, 38119–38124.

Inoue, S., Takewaki, N., Takasuka, T., Mio, T., Adachi, M., Fujii, Y., Miyamoto, C., Arisawa, M., Furuichi, Y., & Watanabe, T. (1995). Characterization and gene cloning of 1,3-β-D-glucan synthase from *Saccharomyces cerevisiae*. *European Journal of Biochemistry*, *231*, 845–854.

Inoue, S. B., Qadota, H., Arisawa, M., Anraku, Y., Watanabe, T., & Ohya, Y. (1996). Signaling toward yeast 1,3-β-glucan synthesis. *Cell Structure and Function*, *21*, 395–402.

Ishihara, S., Hirata, A., Nogami, S., Beauvais, A., Latge, J. P., & Ohya, Y. (2007). Homologous subunits of 1,3-beta-glucan synthase are important for spore wall assembly in *Saccharomyces cerevisiae*. *Eukaryotic Cell*, *6*, 143–156.

Iyer, V., Horak, C., Scafe, C., Botstein, D., Snyder, M., & Brown, P. (2001). Genomic binding sites of the yeast cell-cycle transcription factors SBF and MBF. *Nature*, *409*, 533–538.

Kahn, J., Garcia-Effron, G., Hsu, M., Park, S., Marr, K., & Perlin, D. (2007). Acquired echinocandin resistance in a *Candida krusei* isolate due to modification of glucan synthase. *Antimicrobial Agents and Chemotherapy*, *51*, 1876–1878.

Kang, M., Elango, N., Mattia, E., Au-Young, J., Robbins, P., & Cabib, E. (1984). Isolation of chitin synthetase from *Saccharomyces cerevisiae*. Purification of an enzyme by entrapment in the reaction product. *Journal of Biological Chemistry*, *259*, 14966–14972.

Kang, M. S., & Cabib, E. (1986). Regulation of fungal cell wall growth: a guanine nucleotide-binding, proteinaceous component required for activity of (1->3)-β-D-glucan synthase. *Proceedings of the National Academy of Sciences of the United States of America*, *83*, 5808–5812.

Kelly, R., Register, E., Hsu, M. J., Kurtz, M., & Nielsen, J. (1996). Isolation of a gene involved in 1,3-β-glucan synthesis in *Aspergillus nidulans* and purification of the corresponding protein. *Journal of Bacteriology*, *178*, 4381–4391.

Ketela, T., Green, R., & Bussey, H. (1999). *Saccharomyces cerevisiae* Mid2p is a potential cell wall stress sensor and upstream activator of the *PKC1-MPK1* cell integrity pathway. *Journal of Bacteriology*, *181*, 3330–3340.

Kollár, R., Reinhold, B., Petráková, E., Yeh, H., Ashwell, G., Drgonová, J., Kapteyn, J., Klis, F., & Cabib, E. (1997). Architecture of the yeast cell wall. β(1–>6)-glucan interconnects mannoprotein, β(1–>)3-glucan, and chitin. *Journal of Biological Chemistry*, *272*, 17762–17775.

Kopecká, M., Gabriel, M., & Necas, O. (1974). A method of isolating anucleated yeast protoplasts unable to synthesize the glucan fibrillar component of the wall. *Journal of General Microbiology*, *81*, 111–120.

Kreger, D., & Kopecká, M. (1976). On the nature and formation of the fibrillar nets produced by protoplasts of *Saccharomyces cerevisiae* in liquid media: an electronmicroscopic, x-ray diffraction and chemical study. *Journal of General Microbiology*, *92*, 207–220.

Kupiec, M., Byers, B., Esposito, R. E., & Mitchell, A. P. (1997). Meiosis and sporulation in *Saccharomyces cerevisiae*. In J. R. Pringle, J. R. Broach, & E. W. Jones (Eds.), *The molecular and cellular biology of the yeast Saccharomyces: Cell cycle and cell biology.* (pp. 889–1036) Cold Spring Harbor, New York: Cold Spring Harbor Laboratory Press.

Lagorce, A., Hauser, N., Labourdette, D., Rodriguez, C., Martin-Yken, H., Arroyo, J., Hoheisel, J., & François, J. (2003). Genome-wide analysis of the response to cell wall mutations in the yeast *Saccharomyces cerevisiae*. *Journal of Biological Chemistry*, *278*, 20345–20357.

Laverdière, M., Lalonde, R., Baril, J., Sheppard, D., Park, S., & Perlin, D. (2006). Progressive loss of echinocandin activity following prolonged use for treatment of *Candida albicans* oesophagitis. *Journal of Antimicrobial Chemotherapy*, *57*, 705–708.

Lesage, G., Sdicu, A. M., Menard, P., Shapiro, J., Hussein, S., & Bussey, H. (2004). Analysis of β-1,3-glucan assembly in *Saccharomyces cerevisiae* using a synthetic interaction network and altered sensitivity to caspofungin. *Genetics*, *167*, 35–49.

Lesage, G., Shapiro, J., Specht, C., Sdicu, A., Ménard, P., Hussein, S., Tong, A., Boone, C., & Bussey, H. (2005). An interactional network of genes involved in chitin synthesis in *Saccharomyces cerevisiae*. *BMC Genetics*, *6*, 8.

Lew, D. J., Weinert, T., & Pringle, J. R. (1997). Cell cycle control in *Saccharomyces cerevisiae*. In J. R. Pringle, J. R. Broach, & E. W. Jones (Eds.), *The molecular and cellular biology of the yeast Saccharomyces: cell cycle and cell biology.* (pp. 607–695) Cold Spring Harbor, New York: Cold Spring Harbor Laboratory Press.

Liu, J., Farmer, J. J., Lane, W., Friedman, J., Weissman, I., & Schreiber, S. (1991). Calcineurin is a common target of cyclophilin-cyclosporin A and FKBP-FK506 complexes. *Cell*, *66*, 807–815.

Lussier, M., White, A. M., Sheraton, J., di Paolo, T., Treadwell, J., Southard, S. B., Horenstein, C. I., Chen-Weiner, J., Ram, A. F., Kapteyn, J. C., Roemer, T. W., Vo, D. H., Bondoc, D. C., Hall, J., Zhong, W. W., Sdicu, A. M., Davies, J., Klis, F. M., Robbins, P. W., & Bussey, H. (1997). Large scale identification of genes involved in cell surface biosynthesis and architecture in *Saccharomyces cerevisiae*. *Genetics*, *147*, 435–450.

Mackay, V., & Manney, T. R. (1974). Mutations affecting sexual conjugation and related processes in Isolation and phenotypic characterization of nonmating mutants. *Saccharomycee Cerevisiae I. Genetics*, *76*, 255–271.

Magnelli, P., Cipollo, J., & Abeijon, C. (2002). A refined method for the determination of *Saccharomyces cerevisiae* cell wall composition and β-1,6-glucan fine structure. *Analytical Biochemistry*, *301*, 136–150.

Manners, D., Masson, A., Patterson, J., Björndal, H., & Lindberg, B. (1973). The structure of a β-(1–>6)-D-glucan from yeast cell walls. *Biochemical Journal*, *135*, 31–36.

Markovich, S., Yekutiel, A., Shalit, I., Shadkchan, Y., & Osherov, N. (2004). Genomic approach to identification of mutations affecting caspofungin susceptibility in *Saccharomyces cerevisiae*. *Antimicrobial Agents and Chemotherapy, 48*, 3871–3876.

Matheos, D. P., Kingsbury, T. J., Ahsan, U. S., & Cunningham, K. W. (1997). Tcn1p/Crz1p, a calcineurin-dependent transcription factor that differentially regulates gene expression in *Saccharomyces cerevisiae*. *Genes & Development*, *11*, 3445–3458.

Mazur, P., & Baginsky, W. (1996). In vitro activity of 1,3-β-D-glucan synthase requires the GTP-binding protein Rho1. *Journal of Biological Chemistry*, *271*, 14604–14609.

Mazur, P., Morin, N., Baginsky, W., el-Sherbeini, M., Clemas, J. A., Nielsen, J. B., & Foor, F. (1995). Differential expression and function of two homologous subunits of yeast 1,3-β-D-glucan synthase. *Molecular and Cellular Biology*, *15*, 5671–5681.

Meaden, P., Hill, K., Wagner, J., Slipetz, D., Sommer, S., & Bussey, H. (1990). The yeast KRE5 gene encodes a probable endoplasmic reticulum protein required for (1->6)-β-D-glucan synthesis and normal cell growth. *Molecular and Cellular Biology*, *10*, 3013–3019.

Miller, C., Lomaestro, B., Park, S., & Perlin, D. (2006). Progressive esophagitis caused by *Candida albicans* with reduced susceptibility to caspofungin. *Pharmacotherapy*, *26*, 877–880.

Mio, T., Adachi-Shimizu, M., Tachibana, Y., Tabuchi, H., Inoue, S. B., Yabe, T., Yamada-Okabe, T., Arisawa, M., Watanabe, T., & Yamada-Okabe, H. (1997). Cloning of the *Candida albicans homolog* of *Saccharomyces cerevisiae GSC1/FKS1* and its involvement in β-1,3-glucan synthesis. *Journal of Bacteriology*, *179*, 4096–4105.

Mol, P. C., Park, H. M., Mullins, J. T., & Cabib, E. (1994). A GTP-binding protein regulates the activity of (1–>3)-beta-glucan synthase, an enzyme directly involved in yeast cell wall morphogenesis. *Journal of Biological Chemistry*, *269*, 31267–31274.

Montijn, R., Vink, E., Müller, W., Verkleij, A., Van Den Ende, H., Henrissat, B., & Klis, F. (1999). Localization of synthesis of β1,6-glucan in *Saccharomyces cerevisiae*. *Journal of Bacteriology*, *181*, 7414–7420.

Morris, G. J., Winters, L., Coulson, G. E., & Clarke, K. J. (1986). Effect of osmotic stress on the ultrastructure and viability of the yeast *Saccharomyces cerevisiae. Journal of General Microbiology*, *132*, 2023–2034.

Muhua, L., Karpova, T., & Cooper, J. (1994). A yeast actin-related protein homologous to that in vertebrate dynactin complex is important for spindle orientation and nuclear migration. *Cell*, *78*, 669–679.

Neiman, A. (2005). Ascospore formation in the yeast *Saccharomyces cerevisiae. Microbiology and Molecular Biology Reviews*, *69*, 565–584.

Nyfeler, R., & Keller-Schierlein, W. (1974). Metabolites of microorganisms. 143. Echinocandin B, a novel polypeptide-antibiotic from *Aspergillus nidulans var. echinulatus*: isolation and structural components. *Helvetica Chimica Acta*, *57*, 2459–2477.

O'Keefe, S., Tamura, J., Kincaid, R., Tocci, M., & O'Neill, E. (1992). FK-506- and CsA-sensitive activation of the interleukin-2 promoter by calcineurin. *Nature*, *357*, 692–694.

Ohya, Y., Caplin, B., Qadota, H., Tibbetts, M., Anraku, Y., Pringle, J., & Marshall, M. (1996). Mutational analysis of the β-subunit of yeast geranylgeranyl transferase I. *Molecular and General Genetics*, *252*, 1–10.

Ohya, Y., Goebl, M., Goodman, L., Petersen-Bjørn, S., Friesen, J., Tamanoi, F., & Anraku, Y. (1991). Yeast *CAL1* is a structural and functional homologue to the *DPR1 (RAM)* gene involved in *ras* processing. *Journal of Biological Chemistry*, *266*, 12356–12360.

Ohya, Y., Qadota, H., Anraku, Y., Pringle, J., & Botstein, D. (1993). Suppression of yeast geranylgeranyl transferase I defect by alternative prenylation of two target GTPases, Rho1p and Cdc42p. *Molecular Biology of the Cell*, *4*, 1017–1025.

Onishi, J., Meinz, M., Thompson, J., Curotto, J., Dreikorn, S., Rosenbach, M., Douglas, C., Abruzzo, G., Flattery, A., Kong, L., Cabello, A., Vicente, F., Pelaez, F., Diez, M. T., Martin, I., Bills, G., Giacobbe, R., Dombrowski, A., Schwartz, R., Morris, S., Harris, G., Tsipouras, A., Wilson, K., & Kurtz, M. B. (2000). Discovery of novel antifungal (1,3)-β-D-glucan synthase inhibitors. *Antimicrobial Agents and Chemotherapy*, *44*, 368–377.

Ozaki, K., Tanaka, K., Imamura, H., Hihara, T., Kameyama, T., Nonaka, H., Hirano, H., Matsuura, Y., & Takai, Y. (1996). Rom1p and Rom2p are GDP/GTP exchange proteins (GEPs) for the Rho1p small GTP binding protein in *Saccharomyces cerevisiae. EMBO Journal*, *15*, 2196–2207.

Page, N., Gerard-Vincent, M., Menard, P., Beaulieu, M., Azuma, M., Dijkgraaf, G. J., Li, H., Marcoux, J., Nguyen, T., Dowse, T., Sdicu, A. M., & Bussey, H. (2003). A *Saccharomyces cerevisiae* genome-wide mutant screen for altered sensitivity to K1 killer toxin. *Genetics*, *163*, 875–894.

Parent, S. A., Nielsen, J. B., Morin, N., Chrebet, G., Ramadan, N., Dahl, A. M., Hsu, M. J., Bostian, K. A., & Foor, F. (1993). Calcineurin-dependent growth of an FK506- and CsA-hypersensitive mutant of *Saccharomyces cerevisiae*. *Journal of General Microbiology*, *139*, 2973–2984.

Perlin, D. S. (2007). Resistance to echinocandin-class antifungal drugs. *Drug Resistance Updates*, *10*, 121–130.

Pfaller, M., Boyken, L., Hollis, R., Kroeger, J., Messer, S., Tendolkar, S., & Diekema, D. (2008). In vitro susceptibility of invasive isolates of *Candida* spp. to anidulafungin, caspofungin, and micafungin: Six years of global surveillance. *Journal of Clinical Microbiology*, *46*, 150–156.

Pruyne, D., & Bretscher, A. (2000a). Polarization of cell growth in yeast. *Journal of Cell Science*, *113*(Pt 4), 571–585.

Pruyne, D., & Bretscher, A. (2000b). Polarization of cell growth in yeast I. Establishment and maintenance of polarity states. *Journal of Cell Science*, *113*(Pt 3), 365–375.

Qadota, H., Ishii, I., Fujiyama, A., Ohya, Y., & Anraku, Y. (1992). *RHO* gene products, putative small GTP-binding proteins, are important for activation of the *CAL1/CDC43* gene product, a protein geranylgeranyltransferase in *Saccharomyces cerevisiae*. *Yeast*, *8*, 735–741.

Qadota, H., Python, C. P., Inoue, S. B., Arisawa, M., Anraku, Y., Zheng, Y., Watanabe, T., Levin, D. E., & Ohya, Y. (1996). Identification of yeast Rho1p GTPase as a regulatory subunit of 1,3-β-glucan synthase. *Science*, *272*, 279–281.

Radding, J., Heidler, S., & Turner, W. (1998). Photoaffinity analog of the semisynthetic echinocandin LY303366: identification of echinocandin targets in *Candida albicans*. *Antimicrobial Agents and Chemotherapy*, *42*, 1187–1194.

Rajavel, M., Philip, B., Buehrer, B. M., Errede, B., & Levin, D. E. (1999). Mid2 is a putative sensor for cell integrity signaling in *Saccharomyces cerevisiae*. *Molecular and Cellular Biology*, *19*, 3969–3976.

Ram, A. F., Kapteyn, J. C., Montijn, R. C., Caro, L. H., Douwes, J. E., Baginsky, W., Mazur, P., van den Ende, H., & Klis, F. M. (1998). Loss of the plasma membrane-bound protein Gas1p in *Saccharomyces cerevisiae* results in the release of β1,3-glucan into the medium and induces a compensation mechanism to ensure cell wall integrity. *Journal of Bacteriology*, *180*, 1418–1424.

Ram, A. F., Wolters, A., Ten Hoopen, R., & Klis, F. M. (1994). A new approach for isolating cell wall mutants in *Saccharomyces cerevisiae* by screening for hypersensitivity to calcofluor white. *Yeast*, *10*, 1019–1030.

Ribas, J., Diaz, M., Duran, A., & Perez, P. (1991). Isolation and characterization of *Schizosaccharomyces pombe* mutants defective in cell wall (1-3)β-D-glucan. *Journal of Bacteriology*, *173*, 3456–3462.

Sanchez-Perez, I., Rodriguez-Hernandez, C. J., Manguan-Garcia, C., Torres, A., Perona, R., & Murguia, J. R. (2004). FK506 sensitizes mammalian cells to high osmolarity by modulating p38 MAP kinase activation. *Cellular and Molecular Life Sciences*, *61*, 700–708.

Schimoler-O'Rourke, R., Renault, S., Mo, W., & Selitrennikoff, C. (2003). *Neurospora crassa* FKS protein binds to the (1,3)β-glucan synthase substrate, UDP-glucose. *Current Microbiology*, *46*, 408–412.

Sekiya-Kawasaki, M., Abe, M., Saka, A., Watanabe, D., Kono, K., Minemura-Asakawa, M., Ishihara, S., Watanabe, T., & Ohya, Y. (2002). Dissection of upstream regulatory components of the Rho1p effector, 1,3-β-glucan synthase, in *Saccharomyces cerevisiae*. *Genetics*, *162*, 663–676.

Shahinian, S., & Bussey, H. (2000). β-1,6-Glucan synthesis in *Saccharomyces cerevisiae*. *Molecular Microbiology*, *35*, 477–489.

Shematek, E. M., Braatz, J. A., & Cabib, E. (1980). Biosynthesis of the yeast cell wall. I. Preparation and properties of β-(1->3)glucan synthetase. *Journal of Biological Chemistry*, *255*, 888–894.

Shematek, E. M., & Cabib, E. (1980). Biosynthesis of the yeast cell wall. II. Regulation of β-(1->3)glucan synthetase by ATP and GTP. *Journal of Biological Chemistry*, *255*, 895–902.

Stathopoulos, A., & Cyert, M. (1997). Calcineurin acts through the *CRZ1/TCN1*-encoded transcription factor to regulate gene expression in yeast. *Genes & Development*, *11*, 3432–3444.

Suzuki, M., Igarashi, R., Sekiya, M., Utsugi, T., Morishita, S., Yukawa, M., & Ohya, Y. (2004). Dynactin is involved in a checkpoint to monitor cell wall synthesis in *Saccharomyces cerevisiae*. *Nature Cell Biology*, *6*, 861–871.

Szaniszlo, P. J., Kang, M. S., & Cabib, E. (1985). Stimulation of β(1->3)glucan synthetase of various fungi by nucleoside triphosphates: generalized regulatory mechanism for cell wall biosynthesis. *Journal of Bacteriology*, *161*, 1188–1194.

Takai, Y., Sasaki, T., & Matozaki, T. (2001). Small GTP-binding proteins. *Physiological Reviews*, *81*, 153–208.

Terashima, H., Yabuki, N., Arisawa, M., Hamada, K., & Kitada, K. (2000). Up-regulation of genes encoding glycosylphosphatidylinositol (GPI)-attached proteins in response to cell wall damage caused by disruption of *FKS1* in *Saccharomyces cerevisiae*. *Molecular and General Genetics*, *264*, 64–74.

Travers, K., Patil, C., Wodicka, L., Lockhart, D., Weissman, J., & Walter, P. (2000). Functional and genomic analyses reveal an essential coordination between the unfolded protein response and ER-associated degradation. *Cell*, *101*, 249–258.

Traxler, P., Gruner, J., & Auden, J. (1977). Papulacandins, a new family of antibiotics with antifungal activity I. Fermentation, isolation, chemical and biological characterization of papulacandins A, B, C, D and E. *Journal of Antibiotics (Tokyo)*, *30*, 289–296.

Utsugi, T., Minemura, M., Hirata, A., Abe, M., Watanabe, D., & Ohya, Y. (2002). Movement of yeast 1,3-β-glucan synthase is essential for uniform cell wall synthesis. *Genes to Cells*, *7*, 1–9.

Verna, J., Lodder, A., Lee, K., Vagts, A., & Ballester, R. (1997). A family of genes required for maintenance of cell wall integrity and for the stress response in *Saccharomyces cerevisiae*. *Proceedings of the National Academy of Sciences of the United States of America*, *94*, 13804–13909.

Vink, E., Rodriguez-Suarez, R. J., Gerard-Vincent, M., Ribas, J. C., de Nobel, H., van den Ende, H., Duran, A., Klis, F. M., & Bussey, H. (2004). An *in vitro* assay for (1 -> 6)-β-D-glucan synthesis in *Saccharomyces cerevisiae*. *Yeast*, *21*, 1121–1131.

Waddle, J., Karpova, T., Waterston, R., & Cooper, J. (1996). Movement of cortical actin patches in yeast. *Journal of Cell Biology*, *132*, 861–870.

Zhao, C., Jung, U. S., Garrett-Engele, P., Roe, T., Cyert, M. S., & Levin, D. E. (1998). Temperature-induced expression of yeast *FKS2* is under the dual control of protein kinase C and calcineurin. *Molecular and Cellular Biology*, *18*, 1013–1022.

CHAPTER 3.3.4

# Biochemical and Molecular Properties of Biosynthetic Enzymes for (1,3)-β-Glucans in Embryophytes, Chlorophytes and Rhodophytes


**Lynette Brownfield[1], Monika Doblin[2], Geoffrey B. Fincher[3] and Antony Bacic[2,4]**

[1]*Department of Biology, University of Leicester, University Road, Leicester, LE1 7RH, UK*

[2]*Plant Cell Biology Research Centre, School of Botany, University of Melbourne, VIC, Australia*

[3]*Australian Centre for Plant Functional Genomics, University of Adelaide, Glen Osmond, SA, Australia*

[4]*Australian Centre for Plant Functional Genomics, School of Botany, University of Melbourne, VIC, Australia*


## 1.A Introduction

(1,3)-β-Glucans[1] are widely distributed in higher and lower plants. Whereas many (1,3)-β-glucans are linear, unsubstituted chains of (1,3)-β-glucosyl residues, structural variants include branch-on-branch (1,3;1,6)-β-glucans, side-chain-branched (1,3;1,6)-β-glucans, and cyclic (1,3;1,6)-β-glucans (see Chapter 2.1). The (1,3)-β-glucans of plants fulfil a number of roles and, *inter alia*, may function as specialized components of some cell walls. Certain plant (1,3)-β-glucans are referred to as callose, on the basis of their intense yellow, UV-induced fluorescence in the presence of the Aniline Blue fluorochrome. Here, we will review the characteristics and biology of (1,3)-β-glucans in the embryophytes (land plants), and in selected lower plants, namely the green algae (chlorophytes) and the red algae (rhodophytes).

[1] The prefix D- and L- referring to monosaccharide configurations are omitted throughout except where ambiguities might arise.

## 1.B (1,3)-β-Glucans in Embryophytes

The following section briefly summarizes more extensive reviews of the biology of callose by Kauss (1996), Stone and Clarke (1992), and chapters in this book. Callose is deposited in a number of specialized, extracellular locations during the development of a plant. Callose is usually not considered to be an integral structural component of the mature cell wall of most tissues, but it is nevertheless found in specialized walls and transiently where the wall is developing. For example, callose is found at the cell plate between dividing cells before being degraded as the wall matures (see Chapter 4.4.1). Callose is also found lining the plasmodesmatal canal and pores on sieve plates in the phloem. It is present transiently in cotton fibres (a specialized ovule hair), root hairs, spiral thickenings of tracheids, guard cells, lenticels and abscission zones. The role of callose in wall biology is unclear but it has been suggested to act as a scaffold for the addition of other cell wall components, as a permeability barrier, or as a strengthening substance (Stone and Clarke, 1992).

Callose is also found in various locations in plant reproductive tissues (see Chapter 4.4.3). The pollen mother cell is lined by a callose layer that subsequently surrounds the tetrad of microspores (Worral et al., 1992). Deposition of this layer starts shortly before, and continues through, microspore meiosis. The callose is degraded once the pollen grain exine is complete. Callose is also found in the walls of cells of megasporocytes during embryo-sac formation, and this callose is also degraded during development (Tucker et al., 2001). The pollen-tube cell wall is another important location of callose (see Chapter 4.4.3). The pollen tube grows extracellularly through the transmitting tract of the style to deliver the two sperm cells to the embryo sac, enabling double fertilization to occur. Callose is a major component of the pollen-tube wall (>80% in *Nicotiana*), where it forms a thick, inner wall behind the pollen-tube tip (Rae et al., 1985; Ferguson et al., 1998; Li et al., 1999). The pollen tube is the only known case in which callose acts as the main structural polysaccharide throughout most of the growth and development of a higher plant tissue (Stone and Clarke, 1992; Li et al., 1999). Pollen-tube callose is believed to have a mechanical role as a load-bearing material (Parre and Geitmann, 2005).

Apart from developmentally regulated callose deposition, plants also produce callose in response to wounding, particularly under certain conditions in which the plasma membrane is perturbed (see Chapters 4.4.4 and 4.4.5). This wound-activated callose is rapidly deposited in many cell types in response to physiological stress such as desiccation or metal toxicity (Shaeffer and Walton, 1990; Kartusch, 2003; Hirano et al., 2006), chemical treatment, mechanical wounding or pathogen challenge (Stone and Clarke, 1992). In the latter situation, callose is rapidly laid down at the inside surface of the wall adjacent to the plasma membrane, and although it was originally proposed to provide an additional barrier to penetration of the invading pathogen into the protoplast (Stone and Clarke, 1992), more recent results suggest that its function may be more complex (see Jacobs et al., 2003; Nishimura et al., 2003).

## I.C The Synthesis of (1,3)-β-Glucan in Embryophytes

The polysaccharide synthase responsible for the production of callose is (1,3)-β-glucan synthase (UDP-Glc: 1,3-β-glucan 3-β-glycosyl transferase, EC 2.4.1.34; Stone and Clarke, 1992). It is almost certainly located in the plasma membrane and uses UDP-Glc as a substrate, its binding site lying on the cytoplasmic surface of the plasma membrane (Raymond et al., 1978; Mueller and Maclachlan, 1983; Kudlicka and Brown, 1997; Delmer, 1999). As discussed below, there remain a number of unanswered questions about the identity of callose synthases, but the weight of evidence suggests that they are members of the GT48 family of glycosyltransferases (Coutinho and Henrissat, 1999; see Section 1.C.b.) and are integral membrane, possibly multi-subunit enzyme complexes with molecular sizes greater than 500 kDa (Thelen and Delmer, 1986; Eiberger and Wassermann, 1987; Sloan et al., 1987; Kudlicka and Brown, 1997; Li et al., 2003; Kjell et al., 2004; see Section 1.C.e). Assumed to be included within each complex is a catalytic subunit(s), as well as other components that regulate the activity of the enzyme. Plant extracts from many sources display high levels of callose synthase activity *in vitro*. This activity can be assayed by measuring the incorporation of radiolabelled Glc from UDP-[$^{14}$C]-Glc or UDP-[$^{3}$H]-Glc into ethanol-insoluble products (Stone and Clarke, 1992). The products of the *in vitro* reactions have been characterized both enzymatically and chemically, and shown to be (1,3)-β-glucan.

The identification of the gene family that encodes the presumed plant callose synthase catalytic subunits was facilitated by work on the genes responsible for the synthesis of (1,3)-β-glucan in fungi. In the yeast *Saccharomyces cerevisiae*, mutant analyses suggested that the *FKS* (FK506 supersensitive) genes that encode large (>160 kDa), membrane-bound polypeptides are essential for (1,3)-β-glucan biosynthesis (Douglas et al., 1994a; Douglas et al., 1994b; Eng et al., 1994; Ram et al., 1994; Castro et al., 1995; Garrett-Engele et al., 1995; Inoue et al., 1995; Mazur et al., 1995; Cabib et al., 2001; Dijkgraaf et al., 2002); a conclusion supported by biochemical studies (Kang and Cabib, 1986; Mol et al., 1994; Inoue et al., 1995; Schimoler-O'Rourke et al., 2003; see Chapters 3.3.3 and 4.3). Shin and Brown (1998) first observed that a 170-kDa polypeptide enriched in callose synthase activity from cotton fibres (*Gossypium hirsutum*) yielded a tryptic peptide showing some similarity with the yeast FKS catalytic subunit. Subsequent searches of EST databases for similar plant genes led to the identification of a family of candidate genes for callose synthase catalytic subunits. The family has been denoted the Glucan Synthase-Like (*GSL*) gene family (Saxena and Brown, 2000) and the proteins encoded by these genes are classified in the GT48 family of glycosyltransferases (Coutinho and Henrissat, 1999; http://afmb.cnrs-mrs.fr/CAZY/). Since 2000, considerable evidence supporting a role for GSL proteins in callose synthesis has accumulated through molecular and functional analyses and through biochemical studies. Although no other candidates for the callose synthase catalytic subunit have been identified (see below), proof of catalytic activity remains elusive.

### I.C.a The GSL Proteins

The GSLs are large proteins that vary in size from about 1770 to 1950 amino acids in *Arabidopsis* (Saxena and Brown, 2000; Hong et al., 2001a), and are predicted to contain many transmembrane helices, with the precise number varying between 13 and 19 depending upon the specific GSL family member and the transmembrane helix prediction program used (Cui et al., 2001; Doblin et al., 2001; Verma and Hong, 2001; Østergaard et al., 2002; Li et al., 2003). The predicted topology shown in Fig. 1 is typical of GSL proteins and the related FKS proteins. There is an $NH_2$-terminal region of ~350–500 amino acids that is often predicted to be cytoplasmic although some prediction programs place it in the extracellular space, followed by a membrane spanning region that in most GSL proteins is predicted to contain between six and nine transmembrane helices. The large central 'catalytic' region of ~620–780 amino acids is predicted to be cytoplasmic in all GSL proteins. At the COOH-terminal end there is another membrane-spanning region predicted to contain between eight and 10 transmembrane helices. Differences in the predicted number of transmembrane helices alters the predicted location of the COOH-terminus, although most prediction programs place it in the cytoplasm.

The GSL and FKS proteins are most similar to each other in the central, cytoplasmic domain (Cui et al., 2001; Doblin et al., 2001; Hong et al., 2001a; Østergaard et al., 2002). In this region the GSL and FKS proteins are approximately 30% identical and 50% similar at the amino acid level (Cui et al., 2001) although the identity can be over 80% between particular members of the GSL or FKS family (Douglas, 2001). The high degree of conservation in this region has led to the hypothesis that it may contain the catalytic site. Sequence similarity is lowest in the $NH_2$-terminal region of the GSL and FKS proteins, which led Cui et al. (2001) to suggest that this region may contain regulatory sites, with the sequence divergence being attributable to differences in regulatory mechanisms between different GSL and FKS proteins.

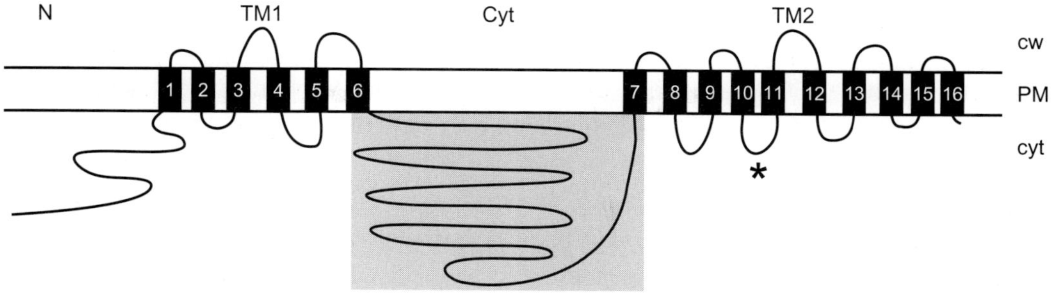


**Fig. 1: Consensus topology model of plant GSL and fungal FKS polypeptides. The plant GSL proteins are predicted to be divided into four regions: the cytoplasmic $NH_2$-terminal domain (N), the first membrane region predicted to contain ~6 transmembrane (TM) domains (TM1), the central cytoplasmic domain (Cyt) and the second membrane region predicted to contain ~10 TM domains (TM2) as indicated above each region. The GSL polypeptide is shown in the plasma membrane (PM) and the cell wall (cw) and cytoplasm (cyt) are indicated. An asterisk indicates the approximate location of the RXTG motif. Grey-shaded box indicates highest region of conservation between plant GSLs and fungal FKS proteins. Adapted from Doblin et al. (2001) (Reproduced with permission of the American Society of Plant Biologists).**

The GSL proteins are located in the plasma membrane, consistent with them having a role in callose synthesis, even though they do not contain a recognised signal peptide. The most likely way in which these polypeptides are targeted to the plasma membrane is that the first transmembrane helix directs the insertion of the nascent polypeptide into the endoplasmic reticulum (ER) for passage through the secretory pathway (Singer et al., 1987; Mothes et al., 1997). A GFP–AtGSL5 fusion protein was located at the plasma membrane when transiently expressed in onion epidermal cells (Østergaard et al., 2002). Antibody labelling also shows a plasma-membrane location using a conserved anti-GSL peptide antibody in sections of mung bean hypocotyls (Nakashima et al., 2003), and NaGSL1 was detected in both isolated plasma membrane preparations and by immunoelectron microscopy in the plasma membrane in sectioned material from *Nicotiana alata* pollen tubes by an anti-NaGSL1 antibody (Brownfield et al., 2007, Brownfield et al, 2008).

One nagging discrepancy in the assignment of callose synthase activity to the products of *GSL* genes has been the inability to identify an active site in the translated amino acid sequences. While there is considerable evidence that the GSL proteins are involved in callose synthesis and that they use UDP-Glc as a donor of Glc residues, they do not contain the UDP-Glc binding motif, 'D, D, D, QxxRW', that is characteristic of GT2 glycosyltransferases such as the cellulose synthase and related cellulose synthase-like (CSL) proteins as well as

CrdS, the *Agrobacterium* curdlan synthase (Saxena et al., 1995; Stasinopoulos et al., 1999; Charnock et al., 2001; Doblin et al., 2002; Karnezis et al., 2003). The question has been raised as to whether the GSL proteins carry the catalytic motif, or whether they are simply a pore-forming unit in a larger callose synthase complex with the catalytic domain on a separate protein (see Section I.C.e.). In this connection, the GSL proteins have some structural similarity to bacterial and eukaryotic transporters (Douglas et al., 1994b; Cui et al., 2001; Hong et al., 2001a; Dijkgraaf et al., 2002), but it nevertheless remains possible that GSL proteins are catalytically active but use an as yet unrecognised group of amino acid residues for UDP-Glc binding. An RXTG motif (Fig. 1) similar to the glycogen synthase substrate-binding motif R/KXGG (Farkas et al., 1990) has been proposed as the UDP-Glc binding consensus sequence within GSL proteins (Østergaard et al., 2002; Brownfield, 2005), but experimental proof that this motif does bind UDP-Glc is yet to be provided.

Affinity labelling and binding experiments have provided some evidence that GSL proteins do indeed bind UDP-Glc. Polypeptides close to the expected molecular weight of GSL proteins (~220 kDa) have been labelled with radioactive UDP-Glc or UDP-Glc analogues in callose synthase preparations from carrot (Lawson et al., 1989), cotton (Delmer et al., 1991) and red beet (Mason et al., 1990) (see Table 1 and Section I.C.d). A 200-kDa polypeptide, which is possibly ScFKS1, from a yeast extract enriched for (1,3)-β-glucan synthase activity, bound radiolabelled UDP-Glc (Inoue et al., 1996) and a 193-kDa polypeptide in product-entrapped material from *Neurospora crassa* labelled with 5-azido-[β-$^{32}$P]-UDP-Glc (Schimoler-O'Rourke et al., 2003). Analysis of tryptic peptides from the 193-kDa polypeptide by peptide mass fingerprinting and tandem mass spectroscopy showed this polypeptide was NcFKS1 (Schimoler-O'Rourke et al., 2003). Together, these data provide strong evidence that FKS and GSL proteins bind UDP-Glc, but do not necessarily mean that UDP-Glc is the substrate for an active callose synthase.

### I.C.b The GSL Gene Family

Plant *GSL* genes are members of multi-gene families (Fig. 2). In the *GSL* family from *Arabidopsis* there are 12 genes (Hong et al., 2001a), and there are 10 in rice (Yamaguchi et al., 2006), more than nine in *Populus trichocarpa* (http://genome.jgi-psf.org/Poptr1/Poptr1.home.html), and at least eight in wheat (Voigt et al., 2006) and seven in barley (Schober et al., 2009). The *GSL* genes can be divided into two distinct groups, based on the arrangements and numbers of introns (Doblin et al., 2001). One group contains three or less introns, while others are highly fragmented, with up to 40 or more introns (Saxena and Brown, 2000; Doblin et al., 2001; Verma and Hong, 2001; Hong et al., 2001a; Jacobs et al., 2003).

**Table 1: Molecular weight of major polypeptides present in callose synthase preparations**

| Plant | Callose synthase enrichment procedure | Molecular weight of major polypeptides in preparations determined by SDS–PAGE (kDa) | Molecular weight of UDP-Glc binding polypeptides (kDa) | Reference |
|---|---|---|---|---|
| Barley (*Hordeum vulgare*) | Sucrose gradient centrifugation and product entrapment | 36, 52, 66 and 170 | – | Pedersen et al., 1993 |
| | Non-denaturing gel electrophoresis | 36, 52, 54, 60, 70 and 94 | | Pedersen et al., 1993 |
| | Sucrose gradient centrifugation and non-denaturing gel electrophoresis | 35, 50, 75, 105 and 250 | – | Li et al., 2003 |
| Carrot (*Daucus carota* L.) | Sucrose gradient centrifugation | – | 43, 57 and 150 | Lawson et al., 1989 |
| Cauliflower (*Brassica oleracea*) | Anion-exchange chromatography and product entrapment | 32, 35, 57, 65 and 66 | – | Fredrikson et al., 1991 |
| Celery (*Apium graveolans* L.) | Anion-exchange, affinity chromatography and product entrapment | 27, 31, 35, 41, 56 and 72 | 44, 54, 71, 80 | Slay et al., 1992 |
| Cotton (*Gossypium hirsutum* L.) | Glycerol gradient | – | 52 and>200 | Delmer et al., 1991 |
| | Immunoprecipitation | – | 34 | Delmer et al., 1991 |
| | Product entrapment | – | 52 | Li et al., 1993 |
| French bean (elicitor treated) (*Phaseolus vulgaris* L.) | Anion-exchange and size-exclusion chromatography | 55 and 65 | – | McCormack et al., 1997 |
| Hybrid aspen (*Populus tremula* × *tremuloides*) | Glycerol gradient centrifugation | 30–32, 55, 62 and 64, (125, minor) | – | Colombani et al., 2004 |
| Mung bean (*Vigna radiata*) | Non-denaturing PAGE | 32, 38, 54, 64 and 78 | 32, 53, 64 and 78 | Kudlicka and Brown, 1997 |

*(Continued)*

**Table 1: (Continued)**

| Plant | Callose synthase enrichment procedure | Molecular weight of major polypeptides in preparations determined by SDS–PAGE (kDa) | Molecular weight of UDP-Glc binding polypeptides (kDa)[a] | Reference |
|---|---|---|---|---|
| Ornamental tobacco (*Nicotiana alata*) | Sucrose gradient centrifugation and product entrapment | 190 (=220 below) | – | Turner et al., 1998 |
| | Sucrose gradient centrifugation and product entrapment | 35, 55, 60, 61, 77, 103 and 220 | | Brownfield et al., 2007 |
| Pea stems | Product entrapment and preparative isoelectric focussing | 22, 30, 70, 80 and 100 | | Dhugga and Ray, 1994 |
| | Glycerol gradient centrifugation | 55 and 70 (30 and 100 minor) | 55 | Dhugga and Ray, 1994 |
| Peanut (*Arachis hypogaea*) | Hydroxylapatite and anion-exchange chromatography | 48 | – | Kamat et al., 1992 |
| Red beet (*Beta vulgaris*) | Product entrapment | 57, 60 and 78 | 57, 83 and 92 | Frost et al., 1990 |
| | Product entrapment | 57 | 200, 76, 60 and 57 | Mason et al., 1990 |
| | Glycerol gradient centrifugation and product entrapment | 27, 31, 42–47 and 57 | – | Qi et al., 1995 |
| | Product entrapment | 27, 31, 35, 43, 57, 67, 70, 83 and 92 | – | Wu et al., 1991 |
| | Product entrapment and protease digestion | 27, 29, 31, 43, 57, 83 and 92 | – | Wu and Wasserman, 1993 |
| Ryegrass (*Lolium multiflorum*) | $CaCl_s$ wash and product entrapment | 29–32, 38, 41 and 50–55 | – | Bulone et al.,1995 |
| | Immunoprecipitation | 30, 31, 54 and 58 | 31 | Meikle et al., 1991 |
| Soybean (*Glycine max*) | Sucrose gradient centrifugation – two rounds | 31 | – | Fink et al., 1990 |

[a]Potential UDP-Glc binding polypeptides have been identified by affinity labelling with radioactive UDP-Glc or photo-reactive analogues.

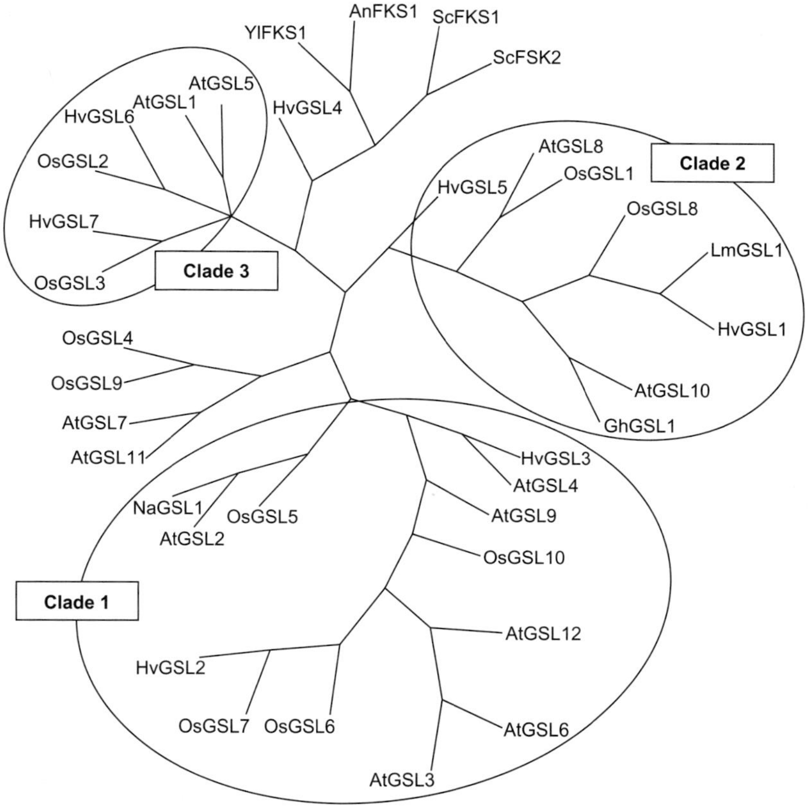


**Fig. 2: A phylogenetic tree graphically representing amino acid sequence homology of publicly available full-length plant GSL sequences and partially cloned fragments of barley GSLs. Following are the accession numbers of the GSLs in the above phylogenetic tree: *Arabidopsis thaliana* AtGSL1 (NM_116736), AtGSL2 (NM_179622), AtGSL3 (NM_179847), AtGSL4 (NM_112317), AtGSL5 (NM_116593), AtGSL6 (AF237733), AtGSL7 (NM_100528), AtGSL8 (NM_179940), AtGSL9 (NM_123045), AtGSL10 (NM_111596), AtGSL11 (NM_115772), AtGSL12 (NM_121303); *Oryza sativa* OsGSL1 (XM_550490, OsGSL2 (NM_191973), OsGSL3 (NM_191270), OsGSL4 (NM_193211), OsGSL5 (AP_007226), OsGSL6 (NM_001065181), OsGSL7 (XM_468556), OsGSL8 (NM_187562), OsGSL9 (NM_001049786) OsGSL10 (NM_187591); *Gossypium hirsutum* GhGSL1 (AF085717); *Lolium multiflorum* LmGSL1 (AY286332); *Nicotiana alata* NaGSL1 (AF304372); *Hordeum vulgare* HvGSL1 (AY177665); *Yarrowia lipolytica* YlGSL1 (AF198090); *Aspergillus nidulans* AnGSL1 (AACD01000061); *Saccharomyces cerevisiae* ScFKS1 (SCU12893), ScFKS2 (SCU16783).**

Over their entire length the plant GSL proteins share between about 40% and 80% identity with each other at the amino acid level (Doblin et al., 2001). Phylogenetic analyses show that GSL proteins are divided into three distinct clades (Fig. 2; Doblin et al., 2001). The largest clade includes six of the 12 *Arabidopsis* proteins (AtGSL2, AtGSL3, AtGSL4, AtGSL6, AtGSL9 and AtGSL12), along with NaGSL1 from *N. alata* and OsGSL5, OsGSL6, OsGSL7 and OsGSL10 from rice (*Orzya sativa*). The second clade contains two more *Arabidopsis* proteins (AtGSL8 and AtGSL10), GhGSL1 from cotton (*Gossypium hirsutum*), two from rice (OsGSL1 and OsGSL8), two from barley (*Hordeum vulgare*) (HvGSL1 and HvGSL5), and LmGSL1 from ryegrass (*Lolium multiflorum*). The two remaining *Arabidopsis* GSL proteins (AtGSL1 and AtGSL5) and two GSL proteins from rice (OsGSL2 and OsGSL3) and two GSL proteins from barley (HvGSL6 and HvGSL7) are in the third clade. Proteins in the third clade are clearly distinct from the other GSLs, being 100–150 amino acids shorter at their $NH_2$-terminus and being encoded by the *GSL* genes that contain less than three exons. In contrast, the other *Arabidopsis GSL* genes in clades 1 and 2 have approximately 40 exons (Doblin et al., 2001; Hong et al., 2001; Verma and Hong, 2001). All 12 *Arabidopsis* genes appear to be functional, based on the observation that no pseudogenes are apparent in the complete genome sequence (Hong et al., 2001a).

Each of the 12 *Arabidopsis* genes is likely to be responsible for callose synthesis in a different location within the plant or at a different stage of development (Hong et al., 2001a). The expression of particular *GSL* genes specifically in tissues that produce callose, such as the expression of *GhGSL1* (Cui et al., 2001) in cotton fibres and *NaGSL1* in pollen tubes (Doblin et al., 2001), supports this suggestion. The *GSL* genes from different species are more similar to each other than *GSL* genes from the same species (Fig. 2), suggesting the *GSL* genes from different species that are closely related may be functional orthologs, encoding callose synthase enzymes that produce callose in the same tissue or developmental stage (Doblin et al., 2001). For example, NaGSL1 and its closest homologues from *Arabidopsis*, AtGSL2 and rice, OsGSL5 (Fig. 2), may be the developmentally regulated callose synthases of pollen tubes, because *NaGSL1*, *AtGSL2* and *OsGSL5* are expressed in mature pollen (Doblin et al., 2001; Becker et al., 2003; Honys and Twell, 2003, 2004, Nishikawa et al., 2005, Yamaguchi et al., 2006). However, it should be noted that there is not an exact correlation between the number and grouping of GSL proteins from different species. Some of the GSL clades shown in Fig. 2 have more GSL members from one species compared to another. For example, clade 1 contains more *Arabidopsis* than rice GSLs. This may indicate that *Arabidopsis* has specialized functions for some GSL proteins in particular tissues and/or developmental stages that are not shared in rice and *vice versa*.

The suggestion that different GSL family members are associated with callose synthesis in different tissues or at different times is further supported by work investigating the *GSL* families in barley and wheat. Northern analysis with a probe that hybridizes to most barley *GSL* transcripts detected expression in developing grain and florets, as well as in roots and coleoptiles, all tissues that produce callose during development (Li et al., 2003). Transcript levels for eight specific barley *GSL* genes have also been monitored in vegetative and floral tissues and in developing endosperm, using quantitative PCR (Schober et al., 2009). Although transcription of each gene was detected in each tissue examined, the levels varied widely with some genes transcribed at high levels only in a particular tissue. In the developing endosperm high transcription levels generally occurred between 2 and 5 days after pollination (Schober et al., 2009) which coincides with the onset of cellularization of the endosperm, a process that requires callose deposition in the nascent cell wall (Brown et al., 1997, Olsen, 2001; Wilson et al., 2006). Similar results were observed with wheat where quantitative PCR analysis showed that although eight wheat *GSL* genes were expressed in stems, leaf blades and spikes, the expression levels of individual genes varied between tissues (Voigt et al., 2006). The level of callose synthase activity and total callose also varied between these tissues.

Furthermore, *GSL* genes other than those contributing to developmentally regulated callose production appear to be involved in generating wound-activate302d callose throughout the plant. For example, when *Arabidopsis* was transformed with double-stranded RNA interference (dsRNAi) constructs designed to individually silence the three putative callose synthase genes, *GSL5*, *GSL6* and *GSL11*, both wound callose and papillary callose were absent in lines transformed with *GSL5* dsRNAi and in a corresponding *AtGSL5* T-DNA insertion line (Jacobs et al., 2003). However, wound and papillary callose were unaffected in *GSL6* and *GSL11* dsRNAi lines (Fig. 3). These data provide strong genetic evidence that the *GSL5* gene of *Arabidopsis* encodes a protein that is essential for wound callose formation. Deposition of callosic plugs, or papillae, at sites of fungal penetration is a widely recognised early response of host plants to microbial attack and has been implicated in impeding entry of the fungus (Stone and Clarke, 1992). Depletion of callose from papillae in *gsl5* plants marginally enhanced the penetration of the grass powdery mildew fungus *Blumeria graminis* on the non-host *Arabidopsis*. Paradoxically, the absence of callose in papillae or haustorial complexes correlated with the effective growth cessation of several normally virulent powdery mildew species and of *Peronospora parasitica* (Nishimura et al., 2003; Jacobs et al., 2003). Similarly, the phenotypic characteristics of the *gsl5* lines are similar to those described for the powdery mildew-resistant *pmr4-1* mutant of *Arabidopsis* (Vogel and Somerville, 2000; Nishimura et al., 2003). The *pmr4* mutation results from changes in the *GSL5* gene (Nishimura et al.,

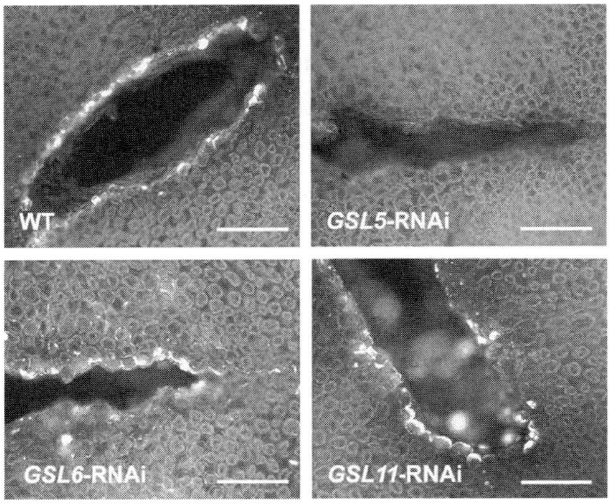


**Fig. 3: Reduced callose accumulation at wound sites in *Arabidopsis GSL5* dsRNAi lines. Leaves of wild-type (WT) plants and *GSL5*, *GSL6* and *GSL11* dsRNAi lines were wounded by cutting with a razor. Aniline Blue fluorochrome staining of leaves revealed reduced callose accumulation at wound sites after 24 h. Bars 200 µm. From Jacobs et al., (2003) (Reproduced with permission of the American Society of Plant Biologists).**

2003; Jacobs et al., 2003). Together, these observations indicate that resistance to wall penetration by the grass powdery mildew fungus is a plant-controlled process to which papillary callose does not contribute to any great extent (Jacobs et al., 2003). They also bring into question the initial notion that callose deposition physically impedes the growth of fungal pathogens and therefore contributes to disease resistance.

### *I.C.c Functional Analysis of GSL Genes*

The silencing or overexpression of genes within the plant provides useful information on the roles of genes and has been employed in the investigation of *GSL* genes. Several studies have shown a loss of specific callose deposits when expression of a single *GSL* gene or pairs of *GSL* genes have been knocked out or down-regulated. Callose is not produced in response to wounding or pathogen challenge in plants lacking *AtGSL5* (Jacobs et al., 2003; Nishimura et al., 2003; see Section I.D.b.). When both *AtGSL5* and *AtGSL1* are mutated, callose deposits between microspores within a tetrad are lacking (Enns et al., 2005; see Chapter 4.4.3). Loss of *AtGSL2* results in a lack of callose in the cell wall of meiocytes, tetrads and microspores (Dong et al., 2005), while weaker alleles lead to a loss of pollen tube callose (Nishikawa et al., 2005). These data show that GSL proteins are required for callose synthesis, but do not indicate if they are the catalytic subunit. Recent work suggests that GSL proteins may also have a regulatory role. Mutations in *AtGSL8* and *AtGSL10* result in defects during pollen development, but specific callose deposits are not missing (Töller et al., 2008). Lack of *AtGSL8* prevents microspores from entering mitosis, where a need for callose has not been described, possibly indicating that the GSL proteins may also have a role in signalling (Töller et al., 2008). Some *AtGSL10* mutants also fail at microspore mitosis, while others display defects in the subsequent cytokinesis, where callose is deposited ectopically, or at the cell plate but in a disrupted and persistent manner (Töller et al., 2008). Thus, a combination of GSL proteins may be required for the correct spatial and temporal deposition of callose.

Hong et al. (2001a) overexpressed *GFP::AtGSL6* under the control of the constitutive CaMV35S promoter in tobacco BY2 cells and observed an increase in Aniline Blue staining at the cell plate, suggesting an increased amount of callose, and increased callose synthase activity. However, the increase in Aniline Blue fluorescence was not quantified and the overexpression of phragmoplastin, which is a protein not directly involved in callose synthesis, also produced an increase in Aniline Blue staining of the cell plate (Geisler-Lee et al., 2002). Thus, it is not clear if *AtGSL6* was responsible for the increased callose synthesis. Expression of *AtGSL5* was found to be increased in the *mpk4* mutant of *Arabidopsis* (Østergaard et al., 2002). This mutant constitutively expresses the systemic acquired resistance defence response (Petersen et al., 2000) and has increased levels of callose deposition (Østergaard et al., 2002). However, the increase in callose deposition and the increased expression of *AtGSL5* have not been directly linked and may be two separate responses to systemic acquired resistance.

Heterologous expression has been used to determine the function of several cell wall polysaccharide synthases, namely members of the CslA and CslH (Dhugga et al., 2004; Liepman et al., 2005; Liepman et al., 2007), CslC (Cocuron et al., 2007) and CslF families (Burton et al., 2006; Doblin et al., 2009). In CslA, CslF and CslH the genes were expressed heterologously in plant cell types that do not usually make the product. This approach is compromised for GSL proteins due to their widespread expression, and the ubiquitous expression of wound-activated callose synthases. Expression studies with *GSL* genes have been conducted using yeast as a heterologous host, but results have been conflicting. Østergaard et al. (2002) transformed *AtGSL5* into a yeast *fks1* mutant that is cyclosporin A-sensitive due to the lesion in the *FKS* gene, and reported partial complementation when the transformed yeast was grown on the drug cyclosporin A. However, there was only a marginal increase in growth in the presence of *AtGSL5*, and no expression studies or callose synthase activity assays were conducted. Brownfield (2005) transformed *NaGSL1* into the yeast *fks1*Δ mutant, the *fks1*Δ*fks2*Δ double mutant, a range of double mutants displaying synthetic interactions with *fks1*Δ, and a range of temperature-sensitive yeast mutants with point mutations in *FKS1*. The NaGSL1 protein

was produced and located, at least partially, in the plasma membrane in the transformed yeast. However, no complementation was observed in any of the mutants and no change in caspofungin sensitivity was observed in the *fks1*Δ mutant. Furthermore, no *in vitro* activity by NaGSL1 could be detected in membrane preparations (Brownfield, 2005). Similarly, *AtGSL6* also did not complement the *fks1*Δ mutant and there was no increase in the (1,3)-β-glucan synthase activity in yeast transformed with *AtGSL6*
(Hong et al., 2001a). The lack of activity of GSL proteins in yeast may relate to incorrect post-translational modifications or the absence of an important co-factor, or may indicate the absence of catalytic activity of these proteins. Thus, yeast does not appear to be a suitable system for the heterologous expression of these complex plant proteins.

### I.C.d Biochemical Identification of GSL Proteins

Biochemical approaches have also been taken to identify components of the callose synthase enzyme, and more recently to verify that GSL proteins are involved in callose synthesis. These strategies have involved attempts to purify the callose synthase enzyme for the identification of protein constituents, but purification of callose synthase enzymes has proved difficult. These difficulties are most likely related to both the membrane location of the enzyme and its size. Because the enzyme is an integral membrane protein, it is generally assumed that a detergent is required to solubilize it. However, only a few detergents are able to extract and preserve callose synthase activity (Lai-Kee-Him et al., 2003) and the best detergent depends on the source of the extract (Wasserman and MacCarthy, 1986; Lawson et al., 1989; Fink et al., 1990; Lai-Kee-Him et al., 2001; Colombani et al., 2004). In addition, two of the detergents commonly used in these procedures, namely digitonin and CHAPS, do not produce true micelles during solubilization, but instead form small vesicles (Lai-Kee-Him et al., 2001). This means that many proteins are likely to co-purify with callose synthase activity through entrapment within the vesicles. Dissociation of any callose synthase complex from membranes could also result in the loss of activity and prevent many common chromatographic techniques from being successfully used during the purification procedure.

Despite these limitations, enriched callose synthase preparations have been made from various plant sources using methods such as glycerol- or sucrose-gradient centrifugation, product entrapment and/or immunoprecipitation. These preparations generally contain a number of polypeptides ranging in size from 25 to 250 kDa depending upon the enrichment procedure and enzyme source (Table 1), consistent with the presence of contaminating polypeptides along with the callose synthase enzyme components. To identify the catalytic subunit of callose synthase from amongst the polypeptides present in callose synthase-enriched fractions, affinity labelling with either radioactive UDP-Glc or a photoreactive substrate analogue has been used. Far from providing a clear candidate polypeptide for the catalytic subunit, a number of polypeptides can be labelled within a single preparation (Table 1), including polypeptides that are not enriched in parallel with the enrichment of callose synthase activity. This suggests that contaminating proteins bind the substrate or its analogues. Given the number of enzymes that use either UDP or Glc as substrates, coupled with the potential for non-specific binding of UDP to proteins in general, it is perhaps not surprising that these results can potentially mask the labelling of the catalytic subunit of the callose synthase enzyme. Nevertheless, a large polypeptide (>150 kDa) was detected in preparations from carrot (Lawson et al., 1989), cotton (Delmer et al., 1991) and red beet (Mason et al., 1990) using this approach (Table 1), and these could be GSL proteins.

Two GSL proteins, HvGSL1 and NaGSL1, have been identified in callose synthase-enriched preparations and evidence provided to link the GSL proteins with wound-activated and developmentally regulated callose synthase activity, respectively. A barley callose synthase was purified more than 60-fold from a microsomal fraction from suspension-cultured cells using CHAPS detergent extraction, $CaCl_2$ treatment, sucrose density gradient centrifugation and non-denaturing gel electrophoresis (Li et al., 2003). Following non-denaturing gel electrophoresis, a single protein band in the gel synthesized Aniline Blue fluorochrome-positive material when provided with UDP-Glc, and this material was shown through specific enzyme hydrolysis to contain mostly (1,3)-β-glucan. Furthermore, the protein band in the gel was recognised by antibodies raised against a 17-kDa protein generated by heterologous expression of a fragment of the *HvGSL1* cDNA. Mass spectrometric peptide mass fingerprinting analyses showed that the mass of tryptic peptides produced by in-gel digestion of the active enzyme matched peptides predicted from the *HvGSL1* gene sequence (Li et al., 2003). Thus, the amino acid sequence predicted from the *HvGSL1* gene was directly linked with the amino acid sequence of the active (1,3)-β-glucan synthase fraction from barley.

NaGSL1 protein was identified in a fraction enriched for callose synthase from *N. alata* pollen tubes (Brownfield et al., 2007). Callose synthase activity was enriched through a process of sucrose density gradient centrifugation followed by solubilization with digitonin and product entrapment. A 220-kDa polypeptide that was specifically detected by an anti-NaGSL1 antibody was enriched. The 220-kDa polypeptide from product-entrapped material was excised from an SDS–PAGE gel and tryptic fragments analysed by mass spectrometry. Peptides detected by peptide mass fingerprinting and amino acid sequence determined by tandem mass spectroscopy both identified the polypeptide as NaGSL1 (Brownfield et al., 2007).

This polypeptide also increased in abundance during pollen tube growth, along with the level of callose synthase activity, and relative levels of this polypeptide in the membrane fractions correlated with the amount of callose synthase activity (Brownfield et al., 2008).

Antibodies have been used to link the *GSL* gene product with callose synthesis in other systems, without the identification of the polypeptide. Cui et al. (2001) raised antibodies against the $NH_2$-terminal region of GhGSL1 and showed that the protein was present in cotton fibres at the time of callose deposition. The anti-GhGSL1 antibodies also bound to a polypeptide of over 200 kDa in material enriched for callose synthase activity by product entrapment (Cui et al., 2001). Anti-GFP antibodies revealed that a GFP-AtGSL6 fusion polypeptide was present in product-entrapped material (Hong et al., 2001a), although there are no data to show if these preparations are also enriched with callose synthase activity.

### *I.C.e Callose Synthase Complexes*

GSL proteins are found in high molecular weight complexes *in vitro.* The HvGSL1 from barley (*Hordeum vulgare*) was reported to be part of a very large callose-synthesizing complex that barely entered the stacking portion of a native electrophoresis gel (Li et al., 2003). Similarly, a GSL protein from spinach (*Spinacea oleracea*) was identified in a large plasma membrane complex by 2D-Blue-native–PAGE (Kjell et al., 2004). Hence, GSL proteins may associate and more than one GSL molecule may be present in each callose synthase complex. Polymers of (1,3)-β-glucan over a certain length associate and can form triple helices of parallel chains, and this has been used as evidence that callose synthesis involves catalytic complexes (Stone and Clarke, 1992; Pelosi et al., 2003). However, this polysaccharide chain aggregation can occur spontaneously and does not necessarily require the presence of multiple catalytic subunits. Lai-Kee-Him et al. (2001) also suggested that there may be several catalytic subunits in the callose synthase complex because the kinetic behavior of callose synthase in *Arabidopsis* plasma membrane preparations suggested positive homotropic cooperativity between the subunits of a complex. However, this behaviour could also be explained if a number of different callose synthase isoforms, each with slightly different kinetic properties, were present in the plasma membrane, or if sub-optimal assay conditions were used.

While there is considerable evidence that GSL proteins are responsible for callose synthesis, it is likely that other proteins are also involved in a callose synthase complex, with roles in regulation of, or providing substrate to, the GSL proteins. The presence and identity of proteins that associate with GSL proteins may vary with different GSL proteins, and between developmentally regulated and wound-activated callose synthesis. In almost every plant species from which the callose synthase enzyme has been partially purified, enriched preparations contain a number of polypeptides (Table 1). It remains to be determined whether such proteins are truly components of a callose synthase complex or are proteinaceous contaminants arising from the purification procedures (see Section I.C.d.) or a combination of both. A number of these proteins have now been identified (Table 2) and how they contribute to an active callose synthase complex is found in a model proposed by Kauss, (1996) (Fig. 4).

**Table 2: Putative identity of polypeptides present in plant callose synthase preparations**

| Plant source | Polypeptide (kDa) | Putative identity | Method of identification | Reference |
|---|---|---|---|---|
| Barley (*Hordeum vulgare*) | 54 | ATPase subunit | Sequence similarity | Bulone et al., 1995 |
| | 250 | HvGSL1 | Antibody labelling and peptide mass fingerprinting | Li et al., 2003 |
| Cotton (*Gossypium hirsutum* L.) | 91 | SuSy | Sequence similarity from affinity-labelled polypeptide | Amor et al., 1995 |
| | 65 | Calnexin | cDNA expression screening with mAb that immunoprecipitates *Vigna radiata* callose synthase | Delmer et al., 1993; Kawagoe and Delmer, 1995 |
| | 34 | Annexin | Sequence similarity | Andrawis et al., 1993 |
| French bean (*Phaseolus vulgaris* L.) | 65 | Protein disulfide isomerase | Sequence similarity | McCormack et al., 1997 |
| Ornamental tobacco (*Nicotiana alata*) | 220 | NaGSL1 | Antibody labelling Peptide mass fingerprinting Tandem mass spectrometry | Brownfield et al., 2007 |
| | 103 | Plasma membrane $H^+$-ATPase | Peptide mass fingerprinting | Brownfield et al., 2007 |
| | 60 | β-subunit of mitochondrial ATPase | Peptide mass fingerprinting | Brownfield et al., 2007 |
| Red beet (*Beta vulgaris*) | 27–31 | PMIP – aquaporin homologue | Sequence similarity and antibody labelling | Qi et al., 1995 |
| | 55 | Tonoplast ATPase subunit | Antibody labelling | Wu et al., 1991 |

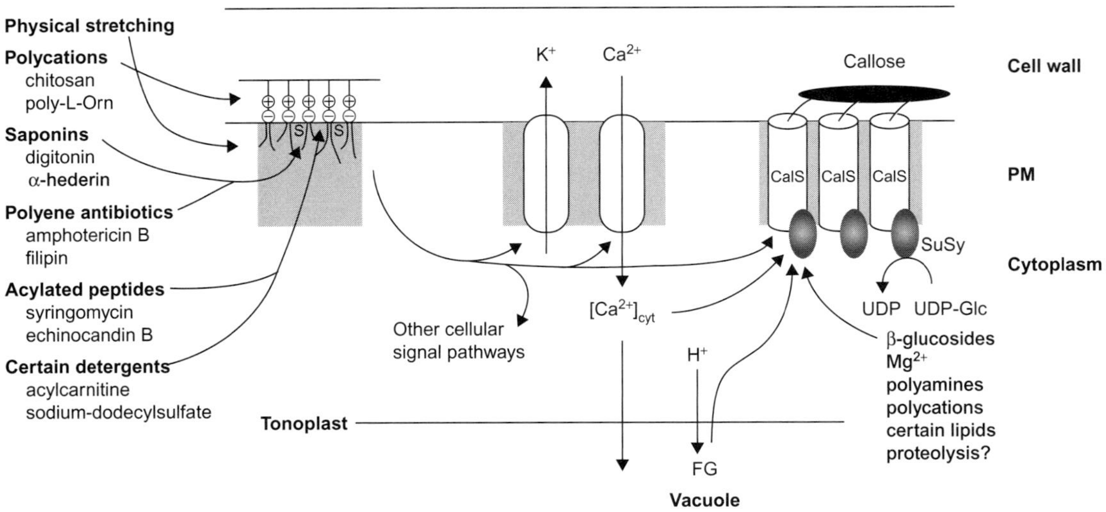


**Fig. 4: Model of the callose synthase enzyme complex and its regulation. A diverse set of elicitors is able to activate the callose synthase enzyme, resulting in callose synthesis and deposition. Elicitors that disturb the integrity of the plasma membrane (left panel: e.g. cationic compounds with multiple positive charges interacting with the negatively charged phospholipids of the plasma membrane, PM; saponins (S) that integrate into the lipid bilayer) create a common but as yet unidentified signal that is able to stimulate ion-transport processes (centre panel), leading to an increase in localized cytoplasmic $Ca^{2+}$ concentration, a potent effector of callose synthase. In addition, the unknown signal either influences callose synthase directly or activates other signal pathways that cause the mobilization of an enzyme activator(s), possibly identical to those previously identified within *in vitro* experiments (right panel: e.g. β-glucosides such as β-furfuryl-β-glucoside (FG) that could be released from the vacuole (Ohana et al., 1993), other compounds such as polyamines and lipids). These three processes are envisaged to occur in close proximity, to explain the localized deposition of callose. The callose synthase (CalS) enzyme is an integral membrane protein comprised of several different subunits with a combined molecular mass of >500 kDa (Eiberger and Wasserman 1987). Their exact identity is not known, although may include regulatory components such as one or more annexins (Andrawis et al., 1993), a pore-forming membrane intrinsic protein, now proposed to be the GSL proteins (Brownfield et al. 2007), and a membrane-associated form of sucrose synthase (SuSy) to channel the substrate UDP-Glc directly into the catalytic subunit(s) (Amor et al., 1995). The catalytic subunit has its substrate binding site located on the cytoplasmic surface of the plasma membrane (Mueller and Maclachlan 1983). This necessitates that there be some type of pore (formed by either the catalytic polypeptide or one or more other components) through which the nascent polysaccharide chain is extruded (Kauss 1985). Once within the apoplastic space, the (1,3)- β-glucan chains aggregate and are deposited as callose. As in the case of cellulose synthesis, the mechanisms which govern chain initiation and termination as well as their deposition, and how this entire process is regulated, are largely unknown. Reproduced from Kauss, (1996) with permission.**

Some identified proteins are most likely to be contaminants. A 103-kDa polypeptide in a callose synthase enriched fraction from *N. alata* pollen tubes identified as a plasma membrane $H^+$-ATPase is most likely a contaminant as it was not enriched with callose synthase activity through product entrapment (Brownfield et al., 2007). Polypeptides between 55 and 60 kDa are commonly observed in callose synthase preparations, and several have been identified as likely to be a subunit of abundant tonoplast, mitochondrial or other ATPases (Wu et al., 1991; Bulone et al., 1995; Brownfield et al., 2007). These polypeptides are also likely to be contaminants as they are probably not involved in callose synthesis and are from locations other than the plasma membrane. Polypeptides of 29–35 kDa are also commonly observed and a 31-kDa polypeptide from red beet storage tissue has sequence similarity to highly abundant plasma membrane intrinsic proteins or aquaporins (Qi et al., 1995). Since aquaporins form pores in the plasma membrane, it has been proposed that they may play a role in callose synthesis by assisting with translocation of polysaccharides across the plasma membrane (Qi et al., 1995), or they too could be contaminants.

Polypeptides that act as molecular chaperones may assist assembly of the enzyme complex. The ~65-kDa polypeptides identified in callose synthase preparations appear to be either protein disulfide isomerase (PDI) or calnexin (McCormack et al., 1997; Delmer et al., 1993; Kawagoe and Delmer, 1995; Table 2). Antibodies against the 65-kDa polypeptide from French bean, with similarity to PDI, label sites of callose deposition, including the cell plate, plasmodesmata and wound papillae (McCormack et al., 1997; Brown et al., 1998), indicating that this polypeptide is co-located with callose synthesis. PDI is an ER protein that catalyses the formation and breakage of disulfide bonds between cysteine residues in proteins during folding and post-translational disulfide exchange. Given the hydrophobic nature of the callose synthase enzyme, it is possible that molecular chaperones would be involved in assisting in the synthesis and/or assembly of the callose synthase polypeptides into an enzyme complex. Similarly, antibodies that recognise calnexin bind to sieve plates of mung bean (*Vigna radiata*) seedlings, as well as to plasmodesmata and wounding sites within onion bulb epidermal cells (Delmer et al., 1993; Kawagoe and Delmer, 1995), and the antibodies interact with callose synthase in a cation-dependent manner. It has been suggested that calnexin, like PDI, acts as a molecular chaperone and therefore could assist in the assembly of the callose synthase complex.

Several studies have identified proteins, a sucrose synthase (SuSy) and a UDP-Glc transferase (UGT1), that may channel UDP-Glc into the callose synthase catalytic subunits to produce (1,3)-β-glucan. Labelling studies identified a 91-kDa polypeptide as the most abundant UDP-Glc binding polypeptide in cotton-fibre plasma membranes (Amor et al., 1995), and polypeptides of this size are also observed in other callose synthase preparations from red beet (Wu et al., 1991; Wu and Wasserman, 1993) and barley (Pedersen et al., 1993) (Table 1). Sequencing of tryptic fragments of the cotton polypeptide showed this to be the membrane-associated form of SuSy, an enzyme that degrades sucrose into fructose and UDP-Glc. Yeast two-hybrid analyses indicated that AtGSL6 interacts with UGT1 and UGT1 is present in product-entrapped material (Hong et al., 2001a; Hong et al., 2001b). AtGSL6 is believed to be involved in callose synthesis at the cell plate and UGT1 also interacts with phragmoplastin and a RHO-like protein called ROP1, but only when ROP1 contains bound GTP (Hong et al., 2001b). On the basis of these results it was proposed that AtGSL6, phragmoplastin and UGT1 form a complex at the cell plate with UGT1 being regulated by ROP1 and channelling UDP-Glc to AtGSL6 (Hong et al., 2001b). UGT1 may also interact with SuSy (Hong et al., 2001b).

Calmodulin may also be part of some callose synthase complexes. Many callose synthases are regulated by $Ca^{2+}$ ions (see Section I.C.f), and calmodulin binds to short peptide sequences in target proteins in response to alterations in intracellular $Ca^{2+}$ concentrations inducing structural changes (Zielinski, 1998). Many of the proteins that calmodulin binds are unable to bind $Ca^{2+}$ themselves and, as such, use calmodulin as a $Ca^{2+}$ sensor and signal transducer. The wound-activated callose synthase from corn (*Zea mays*) coleoptiles is stimulated by calmodulin (Paliyath and Poovaiah, 1988). A putative calmodulin-binding domain was identified in the $NH_2$-terminal region (amino acids 226 to 241) of the GhGSL1 protein from cotton fibres and this region was able to bind calmodulin *in vitro* in the presence of $Ca^{2+}$ (Cui et al., 2001). However, not all GSL proteins appear to interact with calmodulin, as calmodulin-binding domains are not present in all GSL proteins (Brownfield, 2005). Furthermore, the enzyme from soybean (*Glycine max*) does not respond to exogenously supplied calmodulin, although endogenous calmodulin levels may have been sufficient for maximal callose synthase activity in these experiments (Kauss et al., 1983).

A 34-kDa polypeptide from cotton-fibre plasma membrane has been identified as an annexin (Andrawis et al., 1993; Table 2). Annexins interact with membranes in a $Ca^{2+}$- and phospholipid-dependent manner and can specifically be released by treatment of membranes with a $Ca^{2+}$-chelating agent such as EGTA (Buostead et al., 1989). Callose synthase-associated annexins are hypothesized to play a role in localizing and/or anchoring the callose synthase complex to the cytoskeleton (Andrawis et al., 1993). Polypeptides of this size can also be labelled with UDP-Glc analogues, suggesting that UDP-Glc can act as both an effector as well as a substrate of callose synthase, or that annexin plays a role in (1,3)-β-glucan primer synthesis (Delmer et al., 1991; Andrawis et al., 1993). A specific interaction between annexins and callose synthase has been established by reconstitution experiments in which annexin-containing fractions bind to and inhibit the partially purified callose synthase from cotton fibres (Andrawis et al., 1993). However, addition of a recombinant form of the annexin-like protein was later shown to have no effect on callose synthase activity, although this may have been due to the protein being expressed in *Escherichia coli* and thus not having some post-translational modifications that are important for its inhibitory function (Shin and Brown, 1999).

Rather than investigating individual components a few studies have focussed on the observation of putative functional units of callose synthase complexes. At the electron microscope level, Hayashi et al. (1987) observed numerous 20–30-nm doughnut-shaped structures within (1,3)-β-glucan fibrils synthesised *in vitro* and occasionally found that the fibrils protruded from the centre of these structures. Similar ring-shaped structures 40–50 nm in diameter have been observed in the flow-through of an anti-tubulin column loaded with detergent-solubilized plasma membrane preparations of azuki bean (Mizuno, 1998). These structures are sensitive to (1,3)-β-glucanase and dissociate in the presence of $Mg^{2+}$. The callose-synthesising band identified upon separation of detergent-solubilized membrane proteins of mung bean using native PAGE also contained abundant protein complexes 29–38 nm in size that were associated with the (1,3)-β-glucan product (Kudlicka and Brown, 1997). Lai-Kai-Him et al. (2001) and Colombani et al. (2004) were able to obtain preparations that specifically synthesized microfibrillar (1,3)-β-glucan and large protein complexes could be seen attached to the nascent polysaccharide chains (Bulone et al., 1995; Colombani et al., 2004; Pelosi et al., 2006).

Clearly, callose-synthesizing units appear to be variable in form, and no consensus structure is available at present. Given that the *GSL* gene family comprises multiple members (see Section I.C.b) and the complexities introduced by *in vitro* purification approaches, the question of what comprises a functional callose synthase enzyme complex will need to be resolved in the future, perhaps through the use of *in situ* immunoelectron microscopy or fluorescence approaches that have been applied to elucidating the constituents of the cellulose-synthesizing complexes (Taylor et al., 2003; Paredez et al., 2006).

### *I.C.f Regulation of Callose Synthesis*

Regulation of callose synthesis has been studied in many plant species, but most thoroughly in cotton fibres, red beet, legume hypocotyls, mung bean, ryegrass suspension-cultured cells and, more recently, suspension-cultured cells from *Arabidopsis* and aspen. In these cell types it is likely that the major callose synthase activity is attributable to wound activation, but activity from any developmentally regulated enzyme present in these cell types would also be measured. One system where there appears to be no wound-activated callose synthase activity and where callose is deposited developmentally is the pollen tube, for which most studies have been focussed on pollen tubes from *N. alata*. A proposed model of the callose synthesis enzyme complex and its regulation developed by Kauss (1996) is presented in Fig. 4.

In most plant extracts in which *in vitro* callose synthase activity has been reported to date, the enzyme requires $Ca^{2+}$ and a β-glucoside such as cellobiose for maximal activity. Callose synthase activity *in vitro* can also be activated by compounds containing positively charged groups (e.g. polyamines, ruthenium red, chitosan) and amphipathic substances such as digitonin and phospholipids (Kauss, 1985; Kauss and Jeblick, 1985; Köhle et al., 1985; Morrow and Lucas, 1986; Eiberger and Wassermann, 1987; Hayashi et al., 1987; Sloan et al., 1987; Kauss et al., 1989; Fredrikson and Larsson, 1992). Inhibitory compounds include ion chelators, $La^{3+}$ and various unsaturated fatty acids (e.g. arachidonic acid) (Kauss, 1985; Kauss and Jeblick, 1986; Wasserman and MacCarthy, 1986; Sloan et al., 1987; Fredrikson and Larsson, 1992; Kauss, 1996). Information on activators and inhibitors in *in vitro* assays is summarized in Table 3.

The $Ca^{2+}$ dependence of callose synthase activity from most sources is believed to be related to the activation of the wound-activated enzyme upon plasma membrane disruption. The concentration of $Ca^{2+}$ required for maximal activity *in vitro* depends upon the $Mg^{2+}$ concentration, with maximal callose synthase activity occurring at micromolar concentrations of $Ca^{2+}$ in the presence of $Mg^{2+}$ and millimolar concentrations of $Ca^{2+}$ in the absence of $Mg^{2+}$ (Hayashi et al., 1987; McCormack et al., 1997; Colombani et al., 2004). There is little or no activity when all the free $Ca^{2+}$ is complexed by EDTA or EGTA (Kauss, 1996; McCormack et al., 1997). Upon plasma membrane disruption following wounding or pathogen challenge, there is a localized increase in the $Ca^{2+}$ concentration, which is likely to activate callose synthase and lead to a rapid callose deposition in the affected region (Köhle et al., 1985; Fredrikson and Larsson, 1992; Kauss, 1996). $Ca^{2+}$ appears to interact with the callose synthase protein at the cytoplasmic face of the plasma membrane (Fredrikson and Larsson, 1989); however, the precise mechanism of $Ca^{2+}$ stimulation is unknown. It may involve a separate $Ca^{2+}$-binding protein such as calmodulin or an annexin-like protein (see Section I.C.e).

The pollen-tube callose synthase differs from other callose synthases studied in not requiring $Ca^{2+}$, and displays a high level of activity in the presence of EDTA and EGTA. Within pollen tubes there is a very steep, tip-focussed gradient of $Ca^{2+}$ ions (Holdaway-Clarke and Hepler, 2003). Generally, this gradient runs from 3–10 μM $Ca^{2+}$ at the tip to 0.15–0.3 μM, 20 μm behind the tip. As callose is first detected ~30 μm behind the tip in growing pollen tubes (Ferguson et al., 1998), the callose synthase enzyme is presumably active in this location where there is a low $Ca^{2+}$ concentration, necessitating the $Ca^{2+}$ independence of pollen-tube callose synthase.

**Table 3: Parameters involved in callose deposition**

| | | References |
|---|---|---|
| **Critical effectors** | | |
| $Ca^{2+}$, a β-glucoside (often cellobiose is used *in vitro* as it is readily available) | | Kauss, (1985);<br>Hayashi et al., (1987) |
| **Intracellular molecules** | | |
| substrate<br>cations<br>polyamines<br>β-glucosides | UDP-glucose<br>e.g. $Ca^{2+}$, $Mg^{2+}$<br>e.g. spermine, spermidine<br>e.g. cellobiose, laminaribiose, sucrose, glycerol,<br>β-furfuryl-β-glucoside, β-glucosides with hydrophobic aglycones | Delmer, (1987)<br>Kauss, (1985)<br>Kauss and Jeblick, (1985);<br>Hayashi et al., (1987)<br><br>Ohana et al., (1992);<br>Ng et al., (1996) |
| **Molecules involved in physical stretching of the plasma membrane** | | |
| | **Cationic compounds** | |
| | chitosan | Köhle et al., (1985); |
| | poly-L-Orn/Lys/Arg+polyamines | Kauss and Jeblick, (1985); |
| | ruthenium red | Kauss and Jeblick, (1985) |
| | **Saponins** | |
| | Digitonin | Eiberger and Wasserman, (1987);<br>Kauss and Jeblick, (1987) |
| | α-hederin | Kauss, (1996) |
| | **Polyene antibiotics** | |
| | amphotericin B | Kauss, (1996) |
| | filipin | Kauss, (1996) |
| | **Acylated peptides** | |
| | syringomycin | Kauss, (1996) |
| | echinocandin B | Kauss, (1996) |
| | **Detergents** | |
| | acylcarnitine | Kauss and Jeblick, (1986) |
| | sodium dodecylsulfate | Kauss, (1996) |
| | 3-[(3-cholamidopropyl)-dimethylammonio]-1-propane-sulfonate (CHAPS) | Li et al., (1997);<br>Lai Kee Him et al., (2001)<br>Lai Kee Him et al., (2001) |
| | n-octyl-β-D-glucopyranoside (OG) | |

*(Continued)*

Table 3: (Continued)

| | | References |
|---|---|---|
| | zwittergent 3–12, 3–16 | Lai Kee Him et al., (2001); Li et al., (1997) |
| | glycodeoxycholate (GDC) | Lai Kee Him et al., (2001) |
| | Brig 35, Brig 58 | Lai Kee Him et al., (2001) |
| | Mega 10 | Lai Kee Him et al., (2001) |
| | lysophosphatidylcholine | Li et al., (1997) |
| **Partial proteolysis** | | |
| | trypsin | Kauss et al., (1983) Li et al., (1997) |
| **Inhibitors** | | |
| | **Salts/Ions** | |
| | NaF, KF | Kauss, (1996) |
| | $La^{3+}$ | Kauss, (1986) |
| | $Mg^{2+}$ (*Arabidopsis*) | Lai Kee Him et al., (2001) |
| | **Chelators** | |
| | EDTA, EGTA | Kauss et al., (1983) |
| | **Unsaturated fatty acids** | |
| | oleic acid | Kauss and Jeblick, (1986) |
| | linolenic acid | Kauss and Jeblick, (1986) |
| | arachidonic acid | Kauss and Jeblick, (1986) |
| | **Extensive proteolysis** | |
| | Pronase E | Wu and Wasserman, (1993); Kauss et al., (1983) |
| | trypsin | Girard and Maclachlan, (1987) |
| | **Compounds that bind (1,3)-β-glucan** | |
| | Sirofluor (fluorochrome of Aniline Blue) | Morrow and Lucas, (1986) |

Maximal rates of callose synthesis *in vitro* are only achieved in the presence of a β-glucoside, with cellobiose being the most common β-glucoside used in assay mixtures (Morrow and Lucas, 1986; Thelen and Delmer, 1986; Hayashi et al., 1987; Fredrikson and Larsson, 1989; Stone and Clarke, 1992; Pedersen et al., 1993; Dhugga and Ray, 1994; Kauss, 1996). The role of β-glucosides in callose synthesis is unknown. It may act either as a primer, although it is not incorporated into the final product, or as an allosteric activator (MacLachlan, 1982;

Morrow and Lucas, 1986; Hayashi et al., 1987). The β-glucosides appear to act synergistically with $Ca^{2+}$ (Hayashi et al., 1987) and, like $Ca^{2+}$, exert their effect on the cytoplasmic side of the plasma membrane (Fredrikson and Larsson, 1989). The β-glucoside, β-furfuryl-β-glucoside (FG), from mung beans (*Vigna radiata*), activates mung bean callose synthase *in vitro* and is proposed to be an endogenous regulator (Ohana et al., 1991; Ohana et al., 1992; Ohana et al., 1993). Compounds that are either identical or very similar to FG have been identified in peas, sorghum, barley and cotton (Ohana et al., 1992), consistent with the hypothesis that FG has a general role in the stimulation of plant callose synthesis *in vivo*. FG is largely sequestered into the vacuole of barley suspension cultured cells, and physical stresses such as those that activate the wound-activated callose synthase are proposed to activate an FG-carrier protein which results in the elevation of cytoplasmic concentrations of FG, and allows *in vivo* callose synthesis (Ohana et al., 1993). Other naturally occurring β-glucosides with hydrophobic aglycon moieties also have a stimulatory effect on callose synthase activity (Ohana et al., 1992; Ng et al., 1996). These molecules may fulfil the role of enzyme activation in other species in a similar manner to FG and be mimicked by cellobiose in *in vitro* callose synthase assay systems. β-Glucosides may also play a similar role for developmentally regulated callose synthases because the pollen-tube callose synthase also requires a β-glucoside for full activity *in vitro* (Schlüpmann et al., 1993).

Endogenous enzymes such as proteases, phosphatases and phosphodiesterases have been observed to affect callose synthase activity *in vitro* and therefore may also be regulators of callose synthase *in vivo*, although the physiological significance of these regulators is unknown. Callose synthase activity can be influenced by exogenous unsaturated free fatty acids, lysophospholipids, phospholipases and some amphipathic substances including digitonin (Kauss and Jeblick, 1986; Wasserman and MacCarthy, 1986; Saugy et al., 1988; Frost et al., 1990). This suggests that membrane phospholipids are an essential part of the callose synthase environment, and that the properties of this boundary layer are important in maintaining full callose synthase activity. The precise membrane environment required by the callose synthases from different species and tissues may vary slightly, explaining why different detergents, at different concentrations, are required in order to 'solubilize' different callose synthase enzymes while maintaining activity (Wasserman and MacCarthy, 1986; Lawson et al., 1989; Fink et al., 1990; Lai-Kee-Him et al., 2001).

Yeast (1,3)-β-glucan synthase activity is dependent upon GTP with catalytic activity regulated by the small GTP-binding protein Rho1 that binds FKS proteins in its GTP-bound conformation (Mazur and Baginsky, 1996). Although some plant callose synthases are not regulated by GTP (Turner et al., 1998; see Chapters 3.3.3 and 4.3), others may be. A low concentration of

GTP produced a small activation of callose synthase activity from moth bean (*Vigna aconitifolia*) root tips and the Rho-like GTP binding protein Rop1 interacts with UGT1 in its GTP-bound form, and may be present at the cell plate in a complex with UGT1 and AtGSL6 (Hong et al., 2001b).

Treatment of callose synthase preparations with trypsin stimulates activity in cell-free extracts from *N. alata* pollen tubes (Schlüpmann et al., 1993; Li et al., 1999), homogenates of soybean suspension culture cells (Kauss et al., 1983), mung bean microsomal membranes (Kauss, 1996) and, to a limited degree, in pea epicotyl membranes (Girard and MacLachlan, 1987). Recent studies with NaGSL1 show that lower molecular weight species of NaGSL1 were not correlated with increased callose synthase activity when membrane preparations from pollen tubes were treated with trypsin (Brownfield et al., 2008). This suggests that trypsin may act by removing a proteinaceous inhibitor rather than the removal of an auto-inhibitory domain as previously speculated (Li et al., 1999), although small changes in molecular weight may not have been detected. Removal of an inhibitor is consistent with the observation that the detergent CHAPS activates the *N. alata* pollen tube callose synthase in a similar manner to trypsin (Schlüpmann et al., 1993; Li et al., 1997, 1999) and to the auto-activation of callose synthase activity during enrichment of NaGSL1 (Turner et al., 1998). Stimulation of callose synthase activity has also been observed after affinity purification using an immobilized substrate in the presence of detergent (Slay et al., 1992), washing with $CaCl_2$ (Bulone et al., 1995), gradient centrifugation (Colombani et al., 2004) and in the presence of certain detergents (Kamat et al., 1992; Pedersen et al., 1993; McCormack et al., 1997; Lai-Kee-Him et al., 2001) which could all result from the removal of an inhibitory protein during these processes.

Post-translational regulation of NaGSL1 from *N. alata* pollen tubes appears to be important *in vivo*. The 220-kDa NaGSL1 polypeptide is produced after pollen-tube germination and accumulates during pollen-tube growth, as does callose synthase activity (Li et al., 1999; Brownfield et al., 2008). A combination of membrane fractionation and immuno-electron microscopy revealed that NaGSL1 is present predominantly in ER and Golgi membranes in younger pollen tubes when callose synthase was mostly in an inactive (latent) form (Brownfield et al., 2008). A model has been proposed for the regulation callose synthase in *N. alata* pollen tubes where the NaGSL1 protein in ER and Golgi is kept inactive, due to the presence of an inhibitory protein, to prevent ectopic callose deposition (Brownfield et al., 2008). The proteinaceous inhibitor is then removed at the plasma membrane where, in later stages of pollen-tube growth, NaGSL1 is present in both latent and active forms, consistent with the direct deposition of callose into the wall. Removal of the inhibitor could rely on interaction with a specific protease, or further post-translational modification of the NaGSL1 protein.

Evidence is emerging to suggest that post-translational modifications of membrane-bound proteins are important in the regulation of their activity and function in the cell. Glycosylation and phosphorylation and are two common post-translational modifications, but in the case of family GT48 enzymes, of which callose synthases are members, the extent to which glycosylation or phosphorylation might affect activity is largely unknown. An important role for protein lipidation is emerging from studies of mammalian and yeast membrane-bound proteins. Lipid modifications might be involved in the localization and cellular trafficking of proteins, and possibly in their activity (Nadolski and Linder, 2007). Commonly observed lipid modifications include *N*-myristoylation, *S*-palmitoylation and prenylation, as well as the attachment of GPI (glycosylphosphatidylinositol) anchors. For example, palmitolyation of the yeast family GT2 enzyme chitin synthase might be necessary for its transport from the ER to the cell surface (Lam et al., 2006). To date, no evidence for lipid-type modifications has emerged from biochemical studies of callose synthases from plants, and careful examination of GSL protein sequences reveals no GPI anchor signal sequences or other previously characterized motifs that specify lipid attachment to proteins.

## I.D The Synthesis of (1,3)-β-Glucans in Chlorophytes

In the past, investigations of callose biosynthesis have focussed on a relatively small number of embryophytic species (see Table 1). Whereas there are many additional land plant species that could be examined, similar difficulties are likely to be encountered in purifying higher plant callose synthases to homogeneity. Given these challenges, other approaches need to be developed to gain further insight into the process of callose synthesis and its regulation. One possibility is to explore other eukaryotic lineages such as the chlorophytes (green algae) that also contain (1,3)-β-glucan in their cell walls (Astbury and Preston, 1940; Nicolai and Preston, 1952; Stone and Clarke, 1992; see Chapter 2.1). A chlorophytic species that has been the subject of extensive research is the unicellular green alga *Chlamydomonas reinhardtii.* Despite the fact that their cell walls are composed almost exclusively of hydroxyproline-rich glycoproteins (Hicks et al., 2001), their walls do contain some callose, as revealed by Aniline Blue staining (Bai and VanWinkle-Swift, 2000).

*Chlamydomonas* has proven to be an excellent model system for the study of plant cell biology and metabolism (Harris, 2001), and offers a number of advantages over higher plants for the study of cell wall synthesis (Hicks et al., 2001). Its unicellular growth habit and amenability to genetic and molecular biological experimentation permits the use of genetic techniques analogous to those used in yeast for the dissection of biochemical pathways and cellular processes (Tam and Lefebvre, 1993; Asamizu et al., 1999; Hicks et al., 2001). In addition, there are established procedures for the transformation of its nuclear and chloroplast genomes (Kindle, 1990; Boynton and Gillham, 1993; Shimogawara et al., 1998), and functional genomic analysis is possible using RNA interference technology (Rohr et al., 2004; Schroda, 2006). Because *C. reinhardtii* is haploid, mutations produce progeny that do not segregate for the lesion and this facilitates the identification and characterization of any phenotypic variation associated with the phenotypes. The contractile vacuole permits growth in the absence of a normal cell wall structure, allowing most wall-defective mutants to remain viable, obviating the need for selection of conditional phenotypes (Hicks et al., 2001). Thus, saturation mutagenesis and screening for cell wall defects can occur quickly and efficiently, in contrast to higher plants. Importantly, a draft genome sequence for *C. reinhardtii* is now available (Merchant et al., 2007; http://genome.jgi-psf.org/) and its transcriptome has also been mapped on this sequence (Jain et al., 2007). A BLAST search of the *C. reinhardtii* genome (http://genome.jgi-psf.org/) shows that it contains as many as eight *GSL* genes (M. Doblin, unpublished data), although whether these are all functional remains to be verified. Nevertheless, the presence of *GSL* genes supports the finding that this alga synthesizes (1,3)-β-glucan in a manner that might be similar to that in embryophytes and that it could be exploited through comparative genomic, genetic and biochemical means to identify and subsequently test components of the callose synthesis and regulation machinery.

Another notable aspect of unicellular green algae is the potential for reduced functional gene redundancy compared to higher plant genomes. Whereas *C. reinhardtii* (121 Mb) displays a genome complexity approaching that of *Arabidopsis* (140.1 Mb) (Merchant et al., 2007), some recently characterized microalgae have genomes that rival yeast in their simplicity (Courties et al., 1998). The unicellular picophytoplanktonic chlorophytes, *Ostreococcus tauri* and *O. lucimarinus,* belong to the Prasinophyceae, an early diverging class within the green plant lineage. These are the smallest eukaryotes known (0.8 μm in diameter compared to 10 μm for *C. rheinhardtii*), and have genome sizes of 12.6 and 13.2 Mbp, respectively (Derelle et al., 2006; Palenik et al., 2007). Nevertheless, *Ostreococcus* displays all the major features of chlorophytes and other plant cells, having a single chloroplast with a starch granule, a mitochondrion and a Golgi (Courties et al., 1998). While the *Ostreococcus* spp. have not yet been shown to contain callose, BLAST searches have identified at least one *GSL* gene within each genome (http://genome.jgi-psf.org/; M. Doblin, unpublished data). This finding supports the notion of reduced *GSL* redundancy within these unicellular algae. In other species, the occurrence of (1,3)-β-glucan can be inferred by the presence of *GSL* genes, although *Ostreococcus* spp. are devoid of a true cell wall (Courties et al., 1998). This may indicate that the *Ostreococcus* spp. *GSLs* are only induced in response to physical or chemical stress, that they are non-functional, or that essential associated proteins are not present. However, if the *GSL* genes are shown to be functional, it would be of interest to ascertain whether these microalgae produce (1,3)-β-glucan and under what conditions. The complete sequencing of additional chlorophyte genomes such as *Bathycoccus sp.* BAN7, *Dunaliella salina, Micromonas pusilla* and *Volvox carteri f. nagariensis* (http://genome.jgi-psf.org/, http://www.genomesonline.org/) offers further opportunities to exploit comparative genomic technologies (Misumi et al., 2005) for the identification of additional candidate genes that might be involved in (1,3)-β-glucan synthesis.

## I.E The Synthesis of (1,3)-β-Glucans in Rhodophytes

The rhodophytes (red algae) may also present a worthwhile resource to explore (1,3)-β-glucan synthesis. Unicellular red algae of the Cyanidiales have also emerged as model systems for the study of plant processes (Barbier et al., 2005). Recently, the genome sequence of *Cyanidioschyzon merolae* was determined (Matsuzaki et al., 2004) and a relatively large EST dataset from *Galdieria sulphuraria* (Weber et al., 2004) as well as 8 Mb of non-redundant genomic sequence (approximately 70% genome coverage; http://genomics.msu.edu/galdieria) have become available. Both these organisms are small (2 μm in diameter), inhabit sulfate-rich hot springs (pH 0.05–3, temperature >56°C) and have relatively small genomes, similar in size to *Ostreococcus* spp.: 16.5 Mb for *C. merolae* (Matsuzaki et al., 2004) and between 10 and 16 Mb for *Galdieria* spp. (Moreira et al., 1994; Muravenenko et al., 2001). Like *Ostreococcus* spp. but in contrast to *G. sulphuraria*, the cells of *C. merolae* do not have a rigid cell wall. A recent report has demonstrated that *C. merolae* is amenable to transformation by electroporation, and evidence presented for a relatively high rate of homologous recombination (Minoda et al., 2004) suggests that this unicellular alga could be suitable for targeted gene knock-out and gene knock-in approaches. Although most searches of the available *C. merolae* (http://merolae.biol.s.u-tokyo.ac.jp/) and *G. sulphuraria* (http://genomics.msu.edu/galdieria/) sequences failed to detect any genes with similarity to *GSLs* (M. Doblin, unpublished data), this is not an unexpected result given the lack of a wall in *C. merolae* and the fact that it divides by binary fission rather than through formation of a phragmoplast as in higher plants (Barbier et al., 2005). The absence of a wall and thus potentially (1,3)-β-glucan, however, may prove extremely useful experimentally, as such organisms could provide a functional analysis system to test various callose synthase enzyme components and regulators in a background devoid of callose synthesis. Further sequencing of the partially completed *G. sulphuraria* and other rhodophyte genomes of *Chondrus crispus* and *Porphyra purpurea,* together with substantiation of the presence or absence of (1,3)-β-glucan in these species, may provide additional information regarding (1,3)-β-glucan synthesis and regulation that could be applied to higher plant species. Such information would be invaluable for the broader investigation of the origin and evolution of eukaryotic and plant cell walls.

## I.F Future Directions

Now that the characteristics of the gene families encoding higher plant and microbial (1,3)-β-glucan synthases have been described, we can expect the roles of individual genes and enzymes to be defined in detail in the immediate future. The temporal and spatial aspects of individual gene transcription will be increasingly important and this will be linked with information on the transcription factors that regulate the first stages of gene expression and other regulatory processes that might involve small RNA molecules, mRNA turnover and alternative splicing of mRNAs. These advances will be largely driven by functional genomics technologies, linked with new methods in bioinformatics. The potential for the chlorophyte and rhodophyte genomic sequences to contribute to comparative genomics approaches in the study of callose synthesis points to these as fruitful systems to exploit.

In addition, new opportunities to apply biochemical approaches to defining cellular processes associated with callose synthases are likely to be adopted and to make major contributions to our knowledge of callose deposition. For example, one might expect that biochemical approaches will be applied to major areas such as the more detailed description of specific cellular functions of callose synthase gene products, the regulation of enzyme activity through post-translational modifications, and the definition of protein–protein interactions that occur in the various cellular processes in which callose synthases participate. Finally and of central importance, we will need to provide unequivocal evidence that the products of *GSL* genes are indeed the callose synthase catalytic subunits in their own right and, if not, to define their specific function in callose biosynthesis.

## Acknowledgements

The authors wish to thank the Grains Research and Development Corporation (GRDC, Australia) for their grant in the 'Growth and end-use quality of cereals' in the support of this work. GBF also acknowledges the support of the ARC and LB acknowledges the support of an Australian Postgraduate Award (APA).

## References

Amor, Y., Haigler, C. H., Johnson, S., Wainscott, M., & Delmer, D. (1995). A membrane-associated form of sucrose synthase and its potential role in synthesis of cellulose and callose in plants. *Proceedings of the National Academy of Sciences USA*, *92*, 9353–9357.

Andrawis, A., Solomon, M., & Delmer, D. P. (1993). Cotton fibre annexins – A potential role in the regulation of callose synthase. *Plant Journal*, *3*, 763–772.

Asamizu, E., Nakamura, Y., Sato, S., Fukuzawa, H., & Tabata, S. (1999). A large scale structural analysis of cDNAs in a unicellular green alga, *Chlamydomonas reinhardtii*. I. Generation of 3433 non-redundant expressed sequence tags. *DNA Research*, *6*, 369–373.

Astbury, W., & Preston, R. (1940). The structure of the cell wall in some species of the filamentous green alga *Cladophora*. *Proceedings of the Royal Society of London. Series B, Biological Sciences*, *129*, 54–76.

Bai, M., & VanWinkle-Swift, K. (2000). The presence of callose in the primary zygote wall of *Chlamydomonas monoica* and the effects of its degradation on zygote development. *Journal of Phycology*, *36*, 3–4.

Barbier, G., Oesterhelt, C., Larson, M. D., Halgren, R. G., Wilkerson, C., Garavito, R. M., Benning, C., & Weber, A. P. M. (2005). Comparative genomics of two closely related unicellular thermo-acidophilic red algae, *Galdieria sulphuraria* and *Cyanidioschyzon merolae*, reveals the molecular basis of the metabolic flexibility of *Galdieria sulphuraria* and significant differences in carbohydrate metabolism of both algae. *Plant Physiology*, *137*, 460–474.

Becker, J. D., Boavida, L. C., Carneiro, J., Haury, M., & Feijo, J. A. (2003). Transcriptional profiling of *Arabidopsis* tissues reveals the unique characteristics of the pollen transcriptome. *Plant Physiology*, *133*, 713–725.

Boynton, J., & Gillham, N. (1993). Chloroplast transformation in *Chlamydomonas*. *Methods in Enzymology*, *217*, 510–536.

Brown, R. C., Lemmon, B. E., Stone, B. A., & Olsen, O. A. (1997). Cell wall (1-3)- and (1-3, 1-4)-beta-glucans during early grain development in rice (*Oryza sativa* L.). *Planta*, *202*, 414–426.

Brown, I., Trethowan, J., Kerry, M., Mansfield, J., & Bolwell, G. (1998). Localization of components of the oxidative cross-linking of glycoproteins and of callose synthesis in papillae formed during the interaction between non-pathogenic strains of *Xanthomonas campestris* and French bean mesophyll cells. *Plant Journal*, *15*, 333–343.

Brownfield, L. R. (2005). The role of NaGSL1 in callose synthesis in the pollen tubes of *Nicotiana alata*. Melbourne: University of Melbourne.

Brownfield, L. R., Ford, K., Doblin, M. S., Newbigin, E., Read, S., & Bacic, A. (2007). Proteomic and biochemical evidence links the callose synthase in *Nicotiana alata* pollen tubes to the product of the *NaGSL1* gene. *Plant Journal*, *52*, 147–156.

Brownfield, L., Wilson, S., Newbigin, E., Bacic, A., & Read, S. (2008). Molecular control of the glucan-synthase-like protein NaGSL1 and callose synthesis during growth of *Nicotiana alata* pollen tubes. *Biochemical Journal*, 414, 43–52.

Bulone, V., Fincher, G. B., & Stone, B. A. (1995). *In vitro* synthesis of a microfibrillar (1-3)-beta-glucan by a ryegrass (*Lolium multiflorum*) endosperm (1-3)-beta-glucan synthase enriched by product entrapment. *Plant Journal*, *8*, 213–225.

Buostead, C. M., Smallwood, M., Small, H., Bowles, D. J., & Walker, J. H. (1989). Identification of calcium-dependent phospholipid-binding proteins in higher plants. *FEBS Letters*, *244*, 456–460.

Burton, R., Wilson, S., Hrmova, M., Harvey, A., Shirley, N., Medhurst, A., Stone, B., Newbigin, E., Bacic, A., & Fincher, G. (2006). Cellulose synthase-like *CslF* genes mediate the synthesis of cell wall (1, 3; 1, 4)-ß-D-glucans. *Science*, *311*, 1940–1942.

Cabib, E., Roh, D. H., Schmidt, M., Crotti, L. B., & Varma, A. (2001). The yeast cell wall and septum as paradigms of cell growth and morphogenesis. *Journal of Biological Chemistry*, *276*, 19679–19682.

Castro, C., Ribas, J. C., Valdivieso, M. H., Varona, R., Delrey, F., & Duran, A. (1995). Papulacandin-B resistance in budding and fission yeasts – Isolation and characterization of a gene involved in (1,3)-beta-D-glucan synthesis in *Saccharomyces cerevisiae*. *Journal of Bacteriology*, *177*, 5732–5739.

Charnock, S. J., Henrissat, B., & Davies, G. J. (2001). Three-dimensional structures of UDP-sugar glycosyltransferases illuminate the biosynthesis of plant polysaccharides. *Plant Physiology*, *125*, 527–531.

Cocuron, J., Lerouxel, O., Drakakaki, G., Alonso, A., Liepman, A., Keegstra, K., Raikhel, N., & Wilkerson, C. (2007). A gene from the cellulose synthase-like C family encodes a beta-1, 4 glucan synthase. *Proceedings of the National Academy of Sciences USA*, *104*, 8550.

Colombani, A., Djerbi, S., Bessueille, L., Blomqvist, K., Ohlsson, A., Berglund, T., Teeri, T. T., & Bulone, V. (2004). *In vitro* synthesis of (1,3)-β-D-glucan (callose) and cellulose by detergent extracts of membranes from cell suspension cultures of hybrid aspen. *Cellulose*, *11*, 313–327.

Courties, C., Perasso, R., Chretiennot, M., Gouy, M., Guillou, L., & Troussellier, M. (1998). Phylogenetic analysis and genome size of *Ostreococcus tauri* (Chlorophyta, Prasinophyceae). *Journal of Phycology*, *34*, 844–849.

Coutinho, P. M., & Henrissat, B. (1999). Carbohydrate-active enzymes: An integrated database approach. In H. J. Gilbert, G. Davies, B. Henrissat, & B. Svensson (Eds.), *Recent Advances in Carbohydrate Bioengineering* (pp. 3–12). Cambridge: The Royal Society of Chemistry.

Cui, X. J., Shin, H. S., Song, C., Laosinchai, W., Amano, Y., & Brown, R. M. (2001). A putative plant homolog of the yeast beta-1,3-glucan synthase subunit FKS1 from cotton (*Gossypium hirsutum* L.) fibres. *Planta*, *213*, 223–230.

Delmer, D. (1987). Cellulose biosynthesis. *Annual Review of Plant Physiology*, *38*, 259–290.

Delmer, D. (1999). Cellulose biosynthesis: Exciting times for a difficult field of study. *Annual Review of Plant Physiology and Plant Molecular Biology*, *50*, 245–276.

Delmer, D., Solomon, M., & Read, S. (1991). Direct photolabeling with [$^{32}$-P]UDP-glucose for identification of a subunit of cotton fibre callose synthase. *Plant Physiology*, *95*, 556–563.

Delmer, D. P., Volokita, M., Solomon, M., Fritz, U., Delphendahl, W., & Herth, W. (1993). A monoclonal antibody recognizes a 65kDa higher plant membrane polypeptide which undergoes cation-dependent association with callose synthase *in vitro* and co-localizes with sites of high callose deposition *in vivo*. *Protoplasma*, *176*, 33–42.

Derelle, E., Ferraz, C., Rombauts, S., Rouze, P., Worden, A., Robbens, S., Partensky, F., Degroeve, S., Echeynie, S., & Cooke, R. (2006). Genome analysis of the smallest free-living eukaryote *Ostreococcus tauri* unveils many unique features. *Proceedings of the National Academy of Sciences USA*, *103*, 11647.

Dhugga, K. S., Barreiro, R., Whitten, B., Stecca, K., Hazebroek, J., Randhawa, G. S., Dolan, M., Kinney, A. J., Tomes, D., Nichols, S., et al. (2004). Guar seed beta-mannan synthase is a member of the cellulose synthase super gene family. *Science*, *303*, 363–366.

Dhugga, K. S., & Ray, P. M. (1994). Purification of 1,3-beta-D-glucan synthase activity from pea tissue. Two polypeptides of 55 kDa and 70 kDa copurify with enzyme activity. *European Journal of Biochemistry*, *220*, 943–953.

Dijkgraaf, G. J. P., Abe, M., Ohya, Y., & Bussey, H. (2002). Mutations in Fks1p affect the cell wall content of beta-1,3-and beta-1,6-glucan in *Saccharomyces cerevisiae*. *Yeast*, *19*, 671–690.

Doblin, M. S., De Melis, L., Newbigin, E., Bacic, A., & Read, S. M. (2001). Pollen tubes of *Nicotiana alata* express two genes from different beta-glucan synthase families. *Plant Physiology*, *125*, 2040–2052.

Doblin, M. S., Kurek, I., Jacob-Wilk, D., & Delmer, D. P. (2002). Cellulose biosynthesis in plants: From genes to rosettes. *Plant and Cell Physiology*, *43*, 1407–1420.

Dong, X., Hong, Z., Sivaramakrishnan, M., Mahfouz, M., & Verma, D. P. S. (2005). Callose synthase (CalS5) is required for exine formation during microgametogenesis and for pollen viability in *Arabidopsis*. *Plant Journal*, *42*, 315–328.

Douglas, C. M. (2001). Fungal beta(1,3)-D-glucan synthesis. *Medical Mycology*, *39*, 55–66.

Douglas, C. M., Foor, F., Marrinan, J. A., Morin, N., Nielsen, J. B., Dahl, A. M., Mazur, P., Baginsky, W., Li, W. L., Elsherbeini, M., Clemas, J. A., Mandala, S. M., Frommer, B. R., & Kurtz, M. B. (1994a). The *Saccharomyces cerevisiae FKS1* (*ETG1*) gene encodes an integral membrane-protein which is a subunit of 1,3-beta-D-glucan synthase. *Proceedings of the National Academy of Sciences USA*, *91*, 12907–12911.

Douglas, C. M., Marrinan, J. A., Li, W., & Kurtz, M. B. (1994b). A *Saccharomyces cerevisiae* mutant with echinocandin-resistant 1,3-beta-D-glucan synthase. *Journal of Bacteriology*, *176*, 5686–5696.

Eiberger, L. L., & Wassermann, B. P. (1987). Partial purification of digitonin-solubilized beta-glucan synthase from red beet root. *Plant Physiology*, *83*, 982–987.

Eng, W. K., Faucette, L., McLaughlin, M. M., Cafferkey, R., Koltin, Y., Morris, R. A., Young, P. R., Johnson, R. K., & Livi, G. P. (1994). The yeast *FKS1* gene encodes a novel membrane-protein, mutations in which confer Fk506 and cyclosporine-A hypersensitivity and calcineurin-dependent growth. *Gene*, *151*, 61–71.

Enns, L. C., Kanaoka, M. M., Torii, K. U., Comai, L., Okada, K., & Cleland, R. E. (2005). Two callose synthases, GSL1 and GSL5, play an essential and redundant role in plant and pollen development and in fertiliy. *Plant Molecular Biology*, *58*, 333–349.

Farkas, I., Hardy, T. A., DePaoli-Roach, A. A., & Roach, P. J. (1990). Isolation of the *GSY1* gene encoding yeast glycogen synthase and evidence for the existence of a second gene. *Journal of Biological Chemistry*, *265*, 20879–20886.

Ferguson, C., Teeri, T. T., Siika-aho, M., Read, S. M., & Bacic, A. (1998). Location of cellulose and callose in pollen tubes and grains of *Nicotiana tabacum*. *Planta*, *206*, 452–460.

Fink, J., Jeblick, W., & Kauss, H. (1990). Partial purification and immunological characterization of 1,3-beta-glucan synthase from suspension cells of *Glycine max*. *Planta*, *181*, 343–348.

Fredrikson, K., & Larsson, C. (1989). Activation of 1,3-beta-glucan synthase by $Ca^{2+}$, spermine and cellobiose. – Localization of activation sites using inside out plasma membrane vesicles. *Physiologia Plantarum*, *77*, 196–201.

Fredrikson, K., & Larsson, C. (1992). Activators and inhibitors of the plant plasma membrane 1,3-beta-glucan synthase. *Biochemical Society Transactions*, *20*, 710–713.

Fredrikson, K., Kjellbom, P., & Larsson, C. (1991). Isolation and polypeptide composition of a 1,3-beta-glucan synthase from plasma membrane of *Brassica oleracea*. *Physiologia Plantarum*, *81*, 289–294.

Frost, D. J., Read, S. M., Drake, R. R., Haley, B. E., & Wasserman, B. P. (1990). Identification of the UDP-glucose-binding polypeptide of callose synthase from *Beta vulgaris* L. by photoaffinity labeling with 5-azido-UDP-glucose. *Journal of Biological Chemistry*, *265*, 2162–2167.

Garrett-Engele, P., Moilanen, B., & Cyert, M. S. (1995). Calcineurin, the $Ca^{2+}$/calmodulin-dependent protein phosphatase, is essential in yeast mutants with cell integrity defects and in mutants that lack a functional vacuolar $H^+$-ATPase. *Molecular and Cellular Biology*, *15*, 4103–4114.

Girard, V., & MacLachlan, G. (1987). Modulation of pea membrane beta-glucan synthase activity by calcium, polycation, endogenous protease, and protease inhibitor. *Plant Physiology*, *85*, 131–136.

Geisler-Lee, C. J., Hong, Z., & Verma, D. P. S. (2002). Overexpression of the cell plate-associated dynamin-like GTPase, phragmoplastin, results in the accumulation of callose at the cell plate and arrest of plant growth. *Plant Science*, *163*, 33–42.

Harris, E. H. (2001). *Chlamydomonas* as a model organism. *Annual Review of Plant Physiology and Plant Molecular Biology*, *52*, 363–406.

Hayashi, T., Read, S., Bussell, J., Thelen, M. P., Lin, F.-C., Brown, R. M. J., & Delmer, D. (1987). UDP-glucose: (1-3)-beta-glucan synthases from mung bean and cotton. *Plant Physiology*, *83*, 1054–1062.

Hicks, G., Hironaka, C., Dauvillee, D., Funke, R., D'Hulst, C., Waffenschmidt, S., & Ball, S. (2001). When simpler is better. Unicellular green algae for discovering new genes and functions in carbohydrate metabolism. *Plant Physiology*, *127*, 1334–1338.

Hirano, Y., Walthert, L., & Brunner, I. (2006). Callose in root apices of European chestnut seedlings; a physiological indicator of aluminium stress. *Tree Physiology*, *24*, 1270–1283.

Holdaway-Clarke, T. L., & Hepler, P. K. (2003). Control of pollen tube growth: Role of ion gradients and fluxes. *New Phytologist*, *159*, 539–563.

Hong, Z. L., Delauney, A. J., & Verma, D. P. S. (2001a). A cell plate specific callose synthase and its interaction with phragmoplastin. *Plant Cell*, *13*, 755–768.

Hong, Z. L., Zhang, Z. M., Olson, J. M., & Verma, D. P. S. (2001b). A novel UDP-glucose transferase is part of the callose synthase complex and interacts with phragmoplastin at the forming cell plate. *Plant Cell*, *13*, 769–779.

Honys, D., & Twell, D. (2003). Comparative analysis of the *Arabidopsis* pollen transcriptome. *Plant Physiology*, *132*, 553–640.

Honys, D., & Twell, D. (2004). Transcriptome analysis of haploid male gametophyte development in *Arabidopsis*. *Genome Biology*, *5*, R85.13–R85.81.

Inoue, S. B., Qadota, H., Arisawa, M., Watanabe, T., & Ohya, Y. (1996). Signaling toward yeast 1,3-beta-glucan synthesis. *Cell Structure and Function*, *21*, 395–402.

Inoue, S. B., Takewaki, N., Takasuka, T., Mio, T., Adachi, M., Fujii, Y., Miyamoto, C., Arisawa, M., Furuichi, Y., & Watanabe, T. (1995). Characterization and gene cloning of 1,3-beta-D-glucan synthase from *Saccharomyces cerevisiae*. *European Journal of Biochemistry*, *231*, 845–854.

Jacobs, A. K., Lipka, V., Burton, R. A., Panstruga, R., Strizhov, N., Schulze-Lefert, P., & Fincher, G. B. (2003). An *Arabidopsis* callose synthase, GSL5, is required for wound and papillary callose formation. *Plant Cell*, *15*, 2503–2513.

Jain, M., Shrager, J., Harris, E. H., Halbrook, R., Grossman, A. R., Hauser, C., & Vallon, O. (2007). EST assembly supported by a draft genome sequence: An analysis of the *Chlamydomonas reinhardtii* transcriptome. *Nucleic Acids Research*, *35*, 2074–2083.

Kamat, U., Garg, R., & Sharma, C. B. (1992). Purification to homogeneity and characterization of a 1,3-beta-glucan (callose) synthase from germinating *Arachis hypogaea* cotyledons. *Archives of Biochemistry and Biophysics*, *298*, 731–739.

Kang, M. S., & Cabib, E. (1986). Regulation of fungal cell wall growth, a guanine nucleotide-binding, proteinaceous component required for activity of (1,3)-beta-D-glucan synthase. *Proceedings of the National Academy of Sciences USA*, *83*, 5808–5812.

Karnezis, T., Epa, V. C., Stone, B. A., & Stanisich, V. A. (2003). Topological characterization of an inner membrane (1,3)-beta-D-glucan (curdlan) synthase from *Agrobacterium sp* strain ATCC31749. *Glycobiology*, *13*, 693–706.

Kartusch, R. (2003). On the mechanism of callose synthesis induction by metal ions in onion epidermal cells. *Protoplasma*, *220*, 219–225.

Kauss, H. (1985). Callose biosynthesis as a calcium-regulated process and possible relations to the induction of other metabolic changes. *Journal of Cell Science* (Suppl. 2), 89–103.

Kauss, H. (1986). $Ca^{2+}$ dependence of callose synthesis and the role of polyamines in the activation of 1,3-β-glucan synthase by $Ca^{2+}$. In A. J. Trewavas (Ed.), *Molecular and Cellular Aspects of Calcium in Plant Development* (pp. 131–136). Plenum Press.

Kauss, H. (1996). Callose synthesis. In M. Smallwood, J. P. Knox, & D. J. Bowles (Eds.), *Membranes: Specialized Function in Plants* (2nd ed) (pp. 77–92). Oxford: BIOS Scientific Publishers.

Kauss, H., & Jeblick, W. (1985). Activation by polyamines, polycations, and ruthenium red of the $Ca^{2+}$-dependent glucan synthase from soybean cells. *FEBS Letters*, *185*, 226–230.

Kauss, H., & Jeblick, W. (1986). Influence of free fatty acids, lysophosphatidylcholine, platelet-activating factor, acylcarnitine and echinocandin B on 1,3-beta-D-glucan synthase and callose synthesis. *Plant Physiology*, *80*, 7–13.

Kauss, H., & Jeblick, W. (1987). Solubilization, affinity chromatography and $Ca^{2+}$/polyamine activation of the plasma membrane-located 1,3-β-D-glucan synthase. *Plant Science*, *48*, 63–69.

Kauss, H., Jeblick, W., & Domard, A. (1989). The degrees of polymerization and N-acetylation of chitosan determine its ability to elicit callose formation in suspension cells and protoplasts of *Catharanthus roseus*. *Planta*, *178*, 385–392.

Kauss, H., Köhle, H., & Jeblick, W. (1983). Proteolytic activation and stimulation by $Ca^{2+}$ of glucan synthase from soybean. *FEBS Letters*, *158*, 84–88.

Kawagoe, Y., Delmer, D. P. (1995). Cotton sucrose synthase interacts in vitro with calnexin: A possible mechanism for β-glucan synthase assembly. In *Proceedings of the Seventh International Cell Wall Meeting* (p. 188). Spain: Santiago de Compostella.

Kindle, K. (1990). High-frequency nuclear transformation of *Chlamydomonas reinhardtii*. *Proceedings of the National Academy of Sciences USA*, *87*, 1228.

Kjell, J., Rasmusson, A. G., Larsson, H., & Widell, S. (2004). Protein complexes of the plant plasma membrane resolved by Blue Native PAGE. *Physiologia Plantarum*, *121*, 546–555.

Köhle, H., Jeblick, W., Poten, F., Blaschek, W., & Kauss, H. (1985). Chitosan-elicited callose synthesis in soybean cells as a $Ca^{2+}$-dependent process. *Plant Physiology*, *77*, 544–551.

Kudlicka, K., & Brown, R. M. (1997). Cellulose and callose biosynthesis in higher plants. 1. Solubilization and separation of (1,3)- and (1,4)-beta-glucan synthase activities from mung bean. *Plant Physiology*, *115*, 643–656.

Lai-Kee-Him, J., Chanzy, H., Pelosi, L., Putaux, J. L., & Bulone, V. (2003). Recent developments in the field of *in vitro* biosynthesis of plant beta-glucans. *Biocatalysis in Polymer Science*, *840*, 65–77.

Lai-Kee-Him, J., Pelosi, L., Chanzy, H., Putaux, J. L., & Bulone, V. (2001). Biosynthesis of (1,3)-beta-D-glucan (callose) by detergent extracts of a microsomal fraction from *Arabidopsis thaliana*. *European Journal of Biochemistry*, *268*, 4628–4638.

Lam, K., Davey, M., Sun, B., Roth, A., Davis, N., & Conibear, E. (2006). Palmitoylation by the DHHC protein Pfa4 regulates the ER exit of Chs3. *The Journal of Cell Biology*, *174*, 19.

Lawson, S. G., Mason, T. L., Sabin, R. D., Sloan, M. E., Drake, R. R., Haley, B. E., & Wasserman, B. P. (1989). UDP-glucose:(1,3)-beta-glucan synthase from *Daucus carota* L. *Plant Physiology*, *90*, 101–108.

Li, H. J., Bacic, A., & Read, S. M. (1999). Role of a callose synthase zymogen in regulating wall deposition in pollen tubes of *Nicotiana alata* Link et Otto. *Planta*, *208*, 528–538.

Li, J., Burton, R. A., Harvey, A. J., Hrmova, M., Wardak, A. Z., Stone, B. A., & Fincher, G. B. (2003). Biochemical evidence linking a putative callose synthase gene with (1,3)-beta-D-glucan biosynthesis in barley. *Plant Molecular Biology*, *53*, 213–225.

Li, H. J., Turner, A., Doblin, M., Bacic, A., & Read, S. (1997). Localisation and purification of callose synthase from pollen tubes of *Nicotiana*. *Plant Physiology*, *114*, 102.

Li, L., Drake, R. R., Clement, S., & Brown, R. M. (1993). Beta-Glucan synthesis in the cotton fibre. III Identification of UDP-glucose-binding subunits of beta-glucan synthases by photoaffinity labeling with [$\beta$-$^{32}$P]5′-N3-UDP-glucose. *Plant Physiology*, *101*, 1149–1156.

Liepman, A., Nairn, C., Willats, W., Sørensen, I., Roberts, A., & Keegstra, K. (2007). Functional genomic analysis supports conservation of function among *Cellulose Synthase-Like A* gene family members and suggests diverse roles of mannans in plants. *Plant Physiology*, *143*, 1881.

Liepman, A. H., Wilkerson, C. G., & Keegstra, K. (2005). Expression of cellulose synthase-like (*Csl*) genes in insect cells reveals that *CslA* family members encode mannan synthases. *Proceedings of the National Academy of Sciences USA*, *102*, 2221–2226.

MacLachlan, G. A. (1982). Does cellulose synthesis require a primer?. In R. M. J. Brown (Ed.), *Cellulose and Other Natural Polymer Systems* (pp. 327–339) New York: Plenum Press.

Mason, T. L., Read, S. M., Frost, D. J., & Wasserman, B. P. (1990). Inhibition and labelling of red beet uridine 5′ diphospho-glucose:(1,3)-beta-glucan (callose) synthase by chemical modification with formaldehyde and uridine 5′ diphospho-pyridoxal. *Physiologia Plantarum*, *79*, 439–447.

Matsuzaki, M. et al., (2004). Genome sequence of the ultrasmall unicellular red alga *Cyanidioschyzon merolae* 10D. *Nature*, *428*, 653–657.

Mazur, P., & Baginsky, W. (1996). *In vitro* activity of 1,3-beta-D-glucan synthase requires the GTP-binding protein Rho1. *Journal of Biological Chemistry*, *271*, 14604–14609.

Mazur, P., Morin, N., Baginsky, W., Elsherbeini, M., Clemas, J. A., Nielsen, J. B., & Foor, F. (1995). Differential expression and function of 2 homologous subunits of yeast 1,3-beta-D-glucan synthase. *Molecular and Cellular Biology*, *15*, 5671–5681.

McCormack, B. A., Gregory, A. C. E., Kerry, M. E., Smith, C., & Bolwell, G. P. (1997). Purification of an elicitor-induced glucan synthase (callose synthase) from suspension cultures of French bean (*Phaseolus vulgaris* L.): Purification and immunolocation of a probable $M_r$-65 000 subunit of the enzyme. *Planta*, *203*, 196–203.

Meikle, P. J., Ng, K. F., Johnson, E., Hoogenraad, N. J., & Stone, B. A. (1991). The beta-glucan synthase from *Lolium multiflorum*. Detergent solubilization, purification using monoclonal antibodies and photoaffinity labeling with a novel photoreactive pyrimidine analogue or uridine 5′-diphosphoglucose. *Journal of Biological Chemistry*, *266*, 22569–22581.

Merchant, S. S. et al., (2007). The *Chlamydomonas* genome reveals the evolution of key animal and plant functions. *Science*, *318*, 245–250.

Minoda, A., Sakagami, R., Yagisawa, F., Kuroiwa, T., & Tanaka, K. (2004). Improvement of culture conditions and evidence for nuclear transformation by homologous recombination in a red alga, *Cyanidioschyzon merolae* 10D. *Plant and Cell Physiology*, *45*, 667–671.

Misumi, O., Matsuzaki, M., Nozaki, H., Miyagishima, S., Mori, T., Nishida, K., Yagisawa, F., Yoshida, Y., Kuroiwa, H., & Kuroiwa, T. (2005). *Cyanidioschyzon merolae* genome. A tool for facilitating comparable studies on organelle biogenesis in photosynthetic Eukaryotes. *Plant Physiology, 137*, 567–585.

Mizuno, K. (1998). Cellulose synthase-tubulin interactions in *Azuki* bean. In *Proceedings of the Plant Polysaccharides Symposium*. California, U.S.A: University of California-Davis.

Mol, P. C., Park, H. M., Mullins, J. T., & Cabib, E. (1994). A GTP-binding protein regulates the activity of (1-3)-beta-glucan synthase, an enzyme directly involved in yeast-cell wall morphogenesis. *Journal of Biological Chemistry, 269*, 31267–31274.

Moreira, D., Lopez-Archilla, A. I., Amils, R., & Marin, I. (1994). Characterization of two new thermoacidophilic microalgae: Genome organization and comparison with *Galdieria sulphuraria*. *FEMS Microbiology Letters, 122*, 109–114.

Morrow, D. L., & Lucas, W. J. (1986). (1-3)-Beta-D-Glucan synthase from sugar beet. I Isolation and solubilization. *Plant Physiology, 81*, 171–176.

Mothes, W., Heinrich, S. U., Graf, R., Nilsson, I. M., von Heijne, G., Brunner, J., & Rapoport, T. A. (1997). Molecular mechanisms of membrane protein integration into the endoplasmic reticulum. *Cell, 89*, 523–533.

Mueller, S. C., & Maclachlan, G. A. (1983). Radioautographic visualization of beta-glucans formed by pea membranes from UDP-glucose. *Canadian Journal of Botany – Revue Canadienne De Botanique, 61*, 1266–1275.

Muravenenko, O. V., Selyakh, I. O., Kononenko, N. V., & Stadnichuk, I. N. (2001). Chromosome numbers and nuclear DNA contents in the red microalgae *Cyanidium caldarium* and three *Galdieria* species. *European Journal of Phycology, 36*, 227–232.

Nadolski, M., & Linder, M. (2007). Protein lipidation. *FEBS Journal, 274*, 5202–5210.

Nakashima, J., Laosinchai, W., Cui, X. J., & Brown, R. M. (2003). New insight into the mechanism of cellulose and callose biosynthesis: Proteases may regulate callose biosynthesis upon wounding. *Cellulose, 10*, 369–389.

Ng, K., Johnson, E., & Stone, B. A. (1996). Specificity of binding of [beta]-glucoside activators of ryegrass (1-3)-[beta]-glucan synthase and the synthesis of some potential photoaffinity activators. *Plant Physiology, 111*, 1227–1231.

Nicolai, E., & Preston, R. D. (1952). Cell-wall studies in the Chlorophyceae I. A general survey of submicroscopic structure in filamentous species. *Proceedings of the Royal Society of London. Series B, Biological Sciences, 140*, 244–274.

Nishikawa, S., Zinkl, G. M., Swanson, R. J., Maruyama, D., & Preuss, D. (2005). Callose (beta-1,3-glucan) is essential for *Arabidopsis* pollen wall patterning, but not tube growth. *BMC Plant Biology*, *5*.

Nishimura, M. T., Stein, M., Hou, B. H., Vogel, J. P., Edwards, H., & Somerville, S. C. (2003). Loss of a callose synthase results in salicylic acid-dependent disease resistance. *Science*, *301*, 969–972.

Ohana, P., Benziman, M., & Delmer, D. P. (1993). Stimulation of callose synthesis *in vivo* correlates with changes in intracellular distribution of the callose synthase activator [β]-furfuryl-[β]-glucoside. *Plant Physiology*, *101*, 187–191.

Ohana, P., Delmer, D., Volman, G., Steffens, J., Matthews, D., & Benziman, M. (1992). β-Furfuryl-β-glucoside: An endogenous activator of higher plant UDP-glucose:(1-3)-β-synthase. *Plant Physiology*, *98*, 708–715.

Ohana, P., Delmer, D. P., Steffens, J. C., Matthews, D. E., Mayer, R., & Benziman, M. (1991). β-Furfuryl-β-glucoside: An endogenous activator of higher plant UDP-glucose:(1-3)-β-glucan synthase. *Journal of Biological Chemistry*, *266*, 13472–13475.

Olsen, O. A. (2001). Endosperm development: Cellularization and cell fate specification. *Annual Review of Plant Physiology and Plant Molecular Biology*, *52*, 233–267.

Østergaard, L., Petersen, M., Mattsson, O., & Mundy, J. (2002). An *Arabidopsis* callose synthase. *Plant Molecular Biology*, *49*, 559–566.

Palenik, B., Grimwood, J., Aerts, A., Rouze, P., Salamov, A., Putnam, N., Dupont, C., Jorgensen, R., Derelle, E., & Rombauts, S. (2007). The tiny eukaryote *Ostreococcus* provides genomic insights into the paradox of plankton speciation. *Proceedings of the National Academy of Sciences USA*, *104*, 7705.

Paliyath, G., & Poovaiah, B. W. (1988). Promotion of beta-glucan synthase activity in corn microsomal membranes by calcium and protein phosphorylation. *Plant and Cell Physiology*, *29*, 67–73.

Paredez, A., Somerville, C., & Ehrhardt, D. (2006). Visualization of cellulose synthase demonstrates functional association with microtubules. *Science*, *312*, 1491–1495.

Parre, E., & Geitmann, A. (2005). More than a leak sealant. The mechanical properties of callose in pollen tubes. *Plant Physiology*, *137*, 274–286.

Pedersen, L. H., Jacobsen, S., Hejgaard, J., & Rasmussen, S. K. (1993). Characterization and partial purification of beta-1,3-D-glucan (callose) synthase from barley (*Hordeum vulgare*) leaves. *Plant Science*, *91*, 127–138.

Pelosi, L., Imai, T., Chanzy, H., Heux, L., Buhler, E., & Bulone, V. (2003). Structural and morphological diversity of (1,3)-beta-D-glucans synthesized *in vitro* by enzymes from *Saprolegnia monoica*. Comparison with a corresponding *in vitro* product from blackberry (*Rubus fruticosus*). *Biochemistry*, *42*, 6264–6274.

Pelosi, L., Bulone, V., & Heux, L. (2006). Polymorphism of curdlan and (1-3)-β-D-glucans synthesized *in vitro*: A 13C CP-MAS and X-ray diffraction analysis. *Carbohydrate Polymers*, *66*, 199–207.

Petersen, M., Brodersen, P., Naested, H., Andreasson, E., Lindhart, U., Johansen, B., Nielsen, H. B., Lacy, M., Austin, M. J., Parker, J. E., Sharma, S. B., Klessig, D. F., Martienssen, R., Mattsson, O., Jensen, A. B., & Mundy, J. (2000). *Arabidopsis* MAP kinase 4 negatively regulates systemic acquired resistance. *Cell*, *103*, 1111–1120.

Qi, X., Tai, C., & Wasserman, B. (1995). Plasma membrane intrinsic proteins of *Beta vulgaris* L. *Plant Physiology*, *108*, 387.

Rae, A. E., Harris, P. J., Bacic, A., & Clarke, A. E. (1985). Composition of the cell walls of *Nicotiana alata* Link et Otto pollen tubes. *Planta*, *166*, 128–133.

Ram, A. F. J., Wolters, A., Tenhoopen, R., & Klis, F. M. (1994). A new approach for isolating cell-wall mutants in *Saccharomyces cerevisiae* by screening for hypersensitivity to calcofluor white. *Yeast*, *10*, 1019–1030.

Raymond, Y., Fincher, G., & Maclachlan, G. A. (1978). Tissue slice and particulate beta-glucan synthetase activity from *Pisum* epicotyls. *Plant Physiology*, *61*, 938–942.

Rohr, J., Sarkar, N., Balenger, S., Jeong, B., & Cerutti, H. (2004). Tandem inverted repeat system for selection of effective transgenic RNAi strains in *Chlamydomonas*. *Plant Journal*, *40*, 611.

Saugy, M., Farkas, V., & Maclachlan, G. (1988). Phosphatases and phosphodiesterases interfere with 1,3-beta-D-glucan synthase activity in pea epicotyl membrane preparations. *European Journal of Biochemistry*, *177*, 135–138.

Saxena, I. M., & Brown, R. M. J. (2000). Cellulose synthases and related enzymes. *Current Opinion in Plant Biology*, *3*, 523–531.

Saxena, I. M., Brown, R. M. J., Fevre, M., Geremia, R. A., & Henrissat, B. (1995). Multidomain architecture of beta-glycosyl transferases: Implications for mechanism of action. *Journal of Bacteriology*, *177*, 1419–1424.

Schimoler-O'Rourke, R., Renault, S., Mo, W. J., & Selitrennikoff, C. P. (2003). *Neurospora crassa* FKS protein binds to the (1,3)-beta-glucan synthase substrate, UDP-glucose. *Current Microbiology*, *46*, 408–412.

Schlüpmann, H., Bacic, A., & Read, S. M. (1993). A novel callose synthase from pollen tubes of *Nicotiana*. *Planta*, *191*, 470–481.

Schroda, M. (2006). RNA silencing in *Chlamydomonas*: Mechanisms and tools. *Current Genetics*, *49*, 69–84.

Schaeffer, H. J., & Walton, J. D. (1990). Aluminum ions induce oat protoplasts to produce an extracellular (1,3)-beta-D-glucan. *Plant Physiology*, *94*, 13–19.

Shimogawara, K., Fujiwara, S., Grossman, A., & Usuda, H. (1998). High-efficiency transformation of *Chlamydomonas reinhardtii* by electroporation. *Genetics*, *148*, 1821–1828.

Shin, H., & Brown, R.M. (1998). The first biochemical identification of yeast β-1,3-glucan synthase homolog in cotton fibres (*Gossypium hirsutum*). In *Proceedings of the Keystone Symposia on Molecular and Cellular Biology.* Taos, New Mexico. Abstract 145.

Shin, H. S., & Brown, R. M. (1999). GTPase activity and biochemical characterization of a recombinant cotton fiber annexin. *Plant Physiology*, *119*, 925–934.

Singer, S. J., Macher, P. A., & Yaffe, M. P. (1987). On the translocation of proteins across membranes. *Proceedings of the National Academy of Sciences USA*, *84*, 1015–1019.

Slay, R. M., Watada, A. E., Frost, D. J., & Wasserman, B. P. (1992). Characterization of the UDP-glucose: (1,3)-beta-glucan (callose) synthase from plasma membrane of celery: Polypeptide profiles and photolabeling patterns of enriched fractions suggest callose synthase complexes from various sources share a similar structure. *Plant Science*, *86*, 125–136.

Sloan, M. E., Rodis, P., & Wasserman, B. P. (1987). CHAPs solubilisation and functional reconstitution of beta-glucan synthase from red beet (*Beta vulgaris L.*) storage tissue. *Plant Physiology*, *85*, 516–522.

Stasinopoulos, S. J., Fisher, P. R., Stone, B. A., & Stanisich, V. A. (1999). Detection of two loci involved in (1,3)-beta-glucan (curdlan) biosynthesis by *Agrobacterium* sp. ATCC31749, and comparative sequence analysis of the putative curdlan synthase gene. *Glycobiology*, *9*, 31–41.

Stone, B. A., & Clarke, A. E. (1992). *Chemistry and Biology of (1,3)-β-D-glucans*. Melbourne: La Trobe University Press.

Tam, L. W., & Lefebvre, P. A. (1993). Cloning of flagellar genes in *Chlamydomonas reinhardtii* by DNA insertional mutagenesis. *Genetics*, *135*, 375–384.

Taylor, N. G., Howells, R. M., Huttly, A. K., Vickers, K., & Turner, S. R. (2003). Interactions among three distinct CesA proteins essential for cellulose synthesis. *Proceedings of the National Academy of Sciences USA*, *100*, 1450–1455.

Thelen, M. P., & Delmer, D. (1986). Gel-electrophoretic separation, detection and characterization of plant and bacterial UDP-glucose glucosyltransferases. *Plant Physiology*, *81*, 913–918.

Töller, A., Brownfield, L. R., Neu, C., Twell, D., & Schulze-Lefert, P. (2008). Dual function of *Arabidopsis* Glucan Synthase-Like genes *GSL8* and *GSL10* in male gametophyte development and plant growth. *Plant Journal*, *54*, 911–923

Tucker, M. R., Paech, N. A., Willemse, M. T. M., & Koltunow, A. M. G. (2001). Dynamics of callose deposition and beta-1,3-glucanase expression during reproductive events in sexual and apomictic *Hieracium*. *Planta*, *212*, 487–498.

Turner, A., Bacic, A., Harris, P. J., & Read, S. M. (1998). Membrane fractionation and enrichment of callose synthase from pollen tubes of *Nicotiana alata* Link et Otto. *Planta*, *205*, 380–388.

Verma, D. P. S., & Hong, Z. L. (2001). Plant callose synthase complexes. *Plant Molecular Biology*, *47*, 693–701.

Vogel, J., & Somerville, S. (2000). Isolation and characterization of powdery mildew-resistant *Arabidopsis* mutants. *Proceedings of the National Academy of Science USA*, *97*, 1897–1902.

Voigt, C. A., Schäfer, W., & Salomon, S. (2006). A comprehensive view on organ-specific callose synthesis in wheat (*Triticum aestivum* L.): Glucan synthase-like gene expression, callose synthase activity, callose quantification and deposition. *Plant Physiology and Biochemistry*, *44*, 242–247.

Wasserman, B. P., & MacCarthy, K. J. (1986). Regulation of plasma membrane beta-glucan synthase from red beet root by phospholipids. *Plant Physiology*, *82*, 396–400.

Weber, A., Oesterhelt, C., Gross, W., Bräutigam, A., Imboden, L., Krassovskaya, I., Linka, N., Truchina, J., Schneidereit, J., & Voll, H. (2004). EST-analysis of the thermo-acidophilic red microalga *Galdieria sulphuraria* reveals potential for lipid A biosynthesis and unveils the pathway of carbon export from rhodoplasts. *Plant Molecular Biology*, *55*, 17–32.

Wilson, S. M., Burton, R. A., Doblin, M. S., Stone, B. A., Newbigin, E. J., Fincher, G. B., & Bacic, A. (2006). Temporal and spatial appearance of wall polysaccharides during cellularization of barley (*Hordeum vulgare*) endosperm. *Planta*, *224*, 655–667.

Worral, D., Hird, D. L., Hodge, R., Paul, W., Draper, J., & Scott, R. (1992). Premature dissolution of the microsporocyte callose wall causes male sterility in transgenic tobacco. *Plant Cell*, *4*, 759–771.

Wu, A., Harriman, R. W., Frost, D. J., Read, S. M., & Wasserman, B. P. (1991). Rapid enrichment of CHAPS-solubilized UDP-glucose:(1,3)-beta-glucan (callose) synthase from *Beta vulgaris* L. by product entrapment. *Plant Physiology*, *97*, 684–692.

Wu, A., & Wasserman, B. P. (1993). Limited proteolysis of (1,3)-beta-glucan (callose) synthase from *Beta vulgaris* L – Topology of protease-sensitive sites and polypeptide identification using Pronase-E. *Plant Journal*, *4*, 683–695.

Yamaguchi, T., Hayashi, T., Nakayama, K., & Koike, S. (2006). Expression analysis of genes for callose synthases and Rho-type small GTP-binding proteins that are related to callose synthesis in rice anther. *Bioscience, Biotechnology and Biochemistry*, *70*, 639–645.

Zielinski, R. E. (1998). Calmodulin and calmodulin-binding proteins in plants. *Annual Review of Plant Physiology and Plant Molecular Biology*, *49*, 697–725.

CHAPTER 4.1

# Functional Roles of (1,3)-β-Glucans and Related Polysaccharides: Prokaryotes

**Vilma A. Stanisich[1] and Bruce A. Stone[2]**
[1]*Department of Microbiology, La Trobe University, Melbourne, Australia*
[2]*Department of Biochemistry, La Trobe University, Melbourne, Australia*

(1,3)-β-Glucans produced by both Gram-positive and Gram-negative prokaryotes are deposited as extracellular capsules in the case of (1,3)-β-glucan (curdlan) of *Agrobacterium* spp. and *Cellulomonas* spp. and the (1,3;1,2)-β-glucan of serotype 37 strains of *Streptococcus pneumoniae*. Others are secreted to the periplasmic compartment and remain predominantly cell associated; namely, the (1,3;1,6)-β-glucans of *Bradyrhizobium* spp. and other rhizobia. These various glucans exhibit different physico-chemical attributes and have been implicated in a variety of roles that assist in bacterial survival, colonization and cell–cell interactions. The roles include: prevention of desiccation in low-moisture environments due to the water-binding properties of the glucan [(1,3)- and (1,3;1,2)-β-glucans]; protection against predation by soil amoebae [(1,3)-β-glucan] or against phagocytosis by leucocytes [(1,3;1,2)-β-glucan]; enhanced survival in stressful environments [e.g. relief of hypoosmotic stress by the cyclic (1,3;1,6)-β-glucans]; mediation of cell adhesion leading to the formation of flocs and/or biofilms [(1,3)-β-glucans of *Agrobacterium* and *Cellulomonas*]; and modulation of plant defenses enabling infection [(1,3;1,6)-β-glucans]. In *Cellulomonas*, curdlan is used as an extracellularly located nutrient reserve.

## A. Introduction

The prokaryotic (1,3)-β-glucans include the linear glucan, curdlan, the cyclic (1,3;1,6)-β-glucans and the side-chain-branched (1,3;1,2)-β-glucan. These glucans are produced by members of the Gram-negative α-Proteobacteria (i.e. species of *Agrobacterium*, *Bradyrhizobium*, *Azorhizobium* and *Azospirillum*) and the Gram-positive Firmicutes (i.e. species of *Bacillus*,

*Cellulomonas* and *Streptococcus*). Most of the (1,3)-β-glucan-producing organisms are found as free-living inhabitants of the soil that are subject to variable nutritional, chemical and physical stresses. Some can also form close associations with plants, either as colonizers of the surface mucilage of roots or as invaders of plant tissue leading to pathogenic or symbiotic interactions. Survival in the latter situations requires the bacterium to successfully compete against other microbiota in the rhizosphere and/or to overcome the defense mechanisms of the host plant. Only the (1,3;1,2)-β-glucan-producing *Streptococcus pneumoniae* is typically associated with humans either as a commensal or pathogenic component of the nasopharangeal microflora. In this chapter the distribution and functional roles of the three bacterial (1,3)-β-glucans are discussed.

## B. Linear (1,3)-β-Glucan (Curdlan)

### *B.1. Occurrence*

The (1,3)-β-glucan, curdlan, was first detected in a spontaneous mutant of a slime-producing *Agrobacterium* sp. biovar 1 strain that was recovered from soil and was capable of using 10% ethylene glycol as the sole source of carbon (Harada and Harada, 1996). The mutant, 10C3K, produced curdlan in high yield whereas only a small quantity (<7%) was made by the wild-type strain 10C3 (Hisamatsu et al., 1982). The enhanced production of curdlan by 10C3K was accompanied by an inability to produce succinoglycan (Hisamatsu et al., 1980), a water-soluble acidic heteroglycan that is responsible for the slime phenotype. In contrast, both strains produced cyclic (1,2)-β-glucan in similarly low yields. Strain 10C3K is the progenitor of other mutants used to optimize industrial-scale production of curdlan [e.g. NTK-u (Nakanishi et al., 1992); ATCC 31749 (Lee, 2002) and ATCC 31750 (Kim et al., 2003)] or to investigate the genetic basis of curdlan production [*Agrobacterium* sp. LTU50 (Stasinopoulos et al., 1999; Karnezis et al., 2003; McIntosh et al., 2005)] (see Chapter 3.3.1).

Curdlan production appears to be a relatively common characteristic of *Agrobacterium* strains and can be detected readily by the formation of dark-blue-staining colonies on medium containing Aniline Blue, a (1,3)-β-glucan-specific dye (Nakanishi et al., 1974). Nakanishi et al. (1976) used it to screen 687 representative strains from 43 bacterial genera and found that six of 17 *Agrobacterium* strains (i.e. 10C3, 10C3K and various *A. radiobacter* strains) gave rise to stained colonies. A similar study of *Agrobacterium* biovar 1 isolates from garden soils in Australia detected staining in 25 of 40 strains tested (F. Taner and V.A. Stanisich, unpublished data). In each study, curdlan production was confirmed by isolation of the polysaccharide. Some *Agrobacterium* strains that do not produce curdlan may be genetically incapable of doing so whereas others, such as the fully sequenced *A. tumefaciens* strain C58 (Goodner et al., 2001; Wood et al., 2001), have the requisite biosynthetic genes (*crdASC*) (see Chapter 3.3.1) but produce the polymer in low yield. Yet other strains display an unstable phenotype, spontaneously forming curdlan-producing variants when stored on nutrient agar slants for 60 days or on suspension in saline (0.9%) or phosphate buffer (0.02 M) for >5 days (Hisamatsu et al., 1977; Nakanishi et al., 1976). Among representatives of *Rhizobium*, a genus related to *Agrobacterium*, two of 50 strains tested produced curdlan and both were rare Aniline Blue-staining variants recovered from parental non-curdlan producing cultures (Footrakul et al., 1981; Ghai et al., 1981) (Table 1).

Curdlan production has also been detected in soil inhabitants from two Gram-positive genera: *Cellulomonas*, which are noteworthy as cellulolytic bacteria, and *Bacillus*, which are spore-forming bacteria. Six species of *Cellulomonas*, including *C. flavigena* KU isolated from leaf litter (Angelo et al., 1990), all produced curdlan in varying amounts in a mineral salts-glucose liquid culture; however, the colonies they formed did not stain with Aniline Blue (Buller, 1990; Kenyon and Buller, 2002). Likewise, none of the six *Cellulomonas* strains tested by Nakanishi et al. (1976) stained with Aniline Blue. In contrast, Aniline Blue-stained colonies of *Bacillus* were observed by Nakanishi et al. (1976) (5 of 158 strains including *B. brevis* and *B. cereus*) and Gummadi and Kumar (2005) (11 of 50 strains) and, in the latter study, curdlan was isolated from three strains of which *Bacillus* sp. SNC07 produced the highest yield ($3\,g\,L^{-1}$).

Curdlan production thus appears to be confined to soil inhabitants and includes species that share ecological niches (Table 1). *Agrobacterium* and *Rhizobium* are typical members of the nutrient-rich and microbially diverse plant rhizosphere ecosystem that is predominantly populated by Gram-negative bacteria (Garbeva et al., 2004). Members of these two genera are additionally noteworthy because of their close association with plants as overt pathogens and $N_2$-fixing symbionts, respectively. Gram-positive bacteria including some *Bacillus* spp. are also members of the rhizosphere community (Garbeva et al., 2003). Conversely, *Bacillus* and *Cellulomonas* are examples of commonly encountered saprophytes that occur in soil ecosystems which, in comparison to the rhizosphere, are relatively nutrient limited (da Silva and Nahas, 2002). Non-pathogenic/symbiotic strains of *Agrobacterium* and *Rhizobium* are also numerous as soil saprophytes (Krimi et al., 2002; Denison and Kiers, 2004) and *Agrobacterium* spp. can persist in soil that has been continuously fallow for 5 years (Bouzar et al., 1993). Whether curdlan serves significant roles in these bacteria and whether the pathways of production are genetically comparable are intriguing questions yet to be fully resolved.

**Table 1: Occurrence of bacterial (1,3)-β-glucans**

| Glucan type | Bacterial source | Reference |
|---|---|---|
| (1,3)-β-glucan, linear (curdlan) | *Agrobacterium* sp. 10C3 (formerly *Alcaligenes faecalis var myxogenes* strain 10C3) and derivatives (e.g. NTK-u (=IFO13140 =ATCC21680)) | Nakanishi et al. (1976); Harada and Harada (1996) |
| | *Agrobacterium* sp. ATCC31749 and derivatives (LTU50; ATCC31750; R259) | Phillips & Lawford (1983); Stasinopoulos et al. (1999); Lee (2002); Kim et al. (2003) |
| | *A. radiobacter* IFO12607; IFO12665; IFO13127 (=ATCC6466 = NCIM2443); IFO13256; IFO13259[a] | Nakanishi et al. (1976) |
| | *Agrobacterium* sp. CFR-24 | Shivakumar & Vijayendra (2006) |
| | *Rhizobium trifolii* J60B1 | Ghai et al. (1981) |
| | *Rhizobium* sp. TISTR64B | Footrakul et al. (1981) |
| | *Cellulomonas* spp. (e.g. *C. uda* ATCC492; *C. fimi* ATCC15724) | Buller (1990); Kenyon et al. (2005) |
| | *C. flavigena* KU (ATCC53703) | Kenyon & Buller (2002) |
| | *Bacillus* sp. SNC07, SNC10, SNC11 | Gummadi & Kumar (2005) |
| (1,3;1,6)-β-glucans, cyclic | *Bradyrhizobium japonicum* USDA 100; I17; 3I1b71a<br>*Bradyrhizobium* sp. 32H1(=ATCC33848); BR4406; BR8404 | Dudman & Jones (1980); Miller et al. (1990); de Iannino and Ugalde (1993) |
| | *Rhizobium loti* NZP2309 | Estrella et al. (2000) |
| | *Azorhizobium caulinodans* HAMBI216 (=ORS571) | Komaniecka & Choma (2003) |
| | *Azospirillum brasilense* ATCC29710 | Altabe et al. (1994)[b] |
| (1,3)-β-glucan, cyclic (cyclolaminarinose) | *B. japonicum* USDA 110 *ndvC*::Tn*5* mutant (AB-1) | Bhagwat et al. (1999) |
| | *Sinorhizobium meliloti ndvB*::Tn*5* mutant (TY7) with a plasmid-borne *B. japonicum ndv* locus (p5D3) | Pfeffer et al. (1996) |
| (1,3;1,2)-β-glucan, side-chain-branched | *Streptococcus pneumoniae* type 37 (1235/89; 7077/39; 975/96) | Knecht et al. (1970); Llull et al. (1999) |

ATCC = American Type Culture Collection (USA); IFO = Institute Fermentation Osaka (Japan); NCIM = National Collection of Industrial Micoorganisms (India).

[a]Originally named *Radiobacter rhizogenes*; renamed by the IFO.

[b]Reported by Altabe et al. (1994) to be linear glucans; subsequently shown to be cyclic (Altabe et al., 1998).

## B.2. Formation and Functional Roles

Bacterial surface polymers such as loosely associated slimes and cell-associated polysac charides (capsular polymers, lipopolysaccharides and cyclic glycans) have been implicated in significant survival roles (Costerton et al., 1987; Roberts, 1996). These include: cellular recognition, colonization of biotic and abiotic surfaces, and tolerance to toxic agents (e.g. antibiotics, disinfectants) and physical stresses (e.g. desiccation), all of which are properties that would be expected to enhance the persistence of the bacteria in a variety of environmental situations. Scanning electron microscopy on *Agrobacterium* 103CK cultured on agar-solidified medium (0.5% yeast extract; 2% glucose) showed that curdlan is deposited extracellularly as microfibrils that form an envelope (capsule) around the cells (Kako et al., 1989). Curdlan production was detectable at day 2 and increased thereafter even though cell number did not increase. After 8 days, a pellicle composed of enmeshed curdlan and cells was formed and had sufficient tensile strength to be stripped off the bacterial growth as an intact, flexible layer (Fig. 1). Bacterial cells that were not part of the network remained on the surface of the solid medium. The establishment of a pellicle at the air interface presumably reflects the aerobic conditions required for curdlan production (Lee, 2002) and the elasticity of the pellicle, the effects of moisture loss on the viscosity and gelling characteristics of the polymer (Okuyama et al., 1991) (See Chapter 2.2). It is likely that the curdlan pellicle serves to protect cells from physical stresses such as extreme dehydration and the killing effects of UV radiation, as has been demonstrated for the looser cellulosic pellicles formed by *Acetobacter xylinum* and other

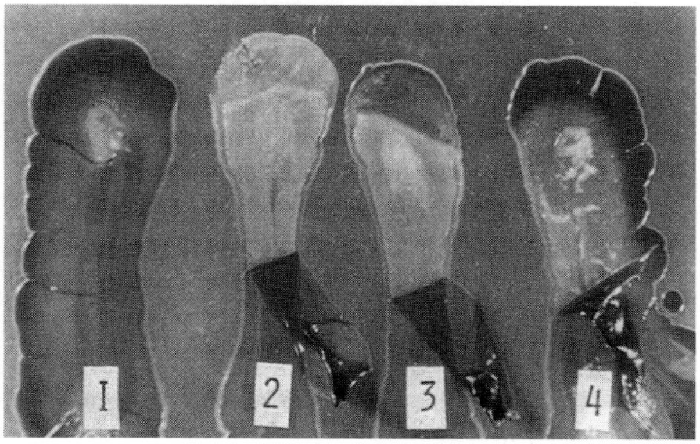


**Fig. 1: Growth of *Agrobacterium* NTK-u on the surface of agar medium containing the (1,3)-β-glucan-specific Aniline Blue dye after 5 days incubation (1) and thereafter for 1 h (2), 3 days (3) and 5 days (4) after stripping off of the curdlan pellicle. From Nakanishi et al. (1976). The colour specifications refer to colours in panels.**

Gram-negative bacteria (Williams and Cannon, 1989). Curdlan has been shown to protect *Agrobacterium* from predation by ubiquitous soil amoebae such as *Dictyostelium discoideum*, *Acanthamoeba castellanii* and *Naegleria* sp. None of these amoebae could feed on *Agrobacterium* LTU50 growing as a confluent layer on agar culture medium if inoculated after the onset of curdlan production, although all could do so if the bacterium was a non-curdlan producing (*crdA*) mutant (Aracic et al., 2008).

Studies on *Agrobacterium* grown in N-limited liquid cultures with excess glucose, as opposed to those above using cultures from solidified medium, have shown that curdlan production occurs during the stationary growth phase on exhaustion of the N-source (Lee, 2002). The curdlan layer surrounding cells of *Agrobacterium* LTU50 from liquid culture appears as a tightly apposed and discrete zone when viewed with fluorescence optics after staining of the sample with Aniline Blue fluorochrome (Fig. 2) or in stained cell sections viewed by transmission electron microscopy (Fig. 3) (McIntosh, 2004). A similarly discrete capsular zone is evident in sections of *Cellulomonas flavigena* KU examined by transmission electron microscopy at low magnification (Angelo et al., 1990; Kenyon and Buller, 2002). In both species, the production of curdlan correlated with the formation of flocs (cell aggregates), suggesting that curdlan assists in cell-to-cell adhesion. If the flocs were depleted of curdlan, such as by treatment of *Agrobacterium* LTU50 cultures with a (1,3)-β-glucan endohydrolase (McIntosh, 2004) or after utilization of the polymer as a nutrient source by *C. flavigena* KU (Voepel and Buller, 1990), they disassociated into planktonic cells. Furthermore, growth of either bacterial species under conditions that failed to

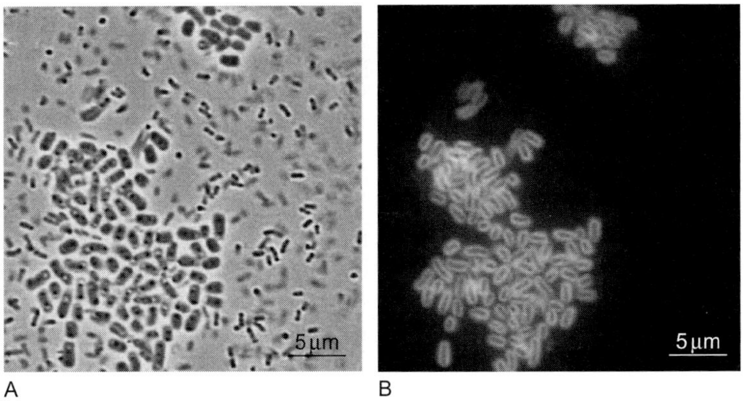


**Fig. 2: Phase contrast image (A) and UV fluorescence image (B) of cells of *Agrobacterium* LTU50 in the same field stained with the (1,3)-β-glucan specific Aniline Blue fluorochrome (Evans et al., 1984), showing small flocs of curdlan-encapsulated cells and curdlan-free, planktonic, cells. From McIntosh (2004). The colour specifications refer to colours in panels.**

elicit curdlan production, such as by culture in a nutritionally complex medium, yielded only capsule-less planktonic cells.

The vast majority of the cells in a shaken liquid culture of *Agrobacterium* LTU50 are in flocs, and it is only these that are encapsulated in curdlan and appear as large rods (1.5 × 3.0 μm) that are easily distinguishable from the less abundant, non-capsulated and slender rods (0.5 × 2.0 μm) of the planktonic form (Fig. 2) (McIntosh, 2004). Floc formation is not exclusive to curdlan-producing strains as both LTU50 and a curdlan non-producing mutant formed small flocs (10–40 μm) that appeared at day 2 and increased in number to day 7. The distinction between the two strains was that small flocs are more numerous in LTU50 cultures and, distinctively, only these cultures contain large flocs (40–100 μm) whose appearance at day 4 coincided with the onset of curdlan production (McIntosh, 2004). At day 7, the large flocs represented ~4% of the total flocs present. These findings suggest that curdlan is not required to initiate floc formation but it is clearly essential for the formation of large flocs. The initiation of floc formation and/or further development may involve other exopolysaccharides produced by the strain, such as cellulose (Matthysse et al., 1995), WSNCE-A (a galactoglucan), WSNCE-B (a glucomannan) (Lee et al., 1997a; 1997b) and a $Mn^{2+}$-inducible, branched galactoglucomannan (Tromp et al., 2008). Cellulose has no obvious role in fully formed flocs since treament of these with cellulase, as opposed to treatment with a (1,3)-β-glucan endohydrolase, had no effect (McIntosh, 2004).

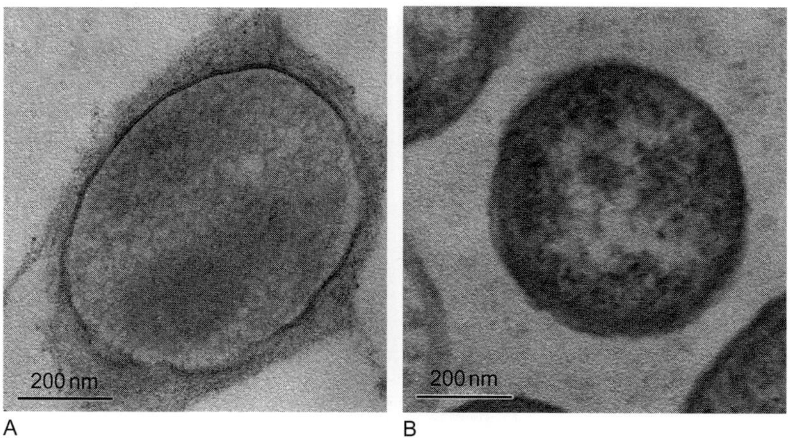


**Fig. 3: Transmission electron microscope images of a section through a curdlan-producing *Agrobacterium* LTU50 cell showing a fibrous capsule (A) and a section through a non-curdlan-producing mutant (B). Cells were fixed, embedded, sectioned and stained with lead citrate-uranyl acetate. Magnification = 96 000×. From McIntosh (2004).**

Curdlan probably facilitates adhesion between adjacent cells through interpenetration of the capsular microfibrils and may further provide mechanical stability to shear forces, enabling the flocs to increase in size. Enlargement of flocs is probably by recruitment of smaller flocs or planktonic cells since curdlan is produced by cells that have ceased to divide (Kako et al., 1989). Indeed, the encapsulated cells within flocs are remarkably homogenous with respect to cell size and extent of the capsular layer (Fig. 2b), and these features are independent of their location within the floc, as assessed using multi-photon confocal imaging (Fig. 4). No dividing cells were detected. Whether curdlan production is elicited by a quorum sensing-type signal (He and Fuqua, 2006) as suggested by its occurrence in aggregated, but not planktonic, cells is not known.

Floc formation is also a characteristic feature of *C. flavigena* KU when it is cultured under N limitation with an excess soluble C source (glucose or cellobiose) or with microcrystalline cellulose or other insoluble substrates (e.g. xylan) (Kenyon et al., 2005). Flocs stained using cationized ferritin and viewed by scanning electron microscopy revealed capsulated cells encased in an interconnected curdlan matrix. In the cellulose-grown cultures, the flocs were closely associated with microscopic particles of cellulose and, at high magnification,

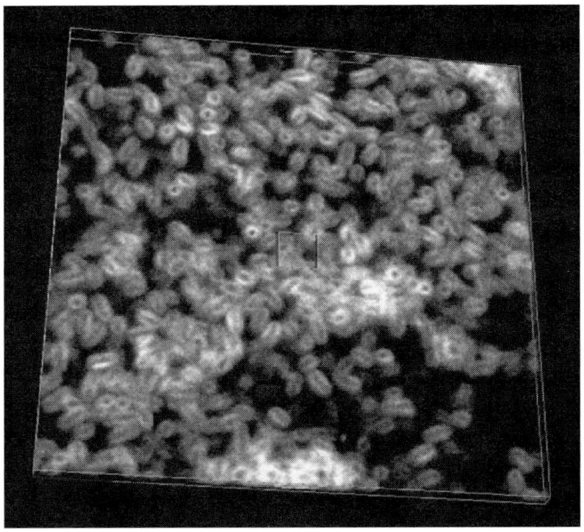

**Fig. 4: Confocal microscope volume graphic image of a representative plane through a floc of *Agrobacterium* LTU50 cells showing the cell arrangement. Image was prepared using VG Studio Max 1.2 (Volume Graphics GmbH, Heidelberg, Germany). Courtesy of Dr Allan Jones (Electron Microscopy Unit, University of Sydney, Australia).**

microcolonies of encapsulated cells were observed attached to the microcrystals. Cells in the microcolonies were covered in numerous surface protrusions that resembled cellulosome complexes, structures that contain the components involved in the multi-step process of cellulose degradation (Bayer et al., 1998). Consistent with this view, most (96%) of the cellulase activity of the culture was associated with the sedimented cellulose floc component and not with the free-floating cells recovered from the supernatant fluid. Thus, *Cellulomonas* curdlan plays a role in cell associations leading to floc formation and to the establishment of microcolonies on insoluble substrates. The biofilm that is formed presumably enhances degradation of the cellulose substrate and the breakdown products appeared to be directly sequestered as very little was detectable in the culture supernatant. Indeed, if excess glucose was present together with cellulose, flocs were formed, but they did not associate with the cellulose particles, so that curdlan *per se* is not responsible for initiating the attachment process.

The likely involvement of curdlan in some aspect of *Cellulomonas* biofilm formation raises the issue of a similar role in *Agrobacterium*, a topic of current interest and in which an uncharacterized, low P-inducible exopolysaccharide has been implicated (Danhorn et al., 2004; Danhorn and Fuqua, 2007). A comparative study of the adhesion of *Agrobacterium* LTU50 and a curdlan-deficient mutant to the roots of tomato and *Arabidopsis* found no difference between the strains, suggesting that the polymer is not essential for these bacterium–plant interactions (Aracic et al., 2005). This is in marked contrast to the role of cellulose in facilitating adhesion of *A. tumefaciens* (Matthysse and McMahan, 1998) and *Rhizobium leguminosarum* (Ausmees et al., 1999; Laus et al., 2005) to root tissue. Curdlan may have a more general role in bacterial persistence as indicated by the significant heat tolerance of *Agrobacterium* LTU50 (55°C for 30 min) compared with the total sensitivity of a curdlan-deficient strain (Aracic et al., 2008). The encapsulated bacteria were effectively insulated from the effects of heat conduction and the physical barrier provided by curdan would presumably extend to other physico-chemical stressors (e.g. UV radiation and high/low pH).

In terms of global cellular metabolism, curdlan production, like that of many other exopolysaccharides, serves a role as a significant sink for fixed carbon when the C/N ratio is severely unbalanced (Sutherland, 2001a). Interestingly, in contrast to other bacteria that are usually unable to utilize the exopolysaccharides that they have synthesized (Cerning 1990; Sutherland, 2001b), *C. flavigena* KU can mobilize the deposited curdlan for use as a nutrient source. This was demonstrated by transfer of aggregated cells to a medium with an excess N source but devoid of a carbon and energy source. Over the initial 18-h incubation period, the reducing sugar content of the culture decreased and was accompanied by the loss of cell aggregation and the disappearance of capsules (Voepel and Buller, 1990; Kenyon and Buller,

2002). Further consistent with this finding was the ability of *C. flavigena* to use the extracted glucan polymer when this was incorporated as the sole carbon and energy source into suitable Aniline Blue-containing agar medium: zones of clearing surrounded the colonies indicating that the growing bacteria synthesized and excreted enzymes capable of hydrolysing the polymer.

In summary, the principal roles that can be ascribed to curdlan are its incorporation of excess carbon during secondary metabolism, its use as a C reserve and its enhancement of cell-to-cell interactions leading to the formation of flocs, biofilms and surface pellicles that can variously provide protection from hydrodynamic shear and heat stress, access to nutrients, and escape from predatory amoebae.

## C. Cyclic (1,3)- and (1,3;1,6)-β-Glucans

### C.1. Occurrence

Low molecular weight periplasmic glucans that have a linear or cyclic backbone structure are intrinsic components of some Gram-negative bacterial cell envelopes (Breedveld and Miller, 1994; Kennedy, 1996; Bohin, 2000). Typical amongst the α-Proteobacteria is the production of structurally diverse cyclic glucans. The best characterized are the solely (1,2)-β-linked cyclic glucans (17–40 sugar residues) produced by species of *Agrobacterium*, *Rhizobium* and *Brucella*, a pathogen of humans and animals (Bundle et al., 1988; Smith and Ficht, 1990). Other, smaller (10–13 sugar residues) cyclic glucans containing both (1,3)- and (1,6)-β-linkages (see Chapter 2.1 for structures) have been identified in $N_2$-fixing root-nodulating or root-associated bacteria, most notably in *Bradyrhizobium japonicum* USDA 110, a slow-growing and well-characterized endosymbiont of soybean (*Glycine max*), and other *Bradyrhizobium* spp. (e.g. BR4406 and BR8404 isolated from different tropical tree legumes) (Table 1). They are also produced by *Rhizobium loti* NZP 2309, a slow-growing strain that is phylogentically closely related to, but distinguishable from, *B. japonicum* USDA 110 (Estrella et al., 2000), and by *Azorhizobium caulinodans*, which forms nodules on the stems and roots of *Sesbania rostrata*, a tropical legume. The free-living diazotroph, *Azospirillum brasilense* strain Cd, which colonizes the sheath layer and intercellular spaces of the root hair surface of non-leguminous plants (Somers et al., 2004), also produces (1,3;1,6)-β-glucans.

Despite the structural diversity of the periplasmic glucans, some serve similar roles in their respective bacteria and in interactions of these bacteria with eukaryotic hosts. This is true of cyclic (1,3;1,6)-β-glucans which have been proposed to be functional analogues of the cyclic

(1,2)-β-glucans (Miller and Gore, 1992). For example, both classes of glucans have been implicated in osmotic regulation and the suppression of plant defence responses leading to nodulation. Nonetheless, the precise roles of each type are not known, and they may have different functions or act at different times in each rhizobium–legume association.

## C.2. Formation and Functional Roles

### C.2.1. Adaptation to hypo-osmotic conditions

Periplasmic glucans are considered to be important intrinsic components of the Gram-negative cell envelope since mutants defective in their synthesis display a variety of pleiotropic phenotypes suggestive of significant structural alteration of the cell envelope (Bohin, 2000; Breedveld and Miller, 1994; Kennedy, 1996). The glucans occur in the periplasm usually only during growth of the bacteria in a medium of very low osmolality (<100 milliosmoles per kilogram $H_2O$) and can accumulate to a high concentration (~10–100 mM), representing 5–10% of the cellular dry weight (Bohin, 2000; Talaga et al., 2002). They may also be released to the extracellular environment and can generally be detected in stationary-phase cultures.

These features also apply to the cyclic (1,3;1,6)-β-glucans of *Bradyrhizobium* spp. and *A. brasilense*, which reach their highest levels in a low osmolality medium (either complex or chemically defined), and the production of which is strongly inhibited in high osmolality media. The latter include media containing a non-ionic solute (e.g. 0.5 M sucrose; 0.3 M fructose; 250 g $L^{-1}$ PEG 6000) or an ionic solute such as 0.25 M NaCl provided that, as in *A. brasilense*, it does not adversely affect cell growth (Tully et al., 1990; Miller and Gore, 1992; Altabe et al., 1994; Pfeffer et al., 1994). A readily permeable, osmotically inactive, solute such as glycerol (0.5 M) does not inhibit cyclic (1,3;1,6)-β-glucan production (Miller and Gore, 1992). Monitoring of *B. japonicum* cyclic glucan production using NMR showed that it was highest during the metabolically active growth phase and that yields depended on the carbon energy source used (Pfeffer et al., 1994). The cell-associated cyclic glucans can be released from the periplasm by treatment with chloroform, freeze–thaw shock or osmotic shock, although the treatments vary in efficiency and, under some growth conditions, more of the glucan may be in the culture medium than is cell associated (Miller and Gore, 1992). In contrast to the situation in *Bradyrhizobium* spp. and *A. brazilense*, production of cyclic (1,3;1,6)-β-glucan by *A. caulinodans* is not osmoregulated (Komaniecka and Choma, 2003), and this is also true of cyclic (1,2)-β-glucan production by *Brucella* spp. (Briones et al., 1997) and a few rhizobial strains (Breedveld et al., 1991; 1993; Soto et al., 1993; de Iannino et al., 2000).

The observation that cyclic (1,3;1,6)-β-glucan production is typically an osmotically regulated property suggests that these glucans provide a mechanism by which the cell can regulate volume and osmotic strength of the periplasmic compartment (Miller et al., 1986). This conclusion is supported by the behaviour of mutants that are defective in cyclic glucan production; these display impaired growth under low osmolality conditions as observed with *B. japonicum ndvB*::Tn*5* or *ndvD*::Tn*5* mutants on an arabinose-gluconate medium (Bhagwat et al., 1996; Chen et al., 2002). Neither class of mutant produces significant *in vivo* amounts of cyclic (1,3;1,6)-β-glucan and growth on low osmolality medium could be assisted by the addition of 100 mM fructose to serve as an osmoprotectant (Bhagwat and Keister, 1995). By contrast, the *B. japonicum ndvC*::Tn*5* mutant, which produces a cyclic (1,3)-β-glucan and not the wild-type glucan (Bhagwat et al., 1999), exhibited near normal growth on low osmolality medium, suggesting that the novel glucan serves as a substitute in osmoprotection (Bhagwat et al., 1996). Consistent with these observations, complementation of the *ndvB*::Tn*5* and *ndvD*::Tn*5* mutations with the respective wild-type gene restored the normal growth phenotype. The conditions affecting (1,3;1,6)-β-glucan production in *B. japonicum* and the behaviour of glucan-defective mutants are comparable to those described for the (1,2)-β-glucans produced by *Agrobacterium* and *Sinorhizobium* (Cangelosi et al., 1990; Dylan et al., 1990). Indeed, the *B. japonicum ndv* locus was initially recognized because of its ability to restore growth of a cyclic (1,2)-β-glucan-deficient *ndvB*::Tn*5* mutant of *S. meliloti* on low osmolality medium (Bhagwat et al., 1993). These data overall support a significant role for cyclic (1,3;1,6)-β-glucans during hypo-osmotic adaptation of free-living *Bradyrhizobium* in soil and in the rhizosphere environment, both of which are osmotically variable (Miller and Wood, 1996).

### *C.2.2. Nodulation and $N_2$-fixation*

The involvement of cyclic β-glucans in roles other than osmoprotection comes from the finding that mutants unable to produce such glucans are defective in virulence (e.g. *A. tumefaciens chv* and *B. abortus cgs* mutants; Arellano-Reynoso et al., 2005; Douglas et al., 1985) or form only ineffective nodules (e.g. *S. meliloti* and *Rhizobium fredii ndv* mutants; Bhagwat et al., 1992; Dylan et al., 1986). Similarly, the cyclic (1,3;1,6)-β-glucans produced by *B. japonicum* USDA 110 are essential for the symbiotic interaction with soybean since glucan-defective mutants (*ndvA*, *ndvB* or *ndvC*) are all impaired in nodule development, although to different degrees. The distinction, mechanistically, between the diverse roles played by the cyclic glucans is perhaps most obvious with the *ndvC*::Tn*5* mutant which produces an altered, cyclic (1,3)-β-linked glucan that is effective in osmoadaptation of *B. japonicum* but is unable to replace the wild-type (1,3;1,6)-β-linked glucans with respect to successful symbiosis of soybean (Bhagwat et al., 1996). Interestingly, a cyclic (1,3)-β-linked glucan that is structurally identical to that of the *B. japonicum ndvC*::Tn*5* mutant is also produced by an *S. meliloti* cyclic (1,2)-β-glucan-deficient *ndvB*::Tn*5* mutant on introduction of a plasmid-borne copy of the *B. japonicum ndv* locus (Bhagwat et al., 1999). In this case, genetic complementation in the *S. meliloti* strain [TY7(p5D3)] resulted both in restored hypo-osmotic adaptation and the ability to form effective, $N_2$-fixing, nodules on alfalfa (Bhagwat et al., 1993; Pfeffer et al., 1996). Although the physiological basis of restored symbiosis by *S. meliloti* remains unknown, the findings suggest that each bacterium–legume interaction has different functional requirements of the glucan, and that the structure of these molecules is less critical in the alfalfa interaction than in that involving soybean (Bhagwat et al., 1999).

Studies on soybean nodule morphogenesis have been conducted using the wild-type *B. japonicum* USDA 110 and its various *ndv*-insertion mutants which are all symbiotically defective (Bhagwat et al. 1996; Chen et al., 2002; Dunlap et al., 1996; Wiedemann and Müller, 2004). All the mutants formed smaller, leghaemoglobin-free (white) ineffective nodules which, in the case of the *ndvB*:: Tn*5* mutant (AB-14), appeared at a similar rate to that of the wild type, but were rare, very small structures (pseudonodules) in the case of the *ndvC*::Tn*5* mutant (AB-1) and also much delayed in development (visible at 14 vs 8 days post infection). The nodules induced by the mutants showed tissue differentiation by day 20 but were morphologically distinguishable from each other and from wild-type nodules. In the case of the wild type, the nodules contained a large number of infected tissue cells, all with numerous bacteroids evenly dispersed in a uniformly stained cytoplasm. The bacteroid content of the mutant-induced nodules was much reduced: 100-fold less for the *ndvD*::Tn*5* mutant (RC-2) and 1000-fold less for the *ndvB*::Tn*5* mutant (Chen et al., 2002). In the *ndvA*- or *ndvB*-infected tissue cells, the bacteroids tended to be clustered in a small area probably coinciding with the point of release from the infection thread; in other nodules, very few, if any, bacteroids were present, and typically the bacteria showed morphological signs of degradation and generally were not surrounded by a symbiosome membrane (Dunlap et al., 1996; Wiedemann and Müller, 2004). In contrast, the bacteroids of the *ndvD*::Tn*5* mutant were evenly dispersed in the tissue cells together with a large number of vesicles containing loosely organized fibrillar material; such vesicles were not present in cells of wild-type nodules (Chen et al., 2002). In *ndvC*-induced nodules, the tissue cells in the central region differentiated into two types: those with a large central vacuole (a feature of uninfected cells) and those without a large vacuole and with an evenly dispersed cytoplasm (a feature of infected cells). Thickened cell walls and a dense cytoplasm were common in both cell types; however, neither bacteria nor infection threads were evident (Dunlap et al., 1996).

Dunlap et al. (1996) also assessed the expression of soybean nodulins in response to infection by the wild-type *B. japonicum* and the *ndvB*::Tn*5* and *ndvC*::Tn*5* mutants. They found that early (ENOD2, ENOD55-1 and *Gm*N93), intermediate (*Gm*N70) and late nodulins (NOD26 and leghaemoglobin) were all expressed in wild-type nodules at the earliest time point (8 days) but that expression in *ndvB*-induced nodules was at reduced levels or was delayed until day 14. By contrast, the early nodulins were not detected at day 14 in the *ndvC*-induced pseudonodules but they, and the other nodulins tested, were present at day 21. These findings show that even when nodulation capacity is severely affected, as with the *ndvC*::Tn*5* mutant, the nodular structures that may be formed show some signs of cellular differentiation and induction of plant nodulins despite the absence of active bacterial invasion. They also demonstrate that cyclic (1,3;1,6)-β-glucan production by *B. japonicum* is an essential component of normal nodule morphogenesis in soybean and cannot be substituted by the cyclic (1,3)-β-glucan formed by the *ndvC*::Tn*5* mutant.

### *C.2.3. Suppression of plant defence response*

The symbiotic interaction between a bacterium and its host plant is ultimately beneficial to both participants. Initially, however, it constitutes an assault that can be perceived by the sophisticated sensory system of the plant and induces a complex biochemical defence response (Dixon et al., 1994; Boller, 1995) (see Chapter 4.4.5). The signaling compounds (elicitors) that initiate the defence response are perceived at low concentration, and many are constituents of the outer layers of the invading organism (Ebel and Mithöfer, 1998). They include oligosaccharides such as the branched (1,3;1,6)-hepta-β-glucoside from the mycelial walls of the soybean pathogen *Phytophthora sojae* (Ayers et al., 1976; Sharp et al., 1984). This fungal β-glucan is a potent elicitor of glyceollins, soybean phytoalexins that have antimicrobial activity and accumulate in response to biotic and abiotic stresses (Parniske et al., 1991). The elicitor-active fungal β-glucan has some structural similarity with the cyclic (1,3;1,6)-β-glucans produced by free-living cells of *B. japonicum* and by bacteroids within soybean root nodules (Rolin et al., 1992; Gore and Miller, 1993). The ability of *B. japonicum* to successfully nodulate soybean implies that it can overcome the plant defence responses, and it may specifically be able to do so by virtue of the role played by its cyclic (1,3;1,6)-β-glucans (Mithöfer, 2002). The use of cDNA microarrays to monitor soybean gene expression has shown that several thousand plant genes are differentially expressed in response to inoculation with *B. japonicum* USDA 110 and that during nodule development the plant defence response is, indeed, reduced (Brechenmacher et al., 2008). For example, the transcript abundance of various flavonoid synthesis genes (such as those leading to production of glyceollins) and of some other defence-related genes is markedly reduced, supporting the view that such repression serves to enhance symbiosis.

In bioassays using wounded soybean cotyledons, the cyclic (1,3;1,6)-β-glucans of *B. japonicum* USDA 110 are only very weak elicitors of glyceollin production compared with the strong response induced by even low concentrations of the *P. sojae* branched (1,3;1,6)-β-glucans (Miller et al., 1994; Mithöfer et al., 1996). Similarly, little if any glyceollin is produced in soybean nodules formed by *B. japonicum* although the levels are elevated in the ineffective nodules formed by *ndvB*::Tn*5* or *ndvC*::Tn*5* mutants (by three-fold and five-fold, respectively) (Bhagwat et al., 1999). More significantly, when tested in combination with the fungal β-glucans, the *B. japonicum* cyclic (1,3;1,6)-β-glucans suppress glyceollin production in soybean cotyledons stimulated by the fungal elicitor (Mithöfer et al., 1996; Bhagwat et al., 1999). In these studies, 35 μM of the wild-type bradyrhizobial glucans were sufficient to cause 50% inhibition of the fungal-induced response, whereas suppression by the cyclic (1,3)-β-glucan produced by the *ndvC*::Tn*5* mutant was relatively ineffective (1 mM glucan caused less than 25% suppression). The suppression effects of the cyclic-β-glucans correlated with their ability to inhibit binding of the fungal β-glucans to receptor (elicitor-binding) proteins in soybean membranes (Mithöfer et al., 2000). Both the cyclic (1,3;1,6)- and the cyclic (1,3)-β-glucans were able to compete with a radioactively labelled ligand, HG-APEA [a 2-(4-aminophenyl) ethylamine conjugate of the fungal β-glucan], for binding to solubilized soybean binding proteins, although the wild-type glucans had a much higher affinity (40-fold) for the receptor proteins than did the cyclic (1,3)-β-glucan (Mithöfer et al., 1996; Bhagwat et al., 1999). In corresponding studies, the cyclic (1,2)-β-glucans from rhizobial species were ineffective in inducing glyceollin production, in suppressing fungal-elicited glyceollin accumulation and in displacing HG-APEA (Miller et al., 1994; Mithöfer et al., 1996), thereby supporting the view that the interaction between the bradyrhizobial cyclic (1,3;1,6)-β-glucans and the soybean β-glucan receptor proteins is specific.

The accumulated findings have lead to the proposal that the bradyrhizobial cyclic (1,3;1,6)-β-glucans are necessary for the development of differentiated and effective nodules due to their role as suppressors of the plant's defence response (Bhagwat et al., 1999; Mithöfer, 2002). Soybean glyceollins have a bacteriostatic effect on *B. japonicum* which is reflected in an increased lag period in liquid cultures (Parniske et al., 1991; Mithöfer et al., 2001). The bacteria can, however, become tolerant to glyceollin (up to 200 μM) and to the inhibitory effects of plant exudates if they are exposed to low concentrations of glyceollin in cultures or *in planta*, as bacterioids are also resistant. In contrast, bacteroids from the cyclic β-glucan-deficient *ndvB*::Tn*5* mutant failed to grow in cultures with 10 μM glyceollin. It was suggested that the arrested development of mutant-induced nodules is due to sensitivity of the bacteroids to the low levels of glyceollin that are present, whereas suppression of glyceollin accumulation in the wild-type nodules enables normal development to proceed (Bhagwat et al., 1999). Other surface polysaccharides (capsular and lipopolysaccharides) also have significant roles in various rhizobia–host symbiotic interactions (Breedveld and Miller, 1998; Mithöfer, 2002), however, the possible contribution of these to the *Bradyrhizobium*–host system remains to be elucidated.

## D. Streptococcal Type 37 (1,3;1,2)-β-Glucan

*Streptococcus pneumoniae* is a leading cause of human morbidity and mortality worldwide causing invasive diseases such as pneumonia, bacteriaemia and meningitis, and less severe infections such as otitis media and sinusitis (Gaillat, 1998). This Gram-positive pathogen produces a variety of virulence factors, however, the capsular polysaccharide (CPS) is considered to be of prime significance since non-encapsulated derivatives are typically avirulent. The CPS is responsible for the characteristic mucoid colonies that clinically isolated pneumococci form on the surface of agar medium and protects the bacterium from complement-mediated opsonophagocytosis during an infection (Moxon and Kroll, 1990). In contrast to *Streptococcus pyogenes* whose pathogenicity is associated with its production of a hyaluronic acid capsule (Yother, 1999), 90 structurally and serologically distinct CPS have been recognised amongst *S. pneumoniae* isolates (Henrichsen, 1995). Of these, only two are neutral polysaccharides and only the type 37 CPS is a homopolymer, consisting of a (1,3)-β-glucan backbone with (1,2)-linked β-glucopyranosyl side branches at each glucosyl residue (i.e. it is composed of (1,3)-β-linked sophorosyl residues) (Adeyeye et al., 1988) (see Chapter 2.1 for structure).

*S. pneumoniae* type 37 isolates, like those of type 3 whose CPS consists of repeated disaccharide residues (Glc-GlcA), produce very large capsules resulting in large, mucoid colonies on solid medium and extreme viscosity at concentrations in excess of 3 mg $ml^{-1}$ water (Knecht et al., 1970). In the case of type 37 strains, three times more of the CPS is recoverable from the culture supernatant than is cell associated. Despite the quantity of CPS that is produced and early descriptions of type 37 strains from infected patients (Kauffmann et al., 1940; Walter et al., 1941), they are in fact infrequently encountered in humans, and are rarely pathogenic (Azoulay-Dupuis et al., 2000). For example, in mouse virulence tests, $10^7$ type 37 bacteria are required to cause lethal infection compared with one type 3 bacterium (Knecht et al., 1970). Consequently, type 37 capsular material is not included in the current anti-pneumococcal vaccine that consists of purified CPS from 23 serotypes commonly associated with invasive disease worldwide.

## E. Conclusion

Functional roles have been ascribed to the three bacterial (1,3)-β-glucans discussed in this chapter but much of the information remains fragmentary. The significant role performed by cyclic (1,3;1,6)-β-glucans in the symbiosis between *Bradyrhizobium japonicum* and its soybean host is not known to extend to the similar glucans produced by plant-associated *Azorhizobium caulinodans* and *Azospirillum brasilense*. Linear (1,3)-β-glucan (curdlan) occurs only in a limited variety of bacteria and assists in the formation of flocs or biofilms in the representative Gram-positive (*Cellulomonas*) and Gram-negative (*Agrobacterium*) strains so far studied. The degree to which these cell–cell associations are vital remains to be determined. The side-chain-branched (1,3;1,2)-β-glucan is unique to *Streptococcus pneumoniae* type 37. Its functional roles and those of the other (1,3)-β-glucans remain fertile areas for future research.

## Acknowledgements

We thank Sanja Aracic, Ferdiye Taner and Danielle Tromp for critically reading the manuscript and Dr Fung Lay for preparation of Figures. Work in our laboratories was supported, in part, by Australian Research Council Grants (AO9925079, LX 0211339).

## References

Adeyeye, A., Jansson, P.-E., Lindberg, B., & Henrichsen, J. (1988). Some biological properties of pneumococcus type 37 and the chemistry of its capsular polysaccharide. *Carbohydrate Research, 180*, 295–299.

Altabe, S. G., de Iannino, N., de Mendoza, D., & Ugalde, R. A. (1994). New osmoregulated β (1-3), β (1-6) glucosyltransferase(s) in *Azospirillum brasilense*. *Journal of Bacteriology*, *196*, 4890–4898.

Altabe, S. G., Talaga, P., Wieruszeski, J.-M., Lippens, G., Ugalde, R., & Bohin, J.-P. (1998). Periplasmic glucans of *Azospirillum brasilense*. In C. Elmerich, A. Kondorosi, & W. E. Newton (Eds.), *Biological nitrogen fixation for the 21st century*, p. 390. Dordrecht: Kluwer.

Angelo, R., Voepel, K. C., & Buller, C. S. (1990). Isolation and characterization of a new strain of *Cellulomonas flavigena*. *Journal of Industrial Microbiology*, *5*, 125–130.

Aracic, S. Stanisich, V. A., Stone, B. A., & Matthysse, A. G. (2005). The role of cellulose and curdlan in biotype 1 *Agrobacterium tumefaciens*. *26th Annual Crown Gall meeting*, Bloomington, Alabama, USA, August 5–7th. Abstract, p. 24.

Aracic, A., Matthysse, A. G., Stone, B. A., & Stanisich, V. A. (2008). The biological role of agrobacterial (1,3)-β-glucan and regulation of its production. *Annual scientific meeting of the Australian Society for Microbiology*, Melbourne, Australia, July 6–10th. Abstract, P04.05.

Arellano-Reynoso, B., Lapaque, N., Salcedo, S., Briones, G., Ciocchini, A. E., Ugalde, R., Moreno, E., Moriyon, I., & Gorvel, J. P. (2005). Cyclic beta-1,2-glucan is a *Brucella* virulence factor required for intracellular survival. *Nature Immunology*, *6*, 618–625.

Ausmees, N., Jonsson, H., Hoglund, S., Ljunggren, H., & Lindberg, M. (1999). Structural and putative regulatory genes involved in cellulose synthesis in *Rhizobium leguminosarum bv. trifolii*. *Microbiology*, *145*, 1253–1262.

Ayers, A. R., Ebel, J., Finelli, F., Berger, N., & Albersheim, P. (1976). Host-pathogen interactions: IX. Quantitative assays of elicitor activity and characterisation of the elicitor present in the extracellular medium of cultures of *Phytophthora megasperma* var. *sojae*. *Plant Physiology*, *57*, 751–759.

Azoulay-Dupuis, E., Rieux, V., Muffat-Joly, M., Bédos, J. P., Vallée, E., Rivier, C., Isturiz, R., Carbon, C., & Moine, P. (2000). Relationship between capsular type, penicillin susceptibility, and virulence of human *Streptococcus pneumoniae* isolates in mice. *Antimicrobial Agents and Chemotherapy*, *44*, 1575–1577.

Bayer, E. A., Shimon, L. J. W., Shoham, Y., & Lamed, R. (1998). Cellulosomes: Structure and ultrastructure. *Journal of Structural Biology*, *124*, 221–234.

Bhagwat, A. A., Tully, R. E., & Keister, D. L. (1992). Isolation and characterisation of an *ndvB* locus from *Rhizobium fredii*. *Molecular Microbiology*, *6*, 2159–2165.

Bhagwat, A. A., Tully, R. E., & Keister, D. L. (1993). Identification and cloning of a cyclic β-(1-3), β-(1-6)-D-glucan synthesis locus from *Bradyrhizobium japonicum*. *FEMS Microbiology Letters*, *114*, 139–144.

Bhagwat, A. A., & Keister, D. L. (1995). Site-directed mutagenesis of the β-(1-3),β-(1-6)-D-glucan synthesis locus of *Bradyrhizobium japonicum*. *Molecular Plant-Microbe Interactions*, *3*, 366–370.

Bhagwat, A. A., Gross, K. C., Tully, R. E., & Keister, D. L. (1996). β-Glucan synthesis in *Bradyrhizobium japonicum*: Characterization of a new locus (*ndvC*) influencing β-(1-6) linkages. *Journal of Bacteriology*, *178*, 4635–4642.

Bhagwat, A. A., Mithöfer, A., Pfeffer, P. E., Kraus, C., Spickers, N., Hotchkiss, A., Ebel, J., & Keister, D. L. (1999). Further studies of the role of cyclic β-glucans in symbiosis: An *ndvC* mutant of *Bradyrhizobium japonicum* synthesizes cyclodecakis-(1-3)-β-glucosyl. *Plant Physiology*, *119*, 1057–1064.

Bohin, J-P. (2000). Osmoregulated periplasmic glucans in Proteobacteria. *FEMS Microbiology Letters*, *186*, 11–19.

Boller, T. (1995). Chemoperception of microbial signals in plant cells. *Annual Review of Plant Physiology and Plant Molecular Biology*, *46*, 189–214.

Bouzar, H., Ouadah, D., Krimi, Z., Jones, J. B., Trovato, M., Petit, A., & Dessaux, Y. (1993). Correlative association between resident plasmids and the host chromosome in a diverse *Agrobacterium* soil population. *Applied and Environment Microbiology*, *59*, 1310–1317.

Breedveld, M. W., Zevenhuizen, L. P., & Zehnder, A. J. B. (1991). Osmotically-regulated trehalose accumulation and cyclic β-(1,2)-glucan excretion by *Rhizobium leguminosarum biovar trifolii* TA-1. *Archives of Microbiology*, *156*, 501–506.

Breedveld, M. W., Dijkema, C., Zevenhuizen, L. P. T., & Zehnder, A. J. B. (1993). Response of intracellular carbohydrates to a NaCl shock in *Rhizobium leguminosarum* TA-1 and *Rhizobium meliloti* SU-47. *Journal of General Microbiology*, *139*, 3157–3163.

Breedveld, M. W., & Miller, K. J. (1994). Cyclic β-glucans of members of the family Rhizobiaceae. *Microbiological Reviews*, *58*, 145–161.

Breedveld, M. W., & Miller, K. J. (1998). Cell-surface β-glucans. In H. P. Spaink, A. Kondorosi, & P. J. J. Hookkaas (Eds.), *The rhizobiaceae: Molecular biology of model plant-associated bacteria* (pp. 81–96). Kluwer Academic Publishers.

Brechenmacher, L., Kim, M-Y., Benitez, M., Li, M., Joshi, T., Calla, B., Lee, M. P., Libault, M., Vodkin, L. O., Xu, D., Lee, S-H., Clough, S. J., & Stacey, G. (2008). Transcription profiling of soybean nodulation by *Bradyrhizobium japonicum*. *Molecular Plant-Microbe Interactions*, *21*, 631–645.

Briones, G., de Iannino, N. I., Steinberg, M., & Ugalde, R. A. (1997). Periplasmic cyclic 1,2-β-glucan in *Brucella* spp. is not osmoregulated. *Microbiology*, *143*, 1115–1124.

Buller, C. S. (1990). Water insoluble polysaccharide polymer and methods thereof. *US patent*, 4,908,310.

Bundle, D. R., Cherwonogrodzky, J. W., & Perry, M. B. (1988). Characterisation of *Brucella* polysaccharide B. *Infection and Immunity*, *56*, 1101–1106.

Cangelosi, G. A., Martinetti, G., & Nester, E. W. (1990). Osmosensitivity phenotypes of *Agrobacterium tumefaciens* mutants that lack periplasmic β-1,2-glucan. *Journal of Bacteriology*, *172*, 2172–2174.

Cerning, J. (1990). Exocellular polysaccharides produced by lactic acid bacteria. *FEMS Microbiology Reviews*, *87*, 113–130.

Chen, R., Bhagwat, A. A., Yaklich, R., & Keister, D. L. (2002). Characterisation of *ndvD*, the third gene involved in the synthesis of cyclic β-(1-3),(1-6)-D-glucans in *Bradyrhizobium japonicum*. *Canadian Journal of Microbiology*, *48*, 1008–1016.

Costerton, J. W., Cheng, K. J., Geesy, G. G., Ladd, T. I., Nickel, J. C., Dasgupta, M., & Marrie, T. J. (1987). Bacterial biofilms in nature and disease. *Annual Review of Microbiology*, *41*, 435–464.

Danhorn, T., Hentzer, M., Givskov, M., Parsek, M. R., & Fuqua, C. (2004). Phosphorus limitation enhances biofilm formation of the plant pathogen *Agrobacterium tumefaciens* through the PhoR-PhoB regulatory system. *Journal of Bacteriology*, *186*, 4492–4501.

Danhorn, T., & Fuqua, C. (2007). Biofilm formation by plant-associated bacteria. *Annual Review of Microbiology*, *61*, 401–422.

da Silva, P., & Nahas, E. (2002). Bacterial diversity in soil in response to different plants, phosphate fertilizing and liming. *Brazilian Journal of Microbiology*, *33*, 304–310.

de Iannino, N. I., & Ugalde, R. A. (1993). Biosynthesis of cyclic β-(1–3),β-(1–6) glucan in *Bradyrhizobium* spp.. *Archives of Microbiology*, *159*, 30–38.

de Iannino, N. I., Briones, G., Iannino, F., & Ugalde, R. A. (2000). Osmotic regulation of cyclic 1,2-β-glucan synthesis. *Microbiology*, *146*, 1735–1742.

Denison, R. F., & Kiers, E. T. (2004). Lifestyle alternatives for rhizobia: Mutualism, parasitism and forgoing symbiosis. *FEMS Microbiology Letters*, *237*, 187–193.

Dixon, R. A., Harrison, M. J., & Lamb, C. J. (1994). Early events in the activation of plant defense responses. *Annual Review of Phytopathology*, *32*, 479–501.

Douglas, C. J., Staneloni, R. J., Rubin, R. A., & Nester, E. W. (1985). Identification and genetic analysis of an *Agrobacterium tumefaciens* chromosomal virulence region. *Journal of Bacteriology*, *161*, 850–860.

Dudman, W. F., & Jones, A. J. (1980). The extracellular glucans of *Rhizobium japonicum* strain 3I1b71a. *Carbohydrate Research*, *84*, 358–364.

Dunlap, J., Minami, E., Bhagwat, A. A., Keister, D. L., & Stacey, G. (1996). Nodule development induced by mutants of *Bradyrhizobium japonicum* defective in cyclic β-glucan synthesis. *Molecular Plant-Microbe Interactions*, *9*, 546–555.

Dylan, T., Helinski, D. R., & Ditta, G. S. (1990). Hypoosmotic adaptation in *Rhizobium meliloti* requires β-(1-2)-glucan. *Journal of Bacteriology*, *172*, 1400–1408.

Dylan, T., Lelpi, L., Stanfield, S., Kashyap, L., Douglas, C., Yanofsky, M., Nester, E., Helinski, D. R., & Ditta, G. (1986). *Rhizobium meliloti* genes required for nodule development are related to chromosomal virulence genes in *Agrobacterium tumefaciens*. *Proceedings of the National Academy of Sciences of the United States of America*, *83*, 4403–4407.

Ebel, J., & Mithöfer, A. (1998). Early events in the elicitation of plant defence. *Planta*, *206*, 335–348.

Estrella, M. J., Pfeffer, P. E., Brouillette, J. N., Ugalde, R. A., & de Iannino, N. I. (2000). Biosynthesis and structure of cell associated glucans in the slow growing *Rhizobium loti* strain NZP 2309. *Symbiosis*, *29*, 173–199.

Evans, N. A., Hoyne, P. A., & Stone, B. A. (1984). Characteristics and specificity of the interaction of a fluorochrome from aniline blue (Sirofluor) with polysaccharides. *Carbohydrate Polymers*, *4*, 215–230.

Footrakul, P., Suyanandana, P., Amemura, A., & Harada, T. (1981). Extracellular polysaccharides of *Rhizobium* from the Bangkok MIRCEN collection. *Journal of Fermentation Technology*, *59*, 9–14.

Gaillat, J. (1998). Epidemiology of systemic *Streptococcus pneumoniae* infections. *Presse Medicale*, *27*(Suppl. 1), 9–16.

Garbeva, P., van Veen, J. A., & van Elsas, J. D. (2003). Predominant *Bacillus* spp. in agricultural soil under different management regimes detected via PCR-DGGE. *Microbial Ecology*, *45*, 302–316.

Garbeva, P., van Veen, J. A., & van Elsas, J. D. (2004). Microbial diversity in soil: Selection of microbial populations by plant and soil type and implications for disease suppressiveness. *Annual Review of Phytopathology*, *42*, 243–270.

Ghai, S. K., Hisamatsu, M., Amemura, A., & Harada, T. (1981). Production and chemical composition of extracellular polysaccharides of *Rhizobium*. *Journal of General Microbiology*, *122*, 33–40.

Gore, R. S., & Miller, K. J. (1993). Cyclic β-1,6-1,3 glucans are synthesized by *Bradyrhizobium japonicum* bacteroids within soybean (*Glyine max*) root nodules. *Plant Physiology*, *102*, 191–194.

Goodner, B., Hinkle, G., Gattung, S., Miller, N., Blanchard, M., Qurollo, B., Goldman, B. S., Cao, Y., Askenazi, M., Halling, C., Mullin, L., Houmiel, K., Gordon, J., Vaudin, M., Iartchouk, O., Epp, A., Liu, F., Wollam, C., Allinger, M., Doughty, D., Scott, C., Lappas, C., Markelz, B., Flanagan, C., Crowell, C., Gurson, J., Lomo, C., Sear, C., Strub, G., Cielo, C., & Slater, S. (2001). Genome sequence of the plant pathogen and biotechnology agent *Agrobacterium tumefaciens* C58. *Science*, *294*, 2323–2328.

Gummadi, S. N., & Kumar, K. (2005). Production of extracellular water insoluble β-1,3-glucan (curdlan) from *Bacillus* sp. SNC07. *Biotechnology and Bioprocess Engineering*, *10*, 546–551.

Harada, T., & Harada, A. (1996). Curdlan and succinoglycan. In S. Dimitrio (Ed.), *Polysaccharides in medical applications* (pp. 21–57). New York: Marcel Dekker.

He, X., & Fuqua, C. (2006). Rhizosphere communication: Quorum sensing by the rhizobia. *Journal of Microbiology and Biotechnology*, *16*, 1661–1677.

Henrichsen, J. (1995). Six newly recognized types of *Streptococcus pneumoniae*. *Journal of Clinical Microbiology*, *33*, 2759–2762.

Hisamatsu, M., Ott, I., Amemura, A., Harada, T., Nakanishi, I., & Kimura, K. (1977). Change in ability of *Agrobacterium* to produce water-soluble and water-insoluble β-glucans. *Journal of General Microbiology*, *103*, 375–379.

Hisamatsu, M., Abe, J. I., Amemura, A., & Harada, T. (1980). Structural elucidation of succinoglycan and related polysaccharides from *Agrobacterium* and *Rhizobium* by fermentation with two special β-D-glycanases and methylation analysis. *Agricultural and Biological Chemistry*, *44*, 1049–1055.

Hisamatsu, M., Amemura, A., Matsuo, T., Matsuda, H., & Harada, T. (1982). Cyclic (1–2)-β-D-glucan and the octasaccharide repeating unit of succinoglycan produced by *Agrobacterium*. *Journal of General Microbiology*, *128*, 1873–1879.

Kako, K., Koreeda, A., & Harada, T. (1989). Electron microscopic studies on curdlan formation by bacteria in solid culture medium. *Bulletin of The Kobe Women's University*, *22*, 183–189.

Karnezis, T., Epa, V. C., Stone, B. A., & Stanisich, V. A. (2003). Topological characterisation of an inner membrane (1–3)-β-D-glucan (curdlan) synthase from *Agrobacterium* sp. strain ATCC31749. *Glycobiology*, *13*, 693–706.

Kauffmann, F., Mørch, E., & Schmith, K. (1940). On the serology of the pneumococcus group. *Journal of Immunology*, *39*, 397–426.

Kennedy, E. P. (1996). Membrane-derived oligosaccharides (periplasmic β-D-glucans) of *Escherichia coli*. In F. C. Neidhardt, R. Curtis, J. L. Ingraham, C. C. Lin, K. B. Low, B. Magasanik, W. S. Reznikoff, M. Riley, M. Schaechter, & H. E. Umbarger, (Eds.)*Escherichia coli and Salmonella typhimurium: Cellular and Molecular Biology: Vol. 1* (pp. 672–679). Washington, D.C.: American Society for Microbiology.

Kenyon, W. J., & Buller, C. S. (2002). Structural analysis of the curdlan-like exopolysaccharide produced by *Cellulomonas flavigena* KU. *Journal of Industrial Microbiology & Biotechnology*, *29*, 200–203.

Kenyon, W. J., Esch, S. W., & Buller, C. S. (2005). The curdlan-type exopolysaccharide produced by *Cellulomonas flavigena* KU forms part of an extracellular glycocalyx involved in cellulose degradation. *Antonie van Leeuwenhoek*, *87*, 143–148.

Kim, M-Y., Ryu, K-E., Choi, W-A., Rhee, Y-H., & Lee, I-Y. (2003). Enhanced production of (1–3)-β-D-glucan by a mutant strain of *Agrobacterium* species. *Biochemical Engineering Journal*, *16*, 163–168.

Knecht, J. C., Schiffman, G., & Austrian, R. (1970). Some biological properties of pneumococcus type 37 and the chemistry of its capsular polysaccharide. *Journal of Experimental Medicine*, *132*, 475–487.

Komaniecka, I., & Choma, A. (2003). Isolation and characterization of periplasmic cyclic β-glucans of *Azorhizobium caulinodans*. *FEMS Microbiology Letters*, *227*, 263–269.

Krimi, Z., Petit, A., Mougel, C., Dessaux, Y., & Nesme, X. (2002). Seasonal fluctuations and long-term persistence of pathogenic populations of *Agrobacterium* spp. in soils. *Applied and Environmental Microbiology*, *68*, 3358–3365.

Laus, M. C., van Brussel, A. A. N., & Kijne, J. W. (2005). Role of cellulose fibrils and exopolysaccharides of *Rhizobium leguminosarum* in attachment to and infection of *Vicia sativa* root hairs. *Molecular Plant-Microbe Interactions*, *18*, 533–538.

Lee, I.-Y. (2002). Curdlan. In E. J. Vandamme, S. De Baets, & A. Steinbuchel (Eds.), *Polysaccharides I: Polysaccharides from prokaryotes. Biopolymers, Vol. 5.* (pp. 135–158). Wiley.

Lee, J. W., Yeomans, W. G. H., Allen, A. L., Gross, R. A., & Kaplan, D. L. (1997a). Microbial production of water-soluble non curdlan type exopolymer B with controlled composition by *Agrobacterium* sp.. *Biotechnology Letters*, *19*, 1217–1221.

Lee, J. W., Yeomans, W. G., Allen, A. L., Kaplan, D. L., Deng, F., & Gross, R. A. (1997b). Exopolymers from curdlan production: Incorporation of glucose-related sugars by *Agrobacterium* sp. strain ATCC 31749. *Canadian Journal of Microbiology*, *43*, 149–156.

Llull, D., Muñoz, R., López, R., & García, E. (1999). A single gene *(tts)* located outside the *cap* locus directs the formation of *Streptococcus pneumoniae* Type 37 capsular polysaccharide: Type 37 pneumococci are natural, genetically binary strains. *Journal of Experimental Medicine*, *190*, 241–251.

Matthysse, A. G., & McMahan, S. (1998). Root colonization by *Agrobacterium tumefaciens* is reduced in *cel*, *attB*, *attD*, and *attR* mutants. *Journal of Bacteriology*, *64*, 2341–2345.

Matthysse, A. G., White, S., & Lightfoot, R. (1995). Genes required for cellulose synthesis in *Agrobacterium tumefaciens*. *Journal of Bacteriology*, *177*, 1069–1075.

McIntosh, M. (2004). *An investigation of the production of curdlan, a (1–3)-β-glucan, by an* Agrobacterium *sp*. PhD thesis, Australia: La Trobe University.

McIntosh, M., Stone, B. A., & Stanisich, V. A. (2005). Curdlan and other bacterial (1–3)-β-D-glucans. *Applied Microbiology and Biotechnology*, *68*, 163–173.

Miller, J. K., & Gore, R. S. (1992). Cyclic beta-1,6–1,3 glucans of *Bradyrhizobium*: Functional analogs of the cyclic beta-1,2-glucans of *Rhizobium*? *Current Microbiology*, *24*, 101–104.

Miller, K. J., & Wood, J. M. (1996). Osmoadaptation by rhizosphere bacteria. *Annual Review of Microbiology*, *50*, 101–136.

Miller, K. J., Hadley, J. A., & Gustine, D. L. (1994). Cyclic β-1,6–1,3-glucans of *Bradyrhizobium japonicum* USDA 110 elicit isoflavenoid production in the soybean (*Glycine max*) host. *Plant Physiology*, *104*, 917–923.

Miller, K. J., Kennedy, E. P., & Reinhold, V. N. (1986). Osmotic adaptation by gram-negative bacteria: Possible role for periplasmic oligosaccharides. *Science*, *231*, 48–51.

Miller, K. J., Gore, R. S., Johnson, R., Benesi, A. J., & Reinhold, V. N. (1990). Cell-associated oligosaccharides of *Bradyrhizobium* spp.. *Journal of Bacteriology*, *172*, 136–142.

Mithöfer, A., Bhagwat, A. A., Feger, M., & Ebel, J. (1996). Suppression of fungal β-glucan-induced plant defense in soybean (*Glycine max. L.*) by cyclic 1,3–1,6-β-glucans from the symbiont *Bradyrhiobium japonicum*. *Planta*, *199*, 270–275.

Mithöfer, A., Fliegmann, J., Neuhaus-Url, G., Schwarz, H., & Ebel, J. (2000). The hepta-β-glucoside elicitor-binding proteins from legumes represent a putative receptor family. *Biological Chemistry*, *381*, 705–713.

Mithöfer, A., Bhagwat, A. A., Keister, D. L., & Ebel, J. (2001). *Bradyrhizobium japonicum* mutants defective in cyclic β-glucan synthesis show enhanced sensitivity to plant defense responses. *Zeitschrift fur Naturforschung*, *56c*, 581–584.

Mithöfer, A. (2002). Suppression of plant defence in rhizobia-legume symbiosis. *Trends in Plant Science*, *7*, 440–444.

Moxon, E. R., & Kroll, J. S. (1990). The role of bacterial polysaccharide capsules as virulence factors. *Current Topics in Microbiology and Immunology*, *150*, 65–85.

Nakanishi, I., Kimura, K., Kusui, S., & Yamazaki, E. (1974). Complex formation of gel-forming bacterial (1-3)-β-D-glucans (curdlan-type polysaccharides) with dyes in aqueous solution. *Carbohydrate Research*, *32*, 47–52.

Nakanishi, I., Kimura, K., Suzuki, T., Ishikawa, M., Banno, I., Sakane, T., & Harada, T. (1976). Demonstration of curdlan-type polysaccharide and some other β-1,3-glucans in microorganisms with aniline blue. *Journal of General and Applied Microbiology*, *22*, 1–11.

Nakanishi, I., Kimura, K., & Kanamaru, T. (1992). Studies on curdlan-type polysaccharide. I. Industrial production of curdlan-type polysaccharide. *Journal of the Takeda Research Laboratory*, *51*, 99–108.

Okuyama, K., Otsubo, A., Fukuzawa, Y., Ozawa, M., Harada, T., & Kasai, N. (1991). Single-helical structure of native curdlan and its aggregation states. *Journal of Carbohydrate Chemistry*, *10*, 645–656.

Parniske, M., Ahlborn, B., & Werner, D. (1991). Isoflavonoid-inducible resistance to the phytoalexin glyceollin in soybean rhizobia. *Journal of Bacteriology*, *173*, 3432–3439.

Pfeffer, P. E., Bécard, G., Rolin, D. B., Uknalis, J., Cooke, P., & Tu, S-I. (1994). *In vivo* nuclear magnetic resonance study of the osmoregulation of phosphocholine-substituted β-1,3;1,6 cyclic glucan and its associated carbon metabolism in *Bradyrhizobium japonicum* USDA 110. *Applied and Environmental Microbiology*, *60*, 2137–2146.

Pfeffer, P. E., Osman, S. F., Hotchkiss, A., Bhagwat, A. A., Keister, D. L., & Valentine, K. M. (1996). Cyclolaminarinose. A new biologically active β-(1-3) cyclic glucan. *Carbohydrate Research*, *296*, 23–37.

Phillips, K. R., & Lawford, H. G. (1983). Curdlan: Its properties and production in batch and continuous fermentations. In M. E. Bushell (Ed.), *Progress in industrial microbiology 18* (pp. 201–229). Amsterdam: Elsevier Scientific Publishing.

Roberts, I. S. (1996). The biochemistry and genetics of capsular polysaccharide production in bacteria. *Annual Review of Microbiology*, *50*, 285–315.

Rolin, D. B., Pfeffer, P. E., Osman, S. F., Szwergold, B. S., Kappler, F., & Benesi, A. J. (1992). Structural studies of a phosphocholine substituted β-(1,3);(1,6) macrocyclic glucan from *Bradyrhizobium japonicum* USDA 110. *Biochimica et Biophysica Acta*, *1116*, 215–225.

Sharp, J. K., Valent, B., & Albersheim, P. (1984). Purification and partial characterization of a β-glucan fragment that elicits phytoalexin accumulation in soybean. *Journal of Biological Chemistry*, *259*, 11312–11320.

Shivakumar, S., & Vijayendra, S. V. N. (2006). Production of exopolysaccharides by *Agrobacterium* sp. CFR-24 using coconut water – a byproduct of food industry. *Letters in Applied Microbiology*, *42*, 477–482.

Smith, L. D., & Ficht, T. A. (1990). Pathogenesis of *Brucella*. *Critical Reviews in Microbiology*, *17*, 209–230.

Somers, E., Vanderleyden, J., & Srinivasan, M. (2004). Rhizosphere bacterial signalling: A love parade beneath our feet. *Critical Reviews in Microbiology*, *30*, 205–240.

Soto, M. J., Lepek, V., Olivares, J., & Toro, N. (1993). Accumulation of cell-associated β (1-2)-glucan in *Rhizobium meliloti* strain GR4 in response to osmotic potential. *Molecular Plant-Microbe Interactions*, *5*, 288–293.

Stasinopoulos, S., Fisher, P. R., Stone, B. A., & Stanisich, V. A. (1999). Detection of two loci involved in (1-3)-β-glucan (curdlan) biosynthesis by *Agrobacterium* sp. ATCC31749, and comparative sequence analysis of the putative curdlan synthase gene. *Glycobiology*, *9*, 31–41.

Sutherland, I. W. (2001a). Microbial polysaccharides from Gram-negative bacteria. *International Dairy Journal*, *11*, 663–674.

Sutherland, I. W. (2001b). The biofilm matrix – an immobilized but dynamic microbial environment. *Trends in Microbiology*, *9*, 222–227.

Talaga, P., Cogez, V., Wieruszeski, J-M., Stahl, B., Lemoine, J., Lippens, G., & Bohin, J-P. (2002). Osmoregulated periplasic glucans of the free-living photosynthetic bacterium *Rhodobacter sphaeroides*. *European Journal of Biochemistry*, *269*, 2464–2472.

Tromp, D., Pettilino, F., Aracic, S., Malcolm, P., Stone, B. A., & Stanisich, V. A. (2008). Production of a novel exopolysaccharide by an *Agrobacterium* strain. *Annual scientific meeting of the Australian Society of Microbiology.* Melbourne, Australia July 6–10th. Abstract, PP07.02.

Tully, R. E., Keister, D. L., & Gross, K. C. (1990). Fractionation of the β-linked glucans of *Bradyrhizobium japonicum* and their response to osmotic potential. *Applied and Environmental Microbiology*, *56*, 1518–1522.

Voepel, K. C., & Buller, C. S. (1990). Formation of an extracellular energy reserve by *Cellulomonas flavigena* strain KU. *Journal of Industrial Microbiology*, *5*, 131–138.

Walter, A. W., Guerin, V. H., Beattie, M. W., Cotler, H. Y., & Bucca, H. B. (1941). Extension of the separation of types among the pneumococci: Description of 17 types in addition to Types 1 to 32 (Cooper). *Journal of Immunology*, *41*, 279–294.

Wiedemann, G., & Müller, P. (2004). Use of Tn*KPK2* for sequencing a 10.6-kb *Pst*I DNA fragment of *Bradyrhizobium japonicum* and for the construction of *aspA* and *ndvA* mutants. *Archives of Microbiology*, *181*, 418–427.

Williams, W. S., & Cannon, R. E. (1989). Alternative environmental roles for cellulose produced by *Acetobacter xylinum*. *Applied and Environmental Microbiology*, *55*, 2448–2452.

Wood, D. W., Setubal, J. C., Kaul, R., Monks, D. E., Kitajima, J. P., Okura, V. K., Zhou, Y., Chen, L., Wood, G. E., Almeida, N. F., Jr., Woo, L., Chen, Y., Paulsen, I. T., Eisen, J. A., Karp, P. D., Bovee, D., Sr., Chapman, P., Clendenning, J., Deatherage, G., Gillet, W., Grant, C., Kutyavin, T., Levy, R., Li, M. J., McClelland, E., Palmieri, A., Raymond, C., Rouse, G., Saenphimmachak, C., Wu, Z., Romero, P., Gordon, D., Zhang, S., Yoo, H., Tao, Y., Biddle, P., Jung, M., Krespan, W., Perry, M., Gordon-Kamm, B., Liao, L., Kim, S., Hendrick, C., Zhao, Z. Y., Dolan, M., Chumley, F., Tingey, S. V., Tomb, J. F., Gordon, M. P., Olson, M. V., & Nester, E. W. (2001) The genome of the natural genetic engineer *Agrobacterium tumefaciens* C58. *Science*, *294*, 2317–2323.

Yother, J. (1999). Common themes in the genetics of streptococcal capsular polysaccharides. In J. B. Goldberg (Ed.), *Genetics of bacterial polysaccharides* (pp. 161–184). Boca Raton, Florida: CRC Press.

CHAPTER 4.2

# Biology of (1,3)-β-Glucans and Related Glucans in Protozoans and Chromistans

**Sverre M. Myklestad [1] and Espen Granum[2]**

[1]*Department of Biotechnology, Norwegian University of Science and Technology (NTNU), Trondheim, Norway*

[2]*Department of Animal and Plant Sciences, University of Sheffield, Sheffield, UK*

Protozoans and chromistans comprise a wide range of organisms, including both phototrophic and heterotrophic species, with very different biologies from those of fungi and higher animals and plants. The storage polysaccharides in the euglenids (paramylon), haptophytes, diatoms and chrysophytes (chrysolaminarin), oomycetes (mycolaminarin) and brown algae (laminarin) are (1,3)- or (1,3;1,6)-β-glucans. Furthermore, (1,3)-β-glucans and related glucans are essential as structural cell wall components in many protozoans and chromistans, including callose ((1,3)-β-glucan), cellulose ((1,4)-β-glucan), chitin ((1,4)-β-*N*-acetylglucosamine) and (1,3:1,4)-β-glucans. There is considerable variation in the chemical structures and properties of different (1,3)-β-glucans, from the fibrillar callose and paramylon to the water-soluble laminarins. The biosynthesis and degradation of the storage (1,3)-β-glucans are particularly dynamic, and cellular quantities vary greatly with growth conditions such as light and nutrient levels and developmental stage. However, detailed molecular studies of (1,3)-β-glucan metabolism in protozoans and chromistans are still scarce.

## Introduction

(1,3)-β-Glucans and related glucans are important both as storage polysaccharides (laminarins and paramylon) and structural cell wall polysaccharides (callose, cellulose, chitin, etc.) in protozoans and chromistans. The nature of such polysaccharides is somewhat correlated with taxonomy (Table 1), but there is considerable variation between, and sometimes even within, classes in each division. Furthermore, the contents and detailed structures of the polysaccharides vary

**Table 1: Major types and structures of polysaccharides in protozoans and chromistans**

| Division<br>Major classes | Storage polysaccharides<br>(cytoplasm/vacuole) | Structural polysaccharides (cell wall)<br>Fibrillar | <br>Matrix | Mucilages<br>(extracellular) |
|---|---|---|---|---|
| Dinophyta<br>Dinophyceae | Starch ((1,4)-α-glucan) | Cellulose ((1,4)-β-glucan) | ? | Galactoglucan |
| Euglenophyta<br>Euglenophyceae | Paramylon ((1,3)-β-glucan) | No cell wall | No cell wall (Euglenophyta matrix) | Complex heteroglycans |
| Cryptophyta<br>Cryptophyceae | Starch ((1,4)-α-glucan) | Cellulose ((1,4)-β-glucan)? | ? | Complex heteroglycans? |
| Haptophyta<br>Haptophyceae | Chrysolaminarin ((1,3)-β-glucan) | Cellulose ((1,4)-β-glucan) | Acidic heteroglycans | Complex heteroglycans |
| Heterokontophyta<br>Bacillariophyceae<br>Chrysophyceae<br>Oomycota<br>Phaeophyceae | (Chryso- or myco-) laminarin ((1,3)-β-glucan) | Callose ((1,3)-β-glucan),<br>Cellulose ((1,4)-β-glucan),<br>Chitin ((1,4)-β-*N*-acetylglucosamine) | Acidic heteroglycans | Complex heteroglycans<br>Proteoglycan |

with the state of growth and development. In this chapter, the occurrence and functional roles of (1,3)-β-glucans and related glucans are treated in a taxonomic framework starting with the protozoan Euglenophyta (Euglenophyceae), followed by the chromistans Haptophyta (Haptophyceae) and Heterokontophyta (Bacillariophyceae, Chrysophyceae, Oomycota and Phaeophyceae).

## 4.2.1. Euglenophyceae

Euglenids (Euglenophyceae, Euglenophyta) are unicellular flagellates, with around 40 genera and 1000 species, most of which are found in fresh waters (van den Hoek et al., 1995). Although the majority are phototrophs with green chloroplasts, several genera are colourless and obligate heterotrophs. The storage polysaccharide in euglenids is paramylon, a fibrillar (1,3)-β-glucan, which is deposited as granules in the cytoplasm. This glucan has attracted considerable interest due to its applications in medicine as an immunostimulant and immunopotentiator (Kondo et al., 1992). Euglenids have no cell wall, but the plasmalemma is covered by a proteinaceous pellicle. In addition, the cells are often surrounded by a thin layer of mucilage.

The storage polysaccharide in euglenids was first discovered in *Euglena gracilis* by Gottlieb (1850), and termed paramylon (or paramylum) since it is isomeric with starch but does not stain with iodine. It occurs as membrane-bound, anisotropic granules of diverse morphology in the

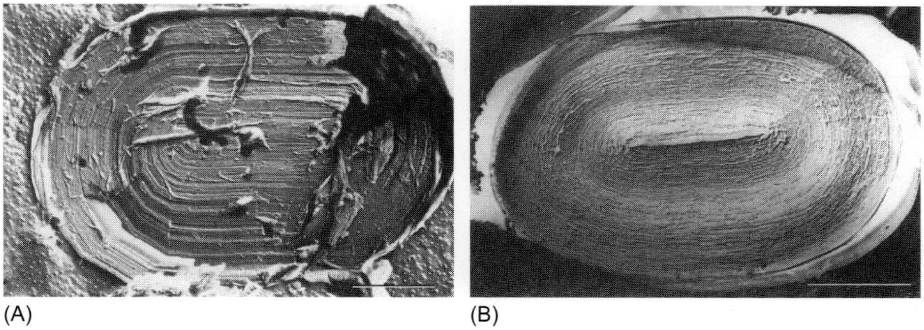

**Fig. 1: Freeze-etch electron micrographs of paramylon granules from *Euglena gracilis*. (A) Mature (wet) paramylon with highly ordered arrays of microfibrils; (B) immature (wet) paramylon with less organized arrays of microfibrils; scale bar = 500 nm (from Kiss et al., 1988a).**

cytoplasmic matrix, and as sheaths around pyrenoids outside the chloroplast envelope (Barras and Stone, 1968; Stone and Clarke, 1992). Structural investigations by X-ray diffraction and periodate oxidation revealed that it is a linear (1,3)-β-glucan (Kreeger and Meeuse, 1952; Clarke and Stone, 1960). Paramylons from the non-photosynthetic euglenids *Peranema trichophorum* (Archibald et al., 1963) and *Astasia ocellata* (Manners et al., 1966) also have the same structure, with degrees of polymerization (DP) of about 80 and 50–55, respectively. Paramylon granules have a very high crystallinity (comparable to that of cellulose) due to higher order aggregates of microfibrils and their interaction with water (Fig. 1; Marchessault and Deslandes, 1979; Booy et al., 1981; Kiss et al., 1987, 1988a; Kiss and Triemer, 1988). The elemental 4-nm microfibrils are composed of triple helices of (1,3)-β-glucan chains.

Paramylon contents in euglenids vary greatly with growth conditions (Barras and Stone, 1968). Although the glucan is produced by photosynthetic carbon assimilation in *E. gracilis*, it accumulates even more during heterotrophic growth on a suitable organic carbon source such as sugars or organic acids (Briand and Calvayrac, 1980; Kiss et al., 1986). In a non-photosynthetic *E. gracilis* mutant, paramylon accumulates up to 90% of cellular dry weight under dark conditions with glucose as carbon source (Barsanti et al., 2001). The polysaccharide is consumed in the dark when exponentially growing cells are transferred to a medium lacking a carbon source (Freyssinet et al., 1972). When dark-grown cells are transferred to the light, paramylon is consumed to support chlorophyll synthesis and chloroplast formation (Fig. 2; Dwyer and Smillie, 1970, 1971; Schwartzbach et al., 1975). Nitrogen starvation in photoheterotrophically grown cells induces disassembly of photosynthetic structures and accumulation of paramylon granules as well as lipid globules in the cytoplasm (García-Ferris et al.,

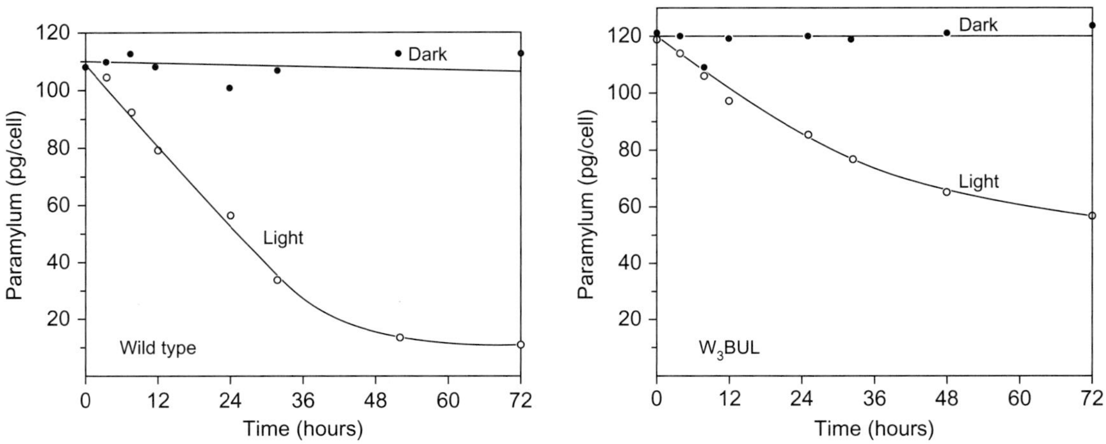


**Fig. 2: Kinetics of light-induced paramylon (paramylum) degradation in wild type and non-photosynthetic mutant ($W_3$BUL) of *Euglena gracilis* (from Schwartzbach et al., 1975).**

1996). Furthermore, pH shifts in the stationary growth phase have dramatic effects on both paramylon and protein contents (Hayashi et al., 1994). The storage polysaccharide may also be important in the development of resting spores; extensive accumulation of paramylon granules was observed in intermediate and mature cysts of *E. gracilis* (Triemer, 1980) and *Eutreptiella gymnastica* (Olli, 1996).

Biosynthesis of paramylon in *E. gracilis* involves a membrane-bound (1,3)-β-glucan synthase using UDP-glucose as substrate (Marechal and Goldemberg, 1964; Dwyer et al., 1970; Tomos and Northcote, 1978; Bäumer et al., 2001; see also Chapter 3.3.2). During photoheterotrophic growth, Calvayrac et al. (1981) observed that mitochondria and chloroplasts develop a common network of vesicles in which paramylon granules are formed. The membrane surrounding the granule has been implicated in the glucan synthesis (Kiss et al., 1988b). In cells returned from anaerobic to aerobic conditions, accumulated wax esters are used as precursors for paramylon synthesis, probably via the glyoxylate cycle (Inui et al., 1992). The polysaccharide is degraded by (1,3)-β-glucan endo- and exo-hydrolases as well as (1,3)-β-glucan phosphorylases (Barras and Stone, 1968; Dwyer et al., 1970; Miyatake and Kitaoka, 1980; see also Chapter 3.1). Photoregulation of paramylon degradation involves both chloroplast and non-chloroplast cell compartments controlled by separate photoreceptors (Schwartzbach et al., 1975). Under osmotic stress, paramylon is the precursor of the osmolyte trehalose (Takenaka et al., 1997).

### 4.2.2. Haptophyceae (Prymnesiophyceae)

The great majority of the Haptophyceae (or Prymnesiophyceae, Haptophyta) are marine, golden-brown, unicellular flagellates, comprising about 75 genera and 500 species (van den Hoek et al., 1995). Studies have particularly been made in the genera *Phaeocystis, Emiliania* and *Chrysochromulina*, because species belonging to these groups are often prominent members of the phytoplankton during blooms in the sea. Chrysolaminarin, a water-soluble (1,3)-β-glucan stored in special vacuoles, is the most important storage polysaccharide, although exceptions are known. The surface of the cells is covered with tiny scales of organic material, which consist mainly of cellulose ((1,4)-β-glucan). In addition, calcified scales (coccoliths) occur in species with coccoid stages. Some species also produce large amounts of extracellular mucopolysaccharides, which are complex heteroglycans.

Chrysolaminarin was observed in vacuoles in different *Phaeocystis* spp. (Kornmann, 1955; Parke et al., 1971; Chang, 1984). A non-colony-forming strain of *Phaeocystis globosa* produces a water-soluble (1,3)-β-glucan with some branching at C(O)6 and DP of about 20, which was classified as chrysolaminarin (Janse et al., 1996). In this strain, the entire polysaccharide could be degraded by a (1,3)-β-glucan endo-hydrolase (from *Trichoderma*), but in a colony-forming strain only 40–70% of the polysaccharide was digested (Alderkamp et al., 2007). The role of the glucan as a storage polysaccharide was indicated by a decrease in cellular concentration during light deprivation (Lancelot and Mathot, 1985a; Janse et al., 1996).

*Pavlova mesolychnon* contains cytoplasmic granules of paramylon-like (1,3)-β-glucan with a DP of more than 50, as revealed by X-ray diffraction (Kreger and van der Veer, 1970). A similar crystalline (1,3)-β-glucan found in *Pavlova prymnesiida* has a unique X-ray diffraction pattern (different from *Euglena*); hence, the authors claim that the term paramylon should be reserved for the euglenoid storage polysaccharide (Kiss and Triemer, 1988). A different storage glucan was discovered in the coccolithophorid *Emiliania huxleyi*, consisting of a (1,6)-β-glucan backbone with branching at C(O)3 (Vårum et al., 1986a). The $^{13}$C-NMR spectrum of the main chain of the glucan (after Smith degradation) provided conclusive evidence for this rather unusual chemical structure (Fig. 3). Recently, a similar water-soluble (1,3;1,6)-β-glucan was reported in the coccolithophorid *Pleurochrysis haptonemofera*, with (1,6)-β-linkages being most abundant (Hirokawa et al., 2008). Furthermore, (1,6)-β-glucan hydrolase activity was detected for the first time in algae in addition to (1,3)-β-glucan hydrolase activity (see also Chapter 3.1).

When nutrients are depleted, *Phaeocystis* blooms enter a stationary phase, during which carbohydrates accumulate (Lancelot, 1984; Alderkamp et al., 2006, 2007). In a study of five

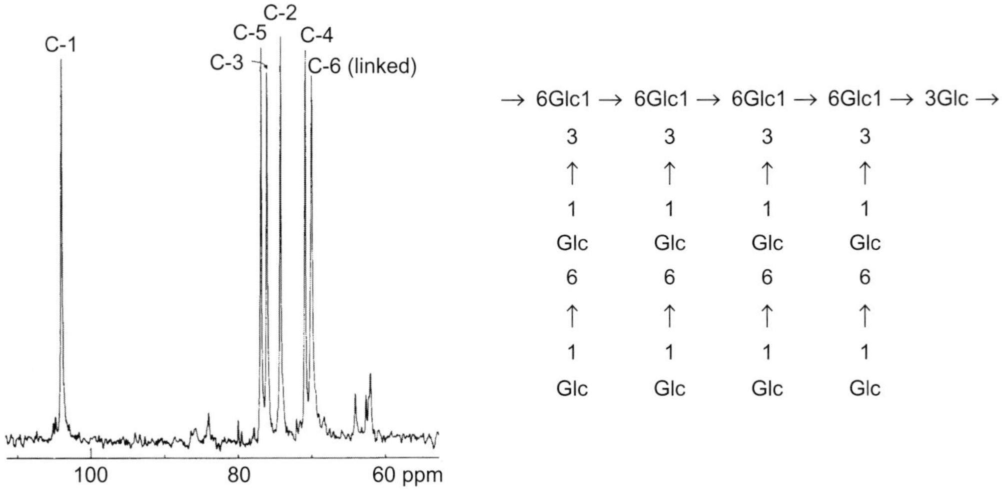


**Fig. 3: $^{13}$C-NMR spectrum of the main chain (after Smith degradation) and proposed chemical structure of (1,3;1,6)-β-glucan from *Emiliania huxleyi* (from Vårum et al., 1986a).**

different strains of *P. globosa*, glucose residues contributed 7–85% of total carbohydrate (van Rijssel et al., 2000). The fraction of glucose units present as chrysolaminarin increased from exponential to stationary growth phase. Strain differences were not detected, but the origin of the sea water used for culturing had a significant effect. During the stationary phase (1,3)-β-glucan contributes up to 50% of particulate organic carbon, whereas mucopolysaccharides contribute 5–60% (Alderkamp et al., 2007). The composition and properties of the *Phaeocystis* colony mucus are markedly different from those of the chrysolaminarin (van Boekel, 1992).

## 4.2.3. Bacillariophyceae (diatoms)

Diatoms (Bacillariophyceae, Heterokontophyta) are unicellular or colonial coccoid, golden-brown algae with siliceous cell walls, comprising around 250 genera and 10 000 species (Round et al., 1990; van den Hoek et al., 1995). They are widespread in the plankton and benthos of the sea as well as fresh waters, and may account for as much as 20% of global primary production (Field et al., 1998; Falkowski and Raven, 2007). The principal storage polysaccharide in diatoms is chrysolaminarin, a water-soluble (1,3)-β-glucan, which is dissolved in special vacuoles. The protoplast is encapsulated in a silica frustule coated by an organic sheath containing acidic and sulphated heteroglycans, and strips of callose, a fibrillar

(1,3)-β-glucan. In addition, diatoms often produce extracellular mucilage, which consists largely of complex heteroglycans. Many planktonic diatoms also have external spines of chitin ((1,4)-β-*N*-acetylglucosamine) fibres.

### 4.2.3.1. Chrysolaminarin

The storage polysaccharide in diatoms was first observed as refractile vacuolar bodies referred to as leucosin (von Stosch, 1951), but was renamed chrysolaminarin as chemical characterization revealed that it is a water-soluble (1,3)-β-glucan similar to those in brown algae and chrysophytes (Beattie et al., 1961). The polysaccharide is easily extracted from the cells by dilute acid or hot water, and can be purified by ethanol precipitation or dialysis and freeze drying (Allan et al., 1972; Myklestad, 1988; Granum and Myklestad, 2002; Chiovitti et al., 2004). Elucidation of detailed structure by methylation and periodate oxidation showed that chrysolaminarins of mixed freshwater diatoms (Beattie et al., 1961) and the marine diatom *Phaeodactylum tricornutum* (Ford and Percival, 1965a) are (1,3)-β-glucans with occasional branch points at C(O)6 and average DP of about 20, whereas in the marine diatom *Skeletonema costatum* (1,3)-β-glucan has branch points both at C(O)2 and C(O)6 and DP 11 (Paulsen and Myklestad, 1978). Chrysolaminarins of the ice diatom *Stauroneis amphioxys* and mixed Antarctic ice diatoms are also predominantly (1,3)-β-glucans with occasional branch points at C(O)2, C(O)4 and C(O)6 and DP 24 (McConville et al., 1986). Similar (1,3)-β-glucan structures were found in the marine diatoms *Coscinodiscus nobilis* (Percival et al., 1980), *Craspedostauros australis*, *Cylindrotheca fusiformis* and *Thalassiosira pseudonana* (Chiovitti et al., 2004). In three other marine diatoms NMR spectrometry revealed (1,3)-β-glucans with some structural variation: in *Chaetoceros debilis* 33% pustulan-like (1,6)-β-linked branches and DP 30; in *Chaetoceros mülleri* only 0.5–0.9% single-unit (1,6)-β-branches and DP 19–24; and in *Thalassiosira weissflogii* no branching and DP 5–13 (Størseth et al., 2004, 2005, 2006). Within each species there is little variation in chemical structure with different growth phases. Chrysolaminarin was localized to the vacuoles in the freshwater diatom *Pinnularia* by alkaline Aniline Blue staining (Waterkeyn and Bienfait, 1987). Furthermore, immunolabelling with a monoclonal (1,3)-β-glucan antibody confirmed that the polysaccharide is located in the vacuoles of the marine diatoms *C. fusiformis*, *P. tricornutum* and *T. pseudonana* (Fig. 4; Chiovitti et al., 2004).

Chrysolaminarin contents in diatoms vary greatly with growth conditions such as light and nutrient levels, indicating a very active and dynamic metabolism (Myklestad, 1988). The (1,3)-β-glucan is produced in the light and consumed in the dark (Handa, 1969; Vårum and

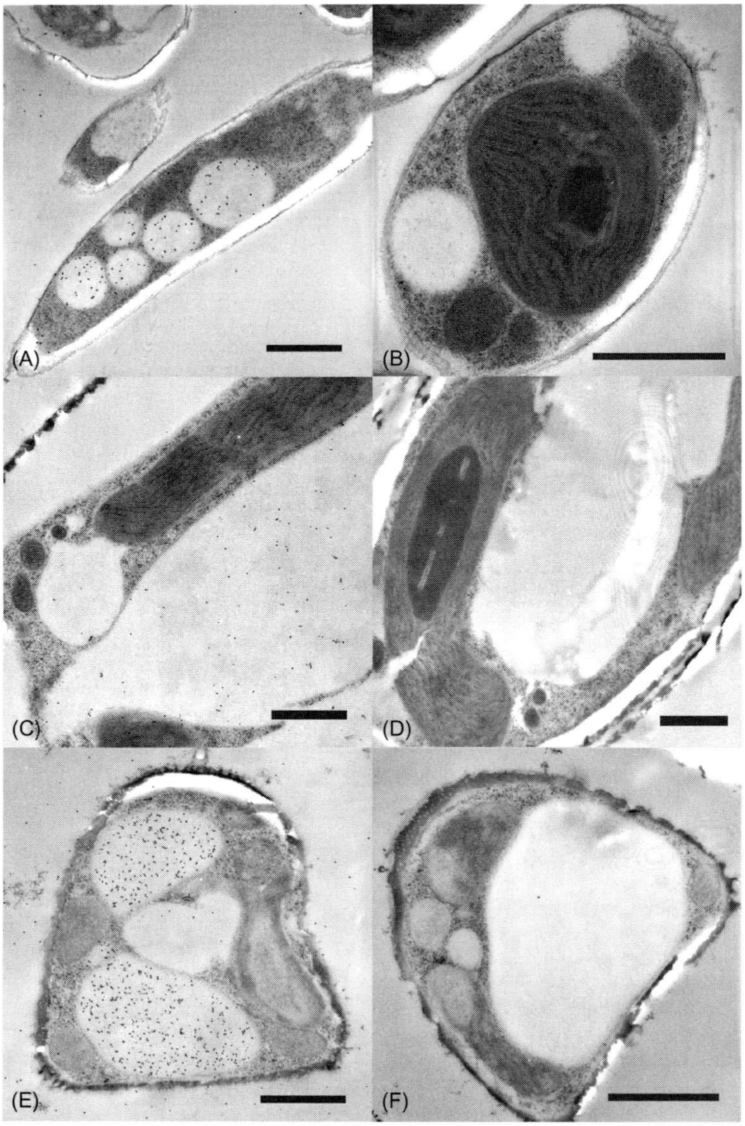


**Fig. 4: Transmission electron micrographs of diatom sections immunolabelled with anti-(1,3)-β-glucan antibody (A, C, E) and corresponding negative controls (B, D, F). (A, B) *Phaeodactylum tricornutum*; (C, D) *Cylindrotheca fusiformis*; (E, F) *Thalassiosira pseudonana*; scale bar = 500 nm (from Chiovitti et al., 2004).**

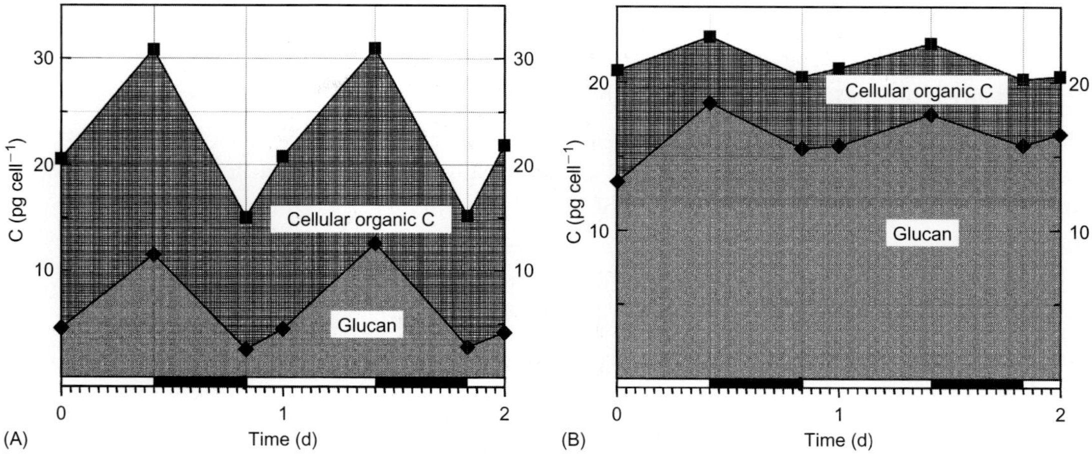


**Fig. 5: Diel changes in cell quotas of chrysolaminarin (glucan) and total organic carbon in *Skeletonema costatum* during exponential (A) and stationary (nitrogen depletion, B) growth phases under a 14:10 h light (white bar):dark (black bar) cycle (from Granum et al., 2002).**

Myklestad, 1984; Granum and Myklestad, 1999, 2001), in accordance with its role as a photosynthetic reserve. In exponentially growing *S. costatum* it increases from a minimum of 5–25% (depending on the length of photo- and scotoperiods) of cellular carbon at the end of the scotophase to a maximum of around 50% at the end of the photophase (Fig. 5; Hitchcock, 1980; Vårum et al., 1986b; Granum et al., 2002). Nutrient limitation often leads to significant accumulation of chrysolaminarin, with nitrogen inducing much higher levels than phosphorus (Myklestad and Haug, 1972; Myklestad, 1974, 1977; Millie, 1986; Guerrini et al., 2000). In *S. costatum* it accumulates up to 80% of cellular carbon under nitrogen deficiency (Fig. 5; Vårum and Myklestad, 1984; Granum et al., 2002). Upon nitrogen replenishment in the dark or low light, the polysaccharide is consumed to support protein synthesis (Vårum and Myklestad, 1984; Granum and Myklestad, 1999, 2001). In contrast, silicon deficiency induces a shift from chrysolaminarin to lipid accumulation (Millie, 1986; Taguchi et al., 1987; Roessler, 1988). Furthermore, (1,3)-β-glucan production is impaired by iron deficiency in the Antarctic diatom *Chaetoceros brevis*, probably due to reduced photosynthetic capacity (van Oijen et al., 2004a). Nevertheless, the evidence suggests that chrysolaminarin functions both as a short-term diurnal reserve and a long-term stockpile reserve in diatoms.

Field investigations of natural marine diatom populations indicate similar diel changes in chrysolaminarin (Morris et al., 1981; Lancelot and Mathot, 1985b; McConville et al., 1985;

Hama and Handa, 1992a, 1992b; van Oijen et al., 2003, 2004b) and accumulation under nutrient deficiency (Haug et al., 1973; Hitchcock, 1978; Sakshaug et al., 1983; Hama and Honjo, 1987; Hama et al., 1988), although its production is impaired by iron deficiency in parts of the Southern Ocean (van Oijen et al., 2005). (1,3)-β-Glucan dynamics may have further implications for cellular buoyancy and vertical migration of diatoms in the sea. Increasing sinking rates with higher carbohydrate contents have been reported for marine planktonic diatoms such as *Coscinodiscus concinnus* (Granata, 1991), *Rhizosolenia* spp. (Villareal et al., 1993) and *T. weissflogii* (Richardson and Cullen, 1995), although some observations are not consistent (Fisher and Harrison, 1996). Hence, the accumulation of chrysolaminarin by nutrient-deficient cells may cause increased cellular density and sinking below the nutricline. Upon nutrient replenishment and (1,3)-β-glucan respiration, the cells may rise toward the surface again, thereby transporting deep nutrients to the euphotic zone. Chrysolaminarin also plays an important role in the development of resting cells and spores (Kuwata et al., 1993), which are means of short- or long-term survival during periods of unfavourable conditions. Diatom blooms often culminate by formation and sinking of resting spores, suggesting that nutrient stress is the primary initiator (Sakshaug and Myklestad, 1973; Garrison, 1981; McQuoid and Hobson, 1996). The marine diatom *Chaetoceros pseudocurvisetus* produces resting spores with extensive accumulation of chrysolaminarin (70% of cellular carbon) as well as lipids (17%) under nitrogen depletion (Kuwata et al., 1993). Formation and germination of resting cells and spores may be important for species succession, dispersal and cycling of nutrients through the water column.

The pathways and regulation of chrysolaminarin metabolism in diatoms have only been investigated to a limited extent. The (1,3)-β-glucan is rapidly formed by photosynthetic carbon assimilation in *S. costatum* (Fig. 6; Myklestad, 1988; Granum and Myklestad, 2001). UDP-glucose pyrophosphorylase and (1,3)-β-glucan synthase activities were measured in the marine diatom *Cyclotella cryptica* (Roessler 1987, 1988; see also Chapter 3.3.2). The glucan synthase activity is significantly reduced by silicon deficiency, which may be related to a shift from chrysolaminarin to lipid accumulation (Roessler, 1988). UDP-glucose pyrophosphorylase activity in the marine diatom *Achnanthes brevipes* is reduced by phosphorus deficiency (Guerrini et al., 2000). Addition of a β-glucan synthesis inhibitor (2,6-dichlorobenzonitrile) in the marine benthic diatom *Cylindrotheca closterium* significantly reduces production of (1,3)-β-glucan as well as exopolymers and colloidal carbohydrate (Underwood et al., 2004). Recent whole-genome sequencing of *T. pseudonana* (Armbrust et al., 2004) and *P. tricornutum* (Bowler et al., 2008) provided evidence of two putative UDP-glucose pyrophosphorylases (one chloroplastic and one cytoplasmic) as well as one (1,3)-β-glucan

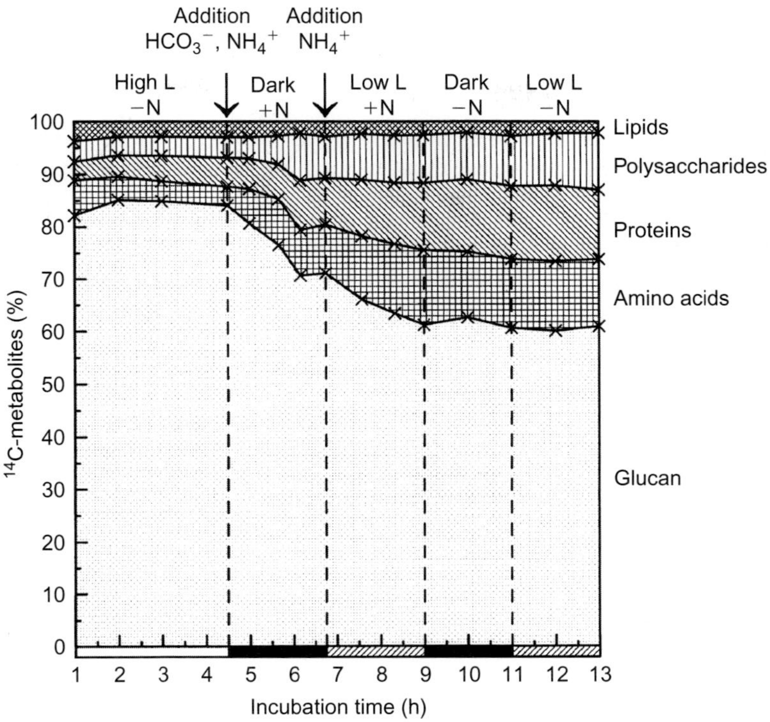


**Fig. 6: Distribution of cellular $^{14}C$-labelled compounds (glucan, chrysolaminarin and sugars; polysaccharides, cell wall polysaccharides and nucleic acids) in *Skeletonema costatum* during a pulse phase with photosynthetic carbon assimilation (white bar) and subsequent chase phases in the dark (black bars) or low light (shaded bars) with (+N) or without (−N) added $NH_4^+$ (from Granum and Myklestad, 2001).**

synthase (probably membrane-located) in each species (Kroth et al., 2008). The genomic data suggest that chrysolaminarin is either synthesized in the tonoplast using UDP-glucose generated in the cytoplasm, or in the chloroplast endoplasmic reticulum *en route* to the vacuole using UDP-glucose formed in the chloroplast.

Mobilization of chrysolaminarin provides precursors for cellular amino acids, proteins, nucleic acids and other polysaccharides in *S. costatum* (Fig. 6; Granum and Myklestad, 2001) and *C. brevis* (van Oijen et al., 2004b), and for exopolymers in *C. closterium* (Underwood et al., 2004). (1,3)-β-Glucan exo-hydrolase activity was demonstrated in some marine diatoms (Myklestad et al., 1982; see also Chapter 3.1), and it increases markedly with increasing nutrient deficiency in *S. costatum* (Vårum and Myklestad, 1984). Addition of a β-glucan hydrolase inhibitor (*p*-nitrophenyl β-glucopyranoside) resulted in glucan accumulation and lower rates of exopolymer and colloidal carbohydrate production in *C. closterium* (Underwood et al., 2004). A suite of putative glycosyl hydrolases were identified in the *P. tricornutum* and *T. pseudonana* genomes (Kroth et al., 2008), including (1,3)-β-glucan exo- and endo-hydrolases and β-glucosidases (most of them probably membrane located, e.g. tonoplast), but there is no evidence of phosphorolytic enzymes. However, both genomes include a putative glucokinase, which may serve to phosphorylate free glucose released by the glycosyl hydrolases. Then glucose-6-phosphate can be further metabolized through the respiratory pathways, providing energy and biosynthetic precursors.

### 4.2.3.2. Callose

Evidence of callose, a fibrillar (1,3)-β-glucan, as a component in the cell wall of *Pinnularia* was provided by alkaline Aniline Blue staining (Waterkeyn and Bienfait, 1987). Microscopic observation revealed three different locations of the (1,3)-β-glucan: (1) as a continuous callosic strip, or joint, between the epitheca and the hypotheca; (2) as minute callosic deposits arranged in a row ('Nebenlinien') on the connecting bands; and (3) as callosic deposits characteristic of the alveoli of the newly formed theca. The callosic strip seems to have a sealing function between the epitheca and the hypotheca. In contrast, the 'Nebenlinien' contain some acidic and sulphated heteroglycans in addition to callose, which have a selective and variable permeability, thus regulating exchanges with the surrounding medium. The metabolism of callose in diatoms has not been investigated, but may involve some of the previously mentioned enzymes identified in the *P. tricornutum* and *T. pseudonana* genomes (see also Chapters 3.1 and 3.3.2). However, a callose-specific (1,3)-β-glucan synthase would be expected to be located in the plasma membrane rather than the tonoplast, in analogy with the *Thalassiosira* chitin synthases (Sugiyama et al., 1999). The hydrolysing enzymes acting on callose are probably also located in the plasma membrane.

## 4.2.4. Chrysophyceae (golden algae)

Chrysophytes (Chrysophyceae, Heterokontophyta) are mainly unicellular or colonial golden-brown algae, which may be flagellate, but there are also some multicellular species (van den Hoek et al., 1995). This class contains about 200 genera and 1000 species, and most of them are found in fresh waters. The storage polysaccharide in chrysophytes is chrysolaminarin, a water-soluble (1,3)-β-glucan, which is dissolved in special vacuoles. In some species the cells are naked (amoeboid), whereas others have cell walls of siliceous or cellulosic scales coated by an organic sheath of acidic heteroglycans. Many chrysophytes produce extracellular mucilages of proteoglycans, or have mucilage bodies below the cell surface which can be discharged.

As in diatoms, the storage polysaccharide in chrysophytes was first observed as refractile vacuolar bodies termed leucosin (Klebs, 1893), but was renamed chrysolaminarin as chemical characterization revealed that it is a water-soluble (1,3)-β-glucan similar to those in brown algae and diatoms (Quillet, 1955). Detailed structural investigation by periodate oxidation showed that chrysolaminarin from *Ochromonas malhamensis* is a (1,3)-β-glucan with occasional branch points at C(O)6 and average DP of about 34 (Archibald et al., 1963). In *Ochromonas tuberculatus* the polysaccharide accumulates in posterior vacuoles, which occupy over half of the cell volume, with older cells containing greater amounts (Hibberd, 1970, 1976).

The biosynthetic pathway of chrysolaminarin in chrysophytes is yet to be resolved, but the (1,3)-β-glucan is rapidly formed by photosynthetic carbon assimilation in *O. malhamensis* (Kauss, 1962). Degradation of the polysaccharide involves a (1,3)-β-glucan phosphorylase, which is regulated allosterically by AMP (Albrecht and Kauss, 1971; see also Chapter 3.1). Furthermore, chrysolaminarin is the precursor of the osmolyte isofloridoside (*O*-α-galactopyranosyl-(1,1)-glycerol) in this chrysophyte (Kauss, 1977). Chrysolaminarin vacuoles isolated from *Poterioochromonas malhamensis* are enriched in β-glucosidase activity (Jochem et al., 1983).

## 4.2.5. Oomycota

Oomycota belongs to the Heterokontophyta, but many have a fungal-like appearance and live as filamentous networks of cells (Deacon, 2006). For many years the oomycetes were regarded as a separate class within the fungi. The oomycetes comprise animal and plant pathogens that can cause significant environmental and economic damage, as well as saprophytes that are beneficial to natural ecosystems by contributing to the recycling of nutrients from organic matter. The principle storage polysaccharide in oomycetes is mycolaminarin, a water-soluble (1,3)-β-glucan, which occurs in the cytoplasm. Furthermore, the cell walls contain (1,3)- or (1,3;1,6)-β-glucan in addition to cellulose and in some cases chitin (Fig. 7).

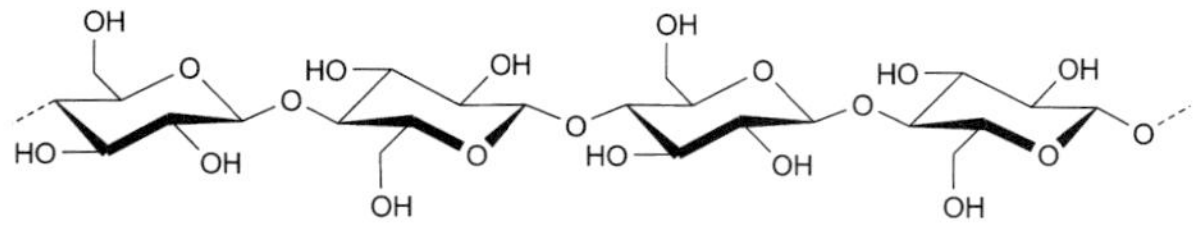

Cellulose ((1,4)-β-glucan)

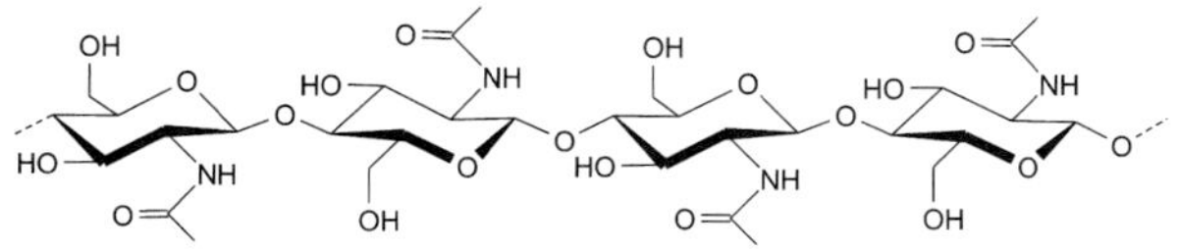

Chitin ((1,4)-β-*N*-acetylglucosamine)

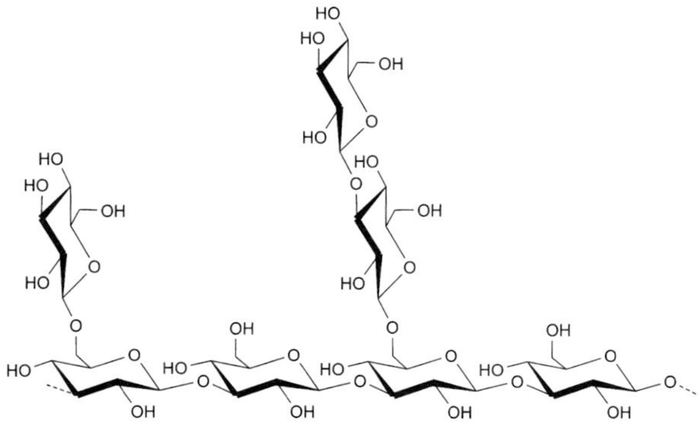

(1,3;1,6)-β-glucan

**Fig. 7: Chemical structures of the major oomycete polysaccharides, cellulose, chitin and (1,3;1,6)-β-glucan.**

### *4.2.5.1. Mycolaminarin*

Glycogen, the usual storage glucan in the fungi, has not been found in the oomycetes, of which *Phytophthora* spp. are the best studied (Stone and Clarke, 1992). The water-soluble (1,3)-β-glucan isolated from *Phytophthora palmivora* (37% of dry weight) was named mycolaminarin because of the structural resemblance to the storage polysaccharide, laminarin, from brown algae (Wang and Bartnicki-Garcia, 1974). It appeared highly homogeneous by gel filtration analysis and contained exclusively glucose units. The mycolaminarin was shown to be a (1,3)-β-glucan with one or two (1,6)-β-branches and average DP 36. Several studies indicate that mycolaminarin functions as a storage polysaccharide in oomycetes (Zevenhuisen and Bartnicki-Garcia, 1970; Wang and Bartnicki-Garcia, 1980; Lee and Mullins, 1994; Du and Mullins, 1998). The cytoplasmic (1,3)-β-glucan of *Achlya* consists of two types: a small neutral and a large phosphorylated polysaccharide (Shapiro and Mullins, 1997). Interestingly, no glucan was found in the apical vesicles, suggesting that mycolaminarin is not directly involved in apical growth.

#### 4.2.5.2. Cellulin

Cellulin is a granular, intracellular component found in some genera of the order Leptomitales (Pringsheim, 1883). It is almost entirely composed of two types of polysaccharides, chitin and (1,3;1.6)-β-glucan, which are in intimate association. In *Apodachlya* sp. the granules consist of 60% chitin and 39% (1,3;1,6)-β-glucan (Lee and Aronson, 1975). The biological function of these intracellular granules is unclear.

#### 4.2.5.3. Cell wall polysaccharides

Cell walls of oomycetes have received much attention because their chemical composition is different from those of the fungi, and the general view that growth and morphogenesis are intricately associated with cell wall metabolism (Bartnicki-Garcia, 1968; Wessels and Sietsma, 1981; Stone and Clarke, 1992). Generally, the cell wall consists of an insoluble inner layer of β-glucans embedded in other polymers, and an alkali-soluble outer layer. The content of total glucans range from 50 to 90% of dry weight (Table 2; Bertke and Aronson, 1980). This fraction occurs primarily as alkali-insoluble (1,3)- or (1,3;1,6)-β-glucan, and comparatively smaller amounts of cellulose (6–10% in four species). In addition, appreciable amounts of chitin (14–18%) were found in *Apodachlya* sp. and *Leptomitus lacteus*. In the fungi and yeasts, chitin is usually closely associated with a branch-on-branch (1,3;1,6)-β-glucan, and there is evidence for covalent linkages between them (Sietsma and Wessels, 1979; Mol and Wessels, 1987; Bartnicki-Garcia, 1999; Sietsma and Wessels, 2006). Both cellulose and (1,3)-β-glucan are thought to play important roles in growing hyphae in the oomycetes. The finding that elongating hyphae in *Achlya bisexualis* lack cellulose seems to require a re-evaluation of hyphal tip growth (Shapiro and Mullins, 2002a, 2002b).

Despite the biological importance of (1,3)-β-glucans both as storage polysaccharides and structural components of the cell walls of oomycetes, and as such playing an important role in morphogenesis and growth, their biosynthesis is poorly understood. (1,3)-β-Glucan synthases from *Saprolegnia monoica* have been purified and characterized (Girard and Fèvre, 1984; Bulone et al., 1990; Girard et al., 1992; Bulone and Fèvre, 1996; Billon-Grand et al., 1997; Pelosi et al., 2003; see also Chapter 3.3.2). Recently, annexin was identified as an activator of (1,3)-β-glucan synthase in *S. monoica* (Bouzenzana et al., 2006). The function of (1,3)-β-glucan endo-hydrolase was studied in three species of the family Saprolegniaceae (Money and Hill, 1997; see also

**Table 2: Chemical composition (as percentage of dry matter) of hyphal cell walls from seven oomycete species**

| Species | Total glucan | Cellulose | Chitin | Protein | Lipid | Ash |
|---|---|---|---|---|---|---|
| *Lagenidium callinectes*[a] | 79 | 7.8 | – | 5.4 | 5.8 | 3.5 |
| *Lagenidium chthamalophilum*[a] | 66 | 7.5 | – | 9.8 | 8.2 | 2.9 |
| *Leptomitus lacteus*[b] | 74 | 6.2–7.5 | 13.8 | 3.8 | 1.8 | 0.3 |
| *Apodachlya* sp.[c] | 50 | 10 | 18 | 6.4 | 1.7 | 0.4 |
| *Apodachlya brachynema*[d] | 80–90 | n.d. | – | 4 | 3–8 | 4 |
| *Dictyuchus sterile*[d] | " | " | – | " | " | " |
| *Pythium* sp.[d] | " | " | – | " | " | " |
| *Saprolegnia ferax*[d] | " | " | – | " | " | " |

– , no evidence for the presence of chitin, although 0.7–3% of glucosamine was found in these hyphal samples; n.d., not determined; ", same as above. The combined action of exo-laminarase, endo-laminarase, cellulase and lipase produced protoplasts from the four oomycete species tested. Both lipase and endo-laminarase were not essential, but aided protoplast formation.
[a]Bertke, C. C. and Aronson, J. M. (1992). Hyphal wall chemistry of *Lagenidium callinectes* and *Lagenidium chthamalophilum*. *Botanica Marina* 35, 147–152.
[b]Aronson, J. M. and Lin, C. C. (1978). Hyphal wall chemistry of *Leptomitus lacteus*. *Mycologia* 70, 363–369.
[c]Lin, C. C., Sicher, R. C. and Aronson, J. M. (1976). Hyphal wall chemistry in *Apodachlya*. *Archives of Microbiology* 108, 85–91.
[d]Sietsma, J. H., Eveleigh, D. E. and Haskins, R. H. (1969). Cell wall composition and protoplast formation of some oomycete species. *Biochimica et Biophysica Acta* 184, 306–317.

Chapter 3.1). Under osmotic stress, extracellular endo-hydrolase activity increased markedly, while the activities of other extracellular enzymes remained relatively constant. This change in endo-hydrolase activity correlated with cell wall strength in oomycete hyphae, consistent with the idea that the enzyme plays a role in controlling the mechanical properties of the cell wall during growth. Putative genes for three (1,3)-β-glucan exo-hydrolases, one (1,3)-β-glucan endo-hydrolase and one (1,3;1,4)-β-glucan endo-hydrolase were cloned and characterized from *Phytophthora infestans* (McLeod et al., 2003).

## 4.2.6. Phaeophyceae (brown algae)

The Phaeophyceae (Heterokontophyta) comprise multicellular brown algae with a wide range of morphologies and sizes (van den Hoek et al., 1995). This class contains 14 orders, 265 genera and 1500–2000 species. With the exception of a few freshwater genera, most brown algae are marine, and the majority grow in the intertidal and upper sublittoral zones. They inhabit rocky shores, with members of the Fucales dominating in the intertidal belt, and members of the Laminariales forming sublittoral kelp forests (Fig. 8). In addition, there are huge submerged beds of the giant kelps *Macrocystis* and *Nereocystis* off the Pacific coast of North America and Southern Australia. Nevertheless, the brown algae are limited to a

**Fig. 8:** ***Laminaria hyperborea*** **kelp forest from Reine, Lofoten, Norway. The colour specifications refer to colours in panels.**

relatively small area compared to phytoplankton, which occupy the euphotic zone of the whole ocean. The principal storage polysaccharide in brown algae is laminarin, a (1,3)-β-glucan which resembles chrysolaminarins in other heterokontophytes and haptophytes. Their cell wall is built of a network of cellulose fibrils (Table 3) strengthened by alginate, an acidic heteroglycan which is the most abundant polysaccharide in brown algae (also found in intercellular space). Some brown algae are of considerable economic interest. Algae for food are produced especially in Japan, while alginate is produced in the USA, Europe and China from *Laminaria*, *Macrocystis*, *Ecklonia*, *Durvillea*, *Ascophyllum* and others. But laminarin has never found a commercial outlet.

Early history and structural work on laminarin has been summarized previously, including a comprehensive list of references (Percival and McDowell, 1967; Stone and Clarke, 1992). While the earliest work was based almost exclusively on classical carbohydrate chemistry, Nelson and Lewis (1974) studied an 'insoluble' laminarin from *Laminaria hyperborea* using a purified (1,3)-β-glucan exo-hydrolase (from a basidiomycete) in addition to chemical methods. The polysaccharide could be separated into three closely related fractions containing (1) a water-soluble, branched, reducing glucan (with a reducing end); (2) an insoluble, unbranched, reducing glucan; and (3) an unbranched, non-reducing glucan with a single mannitol residue at the reducing end. In a more recent study, Read et al. (1996) used

**Table 3: Cellulose contents (as percentages of dry weight) of some brown algae[a]**

| Species | Frond | Whole alga | Stipe |
|---|---|---|---|
| *Laminaria digitata* | 3–5 | | 6–8 |
| *Laminaria saccharina* | 4–5 | | 7–8 |
| *Laminaria hyperborea* | 4–5 | | 8.5–10 |
| *Fucus spiralis* | | 4.4–4.8 | |
| *Fucus serratus* | | 2.0–3.5 | |
| *Fucus vesiculosus* | | 1.3–2.8 | |
| *Ascophyllum nodosum* | | 2.0 | |
| *Pelvetia canaliculate* | | 0.6–1.5 | |

[a]Percival, E. and McDowell, R. H. (1967). Chemistry and Enzymology of Marine Algal Polysaccharides. Academic Press, London.

electrospray-ionization mass spectrometry in combination with chemical methods to analyse structural heterogeneity of laminarin from *Laminaria digitata*. The ratio of laminarin molecules terminating with a reducing (1,3)-β-linked glucose residue (G-chains) to those terminating with a non-reducing (1,1)-β-linked mannitol residue (M-chains) was approximately 1:3. Molecular sizes were in the range of DP 20–30 (average of 25), and the number of (1,6)-β-branches per chain was 0–4 (average of 1.3), of which 75% were single glucosyl residues. Structural analyses by MALDI (matrix-assisted laser desorption/ionization) and FAB (fast atom bombardment) mass spectrometry confirmed the presence of both G- and M-chains in laminarins from *Alaria angusta*, *Chorda filum*, *Laminaria cichorioides*, *L. hyperborea*, *L. digitata* and *Sphaerotrichia divaricata* (Chizhov et al., 1998). The spectra revealed a range of oligomer sizes up to DP 40, peaking at DP 23–26 for most laminarins (except *C. filum* laminarin, peaking at DP 12), and the presence of minor amounts of cyclic oligomers. Detailed characterization of *L. digitata* laminarin by NMR spectrometry indicated an average DP of 33 (polydispersity 1.12) and degree of branching of 0.07 (Kim et al., 2000). Both content and structure of the laminarins in *Fucus evanescens*, *L. cichorioides* and *Laminaria japonica* were shown to vary considerably with age and season, with molecular weights in the range of 10–40 kDa and ratios of (1,3)- to (1,6)-β-linked glucose units from 2:1 to 10:1 (Zvyagintseva et al., 2003).

A different laminarin which is easily soluble in water and devoid of mannitol was first found in *Eisenia bicyclis* (Laminariales). Chemical analyses indicated an essentially linear structure of (1,3)- and (1,6)-β-linked glucose units at a ratio of 3:2 (Handa and Nisizawa, 1961), which was confirmed by NMR spectrometry (Usui et al., 1979). *Eisenia*-type laminarins were also found in *Ishige okamurai* (Maeda and Nisizawa, 1968), *Cystophora scalaris* and *Ecklonia radiata*

(Ram et al., 1981). The laminarins of *Cystoseira barbata* and *Cystoseira crinita* are also mannitol-free, but contain some *N*-acetylhexosamine-terminated chains (Chizhov et al., 1998).

Laminarin contents in brown algae vary markedly with the season (Stone and Clarke, 1992). In *L. hyperborea* fronds collected from different locations along the Norwegian coast (59.5–70.5°N), laminarin increased dramatically from a minimum below 2% in May to a maximum above 30% of algal dry mass in October–November, and then declined during winter and spring (Fig. 9; Jensen and Haug, 1956). Although no nutrient analyses were performed during this investigation, it is well known from other studies that nutrient levels along the Norwegian coast are low from May and throughout the summer (Sakshaug and Myklestad, 1973). Whereas protein synthesis and growth are slow during this period, the synthesis of laminarin is not inhibited by low concentrations of nutrients such as nitrogen and phosphorus, which are not incorporated into the glucan. Even during winter, when photosynthesis is too low for growth, protein is produced using laminarin reserves (Jensen and Haug, 1956; Hellebust and Haug, 1972). Also in *Laminaria longicrusis*, a low growth rate in the summer is associated with nitrogen depletion and laminarin accumulation (Chapman and Craigie, 1977). Upon fertilization with nitrate, the growth rate increases markedly, and laminarin levels decrease. Indeed, growth of a new frond in complete darkness was reported in *L. hyperborea* (Lüning, 1969). Gagne et al. (1982) also observed that carbohydrate accumulated in *L. longicrusis* during spring and summer was utilized during fall and early winter. At a location with year-round supply of nitrogen, the growth followed the seasonal light regime with only small increases in cellular carbon and nitrogen reserves. Sjøtun and Gunnarsson (1995) also found that laminarin in *Laminaria* was utilized when growth increased around mid-winter. In their pioneering work on *Fucus* zygotes, Quatrano and Stevens (1976) suggested that laminarin provides precursors for cell wall biosynthesis and energy for increased rates of respiration triggered by fertilization.

### 4.2.7. Other classes

Dinoflagellates (Dinophyceae, Dinophyta) comprise both photosynthetic and heterotrophic flagellates, with about 130 genera and 2000 species (van den Hoek et al., 1995). The principal storage polysaccharide is starch, a (1,4;1,6)-α-glucan, which occurs in the form of grains outside the chloroplasts. The cells are surrounded by a complex theca and in some cases a thin additional layer, the pellicle. Although the pellicle is usually referred to as cellulosic, linear (1,3;1,4)-β-glucans were found in *Peridinium westii* (Nevo and Sharon, 1969).

Chlorarachniophytes (Chlorarachniophyceae, Chlorarachniophyta) are a small group of amoeboid flagellate algae that were only recently discovered (van den Hoek et al., 1995).

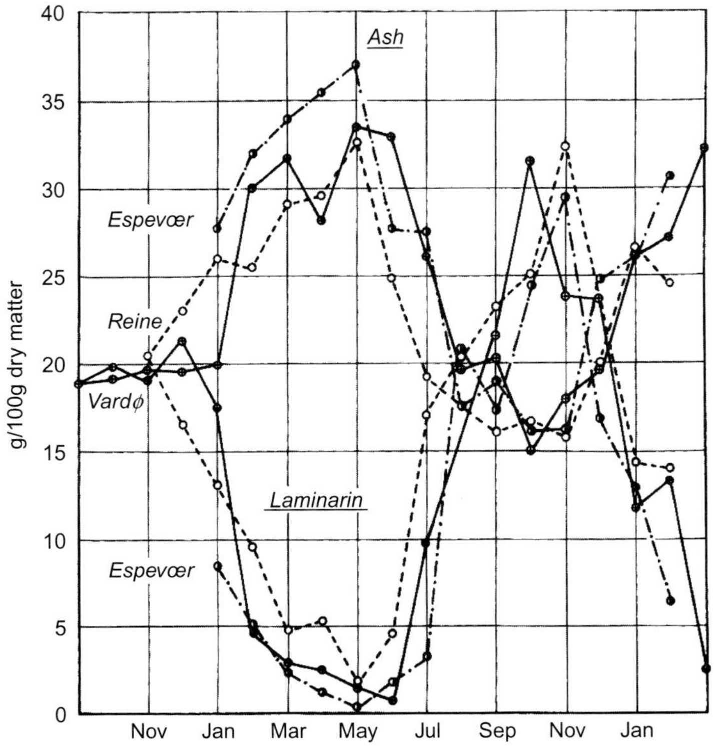


**Fig. 9: Seasonal variation of laminarin and ash in *Laminaria hyperborea* fronds from the Norwegian coast (from Jensen and Haug, 1956).**

Methylation analyses of the storage polysaccharides of two *Chlorarachnion* spp. revealed a long-chain (1,3)-β-glucan (McFadden et al., 1997). Furthermore, the (1,3)-β-glucan was immunolocalized to cytoplasmic vacuoles encapsulating the pyrenoids.

It is believed that other classes within the division Heterokontophyta also store laminarin-like (1,3)-β-glucans, including the Dictyochophyceae, Eustigmatophyceae, Raphidophyceae and Xanthophyceae (van den Hoek et al., 1995), but detailed structural studies are scarce. However, the storage polysaccharide of the raphidophyte *Haramonas dimorpha* was shown to be predominantly (1,3)-β-glucan with a low proportion of side-branching with glucosyl residues and DP 12–16 (Chiovitti et al., 2006). In the eustigmatophyte *Monodus subterraneus*, (1,3;1,4)-β-glucans were the main components of an alkaline extract which may originate from the cell wall (Beattie and Percival, 1962; Ford and Percival, 1965b).

## References

Albrecht, G. J., & Kauss, H. (1971). Purification, crystallization and properties of a β-(1→3)-glucan phosphorylase from *Ochromonas malhamensis*. *Phytochemistry*, *10*, 1293–1298.

Alderkamp, A. C., Nejstgaard, J. C., Verity, P. G., Zirbel, M. J., Sazhin, A. F., & van Rijssel, M. (2006). Dynamics in carbohydrate composition of *Phaeocystis pouchetii* colonies during spring blooms in mesocosms. *Journal of Sea Research*, *55*, 169–181.

Alderkamp, A. C., Buma, A. G. J., & van Rijssel, M. (2007). The carbohydrates of *Phaeocystis* and their degradation in the microbial food web. *Biogeochemistry*, *83*, 99–118.

Allan, G. G., Lewin, J., & Johnson, P. G. (1972). Marine polymers. IV. Diatom polysaccharides. *Botanica Marina*, *15*, 102–108.

Archibald, A. R., Cunningham, W. L., Manners, D. J., Stark, J. R., & Ryley, J. F. (1963). Studies on the metabolism of the protozoa. 10. The molecular structure of the reserve polysaccharides from *Ochromonas malhamensis* and *Peranema trichophorum*. *Biochemical Journal*, *88*, 444–451.

Armbrust, E. V., Berges, J. A., Bowler, C., Green, B. R., Martinez, D., Nicholas, H., Putnam, N. H., Zhou, S., Allen, A. E., Apt, K. E., Bechner, M., Brzezinski, M. A., Chaal, B. K., Chiovitti, A., Davis, A. K., Demarest, M. S., Detter, J. C., Glavina, T., Goodstein, D., Hadi, M. Z., Hellsten, U., Hildebrand, M., Jenkins, B. D., Jurka, J., Kapitonov, V. V., Kröger, N., Lau, W. W. Y., Lane, T. W., Larimer, F. W., Lippmeier, J. C., Lucas, S., Medina, M., Montsant, A., Obornik, M., Parker, M. S., Palenik, B., Pazour, G. J., Richardson, P. M., Rynearson, T. A., Saito, M. A., Schwartz, D. C., Thamatrakoln, K., Valentin, K., Vardi, A., Wilkerson, F. P., & Rokhsar, D. S. (2004). The genome of the diatom *Thalassiosira pseudonana*: ecology, evolution, and metabolism. *Science*, *306*, 79–86.

Barras, D. R., & Stone, B. A. (1968). Carbohydrate composition and metabolism in *Euglena*. In D. E. Buetow (Ed.),*The Biology of Euglena: Vol. 2* (pp. 149–191). New York: Academic Press.

Barsanti, L., Vismara, R., Passarelli, V., & Gualtieri, P. (2001). Paramylon (β-1,3-glucan) content in wild type and WZSL mutant of *Euglena gracilis*: effects of growth conditions. *Journal of Applied Phycology*, *13*, 59–65.

Bartnicki-Garcia, S. (1968). Cell wall chemistry, morphogenesis, and taxonomy of fungi. *Annual Review of Microbiology*, *22*, 87–108.

Bartnicki-Garcia, S. (1999). Glucans, walls, and morphogenesis: on the contributions of J. G. H. Wessels to the golden decades of fungal physiology and beyond. *Fungal Genetics and Biology*, *27*, 119–127.

Bäumer, D., Preisfeld, A., & Ruppel, H. G. (2001). Isolation and characterization of paramylon synthase from *Euglena gracilis* (Euglenophyceae). *Journal of Phycology*, *37*, 38–46.

Beattie, A., & Percival, E. (1962). The polysaccharides synthesized by *Monodus subterraneus* when grown on artificial media under bacteria-free conditions. *Proceedings of the Royal Society of Edinburgh, Section B: Biology*, *68*, 171–185.

Beattie, A., Hirst, E. L., & Percival, E. (1961). Studies on the metabolism of the Chrysophyceae: comparative structural investigations on leucosin (chrysolaminarin) separated from diatoms and laminarin from the brown algae. *Biochemical Journal*, *79*, 531–537.

Bertke, C. C., & Aronson, J. M. (1980). Hyphal wall composition in *Apodachlyella completa*. *Current Microbiology*, *4*, 235–238.

Billon-Grand, G., Marais, M. F., Joseleau, J. P., Girard, V., Gay, L., & Fèvre, M. (1997). A novel (1,3)-β-glucan synthase from the oomycete *Saprolegnia monoica*. *Microbiology*, *143*, 3175–3183.

Booy, F. P., Chanzy, H., & Boudet, A. (1981). An electron diffraction study of paramylon storage granules from *Euglena gracilis*. *Journal of Microscopy*, *121*, 133–140.

Bouzenzana, J., Pelosi, L., Briolay, A., Briolay, J., & Bulone, V. (2006). Identification of the first oomycete annexin as a (1→3)-β-D-glucan synthase activator. *Molecular Microbiology*, *62*, 552–565.

Bowler, C., Allen, A. E., Badger, J. H., Grimwood, J., Jabbari, K., Kuo, A., Maheswari, U., Martens, C., Maumus, F., Otillar, R. P., Rayko, E., Salamov, A., Vandepoele, K., Beszteri, B., Gruber, A., Heijde, M., Katinka, M., Mock, T., Valentin, K., Verret, F., Berges, J. A., Brownlee, C., Cadoret, J.-P., Chiovitti, A., Choi, C. J., Coesel, S., De Martino, A., Detter, J. C., Durkin, C., Falciatore, A., Fournet, J., Haruta, M., Huysman, M. J. J., Jenkins, B. D., Jiroutova, K., Jorgensen, R. E., Joubert, Y., Kaplan, A., Kröger, N., Kroth, P. G., La Roche, J., Lindquist, E., Lommer, M., Martin-Jézéquel, V., Lopez, P. J., Lucas, S., Mangogna, M., McGinnis, K., Medlin, L. K., Montsant, A., Oudot-Le Secq, M.-P., Napoli, C., Obornik, M., Parker, M. S., Petit, J.-L., Porcel, B. M., Poulsen, N., Robison, M., Rychlewski, L., Rynearson, T. A., Schmutz, J., Shapiro, H., Siaut, M., Stanley, M., Sussman, M. R., Taylor, A. R., Vardi, A., von Dassow, P., Vyverman, W., Willis, A., Wyrwicz, L. S., Rokhsar, D. S., Weissenbach, J., Armbrust, E. V., Green, B. R., Van de Peer, Y., & Grigoriev, I. V. (2008). The *Phaeodactylum* genome reveals the evolutionary history of diatom genomes. *Nature*, *456*, 239–244.

Briand, J., & Calvayrac, R. (1980). Paramylon synthesis in heterotrophic and photoheterotrophic *Euglena* (Euglenophyceae). *Journal of Phycology*, *16*, 234–239.

Bulone, V., & Fèvre, M. (1996). A 34-kilodalton polypeptide is associated with 1,3-β-glucan synthase activity from the fungus *Saprolegnia monoica*. *FEMS Microbiology Letters*, *140*, 145–150.

Bulone, V., Girard, V., & Fèvre, M. (1990). Separation and partial purification of 1,3-β-glucan and 1,4-β-glucan synthases from *Saprolegnia*. *Plant Physiology*, *94*, 1748–1755.

Calvayrac, R., Laval-Martin, D., Briand, J., & Farineau, J. (1981). Paramylon synthesis by *Euglena gracilis* photoheterotrophically grown under low $O_2$ pressure: description of a mitochloroplast complex. *Planta*, *153*, 6–13.

Chang, F. H. (1984). The ultrastructure of *Phaeocystis pouchetii* (Prymnesiophyceae) vegetative colonies with special reference to the production of new mucilaginous envelope. *New Zealand Journal of Marine and Freshwater Research*, *18*, 303–308.

Chapman, A. R. O., & Craigie, J. S. (1977). Seasonal growth in *Laminaria longicrusis*: relations with dissolved inorganic nutrients and internal reserves of nitrogen. *Marine Biology*, *40*, 197–205.

Chiovitti, A., Molino, P., Crawford, S. A., Teng, R., Spurck, T., & Wetherbee, R. (2004). The glucans extracted with warm water from diatoms are mainly derived from intracellular chrysolaminaran and not extracellular polysaccharides. *European Journal of Phycology*, *39*, 117–128.

Chiovitti, A., Ngoh, J. E., & Wetherbee, R. (2006). 1,3-β-D-Glucans from *Haramonas dimorpha* (Raphidophyceae). *Botanica Marina*, *49*, 360–362.

Chizhov, A. O., Dell, A., Morris, H. R., Reason, A. J., Haslam, S. M., McDowell, R. A., Chizhov, O. S., & Usov, A. I. (1998). Structural analysis of laminarans by MALDI and FAB mass spectrometry. *Carbohydrate Research*, *310*, 203–210.

Clarke, A. E., & Stone, B. A. (1960). Structure of the paramylon from *Euglena gracilis*. *Biochimica et Biophysica Acta*, *44*, 161–163.

Deacon, J. (2006). *Fungal Biology*. Malden (MA): Blackwell.

Du, X., & Mullins, J. T. (1998). $Ca^{2+}$-induced sporulation in *Achlya bisexualis*: reserve 1,3-β-glucans provides both carbon and phosphorus. *Mycologia*, *90*, 990–994.

Dwyer, M. R., & Smillie, R. M. (1970). A light-induced β-1,3-glucan breakdown associated with the differentiation of chloroplasts in *Euglena gracilis*. *Biochimica et Biophysica Acta*, *216*, 392–401.

Dwyer, M. R., & Smillie, R. M. (1971). β-1,3-Glucan: a source of carbon and energy for chloroplast development in *Euglena gracilis*. *Australian Journal of Biological Sciences*, *24*, 15–22.

Dwyer, M. R., Smydzuk, J., & Smillie, R. M. (1970). Synthesis and breakdown of β-1,3-glucan in *Euglena gracilis* during growth and carbon depletion. *Australian Journal of Biological Sciences*, *23*, 1005–1013.

Falkowski, P. G., & Raven, J. A. (2007). *Aquatic Photosynthesis* (Ed. 2). Princeton (NJ): Princeton University Press.

Field, C. B., Behrenfeld, M. J., Randerson, J. T., & Falkowski, P. (1998). Primary production of the biosphere: integrating terrestrial and oceanic components. *Science*, *281*, 237–240.

Fisher, A. E., & Harrison, P. J. (1996). Does carbohydrate content affect the sinking rates of marine diatoms? *Journal of Phycology*, *32*, 360–365.

Ford, C. W., & Percival, E. (1965a). The carbohydrates of *Phaeodactylum tricornutum*. Part I. Preliminary examination of the organism, and characterisation of low molecular weight material and of a glucan. *Journal of the Chemical Society*, *1298*, 7035–7041.

Ford, C. W., & Percival, E. (1965b). Polysaccharides synthesised by *Monodus subterraneus*. Part II. The cell-wall glucan. *Journal of the Chemical Society*, *551*, 3014–3016.

Freyssinet, G., Heizmann, P., Verdier, G., Trabuchet, G., & Nigon, V. (1972). Influence des conditions nutritionnelles sur la réponse de l'éclairement chez les *Eugléñes* etiolées. *Physiologie Végétale*, *10*, 421–442.

Gagne, J. A., Mann, K. H., & Chapman, A. R. O. (1982). Seasonal pattern of growth and storage in *Laminaria longicrusis* in relation to differing patterns of availability of nitrogen in the water. *Marine Biology*, *69*, 91–101.

García-Ferris, C., de los Ríos, A., Ascaso, C., & Moreno, J. (1996). Correlated biochemical and ultrastructural changes in nitrogen-starved *Euglena gracilis*. *Journal of Phycology*, *32*, 953–963.

Garrison, D. L. (1981). Monterey Bay phytoplankton. II. Resting spore cycles in coastal diatom populations. *Journal of Plankton Research*, *3*, 137–156.

Girard, V., Bulone, V., & Fèvre, M. (1992). Separation and partial peptide characterization of β-(1,3)-glucan synthase from *Saprolegnia*. *Plant Science*, *82*, 145–153.

Girard, V., & Fèvre, M. (1984). β-1-4- and β-1-3-glucan synthases are associated with the plasma membrane of the fungus *Saprolegnia*. *Planta*, *160*, 400–406.

Gottlieb, J. (1850). Über eine neue, mit Stärkemehl isomere Substanz. *Annalen der Chemie und Pharmazie*, *75*, 51–61.

Granata, T. C. (1991). Diel periodicity in growth and sinking rates of the centric diatom *Coscinodiscus concinnus*. *Limnology and Oceanography*, *36*, 132–139.

Granum, E., & Myklestad, S. M. (1999). Effects of $NH_4^+$ assimilation on dark carbon fixation and β-1,3-glucan metabolism in the marine diatom *Skeletonema costatum* (Bacillariophyceae). *Journal of Phycology*, *35*, 1191–1199.

Granum, E., & Myklestad, S. M. (2001). Mobilization of β-1,3-glucan and biosynthesis of amino acids induced by $NH_4^+$ addition to N-limited cells of the marine diatom *Skeletonema costatum* (Bacillariophyceae). *Journal of Phycology*, *37*, 772–782.

Granum, E., & Myklestad, S. M. (2002). A simple combined method for determination of β-1,3-glucan and cell wall polysaccharides in diatoms. *Hydrobiologia*, *477*, 155–161.

Granum, E., Kirkvold, S., & Myklestad, S. M. (2002). Cellular and extracellular production of carbohydrates and amino acids by the marine diatom *Skeletonema costatum*: diel variations and effects of N depletion. *Marine Ecology Progress Series*, *242*, 83–94.

Guerrini, F., Cangini, M., Boni, L., Trost, P., & Pistocchi, R. (2000). Metabolic responses of the diatom *Achnanthes brevipes* (Bacillariophyceae) to nutrient limitation. *Journal of Phycology*, *36*, 882–890.

Hama, J., & Handa, N. (1992a). Diel photosynthetic production of cellular organic matter in natural phytoplankton populations, measured with $^{13}C$ and gas chromatography/mass spectrometry. I. Monosaccharides. *Marine Biology*, *112*, 175–181.

Hama, J., & Handa, N. (1992b). Diel variation of water-extractable carbohydrate composition of natural phytoplankton populations in Kinu-ura Bay. *Journal of Experimental Marine Biology and Ecology*, *162*, 159–176.

Hama, T., & Honjo, T. (1987). Photosynthetic products and nutrient availability in phytoplankton population from Gokasho Bay, Japan. *Journal of Experimental Marine Biology and Ecology*, *112*, 251–266.

Hama, T., Handa, N., Takahashi, M., Whitney, F., & Wong, C. S. (1988). Change in distribution patterns of photosynthetically incorporated C during phytoplankton bloom in controlled experimental ecosystem. *Journal of Experimental Marine Biology and Ecology*, *120*, 39–56.

Handa, N. (1969). Carbohydrate metabolism in the marine diatom *Skeletonema costatum*. *Marine Biology*, *4*, 208–214.

Handa, N., & Nisizawa, K. (1961). Structural investigations of a laminarin isolated from *Eisenia bicyclis*. *Nature*, *192*, 1078–1080.

Haug, A., Myklestad, S., & Sakshaug, E. (1973). Studies on the phytoplankton ecology of the Trondheimsfjord. I. The chemical composition of phytoplankton populations. *Journal of Experimental Marine Biology and Ecology*, *11*, 15–26.

Hayashi, M., Toda, K., Ishiko, H., Komatsu, R., & Kitaoka, S. (1994). Effects of shifting pH in the stationary phase of growth on the chemical composition of *Euglena gracilis*. *Bioscience, Biotechnology, and Biochemistry*, *58*, 1964–1967.

Hellebust, J. A., & Haug, A. (1972). *In situ* studies on alginic acid synthesis and other aspects of the metabolism of *Laminaria digitata*. *Canadian Journal of Botany*, *50*, 177–184.

Hibberd, D. J. (1970). Observations on the cytology and ultrastructure of *Ochromonas tuberculatus* sp. nov. (Chrysophyceae), with special reference to the discobolocysts. *British Phycological Journal*, *5*, 119–143.

Hibberd, D. J. (1976). The ultrastructure and taxonomy of the Chrysophyceae and Prymnesiophyceae (Haptophyceae): a survey with some new observations of the ultrastructure of the Chrysophyceae. *Botanical Journal of the Linnean Society*, *72*, 55–80.

Hirokawa, Y., Fujiwara, S., Suzuki, M., Akiyama, T., Sakamoto, M., Kobayashi, S., & Tsuzuki, M. (2008). Structural and physiological studies on the storage β-polyglucan of haptophyte *Pleurochrysis haptonemofera*. *Planta*, *227*, 589–599.

Hitchcock, G. L. (1978). Labelling patterns of carbon-14 in net plankton during a winter-spring bloom. *Journal of Experimental Marine Biology and Ecology*, *31*, 141–153.

Hitchcock, G. L. (1980). Diel variation in chlorophyll *a*, carbohydrate and protein content of the marine diatom *Skeletonema costatum*. *Marine Biology*, *57*, 271–278.

Inui, H., Miyatake, K., Nakano, Y., & Kitaoka, S. (1992). Synthesis of reserved polysaccharide from wax esters accumulated as the result of anaerobic energy generation in *Euglena gracilis* returned from anaerobic to aerobic conditions. *International Journal of Biochemistry*, *24*, 799–803.

Janse, I., van Rijssel, M., van Hall, P.-J., Gerwig, G. J., Gottschal, J. C., & Prins, R. A. (1996). The storage glucan of *Phaeocystis globosa* (Prymnesiophyceae) cells. *Journal of Phycology*, *32*, 382–387.

Jensen, A., & Haug, A. (1956). Geographical and seasonal variation in the chemical composition of *Laminaria hyperborea* and *Laminaria digitata* from the Norwegian coast. *Report No. 14, Norwegian Institute of Seaweed Research. Oslo: Akademisk trykningssentral.*

Jochem, P., Thomson, K. S., & Schwab, D. (1983). Isolation of vacuoplasts from *Poterioochromonas malhamensis*. *Plant Physiology*, *73*, 418–421.

Kauss, H. (1962). Der Einfluss von Thiamin- und Biotin-mangel auf den Stoffwechsel von *Ochromonas malhamensis*. *Vorträge aus dem Gesamtgebiet der Botanik*, *1*, 129–132.

Kauss, H. (1977). Biochemistry of osmotic regulation. In D. H. Northcote (Ed.), *International Review of Biochemistry: Vol. 13: Plant Biochemistry II* (pp. 119–140). Baltimore: University Park Press.

Kim, Y.-T., Kim, E.-H., Cheong, C., Williams, D. L., Kim, C.-W., & Lim, S.-T. (2000). Structural characterization of β-D-(1→3, 1→6)-linked glucans using NMR spectroscopy. *Carbohydrate Research*, *328*, 331–341.

Kiss, J. Z., & Triemer, R. E. (1988). A comparative study of the storage carbohydrate granules from *Euglena* (Euglenida) and *Pavlova* (Prymnesiida). *Journal of Protozoology*, *35*, 237–241.

Kiss, J. Z., Vasconcelos, A. C., & Triemer, R. E. (1986). Paramylon synthesis and chloroplast structure associated with nutrient levels in *Euglena* (Euglenophyceae). *Journal of Phycology*, *22*, 327–333.

Kiss, J. Z., Vasconcelos, A. C., & Triemer, R. E. (1987). Structure of the euglenoid storage carbohydrate, paramylon. *American Journal of Botany*, *74*, 877–882.

Kiss, J. Z., Roberts, E. M., Brown, R. M., Jr, & Triemer, R. E. (1988a). X-ray and dissolution studies of paramylon storage granules from *Euglena*. *Protoplasma*, *146*, 150–156.

Kiss, J. Z., Vasconcelos, A. C., & Triemer, R. E. (1988b). The intramembranous particle profile of the paramylon membrane during paramylon synthesis in *Euglena* (Euglenophyceae). *Journal of Phycology*, *24*, 152–157.

Klebs, G. (1893). Flagellatenstudien II. *Zeitschrift für Wissenschaftliche Zoologie*, *55*, 353–445.

Kondo, Y., Kato, A., Hojo, H., Nozoe, S., Takeuchi, M., & Ochi, K. (1992). Cytokine-related immunopotentiating activities of paramylon, a β-(1→3)-D-glucan from *Euglena gracilis*. *Journal of Pharmacobiodynamics*, *15*, 617–621.

Kornmann, P. (1955). Beobachtungen an *Phaeocystis*-Kulturen. *Helgoländer Wissenschaftliche Meeresuntersuchungen*, *5*, 218–233.

Kreger, D. R., & Meeuse, B. J. D. (1952). X-ray of *Euglena*-paramylon, of the acid-insoluble glucan of yeast cell walls and of laminarin. *Biochimica et Biophysica Acta*, *9*, 699–700.

Kreger, D. R., & van der Veer, J. (1970). Paramylon in a chrysophyte. *Acta Botanica Neerlandica*, *19*, 401–402.

Kroth, P. G., Chiovitti, A., Gruber, A., Martin-Jezequel, V., Mock, T., Parker, M. S., Stanley, M. S., Kaplan, A., Caron, L., Weber, T., Maheswari, U., Armbrust, E. V., & Bowler, C. (2008). A model for carbohydrate metabolism in the diatom *Phaeodactylum tricornutum* deduced from comparative whole genome analysis. *PLoS One*, *3*, e1426.

Kuwata, A., Hama, T., & Takahashi, M. (1993). Ecophysiological characterization of two life forms, resting spores and resting cells, of a marine planktonic diatom, *Chaetoceros pseudocurvisetus*, formed under nutrient depletion. *Marine Ecology Progress Series*, *102*, 245–255.

Lancelot, C. (1984). Metabolic changes in *Phaeocystis poucheti* (Hariot) Lagerheim during the spring bloom in Belgian coastal waters. *Estuarine, Coastal and Shelf Science*, *18*, 593–600.

Lancelot, C., & Mathot, S. (1985a). Biochemical fractionation of primary production by phytoplankton in Belgian coastal waters during short- and long-term incubations with $^{14}C$-bicarbonate. II. *Phaeocystis poucheti* colonial population. *Marine Biology*, *86*, 227–232.

Lancelot, C., & Mathot, S. (1985b). Biochemical fractionation of primary production by phytoplankton in Belgian coastal waters during short- and long-term incubations with $^{14}C$-bicarbonate. I. Mixed diatom population. *Marine Biology*, *86*, 219–226.

Lee, H. Y., & Aronson, J. M. (1975). Composition of cellulin, the unique chitin-glucan granules of the fungus, *Apodachlya* sp. *Archives of Microbiology*, *102*, 203–208.

Lee, J., & Mullins, J. T. (1994). Cytoplasmic water-soluble β-glucans in *Achlya*: response to nutrient limitation. *Mycologia*, *86*, 235–241.

Lüning, K. (1969). Growth of amputated and dark-exposed individuals of the brown alga *Laminaria hyperborea*. *Marine Biology*, *2*, 218–223.

Maeda, M., & Nisizawa, K. (1968). Laminarin of *Ishige okamurai*. *Carbohydrate Research*, *7*, 97–99.

Manners, D. J., Ryley, J. F., & Stark, J. R. (1966). Studies on the metabolism of the protozoa: the molecular structure of the reserve polysaccharide from *Astasia ocellata*. *Biochemical Journal*, *101*, 323–327.

Marchessault, R. H., & Deslandes, Y. (1979). Fine structure of (1→3)-β-D-glucans: curdlan and paramylon. *Carbohydrate Research*, *75*, 231–242.

Marechal, L. R., & Goldemberg, S. H. (1964). Uridine diphosphate glucose-β-1,3-glucan β-3-glucosyltransferase from *Euglena gracilis*. *Journal of Biological Chemistry*, *239*, 3163–3167.

McConville, M. J., Mitchell, C., & Wetherbee, R. (1985). Patterns of carbon assimilation in a microalgal community from annual sea ice, East Antarctica. *Polar Biology*, *4*, 135–141.

McConville, M. J., Bacic, A., & Clarke, A. E. (1986). Structural studies of chrysolaminarin from the ice diatom *Stauroneis amphioxys* (Gregory). *Carbohydrate Research*, *153*, 330–333.

McFadden, G. I., Gilson, P. R., & Sims, I. M. (1997). Preliminary characterization of carbohydrate stores from chlorarachniophytes (division: Chlorarachniophyta). *Phycological Research*, *45*, 145–151.

McLeod, A., Smart, C. D., & Fry, W. E. (2003). Characterization of 1,3-β-glucanase and 1,3;1,4-β-glucanase genes from *Phytophthora infestans*. *Fungal Genetics and Biology*, *38*, 250–263.

McQuoid, M. R., & Hobson, L. A. (1996). Diatom resting stages. *Journal of Phycology*, *32*, 889–902.

Millie, D. F. (1986). Nutrient-limitation effects on the biochemical composition of *Cyclotella meneghiniana* (Bacillariophyta): an experimental and statistical analysis. *Canadian Journal of Botany*, *64*, 19–26.

Miyatake, K., & Kitaoka, S. (1980). Convenient detection of β-1,3-glucan phosphorylases of *Euglena gracilis* z by polyacrylamide disc gel electrophoresis. *Bulletin of the University of Osaka Prefecture, Series B: Agriculture and Biology*, *32*, 49–54.

Mol, P. C., & Wessels, J. G. H. (1987). Linkages between glucosaminoglycan and glucan determine alkali-insolubility of the glucan in walls of *Saccharomyces cerevisiae*. *FEMS Microbiology Letters*, *41*, 95–99.

Money, N. P., & Hill, T. W. (1997). Correlation between endo-glucanase secretion and cell wall strength in oomycete hyphae: implications for growth and morphogenesis. *Mycologia*, *89*, 777–785.

Morris, I., Smith, A. E., & Glover, H. E. (1981). Products of photosynthesis in phytoplankton off the Orinoco River and in the Caribbean Sea. *Limnology and Oceanography*, *26*, 1034–1044.

Myklestad, S. (1974). Production of carbohydrates by marine planktonic diatoms. I. Comparison of nine different species in culture. *Journal of Experimental Marine Biology and Ecology*, *15*, 261–274.

Myklestad, S. (1977). Production of carbohydrates by marine planktonic diatoms. II. Influence of the N/P ratio in the growth medium on the assimilation ratio, growth rate, and production of cellular and extracellular carbohydrates by *Chaetoceros affinis* var. *willei* (Gran) Hustedt and *Skeletonema costatum* (Grev.) Cleve. *Journal of Experimental Marine Biology and Ecology*, *29*, 161–179.

Myklestad, S., & Haug, A. (1972). Production of carbohydrates by the marine diatom *Chaetoceros affinis* var. *willei* (Gran) Hustedt. I. Effect of the concentration of nutrients in the culture medium. *Journal of Experimental Marine Biology and Ecology*, *9*, 125–136.

Myklestad, S., Djurhuus, R., & Mohus, A. (1982). Demonstration of *exo*-(β-1,3)-D-glucanase activity in some planktonic diatoms. *Journal of Experimental Marine Biology and Ecology*, *56*, 205–211.

Myklestad, S. M. (1988). Production, chemical structure, metabolism, and biological function of the (1→3)-linked, β-D-glucans in diatoms. *Biological Oceanography*, *6*, 313–326.

Nelson, T. E., & Lewis, B. A. (1974). Separation and characterization of the soluble and insoluble components of insoluble laminarin. *Carbohydrate Research*, *33*, 63–74.

Nevo, Z., & Sharon, N. (1969). The cell wall of *Peridinium westii*, a non-cellulosic glucan. *Biochimica et Biophysica Acta*, *173*, 161–175.

Olli, K. (1996). Resting cyst formation of *Eutreptiella gymnastica* (Euglenophyceae) in the northern coastal Baltic Sea. *Journal of Phycology*, *32*, 535–542.

Parke, M., Green, J. C., & Manton, I. (1971). Observations on the fine structure of zoids of the genus *Phaeocystis* (Haptophyceae). *Journal of the Marine Biological Association of the United Kingdom*, *51*, 927–941.

Paulsen, B. S., & Myklestad, S. (1978). Structural studies of the reserve glucan produced by the marine diatom *Skeletonema costatum* (Grev.) Cleve. *Carbohydrate Research*, *62*, 386–388.

Pelosi, L., Imai, T., Chancy, H., Heux, L., Buhler, E., & Bulone, V. (2003). Structural and morphological diversity of (1,3)-β-D-glucans synthesized *in vitro* by enzymes from *Saprolegnia monoica:* comparison with a corresponding *in vitro* product from blackberry (*Rubus fruticosus*). *Biochemistry*, *42*, 6264–6274.

Percival, E., & McDowell, R. H. (1967). *Chemistry and Enzymology of Marine Algal Polysaccharides*. London: Academic Press.

Percival, E., Rahman, M. A., & Weigel, H. (1980). Chemistry of the polysaccharides of the diatom *Coscinodiscus nobilis*. *Phytochemistry*, *19*, 809–811.

Pringsheim, N. (1883). Über Cellulinkörner, eine Modifikation der Cellulose in Körnerform. *Berichte der Deutschen Botanischen Gesellschaft*, *1*, 288–308.

Quatrano, R. S., & Stevens, P. T. (1976). Cell wall assembly in *Fucus* zygotes. *Plant Physiology*, *58*, 224–231.

Quillet, M. (1955). Sur la nature chimique de la leucosine, polysaccharide de réserve caractéristique des Chrysophycees, extraite d'*Hydrurus foetidus*. *Comptes Rendus de l'Académie des Sciences Paris*, *240*, 1001–1003.

Ram, S., Beyer, R., Shepherd, M. G., & Sullivan, P. A. (1981). Isolation and analysis of neutral glucans from *Ecklonia radiata* and *Cystophora scalaris*. *Carbohydrate Research*, *96*, 95–104.

Read, S. M., Currie, G., & Bacic, A. (1996). Analysis of the structural heterogeneity of laminarin by electrospray-ionisation–mass spectrometry. *Carbohydrate Research*, *281*, 187–201.

Richardson, T. L., & Cullen, J. J. (1995). Changes in buoyancy and chemical composition during growth of a coastal marine diatom: ecological and biogeochemical consequences. *Marine Ecology Progress Series*, *128*, 77–90.

Roessler, P. G. (1987). UDPglucose pyrophosphorylase activity in the diatom *Cyclotella cryptica*: pathway of chrysolaminarin biosynthesis. *Journal of Phycology*, *23*, 494–498.

Roessler, P. G. (1988). Changes in the activities of various lipid and carbohydrate biosynthetic enzymes in the diatom *Cyclotella cryptica* in response to silicon deficiency. *Archives of Biochemistry and Biophysics*, *267*, 521–528.

Round, F. E., Crawford, R. M., & Mann, D. G. (1990). *The Diatoms: Biology and Morphology of the Genera*. Cambridge: Cambridge University Press.

Sakshaug, E., & Myklestad, S. (1973). Studies on the phytoplankton ecology of the Trondheimsfjord. III. Dynamics of phytoplankton blooms in relation to environmental factors, bioassay experiments and parameters for the physiological state of the populations. *Journal of Experimental Marine Biology and Ecology*, *11*, 157–188.

Sakshaug, E., Andresen, K., Myklestad, S., & Olsen, Y. (1983). Nutrient status of phytoplankton communities in Norwegian waters (marine, brackish, and fresh) as revealed by their chemical composition. *Journal of Plankton Research*, *5*, 175–196.

Schwartzbach, S. D., Schiff, J. A., & Goldstein, N. H. (1975). Events surrounding the early development of *Euglena* chloroplasts. V. Control of paramylum degradation. *Plant Physiology*, *56*, 313–317.

Shapiro, A., & Mullins, J. T. (1997). Localization of cytoplasmic water-soluble reserve 1,3-β-glucans in *Achlya* with immunostaining. *Mycologia*, *89*, 89–91.

Shapiro, A., & Mullins, J. T. (2002a). Hyphal tip growth in *Achlya bisexualis*. I. Distribution of (1,3)-β-glucan in elongating and non-elongating regions of the wall. *Mycologia*, *94*, 267–272.

Shapiro, A., & Mullins, J. T. (2002b). Hyphal tip growth in *Achlya bisexualis*. II. Distribution of cellulose in elongating and non-elongating regions of the wall. *Mycologia*, *94*, 273–279.

Sietsma, J. H., & Wessels, J. G. H. (1979). Evidence for covalent linkages between chitin and β-glucan in a fungal wall. *Journal of General Microbiology*, *114*, 99–108.

Sietsma, J. H., & Wessels, J. G. H. (2006). Apical wall biogenesis. *Mycota*, *1*, 53–72.

Sjøtun, K., & Gunnarsson, K. (1995). Seasonal growth pattern of an Icelandic *Laminaria* population (section Simplices, Laminariaceae, Phaeophyta) containing solid- and hollow-stiped plants. *European Journal of Phycology*, *30*, 281–287.

Stone, B. A., & Clarke, A. E. (1992). *Chemistry and Biology of (1→3)-β-Glucans.* Melbourne:La Trobe University Press.

Størseth, T. R., Hansen, K., Skjermo, J., & Krane, J. (2004). Characterization of a β-D-(1→3)-glucan from the marine diatom *Chaetoceros mülleri* by high-resolution magic-angle spinning NMR spectroscopy on whole algal cells. *Carbohydrate Research*, *339*, 421–424.

Størseth, T. R., Hansen, K., Reitan, K. I., & Skjermo, J. (2005). Structural characterization of β-D-(1→3)-glucans from different growth phases of the marine diatoms *Chaetoceros mülleri* and *Thalassiosira weissflogii*. *Carbohydrate Research*, *340*, 1159–1164.

Størseth, T. R., Kirkvold, S., Skjermo, J., & Reitan, K. I. (2006). A branched β-D-(1→3,1→6)-glucan from the marine diatom *Chaetoceros mülleri* characterized by NMR. *Carbohydrate Research*, *341*, 2108–2114.

Sugiyama, J., Boisset, C., Hashimoto, M., & Watanabe, T. (1999). Molecular directionality of β-chitin biosynthesis. *Journal of Molecular Biology*, *286*, 247–255.

Taguchi, S., Hirata, J. A., & Laws, E. A. (1987). Silicate deficiency and lipid synthesis of marine diatoms. *Journal of Phycology*, *23*, 260–267.

Takenaka, S., Kondo, T., Nazeri, S., Tamura, Y., Tokunaga, M., Tsuyama, S., Miyatake, K., & Nakano, Y. (1997). Accumulation of trehalose as a compatible solute under osmotic stress in *Euglena gracilis* Z. *Journal of Eukaryotic Microbiology*, *44*, 609–613.

Tomos, A. D., & Northcote, D. H. (1978). A protein-glucan intermediate during paramylon synthesis. *Biochemical Journal*, *174*, 283–290.

Triemer, R. E. (1980). Role of Golgi apparatus in mucilage production and cyst formation in *Euglena gracilis* (Euglenophyceae). *Journal of Phycology*, *16*, 46–52.

Underwood, G. J. C., Boulcott, M., Raines, C. A., & Waldron, K. (2004). Environmental effects on exopolymer production by marine benthic diatoms: dynamics, changes in composition, and pathways of production. *Journal of Phycology*, *40*, 293–304.

Usui, T., Toriyama, T., & Mizuno, T. (1979). Structural investigation of laminaran of *Eisenia bicyclis*. *Agricultural and Biological Chemistry*, *43*, 603–611.

van Boekel, W. H. M. (1992). *Phaeocystis* colony mucus components and the importance of calcium ions for colony stability. *Marine Ecology Progress Series*, *87*, 301–305.

van den Hoek, C., Mann, D. G., & Jahns, H. M. (1995). *Algae: an Introduction to Phycology*. Cambridge: Cambridge University Press.

van Oijen, T., van Leeuwe, M. A., & Gieskes, W. W. C. (2003). Variation of particulate carbohydrate pools over time and depth in a diatom-dominated plankton community at the Antarctic Polar Front. *Polar Biology*, *26*, 195–201.

van Oijen, T., van Leeuwe, M. A., Gieskes, W. W. C., & de Baar, H. J. W. (2004a). Effects of iron limitation on photosynthesis and carbohydrate metabolism in the Antarctic diatom *Chaetoceros brevis* (Bacillariophyceae). *European Journal of Phycology*, *39*, 161–171.

van Oijen, T., van Leeuwe, M. A., Granum, E., Weissing, F. J., Bellerby, R. G. J., Gieskes, W. W. C., & de Baar, H. J. W. (2004b). Light rather than iron controls photosynthate production and allocation in Southern Ocean phytoplankton populations during austral autumn. *Journal of Plankton Research*, *26*, 885–900.

van Oijen, T., Veldhuis, M. J. W., Gorbunov, M. Y., Nishioka, J., van Leeuwe, M. A., & de Baar, H. J. W. (2005). Enhanced carbohydrate production by Southern Ocean phytoplankton in response to in situ iron fertilisation. *Marine Chemistry*, *93*, 33–52.

van Rijssel, M., Janse, I., Noordkamp, D. J. B., & Gieskes, W. W. C. (2000). An inventory of factors that affect polysaccharide production by *Phaeocystis globosa*. *Journal of Sea Research*, *43*, 297–306.

Villareal, T. A., Altabet, M. A., & Culver-Rymsza, K. (1993). Nitrogen transport by vertically-migrating diatom mats in the North Pacific Ocean. *Nature*, *363*, 709–712.

von Stosch, H. A. (1951). Über das Leukosin, den Reservestoff der Chrysophyten. *Naturwissenschaften*, *38*, 192–193.

Vårum, K. M., & Myklestad, S. (1984). Effects of light, salinity and nutrient limitation on the production of β-1,3-D-glucan and exo-D-glucanase activity in *Skeletonema costatum* (Grev.) Cleve. *Journal of Experimental Marine Biology and Ecology*, *83*, 13–26.

Vårum, K. M., Kvam, B. J., Myklestad, S., & Paulsen, B. S. (1986a). Structure of a food-reserve β-D-glucan produced by the haptophyte alga *Emiliania huxleyi* (Lohmann) Hay and Mohler. *Carbohydrate Research*, *152*, 243–248.

Vårum, K. M., Østgaard, K., & Grimsrud, K. (1986b). Diurnal rhythms in carbohydrate metabolism of the marine diatom *Skeletonema costatum* (Grev.) Cleve. *Journal of Experimental Marine Biology and Ecology*, *102*, 249–256.

Wang, M. C., & Bartnicki-Garcia, S. (1974). Mycolaminarans: storage (1,3)-β-D-glucans from the cytoplasm of the fungus *Phytophthora palmivora*. *Carbohydrate Research*, *37*, 331–338.

Wang, M. C., & Bartnicki-Garcia, S. (1980). Distribution of mycolaminarans and cell wall β-glucans in the life cycle of *Phytophthora*. *Experimental Mycology*, *4*, 269–280.

Waterkeyn, L., & Bienfait, A. (1987). Localisation et rôle des β-1,3-glucanes (callose et chrysolaminarine) dans le genre *Pinnularia* (Diatomées). *Cellule*, *74*, 199–226.

Wessels, J. G. H., & Sietsma, J. H. (1981). Fungal cell walls: a survey. In W. Tanner & F. A. Loewus (Eds.), *Encyclopedia of Plant Physiology: Vol. 13B: Plant Carbohydrates II* (pp. 352–394). Berlin: Springer Verlag.

Zevenhuizen, L. P. T. M., & Bartnicki-Garcia, S. (1970). Structure and role of a soluble cytoplasmic glucan from *Phytophthora cinnamomi*. *Journal of General Microbiology*, *61*, 183–188.

Zvyagintseva, T. N., Shevchenko, N. M., Chizhov, A. O., Krupnova, T. N., Sundukova, E. V., & Isakov, V. V. (2003). Water-soluble polysaccharides of some far-eastern brown seaweeds: distribution, structure, and their dependence on the developmental conditions. *Journal of Experimental Marine Biology and Ecology*, *294*, 1–13.

CHAPTER 4.3

# Organization of Fungal, Oomycete and Lichen (1,3)-β-Glucans

**Cecile Clavaud*, Vishukumar Aimanianda* and Jean-Paul Latge**
*Aspergillus Unit, Institut Pasteur, Paris, France*

## Introduction

The cell walls of fungi, oomycetes and lichens osmotically protect the underlying protoplasm, determine and maintain cell shape, participate in colonizing insoluble substrates, and provides resistance to environmental biotic and abiotic threats. In addition, being at the interface with the environment, the cell wall plays an important role in establishing pathogenic and symbiotic relationships with other organisms including invertebrates, vertebrates and higher plants. The presence of β-glucans in the cell wall of water moulds (oomycetes) was shown more than 100 years ago by Mangin (1890a,b) and Tanret (1897) in various ascomycetes and basidiomycetes. Since then, many reports have shown that (1,3)-β-glucans are the major structural components of the wall and play an essential biological role in its interactions with other eukaryotes. For many years, these glucans were considered to contribute to the inert exoskeleton of the cell wall. Data reviewed in this chapter show that their composition is highly dynamic, exhibiting continuous changes in location and structure, partly due to cross-linking with other wall components, during morphogenesis and host infection. Moreover, (1,3)-β-glucans are not only the major structural component of the cell wall but are also found intracellularly and secreted. The structure and role of these (1,3)-β-glucans in cellular organization are discussed in this chapter.

## I. (1,3)-β-Glucans in Fungi

### A. Fungal and Yeast (1,3)-β-Glucan Extracellular Polysaccharides

Secretion of (1,3;1,6)-β-glucans has been reported for many fungal species (Flemming et al., 2007). These secreted glucans are either linear or branched with side chains composed of a

*Cecile Clavaud and Vishukumar Aimanianda contributed equally to the Chapter.

single glucose unit to short (1,3)-β- or (1,6)-β-linked oligosaccharides. The number of glucose units between two branches also varies with fungal species (see Chapter 2.1 for examples of the composition of these secreted glucans).

When grown in liquid culture, secretion of exopolysaccharides results in increased viscosity of the culture medium. This has some biotechnological significance; for example, side-chain-branched (1,3;1,6)-β-glucan in the extracellular matrix of *Botrytis cinerea* produced during grape fermentation must be removed by (1,3)-β-glucanase to facilitate filtration of sweet Bordeaux white wines (Dubourdieu and Ribereau-Gayon, 1980). These extracellular polysaccharides exhibit a triple-helix conformation and are able to form hydrogels in water solution and nanochannel-like structures. Besides playing a role in permeability allowing the transport of water and nutrients, these polysaccharides can also protect fungi against dessication as in sclerotia and serve as a nutrient source during sclerotial germination (in a way similar to intracellular glucan reserves of oomycetes) (Iwamuro et al., 1982; Ohno et al., 1986; Latgé et al., 1988; Ohno et al., 1995; Ko and Lin, 2004). The surface mucilages have been suggested to play a role in attaching the fungus to its plant or animal host (as discussed in Chapters 4.4.5, 4.5.1 and 4.5.2). Recently, (1,3)-β-glucans have been found to be essential components of biofilms. These three-dimensional gel-like structures create a highly charged environment in which the organisms are immobilized (Flemming et al., 2000). The biofilm matrix usually encloses microcolonies, in which the individual cells are separated by water channels allowing nutrient flow (Donlan and Costerton, 2002). Although much is known about bacterial biofilms, very little is understood about biofilms produced by fungi. The pathogenic fungus *Candida albicans* produces biofilms whose matrix consists of about 40% of carbohydrate. The matrix is unaffected by treatment with proteinase K, chitinase and β-*N*-acetylglucosaminidase, whereas lyticase with highly active (1,3)-ß-glucanase resulted in an 85% decrease in optical density, indicating that the biofilm matrix was mainly composed of (1,3)-β-glucans (Al-Fattani and Douglas, 2006; Nett et al., 2007). This (1,3)-β-glucan-rich matrix (structure has not been investigated in detail) limits the penetration of antimicrobials into cell cytoplasm, thus increasing the biofilm-associated resistance to several drug classes (Douglas, 2002). (1,3)-β-Glucans, however, are not a part of all fungal biofilms; for example, biofilm matrices from *Aspergillus fumigatus* and *Cryptococcus neoformans* do not contain (1,3)-β-glucans.

## B. (1,3)-β-Glucans in fungal cell walls

The cell wall contributes 15–25% of the cell dry mass and it is estimated that about 80–95% of yeast and mould cell wall material is composed of neutral sugars (Latgé and Calderone,

2005; Lesage and Bussey, 2006). Branch-on-branch (1,3;1,6)-β-glucans and chitin (a 1,4-*N*-acetyl glucosamine polymer) are the major cell wall polysaccharides, but other polymers of glucose, mannose, galactose, galactosamine, glucosamine, xylose, fucose, hexuronic acids, and *N*-acetylglucosamine can also be found. The (1,3;1,6)-β-glucan–chitin core is covalently associated with other cell wall polysaccharides and their composition varies with the fungal genus or class. The cell wall components can be further separated into fibrillar and amorphous fractions (Kopecka et al., 1995; Osumi, 1998a). Branch-on-branch (1,3;1,6)-β-glucans are found mainly in the fibrillar cell wall fraction as an alkali-insoluble component. However, it should be emphasized that cell wall composition has been examined in detail in very few yeast or mould species (Table 1).

### 1. *Isolation of (1,3)-β-glucans from cell walls*

To analyse different components of the insoluble wall, it is essential to solubilize the wall into different fractions (Fleet, 1991). Since the early work of Tanret (1897), chemical fractionation of the wall structure has involved a series of sequential hot buffer, hot-alkali and/or cold/hot-acid treatments. In *S. cerevisiae*, wall extraction using alkali/acid treatment (Manners and Meyer, 1977) resulted in three fractions: (i) an alkali-soluble fraction containing mannan, (1,3)-β-glucan and a small proportion of (1,6)-β-glucan, (ii) an alkali-insoluble and acid-soluble (1,6)-β-glucan, and (iii) an alkali/acid-insoluble fraction containing (1,3)-β-glucan linked to chitin. In addition, Manners and Meyer (1977) showed that *S. pombe* cell wall, which is devoid of chitin, is sensitive to hot alkali, so that dilute alkali at 4°C could be used to obtain alkali-soluble and -insoluble fractions. The alkali-soluble fraction was a mixture of polysaccharides including galactomannan, (1,3)-α-glucan and a branched (1,3)-β-glucan, whereas the alkali-insoluble fraction consisted of highly branched (1,3;1,6)-β-glucan. Nisini et al. (2007) determined the chemical composition of β-glucan in *C. albicans* after extracting the walls

**Table 1: Chemical composition of the cell wall of some yeasts and fungi**

| Polysaccharide (%) | Saccharomyces cerevisiae[a] | Schizosaccharomyces pombe[b] | Aspergillus fumigatus[c] | Candida albicans[d] |
|---|---|---|---|---|
| (1,3;1,6)-β-Glucan | 50–55 | 48–54 | 20–35 | 40 |
| (1,6)-β-Glucan | 5–10 | 2–4 | – | 20 |
| Chitin | 1–2 | – | 7–15 | 1–2 |
| Mannoprotein/ galactomannan | 35–40 | 9–14 | 20–25 | 35–40 |
| (1,3)-α-Glucan | – | 28–32 | 25–40 | – |

[a]Klis et al., 2002; [b]Manners and Meyer, 1977, Kopecka et al., 1995; [c]Fontaine et al., 2000, Maubon et al., 2006; [d]Klis et al., 2001.

with 1% NaOH at 100°C for 24h followed by a second extraction of the insoluble material with 0.5M acetic acid at 80°C. According to nuclear magnetic resonance (NMR) analysis this β-glucan consisted of (1,3)-β- and (1,6)-β-linked polymers in the ratio of ~1:1, which was further confirmed by the selective enzymatic digestion of the β-glucan.

Cell wall fractionation by chemical methods could cause degradation of the constitutent polymers, a and combination of enzymatic and mild chemical treatments is the most appropriate method for such fractionation (Ip et al., 1992; Hong et al., 1994; Ram et al., 1994; Magnelli et al., 2002; Magnelli et al., 2005). In addition, use of specific hydrolytic enzymes allows direct quantification of the component (1,3)-β-glucans, (1,6)-β-glucans or chitin. Preparations such as Zymolyase 100T (Perez and Ribas, 2004) used in wall analyses are a complex mixture of enzymes that favour complete solubilization of the wall components. More recently, recombinant enzymes have been used for wall analysis. Quantazyme is one such commercial recombinant endo-(1,3)-β-glucanase that binds to pentaoligosaccharide at its catalytic site. LamA (laminarinase), another recombinant (1,3)-β-glucanase from *Thermotoga neapolitana* expressed in *Escherichia coli* host cells, is highly thermostable (Zverlov et al., 1997). An endo-(1,6)-β-glucanase of *Trichoderma harzianum* expressed in *Pichia pastoris* releases gentiooligosaccharides from pustulan as well as *S. cerevisiae* cell wall (Bom et al., 1998). A chitinase from *Serratia marcescens* releases chitobiose as primary product from the cell wall chitin (Vorgias et al., 1992). However, the enzyme is sometimes purified in association with *N*-acetylhexosaminidase and hence the final product released is *N*-acetylglucosamine (Roberts and Cabib, 1982). Recombinant (1,3)-α-glucanase (mutanase) is also available from the mycoparasitic fungus *Trichoderma harzianum* or a strain of *Paenibacillus* (Fuglsang et al., 2000; Grun et al., 2006; Sumitomo et al., 2007). These recombinant enzymes are essential in the stepwise quantification of different polysaccharides in the wall structure. Most often, they are used sequentially; for example, Magnelli et al. (2005) performed a series of three enzymatic digestions – recombinant chitinase, (1,6)- β-glucanase and (1,3)-β-glucanase) – and one-step acid hydrolysis to determine the cell wall composition in *S. pombe*.

The protocol shown in Fig. 1 represents a summary of many cell wall fractionation protocols reported in the literature, and seems the most appropriate to date, with the following steps: (1) removal of non-covalently associated components (mainly proteins) from the cell wall fraction by treatment with a detergent under reducing conditions; (2) separation of alkali-soluble and -insoluble fractions; (3) fractionation of the respective fractions using recombinant enzymes; and (4) characterization of the different soluble oligosaccharides by classical carbohydrate analytical chemistry. Most of the studies have been undertaken with wall extracts from disrupted cells. Undisrupted entire cells can also be used advantageously: upon repeated

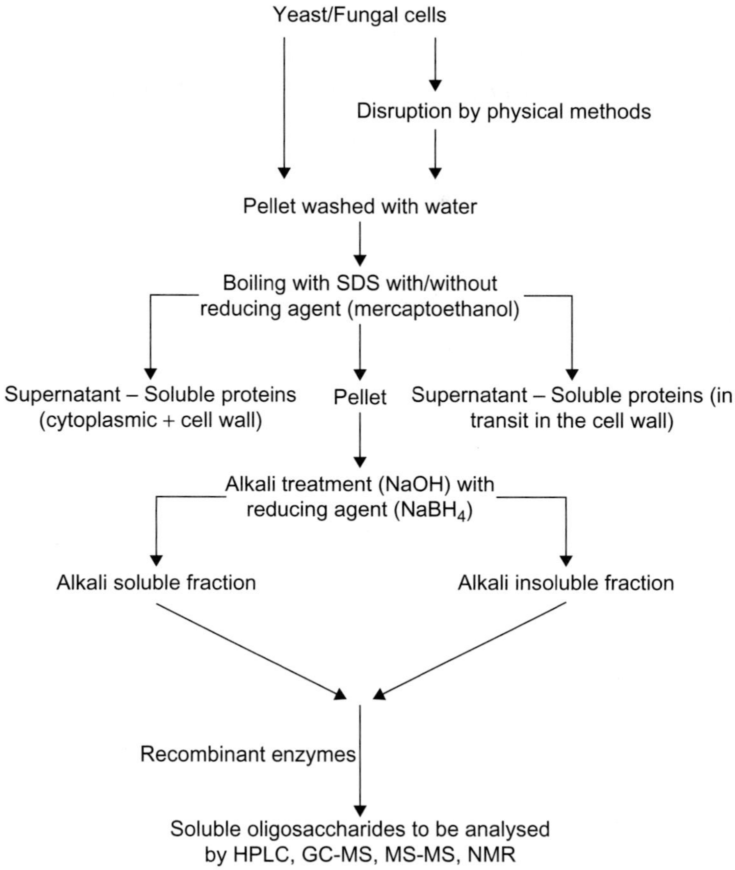


**Fig. 1: General protocol for the extraction of yeast and fungal wall components.**

treatments similar to that outlined in Fig. 1, the whole cells produce 'cell ghosts', maintaining the native structure of the cell wall that could be used further for analysis of the role of wall polysaccharides.

## 2. *Structure and molecular organization of (1,3)-β-glucans in yeast and mould cell walls*

*Cell wall (1,3;1,6)-β-glucan has a branch-on-branch structure* In *S. cerevisiae*, the average length of an individual glucan chain chemically isolated from cell wall is about 1500 glucose residues, corresponding to a molecular mass of $2.4 \times 10^5$ Da and accounting for a fibre length of approximately 600 nm (Muller et al., 1997). The length of (1,3)-β-glucan fibres corresponds to three to six times the wall thickness. Examination of intact yeast cells by solid state NMR confirmed the existence of (1,3)-β-glucans mostly in the helical conformation, either as a single chain or in the form of a triple helix in which individual chains are connected by hydrogen bonds (Williams et al., 1991). Under electron microscopy, the fibres showed a diameter of 20–60 nm, having lateral association with the other chains that were 0.5–1.0 nm long (Kopecka and Kreger, 1986). The (1,3)-β-glucans belong to a family of helical glycans, having a shape comparable to that of a flexible wire spring that can be elongated to different extents depending on the state, which explains the elasticity as well as tensile strength of the wall (Rees et al., 1982; Klis et al., 2002) (see also Chapter 2.2). Physicochemistry X-ray diffractograms of isolated *S. cerevisiae* wall showed low crystallinity of the (1,3)-β-glucan network, which is attributed to the (1,6)-β-branching of the glucans (Manners et al., 1973). In *S. cerevisiae*, the (1,3)-β-glucan of the wall is branched with 3% interchain (1,6)-β-linkages (Kollar et al., 1997), whereas in *A. fumigatus* interchain (1,6)-β-linkages account for nearly 4% (Fontaine et al., 2000), but the crystallinity properties of the *A. fumigatus* β-glucans have not been investigated. In *S. pombe*, 48–54% of the cell wall polysaccharide is (1,3;1,6)-β-glucan, having 2–4% (1,6)-β-branches (Bush et al., 1974; Manners and Meyer, 1977; Kopecka et al., 1995). Lowman et al. (2003) observed higher levels of (1,6)-β-branching in *C. albicans* compared to *S. cerevisiae.* In *C. albicans*, the degree of polymerization of (1,3)-β-glucan was slightly higher in the blastospores (yeast form) compared to the hyphal structure, but in both forms it was significantly lower than in *S. cerevisiae*.

*Branch-on-branch (1,3;1,6)-β-glucans are linked to other polysaccharides* (1,3)-β-Glucans neosynthesized at the growing apex in both yeast and fungi (Beauvais et al., 2001) are alkali soluble and associated to wall flexibility. In *S. cerevisiae*, the catalytic and regulatory subunits of the (1,3)-β-glucan synthase complex, Fks1p and Rho1p, migrate with the cortical actin patches to the cell surface at the site of polarized growth, and any defect in Fks1p movement results in the uneven thickening and thus fragility of the wall, eventually leading to cell lysis (Utsugi et al., 2002). Genetic and chemical/genetic interaction screening studies identified 32 core genes (including *FKS1/FKS2*, the glucan synthesis regulator *KNR4* and a (1,3)-β-glucanosyltransferase *GAS1*) involved in both wall assembly and polarized growth in *S. cerevisiae* (Lesage et al., 2004). Direct genetic interactions were found between genes involved in (1,3)-β-glucan synthesis and polarized growth. In *S. pombe*, (1,3)-β-glucan synthesis, as indicated by location of Bgs4p (one of the (1,3)-β-glucan synthase catalytic subunits), occurs at growing poles. Cortes and coworkers also observed cell lysis upon deletion of *BGS4*, which indicates the essential role of (1,3)-β-glucans at the growing region (Cortes et al., 2002).

Covalent linkages between (1,3)-β-glucans and chitin give the wall rigidity. As a result, fungal cells can withstand turgor pressure changes. It is the associations between

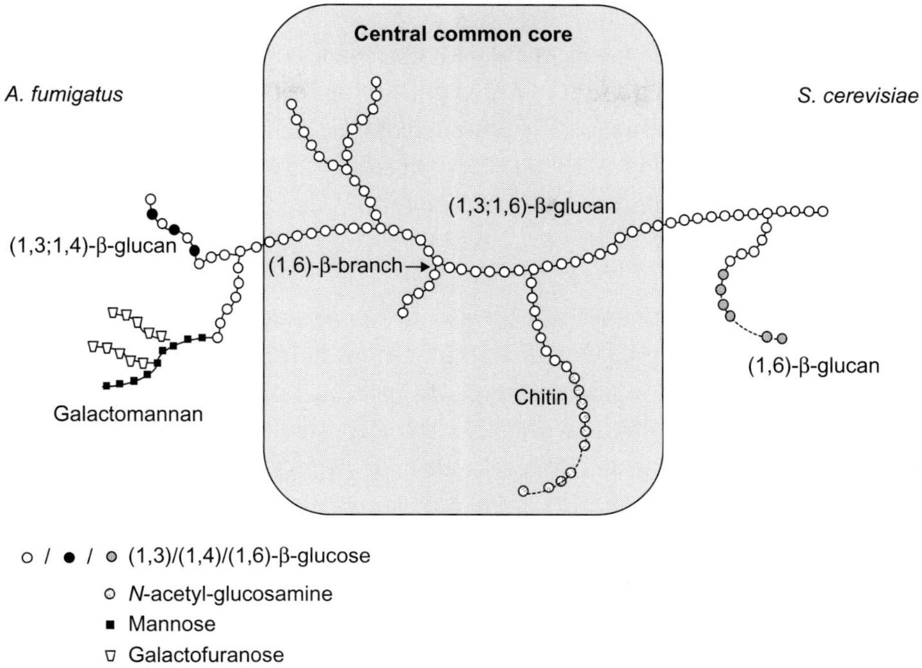


**Fig. 2: Comparison of the wall composition of *Aspergillus fumigatus* and *Saccharomyces cerevisiae*. In both *A. fumigatus*, and *S. cerevisiae* the branch-on-branch (1,3;1,6)-β-glucan central core is covalently attached to chitin through a (1,4)-β-linkage. Associated polysaccharides bound to the glucan–chitin complex differ: (1,3;1,4)-β-glucan and galactomannan are present in *A. fumigatus* whereas (1,6)-β-glucan is present in *S. cerevisiae*.**

(1,3;1,6)-β-glucans and other wall polysaccharides that are responsible for the shape of the cell, and they are markedly different in composition among different fungal and yeast species. In contrast with the site of (1,3)-β-glucan neosynthesis, the location where these cross-linking events occur is unknown. Fig. 2 shows the different linkages occurring in the walls of the yeast *S. cerevisiae* and the filamentous fungus *A. fumigatus*.

In *S. cerevisiae*, (1,3)-β-glucans are linked to chitin (contributing 2–4% of the total cell wall) through (1,4)-β-glycosidic linkages (Kollar et al., 1997). This linkage makes the chitin–glucan complex alkali insoluble. (1,3;1,6)-β-Glucans are also bound to (1,6)-β-glucan chains which account for 12% of the polysaccharides. (1,6)-β-Glucans have an average degree of polymerization of 140–350 glucose residues (Manners et al., 1973; Kollar et al., 1997) and play a critical role in interconnecting the glucan network with all other wall components into a lattice structure (Kapteyn et al., 1999). In *C. albicans*, the alkali/acid-insoluble glucan fraction consisted of hexoses and hexosamine with glucose accounting for ~93% and 3–4% glucosamine; however, the binding pattern of the glucans and chitin has not been investigated in detail, although some studies suggest the presence of specific binding between glucan and chitin (Gopal et al., 1984). Iorio et al. (2008) suggested that the (1,3)-β-glucan chains are branched directly on a (1,6)-β-glucan backbone, the individual (1,3)-β-glucans chains being held together by hydrogen bonds (Klis et al., 2001).

In *Schizophyllum commune* and *Agaricus bisporus*, nearly all wall (1,3;1,6)-β-glucans are insoluble due to their covalent linkage with chitin, whereas in *Neurospora crassa*, *Aspergillus nidulans* and *Coprinus cinereus*, a major portion of (1,3;1,6)-β-glucan in the alkali-insoluble fraction is dimethyl sulfoxide soluble, with only a small residue of resistant (1,3;1,6)-β-glucan bound to chitin (Sietsma and Wessels, 1981). In *A. fumigatus*, as in *S. cerevisiae*, the branch-on-branch (1,3;1,6)-β-glucan central core and chitin are attached covalently through (1,4)-β-linkages (Latgé and Calderone, 2005). This glucan–chitin complex is in turn linked to a (1,3;1,4)-β-glucan (with [3Glcβ1-4Glcβ1] repeating units), which accounts for nearly 10% of the wall β-glucan, and to galactomannan (Fontaine et al., 2000). *A. fumigatus* contains a higher content of chitin (10%) than *S. cerevisiae* but is devoid of (1,6)-β-glucan (Fontaine et al., 2000). In contrast to yeast, mannans do not occur as mannoproteins or peptidomannans but are covalently bound to short chains of (1,3)-β-glucans and galactans (Stokke et al., 1993; Kath and Kulicke, 1999; Fontaine et al., 2000). The galactomannan in *A. fumigatus* is composed of a linear α-mannan having a repeating mannose oligosaccharide unit (6Manα1-2Manα1-2Manα1-2Manα1) with short chains of (1,5)-β-galactofuranose residues (Latgé et al., 1994). Additionly, there is a fibrillar branch-on-branch (1,3;1,6)-β-glucan complex embedded in an amorphous matrix consisting of (1,3)-α-glucans and galactomannan (Beauvais et al., 2007).

***(1,3;1,6)-β-Glucan-anchored proteins*** After removal of non-covalently bound proteins, covalently bound wall proteins can be released either by the degradation of polysaccharides with specific hydrolytic enzymes, such as (1,3)-β-glucanases, (1,6)-β-glucanases and chitinases, mild alkali treatment or anhydrous hydrofluoric acid/pyridine treatment (Fig. 3) (de Groot et al., 2004).

In *S. cerevisiae*, two categories of proteins are covalently bound to the core (1,3;1,6)-β-glucan. In the first group are the PIR (putative proteins with internal repeats) proteins that are linked to the (1,3;1,6)-β-glucan by alkali-labile ester linkages (Mrsa et al., 1997; Mrsa and Tanner, 1999; Ecker et al., 2006). Treatment with 30 mM sodium hydroxide in ice/4°C for

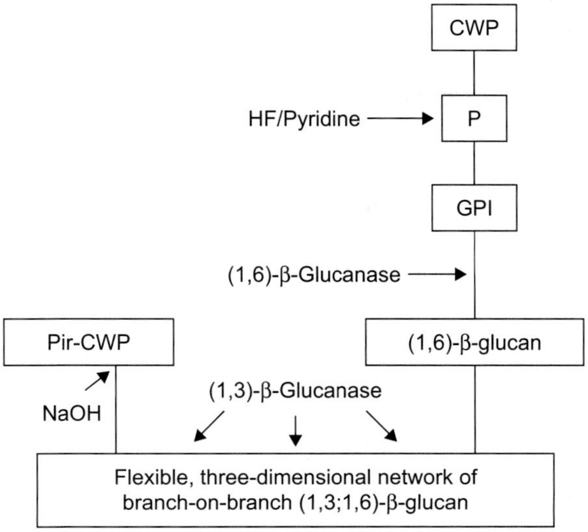


**Fig. 3: Schematic representation of covalently bound wall proteins (CWP) in yeast (de Groot et al., 2004) and methods for their extraction. CWP are anchored in two different ways: GPI-proteins (GPI-P) which are anchored by a glycosylphosphatidyl inositol (GPI) group are immobilized in the cell wall with their C-termini bound to a (1,6)-β-glucan linker coupled to (1,3)-β-glucan; proteins with internal repeat (Pir-P) are directly associated to (1,3:1,6)-β-glucan through an alkali-sensitive linkage. HF=hydrofluoric acid.**

12 h releases such proteins (Mrsa et al., 1997). In *S. cerevisiae* there are five PIR proteins (Pir1p–Pir4p and YJL 160C) with 1–10 repeats, encoded by genes that do not have a glycosylphosphatidylinositol (GPI)-anchor signal at their C-terminus (de Groot et al., 2005). They are released by (1,3)-β-glucanases, but not (1,6)-β-glucanases. The PIR proteins are synthesized as pre-pro-peptides; the pro-part is cleaved by serine protease in the Golgi apparatus. The ester linkage is between the carboxyl group of Gln74 in the repetitive sequence $QIGDGQ_{74}VQ$ of Pir4p and a hydroxyl group on a glucose moiety of (1,3;1,6)-β-glucan (Ecker et al., 2006). The linkage formation between PIR proteins and (1,3;1,6)-β-glucan is an extra-cytoplasmic phenomenon. Ecker et al. (2006) speculated that the energy released during hydrolysis of the amide bond on the Pir4p-Gln74 residue is sufficient for the formation of such an ester linkage. Though there is speculation, direct evidence is lacking to show that PIR proteins participate in linking (1,3;1,6)-β-glucan chains. PIR proteins or proteins extracted with a low concentration of alkali are found in other yeasts but are absent in *A. fumigatus* (Klis et al., 2007).

The second category of proteins bound covalently to the (1,3;1,6)-β-glucan complex are the GPI-anchored proteins which are linked through (1,6)-β-glucan (Montijn et al., 1994; Mrsa et al., 1997; Dijkgraaf et al., 2002). These proteins are initially synthesized as GPI-modified polypeptides that undergo trans-mannosylation involving hydrolysis of the oligomannosyl moiety from the GPI, followed by transfer to (1,6)-β-glucans (Kollar et al., 1997; Fujii et al., 1999). These GPI-anchored cell wall proteins (GPI-CWPs) are immobilized with their C-termini attached to (1,6)-β-glucans, which in turn are coupled to (1,3;1,6)-β-glucan (GPI-CWP→(1,6)-β-glucans→(1,3)-β-glucans) (Kapteyn et al., 1995; Kapteyn et al., 1996; van der Vaart et al., 1996). In *S. cerevisiae, in silico* analysis has identified 60–70 GPI-anchored CWP (de Groot et al., 2003); however, only half of them are supposed to be bound to (1,3;1,6)-β-glucans (Pittet and Conzelmann, 2007). Similar GPI-anchored wall proteins have been identified in many yeasts including *Candida glabrata* (Weig et al., 2004), *Yarrowia lipolytica* (Jaafar and Zueco, 2004), *S. pombe* (de Groot et al., 2007), *C. albicans* (de Groot et al., 2004) and *Paracoccidioides brasiliensis* (da Silva Castro et al., 2008). In yeasts, all GPI-CWPs are adhesins. The expression patterns of many genes encoding (1,3)-β-glucan-bound adhesins are regulated in response to growth or environmental changes (Huang et al., 2003). In *S. cerevisiae*, activation of the mating pathway induces expression of the agglutinin genes *AGA1*, *AGA2* and *SAG1/AGα1*. Pseudohyphae, invasive growth differentiation, formation of colonies, and biofilms modulate the expression of the flocculin genes *FLO1*, *FLO5*, *FLO9*, *FLO10* and *FLO11* (Dranginis et al., 2007). *EPA* gene family members code for a galectin class of adhesins that are found in *C. glabrata* while *ALS* codes for hydrophobic cell surface adhesins found in *C. albicans*.

The situation is different in filamentous fungi such as *Aspergillus*, where no GPI-CWPs have been identified in the mycelium either biochemically or through bioinformatic analysis (Latgé and Calderone, 2005). Absence of GPI-CWPs is consistent with the lack of (1,6)-β-glucan in their wall structures (Bruneau et al., 2001; de Groot et al., 2004; Latgé et al., 2005). This suggests that, in contrast to the current hypothesis, GPI-anchored proteins do not play any linker role in the three-dimensional organization of the fungal wall and are not essential for cell wall integrity (Latgé, 2007). In *A. fumigatus*, only the rodlet proteins encoded by the RodA gene have been found to be covalently bound to the wall but the nature of the anchoring polysaccharide and linkage pattern remains unknown (Aimanianda et al., submitted).

### 3. Structural variations

***Location of (1,3)-β-glucan in the cell wall*** The location of different polysaccharides in the fungal wall, in contrast to plant walls, is poorly characterized. The dye, Calcofluor White, which binds strongly to fibrillar polysaccharides has been used intensively for fluorescence microscopy in yeast studies (Martin-Cuadrado et al., 2003); however, it lacks specificity and binds a number of fibrillar polysaccharides including chitin. Aniline Blue has also been used for identification of the (1,3)-β-glucan synthesis site but not to localize this polysaccharide in a cell wall (Beauvais et al., 2001) (see Chapter 2.1). Localization of the different cell wall layers requires the use of electron microscopy. A simple chemical fixation preceded by different glucanase treatments of the wall can identify different specific structures (Kopecka et al., 1995). Carbon-platinum replicas can also be used to visualize the fibrillar or amorphous nature of the (1,3)-β-glucans (Latgé et al., 1984). A precise location of the different wall components *in situ* requires the use of immunocytochemical methods (Osumi, 1998a). Several antibodies to (1,3)-β-glucans are available and different types of β-glucan in the *S. pombe* wall have been located using colloidal gold-coupled monoclonal antibodies and immuno-electron microscopy (Humbel et al., 2001). Monoclonal antibodies recognising specifically (1,3;1,4)-β-glucan (Meikle et al., 1994), (1,3)-β-glucan (Meikle et al., 1991), (1,3;1,6)-β-glucan (Kondori et al. 2008) and dectin-1 (Adachi et al., 2004), a vertebrate (1,3)-β-glucan-binding protein (see Chapters 3.2 and 4.5.2) have also been used for the location of (1,3)-β-glucans (Fig. 4) (Latgé, unpublished).

In *S. cerevisiae*, in association with chitin, (1,3;1,6)-β-glucans form an inner fibrillar layer and, through (1,6)-β-glucosyl linkages, are attached to the mannoproteins that form the outer layer of the wall (Cabib et al., 2001; Grün et al., 2003), and (1,3)-β-glucans are only present at the surface of the wall at the bud scar (the site of separation of the mother and daughter cells) (Gantner et al., 2005; Latgé, 2007). In the fission yeast, *S. pombe*, the wall consists of an inner layer of (1,3)-α-glucans, a median layer of branch-on-branch (1,3;1,6)-β-glucan and (1,6)-β-glucan, and an electron-dense outer layer of galactomannan and galactomannoprotein (Humbel et al., 2001). In *A. fumigatus*, the (1,3)-β-glucan homogeneously spans the mycelial wall (Latgé, unpublished).

The location of (1,3)-β-glucan also varies with the age of the cell. (1,3)-β-Glucan synthesis occurs at the apex (hyphal tip) in an unorganized form, whereas linking with other polysaccharides occurs in the subapical region and is responsible for the rigidity of the wall (Sietsma and Wessels, 1990). The distribution of (1,3)-β-glucan as well as its structure and linkage with other polysaccharides vary depending on the morphological changes that occur during the yeast/fungal cell cycle; for example, during monopolar or bipolar growth and cell division during asexual growth, mating and sporulation (Perez and Ribas, 2004). The kinetics and chemistry of these changes have not been characterized.

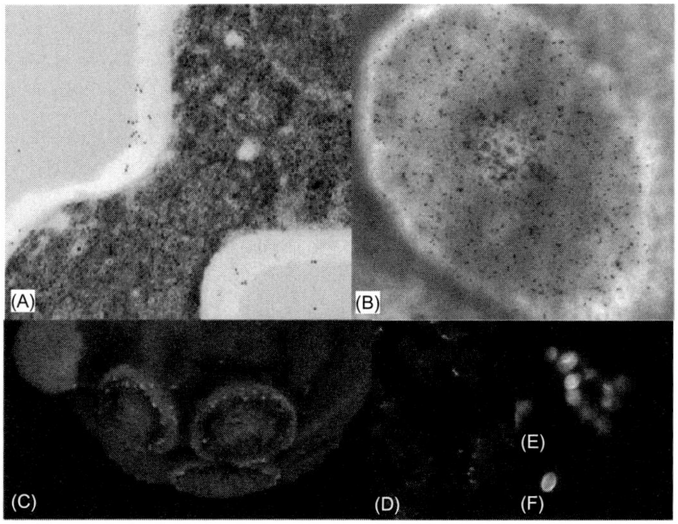


**Fig. 4: Labelling of (1,3)-β-glucans using Dectin1-Fc chimeric protein, post-labelled with anti-human Fc conjugate coupled to fluorescein isothiocyanate (D–F) or to colloidal gold (A–C). Post-embedding labelling on thin sections of *S. cerevisiae* yeasts (A) and zymosan (a yeast cell wall preparation) (B). Note the intense labelling of the zymosan in contrast to the poor labelling of the intact yeast. (C) SEM labelling of yeast bud scars. Note the emerging bud at the left is not labelled. (D–F) Immunofluorescent labeling of intact yeasts (D, poorly reactive), zymosan (E) and NaOH-treated yeast (F) (extremely reactive) (Latgé et al., unpublished).**

*Composition of the septum and the cell wall compared* In yeasts, the septum constitutes a division site. During cytokinesis, daughter cells separate physically and become two independent entities. In yeasts such as *S. pombe* this process is coupled with the synthesis of a separation septum which is further cleaved by septation. In moulds, septa are not associated with cell separation, being rather used as pillars able to substain the long mycelial threads that would collapse without them. Because of its different cellular function, it can be expected that the septum has a composition different from the external cell wall. Accordingly, septa are more brightly stained with Calcofluor White than the cellular wall, in both yeasts and moulds. The composition of the septum has not been analysed chemically since it accounts for a small percentage of the total cell wall mass (4% in *A. fumigatus* for example) and cannot be separated from the main cell wall to undertake a chemical anlysis. Accordingly, the septum could only be investigated using immunocytochemical techniques. *S. pombe* is one of the few fungal species in which the septum has been thoroughly analysed using such techniques (Fig. 5). Humbel et al. (2001) observed three layers in the septum (two electron-dense layers separated by a less dense

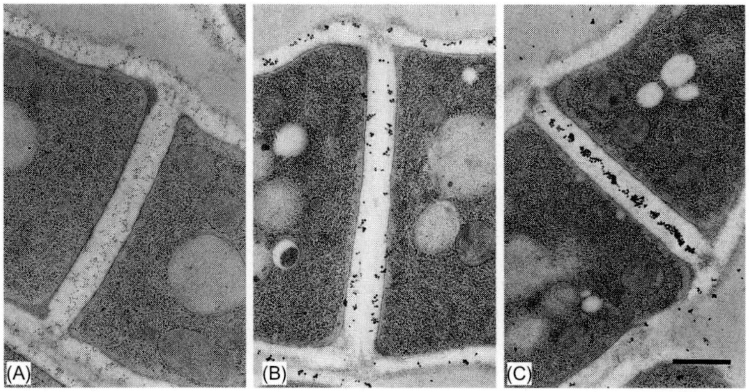


**Fig. 5: Localization by immuno-electron microscopy of (1,3)- and (1,6)-β-glucans in the septum of *S. pombe* cells. Thin sections were labelled for branch-on branch (1,3;1,6)-β-glucan (A), (1,6)-β-glucan (B) and linear (1,3)-β-glucan (C), and detected with 10 nm (A) and silver-enhanced ultra-small gold particles (B,C). The branch-on branch (1,3;1,6)-β-glucan is located in both the primary and secondary septum. (1,6)-β-glucan, however, is present only in the secondary septum, and linear (1,3)-β-glucan is exclusively located in the primary septum. Reproduced from Humbel et al. (2001) by permission of John Wiley & Sons Ltd.**

layer). In the secondary septa they reported the presence of a (1,6)-β-branched-(1,3)-β-glucan and (1,6)-β-glucan. The linear (1,3)-β-glucans were present only at the inner primary septum of *S. pombe* but not in the yeast wall. On the edge of the septum, (1,3)-α-glucans cover the (1,3)-β-glucan layer. Linear (1,3)-β-glucan appears in the early stages of septum formation, persists until completion of the septum and disappears just before cell division, confirming a role played by (1,3)-β-glucan during septum formation and separation of the daughter cells (Alonso-Nunez et al., 2005).

*Comparison of wall composition between Conidia and Hyphae* Transmission electron microscopic studies showed that the morphology of the conidial and mycelial cell wall is usually very different (Schmid et al., 2001; Palleschi et al., 2005; Bocchinfuso et al., 2008), but very few chemical comparative studies between the conidial and mycelial wall have been performed. The (1,3)-β-glucan content in *A. fumigatus* conidia and hyphal walls is similar: 38% compared with 31%, respectively (Maubon et al., 2006). However, labelling with a specific monoclonal antibody or soluble recombinant dectin-1 polypeptide indicated that (1,3)-β-glucans are distributed differently (Hohl et al., 2005; Gersuk et al., 2006): they are present on the surface of germinating conidia and hyphae but not on resting conidia, where they are covered by three different layers, from the inner to the outer, composed of (1,3)-α-glucan, melanin and hydrophobic proteins (Paris et al., 2003).

*Dimorphic forms* Most human fungal pathogens exist both in the unicellular (*in vivo*) and mycelial (*in vitro*) form, and the morphological changes between the two states depend on the environmental conditions. For example, *Histoplasma capsulatum*, *P. brasiliensis*, *Blastomyces dermatitidis*, *Coccidioides immitis* and *Cryptococcus neoformans* show a saprophytic mycelial form at ambient temperature (soil environment) and a parasitic yeast (or spherule) form at mammalian body temperature (Rappleye et al., 2007). In contrast, hypha is the invading form of *Candida albicans*.

In *P. brasiliensis*, dimorphism depends on the thermal conditions: the yeast form exists at 37°C and mycelial form at lower temperatures. The yeast form has a 200–600-nm-thick wall with three layers, (1,3)-β-glucan being present at the inner surface together with chitin, whereas the wall of the mycelial form contains a single layer of 80–150 nm thickness, composed of (1,3)-β-glucan and chitin. (1,3)-α-Glucan accounts for 40% of the wall of the yeast form whereas the mycelial form is richer in (1,3)-β-glucan (Kanetsuna and Carbonell, 1970). In the mycelial form the (1,3;1,6)-β-glucan is a branch-on-branch molecule with a few (1,6)-linkages and is bound to chitin. There is thermal dependency of (1,3)-β-glucan synthesis during transition from the yeast to the mycelial form. On transfer from 37°C to 20°C (yeast to mycelial form), there is a decrease in the incorporation of $^{14}$C-labelled substrate into (1,3)-α-glucan with a concomitant increase in its incorporation into (1,3)-β-glucan (Kanetsuna et al., 1972).

In *B. dermatitidis*, the major glucan in the mycelial form is a (1,3)-α-glucan (60%) with 40% glucan occurring as (1,3)-β-glucan, whereas the yeast form contained 95% (1,3)-α-glucan and only 5% of (1,3)-β-glucans, respectively (Kanetsuna and Carbonell, 1971). Although the level of (1,3)-α-glucan in both forms was lower compared to *P. brasiliensis*, thermal transformation obviously resulted in the increased level of (1,3)-β-glucans even in the mycelial form of *B. dermatitidis*.

The wall structure of *C. albicans* and polysaccharide components at the blastospore (budding yeast) and hyphal stages was analysed. Filamentous form walls contained 10–20% more (1,3)-β-glucan and 20–30% less (1,6)-β-glucan compared to blastospores (Adt et al., 2006). This observation is in accordance with the findings of Lowman et al. (2003) and Miura et al. (2003) who reported 73% (1,6)-β-branching in blastospores compared to 30% branching in hyphae. In *C. albicans* dimorphism is not only associated with changes in wall (1,3)-β-glucan content or composition but changes in the location of the (1,3)-β-glucan layer in the wall. In yeast form, budding and cell separation create permanent scars where β-glucans are exposed to the cell surface, while during filamentous growth no cell separation or subsequent β-glucan exposure occurs (Gantner et al., 2005).

The mycelium-to-yeast conversion in *Sporothrix schenckii* depends on the availability of $CO_2$ in addition to the temperature (Travassos and Lloyd, 1980; Rodriguez-Del Valle et al., 1983). The yeast form appears at 37°C in the presence of 5% $CO_2$, whereas it occurs in the mycelial form at a lower temperature and lower $CO_2$ content. In addition to chitin, glucans having (1,3)-β-, (1,4)-β-, and (1,6)-β-linkages are present in the cell wall of *Sporothrix schenckii* (Travassos and Lloyd, 1980) without any obvious quantitative difference between the dimorphic forms, so there is no evidence that (1,3)-β-glucan influences cell morphology in *S. schenckii*.

*Environmental conditions* Earlier, it was postulated that wall composition would not vary with the nutrients in the environment owing to its function as a skeleton in the wall structure. However, recent studies showed that there was an approximately 1.5-fold decrease in the (1,3)-β-glucan:mannan ratio in the wall when *S. cerevisiae* was grown in a synthetic minimal medium compared to a rich yeast-peptone-dextrose medium. In addition to the effect of nutrients, the physico-chemical parameters of the culture played an essential role. There was a 40% decrease in the (1,6)-β-glucan level when the pH of the culture medium rose from 4.0 to 6.0. Similarly, there was a significant increase in (1,6)-β-glucan when the temperature rose from 22 to 37°C (Aguilar-Uscanga and François, 2003; Aguilar-Uscanga et al., 2007). The molecular size of the (1,3)-β- glucan molecules was also dependent on the environmental conditions such as growth phase and carbon source (Cabib et al., 1998). The origin of the strain is also important; for example, while studying wall polysaccharide composition and sensitivity to (1,3)-β-glucanase, three *S. cerevisiae* strains from different origins (bakers' yeast and two yeasts isolated from agave), when grown in a media containing agava juice, showed changes in (1,3)-β-glucan composition (Aguilar-Uscanga et al., 2007). The ratio of (1,3)-β-glucan/(1,3)-α-glucan also varies when *A. fumigatus* is grown in defined Brian medium or yeast-extract-based medium (Beauvais, unpublished).

*Antifungal compounds* When the structure of cell wall (1,3;1,6)-β-glucan complex is changed, the rigidity of the wall is altered. Such a change could be achieved by using antifungal agents; for example, addition of Congo Red, which blocks interactions between the helical glucan strands, to cultures of *S. cerevisiae* cells results in a loss of rigidity (Kopecka and Gabriel, 1992). Treatment of *Aspergillus* and *Candida* spp. with echinocandins (that inhibit (1,3)-β-glucan synthase activity) induces swelling and lysis of the cell at the site of active wall synthesis (Odds et al., 2003; Nosanchuk, 2006). Addition of subinhibitory concentrations of cell wall polysaccaharide inhibitors to the medium modifies the location and composition of (1,3)-β-glucans. *A. fumigatus* growing in the presence of subinhibitory concentrations of caspofungin (a (1,3)-β-glucan synthase inhibitor) displayed a higher concentration of (1,3)-β-glucan in the outer layer of the mycelial wall (Hohl, 2008). *Candida* isolates that are able to grow in the presence of caspofungin concentrations above the minimum inhibitory concentration showed an 18% decrease in (1,3)-β-glucan content compared to the cell wall of the same strain grown in the absence of antifungal agent (Stevens et al., 2006), and this decrease was accompanied by a 10-fold increase in chitin content.

*Cell wall mutants* Mutant libraries have been very useful to correlate cell wall changes with the putative function of the cell wall genes. For example, though no enzymatic evidence exists, chemical analysis of the cell wall mutants supports the candidacy of two gene products, namely CRH1p and CRH2p, as transglycosidases involved in the cross-linking of (1,3;1,6)-β-glucan and chitin. Crh1p and Crh2p are GPI-anchored proteins (Hamada et al., 1998) localized in the site of polarized growth (Rodriguez-Pena et al., 2000). Cabib et al. (2007) showed that a part of the cell wall chitin is bound to (1,6)-β-glucan. This particular (1,6)-β-glucan-linked chitin disappears in *ΔCRH1* or *ΔCRH2* and *ΔCRH1ΔCRH2* mutants, suggesting a role played by Crh1p and Crh2p in (1,6)-β-glucan–chitin cross-linking.

Mutations in the *KRE* genes result in a decrease in the concentration of (1,6)-β-glucan showing that these genes are involved in (1,6)-β-glucan synthesis in the yeast cell wall (Bussey, 1991). Herrero et al. (2004) compared cell wall composition of wild-type *C. albicans* with that of a *KRE5* null mutant and observed a considerable decrease in (1,6)-β-glucan level in the mutant, indicating involvement of *KRE5* gene products in (1,6)-β-glucan metabolism. They also observed biochemically that, in the *C. albicans* mutant, both (1,3)-β- and (1,6)-β-glucans have significantly fewer branched glucose units than in *S. cerevisiae*. Comparative chemogenomics between fungal genera is of interest since they may show that there is not a direct link between the composition of the cell wall of the mutant and the putative gene function. For example, two orthologs of the yeast *KRE* genes are present in *A. fumigatus*: *KRE2* and *KRE6*. Since no (1,6)-β-glucan occurs in this fungal species, this clearly indicates *KRE2* and *KRE6* are not associated with (1,6)-β-glucan synthesis, in contrast to the studies by Bussey (1991).

Though mutational analysis is of use in monitoring cell wall changes, there exist compensatory mechanisms that pose difficulties in such studies. Cell wall modification can only represent a secondary effect resulting from this compensatory mechanism that has nothing to do with the primary function of the gene mutated (Lussier et al., 1997). Another problem is the multigenic nature of protein families responsible for a single function, and the interconnection between genes responsible for the synthesis or modification of different polysaccharides. There exist five orthologs of the (1,3)-β-glucanosyltransferase *GAS(1-5)* in *S. cerevisiae* (Ragni et al., 2007). *GAS1* is the gene expressed during vegetative growth (Popolo and Vai, 1999; Carotti et al., 2004). However, Gas2p and Gas4p, though expressed in the cells undergoing sporulation and absent during vegetative growth, can replace Gas1p function under specific environmental conditions (Ragni et al., 2007). In addition, inactivation of *GAS1* results in a decreased (1,3)-β-glucan level, with increased chitin and mannoprotein levels in the cell wall (Popolo et al., 1997; Ram et al., 1998). Synthetic lethality is another useful screen to pinpoint interconnecting pathways. The synthetic lethality between *GAS1* and *CHS3* or *KRE6* suggests the involvement of Gasp during glucan–chitin cross-linking or (1,3;1,6)-β-glucan branching. This has, however, not been correlated with the biochemical function demonstrated *in vitro* for these proteins, which were shown to have a (1,3)-β-glucan elongation function (Mouyna et al., 2000).

*Composition of the regenerating protoplast wall* The regenerating yeast/fungal protoplasts provide a very useful way to understand wall polysaccharide assembly. Protoplasts are produced by removing the cell wall enzymatically, including the use of (1,3)-β-glucanases, in an osmotically protected medium. Regeneration occurs in a nutritive medium with a reduced concentration of osmolytes. Reverting protoplasts provide information about the kinetics of assembly and the type of polymer synthesized, which depends on the stage of regeneration. Gene expression analysis of protoplasts undergoing wall regeneration can be an additional tool in the elucidation of wall polysaccharide biosynthesis. At present, the data are too scattered to provide a global picture of the early stages of regeneration and more work is needed to identify the first polysaccharide(s) that anchor the cell wall to the plasma membrane. In yeast and fungi both (1,3)-β-glucan and chitin are synthesized first and seem to be the anchoring structure. Of the two, depending on the species, the first polysaccharide synthesized will be chitin (*C. albicans*; Sentandreu et al., 1983) or (1,3)-β-glucan (*S. pombe*; Osumi, 1998a).

Accumulation of glucans and chitin happens shortly after suspension of the protoplasts of *C. albicans* in regeneration medium (Sentandreu et al., 1983). Initially, a micro-fibrillar skeleton of chitin is deposited and its concentration increases with time. The chitin network allows the spatial arrangement of (1,3)-β-glucan and finally of mannoproteins, as demonstrated by confocal electron microscopy (Elorza et al., 1985; Rico et al., 1997). The wall of *C. albicans* protoplasts becomes apparent after just 1 h of incubation and a normal wall is regenerated a few hours later. The incorporation of radio-labelled glucose in regenerating *C. abicans* spheroplasts has been used to follow the biosynthesis of wall glucans (Gopal et al., 1984); 40% of the total radioactivity incorporated was associated with (1,3)-β-glucan and about 50% was associated with a mixed polymer containing (1,3)-β- and (1,6)-β-glucan linkages as well as chitin. Using enzymatic degradation, the presence of at least two highly branched glucans with predominantly (1,3)-β-and (1,6)-β-glucosidic linkages has been confirmed (Gopal et al., 1984). Electron microscopy of regenerating *C. albicans* protoplasts revealed the synthesis of

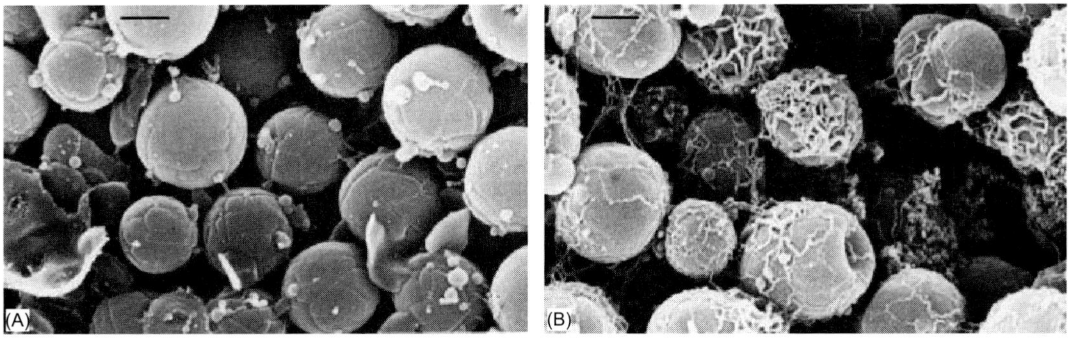


**Fig. 6: SEM of protoplasts of *S. cerevisiae*. (A) Fresh protoplasts appeared as round cells with a smooth surface but irregularly distributed wrinkles. (B) Protoplasts after 2 h in regenerating medium showing thick fibrils. Reproduced from Pardo et al. (1999) by permission of John Wiley & Sons Ltd.**

(1,3)-β-glucan initially as particles, which were then transformed into thin micro-fibrils (Osumi et al., 1998b). The micro-fibrils further aggregated to form interconnected thicker bundles leading to a wide-meshed bundle of long fibres having an approximate diameter of 2.8 nm.

Studies on *S. pombe* protoplast regeneration using low-voltage scanning and transmission electron microscopy in association with computer graphics showed the formation of (1,3)-β-glucan particles in the initial 10 min that subsequently transformed into a fibrillar network on the protoplast surface (Osumi et al., 1998b). In *S. pombe* the micro-fibrils were of 2 nm diameter, twisting over eachother to form an 8-nm fibre after 1 h, and into 16-nm flat bundles after 5 h of incubation, covering the entire protoplast. The inter-fibrillar space was then gradually filled by amorphous α-galactomannan particles and completion of wall regeneration took place in 12 h.

Fresh protoplasts of *S. cerevisiae* under the scanning electron microscope show round cells having a smooth surface. After 2 h of regeneration, protoplasts had a net of thick fibrils irregularly distributed over their surface having all the cell wall components (Fig. 6) (Pardo et al., 1999).

Regenerating protoplasts of *Aspergillus* spp. have also been studied. At 37°C, *A. fumigatus* regeneration starts after 5 min incubation in the appropriate medium, after 30 min (1,3)-β-glucans are produced, and a rigid wall is observed after 6 h (Costachel et al., unpublished data). The kinetics of deposition and composition of the regenerated polysaccharides remain to be analysed.

A more recent application using protoplasting and regeneration techniques was to investigate the variations of mRNA transcripts levels of cell wall-associated genes. Braley and Chaffin (1999) found six GPI-anchored cell wall proteins in *S. cerevisiae* that are overexpressed during protoplast regeneration. Among them was Egt2p, a wall-located cell cycle regulator involved in glucan metabolism. A recent analysis showed that the group of 'structural components of the wall' genes was up-regulated after 2h and remained as such during the process (Castillo et al., 2008). A transcriptome study was also undertaken on *C. albicans* by Castillo et al. (2006). Time-dependent expression divided the genes into 40 clusters. Clusters 1–19 were highly expressed at time 0 and down-regulated following regeneration, whereas clusters 20–40 showed the opposite behaviour, suggesting that the first group is related to cell adaptation to removal of the wall and the second group represents genes implicated in wall regeneration. Among the 6039 genes measured, 407 were up-regulated during the process including 45 that were associated with cell wall organization and biogenesis. For example, 18 genes that were up-regulated code for putative GPI-anchored proteins. Genes coding for (1,3)-β-glucan synthase activity such as *SMI1* and *GSL21*, *GSL22*, as well as those coding for (1,3)-β-glucan-elongating enzymes (*PHR1* and *PHR2*) and glycogen catabolism (*IPF9740* and *GLC3*), were also up-regulated during the first hour of regeneration. After 2h of incubation, as the cell wall became visible, regulation of (1,3)-β-glucan biogenesis occurred and genes coding for glucanase enzymes (*UTR2*, *BGL21*, *EXG2*, *IPF885* and *SCW4*) were overexpressed (Castillo et al., 2006). Proteomic studies have also been undertaken in regenerating protoplasts. For example, Pardo et al. (1999) could identify proteins involved in (1,3)-β-glucan elongation (Gas1-Gas5) and (1,3)-β-glucan hydrolysis (Bgl2, Exg1, Sps2) during cell wall regeneration in *S. cerevisiae*.

### *4. Cell wall (1,3)-β-glucans and fungal taxonomy*

As long ago as 1968, Bartnicki-Garcia pointed out that wall composition could be used as a marker in fungal taxonomy (Bartnicki-Garcia, 1968). As all fungal and yeast species have (1,3)-β-glucans at some point in their life cycle, such molecules cannot be used to define fungal taxons, and with the combination of two major wall polysaccharides Bartnicki-Garcia could classify the whole fungal spectrum into eight categories. Since then, many others have made use of the monosaccharide composition of walls for taxonomic, systematic and phylogenetic purposes (Weijman and Golubev, 1987; Prillinger et al., 1990; Prillinger et al., 1991; Prillinger et al., 1993; Messner et al., 1994). The content of (1,3;1,6)-β-glucan and its branching may be useful in delimiting orders and elaborating fungal evolutionary schemes. As an example, the (1,3;1,6)-β-glucans in the basidiomycetes are more branched than those in the ascomycetes. However, this has been poorly analysed for taxonomic purposes. Carbohydrate-based taxonomy relies more on the polysaccharides associated with the central core, including the branched (1,3;1,6)-β-glucan, rather than the core (1,3;1,6)-β-glucan itself. As pointed out by Leal and Bernabé (1998), alkali-extractable/water-soluble wall polysaccharides are more important in fungal taxonomy than structural components such as (1,3)-β-glucans, because they are more variable.

Bartnicki-Garcia (1968) tentatively considered chytridiomycetes to contain (1,3)-β-glucans along with chitin as the major components of wall. At that time, *Hyaloraphidium curvatum,* was thought to be a chytridiomycete, but is now considered to be an independent member of an evolutionary lineage within the fungi, since its wall contains galactose and mannose but not glucose (Ustinova et al., 2000). In the chromistan oomycetes, (1,3)-β- and (1,6)-β-glucans occur with cellulose, whereas chitin replaces cellulose in Ascomycotina and Basidiomycotina. In contrast to other fungi, the walls of zygomycetes are composed of chitin–chitosan and (1,3)-β-glucans are absent from the walls of the vegetative cells of some species, but are present, as a minor component (<10%), in sporangiospore walls of *Mucor rouxii* and *M. ramannianus* (Dow et al., 1983). However, the concentration of (1,3)-β-glucan can reach 50% in the spore of *Phycomyces blakesleeanus* (Van Laere et al., 1977). A co-polymer of glucose and glucuronic acid has also been identified in Mucorales. The low amount of (1,3)-β-glucan in the spore walls of the chytridiomycetes and Mucorales suggests a most ancient origin of these two orders. However the lack or lower amount of (1,3)-β-glucan is not a general characteristic of all zygomycetes since in the Entomophthorales, the second class of zygomycetes, the walls contains a high amount of (1,3)-β-glucan (Latgé et al., 1984; Latgé and Beauvais, 1987). In the genomic era, it seems now appropriate to revisit a putative correlation between the structure of (1,3)-β-glucans and associated polysaccharides of the cell wall core and the new taxonomy clades revealed by genome sequences. Comparative cell wall studies between yeasts and moulds have already shown interesting results (Latgé, 2007).

## (1,3)-β-Glucans in Lichens

Lichens are symbiotic associations of a fungus with a photosynthetic partner. The fungal partner in lichens is usually an ascomycete but rarely a basidiomycete, whereas the photosynthetic partner is an alga (a green alga or a blue-green alga, sometimes yellow-green/brown alga), cyanobacteria, or both green alga and cyanobacteria. There still exists the argument that this symbiotic association is for mutualism, commensalism or parasitism, as sometimes the fungal hyphae penetrate into the algal cells. As in free-living fungi, lichens contain glucans (either α or β) as the major polysaccharides that serve as the wall material and reserve compounds and arise predominantly from the mycobiont (Gorin et al., 1993). Immunocytochemical location of the (1,3;1,4)-β-glucan lichenan using an anti-(1,3;1,4)-β-glucan antibody has been performed in the lichen-forming ascomycete *Cetraria islandica* (Honegger and Haisch, 2001). Labelling of glucans was intense over the extracellular matrix of the peripheral cortex, as well as in an outer wall layer of medullary hyphae.

Linear (1,3)-β-glucans have been isolated from a number of lichens including *Stereocaulon ramulosum*, *Ramalina usnea* and *Ramalina celastri* (Gorin and Iacomini, 1984; Baron et al., 1988; Stuelp et al., 1999). Recently, using NMR, Carbonero et al. (2001) reported the existence of linear alkali-soluble (1,3)-β-glucan in 15 *Cladonia* sp. Iacomini and co-workers (1987) isolated highly branched (1,3;1,6)-β-glucan, typical of basidiomycetes, from *Dictyonema glabratum* (basidiomycotic lichen formerly known as *Cora pavonia*), while Gorin & Barreto-Berger (1983) showed β-glucans having less than 10% branching in ascomycotic lichens. Stuelp et al. (1999) also isolated alkali-insoluble linear β-glucans from *Ramalina celastri* with regularly distributed (1,3)- and (1,4)-linkages in 1:1 molar ratio. Most glucans, when the mycobiont is an ascomycete, are linear, although Prado et al. described a highly branched β-glucan in the ascolichen *Collema leptosporum* (Prado et al., 1999).

In addition to being a structural component of the mycobiont wall, lichenin-type glucans also function in thallus–water relations (Honegger and Haisch, 2001). The wall polysaccharides are of importance in the taxonomy of lichenized fungi, based on their distribution and diversity in their chemical composition (Carbonero et al., 2001; Carbonero et al., 2002; Sassaki et al., 2002). For example, lichenin-type (1,3;1,4)-β-glucans (lichenan) are the only β-glucans found in Parmeliaceae, whereas the Umbilicariaceae and Cladoniaceae are characterized by the presence of pustulan ((1,6)-β-glucan) and nigeran ((1,3)-α-glucan)-type polysaccharides, respectively (Olafsdottir and Ingolfsdottir, 2001). In *Cetraria islandica*, isolichenan (a (1,3;1,4)-linked β-glucan) is present along with lichenan (Kramer et al., 1995).

## III. (1,3)-β-Glucans in Oomycetes

### *A. Oomycete Cell Wall*

Oomycetes, currently classified as stramenopile (chromistan) eukaryotes (Sogin et al., 1996; Tyler et al., 2006), are a group of filamentous, unicellular protists that physically resemble fungi. However, they differ from fungi in having cellulose instead of chitin in their cell wall and mycelia without septation. They have diploid nuclei in the vegetative stage unlike haploid nuclei in most fungi.

Although the cell wall polysaccharides have not been precisely analysed and studied over the last 20 years, early studies have shown that the major components of the oomycete cell wall are (1,3)-β-glucan, (1,6)-β-glucan and cellulose (Sietsma et al., 1969; Aronson and Lin, 1978; Blaschek et al., 1992). Chitin has been reported in the wall of *Apodachlya* (Leptomitaceae) (Lin et al., 1976). In *Phythium debaryanum*, Yamada & Miyazaki (1976) identified branched (1,3)-β-glucans in the acid extract, a mixture of (1,3)-, (1,6)- and (1,3;1,6)-β-linked glucans in the alkaline extract, and (1,4)-β-glucans fraction extracted by the cuprammonium reagent. They suggested that the wall is composed of cellulosic microfibrils entangled in a matrix of amorphous, branched (1,3)-β- and (1,6)-β-glucan. In *Pythium aphanidermatum*, the mycelial wall consists of 18% cellulose and 82% (1,3;1,6)-β-glucan (Blaschek et al., 1992), and of the two types of branch-on-branch β-glucans identified, one was solubilized by extraction with water at 121° and had a $M_W$ of around 10000 Da and 6% (1,6)-β-linkages. The second fraction was a trifluoroacetic acid-soluble (1,3;1,6)-β-glucan of lower $M_W$ (6000 Da) with 14% (1,6)-β-branches. In *Apodachlya* the wall is composed of (1,3)-β-glucan having (1,6)-β-branches and less than 10% cellulose (Lin et al., 1976). Fabre et al. (1984) isolated and characterized glucans from *Phytophthora parasitica* walls, and separated a mixture of branch-on-branch (1,3;1,6)-β-glucans with relative molecular masses ranging between 9 and 200 kDa. The degree of branching and cross-linking of the (1,3)-β-glucan with other wall components has so far not been investigated in the oomycetes. The (1,3)-β-glucan structure in the oomycetes is however clearly different from plants (embrophytes), where only linear (1,3)-β-glucan (callose) is found.

## B. Storage (1,3)-β-glucans

Among the oomycetes, the intracellular storage polysaccharide, glycogen, is replaced by (1,3)-β-glucans, which is a major difference compared to fungi. In the oomycetes these reserve carbohydrates are known as 'mycolaminarin' and 'cellulin'; the former is a soluble form of (1,3)-β-glucan and the latter an insoluble granular inclusion. Cellulins are the characteristic polysaccharide inclusions in the cytoplasm of the oomycetous Leptomitales. Lee and Aronson (1975) isolated and characterized cellulin granules from an *Apodochlya* sp. which contained 60% chitin and 39% (1,3;1,6 )-β-glucan with less than 0.1% protein. A different composition was reported by Lin et al. (1976) for *Apodochlya* sp., wherein cellulin consisted mainly of β-glucans (50%) with 18% chitin, 10% cellulose and about 6.4% protein. In *Phytophthora palmivora spores*, 'mycolaminarin' is a side-chain-branched (1,3,1,6)-β-glucan with one or two branches at C6 (Wang and Bartnicki-Garcia, 1973). Coulter and Aronson (1977) and Blaschek et al. (1992) observed soluble, side-chain-branched (1,3;1,6)-β-glucans in extracts of *P. aphanidermatum* and *Mindeniella spinospora*. Both these cytoplasmic

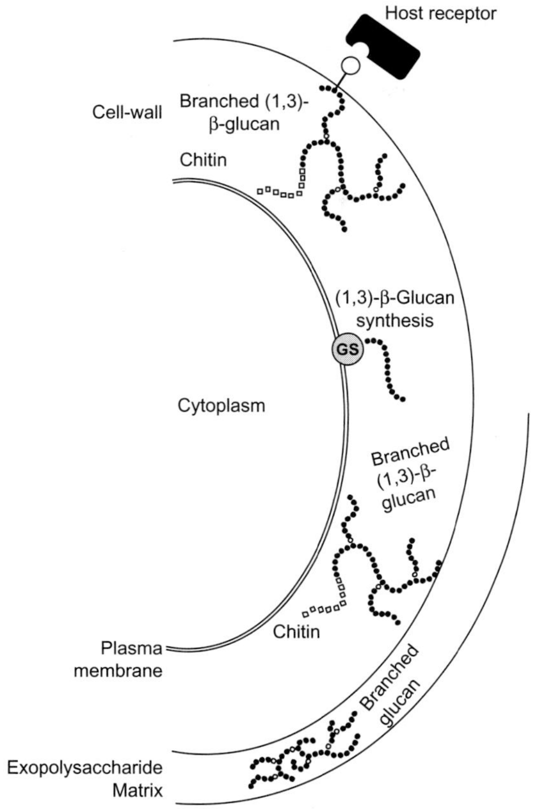


**Fig. 7: Location of (1,3)-β-glucans in Fungi. Linear (1,3)-β-glucans are synthesized at the apex and are remodeled and cross-linked to other polysaccharides in the subapical regions. Exopolysaccharides and GPI-proteins (P) covalently bound to polysaccharides at the surface of the fungal cell wall interact with the external environment.**

glucans are speculated to be utilized for the synthesis of new wall glucans during cellular differentiation (encystment) or carbon starvation (Zevenhuizen and Bartnicki-Garcia, 1970; Woon et al., 1971). In contrast to wall polysaccharides, these reserve polysaccharides remain poorly studied and their function is not well understood.

## Conclusion

Data reviewed in this chapter show the essential role of the (1,3)-β-glucans in fungal life, not only as part of the cell wall but also in the interactions of the fungus with its biotic and abiotic environment (Fig. 7). Data accumulated in the last 20 years have also shown that

(1,3;1,6)-β-glucans can no longer be considered as an inert component of the yeast and mould cell wall, and that the structure is constantly evolving during the fungal growth and cell cycle. The (1,3;1,6)-β-glucans have their own complex cellular life. Synthesized as an amorphous material at the growing apex of the fungal hypha, they become branched and linked to other cell wall polysaccharides. All these structural modifications lead to a different fate for this molecule. The signals and site where the (1,3)-β-glucans are remodelled to be glued in a rigid structure are unknown.

When secreted as an extracellular material, the (1,3)-β-glucan molecule is more branched than the one associated with the cell wall structure. Nobody yet knows what the signals are that govern the secretion of (1,3;1,6)-β-glucans. The site of synthesis of these glucans is also not understood, and the question yet to be answered is: are these glucans produced at the plasma membrane where the cell wall (1,3)-β-glucans are made, or in the Golgi vesicles, and what are the secretion signals?

In addition, it appears that the (1,3)-β-glucans serve as an anchor for surface receptors and adhesins that become bound to these structures to remain at the surface of the cell to fulfil their biological function. During fungal life, the (1,3;1,6)-β-glucan complex can also evolve in terms of location and/or chemical structure due to environmental changes or cell cycle steps. All these changes occuring during fungal growth have not been precisely characterized structurally, and the relationship between chemical modification of the (1,3)-β-glucans and fungal growth remains an exciting challenge in the field of mycology.

## Bibliography

Adachi, Y., Ishii, T., Ikeda, Y., Hoshino, A., Tamura, H., Aketagawa, J., Tanaka, S., & Ohno, N. (2004). Characterization of beta-glucan recognition site on C-type lectin, dectin 1. *Infection and Immunity*, *72*, 4159–4171.

Adt, I., Toubas, D., Pinon, J. M., Manfait, M., & Sockalingum, G. D. (2006). FTIR spectroscopy as a potential tool to analyse structural modifications during morphogenesis of Candida albicans. *Archives of Microbiology*, *185*, 277–285.

Aguilar-Uscanga, B., & François, J. M. (2003). A study of the yeast cell wall composition and structure in response to growth conditions and mode of cultivation. *Letters in Applied Microbiology*, *37*, 268–274.

Aguilar-Uscanga, B., Arrizon, J., Ramirez, J., & Solis-Pacheco, J. (2007). Effect of Agave tequilana juice on cell wall polysaccharides of three Saccharomyces cerevisiae strains from different origins. *Antonie Van Leeuwenhoek*, *91*, 151–157.

Al-Fattani, M. A., & Douglas, L. J. (2006). Biofilm matrix of Candida albicans and Candida tropicalis: chemical composition and role in drug resistance. *Journal of Medical Microbiology*, *55*, 999–1008.

Alonso-Nunez, M. L., An, H., Martin-Cuadrado, A. B., Mehta, S., Petit, C., Sipiczki, M., del Rey, F., Gould, K. L., & de Aldana, C. R. (2005). Ace2p controls the expression of genes required for cell separation in Schizosaccharomyces pombe. *Molecular Biology of the Cell*, *16*, 2003–2017.

Aronson, J. M., & Lin, C. C. (1978). Hyphal Wall Chemistry of Leptomitus lacteus. *Mycologia*, *70*, 363–369.

Braley, R., & Chaffin, W. L. (1999). Variations in mRNA transcript levels of cell wall-associated genes of Saccharomyces cerevisiae following spheroplasting. *FEMS Microbiology Letters*, *181*, 177–185.

Baron, M., Gorin, P. A. J., & Lacomini, M. (1988). Isolation and identification of a linear (1→3)-linked β-image-glucan and other carbohydrate components of the lichen Stereocaulon ramulosum (SW.) Räusch. *Carbohydrate Research*, *177*, 235–239.

Bartnicki-Garcia, S. (1968). Cell wall chemistry, morphogenesis, and taxonomy of fungi. *Annual Revews of Microbioogyl*, *22*, 87–108.

Beauvais, A., Bruneau, J. M., Mol, P. C., Buitrago, M. J., Legrand, R., & Latge, J. P. (2001). Glucan synthase complex of Aspergillus fumigatus. *The Journal of Bacteriology*, *183*, 2273–2279.

Beauvais, A., Schmidt, C., Guadagnini, S., Roux, P., Perret, E., Henry, C., Paris, S., Mallet, A., Prevost, M. C., & Latge, J. P. (2007). An extracellular matrix glues together the aerial-grown hyphae of Aspergillus fumigatus. *Cellular Microbiology*, *9*, 1588–1600.

Blaschek, W., Kasbauer, J., Kraus, J., & Franz, G. (1992). Pythium aphanidermatum: culture, cell-wall composition, and isolation and structure of antitumour storage and solubilised cell-wall (1-->3), (1-->6)-beta-D-glucans. *Carbohydrate Research*, *231*, 293–307.

Bocchinfuso, G., Palleschi, A., Mazzuca, C., Coviello, T., Alhaique, F., & Marletta, G. (2008). Theoretical and experimental study on a self-assembling polysaccharide forming nanochannels: static and dynamic effects induced by a soft confinement. *Journal of Physical Chemistry B*, *112*, 6473–6483.

Bom, I. J., Dielbandhoesing, S. K., Harvey, K. N., Oomes, S. J., Klis, F. M., & Brul, S. (1998). A new tool for studying the molecular architecture of the fungal cell wall: one-step purification of recombinant trichoderma beta-(1-6)-glucanase expressed in Pichia pastoris. *Biochimica et Biophysica Acta*, *1425*, 419–424.

Bruneau, J. M., Magnin, T., Tagat, E., Legrand, R., Bernard, M., Diaquin, M., Fudali, C., & Latge, J. P. (2001). Proteome analysis of Aspergillus fumigatus identifies glycosylphosphatidylinositol-anchored proteins associated to the cell wall biosynthesis. *Electrophoresis*, *22*, 2812–2823.

Bush, D. A., Horisberger, M., Horman, I., & Wursch, P. (1974). The wall structure of Schizosaccharomyces pombe. *Journal of General Microbiology*, *81*, 199–206.

Bussey, H. (1991). K1 killer toxin, a pore-forming protein from yeast. *Molecular Microbiology*, *5*, 2339–2343.

Cabib, E., Drgonova, J., & Drgon, T. (1998). Role of small G proteins in yeast cell polarization and wall biosynthesis. *Annual Review of Biochemistry*, *67*, 307–333.

Cabib, E., Roh, D. H., Schmidt, M., Crotti, L. B., & Varma, A. (2001). The yeast cell wall and septum as paradigms of cell growth and morphogenesis. *Journal of Biological Chemistry*, *276*, 19679–19682.

Cabib, E., Blanco, N., Grau, C., Rodriguez-Pena, J. M., & Arroyo, J. (2007). Crh1p and Crh2p are required for the cross-linking of chitin to beta(1-6)glucan in the Saccharomyces cerevisiae cell wall. *Molecular Microbiology*, *63*, 921–935.

Carbonero, E. R., Sassaki, G. L., Stuelp, P. M., Gorin, P. A., Woranovicz-Barreira, S. M., & Iacomini, M. (2001). Comparative studies of the polysaccharides isolated from lichenized fungi of the genus Cladonia: significance as chemotypes. *FEMS Microbiology Letters*, *194*, 65–69.

Carbonero, E. R., Sassaki, G. L., Gorin, P. A., & Iacomini, M. (2002). A (1-->6)-linked beta-mannopyrananan, pseudonigeran, and a (1-->4)-linked beta-xylan, isolated from the lichenised basidiomycete Dictyonema glabratum. *FEMS Microbiology Letters*, *206*, 175–178.

Carotti, C., Ragni, E., Palomares, O., Fontaine, T., Tedeschi, G., Rodriguez, R., Latge, J. P., Vai, M., & Popolo, L. (2004). Characterization of recombinant forms of the yeast Gas1 protein and identification of residues essential for glucanosyltransferase activity and folding. *European Journal of Biochemistry*, *271*, 3635–3645.

Castillo, L., Martinez, A. I., Garcera, A., Garcia-Martinez, J., Ruiz-Herrera, J., Valentin, E., & Sentandreu, R. (2006). Genomic response programs of Candida albicans following protoplasting and regeneration. *Fungal Genetics and Biology*, *43*, 124–134.

Castillo, L., Martinez, A. I., Gelis, S., Ruiz-Herrera, J., Valentin, E., & Sentandreu, R. (2008). Genomic response programs of Saccharomyces cerevisiae following protoplasting and regeneration. *Fungal Genetics and Biology*, *45*, 253–265.

Cortes, J. C. G., Ishiguro, J., Duran, A., & Ribas, J. C. (2002). Localization of the (1,3)beta-D-glucan synthase catalytic subunit homologue Bgs1p/Cps1p from fission yeast suggests that it is involved in septation, polarized growth, mating, spore wall formation and spore germination. *Journal of Cell Science*, *115*, 4081–4096.

Coulter, D. B., & Aronson, J. M. (1977). Glycogen and other soluble glucans from chytridiomycete and oomycete species. *Archives of Microbiology*, *115*, 317–322.

da Silva Castro, N., Barbosa, M. S., Maia, Z. A., Bao, S. N., Felipe, M. S., Santana, J. M., Soares Mendes-Giannini, M. J., Pereira, M., & de Almeida Soares, C. M. (2008). Characterization of Paracoccidioides brasiliensis PbDfg5p, a cell-wall protein implicated in filamentous growth. *Yeast, 25*, 141–154.

de Groot, P. W. J., Hellingwerf, K. J., & Klis, F. M. (2003). Genome-wide identification of fungal GPI proteins. *Yeast, 20*, 781–796.

de Groot, P. W. J., de Boer, A. D., Cunningham, J., Dekker, H. L., de Jong, L., Hellingwerf, K. J., de Koster, C., & Klis, F. M. (2004). Proteomic analysis of Candida albicans cell walls reveals covalently bound carbohydrate-active enzymes and adhesins. *Eukaryotic Cell, 3*, 955–965.

de Groot, P. W. J., Ram, A. F., & Klis, F. M. (2005). Features and functions of covalently linked proteins in fungal cell walls. *Fungal Genetics and Biology, 42*, 657–675.

de Groot, P. W., Yin, Q. Y., Weig, M., Sosinska, G. J., Klis, F. M., & de Koster, C. G. (2007). Mass spectrometric identification of covalently bound cell wall proteins from the fission yeast Schizosaccharomyces pombe. *Yeast, 24*, 267–278.

Dijkgraaf, G. J. P., Abe, M., Ohya, Y., & Bussey, H. (2002). Mutations in Fks1p affect the cell wall content of beta-1,3-and beta-1,6-glucan in Saccharomyces cerevisiae. *Yeast, 19*, 671–690.

Donlan, R. M., & Costerton, J. W. (2002). Biofilms: survival mechanisms of clinically relevant microorganisms. *Clinical Microbiology Reviews, 15*, 167–193.

Douglas, L. J. (2002). Medical importance of biofilms in Candida infections. *Revista Iberoamericana de Micología, 19*, 139–143.

Dow, J. M., Darnall, D. W., & Villa, V. D. (1983). Two distinct classes of polyuronide from the cell walls of a dimorphic fungus, Mucor rouxii. *J. Bacteriol, 155*, 1088–1093.

Dranginis, A. M., Rauceo, J. M., Coronado, J. E., & Lipke, P. N. (2007). A biochemical guide to yeast adhesins: glycoproteins for social and antisocial occasions. *Microbiology and Molecular Biology Reviews, 71*, 282–294.

Dubourdieu, D., & Ribereau-Gayon, P. (1980). [Occurrence of an extracellular beta-(1 to 3)-glucanase from Botrytis cinerea]. *Comptes Rendus des Seances de l'Académie des Sciences-Série D, Sciences naturelles, 290*, 25–28.

Ecker, M., Deutzmann, R., Lehle, L., Mrsa, V., & Tanner, W. (2006). Pir proteins of Saccharomyces cerevisiae are attached to beta-1,3-glucan by a new protein-carbohydrate linkage. *Journal of Biological Chemistry, 281*, 11523–11529.

Elorza, M. V., Murgui, A., & Sentandreu, R. (1985). Dimorphism in Candida albicans: contribution of mannoproteins to the architecture of yeast and mycelial cell walls. *Journal of General Microbiology, 131*, 2209–2216.

Fabre, I., Bruneteau, M., Ricci, P., & Michel, G. (1984). [Isolation and structural studies of glucans from Phytophthora parasitica]. *European Journal of Biochemistry*, *142*, 99–103.

Fleet, G. H. (1991). The yeast cell wall. In A. H. Rose. & J. S. Harrison. (Eds.), *The Yeasts*, p. 4. London: Academic Press.

Flemming, H. C., Wingender, J., & Griergbe and Mayer, C. (2000). Physico-chemical properties of biofilms. In E. LV. (Ed.), *Biofilms: recent advances in their study and control* (pp. 19–34). Amsterdam: Harwood Academic Publishers.

Flemming, H. C., Neu, T. R., & Wozniak, D. J. (2007). The EPS matrix: the "house of biofilm cells". *The Journal of Bacteriology*, *189*, 7945–7947.

Fontaine, T., Simenel, C., Dubreucq, G., Adam, O., Delepierre, M., Lemoine, J., Vorgias, C. E., Diaquin, M., & Latge, J. P. (2000). Molecular organization of the alkali-insoluble fraction of Aspergillus fumigatus cell wall. *Journal of Biological Chemistry*, *275*, 27594–27607.

Fuglsang, C. C., Berka, R. M., Wahleithner, J. A., Kauppinen, S., Shuster, J. R., Rasmussen, G., Halkier, T., Dalboge, H., & Henrissat, B. (2000). Biochemical analysis of recombinant fungal mutanases. A new family of alpha1,3-glucanases with novel carbohydrate-binding domains. *Journal of Biological Chemistry*, *275*, 2009–2018.

Fujii, T., Shimoi, H., & Iimura, Y. (1999). Structure of the glucan-binding sugar chain of Tip1p, a cell wall protein of Saccharomyces cerevisiae. *Biochimca et Biophysica Acta*, *1427*, 133–144.

Gantner, B. N., Simmons, R. M., & Underhill, D. M. (2005). Dectin-1 mediates macrophage recognition of Candida albicans yeast but not filaments. *The Embo Journal*, *24*, 1277–1286.

Gersuk, G. M., Underhill, D. M., Zhu, L., & Marr, K. A. (2006). Dectin-1 and TLRs permit macrophages to distinguish between different Aspergillus fumigatus cellular states. *The Journal of Immunology*, *176*, 3717–3724.

Gopal, P., Sullivan, P. A., & Shepherd, M. G. (1984). Metabolism of [14C]glucose by regenerating spheroplasts of Candida albicans. *Journal of General Microbiology*, *130*, 325–335.

Gorin, P. A. J., & Barreto-Berger, E. M. (1983). The Chemistry of Polysaccharides of Fungi and Lichens. In G. O. A. Hg (Ed.), *The polysaccharides: Vol. 2* , p. 366. New York: Academic press.

Gorin, P. A., & Iacomini, M. (1984). Polysaccharides of the lichens Cetraria islandica and Ramalina usnea. *Carbohydrate Research*, *128*, 119–132.

Gorin, P. A., Baron, M., Silva, M. L. C., Teixeira, A. Z. A., & Iacomini, M. (1993). Lichen carbohydrates. *Ciencia e Cultura (Brazil)*, *45*, 27–36.

Grun, C. H., Dekker, N., Nieuwland, A. A., Klis, F. M., Kamerling, J. P., Vliegenthart, J. F., & Hochstenbach, F. (2006). Mechanism of action of the endo-(1-->3)-alpha-glucanase MutAp from the mycoparasitic fungus Trichoderma harzianum. *FEBS Letters*, *580*, 3780–3786.

Grün, C. H. (2003). Structure and biosynthesis of fungal a-glucans. **Ph.D. dissertation** Proefschrift Universiteit Utrecht.

Hamada, K., Terashima, H., Arisawa, N., & Kitada, K. (1998). Amino acid sequence requirement for efficient incorporation of glycosylphosphatidylinositol-associated proteins into the cell wall of Saccharomyces cerevisiae. *Journal of Biological Chemistry*, *273*, 26946–26953.

Herrero, A. B., Magnelli, P., Mansour, M. K., Levitz, S. M., Bussey, H., & Abeijon, C. (2004). KRE5 gene null mutant strains of Candida albicans are avirulent and have altered cell wall composition and hypha formation properties. *Eukaryotic Cell*, *3*, 1423–1432.

Hohl, T. M., Van Epps, H. L., Rivera, A., Morgan, L. A., Chen, P. L., Feldmesser, M., & Pamer, E. G. (2005). Aspergillus fumigatus triggers inflammatory responses by stage-specific beta-glucan display. *PLoS Pathogens*, *1*, e30.

Hohl, T. M. (2008). Stage-specificInnate ImmuneRecognition of *Aspergillus fumigatus* and Modulation by Echinocandin Drugs. (3rd ed.). Advances against Aspergillus Miami USA.

Honegger, R., & Haisch, A. (2001). Immunocytochemical location of the (1-->3) (1-->4)-β-glucan lichenin in the lichen-forming ascomycete *Cetraria islandica* (Icelandic moss). *New Phytologist*, *150*, 739–746.

Hong, Z., Mann, P., Shaw, K. J., & Didomenico, B. (1994). Analysis of beta-glucans and chitin in a Saccharomyces cerevisiae cell wall mutant using high-performance liquid chromatography. *Yeast*, *10*, 1083–1092.

Huang, G., Zhang, M., & Erdman, S. E. (2003). Posttranslational modifications required for cell surface localization and function of the fungal adhesin Aga1p. *Eukaryotic Cell*, *2*, 1099–1114.

Humbel, B. M., Konomi, M., Takagi, T., Kamasawa, N., Ishijima, S. A., & Osumi, M. (2001). In situ localization of beta-glucans in the cell wall of Schizosaccharomyces pombe. *Yeast*, *18*, 433–444.

Iacomini, M., Zanin, S. M. W., Fontana, J. D., Hogge, L., & Gorin, P. A. (1987). Isolation and characterization of β-image-glucan, heteropolysaccharide, and trehalose components of the basidiomycetous lichen Cora pavonia. *Carbohydrate Research*, *168*, 55–65.

Iorio, E., Torosantucci, A., Bromuro, C., Chiani, P., Ferretti, A., Giannini, M., Cassone, A., & Podo, F. (2008). Candida albicans cell wall comprises a branched beta-D-(1-->6)-glucan with beta-D-(1-->3)-side chains. *Carbohydrate Research*, *343*, 1050–1061.

Ip, C. C., Miller, W. J., Kubek, D. J., Strang, A. M., van Halbeek, H., Piesecki, S. J., & Alhadeff, J. A. (1992). Structural characterization of the N-glycans of a recombinant hepatitis B surface antigen derived from yeast. *Biochemistry*, *31*, 285–295.

Iwamuro, Y., Aoki, M., Mikami, Y., Obi, Y., & Kisaki, T. (1982). Change in Molecular Weight of Pendulan from Porodisculus pendulus. *Journal of fermentation technology*, *60*, 405–409.

Jaafar, L., & Zueco, J. (2004). Characterization of a glycosylphosphatidylinositol-bound cell-wall protein (GPI-CWP) in Yarrowia lipolytica. *Microbiology*, *150*, 53–60.

Kanetsuna, F., & Carbonell, L. M. (1970). Cell wall glucans of the yeast and mycelial forms of Paracoccidioides brasiliensis. *The Journal of Bacteriology*, *101*, 675–680.

Kanetsuna, F., & Carbonell, L. M. (1971). Cell wall composition of the yeastlike and mycelial forms of Blastomyces dermatitidis. *The Journal of Bacteriology*, *106*, 946–948.

Kanetsuna, F., Carbonell, L. M., Azuma, I., & Yamamura, Y. (1972). Biochemical studies on the thermal dimorphism of Paracoccidioides brasiliensis. *The Journal of Bacteriology*, *110*, 208–218.

Kapteyn, J. C., Montijn, R. C., Dijkgraaf, G. J., Van den Ende, H., & Klis, F. M. (1995). Covalent association of beta-1,3-glucan with beta-1,6-glucosylated mannoproteins in cell walls of Candida albicans. *The Journal of Bacteriology*, *177*, 3788–3792.

Kapteyn, J. C., Montijn, R. C., Vink, E., de la Cruz, J., Llobell, A., Douwes, J. E., Shimoi, H., Lipke, P. N., & Klis, F. M. (1996). Retention of Saccharomyces cerevisiae cell wall proteins through a phosphodiester-linked beta-1,3-/beta-1,6-glucan heteropolymer. *Glycobiology*, *6*, 337–345.

Kapteyn, J. C., Van Den Ende, H., & Klis, F. M. (1999). The contribution of cell wall proteins to the organization of the yeast cell wall. *Biochimica Et Biophysica Acta-General Subjects*, *1426*, 373–383.

Kath, F., & Kulicke, W.-M. (1999). Polymer analytical characterization of glucan and mannan from yeast Saccharomyces cerevisiae. *Die Angewandte Makromolekulare Chemie*, *268*, 69–80.

Klis, F. M., De Groot, P., & Hellingwerf, K. (2001). Molecular organization of the cell wall of Candida albicans. *Medical Mycology*, *39*, 1–8.

Klis, F. M., Mol, P., Hellingwerf, K., & Brul, S. (2002). Dynamics of cell wall structure in Saccharomyces cerevisiae. *FEMS Microbiology Reviews*, *26*, 239–256.

Klis, F. M., de Jong, M., Brul, S., & de Groot, P. W. J. (2007). Extraction of cell surface-associated proteins from living yeast cells. *Yeast*, *24*, 253–258.

Ko, Y. T., & Lin, Y. L. (2004). 1,3-beta-glucan quantification by a fluorescence microassay and analysis of its distribution in foods. *Journal of Agricultural and Food Chemistry*, *52*, 3313–3318.

Kollar, R., Reinhold, B. B., Petrakova, E., Yeh, H. J., Ashwell, G., Drgonova, J., Kapteyn, J. C., Klis, F. M., & Cabib, E. (1997). Architecture of the yeast cell wall. Beta(1-->6)-glucan interconnects mannoprotein, beta(1-->)3-glucan, and chitin. *Journal of Biological Chemistry, 272*, 17762–17775.

Kondori, N., Edebo, L., & Mattsby-Baltzer, I. (2008). A novel monoclonal antibody recognizing β(1-3) glucans in intact cells of Candida and Cryptococcus. *Acta Pathologica, Microbiologica et Immunologica Scandinavica, 116*, 867–876.

Kopecka, M., & Kreger, D. R. (1986). Assembly of microfibrils in vivo and in vitro from (1-->3)-beta-D-glucan synthesized by protoplasts of Saccharomyces cerevisiae. *Archives of Microbiology, 143*, 387–395.

Kopecka, M., & Gabriel, M. (1992). The influence of congo red on the cell wall and (1-->3)-beta-D-glucan microfibril biogenesis in Saccharomyces cerevisiae. *Archives of Microbiology, 158*, 115–126.

Kopecka, M., Fleet, G. H., & Phaff, H. J. (1995). Ultrastructure of the cell wall of Schizosaccharomyces pombe following treatment with various glucanases. *Journal of Structural Biology, 114*, 140–152.

Krämer, P., Wincierz, U., Grübler, G., Tschakert, J., Voelter, W., & Mayer, H. (1995). Rational approach to fractionation, isolation, and characterization of polysaccharides from the lichen Cetraria islandica. *Arzneimittelforschung, 45*, 726–731.

Latgé, J. P., Fournet, B., Cole, G. T., dubourdieu, D., & Tong, N. (1984). Composition chimique et ultrastructurale des parois de corps hyphaux et des azygospores de *Conidiobolus obscurus. Canadian Journal of Microbiology, 30*, 1507–1521.

Latgé, J. P., & Beauvais, A. (1987). Wall composition of the protoplastic Entomophtorales. *Journal of Invertebrate Pathology, 50*, 53–57.

Latgé, J. P., Boucias, D. G., & Fournet, B. (1988). Structure of the extracellular polysaccharide of *Nomuraea rileyi. Carbohydrate Research, 181*, 282–286.

Latgé, J. P., Kobayashi, H., Debeaupuis, J. P., Diaquin, M., Sarfati, J., Wieruszeski, J. M., Parra, E., Bouchara, J. P., & Fournet, B. (1994). Chemical and immunological characterization of the extracellular galactomannan of Aspergillus fumigatus. *Infection and Immunity, 62*, 5424–5433.

Latgé, J. P., & Calderone, R. (2005). The fungal cell wall. In U. Kües & R. Fischer (Eds.), *The Mycoto I. Growth, Differentiation and Sexuality* (pp. 73–104). Berlin, Heidelberg: Springer-Verlag.

Latgé, J. P., Mouyna, I., Tekaia, F., Beauvais, A., Debeaupuis, J. P., & Nierman, W. (2005). Specific molecular features in the organization and biosynthesis of the cell wall of Aspergillus fumigatus. *Medical Mycology, 43*, S15–S22.

Latgé, J. P. (2007). The cell wall: a carbohydrate armour for the fungal cell. *Molecular Microbiology*, *66*, 279–290.

Latgé, J.P. (unpublished).

Leal, J. A., & Bernabé, M. (1998). Chemical Fungal Taxonomy. In (M. Dekker, Ed., J. C. Frisvad, P. D Bridge & D. K. Arora, New York & Hong Kong.

Lee, H. Y., & Aronson, J. M. (1975). Composition of cellulin, the unique chitin-glucan granules of the fungus, Apodachlya sp. *Archives of Microbiology*, *102*, 203–208.

Lesage, G., Sdicu, A. M., Menard, P., Shapiro, J., Hussein, S., & Bussey, H. (2004). Analysis of beta-1,3-glucan assembly in Saccharomyces cerevisiae using a synthetic interaction network and altered sensitivity to caspofungin. *Genetics*, *167*, 35–49.

Lesage, G., & Bussey, H. (2006). Cell wall assembly in Saccharomyces cerevisiae. *Microbiology and Molecular Biology Reviews*, *70*, 317–343.

Lin, C. C., Sicher, R. C., Jr., & Aronson, J. M. (1976) Hyphal wall chemistry in Apodachlya. *Archives of Microbiology*, *108*, 85–91.

Lowman, D. W., Ferguson, D. A., & Williams, D. L. (2003). Structural characterization of (1-->3)-beta-D-glucans isolated from blastospore and hyphal forms of Candida albicans. *Carbohydrate Research*, *338*, 1491–1496.

Lussier, M., White, A. M., Sheraton, J., di Paolo, T., Treadwell, J., Southard, S. B., Horenstein, C. I., Chen-Weiner, J., Ram, A. F., Kapteyn, J. C., Roemer, T. W., Vo, D. H., Bondoc, D. C., Hall, J., Zhong, W. W., Sdicu, A. M., Davies, J., Klis, F. M., Robbins, P. W., & Bussey, H. (1997). Large scale identification of genes involved in cell surface biosynthesis and architecture in Saccharomyces cerevisiae. *Genetics*, *147*, 435–450.

Magnelli, P., Cipollo, J. F., & Abeijon, C. (2002). A refined method for the determination of Saccharomyces cerevisiae cell wall composition and beta-1,6-glucan fine structure. *Analytical Biochemistry*, *301*, 136–150.

Magnelli, P. E., Cipollo, J. F., & Robbins, P. W. (2005). A glucanase-driven fractionation allows redefinition of Schizosaccharomyces pombe cell wall composition and structure: assignment of diglucan. *Analytical Biochemistry*, *336*, 202–212.

Mangin, L. (1890a). Chimie végétale-Sur la callose, nouvelle substance fondamentale existant dans la membrane. *Comptes Rendus de l'Académie des Sciences*, *111*, 644–647.

Mangin, L. (1890b). Chimie végétale – Sur les réactifs colorants des substances fondamentales de la membrane. *Comptes Rendus de l'Académie des Sciences*, *111*, 120–123.

Manners, D. J., Masson, A. J., Patterson, J. C., Bjorndal, H., & Lindberg, B. (1973). The structure of a beta-(1-->6)-D-glucan from yeast cell walls. *Biochemical Journal*, *135*, 31–36.

Manners, D. J., & Meyer, M. T. (1977). The molecular structures of some glucans from the cell walls of Schizosaccharomyces pombe. *Carbohydrate Research*, *57*, 189–203.

Martin-Cuadrado, A. B., Dueñas, E., Vázquez de Aldana, C. R., & del Rey, F. (2003). The endo-beta-1,3-glucanase eng1p is required for dissolution of the primary septum during cell separation in Schizosaccharomyces pombe. *Journal of Cell Science*, *116*, 1689–1698.

Maubon, D., Park, S., Tanguy, M., Huerre, M., Schmitt, C., Prevost, M. C., Perlin, D. S., Latge, J. P., & Beauvais, A. (2006). AGS3, an alpha(1-3)glucan synthase gene family member of Aspergillus fumigatus, modulates mycelium growth in the lung of experimentally infected mice. *Fungal Genetics and Biology*, *43*, 366–375.

Meikle, P. J., Bonig, I., Hoogenraad, N. J., Clarke, A. E., & Stone, B. A. (1991). The location of (1-3)-beta-glucans in the walls of pollen tubes of *Nicotiana alata* using a (1-3)-beta-glucanspecific monoclonal antibody. *Planta*, *185*, 1–8.

Meikle, P. J., Hoogenraad, N. J., Bonig, I., Clarke, A. E., & Stone, B. A. (1994). A (1-->3,1-->4)-beta-glucan-specific monoclonal antibody and its use in the quantitation and immunocytochemical location of (1-->3,1-->4)-beta-glucans. *The Plant Journal*, *5*, 1–9.

Messner, R., Prillinger, H., Altmann, F., Lopandic, K., Wimmer, K., Molnar, O., & Weigang, F. (1994). Molecular characterization and application of random amplified polymorphic DNA analysis of Mrakia and Sterigmatomyces species. *International Journal of Systematic Bacteriology*, *44*, 694–703.

Miura, N. N., Adachi, Y., Yadomae, T., Tamura, H., Tanaka, S., & Ohno, N. (2003). Structure and biological activities of beta-glucans from yeast and mycelial forms of Candida albicans. *Microbiology and Immunology*, *47*, 173–182.

Montijn, R. C., van Rinsum, J., van Schagen, F. A., & Klis, F. M. (1994). Glucomannoproteins in the cell wall of Saccharomyces cerevisiae contain a novel type of carbohydrate side chain. *Journal of Biological Chemistry*, *269*, 19338–193342.

Mouyna, I., Fontaine, T., Vai, M., Monod, M., Fonzi, W. A., Diaquin, M., Popolo, L., Hartland, R. P., & Latge, J. P. (2000). Glycosylphosphatidylinositol-anchored glucanosyltransferases play an active role in the biosynthesis of the fungal cell wall. *Journal of Biological Chemistry*, *275*, 14882–14889.

Mrsa, V., Seidl, T., Gentzsch, M., & Tanner, W. (1997). Specific labelling of cell wall proteins by biotinylation. Identification of four covalently linked O-mannosylated proteins of Saccharomyces cerevisiae. *Yeast*, *13*, 1145–1154.

Mrsa, V., & Tanner, W. (1999). Role of NaOH-extractable cell wall proteins Ccw5p, Ccw6p, Ccw7p and Ccw8p (members of the Pir protein family) in stability of the Saccharomyces cerevisiae cell wall. *Yeast*, *15*, 813–820.

Muller, A., Ensley, H., Pretus, H., McNamee, R., Jones, E., McLaughlin, E., Chandley, W., Browder, W., Lowman, D., & Williams, D. (1997). The application of various protic acids in the extraction of (1-->3)-beta-D-glucan from Saccharomyces cerevisiae. *Carbohydrate Research*, *299*, 203–208.

Nett, J., Lincoln, L., Marchillo, K., Massey, R., Holoyda, K., Hoff, B., VanHandel, M., & Andes, D. (2007). Putative role of beta-1,3 glucans in Candida albicans biofilm resistance. *Antimicrobial Agents and Chemotherapy*, *51*, 510–520.

Nisini, R., Torosantucci, A., Romagnoli, G., Chiani, P., Donati, S., Gagliardi, M. C., Teloni, R., Sargentini, V., Mariotti, S., Iorio, E., & Cassone, A. (2007). beta-Glucan of Candida albicans cell wall causes the subversion of human monocyte differentiation into dendritic cells. *Journal of Leukocyte Biology*, *82*, 1136–1142.

Nosanchuk, J. D. (2006). Current status and future of antifungal therapy for systemic mycoses. *Recent patents on Anti-Infective Drug Discovery*, *1*, 75–84.

Odds, F. C., Brown, A. J., & Gow, N. A. (2003). Antifungal agents: mechanisms of action. *Trends in Microbiology*, *11*, 272–279.

Ohno, N., Suzuki, I., & Yadomae, T. (1986). Structure and antitumor activity of a beta-1,3-glucan isolated from the culture filtrate of Sclerotinia sclerotiorum IFO 9395. *Chemical and Pharmaceutical Bulletin (Tokyo)*, *34*, 1362–1365.

Ohno, N., Terui, T., Chiba, N., Kurachi, K., Adachi, Y., & Yadomae, T. (1995). Resistance of highly branched (1-->3)-beta-D-glucans to formolysis. *Chemical and Pharmaceutical Bulletin (Tokyo)*, *43*, 1057–1060.

Olafsdottir, E. S., & Ingolfsdottir, K. (2001). Polysaccharides from lichens: structural characteristics and biological activity. *Planta Medica*, *67*, 199–208.

Osumi, M. (1998a). The ultrastructure of yeast: cell wall structure and formation. *Micron*, *29*, 207–233.

Osumi, M., Sato, M., Ishijima, S. A., Konomi, M., Takagi, T., & Yaguchi, H. (1998b). Dynamics of cell wall formation in fission yeast, Schizosaccharomyces pombe. *Fungal Genetics and Biology*, *24*, 178–206.

Palleschi, A., Bocchinfuso, G., Coviello, T., & Alhaique, F. (2005). Molecular dynamics investigations of the polysaccharide scleroglucan: first study on the triple helix structure. *Carbohydrate Research*, *340*, 2154–2162.

Pardo, M., Monteoliva, L., Pla, J., Sanchez, M., Gil, C., & Nombela, C. (1999). Two-dimensional analysis of proteins secreted by Saccharomyces cerevisiae regenerating protoplasts: a novel approach to study the cell wall. *Yeast*, *15*, 459–472.

Paris, S., Debeaupuis, J. P., Crameri, R., Carey, M., Charles, F., Prevost, M. C., Schmitt, C., Philippe, B., & Latge, J. P. (2003). Conidial hydrophobins of Aspergillus fumigatus. *Applied and Environmental Microbiology*, *69*, 1581–1588.

Perez, P., & Ribas, J. C. (2004). Cell wall analysis. *Methods*, *33*, 245–251.

Pittet, M., & Conzelmann, A. (2007). Biosynthesis and function of GPI proteins in the yeast Saccharomyces cerevisiae. *Biochimica et Biophysica Acta*, *1771*, 405–420.

Popolo, L., Gilardelli, D., Bonfante, P., & Vai, M. (1997). Increase in chitin as an essential response to defects in assembly of cell wall polymers in the ggp1delta mutant of Saccharomyces cerevisiae. *The Journal of Bacteriology*, *179*, 463–469.

Popolo, L., & Vai, M. (1999). The Gas1 glycoprotein, a putative wall polymer cross-linker. *Biochimica Et Biophysica Acta-General Subjects*, *1426*, 385–400.

Prado, S. R. T., Gorin, P. A., Stuelp, P. M., Honda, N. K., & Iacomini, M. (1999). An unusual juxtaposition of polysaccharide components of Collema leptosporum. *Carbohydrate Polymers*, *40*, 271–276.

Prillinger, H., Dörfler, C., Laaser, G., & Hauska, G. (1990). Ein Beitrag zur Systematik und Entwicklungsbiologie höherer Pilze Hefe-Typen der Basidiomyceten. Teil III: Ustilago-typ. *Zeitschrift für Mykologie*, *56*, 251–278.

Prillinger, H., Dörfler, C., Laaser, G., & Lockau, W. (1991). Ein Beitrag zur Systematik und Entwicklungsbiologie höherer Pilze Hefe-Typen der Basidiomyceten. Teil II: Microbotryum-typ. *Botanica Acta*, *104*, 5–17.

Prillinger, H., Oberwinkler, F., Umile, C., Tlachac, K., Bauer, R., Dörfler, C., & Taufratzhofer, E. (1993). Analysis of cell wall carbohydrate (neutral sugars) from ascomycetes and basidiomycetes yeast with and without derivatization. *The Journal of General and Applied Microbiology*, *39*, 1–34.

Ragni, E., Fontaine, T., Gissi, C., Latge, J. P., & Popolo, L. (2007). The Gas family of proteins of Saccharomyces cerevisiae: characterization and evolutionary analysis. *Yeast*, *24*, 297–308.

Ram, A. F., Wolters, A., Ten Hoopen, R., & Klis, F. M. (1994). A new approach for isolating cell wall mutants in Saccharomyces cerevisiae by screening for hypersensitivity to calcofluor white. *Yeast*, *10*, 1019–1030.

Ram, A. F., Kapteyn, J. C., Montijn, R. C., Caro, L. H., Douwes, J. E., Baginsky, W., Mazur, P., van den Ende, H., & Klis, F. M. (1998). Loss of the plasma membrane-bound protein Gas1p in

Saccharomyces cerevisiae results in the release of beta1,3-glucan into the medium and induces a compensation mechanism to ensure cell wall integrity. *The Journal of Bacteriology*, *180*, 1418–1424.

Rappleye, C. A., Eissenberg, L. G., & Goldman, W. E. (2007). Histoplasma capsulatum alpha-(1,3)-glucan blocks innate immune recognition by the beta-glucan receptor. *Proceedings of the National Academy of Sciences of the United States of America*, *104*, 1366–1370.

Rees, D. A., Morris, E. R., Thom, D., & Madden, J. K. Aspinall G. O. (Ed.). (1982). *The Polysaccharides: Vol. I*. New York: Academic Press.

Rico, H., Carrillo, C., Aguado, C., Mormeneo, S., & Sentandreu, R. (1997). Initial steps of wall protoplast regeneration in Candida albicans. *Research in Microbiology*, *148*, 593–603.

Roberts, R. L., & Cabib, E. (1982). Serratia marcescens chitinase: one-step purification and use for the determination of chitin. *Analytical Biochemistry*, *127*, 402–412.

Rodriguez-Del Valle, N., Rosario, M., & Torres-Blasini, G. (1983). Effects of pH, temperature, aeration and carbon source on the development of the mycelial or yeast forms of Sporothrix schenckii from conidia. *Mycopathologia*, *82*, 83–88.

Rodriguez-Pena, J. M., Cid, V. J., Arroyo, J., & Nombela, C. (2000). A novel family of cell wall-related proteins regulated differently during the yeast life cycle. *Molecular and Cellular Biology*, *20*, 3245–3255.

Sassaki, G. L., Reis, R. A., Gorin, P. A., & Iacomini, M. (2002). The glucans of lichenized fungi. *Mitteilungen aus dem Institut für allgemeine Botanik in Hamburg*, *30*, 195–211.

Schmid, F., Stone, B. A., McDougall, B. M., Bacic, A., Martin, K. L., Brownlee, R. T., Chai, E., & Seviour, R. J. (2001). Structure of epiglucan, a highly side-chain/branched (1-->3;1-->6)-beta-glucan from the micro fungus Epicoccum nigrum Ehrenb. ex Schlecht. *Carbohydrate Research*, *331*, 163–171.

Sentandreu, R., Herrero, E., Elorza, M. V., Rico, H., & Pastor, J. (1983). Synthesis and assembly of wall polymers on regenerating yeast protoplasts. *Experientia Supplementum*, *46*, 187–195.

Sietsma, J. H., Eveleigh, D. E., & Haskins, R. H. (1969). Cell wall composition and protoplast formation of some Oomycete species. *Biochimica Et Biophysica Acta*, *184*, 306–317.

Sietsma, J. H., & Wessels, J. G. (1981). Solubility of (1-->3)-beta-D/(1-->6)-beta-D-glucan in fungal walls: importance of presumed linkage between glucan and chitin. *Journal of General Microbiology*, *125*, 209–212.

Sietsma, J. H., & Wessels, J. G. (1990). The occurrence of glucosaminoglycan in the wall of Schizosaccharomyces pombe. *Journal of General Microbiology*, *136*, 2261–2265.

Sogin, M. L., Morrison, H. G., Hinkle, G., & Silberman, J. D. (1996). Ancestral relationships of the major eukaryotic lineages. *Microbiologia*, *12*, 17–28.

Stevens, D. A., Ichinomiya, M., Koshi, Y., & Horiuchi, H. (2006). Escape of Candida from caspofungin inhibition at concentrations above the MIC (paradoxical effect) accomplished by increased cell wall chitin; evidence for beta-1,6-glucan synthesis inhibition by caspofungin. *Antimicrobial Agents and Chemotherapy*, *50*, 3160–3161.

Stokke, B. T., Elgsaeter, A., Hara, C., Kitamura, S., & Takeo, K. (1993). Physicochemical properties of (1-->6)-branched (1-->3)-beta-D-glucans. 1. Physical dimensions estimated from hydrodynamic and electron microscopic data. *Biopolymers*, *33*, 561–573.

Stuelp, P. M., Carneiro Leão, A. M. A., Gorin, P. A. J., & Iacomini, M. (1999). The glucans of Ramalina celastri: relation with chemotypes of other lichens. *Carbohydrate Polymers*, *40*, 101–106.

Sumitomo, N., Saeki, K., Ozaki, K., Ito, S., & Kobayashi, T. (2007). Mutanase from a Paenibacillus isolate: nucleotide sequence of the gene and properties of recombinant enzymes. *Biochimica Et Biophysica Acta*, *1770*, 716–724.

Tanret, C. (1897). Recherches sur les champignons. *Bulletin de la Société Chimique de France*, *17*, 921–927.

Travassos, L. R., & Lloyd, K. O. (1980). Sporothrix schenckii and related species of Ceratocystis. *Microbiological Reviews*, *44*, 683–721.

Tyler, B. M., Tripathy, S., Zhang, X., Dehal, P., Jiang, R. H., Aerts, A., Arredondo, F. D., Baxter, L., Bensasson, D., Beynon, J. L., Chapman, J., Damasceno, C. M., Dorrance, A. E., Dou, D., Dickerman, A. W., Dubchak, I. L., Garbelotto, M., Gijzen, M., Gordon, S. G., Govers, F., Grunwald, N. J., Huang, W., Ivors, K. L., Jones, R. W., Kamoun, S., Krampis, K., Lamour, K. H., Lee, M. K., McDonald, W. H., Medina, M., Meijer, H. J., Nordberg, E. K., Maclean, D. J., Ospina-Giraldo, M. D.,Morris, P. F., Phuntumart, V., Putnam, N. H., Rash, S., Rose, J. K., Sakihama, Y., Salamov, A. A., Savidor, A., Scheuring, C. F., Smith, B. M., Sobral, B. W., Terry, A., Torto-Alalibo, T. A., Win, J., Xu, Z., Zhang, H., Grigoriev, I. V., Rokhsar, D. S., & Boore, J. L. (2006). Phytophthora genome sequences uncover evolutionary origins and mechanisms of pathogenesis. *Science*, *313*, 1261–1266.

Ustinova, I., Krienitz, L., & Huss, V. A. (2000). Hyaloraphidium curvatum is not a green alga, but a lower fungus; Amoebidium parasiticum is not a fungus, but a member of the DRIPs. *Protist*, *151*, 253–262.

Utsugi, T., Minemura, M., Hirata, A., Abe, M., Watanabe, D., & Ohya, Y. (2002). Movement of yeast 1,3-b-glucan synthase is essential for uniform cell wall synthesis. *Genes to Cells*, *7*, 1–9.

van der Vaart, J. M., van Schagen, F. S., Mooren, A. T., Chapman, J. W., Klis, F. M., & Verrips, C. T. (1996). The retention mechanism of cell wall proteins in Saccharomyces cerevisiae. Wall-bound Cwp2p is beta-1,6-glucosylated. *Biochimica Et Biophysica Acta*, *1291*, 206–214.

Van Laere, A. J., Carlier, A. R., & Van Assche, J. A. (1977). Cell wall carbohydrates in Phycomyces blakeslesanus Burgeff. *Archives of Microbiology*, *112*, 303–306.

Vorgias, C. E., Kingswell, A. J., Dauter, Z., & Oppenheim, A. B. (1992). Crystallization of recombinant chitinase from the cloned chiA gene of Serratia marcescens. *Journal of Molecular Biology*, *226*, 897–898.

Wang, M. C., & Bartnicki-Garcia, S. (1973). Novel phosphoglucans from the cytoplasm of Phytophthora palmivora and their selective occurrence in certain life cycle stages. *Journal Biological Chemistry*, *248*, 4112–4118.

Weig, M., Jansch, L., Gross, U., De Koster, C. G., Klis, F. M., & De Groot, P. W. J. (2004). Systematic identification in silico of covalently bound cell wall proteins and analysis of protein-polysaccharide linkages of the human pathogen Candida glabrata. *Microbiology*, *150*, 3129–3144.

Weijman, A. C. M., & Golubev, W. I. (1987). Carbohydrate patterns and taxonomy of yeast and yeast-like fungi. *Studies in Mycology*, *30*, 361–371.

Williams, D. L., McNamee, R. B., Jones, E. L., Pretus, H. A., Ensley, H. E., Browder, I. W., & Di Luzio, N. R. (1991). A method for the solubilization of a (1-->3)-beta-D-glucan isolated from Saccharomyces cerevisiae. *Carbohydrate Research*, *219*, 203–213.

Woon, F. A., Singh, R. P., & Hodgson, W. A. (1971). *Phytopathology*, *61*, 1006.

Yamada, M., & Miyazaki, T. (1976). Ultrastructure and chemical analysis of the cell wall of *Pythium debaryanum*. *Japanese Journal of Microbiology*, *20*, 83–91.

Zevenhuizen, L. P., & Bartnicki-Garcia, S. (1970). Structure and role of a soluble cytoplasmic glucan from *Phytophthora cinnamomi*. *Journal of General Microbiology*, *61*, 183–188.

Zverlov, V. V., Volkov, I. Y., Velikodvorskaya, T. V., & Schwarz, W. H. (1997). Highly thermostable endo-1,3-beta-glucanase (laminarinase) LamA from Thermotoga neapolitana: nucleotide sequence of the gene and characterization of the recombinant gene product. *Microbiology*, *143*, 1701–1708.

CHAPTER 4.4.1

# Callose in Cell Division

**Roy C. Brown and Betty E. Lemmon**
*Department of Biology, University of Louisiana-Lafayette, Lafayette, LA, USA*

## 1.A. Introduction

Among the numerous functions of callose in land plants, nowhere is the role more fundamental than in cytokinesis. Emerging molecular and comparative information shows that callose is briefly present in new walls of all groups of land plants, in all types of cytokinesis, and that certain genes for callose synthase may be cytokinesis specific.

Plant cytokinesis is achieved by deposition of a cell plate mediated by the phragmoplast. It encompasses at least five distinct phases: 1) assembly of the phragmoplast consisting of opposing arrays of co-aligned microtubules and F-actin; 2) transport of vesicles/membrane components and other molecules to the division site; 3) development and expansion of the cell plate; 4) deposition of callose on luminal surfaces of cell plate membranes; and 5) union of the cell plate with the parental walls. Before callose is replaced by cellulose and other wall constituents of mature walls, the new walls are flexible and easily joined with each other and with parental walls. This is an important feature of the complex process of cytokinesis in plants.

Phragmoplasts are initiated at the interface of opposing microtubule arrays and develop into distinctive structures consisting of highly organized cytoskeletal fibers separated by a midzone where new walls will be deposited. The midzone appears as a dark zone when the cytoskeleton is labelled by immunofluorescence (see Figs 1–3 and 5–7).

In meristematic cells (Figs 1–3), such as those of root and shoot tips, the phragmoplast is initiated at the interface of opposing microtubules in the interzone between telophase nuclei. As the cell plate is deposited (Fig. 4), the phragmoplast expands centrifugally as a ring to join with parental walls at the division site previously marked by the predictive preprophase band (PPB) of cortical microtubules (Fig. 1). Such phragmoplasts have been termed interzonal, primary or conventional.

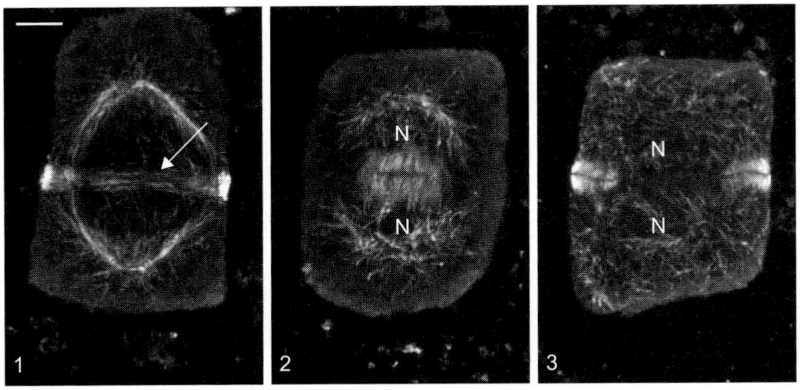


**Figs 1–3: The cytokinetic apparatus in meristematic cells of the daylily, *Hemerocallis fulva*. (1) The PPB (arrow) predicts the future division site. (2) The phragmoplast forms in the interzone between sister nuclei (N). (3) The phragmoplast expands as a ring to the periphery as it guides the forming cell plate to junction with parental walls at the site previously prepared by the PPB. Bar = 5 μm.**

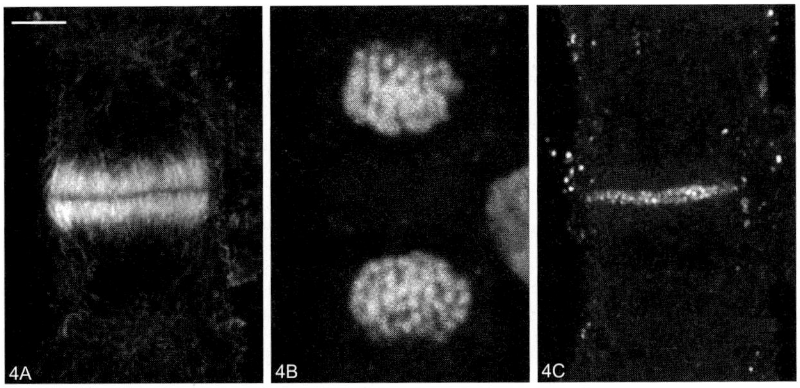


**Fig. 4: The cell plate in a meristematic cell of daylily, *Hemerocallis fulva*, triple stained for microtubules (A), nuclei (B) and callose (C). The cell plate grows as an enlarging disc-shaped structure in association with the phragmoplast that expands as a ring. Callose produced at the luminal membrane surfaces of the cell plate was detected with a (1,3)-ß-glucan antibody. Bar = 3.5 μm. The colour specifications refer to colours in panels.**

In syncytial systems (Figs 5–7) such as nuclear endosperm, many phragmoplasts are initiated simultaneously at the interfaces of microtubules radiating from surfaces of interphase nuclei. These radial microtubule systems (Fig. 5) mark the perimeter of nuclear–cytoplasmic domains (NCDs) and determine placement of walls. Phragmoplasts form (Fig. 6), expand and fuse with

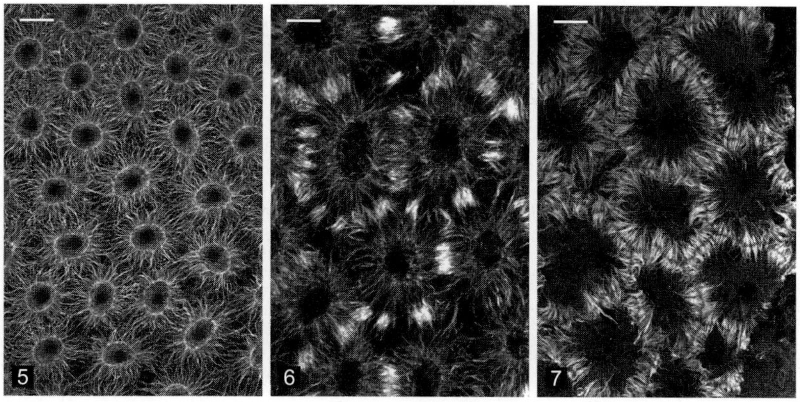


**Figs 5–7: Phragmoplast development in a sheet of syncytial endosperm of a mustard, *Coronopus didymus*, seed seen in face view. (5) Microtubules radiating from interphase nuclei in the undivided cytoplasm organize the cytoplasm into nuclear–cytoplasmic domains (NCDs). Bar = 22 μm. (6) Phragmoplasts (mini-phragmoplasts) are initiated at the boundaries of NCDs. Bar = 16.5 μm. (7) Fusion of phragmoplasts surrounding NCDs results in a complex of phragmoplasts. Bar = 16.5 μm.**

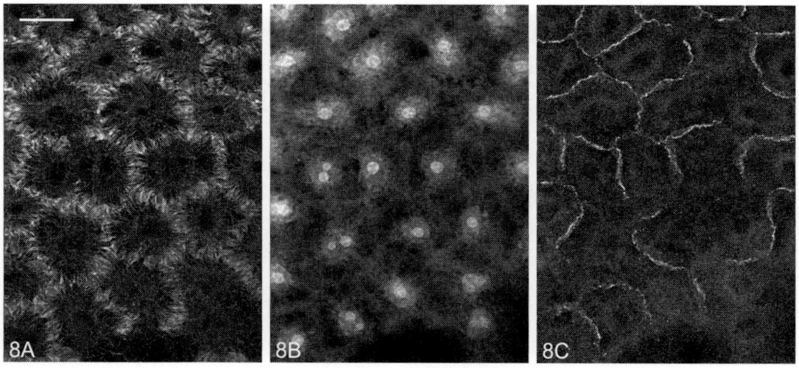


**Fig. 8: Face view of a sheet of cellularizing endosperm of a mustard, *Coronopus didymus*, seed. A complex of fused phragmoplasts surrounding NCDs in the syncytium directs simultaneous deposition of a network of walls as revealed by callose localization with (1,3)-β-glucan antibody. (A) Microtubules, (B) nuclei, (C) developing callosic cell plates. Bar = 22 μm.**

each other to form complexes (Fig. 7) in which a network of walls is deposited (Fig. 8). These phragmoplasts have been termed adventitious (not formed in the interzone), secondary or non-conventional. Both types of phragmoplasts are structurally similar and function in mediating new wall deposition in the midzone where callose is produced at the newly formed membranes.

The major cytoskeletal elements of phragmoplasts are microtubules and actin filaments, both of which are in proper orientation for transport of vesicles to the midzone. It appears that microtubules are of prime importance in the transport of vesicles; if microtubules are perturbed, the cell plate does not form. F-actin appears to be important in directing expansion of the cell plate. Reorganization of microtubules to the perimeter where more vesicles are brought into alignment is controlled in part by cell plate assembly itself (Yasuhara and Shibaoka, 2000). Once the cell plate reaches the stage when callose is deposited, the microtubules are present only at the outer margins.

Cell plate assembly is a complex membrane fusion process (Mayer and Jürgens, 2004; Dhonukshe et al., 2006). As described by Staehelin and coworkers (e.g. Samuels et al., 1995; Otegui and Staehelin, 2000; Seguí-Samarro et al., 2004), it is a continuum beginning with accumulated vesicles associated with a cell plate assembly matrix (CPAM) in the midzone, fusion of vesicles into a tubular–vesicular network (TVN), then to a tubular network that widens and coalesces into a fenestrated sheet. The CPAM is thought to stabilize microtubules and promote membrane fusion (Haas and Otegui, 2007). The KNOLLE protein of Arabidopsis, a syntaxin-related protein, and its binding partner KEULE, a Sec1 homolog, are concentrated in cell plates where they play a role in vesicle fusion (Waizenegger et al., 2000). After initial vesicle fusion, thin tubes (fusion tubes) grow from the fusing vesicles and intertwine to produce the TVN. The formation and stabilization of fusion tubes is mediated by dynamin-like proteins (soybean phragmoplastin and its homologs, ADL1A and ADL1E, from Arabidopsis) (Mayer and Jürgens, 2004).

In meristematic cells, cell plate formation begins in the early phragmoplast occupying the centre of the interzonal region between sister nuclei and grows centrifugally as the phragmoplast expands to join parental walls (Fig. 4). The transition from TVN to tubular network is marked by disappearance of the CPAM and microtubules and a reduction in the number of clatharin-coated budding bodies. It is at this time that callose and matrix polysaccharides are deposited. As the wall is assembled within the lumen of the membranous network, wide tubules expand to produce a fenestrated or almost complete sheet. The fenestrated plate continues to grow centrifugally as microtubule assembly, vesicle delivery and vesicle fusion continues at the margins. Finally, as the cell plate approaches the parental cell wall, numerous finger-like projections reach out to grasp the parental cell membrane. This bracing is followed by maturation of the cell plate, which features closing of the fenestrations and the beginning of cellulose synthesis. As the cellulosic wall is assembled callose disappears except for its association with plasmodesmata. The removal of callose is mediated by specific (1,3)-ß-glucan endohydrolases (Martin and Somers, 2004).

The stages of cell plate formation are similar for meristematic and syncytial types (Mayer and Jürgens, 2004; Otegui, 2007). However, in the formation of anticlinal walls during cellularization in syncytia only one edge of the new wall joins with the parental central cell wall and the opposite edge continues to grow into the centre of the cell. Periclinal walls join with callosic anticlinal walls and both types remain in a juvenile callosic state for a prolonged period.

## 1.B. Alveolation in Syncytial Systems

One of the most dramatic and potentially useful systems for understanding the role of callose in wall formation is the process of alveolation that initiates cellularization of large syncytia. This unusual type of wall formation occurs in the syncytial endosperm (Olsen et al., 1995; Brown and Lemmon, 2007) and the syncytial female gametophytes of gymnosperms (Brown and Lemmon, 2008). In both cases, the syncytium lines the periphery of a large chamber and encloses a central vacuole (Fig. 9). The multinucleate cytoplasm is organized into equidistant hexagonally packed NCDs by microtubules that radiate from the nuclei (Fig. 5). Polarization of the NCDs in axes perpendicular to the outer central cell wall, rearrangement of the microtubules, and vacuolation of the cytoplasm prepare the syncytium for cellularization.

The anticlinal walls are initiated in thin layers of cytoplasm at the perimeters of the vacuolated NCDs. The cell plates begin in a patch-work pattern at the boundaries of NCDs and coalesce into thin wavy walls that join immediately with the nearby central cell wall. The nuclear-based microtubules become distinctly polarized: those emanating from the inner tips of nuclei become arranged into crowns of microtubules in the undivided canopy of cytoplasm adjacent to the central vacuole. Phragmoplasts that form adventitiously (not in interzones) at the interfaces of microtubules from adjacent NCDs (Figs 5–7) mediate continued unidirectional growth of the network of anticlinal walls. This highly asymmetrical cell plate growth is similar to that of 'polarized' cytokinesis in walled vacuolated cells and cambial cells (Cutler and Ehrhardt, 2002).

Meeting no wall on the opposite (vacuolar) side, the walls continue to grow toward the central vacuole forming tightly packed elongated open-ended compartments called alveoli (Fig. 10) because of their resemblance to compartments of a honeycomb. As the anticlinal walls continue to grow simultaneously, the entire canopy of undivided cytoplasm containing the nuclei and complex of phragmoplasts (Fig. 7) adjacent to the central vacuole moves inward. The walls grow in association with adventitious phragmoplasts (Fig. 11). There is considerable transfer of vacuolar contents into the alveoli as they grow and the central vacuole decreases in size.

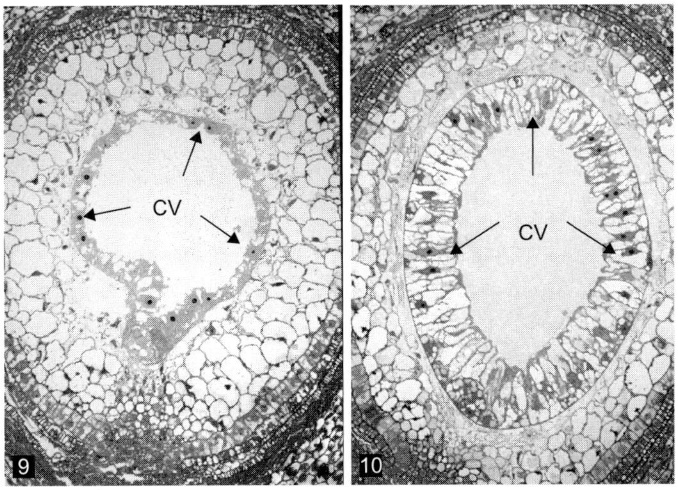


Figs 9–10: Cross-sections of rice (*Oryza sativa*) grains showing early developing endosperm. (9) At the syncytial stage, the multinucleate cytoplasm (arrows) lines the periphery of the central cell and surrounds the large central vacuole (CV). (10) Cellularization of the syncytium is initiated by formation of alveoli (arrows). Anticlinal walls have formed between adjacent nuclei, joined with the central cell wall and continue to grow in association with adventitious phragmoplasts in the canopy of cytoplasm (seen in face view in Fig. 8). The anticlinal walls have formed hexagonally packed open-ended alveoli in which the nuclei are preparing to divide periclinally. Compare their position in the alveoli to that of nuclei in a canopy of cytoplasm shown in Fig. 12. Bar = 50 μm.

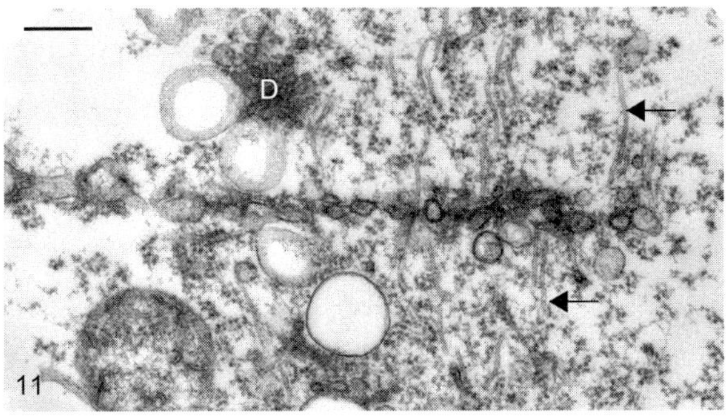


Fig. 11: TEM of the leading edge of an anticlinal wall in the developing endosperm of rice (*Oryza sativa*) seen in side view. The fusion of vesicles is mediated by adventitious phragmoplasts (seen in face view in Fig. 8) in the canopy of cytoplasm adjacent to the central vacuole. Arrows indicate phragmoplast microtubules. D, dictyosome. Bar = 0.3 μm.

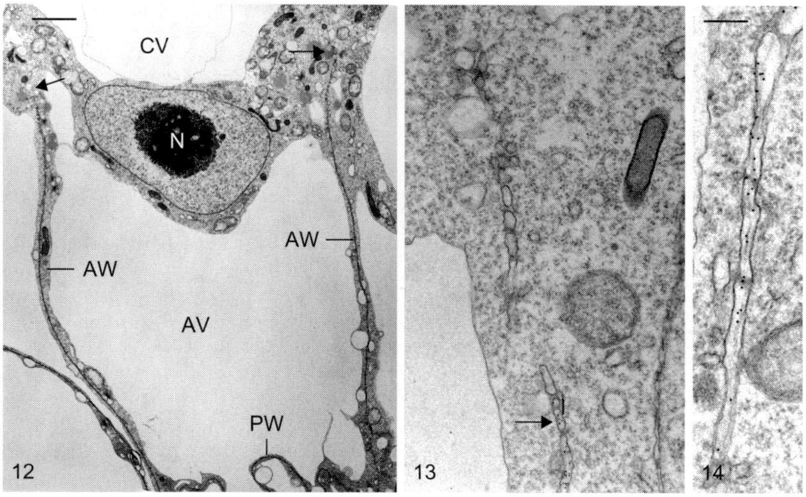


**Figs 12–14: Wall development in endosperm of rice (*Oryza sativa*). TEM of a displaced (second tier) alveolus in longitudinal section. (12) The leading edges (arrows) of the unidirectionally growing anticlinal walls (AW) are assembled in the canopy of cytoplasm adjacent to the central vacuole (CV). AV, alveolar vacuole; PW, periclinal wall; N, nucleus. Bar = 3 μm. (13) Detail of the leading edge of an alveolar wall labelled with (1,3)-ß-glucan antibody gold shows that no callose is deposited until the anticlinal wall becomes lamellar (arrow). Bar = 0.5 μm. (14) Anti-(1,3)-ß-glucan gold-labelled callose in a continuous region of anticlinal wall. Bar = 0.25 μm.**

The continued polar growth of anticlinal walls makes it possible to determine with great accuracy the stages of wall deposition. The appearance of callose in endosperm walls has been studied in cereals (Brown et al., 1997; Wilson et al., 2006; Philippe et al., 2006) and Arabidopsis (Otegui and Staehelin, 2000) using the highly specific antibody to (1,3)-β-glucan developed by Meikle et al. (1991). Callose is present in anticlinal walls after the initial stage of vesicle fusion results in lamellar walls. This is clearly seen in the long stretches of anticlinal wall where the growing tip near the central vacuole is still in the stage of vesicle fusion (Figs 12–14) and lacks callose (Fig. 13) and the older more consolidated portion contains abundant callose (Fig. 14).

Anticlinal wall growth stops and the phragmoplasts are dismantled as a wave of mitosis and cytokinesis occurs in the elongated alveoli. The alveoli shown in Fig. 10 contain nuclei that have moved from the canopy of cytoplasm adjacent to the central vacuole to a more central position in preparation for periclinal division. The periclinal walls form in association with interzonal phragmoplasts that expand to join with the surrounding callosic anticlinal (parental) walls (Brown et al., 1997). Callose is produced in the periclinal wall and the two walls are indistinguishable after junction (Brown et al., 1997; Wilson et al., 2006). The periclinal walls cut off an outer layer of walled cells and an inner layer of displaced alveoli adjacent to the central vacuole. A displaced (interior) alveolus is shown in Fig. 12. Crowns of microtubules again emanate from nuclei in the inner layer of alveoli and phragmoplasts formed adventitiously at the interfaces of the opposing microtubule systems mediate continued unidirectional growth of the anticlinal walls. The new anticlinal cell plate will never join a parental wall. It fuses with the leading edge of the previously deposited anticlinal wall and continues in the midzone of the newly organized phragmoplasts on its path toward the centre. This renewed activity of anticlinal wall deposition may occur two or more times after interruption by periclinal divisions until closure in the centre. The two processes of wall formation, anticlinal between non-sister nuclei and periclinal between sister nuclei, alternate until the endosperm is cellularized.

Both anticlinal and periclinal walls remain in a prolonged juvenile callosic stage during cellularization (Brown et al.,1997; Wilson et al., 2006; Philippe et al., 2006). These walls are flexible and easily join with each other, an important feature of the unusual pattern of wall formation in large syncytia. The presence of callose in endosperm walls could be advantageous in that it provides developmental plasticity that can accommodate the rapidly enlarging seed and also provides a ready source of glucose that can be used either as a source of metabolic energy or substrate in the synthesis of other wall components (Otegui, 2007). For details of the developmental appearance of key polysaccharides during wall maturation in cereal endosperm, see Philippe et al. (2006) and Wilson et al. (2006).

## 1.C. Deposition of Callose

Studies on the role of callose in cell plate development, as in other processes, have been hampered by difficulties in isolating functional callose synthase (CalS). Like cellulose synthase (CelS), CalS is probably a multi-subunit membrane-associated protein making it difficult to extract and purify as the complex disassociates resulting in loss of activity. CalS is often found closely associated with CelS. It was once suggested that CelS and CalS could be the same enzyme that switches linkage of its products in response to interaction with associated proteins, phosphorylation or binding of calcium (Delmer, 1999). However, recent findings demonstrate that CelS and CalS belong to different *gene* families: CelS belongs to family *GT2* (http://afmb.cnrs-mrs.fr/CAZY/) and CalS is in family *GT48* (see also Chapter 3.3.4).

Cloning of genes encoding the 200-kDa catalytic unit of (1,3)-ß-glucan synthase from yeasts and other fungi facilitated the cloning of CalS cDNAs from higher plants including Arabidopsis. These findings show that the conserved D, D,D and QXXRW motifs of the CelS genes are absent in FSK1 homologs from yeast (Ishiguro et al., 1997) and the Arabidopsis CalS1 gene (Hong et al., 2001a). The Arabidopsis genome contains 12 homologous sequences. All seem to be functional and may be expressed either at very low levels or induced under specific growth conditions or stress (see also Chapter 3.3.4).

Of particular interest is GSL 6 (CalS1), which is specifically expressed in cell plate deposition (Hong et al., 2001a). A CalS1::GFP protein under the control of the 35S promoter was expressed in tobacco BY-2. During interphase a pattern of punctate dots was seen throughout the cytoplasm whereas the forming cell plates in cells undergoing cytokinesis were brightly fluorescent. When the Aniline Blue fluorochrome was used to stain callose, the cell plates were brighter in cells overexpressing the CalS::GFP construct than in control cells, and the fluorescence persisted much longer after completion of the cell plate.

In a yeast two-hybrid system, CalS1 interacts with UDP-glucose transferase (UGT1) and phragmoplastin (Hong et al., 2001a). It also associates with ROP1, possibly forming the membrane complex that functions in cell plate formation (Hong et al., 2001b). Another possible component is sucrose synthase (SuSy). UDP-glucose from SuSy could be transferred to CalS1 through UGT1 to form a substrate channel that facilitates the rapid deposition of callose in cell plate formation.

Callose deposition coincides with and is thought to facilitate flattening and stiffening of the developing cell plate. Detailed investigations of the cell plate (Samuels et al., 1995) indicate that callose first appears in the smooth tubular network and coincides with disappearance of phragmoplast microtubules from the region. Caffeine interrupts development of the TVN and the subsequent deposition of callose (Samuels and Staehelin, 1996). Experimental treatment of cells with caffeine before and after entry into the mitotic cycle showed that caffeine neither arrests the accumulation of CalS nor inhibits the enzyme itself (Yasuhara, 2005). Depolymerization of the microtubules in the central region of the phragmoplast is closely related to the start of callose deposition but the exact signalling pathway remains unknown. Experimental evidence implicates the tobacco MAPK kinase kinase, NPK1, in the signalling process but suggests that it may be downstream from the caffeine-sensitive stage (Yasuhara, 2005).

It can be concluded that callose is synthesized at membranes as it is never observed in Golgi stacks or vesicles (e.g. Wilson et al., 2006). CalS may be regulated by calcium, as calcium is known to be required for callose synthesis (Kakimoto and Shibaoka, 1992) and high concentrations of calcium occur in the cell plate. Callose also appears at the surface of the parental wall where the growing cell plate approaches the division site formally occupied by the PPB (Samuels et al., 1995). The importance of the designated division site in cells with PPBs has recently been emphasized by a study demonstrating that the microtubule-associated protein AIR9, which decorates microtubules of the PPB, reappears at cell plate insertion and has a function in new wall maturation (Buschmann et al., 2006). When cell plates are forced to join at sites other than that designated by the PPB, they exhibit prolonged staining for callose indicating that maturation is delayed. In general, any impairment of microtubule dynamics appears to cause prolonged callose production. Experiments on the liverwort *Riella* suggested that cortical microtubules have a role in regulation of callose and cellulose biosynthesis (Scherp et al., 2002). Their putative binding to the CelS site via an unknown binding protein activates cellulose synthesis, and the destruction of this complex results in activation of the nearby CalS, resulting in callose production. Cell plates formed in the presence of microtubule perturbing herbicides are not only abnormal in shape (branching and meandering about in the cell) but are greatly enriched in callose and accumulate very low levels of cellulose (Vaughn, 2006).

## 1.D. The Function of Callose in Cytokinesis

The precise role of callose in cytokinesis is not known. In the forming cell plate, callose is thought to mechanically stabilize the delicate membrane networks and 'create a spreading force' that widens the tubules of the TVN to produce the fenestrated plate (Samuels et al., 1995; Samuels and Staehelin, 1996). Callose may serve as a scaffold or matrix onto or into which the more permanent polysaccharides and proteins of the mature wall may be deposited (Stone and Clarke, 1992). The formation of hydrogels by callose would allow insertion or co-gelation with other polymers in the cell plate. The subsequent removal of callose is presumably mediated by (1,3)-ß-glucan endohydrolases, such as those expressed in oat endosperm (Martin and Sommers, 2004), without affecting the other polymers.

Callose is also involved in the junction of the cell plate with the parental wall, where its role may be related to the general role of callose as part of the wound response cascade. We speculate that the cell may anticipate rupture of the parental membrane/wall that will be necessary for the new cell plate to complete cytokinesis and is able to mobilize the cellular machinery that produces callose for a rapid and pliable repair until the strut and matrix wall can be reconstructed.

Rearrangement of the cortical microtubule system into the PPB in walled cells may reflect membrane remodelling that moves CalS and CelS complexes as well as putative cortical MTOCs ($\gamma$-tubulin complexes) to the future division site. Callose appears both in the cell plate and at the site of parental wall rupture, and cortical microtubules reappear in daughter cells first in the vicinity of the site of cytokinesis. No PPB occurs in the cytokinetic apparatus of cells without cellulose walls, i.e. those of the reproductive lineage (Brown and Lemmon, 2001). This suggests that the PPB is advantageous in preparing the cellulosic walls for junction with the cell plate and is not necessary for junction of non-cellulosic walls. In the periclinal division of alveoli in endosperm, no PPBs are formed and the cell plate joins with parental (anticlinal) walls that are rich in callose. Callosic walls appear to be particularly efficient at fusing with each other, thus facilitating the cellularization of syncytial systems. Although the specific role for callose in the cell plate is still not completely known, its ubiquitous appearance attests to its importance. Callose synthesis at membranes of the forming cell plate may serve as a cytokinetic check point for the redistribution of phragmoplast microtubules into a ring-like structure at the periphery of the cell plate. It is likely that the nature of callose allows for the rapid assembly of a pliable cell plate that can accommodate the strain of adjacent growing walls as well as serve as a scaffold, and possibly substrate, for other polysaccharides in development of the primary wall.

## References

Brown, R. C., & Lemmon, B. E. (2001). Cytoskeleton and the spatial control of cytokinesis in the plant life cycle. *Protoplasma*, *215*, 35–49.

Brown, R. C., & Lemmon, B. E. (2007). The developmental biology of cereal endosperm. *Plant Cell Monographs*, *8*, 1–20.

Brown, R. C., & Lemmon, B. E. (2008). Microtubules in early development of the megagametophyte of *Ginkgo biloba*. *Journal of Plant Research*, *121*, 397–406.

Brown, R. C., Lemmon, B. E., Stone, B. A., & Olsen, O.-A. (1997). Cell wall (1-3) and (1-3, 1-4) ß-glucans during early grain development in rice (*Oryza sativa* L.). *Planta*, *202*, 414–426.

Buschmann, H., Chan, J., Sanchez-Pulido, L., Andrade-Navarro, M. A., Doonan, J. H., & Lloyd, C. W. (2006). Microtubule-Associated AIR9 recognizes the cortical division site at preprophase and cell-plate insertion. *Current Biology*, *16*, 1938–1943.

Cutler, S. R., & Ehrhardt, D. W. (2002). Polarized cytokinesis in vacuolated cells of *Arabidopsis*. *Proceedings of the National Academy of Sciences of the USA*, *99*, 2812–2817.

Delmer, D. P. (1999). Cellulose biosynthesis: Exciting times for a difficult field of study. *Annual Review of Plant Physiology and Molecular Biology*, *50*, 245–276.

Dhonukshe, P., Baluska, F., Schlicht, M., Hlavacka, A., Samaj, J., Friml, J., & Gadella, T. W. J. (2006). Endocytosis of cell surface material mediates cell plate formation during plant cytokinesis. *Developmental Cell*, *10*, 137–150.

Haas, T. J., & Otegui, M. S. (2007). Electron tomography in plant cell biology. *Journal of Integrative Plant Biology*, *49*, 1091–1099.

Hong, Z., Delauney, A. J., & Verma, D. P. S. (2001a). A cell plate-specific callose synthase and its interaction with phragmoplastin. *Plant Cell*, *13*, 755–768.

Hong, Z. L., Zhang, Z. M., Olson, J. M., & Verma, D. P. S. (2001b). A novel UDP-glucose transferase is part of the callose synthase complex and interacts with phragmoplastin at the forming cell plate. *Plant Cell*, *13*, 769–779.

Ishiguro, J., Saitou, A., Duran, A., & Ribas, J. C. (1997). *cps1+*, a *Schizosaccharomyces pombe* gene homolog of *Saccharomyces cerevisiae* FKS genes whose mutation confers hypersensitivity to cyclosporin A and papulacandin B. *Journal of Bacteriology*, *179*, 7653–7662.

Kakimoto, T., & Shibaoka, H. (1992). Synthesis of polysaccharides in phragmoplasts isolated from tobacco BY-2 cells. *Plant and Cell Physiology*, *33*, 353–361.

Martin, D. J., & Somers, D. A. (2004). A (1-3)-ß-glucanase expressed during oat endosperm development. *Journal of Cereal Science*, *39*, 265–272.

Mayer, U., & Jürgens, G. (2004). Cytokinesis: lines of division taking shape. *Current Opinion in Plant Science*, *7*, 599–604.

Meikle, P. J., Bonig, I., Hoogenraad, N. J., Clarke, A. E., & Stone, B. A. (1991). The location of (1,3)-β-glucans in the walls of pollen tubes of *Nicotiana alata* using a (1,3)-β-glucan specific monoclonal antibody. *Planta*, *185*, 1–8.

Olsen, O.-A., Brown, R. C., & Lemmon, B. E. (1995). Pattern and process of wall formation in developing endosperm. *BioEssays*, *17*, 803–812.

Otegui, M. (2007). Endosperm cell walls: Formation, composition, and functions. *Plant Cell Monographs*, *8*, 159–177.

Otegui, M., & Staehelin, L. A. (2000). Syncytial-type cell plates: A novel kind of cell plate involved in endosperm cellularization of Arabidopsis. *Plant Cell*, *12*, 933–947.

Philippe, S., Saulnier, L., & Guillon, F. (2006). Arabinoxylan and (1-3), (1-4)-ß-glucan deposition in cell walls during wheat endosperm development. *Planta*, *224*, 449–461.

Samuels, A. L., & Staehelin, L. A. (1996). Caffeine inhibits cell plate formation by disrupting membrane reorganization just after the vesicle fusion step. *Protoplasma*, *195*, 144–155.

Samuels, A. L., Giddings, T. H., & Staehelin, L. A. (1995). Cytokinesis in tobacco BY-2 and root tip cells: A new model of cell plate formation in higher plants. *Journal of Cell Biology*, *130*, 1345–1357.

Scherp, P., Grotha, R., & Kutschera, U. (2002). Interaction between cytokinesis-related callose and cortical microtubules in dividing cells of the liverwort *Riella helicophylla*. *Plant Biology*, *4*, 619–624.

Seguí-Simarro, J. M., Austin, J. R., II, White, E. A., & Staehelin, L. A. (2004). Electron tomographic analysis of somatic cell plate formation in meristematic cells of Arabidopsis preserved by high-pressure freezing. *Plant Cell*, *16*, 836–856.

Stone, B. A., & Clarke, A. E. (1992). *Chemistry and biology of (1,3)-β-glucans*. Australia: LaTrobe University Press.

Vaughn, K. C. (2006). The abnormal cell plates formed after microtubule disrupter herbicide treatments are enriched in callose. *Pesticide Biochemistry and Physiology*, *84*, 63–71.

Waizenegger, I., Lukowitz, W., Assaad, F., Schwartz, H., Jürgens, G., & Mayer, U. (2000). The Arabidopsis KNOLLE and KEULE genes interact to promote vesicle fusion during cytokinesis. *Current Biology*, *10*, 1371–1374.

Wilson, S. M., Burton, R. A., Doblin, M. S., Stone, B. A., Newbigin, E. J., Fincher, G. B., & Bacic, A. (2006). Temporal and spatial appearance of wall polysaccharides during cellularization of barley (*Hordeum vulgare*) endosperm. *Planta*, *224*, 655–667.

Yasuhara, H. (2005). Caffeine inhibits callose deposition in the cell plate and the depolymerization of the microtubules in the central region of the phragmoplast. *Plant and Cell Physiology*, *46*, 1083–1092.

Yasuhara, H., & Shibaoka, H. (2000). Inhibition of cell-plate formation by brefeldin A inhibited the depolymerization of microtubules in the central region of the phragmoplast. *Plant Cell Physiology*, *34*, 21–29.

CHAPTER 4.4.2

# *Cytology of the (1-3)-β-Glucan (Callose) in Plasmodesmata and Sieve Plate Pores*

**Amit Levy and Bernard L. Epel**
*Department of Plant Sciences, George S. Wise Faculty of Life Sciences, Tel Aviv University, Israel*

## I. Introduction

Plasmodesmata (Pd) are co-axial membranous channels that cross walls of adjacent plant cells, linking the cytoplasm, plasma membranes and endoplasmic reticulum (ER) of cells and allowing direct cytoplasmic cell-to-cell communication of both small molecules and macromolecules (proteins and RNA). Transport through Pd mediates many processes in plants, among them information transfer for coordination of development, movement of photosynthesis products from mature to developing and storage tissues, responses to pathogen infection and systemic gene silencing. In addition, many plant viruses exploit Pd as conduits for spread of infection between cells (reviewed in Beachy and Heinlein, 2000; Citovsky and Zambryski, 2000; Ding et al., 1999; Haywood et al., 2002; Heinlein, 2002; Heinlein and Epel, 2004; Wu et al., 2002; Zambryski and Crawford, 2000). By interconnecting neighbouring plant cells, Pd create a supra-cellular entity termed the 'symplast'. Pd within a specific tissue can be locally or temporally regulated to create a 'symplast domain'. Such a domain acts as an isolated developmental and physiological unit to meet local demands for signalling, transport or isolation (see Lucas et al., 1993; Rinne and van der Schoot, 1998; Rinne and van der Schoot, 2003; Roberts and Oparka, 2003). Additionally, Pd serve as precursors in the development of sieve plate pores which are essential for mass flow through sieve elements. In electron microscopy studies, the observed diameter of Pd is 20–50 nm (Ehlers and Kollmann, 2001), while the pores in sieve plate are 200–400 nm in width, and can even reach 1 μm in some cucurbits (Sjolund, 1997).

A basic structure of simple primary Pd consists of two coaxial membrane tubes. The inner membrane along the Pd axis, termed the desmotubule, is continuous with and connects the

ER of the bordering cells; the outer coaxial membrane is continuous with and connects plasma membranes of adjacent cells (Fig. 1). Between the membranes is a sleeve that interconnects the cytoplasm of the neighbouring cells. Within the cytoplasmic sleeve are particles whose identity is still unknown, but have been interpreted to be cytoskeletal proteins (Overall, 1999; Overall and Blackman, 1996).

Pd conductivity is not a static feature, but one that shows a high degree of plasticity. Changes in Pd can be developmentally regulated. For example, during sink-to-source transition in leaf development, Pd structure changes from a single tunnel (simple Pd) to a branched one (Fig. 1),

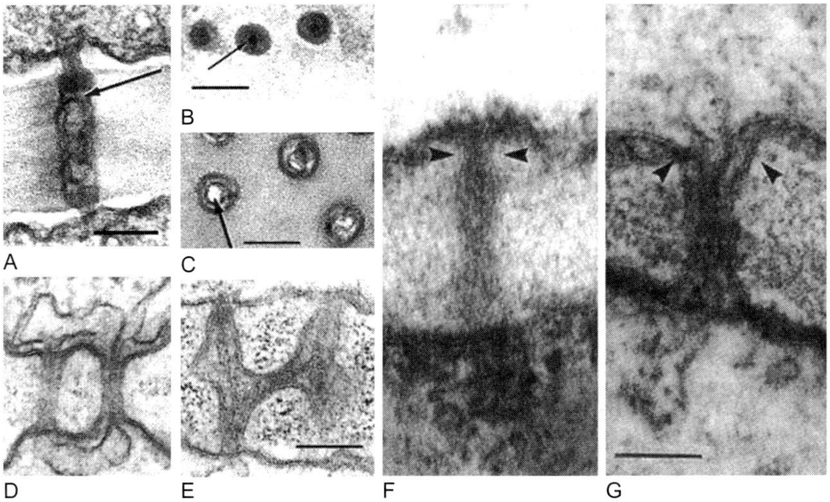


**Fig. 1: Electron micrographs of plasmodesmata. (A–C) Pd between adjacent Krantz mesophyll cells in sugarcane leaf. (A) Longitudinal section of plasmodesma: electron-opaque structures and internal sphincters are present at both ends near orifices of plasmodesma. The desmotubule is open between two sphincters (arrow). (B) Transverse section of Pd at the level of the sphincter; the sphincter is electron dense and the desmotubule is constricted (arrow). (C) Transverse section at mid level showing open ER lumen (arrow). Bars (A, C) = 200 nm, bar (B) = 150 nm. (D, E) Simple plasmodesma (D) in young cell walls in the needles of *Metasequoia glaptostoboides*, and branched plasmodesma (E) in young wall possibly having developed from lateral fusion of two adjacent simple Pd. Bar = 100 nm. (F, G) Longitudinal section of Pd between adjacent onion root cells not treated with callose inhibitor DDG (F) or treated with DDG (G) prior to tissue fixation. Plasmodesma in the non-treated sample shows a raised collar and constricted neck region while Pd in tissue treated with DDG exhibits a funnel-shaped neck configuration (arrowheads). Bar = 100 nm. (A–C from Robinson-Beers and Evert, 1991; D and E from Kollmann and Glockmann, 1999; F and G from Radford et al., 1998. Reproduced with permission of Springer.).**

and the transfer rate of proteins between cells decreases (Liarzi and Epel, 2005; Oparka et al., 1999). Changes can also be local and temporary, and be caused by environmental factors. The control of these localized and transient changes alters Pd conductivity, and plays a role in developmental and defence processes (see for example Bucher et al., 2001; Iglesias and Meins, 2000; Wolf et al., 1989; Zambryski and Crawford, 2000). Different mechanisms may be involved in changes in cell-to-cell transport through Pd changes in area of contact of Pd between cells, changes in number of Pd, increases in the length of the Pd channel, and changes in the cross-sectional area of the cytoplasmic sleeve (Heinlein and Epel, 2004; Liarzi and Epel, 2005). It has also been suggested that structural changes in Pd due to the action of cytoskeletal Pd proteins may alter transport through Pd (Blackman and Overall, 1998; Blackman et al., 1999; Ding et al., 1996; Radford and White, 1998; White et al., 1994). There is little evidence for this suggestion. Much evidence however is available for conductivity changes in Pd due to alteration in the cell wall sheath that surrounds Pd. This second mechanism is suggested to be mediated by the synthesis and hydrolysis of callose (see reviews by Heinlein and Epel, 2004; Overall, 1999; Roberts and Oparka, 2003; Schulz, 1999).

## I.A. Callose Localization in Pd and Sieve Plates

The specialized cell wall sheath which surrounds the Pd is devoid of cellulose is composed from callose, non-esterified pectin and probably other non-cellulosic polysaccharides (Dahiya and Brewin, 2000; Roy et al., 1997; Sutherland et al., 1999). Callose localization associated with Pd was shown by both immunolabelling studies; for example, in tomato (*Lycopersicon esculentum*) (Orfila and Knox, 2000), kiwifruit (*Actinidia deliciosa*) (Sutherland et al., 1999), maize (*Zea mays*) (Balestrini et al., 1994; Baluska et al., 1999; Turner et al., 1994), apple (*Malus domestica*) (Roy et al., 1997), wheat (*Triticum aestivum)* (Sivaguru et al., 2000; Fig. 2A), cotton (*Gossypium hirsutum, Gossypium barbadense* ) (Rodríguez-Gálvez and Mendgen, 1995; Ruan et al., 2004), birch (*Betula pubescens*) (Rinne et al., 2001), bean (*Phaseolus vulgaris*) (Brown et al., 1998; Northcote et al., 1989), barley (*Hordeum vulgare*) (Trethewey and Harris, 2002; Wilson et al., 2006), pea (*Pisum sativum*) (Dahiya and Brewin, 2000), carnation (*Dianthus caryophyllus*) (Trillas et al., 2000), tobacco (*Nicotiana tabacum*) (Oparka et al., 1997), and *Arabidopsis thaliana* suspension cells (Bayer et al., 2004); and by cytochemical Aniline Blue staining; for example, in *Nicotiana benthamiana* (Gorshkova et al., 2003; Guenoune-Gelbart et al., 2008), tobacco (*Nicotiana tabacum*) (Levy et al., 2007a; Sagi et al., 2005; Fig. 2C), *Sporobolis africanus* (Botha and Cross, 2000), onion (*Allium cepa*) (Radford et al., 1998), and Arabidopsis (Levy et al., 2007a). Association of callose with sieve areas and sieve plates in angiosperms (Fig. 2B,D), and with equivalent structures in

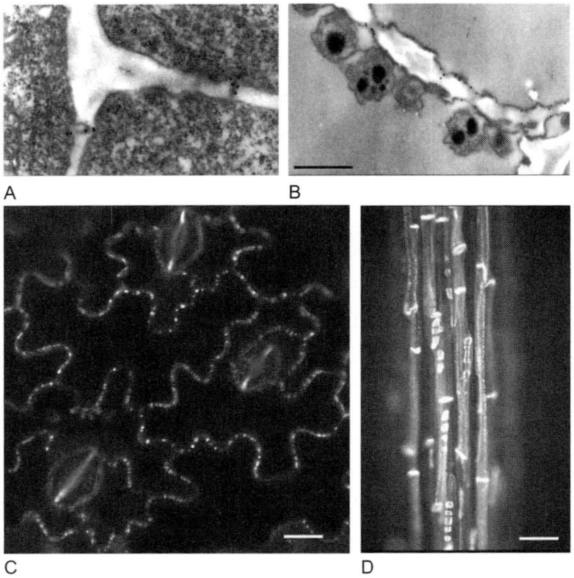


**Fig. 2: Callose localization in Pd and sieve plates. (A, B) Immunogold labelling of callose in wheat root cells, showing presence of callose at Pd regions (A) and sieve plates (B). Bar = 300 nm. (From Sivaguru et al., 2000. Reproduced with permission of American Society of Plant Biologists.) (C, D) Callose staining with Aniline Blue fluorochrome of epidermal cell walls of tobacco showing punctuate callose labelling in the Pd (C) and massive callose accumulation in the sieve elements pores of *Dahlia pinnata* (D). Bar (C) = 20 μm. Bar (D) = 100 μm. (C from Levy, unpublished results. D from Aloni and Peterson, 1990; reproduced with permission of Springer.).**

gymnosperms, pteridophytes, bryophytes, lycopods, polypods, horsetails and certain brown macroalgae, has been widely demonstrated (see Behnke and Sjolund, 1990; Stone and Clarke, 1992).

Raised collar structures have been observed to exist in the cell wall surrounding Pd orifices (Badelt et al., 1994; Cook et al., 1997; Heinlein and Epel, 2004; Olesen, 1979; Olesen and Robards, 1990; Overall, 1999; Rinne and van der Schoot, 1998; Robinson-Beers and Evert. 1991) (Fig. 1F). It was suggested that these might form sphincter-like structures largely composed of callose (Olesen and Robards, 1990). Alternatively, Turner et al. (1994) concluded that in maize root tips the wall sheath around Pd is subdivided, and callose is deposited at the peripheral zone surrounding the collar, rather than in the collar itself. During dormancy, callose was also seen deposited inside the Pd channel, creating an inner plug (Rinne et al., 2001).

The localization of callose at Pd and sieve plates suggests the presence of enzymes for callose formation and depolymerization. Callose synthases in plants are suggested to consist of complexes containing a number of proteins. They are likely encoded in part by the *GSL* (Glucan Synthase-Like) genes which were identified based on their similarity to fungal *FKS* genes, that are believed to encode the fungal (1,3)-β-glucan synthases (Farrokhi et al., 2006). The GSL proteins are large (200–220 kDa), integral membrane proteins predicted to contain 14–16 transmembrane domains, divided into two membranal regions separated by a large, central cytoplasmic region, and an N-terminal cytoplasmic extension (Brownfield et al., 2007; Farrokhi et al., 2006). The *GSL* genes exist in multigene families (for example, 12 in Arabidopsis and 13 in rice), and it was proposed that each gene is responsible for callose synthesis at a different location; for example, the expression of *GhGSL1* in cotton fibres and *NaGSL1* in pollen tubes (Brownfield et al., 2007; Cui et al., 2001; Doblin et al., 2001; Hong et al., 2001). The identity of the GSL proteins forming the callose synthase complex in Pd and sieve plates is still unknown. However, a non-catalytic 65-kDa protein that is associated with callose synthesis in french bean (*Phaseolus vulgaris*) was identified. This protein was identified and purified following fungal elicitor treatment of suspension-cultured cells, which induce callose synthase activity (McCormack et al., 1997). Its N-terminal sequence differs from that of calreticulin, a known Pd-associated protein which also has a size of 65 kDa and runs similarly on gels. Immunolocalization of this 65-kDa protein in French bean tissues showed its association with the Golgi and plasmalemma in newly synthesized cell plate walls, and with the plasmalemma–wall interface in wounded cells. It was also found localized at sieve plates of sieve elements after wounding (McCormack et al., 1997) and with Pd between mesophyll cells and in pits of developing xylem vessels (Brown et al., 1998; Gregory et al., 2002). These results suggest that this 65-kDa protein may be a Pd callose synthase-associated protein.

(1,3)-β-Glucan endo-hydrolases, enzymes that catalyse callose degradation, are also present in large gene families in plants (e.g. 50 genes in Arabidopsis), and are involved in various developmental processes such as germination, pollen tube growth regulation and microspore maturation (reviewed in Leubner-Metzger, 2003; Simmons, 1994), as well as in defence processes through their activity as pathogenesis-related (PR) proteins (reviewed in Leubner-Metzger and Meins, 1999). It is still unknown which members of this large gene family function under which circumstances and in which tissues; however, based on microarray data, 44 genes from this family were grouped into 13 expression clusters (Doxey et al., 2007). Genes specific to leaves, roots and flower organs, or exhibiting a PR response were clustered. The largest identified group contains 13 genes which are expressed in a variety of tissues and exhibit minimal responses to stresses and hormones (Doxey et al., 2007). A (1,3)-β-glucan endo-hydrolase from this group, termed AtBG_pap (plasmodesmal associated protein), was identified by MS/MS proteomic analysis of an Arabidopsis Pd-enriched fraction (Levy et al., 2007a). Expression of a green fluorescent protein (GFP) fusion of this protein in tobacco showed that it is located in the cell membrane and is highly enriched at Pd sites. In spongy mesophyll cells, a punctuate fluorescence was obtained only in walls that connect adjacent cells. As AtBG_pap was also shown to be a glycosylphosphatidylinositol-anchored protein (Elortza et al., 2006; Elortza et al., 2003), it seems that this (1,3)-β-glucan endo-hydrolase is anchored to the plasma membrane facing the cell wall sheath surrounding the Pd.

## I.B. Pd Regulation by Callose Turnover

Movement of cytoplasmic proteins through Pd is assumed to be through the cytoplasmic sleeve of the Pd. Plasma membrane and ER membrane proteins also may move cell to cell via Pd membranes if not anchored (Guenoune-Gelbart et al., 2008). Reversible callose accumulation apparently plays a role in regulating cell-to-cell transport through sieve pores and Pd. Olesen and Robards (1990) suggested that the sphincter complex surrounding Pd might contain glucan synthase molecules that are involved in the deposition of callose. Callose accumulation in the wall surrounding a plasmodesma will result in a decrease in pore area available for transport. Callose hydrolysis, in contrast, would release closure and allow increased passage through the cytoplasmic sleeve (Fig. 3). The closure of the cytoplasmic sleeve might also result in the inhibition of ER membrane protein trafficking if there are protein domains protruding into the cytoplasmic sleeve (Guenoune-Gelbart et al., 2008). Callose deposition is associated with the neck regions of Pd, the areas surrounding both ends of the channels (Radford et al., 1998) (Fig. 1). Neck constrictions are assumed to be Pd 'bottle necks' restricting the size of the cytoplasmic sleeve at the site of constriction, and limiting the size exclusion limit (SEL, the size of the largest molecule that will move cell to cell) for molecular movement between the cells (Badelt et al., 1994; Blackman and Overall, 2001; Ding et al., 1992; Heinlein and Epel, 2004; Overall, 1999; Radford et al., 1998) (Fig. 3).

These suggested mechanisms gain strength from various reports showing that the accumulation of callose in the wall sheath around the channel results in decreased cell-to-cell movement of florescent dyes (Radford and White, 2001; Rinne et al., 2005; Rinne and van der Schoot, 1998; Sivaguru et al., 2000), whereas treatments inhibiting callose deposition resulted in increased diameter of Pd orifices and higher Pd SEL (Radford et al., 1998; Radford and White, 2001; Wolf et al., 1991). Similar results were shown in sieve elements. Induced phloem callose led to a decrease in the lateral movement of $C^{14}$-assimilates and auxin, while treatments that stimulate breakdown of sieve plate callose led to increased movement of fluorescein through the sieve tubes (Aloni et al., 1991; Hollis and Tepper, 1971; Maeda et al., 2006; McNairn, 1972; McNairn and Currier, 1968; Webster and Currier, 1965).

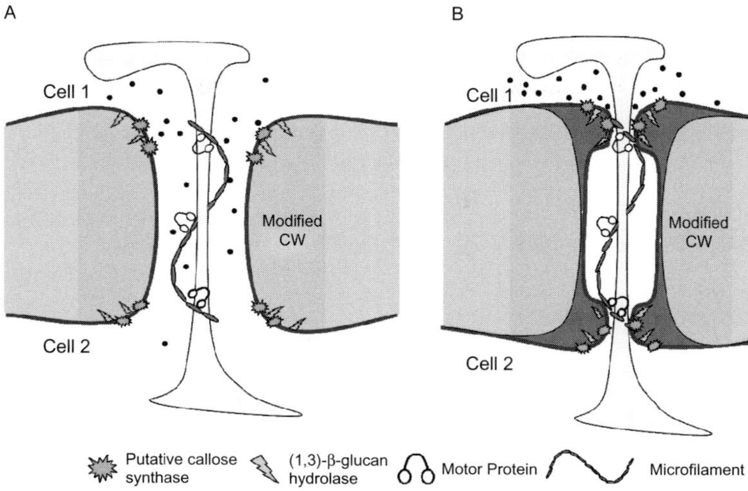


**Fig. 3: A schematic model for non-selective regulation of plasmodesmal permeability by callose. In the open state, the cell wall (CW) sheath immediately surrounding the Pd is devoid of cellulose and is composed in part of non-esterified pectin, callose, other non-cellulosic polyglucans and as yet uncharacterized proteins. Embedded in the Pd plasma membrane are callose synthase and (1,3)-β-glucan hydrolase. Cytoplasmic sleeve is open and non-targeted macromolecules (black spheres) diffuse according to electrochemical gradient and Stokes radius. (B) Deposition of callose in the apoplast (dark wall area) around the cytoplasmic sleeve, and especially at the cytoplasmic annulus, results in the formation of a sphincter that reduces the size of the annulus, creating a bottle neck for diffusion. The colour specifications refer to colours in panels.**

Additional support for this proposed model comes from research employing (1,3)-β-glucan endo-hydrolases. In studies with a tobacco mutant that had decreased levels of class I (1,3)-β-glucan endo-hydrolase (TAG4.4), generated by antisense transformation, callose accumulation increased and the SEL of Pd, measured by the cell-to-cell movement of dextrans and peptides, was reduced from 2300 to 1960 Da, and from 1.0 to 0.85 nm, respectively (Iglesias and Meins, 2000). In Arabidopsis AtBG_pap T-DNA knockout mutants, Pd sites contained higher amounts of callose after wounding, and the cell-to-cell conductivity coefficient for GFP decreased from 0.7 in the wild type to 0.59 and 0.52 in the two independent mutant lines (Levy et al., 2007a).

These results indicate that the steady-state level of callose in the cell wall sheath around Pd regulates the non-selective Pd conductivity, and suggest that the amount of callose at Pd sites is a result of the activity of (1,3)-β-glucan synthases and (1,3)-β-glucan endo-hydrolases. It has been questioned whether this mechanism can completely close Pd, and whether it operates alone (Schulz, 1999). In sieve elements, the shutting off of sieve pores usually involves both callose deposition and the plugging of the pore lumen by filamentous P-proteins and in Fabaceae by forisomes (Furch et al., 2007; Sjolund, 1997). Cellular cytoskeleton proteins, actin, myosin and centrin, may also be involved in the closing of Pd. These proteins were found to be localized at the Pd, and were also shown to affect permeability of Pd (reviewed in Aaziz et al., 2001; Heinlein and Epel, 2004).

## I.C. Pd and Sieve Plate Callose in Abiotic Stresses

### *I.C.1. Wound response: is callose an artifact?*

Callose is known to be deposited in sieve plates to seal sieve plate pores upon sieve tube severing (Engleman and Esau, 1964; Furch et al., 2007; Sjolund, 1997; Van Bel, 2003), and even in response to low-intensity ultrasound which causes only minimal injury (Currier and Webster, 1964). Since callose accumulation is induced by stresses, it has been repeatedly questioned whether association of callose with Pd or sieve plates may only be an artifact due to injury caused during the preparation of the material for microscopy. In studies with *Lemna minor*, a vascular plant small enough to allow fixation for electron microscopy without cutting, no callose was found during the development of *Lemna* sieve element pores unless plants were cut prior to fixation, leading to the suggestion that callose is not naturally associated with the development of a sieve pore (Walsh and Melaragno, 1976). On the other hand, Ehlers et al. (2000) in studies with *Vicia faba* and tomato (*Lycopersicum esculentum*) found callose collars even when employing a gentle method for electron microscopy preparation which minimized damage to the sieve elements. A similar conclusion that callose is naturally present was reached by Engleman and Esau (1964) who measured callose levels after rapid killing of the phloem by immersing in boiling water or liquid nitrogen, or by crashing tissues between blocks of solid carbon dioxide or blocks of aluminum heated to 120–140°C. Whereas the amount of callose was dramatically increased as a result of regular wounding, it was still present, albeit in lesser amounts, in tissues that were rapidly killed. These results suggest that callose is present around sieve pores in uninjured sieve tubes.

The question was also studied in the wall sheath surrounding Pd of onion roots. Callose synthesis was inhibited by incubating the plant tissue with 2-deoxy-D-glucose (DDG), an inhibitor of callose formation, prior to fixation. Untreated tissues exhibited constricted Pd neck regions, while tissues treated with DDG did not exhibit a constricted, but a funnel-shaped configuration (Fig. 1F,G). Dissection of tissues further increased the frequency of constrictions. From these results it was concluded that the constricted neck region is a result of callose formation, and that this formation is induced by both initial dissection and fixation (Radford et al., 1998).

Overall, these results suggests that the presence of callose seen in many, but not all, studies of Pd may be an artifact, and emphasize both the need for non-invasive techniques for the measurements of 'natural' Pd callose, and the requirement for a valid control whenever callose deposition/hydrolysis events are analysed. For example, we have found that immersing whole uncut leaves immediately after dissection in 85% ethanol and incubating overnight dramatically reduces the production of wound callose at Pd (see Levy et al., 2007a).

### I.C.2. Other abiotical stresses induce Pd callose

It has long been observed that callose formation is induced at Pd and sieve elements by heating (McNairn, 1972; McNairn and Currier, 1968; Webster and Currier, 1965; Wolf et al., 1991). In a recent study, Furch et al. (2007) identified a difference between the deposition of callose at distant sieve elements as a result of wounding, and as a result of burning. Burning of the leaf tip of both *V. faba* and tomato initiated a callose response at a distant of 3–4 cm while cutting of the tip did not induce such a distal response and callose deposition was restricted to a region within 0.5 cm of the site of cutting. They also showed that burning triggers a longitudinal propagation wave of an electric potential along the phloem, but that cutting only triggers a negligible depolarization wave. As micromolar changes in intercellular calcium concentrations are known to induce callose deposition (Kauss, 1996), it was suggested that the burning-induced action potentials can induce callose synthesis by releasing a calcium influx into the sieve element lumen. In the case of cutting, it was suggested that only a small amount of calcium is released into the lumen, which may be insufficient to induce callose production.

Chemical stresses also induce callose accumulation. Exposure to excess calcium, boron, nickel, cobalt and zinc all led to callose deposition in sieve plates (Peterson and Rauser, 1979). However, although exposure to cobalt, nickel and zinc caused a drastic reduction in $^{14}C$ translocation, no correlation was found between the amount of callose deposited and the reduction of $^{14}C$ translocation, suggesting the involvement of other mechanisms in inhibition of phloem translocation (Peterson and Rauser, 1979). In wheat, exposure of the plant to aluminum induced deposition of Pd callose in roots (Sivaguru et al., 2000). Microinjection of Lucifer yellow carbohydrazide into both epidermal and cortex root cells, and at different locations along the root, showed a reduction in cell-to-cell dye coupling in aluminium-treated plants. The dye coupling was dramatically recovered when aluminum treatments were administered after DDG treatments. Similar results were also obtained when microinjection was performed in tobacco mesophyll leaf cells. Aluminum exposure also resulted in increased expression of calreticulin and myosin VIII, Pd-associated proteins that co-localize with callose. These results provide evidence that aluminium-induced callose is, in part, implicated in the blocking of cell-to-cell communication through Pd. However, as in Peterson and Rauser (1979), it was suggested that other events are involved in the closure of Pd (Sivaguru et al., 2000).

## I.D. Pd and Sieve Plate Callose During Development

### *I.D.1. Sieve plate pore development*

In general, the sequence of sieve plate pore development in angiosperms begins with Pd connection between sieve elements. Callose platelets are then deposited beneath the cell membrane either in addition to, or in place of, some of the cell wall material, to form cone-shaped pads which penetrate deeper into the wall. Later, opposing callose masses fuse to form cylinders surrounding the Pd connections. Pd begin to enlarge by dissolution of surrounding callose until a large intercellular pore is formed. After pore formation is complete, there is still a thin residual cylinder of callose lining the pore (Evert, 1977; Evert, 1990). The probable role of callose platelets is to prevent deposition of more permanent cellulose in sites of future pores, and enable the eventual opening of the pores by callose digestion (Evert, 1990). While ER cisternae have been implicated in deposition of callose at pore sites (Evert, 1990), the enzymes involved in this process and how they target the developing sieve plate remain unknown. This lack of knowledge is probably due to the complexity of vascular tissues and the limited population of vascular cells in plants, making it difficult to analyse the molecular mechanisms that direct the differentiation of the vascular system and the phloem in particular.

In other plants, callose is not always seen during phloem development. For example, in gymnosperms, such as the lycopods *Selaginella kraussiana* (Burr and Evert, 1973), *Lycopodium lucidulum* (Warmbrodt and Evert, 1974) and *Isoetes muricata* (Kruatrachue and Evert, 1974), the horsetail *Equisetum hyemale* (Dute and Evert, 1977), the fork-fern *Psilotum nudum* (Perry and Evert, 1975), and the fern *Botrychium virginianum* (Evert, 1976), callose may be associated with fully formed sieve areas, but does not appear to be associated with developing pores. In contrast, in the fern *Polypodium vulgare* (Liberman-Maxe, 1978), callose was found during pore formation.

Whether callose seen during development is an artifact of handling the plant, or a true characteristic of phloem development is under debate (see Section I.C.1). However, since callose appearance and later disappearance is a timed, ordered and consistent process, it seems likely that the participation of callose is a general characteristic of the process (Esau and Thorsch, 1985).

### *I.D.2. Definitive and dormancy callose*

At the end of the operational life of a phloem sieve element, sieve areas are covered with prominent masses of callose. This callose indicates the cessation of activity of the sieve element and is thereby called 'definitive' callose. In members of the Pinaceae and Cupressaceae, some or all of the sieve cells of a current season over-winter in a functional state. As cambial differentiation begins in the new season, definitive callose forms in the over-wintered sieve cells (Evert, 1990; Hollis and Tepper, 1971; Stone and Clarke, 1992).

In plants such as grapevine (*Vitis vinfera*) (Aloni et al., 1991) and *Grewia tiliaefolia* (Deshpande and Rajendrababu, 1985), in which the sieve elements normally function for two growing seasons, the cambium becomes dormant at the end of the first season and variable amounts of 'dormancy' callose accumulate on each face of the sieve plate and in lateral sieve areas. Shortly after bud break in spring this dormancy callose is gradually digested and the sieve tube assumes the appearance of fully active conduits (Evert, 1990). In grapevine and white ash it was shown that the disappearance of dormancy callose correlates with the resumption of translocation in the sieve tubes (Aloni et al., 1991; Hollis and Tepper, 1971). Both cytokinin and auxin were found to influence this process. Cytokinin promoted callose production in sieve plates of *Coleus blumei* (Aloni et al., 1990) while auxin applied to both grapevine and magnolia (*Magnolia kobus*) had an opposite influence, reducing the amount of callose in the sieve plates (Aloni and Peterson, 1997; Aloni et al., 1991). It is, therefore, possible that at the end of the growing season, when leaves abscise and auxin levels decline, cytokinin is prominent, leading to callose production which plugs sieve tubes. In the spring, leaves synthesize auxin which can move through the phloem, and may contribute to callose dormancy removal. In contrast to grapevine and magnolia, sieve tubes of oak (*Quercus robur*) are almost free of dormancy callose just before bud break, and application of auxin before bud break results in callose accumulation (Aloni and Peterson, 1997). Aloni et al. suggested that this phenomena is an ecological adaptation of oak (a ring-porous species), which increased the sensitivity of the cambium to extremely low levels of auxin. The cambium of these trees is activated by very low auxin levels produced in dormant-looking buds several weeks before bud break. Artificial auxin application to sieve tubes in the experiment was probably an overdose (Aloni and Peterson, 1997). In the perennials goat willow (*Salix caprea*), sycamore maple (*Acer pseudoplatanus*), and European ash (*Fraxinus excelsior*), the (1,3)-β-glucan endo-hydrolase activity in the cambium and secondary phloem cells rises in early spring just before bud breaks, consistent with (1,3)-β-glucan endo-hydrolase playing a role in hydrolysis of callose for phloem reactivation (Krabel et al., 1993).

A parallel mechanism of 'dormancy callose' was discovered for Pd in birch (Rinne et al., 2001; Rinne and van der Schoot, 1998). During dormancy development induced by exposure to short days, the Pd of the shoot apical meristem are firmly closed by intra- and extracellular structures containing callose (Rinne and van der Schoot, 1998) (Fig. 4). Dye-coupling measurements in the meristem with Lucifer yellow carbohydrazide showed that this closure results in symplasmic isolation of the meristem cells and blockage of signalling networks (Rinne et al., 2001; Rinne and van der Schoot, 1998). The breakage of dormancy, which is induced by exposure to cold, restored the capacity of Pd to transport fluorescent dyes between cells.

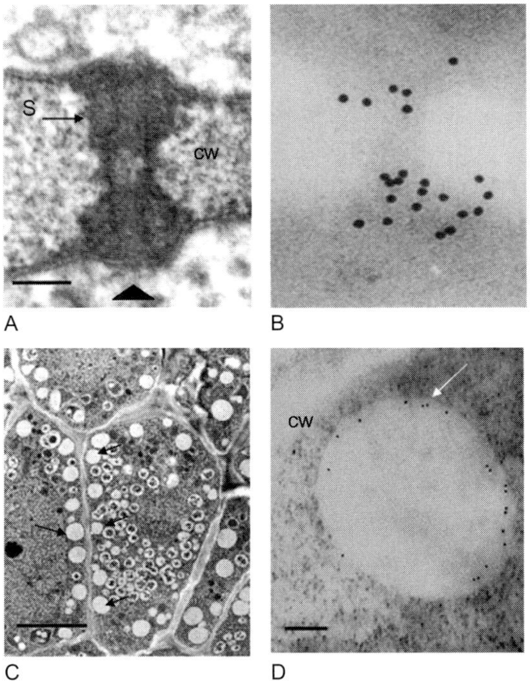


**Fig. 4: Callose deposition and (1,3)-β-glucan hydrolase activation during dormancy cycle in birch. (A, B) Pd of deactivated meristem exposed to short days. Tannic acid staining (A) and callose immunolabelling (B) detect the presence of sphincters (arrow in A) and callose in the meristem Pd, respectively. (C, D) Chilling-induced dormancy release induces the formation of (1,3)-β-glucan hydrolase-containing vesicles at the cell periphery. Small vacuoles localize at the peripheral part of the cytoplasm in contact with cell wall (arrows, C). Immunolabelling of (1,3)-β-glucan hydrolase indicates that the enzyme is present in the peripheral part of vacuoles allowing contact with cell wall and Pd (arrow, D). (From Rinne and van der Schoot, 2004. Reproduced with permission of Haworth Press, Inc.) S, sphincter; CW, cell wall. Bar (A, B) = 100 nm, (C) = 5 μm, (D) = 200 nm.**

During this process, callose-containing sphincters in the shoot apical meristem disappear, and (1,3)-β-glucan endo-hydrolase proteins are upregulated (Rinne et al., 2001). (1,3)-β-Glucan endo-hydrolases were localized in spherosome-like vacuoles or lipid bodies that during chilling became aligned with the plasma membrane and often associated with Pd (Rinne et al., 2001) (Fig. 4). As in phloem dormancy, these enzymes are suggested to have a major role in the implementation of this 'dormancy cycle'. The genes encoding the enzymes involved in this process are as yet uncharacterized.

### *I.D.3. Cotton fibre elongation*

Cotton fibres typically elongate over 2000–3000-fold within 3 weeks of initiation, while adjacent cells expand by 5–10-fold during this period. This indicates a high degree of cell autonomy of the fibres. The symplasmic continuity between fibres and epidermis was tested by the ability of carboxyfluorescein to move into the fibres in three different cotton genotypes (Ruan et al., 2004). At the onset of the rapid phase of elongation Pd close, and then at the end of elongation they reopen. Both immunolabelling and Aniline Blue staining showed that callose is present at Pd of the fibre base when Pd are closed, but not when Pd are open (Ruan et al., 2004). Expression of a fibre-specific (1,3)-β-glucan endo-hydrolase, *GhGluc1*, was strong during callose degradation whereas it was undetectable during callose deposition. It has been suggested that Pd closure by callose may provide a mechanism for fibre cells to generate and maintain a high turgor to drive elongation. By using transporters for solute uptake across the plasma membrane into the fibres, Pd closure can help maintain high turgor and the attraction of water into the fibre (Ruan, 2007).

## I.E. Pd and Sieve Plate Callose in Biotic Stresses

### *I.E.1. Herbivorous attack*

Since any damage to sieve tubes results in callose formation and blockage, phloem feeder insects, to successfully feed on sieve element sap, have developed special mechanisms to overcome stress callose deposition. Upon insertion of the aphid's styles into the sieve tubes, sheath saliva is secreted, preventing influx of calcium from the wall through the puncture site (Will and van Bel, 2006). Moreover, since the salivary proteins contain calcium-binding domains (Will et al., 2007), it was suggested that aphid saliva can act as a chemical calcium scavenger to prevent an increase in calcium concentration and hence the sieve tube sealing response.

In contrast, the brown plant hopper (*Nilaparvata lugens*) uses a different mechanism to overcome sieve tube callose. While genes encoding (1,3)-β-glucan synthases are upregulated and sieve tube callose is deposited upon stylet insertion in both resistant and susceptible rice plants, genes encoding (1,3)-β-glucan endo-hydrolases are only upregulated in susceptible plants (Hao et al., 2008). This suggests that the limited or absent expression of these (1,3)-β-glucan endo-hydrolase genes allows sieve tube occlusion to be maintained in the resistant plants, and is probably the key reason for their resistance (Hao et al., 2008). Thus, instead of preventing the callose synthesis as in aphids, the brown plant hoppers employ the plant (1,3)-β-glucan endo-hydrolases to hydrolyse sieve plate callose.

### I.E.2. Viral infection

Many viruses move cell to cell with the help of one or more viral encoded movement pro teins (MPs) that target to and purportedly dilate Pd (Beachy and Heinlein, 2000). However, the mechanism by which MP increases Pd SEL is still unknown. Measurements of callose levels around Pd revealed that infection of *N. tabacum* with a minimal replicon of tobacco mosaic virus, lacking coat protein and MP ($TMV^{\Delta CP \Delta MP}$), led to the deposition of callose around the channels. Expression of the TMV MP alone had no effect on callose levels and Pd opening in MP transgenic *N. tabacum*. However, infection of the MP transgenic plants with the $TMV^{\Delta CP \Delta MP}$ replicon led to a reduction in callose accumulation (Guenoune-Gelbart et al., 2008). This suggests that callose deposition arises largely as a result of a plant defence mechanism aimed at restricting virus spread, and that MP functions synergistically with viral replicase in mediating the degradation of the callose and opening Pd. To achieve this, the virus must somehow recruit and/or activate (1,3)-β-glucan endo-hydrolase at Pd and/or inactivate callose synthase.

The importance of (1,3)-β-glucan endo-hydrolase in cell-to-cell movement of viruses was demonstrated in the TAG4.4 tobacco mutant that had decreased levels of a class I (1,3)-β-glucan endo-hydrolase (see Section I.B). In this mutant line, susceptibility to TMV infection was decreased, and cell-to-cell movement of the cucumber mosaic virus 3a MP:GFP fusion was delayed as well (Beffa et al., 1996; Iglesias and Meins, 2000). Moreover, when the (1,3)-β-glucan endo-hydrolase coding sequence was cloned into TMV, the virus spread faster, while expression of the gene in an antisense formation led to opposite results (Bucher et al., 2001).

Based on these results it was speculated that MP, together with replicase, recruits and/or activates (1,3)-β-glucan endo-hydrolase at the Pd, thus decreasing the deposition of callose associated with Pd and increasing cell-to-cell viral trafficking (Beffa et al., 1996; Iglesias and Meins, 2000; Guenoune-Gelbart et al., 2008).

The plasmodesmal associated (1,3)-β-glucan endo-hydrolase AtBG_pap (Levy et al., 2007a) is apparently not involved in virus spread. RNA measurements following infection of Arabidopsis with cucumber mosaic virus showed no induction or repression of transcription of the *AtBG_pap* gene, again suggesting that it is not involved in virus spread (unpublished data). In contrast, infection of tobacco with TMV led to a 20-fold increase in class I (1,3)-β-glucan endo-hydrolase protein levels (Vogelilange et al., 1988); however, since class I (1,3)-β-glucan endo-hydrolase enzymes are targeted to the ER lumen and the vacuole (Keefe et al., 1990), their potential activity at Pd requires the presence of a still unknown targeting mechanism.

Which of the multitude of (1,3)-β-glucan endo-hydrolases found in higher plants (in Arabidopsis there are 50 (1,3)-β-glucan endo-hydrolases) facilitates viral spread? How do they target to Pd? Recently, we proposed a mechanism for (1,3)-β-glucan endo-hydrolase targeting to Pd by viruses (Levy et al., 2007b): ER-derived vesicles containing MP as intrinsic membrane protein, and which have within the vesicle lumen (1,3)-β-glucan endo-hydrolase, target to the plasma membrane adjacent to Pd delivering their lumen content to the cell wall (see Fig. 4C,D). We speculated that the targeting to the wall is performed by cytoplasmic proteins. Based on the reported interaction between potato virus x (PVX) TGB12K MP and members of the 'ankyrin repeats' (AKR) family termed TIPs (TGB12K interacting proteins; Fridborg et al., 2003), we suggested that AKR can direct MP+(1,3)-β-glucan endo-hydrolase-containing vesicles to the cell periphery. However, by using the bimolecular fluorescence complementation technique to identify an interaction between two Arabidopsis TIPs homologues and the MP of both TMV and TMV(Cg), we found no such interaction (Levy, unpublished results).

As interactions between MP and microtubule and microfilaments have been described (Beachy and Heinlein, 2000; Heinlein and Epel, 2004; Heinlein et al., 1995; Zambryski, 1995), it was suggested that targeting of MP to the Pd occurs through association with one of the cytoskeletal proteins. However, recent evidence suggests that targeting of MP to Pd may not require an intact microtubule cytoskeleton (Wright et al., 2007). We therefore suggest that the (1,3)-β-glucan endo-hydrolase-containing vesicles are transported to the cell wall through MP–microfilament interactions where the glucanase is secreted into the Pd wall sleeve hydrolysing the callose barrier, while viral RNA and viral replicase are carried on MP rafts floating in the ER lipids (Fig. 5). Under the influence of a concentration gradient between infected and non-infected cell, the raft and its cargo diffuse in ER to the adjacent cell (Guenoune-Gelbart et al., 2008).

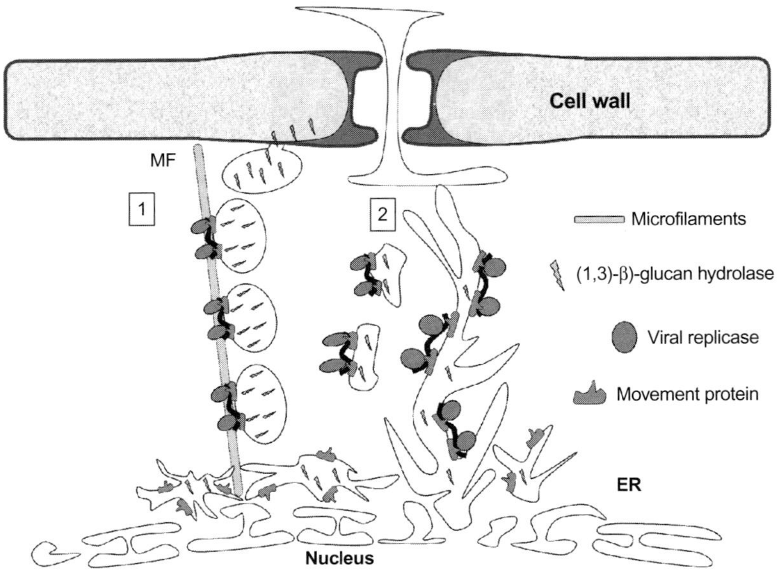


**Fig. 5: A model for the hydrolysis of callose and the gating of Pd during viral infection. Viral infection results in accumulation of callose around Pd. Viral-encoded MP inserts into ER membrane and functions as protein raft for a viral replication complex containing viral RNA and replicase. ER vesicles are formed containing the membrane-associated MP raft and associated cargo and (1,3)-β-glucan hydrolase enzymes in their lumen. The vesicles target to the cell periphery adjacent to Pd by the microfilaments. The vesicles fuse to the plasma membrane and deliver their (1,3)-β-glucan hydrolase cargo to the cell wall (1). (1,3)-β-Glucan hydrolase hydrolyses callose allowing Pd to dilate, and enables the viral replication complex to diffuse in ER to adjacent cell (2). (Modified from Levy et al., 2007b).**

## Acknowledgements

This article was written with financial support from Israel Science Foundation and The Manna Center for Plant Biosciences. We thank Roni Aloni, Chris van der Schoot, Robyn Overall, Ray Evert and Josef Samaj for kindly providing us with figures.

## References

Aaziz, R., Dinant, S., & Epel, B. L. (2001). Plasmodesmata and plant cytoskeleton. *Trends in Plant Science*, *6*, 326–330.

Aloni, R., Baum, S. F., & Peterson, C. A. (1990). The role of cytokinin in sieve tube regeneration and callose production in wounded Coleus internodes. *Plant Physiology*, *93*, 982–989.

Aloni, R., & Peterson, C. A. (1990). The functional significance of phloem anastomoses in stems of Dahlia-pinnata Cav. *Planta*, *182*, 583–590.

Aloni, R., & Peterson, C. A. (1997). Auxin promotes dormancy callose removal from the phloem of Magnolia kobus and callose accumulation and earlywood vessel differentiation in *Quercus robur*. *Journal of Plant Research*, *110*, 37–44.

Aloni, R., Raviv, A., & Peterson, C. A. (1991). The role of auxin in the removal of dormancy callose and resumption of phloem activity in *Vitis-vinifera*. *Canadian Journal of Botany-Revue Canadienne De Botanique*, *69*, 1825–1832.

Badelt, K., White, R. G., Overall, R. L., & Vesk, M. (1994). Ultrastructural specializations of the cell-wall sleeve around plasmodesmata. *American Journal of Botany*, *81*, 1422–1427.

Balestrini, R., Romera, C., Puigdomenech, P., & Bonfante, P. (1994). Location of a cell-wall hydroxyproline-rich glycoprotein, cellulose and beta-1,3-glucans in appeal and differentiated regions of maize mycorrhizal roots. *Planta*, *195*, 201–209.

Baluska, F., Samaj, J., Napier, R., & Volkmann, D. (1999). Maize calreticulin localizes preferentially to plasmodesmata in root apex. *Plant Journal*, *19*, 481–488.

Bayer, E., Thomas, C. L., & Maule, A. J. (2004). Plasmodesmata in *Arabidopsis thaliana* suspension cells. *Protoplasma*, *223*, 93–102.

Beachy, R. N., & Heinlein, M. (2000). Role of P30 in replication and spread of TMV. *Traffic*, *1*, 540–544.

Beffa, R. S., Hofer, R. M., Thomas, M., & Meins, F. (1996). Decreased susceptibility to viral disease of beta-1,3-glucanase-deficient plants generated by antisense transformation. *Plant Cell*, *8*, 1001–1011.

Behnke, H. D., & Sjolund, R. D. (1990). Sieve elements. Berlin, Heidelberg, New York: Springer-Verlag.

Blackman, L. M., & Overall, R. L. (1998). Immunolocalization of the cytoskeleton to plasmodesmata of *Chara corallina*. *Plant Journal*, *14*, 733–741.

Blackman, L. M., & Overall, R. L. (2001). Structure and function of plasmodesmata. *Australian Journal of Plant Physiology*, *28*, 709–727.

Blackman, L. M., Harper, J. D. I., & Overall, R. L. (1999). Localization of a centrin-like protein to higher plant plasmodesmata. *European Journal of Cell Biology*, *78*, 297–304.

Botha, C. E. J., & Cross, R. H. N. (2000). Towards reconciliation of structure with function in plasmodesmata – Who is the gatekeeper? *Micron*, *31*, 713–721.

Brown, I., Trethowan, J., Kerry, M., Mansfield, J., & Bolwell, G. P. (1998). Localization of components of the oxidative cross-linking of glycoproteins and of callose synthesis in papillae formed during the interaction between non-pathogenic strains of *Xanthomonas campestris* and French bean mesophyll cells. *Plant Journal*, *15*, 333–343.

Brownfield, L., Ford, K., Doblin, M. S., Newbigin, E., Read, S., & Bacic, A. (2007). Proteomic and biochemical evidence links the callose synthase in *Nicotiana alata* pollen tubes to the product of the NaGSL1 gene. *Plant Journal*, *52*, 147–156.

Bucher, G. L., Tarina, C., Heinlein, M., Di Serio, F., Meins, F., Jr., & Iglesias, V. A. (2001) Local expression of enzymatically active class I beta-1, 3-glucanase enhances symptoms of TMV infection in tobacco. *Plant Journal*, *28*, 361–369.

Burr, F. A., & Evert, R. F. (1973). Some aspects of sieve-element structure and development in Selaginella kraussiana. *Protoplasma*, *78*, 81–97.

Citovsky, V., & Zambryski, P. (2000). Systemic transport of RNA in plants. *Trends in Plant Science*, *5*, 52–54.

Cook, M. E., Graham, L. E., Botha, C. E. J., & Lavin, C. A. (1997). Comparative ultrastructure of plasmodesmata of Chara and selected bryophytes: Toward an elucidation of the evolutionary origin of plant plasmodesmata. *American Journal of Botany*, *84*, 1169–1178.

Cui, X. J., Shin, H. S., Song, C., Laosinchai, W., Amano, Y., & Brown, R. M. (2001). A putative plant homolog of the yeast beta-1,3-glucan synthase subunit FKS1 from cotton (*Gossypium hirsutum* L.) fibers. *Planta*, *213*, 223–230.

Currier, H. B., & Webster, D. H. (1964). Callose formation and subsequent disappearance: Studies in ultrasound stimulation. *Plant Physiology*, *39*, 843–847.

Dahiya, P., & Brewin, N. J. (2000). Immunogold localization of callose and other cell wall components in pea nodule transfer cells. *Protoplasma*, *214*, 210–218.

Deshpande, B. P., & Rajendrababu, T. (1985). Seasonal-changes in the structure of the secondary phloem of *Grewia-tiliaefolia*, a deciduous tree from India. *Annals of Botany*, *56*, 61–71.

Ding, B., Itaya, A., & Woo, Y. M. (1999). Plasmodesmata and cell-to-cell communication in plants Vol 190. *International Review of Cytology – A Survey of Cell Biology*, *190*, 251–316.

Ding, B., Kwon, M. O., & Warnberg, L. (1996). Evidence that actin filaments are involved in controlling the permeability of plasmodesmata in tobacco mesophyll. *Plant Journal*, *10*, 157–164.

Ding, B., Turgeon, R., & Parthasarathy, M. V. (1992). Substructure of freeze-substituted plasmodesmata. *Protoplasma*, *169*, 28–41.

Doblin, M. S., De Melis, L., Newbigin, E., Bacic, A., & Read, S. M. (2001). Pollen tubes of *Nicotiana alata* express two genes from different beta-glucan synthase families. *Plant Physiology*, *125*, 2040–2052.

Doxey, A. C., Yaish, M. W. F., Moffatt, B. A., Griffith, M., & McConkey, B. J. (2007). Functional divergence in the Arabidopsis -1,3-glucanase gene family inferred by phylogenetic reconstruction of expression states. *Molecular Biology and Evolution*, *24*, 1045–1055.

Dute, R. R., & Evert, R. F. (1977). Sieve-element ontogeny in the root of *Equisetum hyemale*. *American Journal of Botany*, *64*, 421–438.

Ehlers, K., Knoblauch, M., & van Bel, A. J. E. (2000). Ultrastructural features of well-preserved and injured sieve elements: Minute clamps keep the phloem transport conduits free for mass flow. *Protoplasma*, *214*, 80–92.

Ehlers, K., & Kollmann, R. (2001). Primary and secondary plasmodesmata: Structure, origin, and functioning. *Protoplasma*, *216*, 1–30.

Elortza, F., Mohammed, S., Bunkenborg, J., Foster, L. J., Nuhse, T. S., Brodbeck, U., Peck, S. C., & Jensen, O. N. (2006). Modification-specific proteomics of plasma membrane proteins: Identification and characterization of glycosylphosphatidylinositol-anchored proteins released upon phospholipase D treatment. *Journal of Proteome Research*, *5*, 935–943.

Elortza, F., Nuhse, T. S., Foster, L. J., Stensballe, A., Peck, S. C., & Jensen, O. N. (2003). Proteomic analysis of glycosylphosphatidylinositol-anchored membrane proteins. *Molecular and Cellular Proteomics*, *2*, 1261–1270.

Engleman, E. M., & Esau, K. (1964). The problem of callose deposition in phloem. *Science*, *144*, 562.

Esau, K., & Thorsch, J. (1985). Sieve plate pores and plasmodesmata, the communication channels of the symplast – Ultrastructural aspects and developmental relations. *American Journal of Botany*, *72*, 1641–1653.

Evert, R. F. (1976). Some aspects of the sieve-element structure and development in Botrychium virginianum. *Israelian Journal of Botany*, *25*, 101–126.

Evert, R. F. (1977). Phloem structure and histochemistry. *Annual Review of Plant Physiology and Plant Molecular Biology*, *28*, 199–222.

Evert, R. F. (1990). Dicotyledons. In H. D. Behnke & R. D. Sjolund (Eds.), *Sieve elements* (pp. 103–137). Berlin, Heidelberg, New York: Springer-Verlag.

Farrokhi, N., Burton, R. A., Brownfield, L., Hrmova, M., Wilson, S. M., Bacic, A., & Fincher, G. B. (2006). Plant cell wall biosynthesis: Genetic, biochemical and functional genomics approaches to the identification of key genes. *Plant Biotechnology Journal*, *4*, 145–167.

Fridborg, I., Grainger, J., Page, A., Coleman, M., Findlay, K., & Angell, S. (2003). TIP, a novel host factor linking callose degradation with the cell-to-cell movement of Potato virus X. *Molecular Plant-Microbe Interactions*, *16*, 132–140.

Furch, A. C. U., Hafke, J. B., Schulz, A., & van Bel, A. J. E. (2007). Ca2+-mediated remote control of reversible sieve tube occlusion in *Vicia faba*. *Journal of Experimental Botany*, *58*, 2827–2838.

Guenoune-Gelbart, D., Elbaum, M., Sagi, G., Levy, A., & Epel, B. L. (2008). *Tobacco mosaic virus* (TMV) replicase and movement protein function synergistically in facilitating TMV spread by lateral diffusion in the plasmodesmal desmotubule of Nicotiana benthamiana. *Molecular Plant-Microbe Interactions*, *21*, 335–345.

Gorshkova, E. N., Erokhina, T. N., Stroganova, T. A., Yelina, N. E., Zamyatnin, A. A., Kalinina, N. O., Schiemann, J., Solovyev, A. G., & Morozov, S. Y. (2003). Immunodetection and fluorescent microscopy of transgenically expressed hordeivirus TGBp3 movement protein reveals its association with endoplasmic reticulum elements in close proximity to plasmodesmata. *Journal of General Virology*, *84*, 985–994.

Gregory, A. C. E., Smith, C., Kerry, M. E., Wheatley, E. R., & Bolwell, G. P. (2002). Comparative subcellular immunolocation of polypeptides associated with xylan and callose synthases in French bean (*Phaseolus vulgaris*) during secondary wall formation. *Phytochemistry*, *59*, 249–259.

Hao, P., Liu, C., Wang, Y., Chen, R., Tang, M., Du, B., Zhu, L., & He, G. (2008). Herbivore-induced callose deposition on the sieve plates of rice: An important mechanism for host resistance. *Plant Physiology*, *146*, 1810–1820.

Haywood, V., Kragler, F., & Lucas, W. J. (2002). Plasmodesmata: Pathways for protein and ribonucleoprotein signaling. *Plant Cell*, *14*(Suppl.), S303–S325.

Heinlein, M. (2002). Plasmodesmata: Dynamic regulation and role in macromolecular cell-to-cell signaling. *Current Opinion in Plant Biology*, *5*, 543.

Heinlein, M., & Epel, B. L. (2004). Macromolecular transport and signaling through plasmodesmata. *International Review of Cytology – A Survey of Cell Biology*, *235*, 93–164.

Heinlein, M., Epel, B. L., Padgett, H. S., & Beachy, R. N. (1995). Interaction of tobamovirus movement proteins with the plant cytoskeleton. *Science*, *270*, 1983–1985.

Hollis, C. A., & Tepper, H. B. (1971). Auxin transport within intact dormant and active white ash shoots. *Plant Physiology*, *48*, 146–149.

Hong, Z. L., Delauney, A. J., & Verma, D. P. S. (2001). A cell plate specific callose synthase and its interaction with phragmoplastin. *Plant Cell*, *13*, 755–768.

Iglesias, V. A., & Meins, F., Jr. (2000). Movement of plant viruses is delayed in a beta-1,3-glucanase-deficient mutant showing a reduced plasmodesmatal size exclusion limit and enhanced callose deposition. *Plant Journal*, *21*, 157–166.

Kauss, H. (1996). Callose synthesis. In M. Smallwood, J. P. Knox, & D. J. Bowles (Eds.), *Membranes: Specialized functions in plants* (pp. 77–92). Oxford: BIOS.

Keefe, D., Hinz, U., & Meins, F. (1990). The effect of ethylene on the cell-type-specific and intracellular-localization of beta-1,3-glucanase and chitinase in tobacco-leaves. *Planta*, *182*, 43–51.

Kollmann, R., & Glockmann, C. (1999). Multimorphology and nomenclature of plasmodesmata in higher plants. In A. J. van Bel & W. J. P. van Kesteren (Eds.), *Plasmodesmata, structure, function, role in cell communication* (pp. 149–172). Berlin: Springer-Verlag.

Krabel, D., Eschrich, W., Wirth, S., & Wolf, G. (1993). Callase-(1,3-beta-D-glucanase) activity during spring reactivation in deciduous trees. *Plant Science*, *93*, 19–23.

Kruatrachue, M., & Evert, R. F. (1974). Structure and development of sieve elements in the leaf of *Isoetes muricata*. *American Journal of Botany*, *61*, 253–266.

Leubner-Metzger, G. (2003). Functions and regulation of beta-1,3-glucanases during seed germination, dormancy release and after-ripening. *Seed Science Research*, *13*, 17–34.

Leubner-Metzger, G., & Meins, F. (1999). Functions and regulation of plant beta-1,3-glucanases (PR-2). In S. K. Datta & S. Muthukrishnan (Eds.), *Pathogenesis-related proteins in plants* (pp. 49–76). Florida: CRC Press.

Levy, A., Erlanger, M., Rosenthal, M., & Epel, B. L. (2007a). A plasmodesmata-associated beta-1, 3-glucanase in Arabidopsis. *Plant Journal*, *49*, 669–682.

Levy, A., Guenoune-Gelbart, D., & Epel, B. L. (2007b). Beta-1,3-glucanases: Plasmodesmal gate keepers for intercellular communication. *Plant Signaling and Behavior*, *2*, 404–406.

Liarzi, O., & Epel, B. L. (2005). Development of a quantitative tool for measuring changes in the coefficient of conductivity of plasmodesmata induced by developmental, biotic, and abiotic signals. *Protoplasma*, *225*, 67–76.

Liberman-Maxe, M. (1978). Sieve cell-wall in phloem of a polypodiaccous fern. *Biologie Cellulaire*, *31*, 201–210.

Lucas, W. J., Ding, B., & van der Schoot, C. (1993). Tansley Review No. 58 Plasmodesmata and the supracellular nature of plants. *New Phytologist*, *125*, 435–476.

Maeda, H., Song, W., Sage, T. L., & Della Penna, D. (2006). Tocopherols play a crucial role in low-temperature adaptation and phloem loading in Arabidopsis. *Plant Cell*, *18*, 2710–2732.

McCormack, B. A., Gregory, A. C. E., Kerry, M. E., Smith, C., & Bolwell, G. P. (1997). Purification of an elicitor-induced glucan synthase (callose synthase) from suspension cultures of French bean (*Phaseolus vulgaris* L.): Purification and immunolocation of a probable M-r-65 000 subunit of the enzyme. *Planta*, *203*, 196–203.

McNairn, R. B. (1972). Phloem translocation and heat-induced callose formation in field-grown *Gossypium-hirsutum*-L. *Plant Physiology*, *50*, 366–370.

McNairn, R. B., & Currier, H. B. (1968). Translocation blockage by sieve plate callose. *Planta*, *82*, 369–380.

Northcote, D. H., Davey, R., & Lay, J. (1989). Use of antisera to localize callose, xylan and arabinogalactan in the cell-plate, primary and secondary walls of plant-cells. *Planta*, *178*, 353–366.

Olesen, P. (1979). Neck constriction in plasmodesmata – Evidence for a peripheral sphincter-like structure revealed by fixation with tannic-acid. *Planta*, *144*, 349–358.

Olesen, P., & Robards, A. W. (1990). The neck region of the plasmodesmata: General architecture and some functional aspects. In A. W. Robards, W. J. Lucas, J. D. Pitts, H. J. Jongsma, & D. C. Spray (Eds.), *Parallels in cell to cell junctions in plants and animals* (pp. 145–170). Berlin, Heidelberg, New York: Springer.

Oparka, K. J., Prior, D. A., Santa Cruz, S., Padgett, H. S., & Beachy, R. N. (1997). Gating of epidermal plasmodesmata is restricted to the leading edge of expanding infection sites of tobacco mosaic virus (TMV). *Plant Journal*, *12*, 781–789.

Oparka, K. J., Roberts, A. G., Boevink, P., Santa Cruz, S., Roberts, L., Pradel, K. S., Imlau, A., Kotlizky, G., Sauer, N., & Epel, B. (1999). Simple, but not branched, plasmodesmata allow the nonspecific trafficking of proteins in developing tobacco leaves. *Cell*, *97*, 743–754.

Orfila, C., & Knox, J. P. (2000). Spatial regulation of pectic polysaccharides in relation to pit fields in cell walls of tomato fruit pericarp. *Plant Physiology*, *122*, 775–781.

Overall, R. L. (1999). Structure of plasmodesmata. In A. J. vanBel & W. J. P. van Kestern (Eds.), *Plasmodesmata, structure, function, role in cell communication* (pp. 129–148). Berlin, Heidelberg, New York: Spring-Verlag.

Overall, R. L., & Blackman, L. M. (1996). A model of the macromolecular structure of plasmodesmata. *Trends in Plant Science*, *1*, 307–311.

Perry, J. W., & Evert, R. F. (1975). Structure and development of the sieve elements in Psilotum nudum. *American Journal of Botany*, *62*, 1038–1052.

Peterson, C. A., & Rauser, W. E. (1979). Callose deposition and photoassimilate export in Phaseolus-vulgaris exposed to excess Cobalt, Nickel, and Zinc. *Plant Physiology*, *63*, 1170–1174.

Radford, J. E., Vesk, M., & Overall, R. L. (1998). Callose deposition at plasmodesmata. *Protoplasma, 201*, 30–37.

Radford, J. E., & White, R. G. (1998). Localization of a myosin-like protein to plasmodesmata. *Plant Journal, 14*, 743–750.

Radford, J. E., & White, R. G. (2001). Effects of tissue-preparation-induced callose synthesis on estimates of plasmodesma size exclusion limits. *Protoplasma, 216*, 47–55.

Rinne, P. L. H., Kaikuranta, P. M., & van der Schoot, C. (2001). The shoot apical meristem restores its symplasmic organization during chilling-induced release from dormancy. *Plant Journal, 26*, 249–264.

Rinne, P. L. H., van den Boogaard, R., Mensink, M. G. J., Kopperud, C., Kormelink, R., Goldbach, R., & van der Schoot, C. (2005). Tobacco plants respond to the constitutive expression of the tospovirus movement protein NSM with a heat-reversible sealing of plasmodesmata that impairs development. *Plant Journal, 43*, 688–707.

Rinne, P. L. H., & van der Schoot, C. (1998). Symplasmic fields in the tunica of the shoot apical meristem coordinate morphogenetic events. *Development, 125*, 1477–1485.

Rinne, P. L. H., & van der Schoot, C. (2003). Plasmodesmata at the crossroads between development, dormancy, and defense. *Canadian Journal of Botany-Revue Canadienne De Botanique, 81*, 1182–1197.

Rinne, P. L. H., & van der Schoot, C. (2004). Cell-cell communication as a key factor in dormancy cycling. *Journal of Crop Improvement, 10*, 113–156.

Roberts, A. G., & Oparka, K. J. (2003). Plasmodesmata and the control of symplastic transport. *Plant Cell and Environment, 26*, 103–124.

Robinsonbeers, K., & Evert, R. F. (1991). Fine-structure of plasmodesmata in mature leaves of sugarcane. *Planta, 184*, 307–318.

Rodríguez-Gálvez, E., & Mendgen, K. (1995). Cell wall synthesis in cotton roots after infection with *Fusarium oxysporum. Planta, 197*, 535–545.

Roy, S., Watada, A. E., & Wergin, W. P. (1997). Characterization of the cell wall microdomain surrounding plasmodesmata in apple fruit. *Plant Physiology, 114*, 539–547.

Ruan, Y. L. (2007). Rapid cell expansion and cellulose synthesis regulated by plasmodesmata and sugar: Insights from the single-celled cotton fibre. *Functional Plant Biology, 34*, 1–10.

Ruan, Y. L., Xu, S. M., White, R., & Furbank, R. T. (2004). Genotypic and developmental evidence for the role of plasmodesmatal regulation in cotton fiber elongation mediated by callose turnover. *Plant Physiology, 136*, 4104–4113.

Sagi, G., Katz, A., Guenoune-Gelbart, D., & Epel, B. L. (2005). Class 1 reversibly glycosylated polypeptides are plasmodesmal-associated proteins delivered to plasmodesmata via the Golgi apparatus. *Plant Cell*, *17*, 1788–1800.

Schulz, A. (1999). Physiological control of plasmodesmal gating. In A. J. E. van Bel & W. J. P. van Kesteren (Eds.), *Plasmodesmata structure, function, role in cell communication*. Berlin, Heidelberg, New York: Springer-Verlag.

Simmons, C. R. (1994). Physiology and molecular-biology of plant 1,3-beta-D-glucanases and 1,31, 4-beta-D-glucanases. *Critical Reviews in Plant Sciences*, *13*, 325–387.

Sivaguru, M., Fujiwara, T., Samaj, J., Baluska, F., Yang, Z. M., Osawa, H., Maeda, T., Mori, T., Volkmann, D., & Matsumoto, H. (2000). Aluminum-induced 1 -> 3-beta-D-glucan inhibits cell-to-cell trafficking of molecules through plasmodesmata. A new mechanism of aluminum toxicity in plants. *Plant Physiology*, *124*, 991–1005.

Sjolund, R. D. (1997). The phloem sieve element: A river runs through it. *Plant Cell*, *9*, 1137–1146.

Stone, B. A., & Clarke, A. E. (1992). Chemistry and physiology of higher plants (1->3)-beta-glucans. In B. A. Stone & A. E. Clarke (Eds.), *Chemistry and Biology of (1->3)-beta-glucans* (pp. 365–429). Victoria Australia: La Trobe University Press.

Sutherland, P., Hallett, L., Redgwell, R., Benhamou, N., & MacRae, E. (1999). Localization of cell wall polysaccharides during kiwifruit (*Actinidia deliciosa*) ripening. *International Journal of Plant Sciences*, *160*, 1099–1109.

Trethewey, J. A. K., & Harris, P. J. (2002). Location of (1 -> 3)- and (1 -> 3),(1 -> 4)-beta-D-glucans in vegetative cell walls of barley (*Hordeum vulgare*) using immunogold labelling. *New Phytologist*, *154*, 347–358.

Trillas, M. I., Cotxarrera, L., Casanova, E., & Cortadellas, N. (2000). Ultrastructural changes and localization of chitin and callose in compatible and incompatible interactions between carnation callus and *Fusarium oxysporum*. *Physiological and Molecular Plant Pathology*, *56*, 107–116.

Turner, A., Wells, B., & Roberts, K. (1994). Plasmodesmata of maize root tips: Structure and composition. *Journal of Cell Science*, *107*, 3351–3361.

Van Bel, A. J. E. (2003). The phloem, a miracle of ingenuity. *Plant Cell and Environment*, *26*, 125–149.

Vogelilange, R., Hansengehri, A., Boller, T., & Meins, F. (1988). Induction of the defense-related glucanohydrolases, beta-1,3-glucanase and chitinase, by tobacco mosaic-virus infection of tobacco-leaves. *Plant Science*, *54*, 171–176.

Walsh, M. A., & Melaragno, J. E. (1976). Ultrastructural features of developing sieve elements in *Lemna-minor*-L-sieve plate and lateral sieve areas. *American Journal of Botany*, *63*, 1174–1183.

Warmbrodt, R. D., & Evert, R. F. (1974). Structure and development of the sieve element in the stem of *Lycopodium lucidulum*. *American Journal of Botany*, *61*, 267–277.

Webster, D. B., & Currier, H. H. (1965). Callose-lateral movement of assimilates from phloem. *Science*, *150*, 1610–1611.

White, R. G., Badelt, K., Overall, R. L., & Vesk, M. (1994). Actin associated with plasmodesmata. *Protoplasma*, *180*, 169–184.

Will, T., Tjallingii, W. F., Thonnessen, A., & van Bel, A. J. E. (2007). Molecular sabotage of plant defense by aphid saliva. *Proceedings of the National Academy of Sciences of the United States of America*, *104*, 10536–10541.

Will, T., & van Bel, A. J. E. (2006). Physical and chemical interactions between aphids and plants. *Journal of Experimental Botany*, *57*, 729–737.

Wilson, S. M., Burton, R. A., Doblin, M. S., Stone, B. A., Newbigin, E. J., Fincher, G. B., & Bacic, A. (2006). Temporal and spatial appearance of wall polysaccharides during cellularization of barley (*Hordeum vulgare*) endosperm. *Planta*, *224*, 655–667.

Wolf, S., et al. (1991). Plasmodesmatal function is probed using transgenic tobacco plants that express a virus movement protein. *Plant Cell*, *3*, 593–604.

Wolf, S., Deom, C. M., Beachy, R., & Lucas, W. J. (1989). Movement protein of tobacco mosaic-virus modifies plasmodesmatal size exclusion limit. *Science*, *246*, 377–379.

Wright, K. M., Wood, N. T., Roberts, A. G., Chapman, S., Boevink, P., MacKenzie, K. M., & Oparka, K. J. (2007). Targeting of TMV movement protein to plasmodesmata requires the actin/ER network: Evidence from FRAP. *Traffic*, *8*, 21–31.

Wu, X. L., Weigel, D., & Wigge, P. A. (2002). Signaling in plants by intercellular RNA and protein movement. *Genes & Development*, *16*, 151–158.

Zambryski, P. (1995). Plasmodesmata: Plant channels for molecules on the move. *Science*, *270*, 1943–1944.

Zambryski, P., & Crawford, K. (2000). Plasmodesmata: Gatekeepers for cell-to-cell transport of developmental signals in plants. *Annual Review of Cell and Developmental Biology*, *16*, 393–421.

CHAPTER 4.4.3

# Callose and its Role in Pollen and Embryo Sac Development in Flowering Plants

**Ed Newbigin[1], Antony Bacic[1,2] and Steve Read[3]**
[1]*School of Botany, University of Melbourne, VIC, Australia*
[2]*Australian Centre for Plant Functional Genomics, School of Botany, University of Melbourne, VIC, Australia*
[3]*Forestry Tasmania, Hobart, TAS, Australia*

Flowering plant (angiosperm) gametes develop by mitosis from haploid gametophyte cells that are formed by meiosis from diploid sporocytes. Male gametes are contained within multicellular pollen grains and female gametes within a multicellular embryo sac. Transient callose walls are a conspicuous feature of the processes that lead to the production of both types of gametes. Mature pollen grains germinate and produce a tube that conveys the male gametes to the embryo sac where fertilization takes place. Flowering plant pollen tubes are characterized by a thick, inner wall of callose and regular callosic cross walls. Here, we review gamete development and the interval between pollination and fertilization (progamic phase) in flowering plants, using *Arabidopsis thaliana* as our example, and discuss the various hypotheses that have been put forward as to the functions of the callose walls.

## I. Introduction

Reproductive development in flowering plants (angiosperms) requires the formation of haploid cells or gametophytes by meiosis. Gametes are then produced from these cells by mitosis. A clear distinction can thus be drawn between reproductive development in animals, where meiosis gives rise to gametes directly, and the same process in flowering plants where the products of meiosis undergo further mitotic divisions prior to the development of gametes. Thus, the products of male gamete development are male gametes (sperm cells) housed within the male gametophyte (pollen grain), and the products of female gamete development are the female gamete (egg cell) housed within the female gametophyte (embryo sac).

Male sperm cells and pollen grains are often referred to as microgametes and microgametophytes, respectively. Similarly, the terms megagamete and megagametophyte are often used to refer to egg cells and embryo sacs.

A feature of both male and female gamete development is the transient appearance of callose, a linear (1,3)-β-glucan, in the walls that surround the gametophytes and gametes. Because it is readily detected with the Aniline Blue fluorochrome, these callose deposits act as useful markers of various histological stages in the developmental sequences that lead to the formation of viable pollen grains (pollen) or embryo sacs (e.g. Tucker et al., 2001; Sannier et al., 2006). Callose deposits have also been argued to serve specific biological functions. Here, we will review these proposed functions, taking as our starting point the summaries of this subject found in Stone and Clarke (1992). As embryo sac and pollen development in angiosperms is already well documented, we will only briefly sketch these processes. Readers interested in more detailed ultrastructural descriptions, or in thorough discussions of the genes that regulate pollen and embryo sac development, are referred to the following selection of recently published reviews (Grossniklaus and Schneitz, 1998; Scott et al., 2004; Skinner et al., 2004; Yadegari and Drews 2004; Ma, 2005; Boavida et al., 2005; Blackmore et al., 2007).

## I.A. Overview of Microgamete Development

Technically, microsporogenesis is the term used to refer to the initiation and formation of microspores from microsporocytes within the male reproductive organ (androecium or stamen) of the flower; and microgametogenesis to the subsequent rounds of mitosis that occur within the microspore and ultimately lead to the formation of a male gametophyte consisting of two microgametes, or sperm cells, enclosed within a large vegetative cell. For the sake of simplicity, however, we will refer to this entire process as microgamete development.

Microgamete development can accordingly be thought of as having two distinct phases: an initial phase during which the microspores are in contact with each other and with cells of the stamen, and a final phase that occurs entirely within the microspore after contact with the stamen has been severed (Goldberg et al., 1993). Indeed, in many species the mitotic division to produce the two microgametes occurs after pollen has been shed by the anther. These species therefore release pollen that is bicellular, made up of a vegetative cell within which is a sperm cell precursor called the generative cell. Other species release pollen grains that are tricellular, composed of a large vegetative cell inside which are two sperm cells.

The following section briefly describes microgamete development in *Arabidopsis,* with an emphasis on times when callose is present. This description is based on that by Owen and

Makaroff (1995) and, unless otherwise stated, uses Boavida et al. (2005) and Blackmore et al. (2007) as references for information on cell wall compositions and changes. The 12 stages referred to here are taken from Owen and Makaroff (1995) and are shown in diagrammatic form in Fig. 1. Light micrographs of *Arabidopsis* microspores at stages of development when callose is present, along with the corresponding Aniline Blue fluorochrome-stained image, are shown in Fig. 2. Although this description specifically refers to *Arabidopsis*, a species with tricellular pollen grains, these stages are typical of microgamete development in the majority of angiosperms. Subsequent sections will describe some of the variant forms of this typical pattern and the locations and proposed roles of callose throughout this developmental sequence.

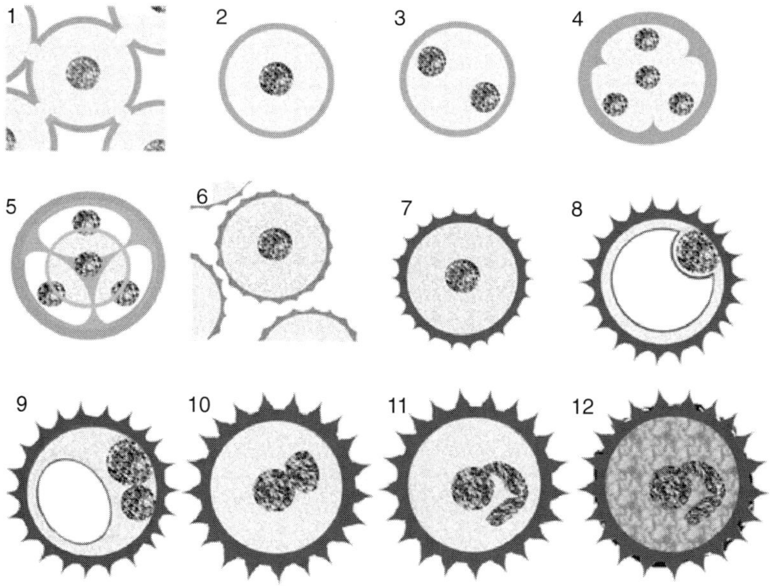


**Fig. 1: Stages of pollen development, following the system of Owen and Makaroff (1995). (1) Premeiosis I: microsporocytes connected by cytoplasmic channels. (2) Premeiosis II: a microsporocyte surrounded by the callose special cell wall. (3) Meiosis: division underway in a microsporocyte. (4) Meiosis complete: before cytokinesis. (5) Tetrad: callose special cell walls (SCWs) are present around the microspores. (6) Released microspore I: microspores are surrounded by differentiating exine. (7) Released microspore II: further differentiation of the exine. (8) Ring-vacuolate microspore: with a large vacuole causing the characteristic signet ring appearance. (9) Bicellular pollen I: asymmetric mitosis gives rise to the vegetative cell surrounding the peripheral generative cell. (10) Bicellular pollen II: with the generative cell central. (11) Second mitotic division: forming the male germ unit. (12) Mature pollen: with storage products accumulated in the cytoplasm. (Reproduced with permission of *New Phytologist* from Blackmore et al., 2007.)**

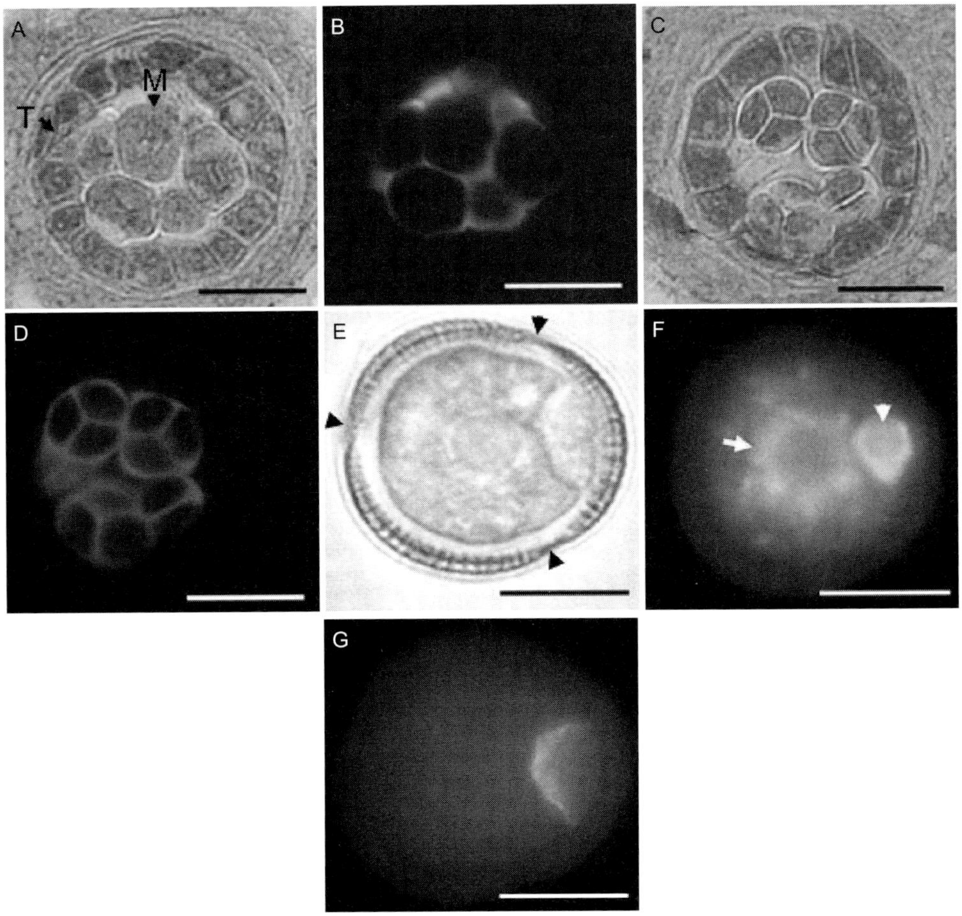


Fig. 2: The appearance of callose during microgamete development in a wild-type *Arabidopsis* plant. (A, C) Toluidine blue-stained cross-section. (E) Bright-field image. (B, D, F and G) Corresponding images viewed with UV illumination after Aniline Blue fluorochrome-staining (B, D and G) or 4′,6-diamidino-2-phenylindole (DAPI) staining (F). (A, B) Microsporocytes (M) undergoing meiosis (stage 3). The surrounding tapetal cell layer (T) is also evident. Callose in the common SCW surrounding each microsporocyte fluoresces when stained with Aniline Blue fluorochrome (B). (C, D) Microspores still in tetrads (stage 5) and surrounded by their individual SCWs. (E–G) *Arabidopsis* microspore after mitosis I (E) Arrowheads indicate areas where the wall is thin and colpi will later form. (F) The arrowhead and arrow indicate the centrally placed vegetative nucleus and peripheral generative nucleus, respectively. (G) The callosic wall separating the generative and vegetative cells is clearly seen. Scale bars: (A–D) 10 μm; (E-G) 5 μm. (A–D reproduced from Figure 8 of Enns et al. (2005), with kind permission from Springer Science+Business media; E–G kindly provided by Professor David Twell, University of Leicester UK.) The colour specifications refer to colours in panels.

### I.A.1. Microgamete Development in Arabidopsis

A mature *Arabidopsis* flower has six stamens, each consisting of a stalk or filament comprised of connective tissue that allows nutrients and water to reach a terminal four-lobed anther. Stamens are arranged in a single ring of organs. Outside the ring of stamens is a ring of four petals and within the ring, at the flower's centre, is the female reproductive organ, the gynoecium. Contained within the gynoecium are numerous ovules. The pollen grains in a fully developed flower are contained in the middle of each lobe of the anthers, in cavities called locules. Mature *Arabidopsis* pollen grains are spheroidal to ovoid in shape, with three slit-like apertures or colpi arranged around the equator (tricolpate). Each pollen grain is tricellular, containing a large vegetative cell inside which are two small sperm cells. At anthesis, the period during which anthers release their pollen grains for dispersal, the anther ruptures along two furrows that run the length of either side.

#### I.A.1.a. Premeiosis I and II (stages 1 and 2)

By the time the anther has developed its characteristic four-lobed appearance, five different cell layers are evident. The innermost cells are diploid microsporocytes, identifiable by their large, centrally located nucleus and prominent nucleolus. Surrounding the microsporocytes is a layer of secretory tapetal cells (Fig. 2A), which have important nutritive functions during pollen development and also deposit sporopollenin and other coatings onto the pollen wall. The tapetal cells degenerate later during anther development. The three cell layers surrounding the tapetal cell layer are the middle layer, endothecium and epidermis. At this stage, microsporocytes are connected to each other and to the tapetal cells by numerous plasmodesmata. Callose accumulates between the plasma membrane and the original primary cell wall of the microsporocytes, which appears to contain both cellulose and methyl-esterified pectin, causing the wall to thicken (Owen and Makaroff, 1995; Rhee and Somerville, 1998). This thickening breaks the cytoplasmic connections that link microsporocytes to each other and to the tapetal cells. Instead of plasmodesmata, microsporocytes become interconnected by large (approx. 0.5 μm in diameter) cytoplasmic channels.

Prior to the onset of meiosis, callosic deposits further thicken the wall around each microsporocyte, forming what is referred to as the common special cell wall (SCW; Blackmore et al., 2007). Because of its high callose content, the common SCW appears electron opaque when uranyl acetate-stained sections are viewed by transmission electron microscopy.

#### I.A.1.b. Meiosis to cytokinesis (stages 3–5)

Meiosis in angiosperms proceeds through the same stages as other eukaryotes. After completion of meiosis II in *Arabidopsis* and most flowering plants, the microsporocytes are coenocytic: each cell contains four haploid nuclei that are located at the cell periphery in an approximate tetrahedral arrangement. Separation of the coenocytic microspore into four haploid daughter cells (cytokinesis) occurs simultaneously (i.e. cytokinesis only occurs when meiosis II is complete). The first evidence of cytokinesis is the appearance of callosic ingrowths in regions of the parental wall lying between the haploid nuclei. These intersporal walls grow centripetally (inwards from the outside) until the four haploid microspores are fully separated from each other. Cytoplasmic connections that previously linked the microsporocytes have been lost by this stage. The four microspores derived from each microsporocyte are joined together in a structure called a tetrad that is completely encased in the thick callosic wall composed of the original common SCW and the walls laid down during cytokinesis (Fig. 2D). The wall around each microspore is called an individual SCW, to distinguish it from the common SCW that surrounded the microsporocyte and now surrounds the tetrad.

#### *I.A.1.c. Released microspore I (stages 6 and 7)*

Wall synthesis in microspores now switches from the deposition of callose to the production of a new fibrillar wall called the primexine composed largely of cellulose. The primexine is the foundation on which the sculpted outer wall of the pollen grain, the exine, is a later elaboration. Primexine is absent from areas where the three colpi will later form on each pollen grain (Fig. 2E). The placement of and number of areas where there is thinning is determined by the geometry of the tetrad produced by meiosis (see below).

Slightly after the onset of primexine synthesis numerous regular undulations appear in the microspore plasma membrane. Fibrous material accumulates at low points in these undulations, eventually resulting in a regular pattern of localized plaques of this material (Fitzgerald and Knox, 1995). On peaks of the undulating membrane, rod-shaped probacula are extruded, giving the primexine a reticulate pattern. During this stage the walls surrounding the tapetal cells become completely degraded and small amounts of fine fibrillar material are released into the anther locule. Callase, a mixture of (1,3)-β-glucan hydrolases secreted by the tapetum, degrades the callose wall surrounding the microspore tetrad and the intersporal cell walls that holds the microspore tetrad together; that is, callase degrades the common and individual SCWs. This mixture of hydrolytic enzymes presumably includes the (1,3)-β-glucanase specifically expressed at this time by the tapetum (Stone and Clarke, 1992; Hird et al., 1993; Bucciaglia and Smith, 1994).

#### *I.A.1.d. Released microspore II to bicellular pollen grains I (stages 8–10)*

The second phase of microgamete development is characterized by male gamete formation within the isolated microspores. Microspores are more rounded than previously and contain several large vacuoles that ultimately coalesce into a single vacuole. The large vacuole repositions the nucleus to one side of the microsporocyte. The polarized microspore with a large central vacuole and a peripheral nucleus has been called the signet ring stage (Knox, 1984). The cytoplasm of tapetal cells is now highly active and contains many vesicles full of osmiophilic material that fuse with the plasma membrane and deposit their contents into the locule, where it is incorporated into the growing exine layer of the microspore cell wall. An inner microspore wall containing unesterified pectins (Van Aelst and Van Went, 1992) called the intine layer also appears at this stage. The intine is synthesized from material produced by the microspore.

The microspore nucleus undergoes an asymmetric first mitotic division to produce a larger vegetative cell that encloses a smaller generative cell (Fig. 2E–G). This generative cell is the precursor of the two sperm cells. The generative cell is separated from the vegetative cell by a thin callose wall that is contiguous with the intine (Fig. 2G; Oh et al., 2005). Degradation of this transient callose wall, presumably by a (1,3)-β-glucanase(s), allows the generative cell to detach from the intine and take up a central position within the vegetative cell.

#### *I.A.1.e. Second mitotic division to mature pollen grains (stages 11 and 12)*

The second mitosis of the generative cell produces two sperm cells that are elongated and linked together. The entire structure (vegetative cell enclosing two sperm cells) is in its final tricellular state and called a pollen grain. The sperm cells have little cytoplasm and are easily distinguished from the nucleus of the vegetative cell, which is lobed and surrounded by a cytoplasm that is highly vacuolated and contains many small storage bodies. Clear physical associations exist between the vegetative nucleus and two sperm cells, an assemblage known as the male germ unit (Lalanne and Twell, 2002; McCormick, 2004). Indeed, a cytoplasmic extension from one sperm cell appears to partially encircle the vegetative nucleus, but whether there is a preference for this sperm cell to fuse with one of the two fusion partners (egg and central cell; see below) at fertilization is unknown. Coordinated transport of the male germ unit down the pollen tube (see below) is essential for fertilization. The tapetal layer and middle layer are no longer present, meaning that the endothecium now forms the inner surface of the locule. Pollen are shed from the anther at anthesis. Callose is not detectably present in mature *Arabidopsis* pollen grains.

#### *I.A.1.f. Progamic phase: Pollen germination and pollen tube growth*

*Arabidopsis* is a self-compatible species and so pollen grains of this species will germinate on the receptive surface (stigma) of the gynoecium of any *Arabidopsis* plant. Most typically, however, this will be the stigma of the flower that produced the pollen grain. When released by the anther, *Arabidopsis* pollen grains are metabolically quiescent and highly desiccated, and must first hydrate for germination to occur. Hydration is not a passive process, mediated by the osmotic gradient that exists between the stigmatic cell and the pollen grain, but a regulated process that requires specific recognition and adhesion of the two cells to each other to occur before hydration can commence (Edlund et al., 2004). Hydration is rapid and is complete within minutes of a pollen grain's attachment to a receptive stigma.

After hydration is complete, the pollen grain germinates to produce a pollen tube that emerges from one of the three colpi, generally the one facing the stigmatic surface (Pruitt and Hülskamp, 1994). The pollen tube is produced by the vegetative cell and conveys the sperm cells to an ovule for fertilization. The pollen tube penetrates the cuticle and underlying cell wall near where the pollen grain has attached to the stigmatic cell, and then elongates by depositing new wall material at its tip (tip growth) through the stigmatic cell wall towards the base of the gynoecium, eventually exiting the stigma and entering the transmitting tract of the style. From here the pollen tube grows along the surface of the ovary towards an ovule.

The pollen tube wall is normally observed as having two distinct layers, an outer fibrillar layer and an inner non-fibrillar and electron-lucent layer (Ferguson et al., 1998, 1999; Lennon and Lord, 2000; see Fig. 3). The growing tube tip is covered solely by the primary fibrillar wall and is largely composed of methyl-esterified pectins that become increasingly de-esterified in older parts of the wall further from the tip, presumably through the action of a wall-bound pectin methylesterase (Li et al., 1994). The inner callose-rich layer is a secondary cell wall that is first visible some distance behind the tip (Ferguson et al., 1998; Lennon and Lord, 2000; see Fig. 3). Cellulose colocalizes with callose in this inner wall, although its deposition appears to commence closer to the tube tip than the deposition of callose. Transverse plugs of callose and cellulose are deposited at regular intervals along the length of the tube, giving pollen tubes a characteristic ladder-like appearance when stained with the Aniline Blue fluorochrome (Pruitt and Hülskamp, 1994). The proposed role of the callose plugs is to seal off the dead portions of the pollen tube from the living vegetative cell, and thus keep the two enclosed sperm cells at the growing tip. Callose plugs also help maintain turgor pressure in the growing tip.

The callosic nature of the wall and the presence of regular callosic plugs are features that are unique to angiosperm pollen tubes. Pollen tubes of gymnosperms (a group of plants that includes conifers, *Ginkgo* and the cycads) have a thick primary wall of cellulose and lack the inner callosic wall and callosic cross walls (Derksen et al., 1999). Angiosperm pollen tubes

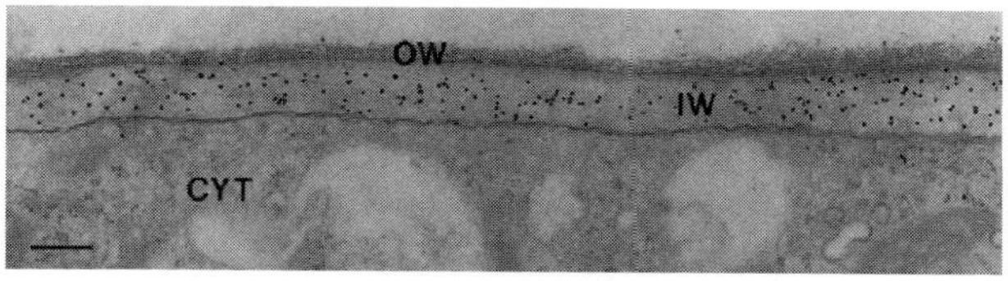


**Fig. 3: Longitudinal section of a *Nicotiana tabacum* pollen tube. The figure shows a region about 200 μm behind the tip and has been labelled with anti-(1,3)-β-glucan monoclonal antibody conjugated to colloidal gold. OW, outer, fibrillar wall layer; IW, inner, electron-lucent wall layer; CYT, cytoplasm. Scale bar: 0.3 μm. (Reproduced from Fig. 7 of Ferguson et al. (1998), with kind permission from Springer Science+Business media.)**

also universally grow at rates that are much faster than those of gymnosperm pollen tubes, with the production of callose walls thought to be the reason for this accelerated growth rate. Faster growing pollen tubes in concert with the callose modifications also resulted in longer pollen tubes, and it is argued that this innovation allowed the truly spectacular diversification of flower shape and form that is such a notable feature of this group of plants (Williams, 2008).

The callose in pollen tubes differs from the callose seen at other times during microgamete development in that the accumulations are not transient. This, and the ability to grow pollen tubes in culture, means that pollen tube callose and the synthase that makes it are amenable to biochemical characterization (Schlüpmann et al., 1993; Brownfield et al., 2007). Pollen tube callose biosynthesis and regulation are reviewed in Chapter 3.3.4.

## I.B. Functions for Callose During Microgamete Development

Ephemeral callosic walls are seen at three different times during microgamete development (see Fig. 2): callose is the major component of the common SCW surrounding the microsporocyte during stages 2–5 prior to meiosis, callose also forms the intersporal walls laid down during cytokinesis as part of the individual SCW (stages 4 and 5), and a temporary callose wall secures the generative cell to the intine after mitosis I (stage 9). The following sections discuss the possible functions of each of these callose walls. Unlike the callose of pollen tubes (see Chapter 3.3.4) the callose that appears during microgamete development has not been subjected to biochemical investigation. Functions are instead inferred by studying pollen development in plants where callose deposition has been altered by mutation or transgenic means, or by studying natural exceptions, species in which callose deposition differs from that seen in the 'typical' *Arabidopsis* pattern described above.

### I.B.1. Role of Callose in the Special Cell Wall

While several possible functions for the common SCW around the microsporocyte have been proposed, the function most often discussed is that callose acts as a physical seal and separates the microsporocyte from the surrounding sporophytic tissues. This is seen as a prerequisite for initiation of the haploid phase of the lifecycle (e.g. see Scott et al., 2004; Blackmore et al., 2007). Callose has often been proposed to act as a sealant in plants largely because callose synthesis is part of a plant cell's response to pathogen attack or physical injury (e.g. see Currier, 1957; Delmer and Amor, 1995; Chapter 4.4.5). The idea that the sporophyte and gametophyte generations need to be physically isolated from each other is based on the observation of extensive degradation of many macromolecules (proteins, ribosomal and messenger RNA) in the microsporocyte prior to meiosis, which led to the suggestion that there is a need for all sporophytic information to be purged from the cell before meiosis can occur (Mackenzie et al., 1967; Dickinson, 1987).

Although widely asserted in the literature, the impermeability of callosic walls has never been conclusively established and in fact may not be very different to that of other primary plant cell walls, where porosity is determined by a pectin-based matrix (Read and Bacic, 1996). So the ability of the common SCW to provide an effective barrier between the microsporocyte and surrounding sporophytic tissues is doubtful. Moreover, there are several examples of plants that lack a SCW, indicating that callose is not always needed for microspore development. For instance, callose is not detectably present in tetrad walls of the perennial herb *Pergularia daemia* (Apocynaceae; Vijayaraghavan and Avdhesh 1977) or the arum lily *Arum alpinum* (Araceae; Anger and Weber, 2006). The palm *Pandanus odoratissimus* (Arecaceae) reportedly also does not deposit callose around its microsporocytes, yet dyads separate after the first meiotic division, microspores after second meiotic division and pollen is normally viable (Periasamy and Amalathas 1991). Curiously, though, microspore tetrads enclosed by a callose wall have been observed in the related species *P. parvus* (Cheah and Stone 1975), so re-examining the early stages of microgamete development in this family will be necessary to confirm whether or not a SCW is present (Furness and Rudall, 2006). Callose is also absent in SCWs of the seagrasses *Amphibolis griffithi* and *A. antarctica* (Cymodoceaceae; McConchie et al., 1982; Pettitt et al., 1984).

Production of non-viable pollen (male sterility) linked to the absence of SCW callose is often cited as evidence of an important role for callose in the SCW. For instance, tobacco plants engineered to secrete (1,3)-β-glucanase into the anther locule before the endogenous 'callases' are present have reduced male fertility associated with the premature disappearance of the SCW soon after the start of meiosis (Worrall et al., 1992). Male sterility is also associated with premature dissolution of the common SCW in non-transgenic lines of petunia and sorghum (Izhar and Frankel, 1971; Warmke and Overman, 1972) and in other species transformed with (1,3)-β-glucanase transgenes expressed in the tapetum (Scott et al., 2004). However, in these plants meiosis and cell division both occur normally, resulting in tetrads that are presumably held together by material other than callose (Worrall et al., 1992). Curiously, down-regulation of the tobacco gene *TAG1*, which encodes the major (1,3)-β-glucanase expressed by tapetal cells, does not affect tetrad dissolution or pollen development (Bucciaglia et al., 2003).

Although loss of the SCW in species where this is present does not prevent meiosis, it reproducibly reduces pollen viability, possibly because the SCW is required for production of the exine layer that surrounds pollen grains (see below). However, mutations in the *Arabidopsis* glucan synthase-like gene *AtGSL2* (also known as *AtCALS5*), which encodes a male-specific (1,3)-β-glucan synthase, prevent production of the common SCW but not the ability of microsporocytes to undergo meiosis and produce tetrads (Nishikawa et al., 2005). While mutations in *AtGSL2* disrupt synthesis of the common SCW, intersporal cell wall formation is not affected and pollen viability is not impaired.

The observation that the inner surface of individual SCWs has the same arrangement of taxon-specific ornamentation as the developing primexine, except in reverse, leads to another proposed function for the SCW: that it acts as a template or mould that fills with primexine to generate its characteristic sculpted pattern of primexine and consequently the pattern of the exine (Waterkeyn and Bienfait, 1970). It is, however, difficult to see how a callose layer on its own could impart the diverse array of patterns seen in the exines of different species. It is more likely that the individual SCW and primexine patterns require both the regulated spatio-temporal placement of polysaccharide synthases, predominantly cellulose synthases, into the plasma membrane, and the specific trafficking of vesicles containing pectic and other non-cellulosic polysaccharides to regions of the plasma membrane, where fusion with this membrane will result in their contents being deposited into the extracellular space. Functional associations between polysaccharide synthase complexes and individual elements of the cortical microtubule array are known to control the former (e.g. see Paredez et al., 2006; DeBolt et al., 2007) while the dynamic organization of the actin cytoskeleton is an essential role for the latter (Šamaj et al., 2006). Consistent with the plasma membrane being a determining factor in exine patterning, the *Arabidopsis dex1* (defective in exine pattern formation) mutation blocks normal invagination of the microspore plasma membrane and delays and significantly reduces exine formation (Paxson-Sowders et al., 2001). The *DEX1* gene encodes a novel calcium-binding protein predicted to be in the plasma membrane. How exine patterns are established and maintained is currently unknown, although centrifugation experiments with developing lily microsporocytes suggest the pattern determinants are already present in the cytoplasm at the start of meiosis (Sheldon and Dickinson, 1983).

An absence of callose around the microspore can have significant effects on exine patterning and thickness. This is seen both in species naturally lacking a SCW and in plants where the SCW is absent, either because of an uncharacterized mutation or as the result of transgenic manipulation. The pollen grains of *Pergularia daemia*, *Amphiobolus* and *Arum alpinum*, for example, mentioned above, are all characterized by absence of a SCW and by an irregularly thin and sparsely deposited exine (Anger and Weber, 2006). The exception to this is *Pandanus odoratissimus*, also mentioned above, which produces pollen with a perfectly formed exine wall (Periasamy and Amalathas, 1991). Likewise, the absence of a SCW in transgenic tobacco leads to the formation of pollen with abnormally thin exine walls that lack regular sculpting but instead have an unusual multilaminate structure with apparently random sporopollenin deposits (Worrall et al., 1992). This phenotype is similar to that seen in the pollen of *Arabidopsis* plants with mutations in *AtGLS2*, one of the genes required for SCW callose (Nishikawa et al., 2005).

### I.B.2. Callose and its Role during Cytokinesis

While the role of the callose SCW during microgamete development remains unclear, less uncertainty surrounds the role of callose in intersporal walls, where initial separation of the individual microspores provides one obvious function. The provision of a subsequently degradable temporary wall, to allow release of microspores from the tetrad, is a possible second function (e.g. see Echlin and Godwin, 1968). However, more significant in terms of its impact on angiosperm evolution is where the intersporal callose walls form and when, because this will determine how many germination pores the mature pollen grain will have. Pore number is arguably the most important character in flowering plant taxonomy as the major division of angiosperms is not monocot versus dicot but one based on pollen type. Basal angiosperms are characterized by having one-pored or uniaperturate pollen, whereas the more highly derived angiosperms (eudicots) have three-pored or triaperturate pollen (Angiosperm Phylogeny Group, 1998). The tricolpate pollen of *Arabidopsis* is an example of one type of triaperturate pollen.

Figure 4 illustrates the two general patterns for the appearance of callose walls after microspore meiosis seen in flowering plants (Furness et al., 2002; Furness and Rudall, 2004).

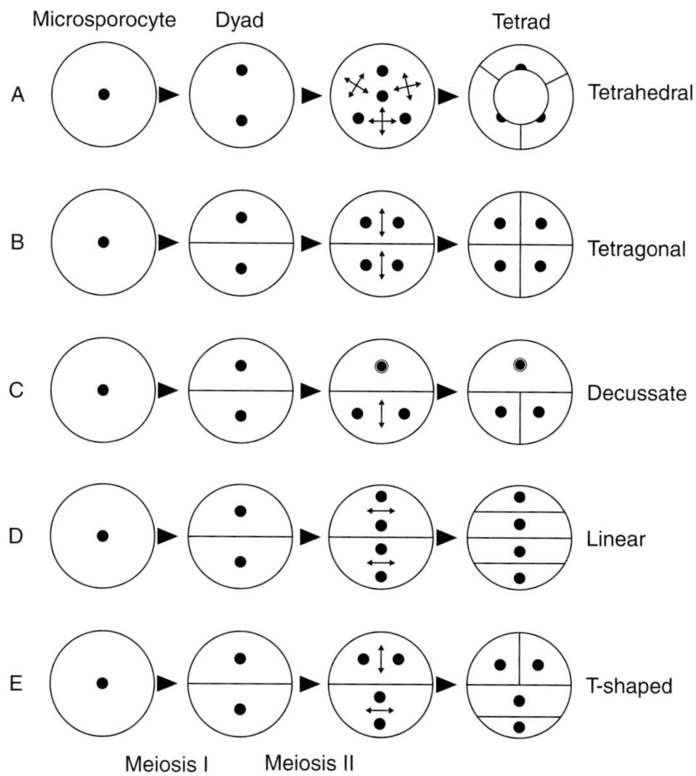


**Fig. 4: Diagram of the development of common tetrad types. (A) Simultaneous division. (B–E) Successive division. Arrowheads indicate equatorial plane of spindle; solid circle indicates nucleus; solid circle inside a small circle indicates nucleus directly below the one above. (Reproduced with permission of the University of Chicago Press from Furness et al., 2002.)**

As already mentioned, *Arabidopsis* microspores undergo simultaneous cytokinesis at the end of meiosis II. In simultaneous cytokinesis, the spindles of the first and second meiotic divisions appear to interact to form a tetrahedral tetrad (Fig. 4A; Furness et al., 2002). In the alternative pattern, successive cytokinesis, a dyad stage corresponding to the first cytoplasmic division, occurs after meiosis I. The second meiotic division follows. But because the spindles can adopt various orientations with respect to each other, a variety of different arrangements of the tetrad are produced (Fig. 4B–E).

Microspore cytokinesis is highly variable among the basal angiosperms, a group of plants that includes species such as the familiar water-lilies (Nymphaeales) and the monocots, of which the grasses (Poaceae) are a well-known example (Angiosperm Phylogeny Group, 1998). In these plants both simultaneous and successive forms of cytokinesis are observed, as well as a number of modified versions where, for instance, an ephemeral cell plate forms after meiosis I and subsequently disperses, with simultaneous cytokinesis occurring after meiosis II (Furness et al., 2002). In palms (Arecaceae), the simultaneous and successive types of cytokinesis can even occur within the same stamen (Sannier et al., 2006). Regardless of the type of cytokinesis, most basal angiosperms produce pollen with a single pore at one pole or an aperture pattern based on this arrangement.

In the more highly derived eudicots, a group that includes ~75% of present day species and most of the plants traditionally classified as dicots (*Arabidopsis* is an example), microspores are formed by simultaneous cytokinesis and pollen grains are triaperturate or have a related aperture pattern (Furness and Rudall, 2004; Angiosperm Phylogeny Group, 1998). Those basal angiosperms that do have triaperturate pollen – for example members of the family Illiciaceae such as star-anise – also have simultaneous cytokinesis, just as in *Arabidopsis* (Sampson, 2000; Furness et al., 2002). It thus appears that triaperturate pollen grains can only be formed by simultaneous cytokinesis, whereas uniaperturate pollen grains can be formed by either simultaneous or successive cytokinesis.

What's the basis for this relationship between callose wall formation and the number of germination pores? Pollen apertures first appear during ontogeny of the exine layer of the pollen wall, but their locations are defined by the geometry of the tetrad and are controlled by the meiotic spindle (Sheldon and Dickinson, 1986; Ressayre et al., 1998, 2005). Combinations of different variable elements during meiosis can account for most of the widespread patterns. These elements include the timing of cytokinesis (successive, simultaneous or intermediate), the orientation of the meiotic axes (tetrahedral, tetragonal, decussate, linear or T-shaped; Fig. 4), and the way callose is deposited to form the intersporal walls (which can be either centripetally as in *Arabidopsis* or centrifugally as in most monocots (Nadot et al., 2006). Generally, apertures are formed at the last points of cytoplasmic contact between meiotic products, with polar apertures being additionally defined by the position of the spindle pole at second meiosis (Ressayre et al., 2005). As an example, in a eudicot tetrad the 12 apertures (4 microspores $\times$ 3 apertures per microspore) form in pairs at midpoints (equatorially) along the edges of the tetrahedron created by the six bipolar spindles (Fig. 4A). Places where apertures will later form are initially marked by patches of microtubules and endomembranes that appear to be involved in blocking primexine template synthesis at these sites (Dickinson and Sheldon, 1984; Muñoz et al., 1995; Ressayre et al., 2002).

There is great interest in the genes and proteins that are involved in male meiotic cytokinesis. By their very nature, mutations that affect the function of these genes are also likely to reduce pollen viability, so genetic screens for male sterility have been extensively used in their identification (e.g. see Caryl et al., 2003; Johnson-Brousseau and McCormick, 2004).

As expected, some male sterility mutations do specifically affect cytokinesis and are linked to formation or dissolution of callose. In *Arabidopsis*, the *GLS* genes, *AtGSL1* and *AtGSL5*, appear to play essential and overlapping roles in synthesizing the intersporal walls (Enns et al., 2005). Microgamete development in mutant *atgsl1* or *atgsl5* plants proceeds normally up to the completion of meiosis I (i.e. a bicellular microsporocyte surrounded by a callosic SCW), suggesting that GSL1 and GSL5 are not required during these early stages. However, while plants homozygous for *gsl1* produce normal pollen, 15% of the pollen from *gsl5*-homozygous plants and 30% of the pollen from *gsl1*/+ *gsl5*/*gls5* plants (i.e. plants homozygous for the *gsl1* mutation and heterozygous for the *gsl5* mutation) and *gsl1*/*gls1* *gsl5*/+ plants are small and shrivelled with misplaced and misshapen apertures. As these pollen grains have completed meiosis, this indicates that a deficiency of GSL1 and GSL5 has effects on later stages of pollen development when callose is not normally present. Significantly, *gls1*/+ *gsl5*/*gsl5* plants also produce enlarged multinucleate pollen grains with more than three pores: the absence of intersporal walls in these pollen grains indicates that GSL1 and GSL5 are required for the synthesis of the intersporal walls but not the callosic SCW (Enns et al., 2005). This phenotype can be contrasted with the effect of mutations in *ATGSL2*, where there is no common SCW but cytokinesis proceeds normally (Nishikawa et al., 2005).

Failure to form intersporal walls is also seen in *Arabidopsis* plants carrying the allelic mutations *tetraspore* and *stud* (Hűlskamp et al., 1997; Spielman et al., 1997). Because microspore cytokinesis does not take place in these mutants, large, multinucleate coenocytic pollen grains with aberrant numbers of misplaced apertures are produced (Spielman et al., 1997). Consistent with the proposed role of microtubules in regulating both the placement of cleavage planes and the sites of future pore formation, the product of *TETRASPORE* is a kinesin, a motor protein that binds to microtubules and is involved in vesicle trafficking (Yang et al., 2003).

Once cytokinesis is complete, breakdown of the individual SCWs generally results in microspore release. Some species, however, naturally release their pollen as a tetrad of fused pollen grains. This is seen in members of the mountain pepper family (Winteraceae), as well as in the water lilies (Nymphaeales), bulrushes (Typhaceae), heaths (Ericaceae), evening primroses

(Onagraceae) and acacias (Fabaceae) (Smyth, 1994; Copenhaver et al., 2000). In these families there is often little or no callose in the intersporal cross walls of the tetrad, which allows the exine layers of adjacent microspores to fuse and prevents the pollen grains from being separated (Prakash et al., 1992; Scott et al., 2004). *Arabidopsis* plants with one of three non-allelic *quartet* (*qrt*) mutations (*qrt1-3*) also produce fused tetrads of pollen grains (Preuss et al., 1994; Copenhaver et al., 2000; Rhee et al., 2003). Although defective or delayed callose degradation has been suggested as a reason why *qrt* microspores fail to separate (Echlin and Godwin, 1968; Izhar and Frankel, 1971), analysis of the *qrt* mutants suggests that the tetrads are instead held together by changes in pectic components in the primary cell wall around the microsporocyte (Rhee and Somerville, 1998). Consistent with this, *QRT1* has recently been shown to encode a pectin methylesterase and *QRT3* an endo-polygalacturonase (Rhee et al., 2003; Francis et al., 2006).

### I.B.3. Role of Callose in Post-Meiotic Male Gamete Development

On the completion of male meiosis, the microspore undergoes two rounds of mitosis (mitosis I and II) to form a three-celled gametophyte composed of a vegetative cell and two sperm cells. In *Arabidopsis* and grasses (Poaceae), mitosis I and II occur before anthesis and mature pollen grains are released in a tricellular state. As noted earlier, in many other species mitosis II occurs within the pollen tube after the grain has germinated. At anthesis, pollen grains from these species are thus in a bicellular state.

Mitosis I is described as an asymmetric division because the two daughter cells are unequal in size and have nuclei that react very differently to DNA-binding stains (see Fig. 2E–G). In addition, these cells have very different developmental fates (Twell et al., 1998). The large vegetative cell constitutes the bulk of the pollen grain and the nuclear chromatin is relatively dispersed. This cell will grow to become the pollen tube upon germination, but does not undergo further division. The smaller generative cell is initially physically isolated from the vegetative cell by a thin, temporary wall of callose attached to the intine (Fig. 2G). The generative cell contains very little cytoplasm and its nuclear DNA is more condensed than that of the vegetative cell (Fig. 2E). This cell undergoes a further symmetrical division to produce the two sperm cells. By the time this has happened, the callosic wall that had earlier separated the vegetative and generative cells is no longer evident. This marked dimorphism of mitosis I products is also evident at the molecular level, with the vegetative cell expressing several genes that the generative cell does not (Twell et al., 1998). Because expression of vegetative cell-specific genes is repressed in the generative cell, it was thought that the generative cell is transcriptionally relatively inactive (McCormick, 2003). However there is recent evidence for extensive generative and sperm cell gene expression in maize and lily (Engel et al., 2003; Okada, et al., 2006) and several generative and sperm cell-specific genes have been characterized in *Arabidopsis* (Engel, et al., 2005; Okada, et al., 2005; Mori et al., 2006; von Besser et al., 2006).

The *Arabidopsis GLS* genes *AtGSL8* and, to a lesser extent, *AtGSL10* appear to play important roles in the asymmetrical division at pollen mitosis I. In pollen containing a mutated version of *AtGSL8*, and in some of the pollen with a mutated version of *AtGSL10*, microspores develop normally in callose-enclosed tetrads, the callose is degraded and the microspore has the characteristic signet ring appearance in preparation for the asymmetric division (stage 8 in Fig. 1). However, these microspores fail to enter mitosis and subsequently collapse (Töller et al., 2008). As it is widely assumed that the GSLs are involved in callose biosynthesis, this is a surprising finding because no callose is detectable by Aniline Blue fluorochrome-staining of wild-type pollen immediately before pollen mitosis I. Possibly very small amounts of callose are made and are required for entry into mitosis. Alternatively, in addition to callose synthesis, the AtGSL8 and AtGSL10 proteins may have roles in signalling.

Development of the generative cell depends on mitosis I being asymmetric, as treatment of cultured microspores with colcemid, a chemical that limits microtubule formation, results in the formation of two equal-sized cells with nuclei that have identical reactions to DNA-binding stains (Tanaka and Ito, 1981). Similarly, mutations that result in a symmetrical division also result in the formation of two cells of equal size that both express vegetative cell-specific genes (Twell et al., 1998). The reason why the generative and vegetative cells adopt such different fates is thought to be the polarized distribution of some gametophytically expressed regulatory factor within the cytoplasm, so that at cytokinesis only the generative cell receives this factor (if the factor actively represses vegetative gene expression) or only the vegetative cell receives this factor (if the factor actively promotes vegetative gene expression). Generative cell development thus critically depends on the callosic wall laid down on the completion of mitosis I isolating one of the daughter nuclei within a region of cytoplasm (the generative pole). As the following *Arabidopsis* mutants demonstrate, anything that alters the placement, synthesis or degradation of this wall also compromises male gametophyte development.

The *sidecar* mutant shows the effects of a misplaced wall (Chen and McCormick, 1996). Mutant plants produce pollen grains with an extra vegetative cell because the first mitotic division in these microspores is symmetrical and produces two equal-sized cells, one of which then undergoes an asymmetric division (equivalent to mitosis I) to form a vegetative cell and a generative cell that then divides to produce two sperm cells (mitosis II). By contrast, the *limpet pollen* (*lip*), *atgsl10*, *gemini pollen 1* (*gem1*) and *two-in-one pollen* (*tio*) mutants illustrate how defects that alter synthesis or degradation of the callosic wall also affect generative cell development. The *lip* mutation results in mature pollen in which the generative or sperm cells remain closely associated with the pollen wall. This association arises because at mitosis I the generative cell fails to migrate inwards toward the centre of the microspore (Howden et al., 1998). Residual wall material around the peripheral generative cell in mutant pollen grains suggests it is trapped in position behind a persistent callosic wall. Failure of the transient callose wall to be degraded, presumably by a (1,3)-β-glucanase(s), is thus one conceivable cause of this mutation (Howden et al., 1998). Mutations in the putative callose synthase *AtGSL10* (see Chapter 3.4.4) also result in some pollen in which the generative or sperm cells remains associated with the wall (Töller et al., 2008). In this case a callose wall is formed between the vegetative and generative cells; however, as the wall is slightly misshapen and persists longer than in wild type, there may be changes to its precise structure that affect function or degradation. In other pollen carrying the *atgsl10* mutation, an ectopic callose wall forms in the middle of the microspore, producing two cells with nuclei that stain equally. The *gem1* and *tio* mutations affect phragmoplast formation and the correct deposition of callose at pollen mitosis I. GEM1 (MOR1), a member of the MAP215 family of microtubule-associated proteins, is involved in establishing interphase arrays of microtubules and is important in formation of the phragmoplast (Whittington et al., 2001, Twell et al., 2002). In *gem1*, microspores either fail to establish a cell plate at pollen mitosis I or form an irregular, often incomplete, branching callosic structure that results in either a single binucleate cell or a bicellular pollen where the two cells are similar in size (Park and Twell 2001; Twell et al., 2002). In the *tio* mutant, pollen grains contain two free vegetative nuclei within a single cell because of a failure of cytokinesis at mitosis I (Oh et al., 2005). Although the callosic wall formed at mitosis I is correctly positioned at the generative pole, it is incomplete and quickly degraded, suggesting that TIO (a protein kinase belonging to the FUSED family) has a specific role in expansion of the cell plate once it has been established (Oh et al., 2005).

## I.C. Overview of Megagamete Development

As with microgamete development, the term megasporogenesis refers to the initiation and formation of megaspores from the megasporocyte within the female reproductive organ (gynoecium) of the flower, and megagametogenesis to the later development of the megagametophyte, or embryo sac, within the ovule. For convenience we will use the terms megagamete or embryo sac development to encompass megasporogenesis and megagametogenesis.

Megagamete development can proceed either by a sexual route, involving meiotic division of a megasporocyte, or by an asexual route (apomixis), involving only mitotic divisions and producing an unreduced embryo sac with an egg cell that develops into an embryo without the need for fertilization (Koltunow, 1993). *Arabidopsis* will be used to illustrate the process of megagamete development by sexual means. Comprehensive reviews of embryo sac development in flowering plants include those by Reiser and Fischer (1993), Russell (1993), Grossniklaus and Schneitz (1998) and Yadegari and Drews (2004), with the following description of this process in *Arabidopsis* being based on the papers of Webb and Gunning (1991), Schneitz et al. (1995), Christensen et al. (1997) and Bajon et al. (1999). The various pathways by which apomixis can produce an embryo sac are described in reviews by Koltunow (1993) and Bicknell and Koltunow (2004).

### *I.C.1. Megagamete Development in* Arabidopsis

Megagamete development in *Arabidopsis* occurs within the ovule, a specialized structure within the gynoecium composed of three fundamental units: a nucellus within which the megasporocyte and subsequently the megagametophyte develop, two integuments that at maturity will enclose the embryo sac, and a funiculus that attaches the ovule to the gynoecium and contains connective tissue through which nutrients and water will pass. At maturity, the *Arabidopsis* embryo sac is a seven-celled, eight-nucleate structure; each cell has a different form of structural specialization, and the whole embryo sac is polarized from the basal (or chalazal) end to the apical (or micropylar) end (Mansfield et al., 1991). At the micropylar end are three cells – a highly vacuolated egg cell and two synergids – that together are called the egg apparatus. In the middle of the ovule is the central cell, a large, vacuolated cell containing many organelles and two nuclei (polar nuclei) that fuse fully or partially prior to fertilization (Grossniklaus and Schneitz, 1998). At the chalazal end of the ovule are three antipodal cells. Fertilization involves a pollen tube entering the ovule through the micropylar opening and releasing its sperm cells. An attractant released by the synergid cells guides the pollen tube to the female gametophyte (Escobar-Restrepo et al., 2007). Fusion of one sperm cell with the egg cell forms a diploid zygote, and fusion of the other sperm cell with the central cell forms a triploid endosperm (double fertilization).

Figure 5 shows an *Arabidopsis* ovule and outlines the main steps in its development, and Fig. 6 shows bright-field and Aniline Blue fluorochrome-stained images of megaspore

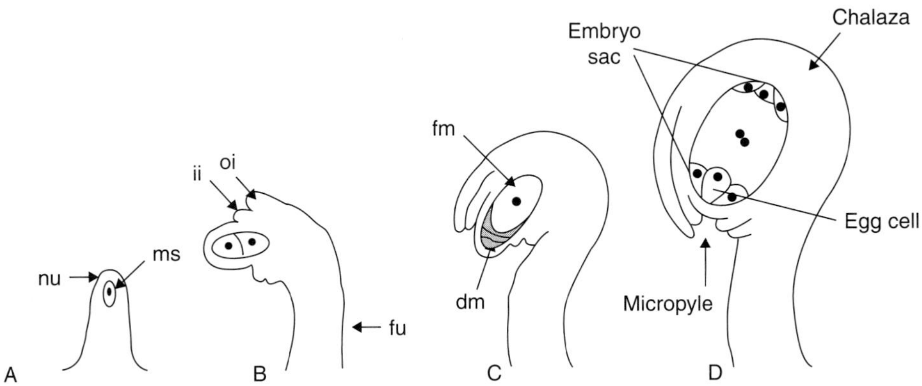


**Fig. 5: *Arabidopsis* ovule development. (A) Ovule shortly after initiation, showing a single megasporocyte (ms); nu, nucellus. (B) Ovule after both integuments have been initiated. At this time, the megasporocyte has undergone the first meiotic division. The axis of the nucellus is transiently perpendicular to the axis of the funiculus (fu); ii, inner integument; oi, outer integument. (C) Ovule after meiosis. The functional megaspore (fm) at the chalazal end has expanded, and the non-functional megaspores are degenerated. The axis of the nucellus is now parallel to the funiculus due to unequal growth, primarily of the integuments; dm, degenerate megaspores. (D) Ovule at maturity. The mature embryo sac contains seven cells and eight nuclei. (Reproduced with permission of the American Society of Plant Biologists from Reiser and Fischer (1993); permission conveyed through Copyright Clearance Centre, Inc.)**

development in *Arabidopsis*. From these figures it can be seen that, unlike in the case of pollen, the megaspore (and later, after fertilization, the developing embryo) maintains continuous contact with the parent plant. The megasporocyte is first apparent as a single sub-dermal cell (called an archesporium) within the nucellus that enlarges more than three-fold to be about 17 μm long and 10 μm wide before undergoing meiosis (Fig. 5A and 6A). Infrequently two megasporocytes form in the same ovule (Fig. 6B). Two meiotic divisions form four megaspores that in *Arabidopsis* are sometimes in a linear tetrad but more commonly have a tetrahedral or decussate arrangement (see Fig. 6C, D, H, I; Webb and Gunning, 1991; Schneitz et al., 1995). The megasporocyte undergoes cytokinesis after meiosis I (Fig. 6G) and again after meiosis II (Fig. 6H, I). Immediately after meiosis is complete, three of the megaspores undergo programmed cell death while the megaspore at the chalazal end of the ovule (the functional megaspore) begins to enlarge (Fig. 5C and 6H, I). All cells in the embryo sac are derived from this megaspore.

Megagametophytes that develop from a single megaspore in this way are classified as monosporic; the term "*Polygonum* type" is used to describe those embryo sacs that, like

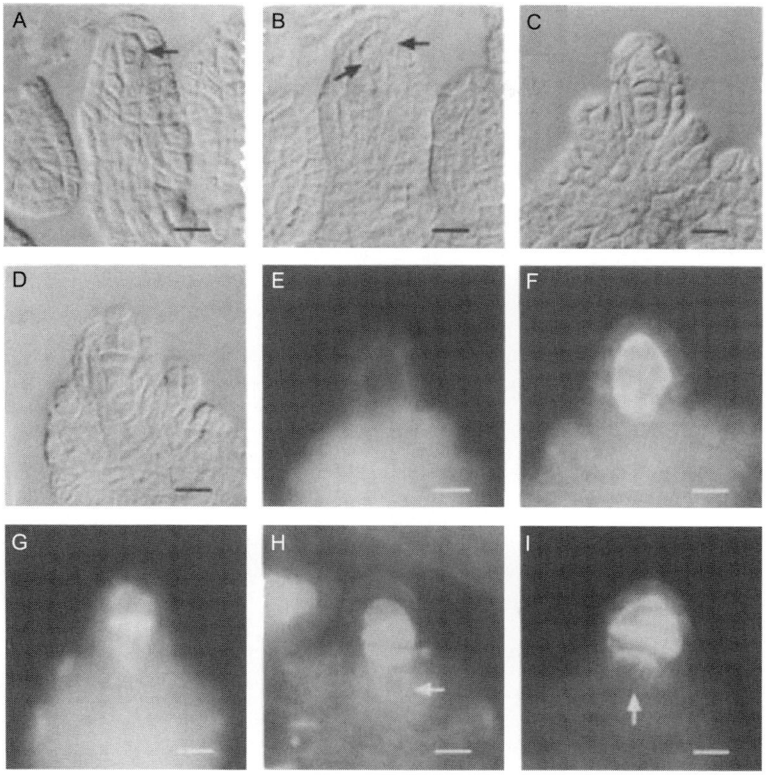


**Fig. 6: Different stages of megagamete development in wild-type *Arabidopsis*. (A–D) Bright-field views of whole-mount ovules. (E–I) Corresponding fluorescent images of Aniline Blue fluorochrome-stained ovules. Scale bars=10 μm. (A) Megaspore before the first nuclear division of meiosis (arrow). Note the large overall size relative to other cells in the nucellus. (B) The two arrows indicate two neighbouring megasporocytes in the same ovule primordium. This occurs rarely and does not lead to the development of two adjacent embryo sacs. (C) A linear tetrad. (D) A multiplanar tetrad. (E) A megasporocyte before the onset of cytokinesis. (F) The appearance of callose in the wall of the megasporocyte is the first sign of the onset of cytokinesis. (G) At meiosis I, a bright disk of callose divides the megasporocyte into two cells of roughly equal size. (H and I) Linear (H) and multiplanar (I) tetrads. Arrows indicate the functional megaspore. (Reproduced with permission of Blackwell Publishing Ltd from Schneitz et al., 1995.)**

*Arabidopsis*, develop from the chalazal megaspore. Approximately 70% of flowering plant species have *Polygonum*-type embryo sacs (Reiser and Fischer, 1993). In an alternative pattern of monosporic development (*Oenothera* type), all cells in the embryo sac are derived from the megaspore at the micropylar end. In still other patterns, multiple megaspores participate in the formation of the embryo sac (bispory, tetraspory).

Small patches of callose are detectable in walls of the megasporocyte before meiosis, but callose is much more evident at the onset of cytokinesis and in the tetrad after meiosis (Fig. 6E–I). Callose is deposited in walls of the functional megaspore where it is in contact with the degenerating ones, but is generally absent elsewhere in the wall of this cell, and to varying extents in the walls between degenerating cells (Webb and Gunning, 1991). By ascribing a sealing function to callose, various authors have suggested that the placement of these walls represents a physical barrier that suppresses the development of the degenerating megaspores by restricting the flow of nutrients or growth factors from the parent plant so only the functional megaspore will continue to develop (e.g. Webb and Gunning, 1991). The way callose is deposited in walls of the functional megaspores of bisporic and tetrasporic embryo sacs, and in monosporic embryo sacs of the *Oenothera* type, is broadly consistent with this hypothesis (Reiser and Fischer, 1993; Grossniklaus and Schneitz, 1998). However, in those forms of *Hieracium* (Asteraceae) capable of producing an embryo sac by sexual means, callose is present in the chalazal wall of the functional megaspore at the chalazal end of the tetrad but is absent from the wall of the megaspore at the micropylar end, a pattern of callose distribution not typical of megaspores undergoing *Polygonum*-type development (Tucker et al., 2001). This suggests that the way callose is deposited is not always essential to selection of the functional megaspore. Transgenic experiments to test whether the loss of the callose barrier affects embryo sac development have not been done.

At maturity, the functional megaspore of *Arabidopsis* is teardrop shaped and rich in organelles. Its nucleus divides thrice mitotically without cytokinesis to produce an eight-nucleated syncitial cell. Four of the nuclei are at the chalazal pole and the other four at the micropylar pole, with a large vacuole separating the two poles. Cellularization produces three antipodal cells at the chalazal pole, and an egg cell and two synergid cells at the micropylar pole. Migration of the two remaining nuclei, one at the chalazal pole and the other at the micropylar pole, to the centre of the ovule after cellularization forms the central cell and completes development of the embryo sac (Fig. 5D).

Cell wall biogenesis during later stages of megagamete development is poorly recorded and the types of polysaccharides present in these walls and how they are distributed is largely unknown. In *Arabidopsis*, as in other species, the cell walls of the egg apparatus are discontinuous and 'bead-like', consisting of regions of cell wall material separated by regions where the plasma membranes of adjacent cells are in direct contact (Cass et al., 1986; Mansfield et al., 1991; Jane, 1997). Membrane-to-membrane contacts are particularly evident at the boundary between the chalazal ends of egg apparatus cells and the central cell, a feature that may be important in the fusion of male and female gametes (Russell, 1993). By contrast, at their micropylar ends synergid cells develop an extensive labyrinth of wall projections (the filiform apparatus) that extends deep into the synergid cytoplasm. In *Arabidopsis* the filiform apparatus stains intensely with the periodic acid and Schiff reagent, indicating that polysaccharides are present that have vicinal hydroxyls in the monosaccharide units (Mansfield et al., 1991). In other species, cell walls of the embryo sac are generally described as being cellulosic and/or non-cellulosic in nature, with occasional deposits of an unknown electron-opaque material (e.g. see Ross and Sumner, 2004). Callose is not reported to be present in embryo sac walls although there is presumably some associated with plasmodesmata (Mansfield et al., 1991).

## I.D. Callose in Apomictic Embryo Sacs

Two different modes of apomixis are recognized that differ in the way the embryo sac originates. In diplospory, a cell at a position similar to that of the megasporocyte in a sexual ovule forms the embryo sac by mitosis, whereas in apospory the embryo sac is formed from one or more somatic cells located elsewhere in the ovule (aposporous initials; Bicknell and Koltunow, 2004). In aposporous apomixis, initial stages of the sexual and apomictic pathways often occur concurrently within the one ovule, although the more rapid development of the apomictic embryo sac usually terminates the development of the sexual embryo sac (Fig. 7; Koltunow, 1993). Detailed descriptions of the cellular events associated with the formation of diplosporous and aposporous embryo sacs are beyond the scope of this review.

In contrast to megagamete development by the sexual pathway, callose is not a feature of the development of an apomictic embryo sac. For instance, in diplosporous apomictic grasses from the genera *Tripsacum* and *Elymus*, callose is absent from the walls of the megasporocyte-equivalent cell, whereas it accumulates normally in related sexual species, suggesting some association in these taxa between callose and meiosis (Carman et al., 1991). In grasses with aposporous apomixis, such as *Poa* and *Pennisetum*, callose is absent from the walls of the aposporous initials but is present in the walls of the nearby megasporocyte (Peel et al., 1997). The aberrant deposition of callose in the megasporocyte in these plants relative to related sexual species is thought to be associated with a failure of the sexual pathway. In *Hieracium*, callose is not deposited in walls of the aposporous initials but is deposited normally in the walls of nearby cells undergoing megagamete development (Fig. 7). If apomixis is thought of as an anomalous form of sexual reproduction, then the absence of callose in cells produced by apospory and diplospory implies a lack of identity between these cells and those produced during normal embryo sac development (Bicknell and Koltunow, 2004).

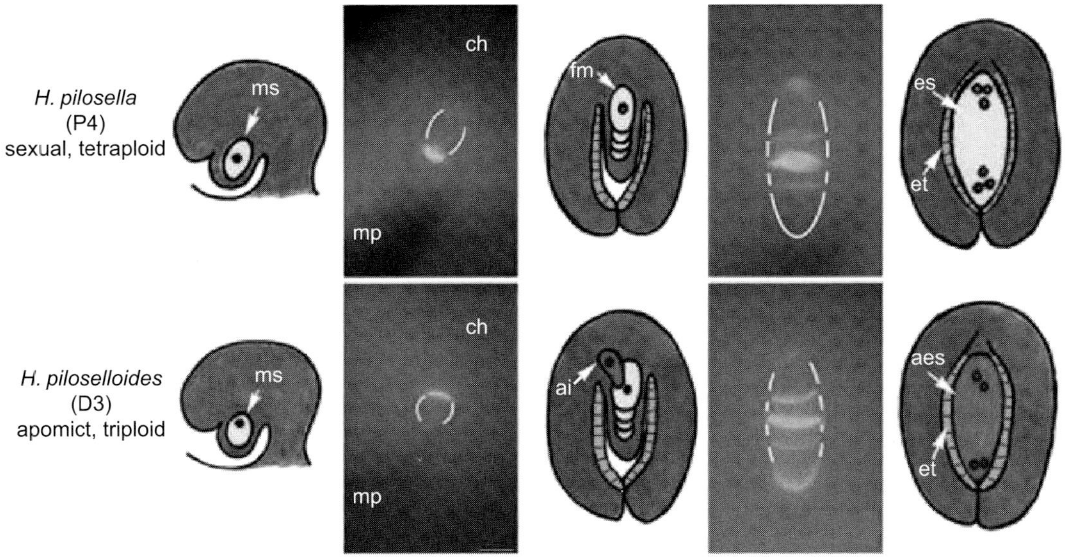


**Fig. 7: Early events of embryo sac formation in ovules of a sexual and an apomictic variety of *Hieracium*. In the colour version of this figure, the early events of reduced embryo sac formation in a sexual plant and an apomict are coloured yellow, and aposporous embryo sac formation is coloured red. The presence of callose in the walls of the megasporocyte (ms) in ovules prior to meiosis and in ovules after meiosis is shown by Aniline Blue fluorochrome-staining viewed with UV light. aes, aposporous embryo sac; ai, aposporous initial; es, embryo sac; ch, chalazal end; et, endothelium; fm, functional megaspore; mp, micropylar end; ms, megasporocyte. (Reproduced with permission of American Society of Plant Biologists from Bicknell and Koltunow (2004); permission conveyed through Copyright Clearance Centre, Inc.) The colour specifications refer to colours in panels.**

## I.E. Conclusions and Future Prospects

Since the publication of Stone and Clarke (1992), the most obvious difference in the ways the roles of callose in microgamete and megagamete development have been explored has been the use of *Arabidopsis* mutants. This approach has not only allowed various hypotheses regarding callose function to be tested, but has also begun to reveal the genes and gene products that are involved in synthesis, deposition and eventual turnover of this callose. Some of these mutants, such as those in *ATGSL8* and *ATGSL10* that affect microgamete development, give rise to phenotypic changes that are first apparent at stages when callose is not present at levels detectable by Aniline Blue fluorochrome staining (Töller et al., 2008). While there are many reason why this should be the case, it does highlight the problems that can arise with over-reliance on this stain to detect callose. At one extreme, a lack of Aniline Blue fluorochrome staining may be taken as evidence that callose is absent, which may not in fact be the case. Callose is often a transient feature of walls and small but functionally significant amounts of it may be present that the fluorochrome fails to detect. At the other extreme, a wall that stains intensely with the Aniline Blue fluorochrome is often defined as being a callose wall. As can been seen in studies on the SCW and dissolution of the microspore tetrad, this can be quite misleading, as other polysaccharides may also be present that, in functional terms, are just as important as callose.

Although a range of polysaccharide-specific antibodies are now available (Knox et al., 1990; Meikle et al., 1991; McCann et al., 1992; Williams et al., 1996), these tools have received limited use in studies of microgamete and megagamete development (for exceptions see Van Aelst and Van Went, 1992; Rhee and Somerville, 1998; Rhee et al., 2003). Most notable is the lack of attention paid so far to the composition of walls in the embryo sac, where there has been no change to the rudimentary descriptions of this structure that were available when Stone and Clarke (1992) was published. This is in spite of the recent interest paid to other aspects of female gametophyte development and function (Grossniklaus and Schneitz, 1998) and suggests that there is now a need for megagamete development to be re-examined using these reagents. A better description of the walls in the megaspore tetrad may, for instance, lead to a better understanding of how the functional megaspore is selected and what role callose plays in this process.

Finally, while *Arabidopsis* can be used as an example of microgamete and megagamete development in a 'typical' plant, as we have already noted there are many exceptions to this example. Examining wall formation in a range of relevant *Arabidopsis* mutants and transgenic plants represents one way of gaining further insight into the role of a transient callose wall associated with a particular developmental stage. An alternative approach is to examine the same stage in plants with atypical forms of microgamete or megagamete development. These methods can be used both as a means of generating and testing various hypotheses regarding callose function, and as a way of gaining a deeper understanding of the biology of pollen and embryo sac development.

## Acknowledgements

This chapter was written with the support of a sabbatical grant from the University of Melbourne to EN, who acknowledges the hospitality he received while a visitor in the School of Cell and Systems Biology, University of Toronto, and who is especially indebted to his host, Professor Daphne Goring, and the members of her lab for their kindness and generosity. The authors also thank Assistant Professor Joe Williams (University of Tennessee) for his comments, and Dr Lynette Brownfield and Professor David Twell (University of Leicester, UK) for their comments and for providing some of the images used in this work.

## References

Anger, E. M., & Weber, M. (2006). Pollen-wall formation in *Arum alpinum*. *Annals of Botany*, *97*, 239–244.

Angiosperm Phylogeny Group. (1998). An ordinal classification for the families of flowering plants. *Annals of the Missouri Botanical Gardens*, *85*, 531–553.

Bajon, C., Horlow, C., Motamayor, J. C., Sauvanet, A., & Robert, D. (1999). Megasporogenesis in *Arabidopsis thaliana* L.: An ultrastructural study. *Sexual Plant Reproduction*, *12*, 99–109.

Bicknell, R. A., & Koltunow, A. M. (2004). Understanding apomixis: Recent advances and remaining conundrums. *Plant Cell*, *16*, S228–S245.

Blackmore, S., Wortley, A. H., Skvarla, J. J., & Rowley, J. R. (2007). Pollen wall development in flowering plants. *New Phytologist*, *174*, 483–498.

Boavida, L. C., Becker, J. D., & Feijó, J. A. (2005). The making of gametes in higher plants. *International Journal of Developmental Biology*, *49*, 595–614.

Brownfield, L., Ford, K., Doblin, M. S., Newbigin, E., Read, S., & Bacic, A. (2007). Proteomic and biochemical evidence links the callose synthase in *Nicotiana alata* pollen tubes to the product of the *NaGSL1* gene. *Plant Journal*, *52*, 147–156.

Bucciaglia, P. A., & Smith, A. G. (1994). Cloning and characterization of *TAG-1* a tobacco anther beta-1,3-glucanase expressed during tetrad dissolution. *Plant Molecular Biology*, *24*, 903–904.

Bucciaglia, P. A., Zimmermann, E., & Smith, A. G. (2003). Functional analysis of a beta-1,3-glucanase gene (*Tag1*) with anther-specific RNA and protein accumulation using antisense RNA inhibition. *Journal of Plant Physiology*, *160*, 1367–1373.

Carman, J. G., Crane, C. F., & Riera-Lizarazu, O. (1991). Comparative histology of cell walls during meiotic and apomeiotic megasporogenesis in two hexaploid Australasian *Elymus* species. *Crop Science*, *31*, 1527–1532.

Caryl, A. P., Jones, G. H., & Franklin, F. C. H. (2003). Dissecting plant meiosis using *Arabidopsis thaliana* mutants. *Journal of Experimental Botany*, *54*, 25–38.

Cass, D. D., Peteya, D. J., & Robertson, B. L. (1986). Megagametophyte development in *Hordeum vulgare*. 2. Late stages of wall development and morphological aspects of megagametophyte cell differentiation. *Canadian Journal of Botany*, *64*, 2327–2336.

Cheah, C. H., & Stone, B. C. (1975). Embryo sac and microsporangium development in *Pandanus* (Pandanaceae). *Phytomorphology*, *25*, 228–238.

Chen, Y. C., & McCormick, S. (1996). *Sidecar* pollen, an *Arabidopsis thaliana* male gametophyte mutant with aberrant cell divisions during pollen development. *Development*, *122*, 3243–3253.

Christensen, C. A., King, E. J., Jordan, J. R., & Drews, G. N. (1997). Megagametogenesis in *Arabidopsis* wild type and the *Gf* mutant. *Sexual Plant Reproduction*, *10*, 49–64.

Copenhaver, G. P., Keith, K. C., & Preuss, D. (2000). Tetrad analysis in higher plants. A budding technology. *Plant Physiology*, *124*, 7–16.

Currier, H. B. (1957). Callose substance in plant cells. *American Journal of Botany*, *44*, 478–488.

DeBolt, S., Gutierrez, R., Ehrhardt, D. W., Melo, C. V., Ross, L., Cutler, S. R., Somerville, C., & Bonetta, D. (2007). Morlin, an inhibitor of cortical microtubule dynamics and cellulose synthase movement. *Proceedings of the National Academy of Sciences of the United States of America*, *104*, 5854–5859.

Delmer, D. P., & Amor, Y. (1995). Cellulose biosynthesis. *Plant Cell*, *7*, 987–1000.

Derksen, J., Li, Y. Q., Knuiman, B., & Geurts, H. (1999). The wall of *Pinus sylvestris* L. pollen tubes. *Protoplasma*, *208*, 26–36.

Dickinson, H. G. (1987). The physiology and biochemistry of meiosis in the anther. *International Review of Cytology*, *107*, 79–109.

Dickinson, H. G., & Sheldon, J. S. (1984). A radial pattern of microtubules extending between the nuclear envelope and the plasma membrane during early male haplophase in flowering plants. *Planta*, *161*, 86–90.

Echlin, P., & Godwin, H. (1968). The utrastructure and ontogeny of pollen in *Heleborus foetidus* L. II. Pollen grain development through callose special wall stage. *Journal of Cell Science*, *3*, 175–186.

Edlund, A. F., Swanson, R., & Preuss, D. (2004). Pollen and stigma structure and function: The role of diversity in pollination. *Plant Cell*, *16*, S84–S97.

Engel, M. L., Chaboud, A., Dumas, C., & McCormick, S. (2003). Sperm cells of *Zea mays* have a complex complement of mRNAs. *Plant Journal*, *34*, 697–707.

Engel, M. L., Davis, R. H., & McCormick, S. (2005). Green sperm. Identification of male gamete promoters in *Arabidopsis*. *Plant Physiology*, *138*, 2124–2133.

Enns, L. C., Kanaoka, M. M., Torii, K. U., Comai, L., Okada, K., & Cleland, R. E. (2005). Two callose synthases, GSL1 and GSL5, play an essential and redundant role in plant and pollen development and in fertility. *Plant Molecular Biology*, *58*, 333–349.

Escobar-Restrepo, J.-M., Huck, N., Kessler, S., Gagliardini, V., Gheyselinck, J., Yang, W.-C., & Grossniklaus, U. (2007). The FERONIA receptor-like kinase mediates male-female interactions during pollen tube reception. *Science*, *317*, 656–660.

Ferguson, C., Teeri, T. T., Siika-aho, M., Read, S. M., & Bacic, A. (1998). Location of cellulose and callose in pollen tubes and grains of *Nicotiana tabacum*. *Planta*, *206*, 452–460.

Ferguson, C., Bacic, A., Anderson, M. A., & Read, S. M. (1999). Subcellular distribution of arabinogalactan proteins in pollen grains and tubes as revealed with a monoclonal antibody raised against stylar arabinogalactan proteins. *Protoplasma*, *206*, 105–117.

Fitzgerald, M. A., & Knox, R. B. (1995). Initiation of primexine in freeze-substituted microspores of *Brassica campestris*. *Sexual Plant Reproduction*, *8*, 99–104.

Francis, K. E., Lam, S. Y., & Copenhaver, G. P. (2006). Separation of Arabidopsis pollen tetrads is regulated by *QUARTET1*, a pectin methylesterase gene. *Plant Physiology*, *142*, 1004–1013.

Furness, C. A., Rudall, P. J., & Sampson, F. B. (2002). Evolution of microsporogenesis in Angiosperms. *International Journal of Plant Science*, *163*, 235–260.

Furness, C. A., & Rudall, P. J. (2004). Pollen aperture evolution – a crucial factor in eudicot success? *Trends in Plant Science*, *9*, 154–158.

Furness, C. A., & Rudall, P. J. (2006). Comparative structure and development of pollen and tapetum in Pandanales. *International Journal of Plant Science*, *167*, 331–348.

Goldberg, R. B., Beals, T. P., & Sanders, P. M. (1993). Anther development: Basic principles and practical applications. *Plant Cell*, *5*, 1217–1229.

Grossniklaus, U., & Schneitz, K. (1998). The molecular and genetic basis of ovule and megagametophyte development. *Seminars in Cell and Developmental Biology*, *9*, 227–238.

Hird, D. L., Worrall, D., Hodge, R., Smartt, S., Paul, W., & Scott, R. (1993). The anther-specific protein encoded by the *Brassica napus* and *Arabidopsis thaliana* A6 gene displays similarity to beta-1,3-glucanases. *Plant Journal*, *4*, 1023–1033.

Hűlskamp, M., Parekh, N. S., Grini, P., Schneitz, K., Zimmerman, I., Lolle, S. J., & Pruitt, R. E. (1997). The *STUD* gene is required for male-specific cytokinesis after telophase II of meiosis in *Arabidopsis thaliana*. *Developmental Biology*, *187*, 114–124.

Howden, R., Park, S. K., Moore, J. M., Orme, J., Grossniklaus, U., & Twell, D. (1998). Selection of T-DNA-tagged male and female gametophytic mutants by segregation distortion in *Arabidopsis*. *Genetics*, *149*, 621–631.

Izhar, S., & Frankel, R. (1971). Mechanism of male sterility in *Petunia*: The relationship between pH, callase activity in the anthers, and the breakdown of the microsporogenesis. *Theoretical and Applied Genetics*, *41*, 104–108.

Jane, W. N. (1997). Ultrastructure of the maturing egg apparatus in *Arundo formosana* Hack. (Poaceae). *International Journal of Plant Science*, *158*, 713–726.

Johnson-Brousseau, S. A., & McCormick, S. (2004). A compendium of methods useful for characterizing *Arabidopsis* pollen mutants and gametophytically- expressed genes. *Plant Journal*, *39*, 761–775.

Knox, R. B. (1984). The pollen grain. In B. M. John (Ed.), *Embryology of Angiosperms* (pp. 197–271). Berlin: Springer-Verlag.

Knox, J. P., Linstead, P. J., King, J., Cooper, C., & Roberts, K. (1990). Pectin esterification is spatially regulated both within cell walls and between developing tissues of root apices. *Planta*, *181*, 512–521.

Koltunow, A. M. (1993). Apomixis: Embryo sacs and embryos formed without meiosis or fertilization of ovules. *Plant Cell*, *5*, 1425–1437.

Lalanne, E., & Twell, D. (2002). Genetic control of male germ unit organization in Arabidopsis. *Plant Physiology*, *129*, 865–875.

Lennon, K. A., & Lord, E. M. (2000). In vivo pollen tube cell of *Arabidopsis thaliana* I. Tube cell cytoplasm and wall. *Protoplasma*, *214*, 45–56.

Li, Y.-G., Chen, F., Linskens, H. F., & Cresti, M. (1994). Distribution of unesterified and esterified pectins in cell walls of pollen tubes of flowering plants. *Sexual Plant Reproduction*, *7*, 145–152.

Ma, H. (2005). Molecular genetic analyses of microsporogenesis and microgametogenesis in flowering plants. *Annual Reviews of Plant Biology*, *56*, 393–434.

Mackenzie, A., Heslop-Harrison, J., & Dickinson, H. G. (1967). Elimination of ribosomes during meiotic prophase. *Nature*, *215*, 997–999.

Mansfield, S. G., Briarty, L. G., & Erni, S. (1991). Early embryogenesis in *Arabidopsis thaliana* I. The mature embryo sac. *Canadian Journal of Botany*, *69*, 447–460.

McCann, M. C., Wells, B., & Roberts, K. (1992). Complexity in the spatial localization and length distribution of plant cell-wall matrix polysaccharides. *Journal of Microscopy*, *166*, 123–136.

McConchie, C. A., Knox, R. B., Ducker, S. C., & Pettitt, J. M. (1982). Pollen wall structure and cytochemistry in the seagrass *Amphibolis griffithii* (Cymodoceaceae). *Annals of Botany*, *50*, 729–732.

McCormick, S. (2004). Control of male gametophyte development. *Plant Cell*, *16*, S142–S153.

Meikle, P. J., Bönig, I., Hoogenraad, N. J., Clarke, A. E., & Stone, B. A. (1991). The location of (1,3)-β-glucans in the walls of pollen tubes of *Nicotiana alata* using a (1,3)-β-glucan specific monoclonal antibody. *Planta*, *185*, 1–8.

Mori, T., Kuroiwa, H., Higashiyama, T., & Kuroiwa, T. (2006). *GENERATIVE CELL SPECIFIC 1* is essential for angiosperm fertilization. *Nature Cell Biology*, *8*, 64–71.

Muňoz, C. A. P., Webster, B. D., & Jernstedt, J. A. (1995). Spatial congruence between exine pattern, microtubules in *Vigna* pollen. *Sexual Plant Reproduction*, *8*, 147–151.

Nadot, S., Forchioni, A., Penet, L., Sannier, J., & Ressayre, A. (2006). Links between early pollen development and aperture pattern in monocots. *Protoplasma*, *228*, 55–64.

Nishikawa, S., Zinkl, G. M., Swanson, R. J., Maruyama, D., & Preuss, D. (2005). Callose (β-1,3 glucan) is essential for *Arabidopsis* pollen wall patterning, but not tube growth. *BMC Plant Biology*, *5*, 22.

Oh, S. A., Johnson, A., Smertenko, A., Rahman, D., Park, S. K., Hussey, P. J., & Twell, D. (2005). A divergent cellular role for the FUSED kinase family in the plant-specific cytokinetic phragmoplast. *Current Biology*, *15*, 2107–2111.

Okada, T., Endo, M., Singh, M. B., & Bhalla, P. L. (2005). Analysis of the histone H3 gene family in *Arabidopsis* and identification of the male-gamete-specific variant *AtMGH3*. *Plant Journal*, *44*, 557–568.

Okada, T., Bhalla, P. L., & Singh, M. B. (2006). Expressed sequence tag analysis of *Lilium longiflorum* generative cells. *Plant and Cell Physiology*, *47*, 698–705.

Owen, H. A., & Makaroff, C. A. (1995). Ultrastructure of microsporogenesis and microgametogenesis in *Arabidopsis thaliana* (L.) Heynh. ecotype Wassilewskija (Brassicaceae). *Protoplasma*, *185*, 7–21.

Paredez, A. R., Somerville, C. R., & Ehrhardt, D. W. (2006). Visualization of cellulose synthase demonstrates functional association with microtubules. *Science*, *312*, 1491–1495.

Park, S. K., & Twell, D. (2001). Novel patterns of ectopic cell plate growth and lipid body distribution in the *Arabidopsis gemini pollen1* mutant. *Plant Physiology*, *126*, 899–909.

Paxson-Sowders, D. M., Dodrill, C. H., Owen, H. A., & Makaroff, C. A. (2001). DEX1, a novel plant protein, is required for exine pattern formation during pollen development in Arabidopsis. *Plant Physiology*, *127*, 1739–1749.

Peel, M. D., Carmen, J. G., & Leblanc, O. (1997). Megasporocyte callose in apomictic buffelgrass, Kentucky bluegrass, *Pennisetum squamulatum* Fressen, *Tripsacum* L. and weeping lovegrass. *Crop Science*, *37*, 724–732.

Periasamy, K., & Amalathas, J. (1991). Absence of callose and tetrad in the microsporogenesis of *Pandanus odoratissimus* with well-formed pollen exine. *Annals of Botany*, *67*, 29–33.

Pettitt, J. M., McConchie, C. A., Ducker, S. C., & Knox, R. B. (1984). Reproduction in seagrasses: Pollen wall morphogenesis in *Amphibolis antarctica* and wall structure in filiform grains. *Nordic Journal of Botany*, *4*, 199–216.

Prakash, N., Lim, A. L., & Sampson, F. B. (1992). Anther and ovule development in *Tasmannia* (Winteraceae). *Australian Journal of Botany*, *40*, 877–885.

Preuss, D., Rhee, S. Y., & Davis, R. W. (1994). Tetrad analysis possible in *Arabidopsis* with mutation of the *QUARTET (QRT)* genes. *Science*, *264*, 1458–1460.

Pruitt, R. E., & Hülskamp, M. (1994). From pollination to fertilization in *Arabidopsis*. In E. M. Meyerowitz & C. R. Somerville (Eds.), *Arabidopsis* (pp. 457–483). Plainview, NY: Cold Spring Harbor Laboratory Press.

Read, S. M., & Bacic, A. (1996). Cell wall porosity and its determination. *Modern Methods of Plant Analysis*, *17*, 63–80.

Reiser, L., & Fischer, R. L. (1993). The ovule and embryo sac. *Plant Cell*, *5*, 1291–1301.

Ressayre, A., Godelle, B., Mignot, A., & Gouyon, P. H. (1998). A morphogenetic model accounting for pollen aperture pattern in flowering plants. *Journal of Theoretical Biology*, *193*, 321–334.

Ressayre, A., Raquin, C., Mignot, A., Godelle, B., & Gouyon, P.-H. (2002). Correlated variation in microtubule distribution, callose deposition during male post-meiotic cytokinesis, and pollen aperture number across *Nicotiana* species (Solanaceae). *American Journal of Botany*, *89*, 393–400.

Ressayre, A., Dreyer, L., Triki-Teurtroy, S., Forchioni, A., & Nadot, S. (2005). Post-meiotic cytokinesis and pollen aperture pattern ontogeny: Comparison of development in four species differing in aperture pattern. *American Journal of Botany*, *92*, 576–583.

Rhee, S. Y., & Somerville, C. R. (1998). Tetrad pollen formation in *quartet* mutants of *Arabidopsis thaliana* is associated with persistence of pectic polysaccharides of the pollen mother cell wall. *Plant Journal*, *15*, 79–88.

Rhee, S. Y., Osborne, E., Poindexter, P. D., & Somerville, C. R. (2003). Microspore separation in the *quartet 3* mutants of *Arabidopsis* is impaired by a defect in a developmentally regulated polygalacturonase required for pollen mother cell wall degradation. *Plant Physiology*, *133*, 1170–1180.

Ross, C. M., & Sumner, M. J. (2004). Development of the unfertilized embryo sac and pollen tubes in the dwarf mistletoe *Arceuthobium americanum* (Viscaceae). *Canadian Journal of Botany*, *82*, 1566–1575.

Russell, S. D. (1993). The egg cell: Development and role in fertilization and early embryogenesis. *Plant Cell*, *5*, 1349–1359.

Šamaj, J., Müller, J., Beck, M., Bohm, N., & Menzel, D. (2006). Vesicular trafficking, cytoskeleton and signalling in root hairs and pollen tubes. *Trends in Plant Science*, *11*, 594–600.

Sampson, F. B. (2000). Pollen diversity in some modern magnoliids. *International Journal of Plant Science*, *161*(Suppl.), S193–S210.

Sannier, J., Nadot, S., Forchioni, A., Harley, M., & Albert, B. (2006). Variations in the microsporogenesis of monosulcate palm pollens. *Botanical Journal of the Linnean Society*, *151*, 93–102.

Schlüpmann, H., Bacic, A., & Read, S. M. (1993). A novel callose synthase from pollen tubes of *Nicotiana*. *Planta*, *191*, 470–481.

Schneitz, K., Hülskamp, M., & Pruitt, R. E. (1995). Wild-type ovule development in *Arabidopsis thaliana*: A light microscope study of cleared whole-mount tissue. *Plant Journal*, *7*, 731–749.

Scott, R. J., Spielman, M., & Dickinson, H. G. (2004). Stamen structure and function. *Plant Cell*, *16*, S46–S60.

Sheldon, J. M., & Dickinson, H. G. (1983). Determination of patterning in the pollen wall of *Lilium henryi*. *Journal of Cell Science*, *63*, 191–208.

Sheldon, J. M., & Dickinson, H. G. (1986). Pollen wall formation in *Lilium*: The effect of chaotropic agents, and the organization of the microtubular cytoskeleton during pattern development. *Planta*, *168*, 11–23.

Skinner, D. J., Hill, T. A., & Gasser, C. S. (2004). Regulation of ovule development. *Plant Cell*, *16*, S32–S45.

Smyth, D. R. (1994). Pollen clusters. *Current Biology*, *4*, 851–853.

Spielman, M., Preuss, D., Li, F.-L., Browne, W. E., Scott, R. J., & Dickinson, H. G. (1997). TETRASPORE is required for male meiotic cytokinesis in *Arabidopsis thaliana*. *Development*, *124*, 2645–2657.

Stone, B. A., & Clarke, A. E. (1992). *The Chemistry and Biology of (1-3)-β-Glucans*. Bundoora: La Trobe University Press.

Tanaka, I., & Ito, M. (1981). Control of division patterns in explanted microspores of *Tulipa gesneriana*. *Protoplasma*, *108*, 329–340.

Töller, A., Brownfield, L. R., Neu, C., Twell, D., & Schulze-Lefert, P. (2008). Dual function of *Arabidopsis* Glucan Synthase-Like genes *GSL8* and *GSL10* in male gametophyte development and plant growth. *Plant Journal*, *54*, 911–923.

Tucker, M. R., Paech, N. A., Willemse, M. T. M., & Koltunow, A. M. (2001). Dynamics of callose deposition and β-1,3-glucanase expression during reproductive events in sexual and apomictic *Hieracium*. *Planta*, *212*, 487–498.

Twell, D., Park, S. K., & Lalanne, E. (1998). Asymmetric division and cell-fate in developing pollen. *Trends in Plant Science*, *3*, 305–310.

Twell, D., Park, S. K., Hawkins, T. J., Schubert, D., Schmidt, R., Smertenko, A., & Hussey, P. J. (2002). MOR1/GEM1 has an essential role in the plant-specific cytokinetic phragmoplast. *Nature Cell Biology*, *4*, 711–714.

Van Aelst, A. C., & van Went, J. L. (1992). Ultrastructural immunolocalization of pectin and glycoproteins in *Arabidopsis thaliana* pollen grains. *Protoplasma*, *168*, 14–19.

von Besser, K., Frank, A. C., Johnson, M. A., & Preuss, D. (2006). *Arabidopsis HAP2* (*GCS1*) is a sperm-specific gene required for pollen tube guidance and fertilization. *Development*, *133*, 4761–4769.

Vijayaraghavan, M. R., & Avdhesh, K. S. (1977). Absence of callose around the microspore tetrad and poorly developed exine in *Pergularia daemia*. *Annals of Botany*, *41*, 923–926.

Warmke, H. E., & Overman, W. A. (1972). Cytoplasmic male sterility in sorghum. 1. Callose behavior in fertile and sterile anthers. *Journal of Heredity*, *63*, 103–108.

Waterkeyn, L., & Bienfait, A. (1970). On a possible function of the callosic special wall in *Ipomea purpurea* (L.) Roth. *Grana*, *10*, 13–20.

Webb, M. C., & Gunning, B. E. S. (1991). Embryo sac development in *Arabidopsis thaliana*. *Sexual Plant Reproduction*, *3*, 244–256.

Whittington, A. T., Vugrek, O., Wei, K. J., Hasenbein, N. G., Sugimoto, K., Rashbrooke, M. C., & Wasterneys, G. O. (2001). MOR1 is essential for organizing cortical microtubules in plants. *Nature*, *411*, 610–613.

Williams, J. H. (2008). Novelties of the flowering plant pollen tube underlie diversification of a key life history stage (in press). *Proceedings of the National Academy of Sciences of the United States of America*.

Williams, M. N. V., Freshour, G., Darvill, A. G., Albersheim, P., & Hahn, M. G. (1996). An antibody Fab selected from a recombinant phage display library detects de-esterified pectic polysaccharide rhamnogalacturonan II in plant cells. *Plant Cell*, *8*, 673–685.

Worrall, D., Hird, D.l., Hodge, R., Paul, W., Draper, J., & Scott, R. (1992). Premature dissolution of the microsporocyte callose wall causes male-sterility in transgenic tobacco. *Plant Cell*, *4*, 759–771.

Yadegari, R., & Drews, G. N. (2004). Female gametophyte development. *Plant Cell*, *16*, S133–S141.

Yang, C. Y., Spielman, M., Coles, J. P., Li, Y., Ghelani, S., Bourdon, V., Brown, R. C., Lemmon, B. E., Scott, R. J., & Dickinson, H. G. (2003). TETRASPORE encodes a kinesin required for male meiotic cytokinesis in *Arabidopsis*. *Plant Journal*, *34*, 229–240.

CHAPTER 4.4.4

# Callose in Abiotic Stress

**Angelika Stass and Walter J. Horst**
*Institute of Plant Nutrition, Leibniz University of Hannover, Hannover, Germany*

Callose, a linear (1,3)-β-glucan, occurs constitutively in cell walls of intact tissues such as sieve plates, cell plates of newly dividing cells, plasmodesmata and in reproductive organs during sporo- and gametogenesis and pollen tube development (Northcote et al., 1989). Its accumulation is usually transitory. Apart from its developmentally determined formation, callose is synthesized rapidly and deposited in a localized manner in response to abiotic stress, wounding, mechanical stress and pathogen attack (Kauss 1987; Bolwell 1993; Benhamou 1995). Therefore, it is an important component of the responses to environmental stresses that are perceived at the plant-cell surfaces and lead to modification of the extracellular matrix (McCormack et al., 1997). It appears that the high capacity for callose synthesis of most plant tissues is related to plant-defence reactions even though callose is not a constitutive component of most cell walls. To assure protection against abiotic and biotic stress, plant cells may need to maintain high levels of callose synthase in the plasma membrane at all times; the synthase is latent in intact cells, but it is available to provide for rapid deposition of callose upon stress.

## Regulation of Callose Formation in Response to Abiotic Stress

It has long been thought that callose could be a bypass activity of the cellulose synthase complex. It was proposed that the same enzyme acts as cellulose synthase or callose synthase depending on the modulation of its conformation and, thereby, its activity through phosphorylation or $Ca^{2+}$ (Delmer, 1999). This hypothesis was supported by Lukowitz et al. (2001), who showed that in *Arabidopsis* mutants defective in the *CYT1* gene, which causes a deficiency in N glycosylation, a 5-fold decrease in cellulose content was accompanied by an accumulation of callose. It was assumed that the mutations in the *CYT1* gene affect the binding of core glucan chains to glycoproteins, which could be required for the proper folding of the cellulose synthase.

However, by now there is convincing evidence that cellulose synthase and callose synthase are expressions of two different genes/gene families (Hong et al., 2001, Scheible and Pauly, 2004, Farrokhi et al., 2006). Using antibodies against subunits of the cellulose synthase or callose synthase, Nakashima et al. (2003) could show that the (1,4)-β-glucan synthase which is localized in the plasma membrane completely disappeared 5 min after wounding, and the (1,3)-β-glucan synthase became labelled. The immuno-location of the (1,3)-β-glucan synthase was in good agreement with the region where the (1,4)-β-glucan synthase was localized before wounding. This shows that the two enzymes do share similar topologies in the membrane, even though they differ in their amino acid sequence.

Biochemically, callose synthase (GSII) can be distinguished from 1,4-β-glucan synthase (GSI) activity by distinct assay conditions. The assay for GSII differs from that for GSI by the inclusion of $Ca^{2+}$ and a higher substrate concentration (Ray 1980). Kudlicka et al. (1997) were able to separate the activities of these two enzymes electrophoretically. The two multienzyme complexes share some polypeptides, but also contain specific polypeptides for the different enzyme activities. Based on the sedimentation of product-entrapped purified callose synthase, it seems that several other proteins may be associated with this complex (Hong et al., 2001b). The functions of some of these proteins may be involved in controlling callose-synthase activity, particularly in response to biotic/abiotic signals (Verma and Hong, 2001).

Kauss (1996) extensively investigated the induction and regulation of callose biosynthesis. Experiments with suspension cells as a model system showed that even though in intact plants callose is deposited in a localized manner, the whole cell surface is able to produce callose, and callose synthesis can be induced with various biochemically unrelated substances such as polycations and amphipathic compounds. These substances alter the integrity of the plasma membrane, recognised for example as leakage of cellular electrolytes such as $K^+$ (Waldmann et al., 1988, Kauss et al., 1990). However, there is no quantitative correlation between leakage of electrolytes and the formation of callose. Induction of callose in suspension cells by aluminium (Al) even reduced the $K^+$ efflux (Staß and Horst, 1995). Also, chemically similar elicitors induced different levels of callose synthesis, suggesting that the induction of callose is not directly linked to plasma-membrane integrity alone. Kauss et al. (1990) could show that the elicitation of callose synthesis requires a lag phase of up to 10 min suggesting signal transduction between the site of perception and response of the callose synthase. However, it cannot be decided yet whether this lag phase is just due to the detection limits of callose quantification.

Callose is synthesized by callose synthase located in the plasma membrane in such a way that the substrate UDP-glucose arrives from the cytoplasm, probably provided by sucrose synthase

(SuSy), which is in close vicinity to the callose and cellulose synthase, and regulated by phosphorylation. Based on their results Amor et al. (1995) and Salnikov et al. (2003) conclude that SuSy exists in a complex with the (1,3)-β-glucan synthase and serves to channel glucose from sucrose to the glucan.

The fact that all plants including algae produce abiotic stress-induced callose (Scherp et al., 2001) suggests a highly conserved signalling pathway that regulates this type of callose-synthase complex. The regulation of wound-induced callose-synthase activity could be mediated by interaction with G-proteins for which a G-protein-binding signature exists in the *Arabidopsis thaliana* callose-synthase gene (*AtCalS1*) sequence (Hong et al., 2001). The enzyme activity also seems to depend on the phospholipids in the membrane surrounding the enzyme (Kauss et al., 1983; Kauss and Jeblick, 1986). This is supported by the fact that mutations causing an altered sterol composition of the plasma membrane in *Arabidopsis* lead to defects in cellulose synthesis and callose deposition (Schrick et al., 2004).

One necessary signal for stress-induced callose synthesis *in vivo* is an increase in the cytosolic $Ca^{2+}$ concentration. However, this alone is not sufficient to trigger callose formation, because elevating cytosolic $Ca^{2+}$ concentration using the $Ca^{2+}$ ionophore A 23187 (Kauss et al., 1990) or ionomycin (Waldmann et al., 1988) did not lead to a substantial formation of callose. This suggests that, in addition to cytosolic $Ca^{2+}$, other endogenous signals are essential for triggering callose synthesis. Ohana et al. (1992) showed that β-furfuryl-β-glycoside is an endogenous activator of (1,3)-β-glucan synthase and the enzyme activation coincides with the transfer of β-furfuryl-β-glycoside from the vacuole to the cytoplasm (Ohana et al., 1993). This glycoside is proposed to be an allosteric effector, binding at a different site to that of $Ca^{2+}$, and thus inducing conformational changes in the enzyme: lowering the $K_m$ for UDP-glucose and raising $V_{max}$. In addition, a further prerequisite for abiotic stress-induced callose formation appears to be the modification of plasma membrane properties such as fluidity and permeability (Kauss et al., 1989).

The current understanding of the events leading to callose induction under abiotic stress and the regulation of callose synthase is summarized in Fig. 1.

Callose once deposited appears to be subject to rapid depolymerization suggesting that callose metabolism is shifted towards callose catabolism. This is supported by observations that (1,3)-β-glucan hydrolase activities are enhanced under conditions promoting callose accumulation. Zabotin et al. (2002) could inhibit the synthesis but also the catabolism of callose in suspension cells with the protein synthesis inhibitor cycloheximide. Collet (2001) induced callose synthesis with different elicitors in two maize genotypes and found differing patterns of

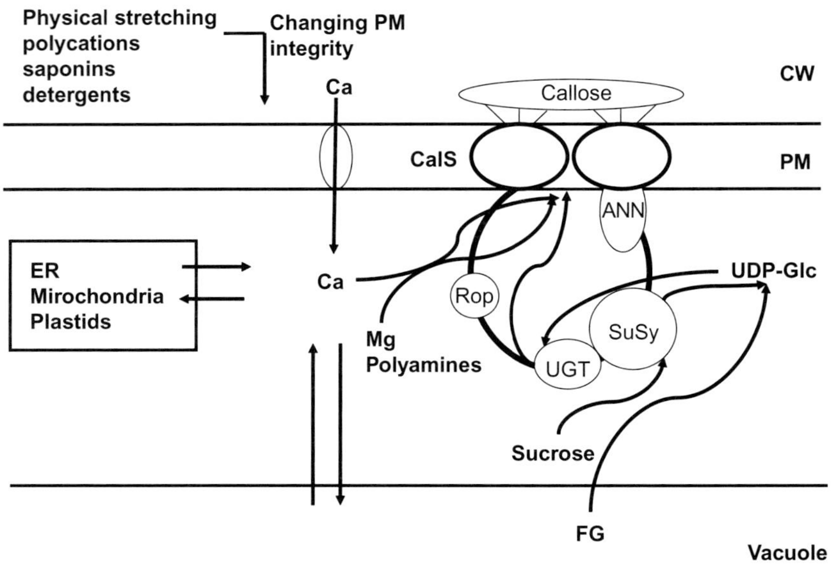


**Fig. 1: Hypothetical model for the induction of callose synthesis, through disruption of plasma membrane (PM) integrity, increased cytosolic $Ca^{2+}$ activity, other possible signals and inductors like $Mg^{2+}$ and polyamines, and regulation of the callose synthase (CalS) composed of transmembrane domains and a hydrophilic loop interacting with an annexin-like protein (ANN), Rop1, a Rho7-like protein, which might regulate CalS by interacting with UGT1 (UDP-glucose transferase) and SuSy (sucrose synthase), which provides the substrate UDP-glc for the enzyme. CW, cell wall; ER, endoplasmic reticulum; FG, β-furfuryl-β-glycoside. (Compiled from Kauss, 1987; Kauss et al., 1990, Verma and Hong, 2001; Ohana et al., 1992).**

callose formation depending on elicitor and genotype mainly affected by callose degradation (Fig. 2).

Levy et al. (2007) demonstrated the cooperation of (1,3)-β-glucan synthase and (1,3)-β-glucanase in the regulation of cell-to-cell transport processes through plasmodesmata. Tobacco lines not transcribing the gene coding for the glucanase accumulated higher callose contents induced by wounding.

## Metal Toxicity and Callose Formation

### *General Aspects*

Various metal ions lead to the induction of callose synthesis when applied in phytotoxic concentrations (Wissemeier et al., 1992, Staß and Horst, 1995; Samardakiewics et al., 1996;

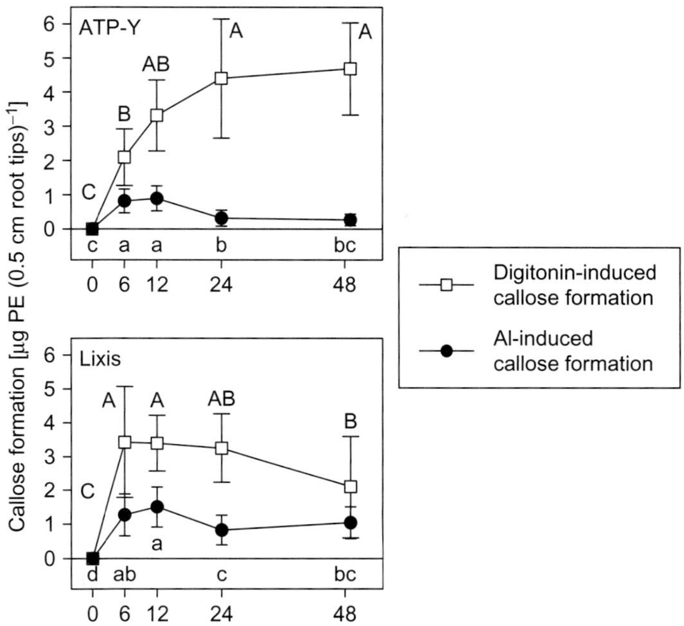


**Fig. 2: Kinetics [h] of Al (25 µM) and digitonin-induced (10 µM) callose formation in root tips of maize in nutrient solution at pH = 4.3, *n* = 15. Means with different letters are significantly different at *P* < 0.05, Tukey test. From Collet (2001).**

Kartusch, 2003). The picture emerging of the ability and effectiveness of different metals as elicitors of callose synthesis is not uniform because the physiological effect of the metals is highly dependent on their speciation. Kartusch (2003) using *Allium cepa* epidermal cells as a model system found that among the metals tested (Cu, Co, Cr, Cd, V, Pb, As, Sr, Li, Mn, Ni, Al) Cu was the most toxic and the induction of callose was not dose dependent beyond a certain threshold concentration. In contrast, Wissemeier et al. (1992) found a dose-dependent induction of callose formation by Mn and Al in leaves of *Vigna unguiculata* and roots of *Glycine max*, respectively. In an investigation of the effect of trivalent metal cations on *Avena sativa* protoplasts, Schaeffer and Walton (1990) found an induction of callose by Y, Yb, Gd and In, but little induction by La, Sc, Fe, Ga and Cr. Staß (1995) found La, Sc and Al to be good inducers of callose synthesis in suspension cells of *Glycine max*. These contradictions arise on the one hand from the different plant systems used and on the other hand from the composition of the treatment solutions. The applied concentration of the cations is not critical, but rather the activity of the phytotoxic form of the cation in solution, which strongly depends on the pH and ionic strength of the treatment solution. Also, the activity of competing cations will strongly influence the cation toxicity and callose formation, as cations can reciprocally ameliorate their toxicity (Kinraide, 1993). Thus, the comparative effectiveness in eliciting callose synthesis by metals can only be assessed if the activities of the metals are considered, which is not the case in many studies.

### Aluminium

One of the most rapid primary manifestations of Al toxicity is an inhibition of root elongation, which can be measured less than 1 h after exposure of the roots to excess Al. An even more sensitive indicator of Al rhizotoxicity is the induction of callose synthesis (Wissemeier et al., 1987; Horst et al., 1992). The most phytotoxic species of Al at low pH is $Al^{3+}$ rather than $Al(OH)^{2+}$ and $Al(OH)_2{}^{+}$ (Kinraide and Parker, 1989). Although the mechanism of Al toxicity at high pH with $Al(OH)_4{}^{-}$ as the predominant Al species has not yet been fully clarified, it is intriguing that inhibition of root elongation was not accompanied by Al-induced callose formation (Wang et al., 2004). It can thus be assumed that the induction of callose synthesis by $Al^{3+}$ is similar to the induction by polycations, which includes both modification of plasma membrane properties and an increase in cytosolic $Ca^{2+}$(see above).

For Al an effect on membrane properties could be shown in *Thermoplasma acidophilum* (Vierstra and Haug, 1978), in artificial membrane vesicles (Deleers et al., 1987) and in root cells of *Quercus rubra* (Chen et al., 1991). There is also circumstantial evidence that plasma membrane composition is a factor determining Al resistance (Staß and Delhaize, 2001, Stival et al., 2006; Ryan et al., 2007).

The additional requirement for stress-induced elicitation of callose formation appears also to be met by Al. An increase in cytosolic $Ca^{2+}$ has been clearly shown (Zhang and Rengel, 1999, Rengel and Zhang, 2003, Bhuja et al., 2004). However La, which induces callose to the same level as Al, is a known Ca-channel blocker (Quist and Roufogalis, 1975, Lansman, 1990). Therefore, $Ca^{2+}$ has to come from internal stores, such as the mitochondria or the vacuole, as has been shown for callose induction under anoxia (see below).

If plants are subjected to Al stress, callose synthesis is a quick and sensitive reaction of the plant root (Horst et al., 1992). In root tips, increased callose contents were found 30 min after Al treatment (Wissemeier et al., 1987, Zhang et al., 1994). In suspension cells as a model system, enhanced callose synthesis could be found 5–10 min after Al application (Staß and Horst, 1995). It thus can be assumed that callose synthesis is switched on instantaneously upon Al application. Callose is confined to the root apex (Wissemeier and Horst, 1995, Jones et al., 2006), and even within the first 10 mm of the root apex callose induction shows a distinct spatial distribution. The distal part of the transition zone (1–2 mm) is the most Al-sensitive apical root zone in maize (Sivaguru and Horst, 1998), showing the highest Al-induced callose formation and Al accumulation. Within the root, callose is first found in the rhizodermis and the outer cell layers of the cortex (Fig. 3A). Formation of callose progresses slowly into the cortex with time (Stass et al., 2006). The longitudinal and radial spatial pattern of Al-induced callose formation appears to be closely related to the distribution of Al in the root apex (Jones et al., 2006).

Aluminium-induced callose formation appears to be a stress indicator and not involved in Al resistance. This conclusion evolves from the negative correlation between the Al resistance of a genotype and its Al-induced callose formation. Thus, Al-induced callose contents in root tips quantified using Aniline Blue fluorescence staining after alkaline extraction (Köhle et al., 1985) has been used to assess genotypic differences in Al resistance in *Zea mays* (Llugany et al., 1994; Horst et al., 1997; Collet et al., 2002; Eticha et al., 2005), *Triticum aestivum* (Schreiner et al., 1994), *Glycine max* (Yang et al., 2000), *Phaseolus vulgaris* (Massot et al., 1999, Rangel et al., 2005) and Arabidopsis (Larsen et al., 1996) grown in hydroponics. This

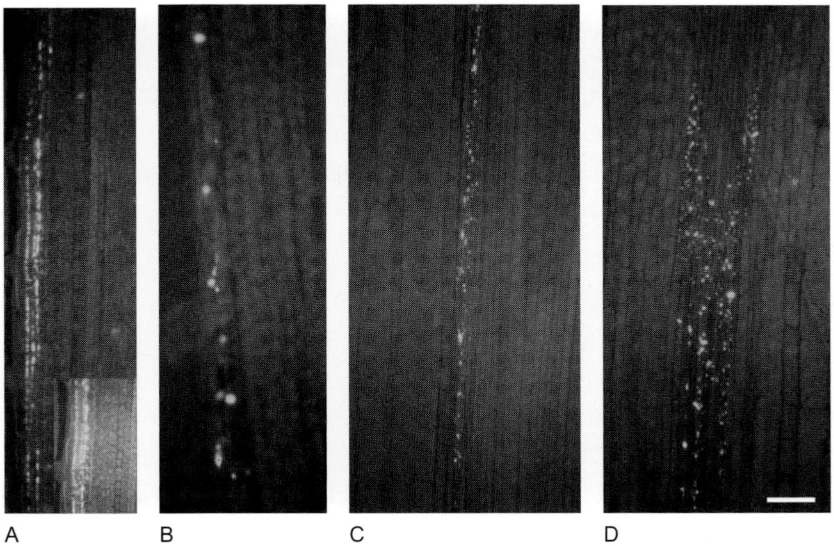


**Fig. 3: Localization of callose by staining with sirofluor in longitudinal sections of root tips of soybean (*Glycine max*). Roots were treated with (A) 25 μM Al for 4 h, (B) 18 μM Pb for 8 h, (C) 1 μM Hg for 4 h, (D) 10 μM Cd for 8 h. Insert in (A) shows a magnification of the rhizodermis and outer cortex. Scale bar = 100 μm. (A. Diening, A. H. Wissemeier, W. J. Horst, unpublished results.) The colour specifications refer to colours in panels.**

technique requires proper standardization of the experimental conditions in order to ensure that Al is the main factor eliciting callose formation (other stress factors induce callose formation as well, see below). However, callose formation has also been used successfully for the characterization of Al stress in trees growing on strongly acid soils (Wissemeier et al., 1998). This, however, required the optimization of the standard protocol of callose quantification to accommodate for the high autofluorescence in spruce roots due to high phenol contents (Hirano and Brunner, 2006). Using Al-induced callose formation for the screening of genotypes for Al resistance has to meet at least two requirements: (i) the genotypic differences in Al resistance have to be expressed without a lag phase, typical for plant species which have an inducible resistance mechanism of the release of organic acid anions (Ma, 2000; Rangel et al., 2007), because in short-term response these resistant genotypes will be sensitive to Al; (ii) the genotypes must not greatly differ in their constitutive capacity for callose production. The latter might have been the case for some *Arabidopsis* mutants characterized for Al resistance by Larsen et al. (1996). In the latter case it is necessary to relate Al-induced callose formation to callose formation induced by another elicitor, which does not interfere with Al resistance mechanisms. The callose induced by this elicitor is set to 100% and Al-induced callose expressed in relation to this value (relative callose formation) as proposed by Schmohl and Horst (2000) and Schmohl et al. (2000) in their work with maize cell-suspension cultures. The elicitor of choice is digitonin, a steroid saponin, which is a very potent elicitor of callose formation and is supposed to induce callose not primarily through ionic properties like Al (see above) but through interaction with sterols of the plasma membrane. Owing to its different way of action, Al-resistance mechanisms should not be affected and the potential of a plant tissue for callose formation can be assessed.

A negative correlation between genotypic Al resistance and callose formation was also found when genotypes were investigated at the protoplast level, even though the correlation was less close (Horst et al., 1997). Thus, callose formation reflects the effect of Al at the plasma membrane directly and supports the view that plasma-membrane properties, such as its negativity and thus ability to bind Al, are properties that contribute to explaining differences in Al resistance of plant species (Wagatsuma et al., 2005). Aluminium may also influence plasma-membrane properties, triggering callose synthesis, indirectly through binding to and modification of the cell-wall network interacting with the plasma membrane (Horst et al., 1999). However, if a genotype has developed resistance mechanisms expressed primarily in the cell wall, callose formation of protoplasts and Al resistance on a whole plant level may not correlate well.

Sivaguru et al. (2000) showed that Al treatment induced callose deposition around plasmodesmata and postulated that this contributed to Al-induced inhibition of root elongation by blocking cell-to-cell solute trafficking in the root tip. This is in line with a prominent role of callose formation and degradation in the control of intercellular communication through plasmodesmata (Levy et al., 2007; see also Chapter 4.4.2). Sivaguru et al. (2006) tested the hypothesis that apoplastic deposition of callose also inhibited apoplastic bypass flow of high molecular weight solutes. However, the inhibition of callose synthesis prevented neither Al-inhibited solute flow nor root elongation. In contrast to these results, de Cnodder et al. (2005) suggested that the 1-aminocyclopropane-1-carboxylic acid (the ethylene precursor) induced callose deposition in the cell walls of the epidermis and cortex cells of the elongation and differentiation zone contributed to the inhibition of root elongation in *Arabidopsis*.

In conclusion, there is little evidence that Al-induced callose formation contributes either to Al toxicity or Al resistance. Thus, it appears that Al-induced callose formation is just an indicator and measure of the stress perceived by the plant-root cells.

### *Manganese and Other Heavy Metals*

Manganese (Mn) induces callose formation in roots, but it is among the least effective of the tested metals. However, Mn-induced callose formation in leaves is a sensitive marker of Mn toxicity in cowpea (Wissemeier et al., 1992). Callose is deposited around the brown spots appearing first on old leaves which are typical Mn-toxicity symptoms in cowpea (*Vigna unguiculata* [L.] Walp.) (Wissemeier and Horst, 1987) as well as other plant species (Horst and Marschner, 1978, Wissemeier et al., 1992). Callose formation in the leaf proved to be a more sensitive indicator of Mn toxicity than the appearance of macroscopic symptoms or the Mn concentration in the leaf (Horst et al., 1999, Fecht-Christoffers et al., 2003). The reasons for the low responsiveness of callose synthesis to Mn in roots compared to leaves are not understood. Three reasons are proposed. (i) Root cortical cells are exposed to micromolar (nutrient solution) but leaf cells to millimolar $Mn^{2+}$ concentrations (apoplastic fluid). Such high apoplastic $Mn^{2+}$ concentrations may lead to an increase in the constitutively low cytosolic $Mn^{2+}$ concentration (Clarkson, 1988) thus triggering callose synthase in a way similar to $Ca^{2+}$ (Morrow and Lucas, 1986). Maintenance of low cytosolic Mn concentrations by enhanced transport of Mn into other cell compartments appears to be an important mechanism of Mn tolerance in some plant species (Hirschi et al., 2000; Delhaize et al., 2003; Peiter et al., 2007). (ii) Manganese-induced oxidative stress in the apoplast (Wissemeier und Horst, 1990; Fecht-Christoffers et al., 2003) could be responsible for callose formation as has been shown for oxidative stress induced by ozone fumigation (Gravano et al., 2004; Bussotti et al., 2005) and as part of the hypersensitive reaction in response to pathogen infection (Li et al., 2008; see Chapter 4.4.5). (iii) Manganese toxicity-induced disturbance of the integrity of the photosynthetic apparatus and photosynthetic carbon fixation could lead to oxidative stress (Houtz et al., 1988, Gonzales et al., 1998; Führs et al., 2008. (iv) Manganese toxicity-induced changes in metabolite composition (Fecht-Christoffers et al., 2007; Führs et al., 2009) and/or compartmentalization could elicit callose synthase, as has been reported by Ohana et al. (1992).

The expression of Mn toxicity (and thus callose synthesis) is not strictly related to the tissue concentration of Mn. Manganese leaf-tissue tolerance is rather dependent on leaf age, genotype, temperature and silicon concentration (Horst et al., 1999). It thus appears that it is not the total Mn concentration but the $Mn^{2+}$ concentration and or physiological/metabolic changes in the apoplast and or symplast triggered by $Mn^{2+}$ that are decisive for the induction of callose synthesis in leaves (Fecht-Christoffers et al., 2007). There is no evidence that the induction of callose formation by Mn is causally related to Mn toxicity or Mn tolerance.

Apart from Mn, other heavy metals are known to induce callose formation. For Pb, Cd and Hg, a distinct pattern of callose formation in roots could be found (Fig. 3). These different patterns probably reflect the different mobility, binding forms, and distribution of the investigated metals, as has been shown for Al (see above). In line with these results, Samardakiewics et al. (1996) found that treatment with $Pb(NO_3)_2$ lead to the deposition of callose in the rhizodermis, but also in the centre of the stele in the root tip. Ueki and Citovsky (2005) showed that Cd induced callose in the plant leaf vascular tissue.

Akhitar et al. (2005) did not quantify callose formation, but found an accumulation of transcripts that encode a callose synthase after *Lemna gibba* had been treated with toxic concentrations of Cu. It is not clear which part of the plant reacted with an increase in transcripts, because they isolated the RNA from the whole plant tissue.

## Wounding, Osmotic Stress and Callose Formation

Callose deposition has been described as a typical plant response to stress aimed at locally isolating the impact of the stress in the plant tissue through the deposition of a physical barrier (Kauss, 1989; Farrokhi et al., 2006). The formation of callose deposits in mechanically wounded plant cells in response to cutting of active sieve tubes during specimen preparation was first described by Fischer (1886). Many subsequent studies have shown that localized deposits of callose develop at the periphery of a variety of plant cells subjected to mechanical stress (Aist 1976, Fincher and Stone, 1981, Kauss, 1985). In damaged tissues callose deposition in the walls of surviving cells is the first event in wound healing according to Galway and

McCully (1987). Currier and Webster (1964) showed that different types of callose deposits were synthesized in cotton petioles at different times post wounding, and also that these different deposits remained for variable times.

In spite of this established knowledge, the underlying physiological and molecular mechanisms are not well understood. Wound callose is first deposited at intercellular connections (i.e. sieve plate pores and pit fields) and then spreads to surrounding wall regions (Currier, 1957). In pea roots, callose was detected in the phloem close to the site of injury 1 min post wounding and synthesis lasted for over 3 h, whereas in cortical parenchyma cells callose was first detected 10 min post wounding and synthesis lasted for about 20 h (Gallway and McCully, 1987). Cells varied in their capacity to synthesize and resorb callose in response to severe disruption of parenchyma cells, but callose in the phloem was not degraded probably because it had ceased functioning. On the other hand, plasmolysis-induced pit callose described by Drake et al. (1978) was resorbed within 4–8 h after recovery of turgor, perhaps reflecting the mildness of the inducing stimulus and the deposition of relatively small amounts of callose.

Jacobs et al. (2003) demonstrated that the silencing of the callose synthase-coding gene *GSL5* substantially reduced callose formation induced by mechanical wounding at the margin of the cut, as well as by pathogen infection as plugs at the site of fungal penetration in Arabidopsis. Callose formation is also induced through Plant-tissue injury by herbivores (Hao et al., 2008). It thus appears that pathogen infection and herbivore-induced callose formation reflect the reaction of plant cells to mechanical injury and are initiated by the same signalling pathway.

Furch et al. (2007) investigated in detail the sequence of events following a mechanical injury to the main vein of *Vicia faba* leaves. They injured the leaf tip and observed the subsequent reaction 3 cm away from this site. Burning the leaf tip caused electro-potential waves 15–20 s after the injury, followed by the dispersion of proteins, so called furisomes, which plugged the sieve pores (15–45 s); this was followed by the synthesis of callose. The callose formation reached its maximum after 20 min. Both of these reactions, which blocked the sieve pores, were reversible with the proteins contracting after 7–15 min, and callose being degraded after 1–2 h. When the injury was induced by cutting the leaf tip, no electro-potential wave was generated. Nevertheless, the furisomes dispersed, but no callose was formed in the observation window 3 cm away from the site of injury. Furch et al. (2007) suggested that the electro-potential wave triggers the release of $Ca^{2+}$ into the lumen of the sieve elements and thus induces the formation of callose, and that the degradation of callose starts as soon as the $Ca^{2+}$ concentration has returned to its initial level.

Paris et al., (2007) showed that the deposition of callose was induced by application of the NO donor sodium nitroprusside, and this induction was additive to the wound-induced callose production. They assumed that increased levels of NO could support the healing responses in plants leading to a rapid restoration of the damaged tissue. A role of NO in the plant-tissue response to wounding and mechanical stress through an increase in cytosolic free $Ca^{2+}$, a necessary signal for the induction of callose synthesis (see above), is well established (Wendehenne et al., 2004),

Callose has also been described to be part of the newly formed cell wall of protoplasts. According to several reports (Asamizu and Nishi 1980; Blaschek et al., 1981; Pilet et al., 1984; Gould et al., 1986) (1,3)-β-glucan (callose) is synthesized by wall-regenerating protoplasts, rather than (1,4)-β-glucan (cellulose). Later investigations by Shea et al. (1989) found that callose was not a component of the initial wall of carrot protoplasts but that intentional wounding by rapid shaking or treatment with dimethyl sulfoxide initiated synthesis of callose. Callose found in earlier studies could be due to impurities in the enzymes used for protoplast preparation leading to proteolytic damage of the plasma membrane or affecting metabolism and even viability. Therefore, callose found in the newly formed cell wall seems to be initiated by mechanical or chemical stress since it is not a normal component of the undisturbed cell wall.

## Callose and Anoxia

The root tip is especially sensitive to anoxia due to its high metabolic activity and thus high demand for oxygen. One of the symptoms that are found to precede cell death in an $O_2$-deprived root tip is induction of callose which could be measured as early as 2 h after the beginning of anoxic stress and was most intense after 24 h. The intensity of callose deposition decreases from the root tip towards the root base, to very low amounts above the root-hair zone (Subbaiah and Sachs, 2001). Under oxygen limitation the activity of invertase responsible for sucrose hydrolysis in well aerated roots is rapidly down-regulated, which leads to a significant reduction in sucrose cleavage. As a consequence, the flux of sucrose to SuSy is favoured (Albrecht and Mustroph, 2003). SuSy, which is supposed to be part of the enzyme complex which forms cellulose and callose (Amor et al., 1995), could then channel glucose from sucrose to callose. The synthesis of callose would still be possible even with reduced ATP supply under limiting oxygen tension, because SuSy does not need ATP to build UDP-Glc, the substrate for callose synthase. A necessary signal for the induction of callose synthesis, an increase in cytosolic free $Ca^{2+}$ activity, has been shown to occur immediately after the onset of anoxia (Subbaiah et al., 1994a and 1994b). This increase in cytosolic $Ca^{2+}$ concentration is due to the release of $Ca^{2+}$ from the mitochondria (Subbaiah et al., 1998). In addition to this necessary signal for callose induction, NO is also suggested to be one of the primary signals of anoxia (Dordas et al., 2003).

Even though callose formation appears to be a rapid response of the plant root apex to anoxia its role in anoxia tolerance is not clear. Albrecht and Mustroph (2003), assuming carbon limitation under anoxia, found that in wheat root tips high levels of callose and cellulose were sustained whereas soluble carbohydrates were depleted. Callose was predominantly localized in the outer cortex whereas cellulose accumulated in the pericycle and endodermis. They concluded that the increase in these structural carbohydrates is necessary for secondary cell wall thickening thus maintaining mechanical stability during long-term anoxia stress.

According to Subbaiah and Sachs (2001) channelling of the carbon resources into the synthesis of callose could facilitate the death of the root tip. Rapid elimination of the tip, a metabolically demanding sink, could prolong the survival of the root axis and the shoot under anoxic conditions. Subbaiah et al. (2000) showed that removal of the root tip prior to submergence improves the seedling's anoxia tolerance (see also Zeng et al., 1999). However, in most maize genotypes studied, root-tip death was a prolonged process during which the necrosis spread into the root axis and led to seedling mortality.

Biemelt et al. (1999) investigating SuSy activity in submerged potato roots interpreted the results differently. Because the flux of carbon through glycolysis is reduced, despite the almost unaltered activity of glycolytic enzymes, it could be limited by end-product utilization. SuSy, which according to Amor et al. (1995) is associated with cellulose and callose synthase and provides the substrate UDP-Glc for these enzymes, might serve to channel glucose directly from sucrose to cellulose and/or callose synthase in the plasma membrane. Therefore, they speculated that glucose from sucrose is partly converted into callose to allow, on the one hand, UDP to be recycled and, on the other hand, carbohydrates to be stored. This explanation is similar to the observed accumulation of fructans (Albrecht et al. 1993, 1997); thus, formation of callose in roots under oxygen deficiency would serve to retain carbohydrates, maintain the sink function of the organ, and store excess carbohydrates for later consumption rather than fuelling glycolysis.

## Callose and Temperature Stress

Even mild changes to supra/sub-optimal temperatures ($\Delta 6°C$) can lead to callose formation (Smith and McCully, 1977). Large deposits of Aniline Blue fluorescent material, indicative of callose, were found in the walls of elongating cells around the shoot apex and in root-cap cells, and appeared to be closely associated with a few pit fields. The plant species investigated differed in their response to changes in temperature: seedlings of *Zea mays, Glycine max* and *Phaseolus vulgaris* showed more callose at 20°C than at 26°C, and seedlings of *Pisum sativum* and *Solanum lycopersicum* showed more callose at 26°C than at 20°C. Under this mild temperature stress no temperature-induced differences could be observed in phloem callose. This is in contrast to results of Currier and co-workers (Webster and Currier, 1965; McNairn and Currier, 1968), who found that the amount of callose increased in cotton hypocotyls heated to 40–45°C, and this accumulation was correlated with inhibition of both basipetal (McNair and Currier, 1968) and lateral (Webster and Currier, 1968) phloem translocation. Treatment for 15 min was sufficient to induce callose formation and inhibition of translocation for several hours. Both callose and phloem translocation returned to near normal levels within 6 h after removal of the heat stress (McNairn and Currier, 1968), suggesting that increased flow resistance due to constriction of plasmodesmata and sieve plate pores by callose might be responsible for this effect of heat stress. This hypothesis is supported by electron micrographs (Shih and Currier, 1969) showing almost complete constriction of sieve plate pores by heat-induced callose in intact cotton stems. In contrast to wound-induced callose in the phloem (Gallway and McCully, 1987), heat-induced callose formation in the phloem was reversible. Callose deposited on sieve plates began to be removed 6 h after heating and was reduced to normal levels within 2 days. Also, Musolan et al. (1975) investigating enzyme activities of cellulose and callose synthases of primary leaves of *P. vulgaris* and *Vigna sinensis* found a severe inhibition of the synthesis of cellulose and a stimulation of callose synthesis in response to heat shock. Callose formed in the phloem and inhibited phloem transport in response to heat but also to cold stress (Majumder and Leopold, 1967).

Callose induction by heat stress was also found under field conditions (McNairn, 1972). Phloem translocation rates in field-grown cotton dropped from morning to afternoon and continued to decline toward the evening. Even though the contribution of water deficit cannot be excluded, heat stress was regarded as the main factor inducing callose formation. Greatly reduced phloem translocation was observed when about 50% or more of the hypocotyl sieve plates were blocked by callose. Callose breakdown appeared to be slower than heat-induced synthesis because inhibition of phloem transport persisted into the evening in spite of decreasing temperatures,

The physiological and molecular mechanisms of temperature stress-induced callose formation have so far not been addressed. It appears attractive to speculate that, based on the model described in (Fig. 1), temperature-induced callose formation is triggered by increasing cytosolic $Ca^{2+}$ activity and modification of plasma-membrane properties typical also for cold as well as heat stress. The most sensitive response of the phloem sieve-plates could be related to their particular, possibly constitutive, capacity for callose synthesis and a need to control phloem transport in response to stress.

## Callose and the Gravitropic Response

The perception of a gravitropic signal and the subsequent bending of the shoot or root are still not understood in detail, but it seems to be clear that the gravity-signalling process is controlled by a network of signalling transduction pathways (Perrin et al., 2005). If there is a function of callose formation in gravitropism the nature of its participation is presently not clear, but it does have the characteristics of a biochemical change in the tissue which precedes gravity-induced curvature.

Jaffe and Leopold (1984) showed that an increase in callose deposition occurs in gravitropically responding coleoptiles of maize and pea plants. The pea mutant ‘Ageotropum’, which does not respond to gravity when etiolated, also fails to produce callose in response to a gravitropic stimulus. These correlations indicate that callose deposition may be a biochemical component of the gravitropic response in plant shoots.

Callose deposition occurred before any detectable gravitropic bending. Thus, Jaffe and Leopold (1984) speculated that callose formation could be part of the transduction between the movement of the statoliths and the final asymmetric distribution of phytohormones, which leads to asymmetric growth.

Callose formation could also be a consequence of gravistimulation. $Ca^{2+}$ is assumed to function as a second messenger in gravity signal transduction within the root-cap statocytes (Perrin et al., 2005). In agreement with this model, the levels of inositol 1,4,5-trisphosphate (IP3), a second messenger known to activate $Ca^{2+}$ release into the cytosol from intracellular stores, fluctuate early in response to gravistimulation on both sides of maize and oat pulvini, before eventually stabilizing with increased concentration at the lower pulvinus half (Perera et al., 2001). As an increased $Ca^{2+}$ concentration is also a necessary signal for callose induction, callose formation could be an effect of gravistimulation. As Delmer (1987) stated this would be an example of ‘accidental’ callose formation. In these situations callose formation occurs, when a localized concentration of $Ca^{2+}$ is needed for other cellular functions; for example, to establish the location of a cell plate or to direct the movement and fusion of vesicles to the tips of growing pollen tubes. In these cases, callose may be deposited only as a consequence of other events and of itself may have no function.

## Concluding Remarks

The constitutive capacity to quickly synthesize callose upon wounding provides cells with the ability to generate a new physical barrier thus sealing the injured plant tissue. The physiological machinery involved in the induction and maintenance of callose deposition can be triggered also by metal toxicity, without conferring any apparent protection against metal toxicity. Aluminium-induced callose formation has been studied in detail and used for the screening of plant genotypes for Al sensitivity, because it is a sensitive and reliable indicator and measure of the level of stress perceived by the plant tissue provided the dynamics of callose turnover and constitutive synthesis capacity are taken into account.

Among other metals inducing callose formation, only Mn has been studied in some detail. The potential of metal toxicity-induced callose formation to increase understanding of the dynamics and spatial perception of stress within plant tissues has not yet been exploited.

Progress in the characterization of the genes coding for callose synthases and hydrolases calls for expression studies of these genes under different abiotic stresses. These could provide spatial and temporal resolution of the site of stress perception, which has not been possible up to now. Reverse genetic approaches offer new prospects for the investigation not only of the role of callose in abiotic stress, but also of the development of abiotic stress itself.

## References

Aist, J. R. (1976). Papillae and related wound plugs of plant cells. *Review in Phytopathology*, *14*, 145–163.

Akhitar, T. A., Lampi, M. A., & Greenberg, B. M. (2005). Identification of six differentially expressed genes in response to copper exposure in the aquatic plant *Lemna gibba* (duckweed). *Environmental Toxicology and Chemistry*, *24*, 1705–1715.

Albrecht, G., & Mustroph, A. (2003). Sucrose utilization via invertase and sucrose synthase with respect to accumulation of cellulose and callose synthesis in wheat roots under oxygen deficiency. *Russian Journal of Plant Physiology*, *50*, 813–820.

Albrecht, G., Biemelt, S., & Baumgarten, S. (1997). Accumulation of fructans following oxygen deficiency stress in related plant species with different flooding tolerance. *New Phytologist*, *136*, 137–144.

Albrecht, G., Kammerer, S., Praznik, W., & Wiedenroth, E. M. (1993). Fructan content of wheat seedlings (*Triticum aestivum* L.) under hypoxia and following re-aeration. *New Phytologist*, *123*, 471–476.

Amor, Y., Haigler, C. H., Johnson, S., Wainscott, M., & Delmer, D. P. (1995). A membrane-associated form of sucrose synthase and its potential role in synthesis of celllulose and callose in plants. *Proceedings of the National Academy of Sciences of the United States of America*, *92*, 9353–9357.

Asamizu, T., & Nishi, A. (1980). Regenerated cell wall of carrot protoplasts isolated from suspension-cultured cells. *Physiologia Plantarum*, *48*, 207–212.

Benhamou, N. (1995). Immunocytochemistry of plant defence mechanisms induced upon microbial attack. *Microscopic Research Techniques*, *31*, 63–78.

Bhuja, P., McLachlan, K., Stephens, J., & Taylor, G. (2004). Accumulation of 1,3-β-D-glucans, in response to aluminum and cytosolic calcium in *Triticum aestivum*. *Plant Cell Physiology*, *45*, 543–549.

Biemelt, S., Hajirezaei, M. R., Melzer, M., Albrecht, G., & Sonnewald, U. (1999). Sucrose synthase activity does not restrict glycolysis in roots of transgenic potato plants under hypoxic conditions. *Planta*, *210*, 41–49.

Blaschek, W., Haass, D., Koehler, H., & Franz, G. (1981). Cell wall regeneration by *Nicotiana tabacum* protoplasts: Chemical and biochemical aspects. *Plant Science Letters*, *22*, 47–57.

Bolwell, G. P. (1993). Dynamic aspects of the plant extracellular matrix. *International Review of Cytology*, *146*, 261–323.

Bussotti, F., Agati, G., Desotgiu, R., Matteini, P.& Tanic, C. (2005). Ozone foliar symptoms in woody plant species assessed with ultrastructural and fluorescence analysis. *New Phytologist*, *166*, 941–955.

Chen, J., Sucoff, E. I., & Stadelmann, E. J. (1991). Aluminum and temperature alteration of cell membrane permeability of *Quercus rubra*. *Plant Physiology*, *96*, 644–649.

Clarkson, D. T. (1988). The uptake and translocation of manganese by plant roots. In R. D. Graham, R. J. Hannam, N. C. Uren (Eds.), *Manganese in soils and plants* (pp 101–111). ISBN 90 247 37583

de Cnodder, T., Vissenberg, K., van der Straeten, D., & Verbelen, J. P. (2005). Regulation of cell length in the *Arabidopsis thaliana* root by the ethylene precursor 1-aminocyclopropane-1-carboxylic acid: A matter of apoplastic reactions. *New Phytologist*, *168*, 541–550.

Collet, L. (2001). *Screening maize (Zea mays) for aluminium resistance-Contribution to the selection for adaptation to acid soils*. PhD-Thesis, University of Hannover.

Collet, L., de Leon, C., Kollmeier, M., Schmohl, N., & Horst, W. J. (2002). Assessment of aluminum sensitivity of maize cultivars using roots of intact plants and excised root tips. *Journal of Plant Nutrition and Soil Science*, *165*, 357–365.

Currier, H. B. (1957). Callose substance in plant cells. *American Journal of Botany*, *44*, 478–488.

Currier, H. B., & Webster, D. H. (1964). Callose formation and subsequent disappearance: Studies in ultrasound stimulation. *Plant Physiology*, *39*, 843–847.

Deleers, M., Servais, J. P., & Wülfert, E. (1987). Aluminium-induced lipid phase separation and membrane fusion does not require the presence of negatively charged phospholipids. *BiochemistryIntern ational*, *14*, 1023–1034.

Delhaize, E., Kataoka, T., Hebb, D. M., White, R. G., & Ryan, R. (2003). Genes encoding proteins of the Cation Diffusion Facilitator family that confer manganese tolerance. *Plant Cell*, *15*, 1131–1142.

Delmer, D. P. (1987). Cellulose biosynthesis. *Annual Review of Plant Physiology*, *38*, 259–290.

Delmer, D. P. (1999). Cellulose Biosynthesis: Exciting times for a difficult field of study. *Annual Review of Plant Physiology and Plant Molecular Biology*, *50*, 245–276.

Dordas, C., Rivoal, J., & Hill, R. D. (2003). Plant haemoglobins, nitric oxide and hypoxic stress. *Annals of Botany*, *91*, 173–178.

Drake, G. A., Carr, D. J., & Anderson, W. P. (1978). Plasmolysis, plasmodesmata and the electrical coupling of oat coleoptile cells. *Journal of Experimental Botany*, *29*, 1205–1214.

Eticha, D., Thé, C., Welcker, C., Narro, L., Staß, A., & Horst, W. J. (2005). Aluminium-induced callose formation in root apices: Inheritance and selection trait for adaptation of tropical maize to acid soils. *Field Crop Research*, *93*, 252–263.

Farrokhi, N., Braun, R. A., Brownfield, L., Hrmova, M., Wilson, S. M., Bacic, A., &Fincher, G. B. (2006). Plant cell wall biosynthesis: genetic, biochemical and functional genomicy approaches to the identification of key genes. *Plant Biotechnology Journal*, *4*, 145–167.

Fecht-Christoffers, M. M., Braun, H. P., Lemaitre-Guillier, C., & VanDorsselaer, A. (2003). Effect of manganese toxicity on the proteome of the leaf apoplast in cowpea. *Plant Physiology*, *133*, 1935–1946.

Fecht-Christoffers, M. M., Maier, P., Iwasaki, K., Braun, H.-P., & Horst, W. J. (2007). The role of the leaf apoplast in manganese toxicity and tolerance in cowpea (*Vigna unguiculata* L., Walp). ISBN: 978-1-4020-5842-4. In B. Sattelmacher & W. J. Horst (Eds.), *The apoplast of higher plants: Compartment of storage, transport, and reactions* (pp. 307–321). Springer Verlag.

Fincher, G. B., Stone, B. A. (1981). Mechanism of non-cellulosic polysaccharides. In W. Tanner, F. A. Loewes (Eds.), *Plant carbohydrates II encyclopedia of plant physiology* (pp. 68–136).

Fischer, A. (1886). Neue Beiträge zur Kenntnis der Siebröhren, Berichte der königlich-sächsischen Gesellschaft der Wissenschaften, mathematisch-physische Klasse. *1*(38), 291–336.

Furch, A. C. U., Hafke, J. B., Schulz, A., & van Bel, A. J. E. (2007). $Ca^{2+}$-mediated remote control of reversible sieve tube occlusion in *Vicia faba*. *Journal of Experimental Botany*, *58*, 2827–2838.

Führs, H., Götze, S., Specht, A., Erban, A., Gallien, S., Heintz, D., Van Dorsselaer, A., Kopka, J., Braun, H.-P., & Horst, W.J. (2009). Characterization of leaf apoplastic peroxidases and metabolites in *Vigna unguiculata* in response to toxic manganese supply and silicon. *Journal of Experimental Botany*, *60*, 1663–1678.

Galway, M. E., & McCully, M. E. (1987). The time course of the induction of callose in wounded pea roots. *Protoplasma*, *139*, 77–91.

Gould, J. H., Palmer, R. L., & Dugger, W. M. (1986). Isolation and culture of cotton ovule epidermal protoplasts (prefiber cells) and analysis of the regenerated wall. *Plant Cell Tissue and Organ Culture*, *6*, 47–59.

Gravano, E., Bussotti, F., Strassser, R. J., Schaub, M., Novak, K., Skell, J., & Tani, C. (2004). Ozone symptoms in leaves of woody plants in open-top chambers: Ultrastructural and physiological characteristics. *Physiologia Plantarum*, *121*, 620–633.

Hao, P., Liu, C., Wang, Y., Chen, R., Tang, M., Du, B., Zhu, L., & He, G. (2008). Herbivore-induced callose deposition on the sieve plates of rice: An important mechanism. *Plant Physiololgy*, *146*, 1810–1820.

Hirano, Y., & Brunner, I. (2006). Quantitative determination of callose in tree roots. *Journal of Plant Physiology*, *163*, 1333–1336.

Hirschi, K. D., Korenkov, V. D., Wilganowski, N. L., & Wagner, G. J. (2000). Expression of Arabidopsis *CAX2* in tobacco. Altered metal accumulation and increased manganese tolerance. *Plant Physiology*, *124*, 125–133.

Hong, Z., Zhang, Z., Olson, J. M., & Verma, D. P. S. (2001). A novel UDP-glucose transferase is part of the callose synthase complex and interacts with phragmoblastin at the forming cell plate. *Plant Cell*, *13*, 769–779.

Horst, W. J., & Marschner, H. (1978). Symptome von Manganüberschuß bei Bohnen (*Phaseolus vulgaris* L.). *Zeitschrift für Pflanzenernährung und Bodenkunde*, *141*, 129–142.

Horst, W. J., Asher, C. J., Cakmak, I., Szulkiewicz, P., & Wissemeier, A. H. (1992). Short-term responses of soybean roots to aluminium. *Journal of Plant Physiology*, *140*, 174–178.

Horst, W. J., Fecht, M., Naumann, A., Wissemeier, A. H., & Maier, P. (1999). Physiology of manganese toxicity and tolerance in *Vigna unguiculata* (L.) Walp. *Journal of Plant Nutrition and Soil Science*, *162*, 263–274.

Horst, W. J., Püschel, A.-K., & Schmohl, N. (1997). Induction of callose formation is a sensitive marker for genotypic aluminium sensitivity in maize. *Plant and Soil*, *192*, 23–30.

Houtz, R. L., Nable, R. O., & Cheniae, G. M. (1988). Evidence for effects on the *in vivo* activity of ribulose-bisphosphate carboxylase/oxygenase during development of Mn toxicity in tobacco. *Plant Physiology*, *86*, 1143–1149.

Jacobs, A. K., Lipka, V., Burton, R. A., Panstruga, R., Strizhov, N., Schulze-Lefert, P., & Fincher, G. B. (2003). An *Arabidopsis* callose synthase, GSL5, is required for wound and papillary callose formation. *Plant Cell*, *15*, 2503–2513.

Jaffe, M. J., & Leopold, A. C. (1984). Callose deposition during gravitropism of *Zea mays* and *Pisum sativum* and its inhibition by 2-deoxy-D-glucose. *Planta*, *161*, 20–26.

Jones, D. L., Blancaflor, E. B., Kochian, L. V., & Gilroy, S. (2006). Spatial coordination of aluminium uptake, production of reactive oxygen species, callose production and walll rigidification in maize roots. *Plant Cell and Environment*, *29*, 1309–1318.

Kartusch, R. (2003). On the mechanism of callose synthesis induction by metal ions in onion epidermal cells. *Protoplasma*, *220*, 219–225.

Kauss, H. (1985). Callose biosynthesis as a $Ca^{2+}$-regulated process and possible relations to the induction of other metabolic changes. *Journal of Cell Science Supplement*, *2*, 89–103.

Kauss, H. (1987). Callose-Synthese. Regulation durch induzierten $Ca^{2+}$-Einstrom in Pflanzenzellen. *Naturwissenschaften*, *74*, 275–281.

Kauss, H. (1989). Fluorometric measurement of callose and other 1,3-β-glucans. In H. F. Linskens, & J. F. Jackson, (Eds.). *In Modern Methods in Plant Analysis: 10* (pp. 127–137). Springer Verlag.

Kauss, H. (1996). Callose synthesis. Membranes: Specialized functions in plants. In M. Smallwood, J. P. Knox, D. J. Bowles (Eds.), (pp. 77–92). Bios Scientific Publishers.

Kauss, H., & Jeblick, W. (1986). Influence of free fatty acids, lysophosphatidylcholine, platelet-activating factor, acylcarnitine, and echinocandin B on 1,3-β-D-glucan synthase and callose synthesis. *Plant Physiology*, *80*, 7–13.

Kauss, H., Köhle, H., & Jeblick, W. (1983). Proteolytic activation and stimulation by $Ca^{2+}$ of glucan synthaase from soybean cells. *FEBS Letters*, *158*, 84–88.

Kauss, H., Waldmann, T., & Quader, H. (1990). $Ca^{2+}$ as a signal in the induction of callose synthesis. In R. Ranjeva, A. M. Boudet (Eds.), *Signal perception and transduction in higher plants* (pp. 117–131). NATO ASI Series H: Cell Biology Vol. 47. ISBN 3 540 51772 3

Kauss, H., Waldmann, T., Jeblick, W., Euler, G., Ranjeva, R., & Domard, A. (1989). $Ca^{2+}$ is an important but not the only signal in callose synthesis induced by chitosan, saponins and polyene antibiotics. In B. J. J. Lugtenberg (Ed.), *Signal molecules in plants and plant-microbe interactions* (pp. 107–116). NATO ASI Series Vol. H 36.

Kinraide, T. B. (1993). Aluminum enhancement of plant growth in acid rooting media. A case of reciprocal alleviation of toxicity by two toxic cations. *Physiologia Plantarum*, *88*, 619–625.

Kinraide, T. B., & Parker, D. R. (1989). Assessing the phytotoxicity of mononuclear hydroxy-aluminum. *Plant, Cell and Environment*, *12*, 479–487.

Köhle, H., Jeblick, W., Poten, F., Blaschek, W., & Kauss, H. (1985). Chitosan-elicited callose synthesis in soybean cells as a $Ca^{2+}$-dependent process. *Plant Physiology*, *77*, 544–551.

Kudlicka, K., & Brown, R. M. (1997). Cellulose and callose biosynthesis in higher plants. I. Solubilization and separation of (1-3) and (1-4)-β-glucan synthase activities from mung bean. *Plant Physiology*, *115*, 643–656.

Lansman, J. B. (1990). Blockade of current through single calcium channels by trivalent lanthanide cations. Effect of ionic radius on the rates of ion entry and exit. *Journal of General Physiology*, *95*, 679–696.

Larsen, P. B., Tai, C. Y., Kochian, L. V., & Howell, S. H. (1996). Arabidopsis mutants with increased sensitivity to aluminum. *Plant Physiology*, *110*, 743–751.

Levy, A., Erlanger, M., Rosenthal, M., & Epel, B. L. (2007). A plasmodesmata-associated β-1-3-glucanase in *Arabidopsis*. *Plant Journal*, *49*, 669–682.

Llugany, M., Massot, N., Wissemeier, A. H., Poschenrieder, C., Horst, W. J., & Barcelo, J. (1994). Aluminium tolerance of maize cultivars as assessed by callose production and root elongation. *Zeitschrift für Pflanzenernährung und Bodenkunde*, *157*, 447–451.

Lukowitz, W., Nickle, T. C., Meinke, D. W., Last, R. L., & Conklin, P. L. (2001). Arabidopsis cyt1 mutants are deficient in a mannose-1-phosphate guanylyltransferase and point to a requirement of N-linked glycosylation for cellulose biosynthesis. *PNAS*, *98*, 2262–2267.

Ma, J. F. (2000). Role of organic acids in detoxification of aluminum in higher plants. *Plant and Cell Physiology*, *41*, 383–390.

Majumder, S. K., & Leopold, A. C. (1967). Callose formation in response to low temperature. *Plant and Cell Physiology*, *8*, 775–778.

Massot, N., Llugany, M., Poschenrieder, C., & Barcelo, J. (1999). Callose production an indicator of aluminum toxicity in bean cultivars. *Journal of Plant Nutrition*, *22*, 1–10.

McCormack, B. A., Gregory, A. C. E., Kerry, M. E., Smith, C., & Bolwell, G. P. (1997). Purification of an elicitor-induced glucan synthase (callose synthase) from suspension cultures of french bean (*Phaseolus vulgaris* L.): Purification and immunolocation of a probable Mr -65000 subunit of the enzyme. *Planta*, *203*, 196–203.

McNairn, R. B. (1972). Phloem translocation and heat-induced callose formation in field-grown *Gossypium hirsutum* L.. *Plant Physiology*, *50*, 366–370.

McNairn, R. B., & Currier, H. B. (1968). Translocation blockage by sieve plate callose. *Planta*, *82*, 369–380.

Morrow, D. L., & Lucas, W. J. (1986). 1-3-β-D-glucan synthase from sugar beet: I Isolation and solubilization. *Plant Physiology*, *81*, 171–176.

Musolan, C., Ordin, L., & Kindinger, J. I. (1975). Effects of heat shock on growth and on lipid and β-glucan synthases in leaves of *Phaoseolus vulgaris* and *Vigna sinensis*. *Plant Physiology*, *55*, 328–332.

Nakashima, J., Laosinchai, W., Cui, X., & Brown, R. M., Jr. (2003) New insight into the mechanism of cellulose and callose biosynthesis: Proteases may regulate callose biosynthesis upon wounding. *Cellulose*, *10*, 369–389.

Ohana, P., Benziman, M., & Delmer, D. P. (1993). Stimulation of callose synthesis in vivo correlates with changes in intracellular distribution of the callose synthase activator β-furfuryl-β-glucoside. *Plant Physiology*, *101*, 187–191.

Ohana, P., Delmer, D. P., Volman, G., Steffens, J. C., Matthews, D. E., & Benziman, M. (1992). β-furfuryl-β-glucoside: An endogenous activator of higher plant UDP-glucose synthase. Biological activity, distribution, and *in vitro* synthesis. *Plant Physiology*, *98*, 708–715.

Paris, R., Lamattina, L., & Casalongue, C. A. (2007). Nitric oxide promotes the wound-healing response of potato leaflets. *Plant Physiology and Biochemistry*, *45*, 80–86.

Peiter, E., Montanini, B., Gobert, A., Pedas, P., Husted, S., & Maathuis, F. J. M. (2007). A secretory pathway-localized cation diffusion facilitator confers plant manganese tolerance. *PNAS*, *104*, 8532–8537.

Perera, I., Heilmann, I., Chang, S., Boss, W., & Kaufman, P. (2001). A role for inositol 1,4,5-trisphosphate in gravitropic signaling and the retention of cold-perceived gravistimulation of oat shoot pulvini. *Plant Physiology*, *125*, 1499–1507.

Perrin, R. M., Yound, L. S., Murthy, N. U. M., Harrison, B. R., Wang, Y., Will, J. L., & Masson, P. H. (2005). Gravity signal transduction in primary roots. *Annals of Botany*, *96*, 737–743.

Pilet, P. E., Blaschek, W., Senn, A., & Franz, G. (1984). Comparison between maize root cells and their respective regenerating protoplasts: Wall polysaccharides. *Planta*, *161*, 465–469.

Quist, E. E., & Roufogalis, B. D. (1975). Determination of the stoichiometry of the calcium pump in human erytrhocytes using lanthanum as a selective inhibitor. *FEBS Letters*, *50*, 135–139.

Rangel, A. F., Rao, I. M., & Horst, W. J. (2005). Proton toxicity interferes with the screening of common bean (Phaseolus vulgaris L.) genotypes for aluminium resistance in nutrient solution. *Journal of Plant Nutrition and Soil Science*, *168*, 607–616.

Rangel, A. F., Rao, I. M., & Horst, W. J. (2007). Spatial aluminium sensitivity of root apices of two common bean (*Phaseolus vulgaris* L.) genotypes with contrasting aluminium resistance. *Journal of Experimental Botany*, *58*, 3895–3904.

Ray, P. (1980). Cooperative action of β-glucan synthetase and UDP-xylose xylosyl transferase of golgi membranes in the synthesis of xyloglucan-like polysaccharide. *Biochimica et Biophysica Acta*, *629*, 431–444.

Rengel, Z., & Zhang, W. H. (2003). Role of dynamics of intracellular calcium in aluminium-toxicity syndrome. *New Phytologist*, *159*, 295–314.

Ryan, P. R., Liu, Q., Sperling, P., Dong, B., Franke, S., & Delhaize, E. (2007). A higher plant Δ8 sphigolipid desaturase with a preference for (Z)-isomer formation confers aluminum tolerance to yeast and plants. *Plant Physiology*, *144*, 1968–1977.

Salnikov, V. V., Grimson, M. J., Seagull, R. W., & Haigler, C. H. (2003). Localization of sucrose synthase and callose in freeze-substituted secondary wall-stage cotton fibers. *Protoplasma*, *221*, 175–184.

Samardakiewicz, S., Strawinski, P., & Wozny, A. (1996). The influence of lead on callose formation in roots of *Lemna minor L. Biologica Plantarum*, *38*, 463–467.

Schaeffer, H., & Walton, J. D. (1990). Aluminum ions induce oat protoplasts to produce an extracellular (1-3)-β-D-Glucan. *Plant Physiology*, *94*, 13–19.

Scheible, W. R., & Pauly, M. (2004). Glycosyltransferases and cell wall biosynthesis: Novel players and insights. *Current Opinion in Plant Biology*, *7*, 285–295.

Scherp, P., Grotha, R., & Kutschera, U. (2001). Deposition of cytokinesis-related callose in *Riella helicophylla* and *Arabidopsis thaliana*. Effects of photolytically altered nifedipine. *Plant Biology*, *3*, 311–318.

Schmohl, N., & Horst, W. J. (2000). Cell wall pectin content modulates aluminium sensitivity of *Zea mays* (L.) cells grown in suspension culture. *Plant, Cell and Environment*, *23*, 735–742.

Schmohl, N., Pilling, J., Fisahn, J., & Horst, W. J. (2000). Pectin methylesterase modulates aluminium sensitivity in *Zea mays* and *Solanum tuberosum*. *Physiologia Plantarum*, *109*, 419–427.

Schreiner, K. A., Hoddinott, J., & Taylor, G. J. (1994). Aluminum-induced deposition of (1,3)-β-glucans (callose) in *Triticum aestivum* L. *Plant and Soil*, *162*, 273–280.

Schrick, K., Fujioka, S., Takatsuto, S., Stierhof, Y. D., Stransky, H., Yoshida, S., & Jürgens, G. (2004). A link between sterol biosynthesis, the cell wall, and cellulose in Arabidopsis. *Plant Journal*, *38*, 227–243.

Shea, E. M., Gibeaut, D. M., & Carpita, N. C. (1989). Structural analysis of the cell walls regenerated by carrot protoplasts. *Planta*, *179*, 293–308.

Shih, C. Y., & Currier, H. B. (1969). Fine structure of phloem cells in relation to translocation in the cotton seedling. *American Journal of Botany*, *56*, 464–472.

Sivaguru, M., & Horst, W. J. (1998). The distal part of the transition zone is the most aluminum-sensitive apical root zone of maize. *Plant Physiology*, *116*, 155–163.

Sivaguru, M., Horst, W. J., Eticha, D., & Matsumoto, H. (2006). Aluminum inhibits apoplastic flow of high-molecular weight solutes in root apices of *Zea mays* L. *Journal of Plant Nutrition and Soil Science*, *169*, 679–690.

Sivaguru, M., Matsumoto, H., & Horst, W. J. (2000). Control of the response to Al stress. *Plant Microtubules*, 103–120.

Smith, M. M., & McCully, M. E. (1977). Mild temperature "stress" and callose synthesis. *Planta*, *136*, 65–70.

Staß, A. (1995). *Kurzfristige Wirkungen von aluminium auf suspensionszellen von glycine max*. Stuttgart: Verlag Ulrich Grauer. ISBN 3-86186-091-0.

Staß, A., & Horst, W. J. (1995). Effect of aluminium on membrane properties of soybean (*Glycine max*) cells in suspension culture. *Plant and Soil*, *171*, 113–118.

Staß, A., & Delhaize, E., et al. (2001). Effect of fatty acid composition of the plasma membrane on Al-tolerance in *Saccharomyces cerevisiae*. In Horst. (Ed.), *Plant Nutrition – Food Security and sustainability of agro-ecosystems through basic and applied research* (pp. 42–43). Dordrecht, The Netherlands: Kluwer Academic Publishers.

Stival da Silva, A. L., Sperling, P., Horst, W. J., Franke, S., Ott, C., Becker, D., Stab, A., Lörz, H., & Heinz, E. (2006). A possible role of sphingolipids in the aluminium resistance of yeast and maize. *Journal of Plant Physiology*, *163*, 26–38.

Subbaiah, C. C., & Sachs, M. M. (2001). Altered patterns of sucrose synthase phosphorylation and localization precede callose induction and root tip death in anoxic maize seedlings. *Plant Physiology*, *125*, 585–594.

Subbaiah, C. C., Bush, D. S., & Sachs, M. M. (1994a). Elevation of cytosolic calcium precedes anoxic gene expression in maize suspension cultured cells. *Plant Cell*, *6*, 1747–1762.

Subbaiah, C. C., Bush, D. S., & Sachs, M. M. (1998). Mitochondrial contribution to the anoxia $Ca^{2+}$ signal in maize suspension-cultured cells. *Plant Physiology*, *118*, 759–771.

Subbaiah, C. C., Kollipara, K., & Sachs, M. M. (2000). A $Ca^{2+}$-dependent cysteine protease is associated with anoxia induced root tip death in maize. *Journal of Experimental Botany*, *51*, 721–730.

Subbaiah, C. C., Zhang, J., & Sachs, M. M. (1994b). Involvement of intracellular calcium in anaerobic gene expression and survival of maize seedlings. *Plant Physiology*, *105*, 368–376.

Ueki, S., & Citovsky, V. (2005). Identification of an interactor of cadmium ion-induced glycine-rich protein involved in regulation of callose levels in plant vasculature. *PNAS*, *102*, 12089–12094.

Verma, D. P. S., & Hong, Z. (2001). Plant callose synthase complexes. *Plant Molecular Biology*, *47*, 693–701.

Vierstra, R., & Haug, A. (1978). The effect of $Al^{3+}$ on the physical properties of membrane lipids in *Thermoplasma acidophilum*. *Biochemical and Biophysical Research Communication*, *84*, 138–143.

Wagatsuma, T., Uemura, M., Maeshima, M., Ishikawa, S., Kawamura, T., Murayama, T., Shiono, Y., Khan, Md. S. H., & Tawaraya, K. (2005). Methylene blue stainability of root-tip protoplasts as an indicator of aluminum tolerance in a wide range of plant species, cultivars and lines. *Soil Science and Plant Nutrition*, *51*, 135–139.

Waldmann, T., Jeblick, W., & Kauss, H. (1988). Induced net $Ca^{2+}$ uptake and callose biosynthesis in suspension-cultured plant cells. *Planta*, *173*, 88–95.

Wang, Y., Stass, A., & Horst, W. J. (2004). Apoplastic binding of Aluminum is involved in Silicon-induced amelioration of Aluminum toxicity in maize. *Plant Physiology*, *136*, 3762–3770.

Webster, D. B., & Currier, H. B. (1965). Callose: Lateral movement of assimilates from phloem. *Science*, *150*, 1610–1611.

Webster, D. H., & Currier, H. B. (1968). Heat-induced callose and lateral movement of assimilates in the phloem. *Canadian Journal of Botany*, *46*, 1215–1220.

Wendehenne, D., Durner, J., & Klessig, D. F. (2004). Nitric oxide: A new player in plant signalling and defence responses. *Current Opinion in Plant Biology*, *7*, 449–455.

Wissemeier, A. H., & Horst, W. J. (1987). Callose deposition in leaves of cowpea (*Vigna unguiculata* (L.) as a sensitive response to high Mn supply. *Plant and Soil*, *102*, 283–286.

Wissemeier, A. H., & Horst, W. J. (1990). Manganese oxidation-capacity of homogenates of cowpea (*Vigna unguiculata* (L.) Walp.) leaves differing in manganese tolerance. *Journal of Plant Physiology*, *136*, 103–109.

Wissemeier, A. H., & Horst, W. J. (1995). Effect of Calcium supply on aluminium-induced callose formation, its distribution and persistence in roots of soybean (*Glycine max* (L.) Merr.). *Journal of Plant Physiology*, *145*, 470–476.

Wissemeier, A. H., Diening, A., Hergenröder, A., Horst, W. J., & Mix-Wagner, G. (1992). Callose formation as parameter for assessing genotypical plant tolerance of aluminium and manganese. *Plant and Soil*, *146*, 67–75.

Wissemeier, A. H., Hahn, G., & Marschner, H. (1998). Callose in roots of Norway spruce (*Picea abies* (L.) Karst.) is a sensitive parameter for aluminium supply at a forest site (Hoglwald). *Plant and Soil*, *199*, 53–57.

Wissemeier, A. H., Klotz, F., & Horst, W. J. (1987). Aluminium induced callose synthesis in roots of soybean (*Glycine max* L.). *Journal of Plant Physiology*, *129*, 487–492.

Yang, Z. M., Sivaguru, M., Horst, W. J., & Matsumoto, H. (2000). Aluminium tolerance is achieved by exudation of citric acid from roots of soybean (*Glycine max*). *Physiologia Plantarum*, *110*, 72–77.

Zabotin, A. I., Barysheva, T. S., Trofimova, O. I., Lozovaya, V. V., & Widholm, J. (2002). Regulation of callose metabolism in higher plant cells *in vitro*. *Russian Journal of Plant Physiology*, *49*, 792–798.

Zhang, G., Hoddinott, J., & Taylor, G. J. (1994). Characterization of 1,3-b-D-glucan (callose) synthesis in roots of *Triticum aestivum* in response to aluminum toxicity. *Journal of Plant Physiology*, *144*, 229–234.

Zhang, W. H., & Rengel, Z. (1999). Aluminium induces an increase in cytoplasmic calcium in intact wheat root apical cells. *Australian Journal of Plant Physiology*, *26*, 401–409.

CHAPTER 4.4.5

# Callose in Biotic Stress (Pathogenesis)

## Biology, biochemistry and molecular biology of callose in plant defence: callose deposition and turnover in plant–pathogen interactions

Christian A. Voigt[1,2] and Shauna C. Somerville[1,2]
[1]Department of Plant Biology, Carnegie Institution of Science, Stanford, CA, USA
[2]Energy Biosciences Institute, University of California, Berkeley, CA, USA

The deposition of the linear (1,3)-β-glucan callose is involved in several fundamental processes of plant development. However, particular attention has been focused on the formation of callose-containing papillae (cell wall thickenings of plants) in response to microbial attack. Papillae are not regarded as a defence response that can completely stop pathogens, rather they are thought to act as a physical barrier to slow pathogen invasion. However, callose-rich papillae are deposited at sites of penetration whether or not penetration is successful and the pathogen gains entry into the host cells, bringing into question the importance of papillae in plant defences. *Arabidopsis thaliana* is one of the best studied models for plant defence responses. Mutants of this plant lacking callose-rich papillae revealed a potential function for callose in modulating signaling via the salicylic acid pathway. Mechanisms regulating callose synthesis are largely unknown, and most data about regulation of callose synthesis are based on correlations rather than direct evidence. Thus, a precise function of callose in plant–microbe interactions has not been demonstrated unequivocally.

Higher plants are exposed to a wide range of potential pathogens. These pathogens are from diverse phyla, including bacteria, fungi, oomycetes, nematodes and plants, and also viruses. Unlike animals, plants lack a somatic adaptive immune system. Rather, the success of plants is dependent on sophisticated defence mechanisms developed during plant–pathogen co-evolution. A further decisive factor of plant defence is the speed with which plants combat invading pathogens. Early and rapid recognition of pathogens and mobilization of biochemical and structural defences are critical for plant protection. As a result, successful infection of plants by pathogens is the exception rather than the rule (Bailey, 1983; Deverall, 1977; Smith, 1978; Thordal-Christensen, 2003). The diversity of specialized plant defence responses reflects in part the high degree of host specialization by many pathogens. Relatively few pathogens have broad host ranges. The strategy of cosmopolitan pathogens is rapid maceration and death of plant tissues in advance of propagation (Sequeira, 1979). In contrast, other pathogens exhibit extreme host specificity, interacting with only selected genotypes within a host plant species; this type of race-cultivar-specific interaction has been explained by the 'gene-for-gene' concept developed by Flor (1947). Such resistance can be seen as an induced protection mechanism activated during pathogen attack, which supplements pre-formed physical and chemical barriers. This induced protection is controlled by plant disease resistance (*R*) genes (Flor, 1971). Upon recognising the presence of corresponding avirulence (avr) proteins or effectors of pathogen origin, R proteins initiate plant defences. Numerous R proteins have been identified and most contain a nucleotide binding site (NBS) and leucine-rich repeats (LRRs), and are localized intracellularly (Dangl and Jones, 2001; Martin et al., 2003). NBS-LRR proteins are broadly related in domain structure to animal CATERPILLER (Ting and Davis, 2005)/NOD (Fritz et al., 2006)/NLR (Wilmanski et al., 2008) proteins and STAND ATPases (Leipe et al., 2004). Plant NBS-LRR-mediated disease resistance is commonly effective against pathogens that can grow only on living host tissue, such as obligate biotrophic or hemi-biotrophic (with an initial biotrophic phase followed by a necrotrophic phase) pathogens. This branch of the plant innate immune system does not generally operate against necrotrophic pathogens, which destroy host tissue during colonization (Glazebrook, 2005). An extension to the 'gene-for-gene' concept, the 'guard hypothesis', implies that R proteins indirectly recognise avirulence proteins or effectors secreted by pathogens. This is achieved by monitoring the integrity of host cellular targets of pathogen effector action (Dangl and Jones, 2001; Kim et al., 2005b; van der Biezen and Jones, 1998).

The second branch of the active immune system uses transmembrane pattern recognition receptors (PRR) to recognise and respond to challengers (Jones and Takemoto, 2004; Nimchuk et al., 2003; Nürnberger et al., 2004). PRR recognise microbe- or pathogen-associated molecular patterns (MAMPs) (Ausubel, 2005). These are characterized by their broad occurrence in all members of a class of pathogens and are often required for pathogen viability. Therefore, they represent ideal targets for plant 'non-self' surveillance systems (Kim et al., 2005b). A variety of MAMPs (e.g. chitin, ergosterol and a transglutaminase from fungi; lipopolysaccharide, flagellin and the elongation factor EF-Tu from bacteria) have been described to date (Nürnberger et al., 2004). Several genes encoding plant PRR have also been identified (Chinchilla et al., 2007; Gomez-Gomez and Boller, 2000; Kaku et al., 2006; Miya et al., 2007; Ron and Avni, 2004; Zipfel et al., 2006).

A current view of the plant immune system that encompasses both NBS-LRR R proteins and PRR is presented in a four-phased Zig-Zag model (Jones and Dangl, 2006). (1) Plants detect MAMP(s) via transmembrane PRR to initiate MAMP-triggered immunity. (2) In successful infections, pathogens release effectors that interfere with MAMP-triggered immunity, resulting in effector-triggered susceptibility. (3) R proteins recognise specific effectors (e.g. avr proteins) and activate effector-triggered immunity that can include hypersensitive cell death. (4) Mechanisms operating on components active in the first three phases select pathogens that have lost specific effectors and plants with new R alleles. The adapted pathogens can suppress or may fail to activate effector-triggered plant immunity, whereas the selected plant isolates recognise different effectors, re-establishing effector-triggered immunity.

In incompatible interactions, plants stop pathogens or limit their spread and multiplication by one or more of the following mechanisms: (1) hypersensitive response that leads to a rapid collapse of host cells in the immediate vicinity of the pathogen; (2) production of phytoalexins, which are low molecular weight stasis-inducing or toxic chemicals of plant origin; (3) biosynthesis of degradative enzymes (e.g. (1,3)-β-glucan and chitin hydrolases to decompose fungal cell walls); and (4) plant cell wall modifications, notably the deposition of callose-containing papillae (Ausubel et al., 1995; Benhamou, 1995; Hammond-Kosack and Jones, 1996; Hardham et al., 2007; Mittler and Lam, 1996). Papillae are cell wall thickenings of plants that are formed at sites of microbial attack (Fig. 1) (reviewed by Stone and Clarke, 1992). An oxidative microburst accompanies papillae formation at the plasma membrane (Thordal-Christensen et al., 1997). In addition to these defences, polarized vesicle-associated secretory processes are likely to be involved in directing the delivery of small antimicrobial molecules, proteins and cell wall components to pathogen contact sites at plant cell surfaces (Hardham et al., 2007; Schulze-Lefert, 2004).

In addition to R protein-triggered, and race- and isolate-specific resistance, plants exhibit a form of non-host resistance that is operationally defined as immunity displayed by an entire plant species against all genetic variants of a pathogen species. An example would be the resistance exhibited by soybean plants against a specialized maize pathogen (Heath, 2000). Non-host resistance is durable and common in nature but still poorly understood. However, it appears that this type of resistance includes both passive and active responses to pathogen attack, including the deposition of callose in papillae and encasing attacked host epidermal cells (Thordal-Christensen, 2003).

Even though callose deposition is a near ubiquitous plant response to pathogen attack, its importance is still controversial. At first glance, this is unexpected, since papillae were

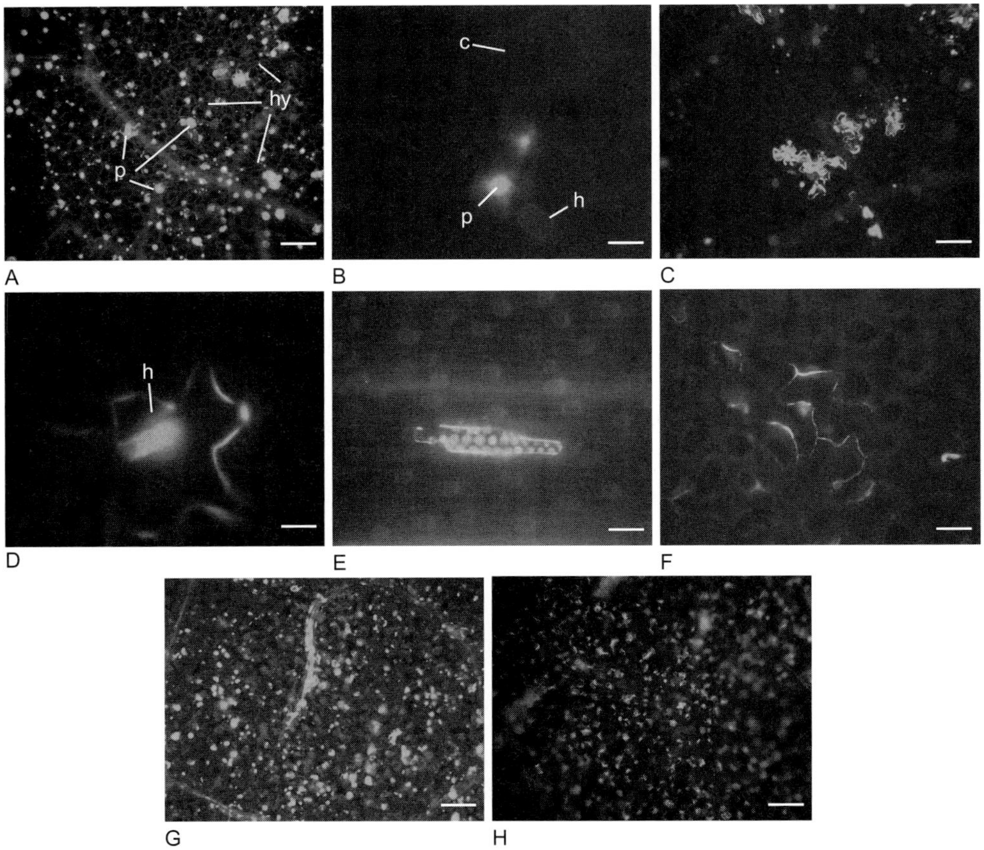


Fig. 1: Callose deposition after microbial attack. Fluorescence microscopy of treated plant leaf tissue was used to visualize callose deposits after staining with Aniline Blue (Ramming et al., 1973). (A) Callose-rich papillae (p) in Col-0 wild-type Arabidopsis leaves are deposited in response to attack with the virulent Arabidopsis powdery mildew, *Golovinomyces cichoracearum*. Picture was taken 3 days post inoculation. Fungal hyphae (hy) stained dark blue. Scale bar, 50 μm. (B) Magnification of a penetration site of a single conidium (c) illustrating callose deposition in papillae (p). Haustorium (h) formation has already started 1 day post inoculation of Col-0 wild-type Arabidopsis leaves with *G. cichoracearum*. Scale bar, 1 μm. (C) Callose deposits outlining attacked cells and papillae in Col-0 wild-type Arabidopsis leaves inoculated with the inappropriate barley powdery mildew, *Blumeria graminis* f. sp. *hordei*. Image was collected 3 days post inoculation. Scale bar, 50 μm. (D) Magnification of a single epidermal leaf cell ringed with callose 3 days post inoculation with *B. graminis* f. sp. *hordei*. The fungal haustorium (h) is also encased with callose. Scale bar, 1 μm. (E) Callose deposition in a leaf epidermal cell of the grass *Brachypodium distachyon* as a defence response in an incompatible interaction. Picture was taken 1 day post inoculation with *G. cichoracearum*. Scale bar, 2.5 μm. (F) Callose deposition in leaf epidermal cells of Arabidopsis

discovered in plant cell walls at sites of fungal penetration by deBary (1863) over 140 years ago. Mangin (1895) reported that callose, a (1,3)-β-glucan with some (1,6)-branches, commonly occurred in papillae (Aspinall and Kessler, 1957). Since then, callose has been identified as the most common constituent in papillae, which may also include proteins (e.g. peroxidases, antimicrobial thionins), phenolics and other constituents (Aist and Williams, 1971; Mercer et al., 1974; Mims et al., 2000). Based on observations of a wide variety of plant–pathogen interactions, papillae are not regarded as a defence mechanism that can completely stop pathogens; rather, they are thought to act as a physical barrier to slow pathogen invasion (reviewed by Stone and Clarke, 1992). The host plant can thus gain time to initiate defence reactions that require gene activation and expression, such as the hypersensitive response, phytoalexin production, and the synthesis and export of pathogenesis-related proteins (Brown et al., 1998; Lamb and Dixon, 1997). However, a general role of callose in preventing or slowing pathogen ingress is not supported by the observation that callose-rich papillae are deposited at sites where the pathogen is successful (Fig. 1A, B) (Aist, 1976). Thus, even though the involvement of callose in plant defence has been investigated for well over a century, its proposed role as a barrier to pathogen entry has not been demonstrated unequivocally. Apart from microbial stress, the local deposition of callose is also induced by abiotic stress and wounding (Jacobs et al., 2003; Mauch-Mani and Mauch, 2005; Ryals et al., 1996; Wheeler, 1974). In this article, we will focus on callose deposition by plants in response to pathogen attack.

Due to the central importance of callose deposition in several key plant processes, including pathogenesis, many attempts have been made to purify and characterize callose synthases and their corresponding genes from plants (Bulone et al., 1995; Dhugga and Ray, 1994; Him et al., 2001; Jacobs et al., 2003; Kudlicka and Brown 1997; Kudlicka et al., 1995; McCormack et al., 1997; Meikle et al., 1991; Schlupmann et al., 1993; Turner et al., 1998).

**Fig. 1: (Continued) wild type in response to chitin treatment. Chitin is a fungal 'microbe-associated molecular pattern' (MAMP). Picture was taken 1 day post spraying with a chitin solution. Scale bar, 2.5 μm. (G) Callose deposition in an Arabidopsis leaf 24 h after syringe infiltration of an attenuated bacterial *Pseudomonas syringae* strain into the apoplast. Callose deposits occur in mesophyll cells. Scale bar, 50 μm. (H) Callose deposition in Arabidopsis mesophyll cells 24 h after syringe infiltration of a solution containing a 22-amino-acid peptide (flg22) from a conserved bacterial flagellin. Flagellin is an important bacterial MAMP. Scale bar, 50 μm. (G and H courtesy of Bill Underwood, Carnegie Institution of Science, Stanford, CA, USA.) The colour specifications refer to colours in panels.**

The molecular mass and subunit composition were determined in various partially purified enzyme preparations (Delmer et al., 1991; Frost et al., 1990; Li and Brown, 1993; Read and Delmer, 1987). Newer results have shown that the callose synthase activity is associated with ~200-kDa polypeptides (Hong et al., 2001; Li et al., 2003; Turner et al., 1998). Callose synthase is likely to be a multi-subunit, membrane-associated enzyme complex (Verma and Hong, 2001). Detergents that are normally used for extraction are likely to dissociate the enzyme complex, destroying activity. Thus, the purification of active callose synthase to homogeneity has not been achieved to date. Despite these difficulties, a family of glucan synthase-like (*GSL*) genes in higher plants has been identified as encoding essential components of callose synthases (Cui et al., 2001; Doblin et al., 2001; Hong et al., 2001; Kauss et al., 1983; Østergaard et al., 2002). The hypothesized function of *GSL* genes is supported by their sequence homology with the *FKS* genes, which are required for callose synthesis in yeast (Cabib et al., 2001; Dijkgraaf et al., 2002; Douglas et al., 1994). The *Arabidopsis thaliana* (Arabidopsis) *GSL5* gene was shown to partially complement a yeast *fks1* mutant (Østergaard et al., 2002). Additionally, the predicted proteins encoded by the *GSL* genes correlate in size with the molecular mass of an ~200-kDa catalytic subunit of putative callose synthases. Finally, Li et al. (2003) showed that the amino acid sequence predicted from a *GSL* gene in barley (*HvGSL1*) resembles the amino acid sequence of an active callose synthase fraction. Recently, similar results were shown for a callose synthase encoded by the *NaGSL1* gene in *Nicotiana alata* (Brownfield et al., 2007).

In most plants, the *GSL* genes are members of a moderately sized gene family. Most information about distinct roles of various *GSL* gene family members is known from Arabidopsis, which has 12 *GSL* genes (Richmond and Somerville, 2000; Verma and Hong, 2001). Seven *GSL* genes have been characterized. *GSL6* (=*CalS1*) encodes a callose synthase subunit that appears to be active at the cell plate of dividing cells (Hong et al., 2001). Additionally, transcript levels for this gene increased slightly after inoculation of Arabidopsis leaves with barley powdery mildew, *Blumeria graminis* f sp. *hordei*, spores (Jacobs et al., 2003). Under these conditions, a modest increase in expression was also observed for *GSL5* and *GSL11*. Moreover, *GSL5* as well as *GSL1* were shown to play an essential and redundant role in plant and pollen development and fertility (Enns et al., 2005), whereas *GSL8* and *GSL10* are independently required for male gametophyte development and plant growth (Töller et al., 2008). A detailed expression profile is available for *GSL2* (=*CalS5*) revealing its importance in exine formation in the pollen wall (Dong et al., 2005). In the context of biotic stress, the GSL5 (=PMR4) callose synthase is required for wound- and pathogen-induced callose deposition (Jacobs et al., 2003; Kim et al., 2005b; Nishimura et al., 2003). Overlapping expression patterns have been observed for the *GSL* gene family members in Arabidopsis. For example, *GSL5*, *GSL6* and *GSL11* transcript levels are elevated following fungal infection of Arabidopsis (Jacobs et al., 2003), but only *GSL5* is responsible for the deposition of callose in papillae and at wound sites. Roles for the remaining Arabidopsis *GSL* genes have not been assigned as yet, but some may contribute to callose deposition at plasmodesmata or in the phloem.

Even though callose is commonly used as marker of plant defence responses in the study of various bacterial virulence mutants and of diverse chemical compounds, mechanisms that directly activate callose synthesis are largely unknown. Callose synthase activity might be partly regulated directly by *GSL* gene expression as modest increases in *GSL* transcript levels are observed following pathogen attack (Donofrio and Delaney, 2001; Jacobs et al., 2003; Østergaard et al., 2002). However, the regulation of callose synthase enzyme activity as proposed by Jacobs et al. (2003) is likely. Thus, post-transcriptional and post-translational regulation seems to be required for proper callose synthesis and deposition. The level of free $Ca^{2+}$ is one mechanism thought to regulate callose synthase activity (Kauss et al., 1983; Köhle et al., 1985). Membrane receptors, activated by pathogen attack or wounding, are thought to mediate local increases in $Ca^{2+}$ concentration (Blume et al., 2000; Grant et al., 2000; Leon et al., 2001; Leon et al., 1998), presumably by activating membrane-bound callose synthases (Amor et al., 1995; Delmer and Amor, 1995; Kauss, 1991). In addition to $Ca^{2+}$ concentration, studies indicate that phosphorylation may play a role in regulating of callose synthases. In yeast, the activities of FKS1 and FKS2 are regulated by Rho1 GTPase, altering the phosphorylation status of the FKS proteins (Calonge et al., 2003; Qadota et al., 1996). A similar mechanism was suggested by Hong et al. (2001): the regulation of an Arabidopsis callose synthase through interaction of the Rho-like GTPase Rop1 with UDP-glucose transferase, a putative subunit of callose synthase complexes. In contrast to $Ca^{2+}$, the regulation of callose synthases via GTPase-mediated phosphorylation would facilitate a specific activation of distinct callose synthase complexes. An additional putative mechanism that could regulate callose synthase activity is the availability of substrate. However, no experimental support for this form of regulation exists. Because the deposition of callose in papillae is highly localized at penetration sites in attacked epidermal cells (Koh et al., 2005; Zimmerli et al., 2004), the callose synthase protein is likely to be either similarly highly localized within the plasma membrane at penetration sites or the enzyme is likely to be selectively activated at these sites. The post-translational modifications of callose synthase (e.g., phosphorylation) may affect either the selective localization or the activation of stress-induced callose synthases at penetration sites.

## Callose in Plant–Fungi Interactions

Many fungal pathogens gain access to plant tissues by penetrating cell walls directly. Hence, the cell wall can be considered the first line of defence against such pathogens. Not surprisingly, plants have evolved means to perceive and respond defensively to physical or chemical events associated with penetration. For example, local mechanical pressure imitating a fungal penetration attempt can elicit the translocation of the nucleus and cytoplasm to the site of stimulation, the intracellular generation of reactive oxygen species, and the transcription of defence-related genes, but not papilla deposition (Gus-Mayer et al., 1998). Mechanical penetration of a cell with a needle also induces the deposition of callose around its tip, reminiscent of papilla formation elicited by attempted fungal penetration (Russo and Bushnell, 1989). The formation of a callose-containing papilla during fungal infection often correlates with failure of the pathogen to cause disease (Aist, 1976; Aist, 1983; Skou et al., 1984; Stolzenburg et al., 1984). Arabidopsis mutants that are unable to respond to or accumulate salicylic acid, an important signal molecule in plant defence, showed a higher susceptibility to the adapted oomycete *Peronospora parasitica* and deposited less callose around haustoria (feeding structures of the pathogen) than wild-type plants (Donofrio and Delaney, 2001). Supporting linkage between callose accumulation and salicylic acid, Østergaard et al. (2002) found high *GLS5* transcript levels, callose synthase activity, callose deposition and salicylic acid levels in a mitogen-activated protein kinase 4 (*mpk4*) mutant; while *mpk4 NahG* plants, unable to accumulate salicylic acid, expressed *GLS5* and accumulated callose at low to non-detectable levels, similar to wild-type plants. Incompatible interactions between Arabidopsis and various non-adapted *Colletotrichum* species elicit the accumulation of callose in plant cell walls beneath fungal appressoria, while compatible interactions with the adapted pathogen, *C. higginsianum*, induced callose deposition at much lower frequency (Shimada et al., 2006). Thus, the frequency of papillary callose deposited beneath fungal appressoria was correlated with failed fungal entry attempts (Fig. 2). Chemical treatment of barley (Bayles et al., 1990) and reed canarygrass (Vance and Sherwood, 1976) resulted in reduced papillae formation and increased susceptibility to fungal penetration, but putative side effects of these chemicals on additional plant defence responses were not considered in these experiments. The barley (*Blumeria graminis* f. sp. *hordei*), but not the Arabidopsis (*Golovinomyces cichoracearum*) powdery mildew, elicits widespread callose deposition in Arabidopsis mesophyll cells subtending infected epidermal cells (Zimmerli et al., 2004). This extensive callose deposition in response to the inappropriate barley pathogen correlated with successful prevention of invasion by this fungus. However, callose deposition is probably not the proximal cause of resistance in Arabidopsis to the barley powdery mildew fungus (Assaad et al., 2004;

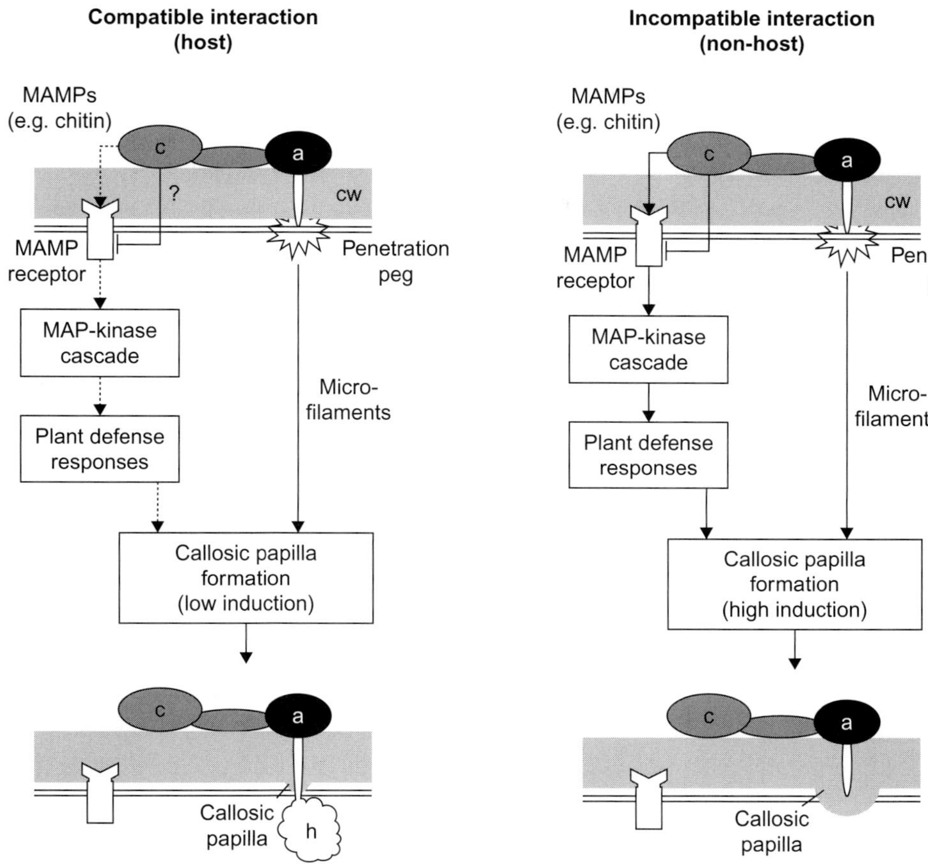


**Fig. 2: Model of pre-invasion resistance and callose-containing papilla formation in plant–fungus interactions. Penetration pegs are observed in both compatible and incompatible plant–fungus interactions. Both types of interaction can trigger callose-rich papilla formation possibly by perception of fungal MAMP by transmembrane PRR. Subsequent signal transduction is thought to involve mitogen-activated protein (MAP) kinase cascades and the activation of callose synthesis at sites of fungal penetration. In compatible interactions, development of callose-rich papillae may be reduced, possibly due to suppression of MAMP recognition by an unknown factor produced by an adapted pathogen. In compatible and incompatible interactions, penetration peg formation from an appressorium triggers reorganization of microfilaments (Hardham et al., 2007; Kobayashi et al., 1992; Shimada et al., 2006; Takemoto et al., 2003) and callose-rich papilla formation. a = appressorium; c = conidium; cw = cell well; h = haustorium. (adapted from Shimada et al., 2006.)**

Collins et al., 2003; Lipka et al., 2005; Stein et al. 2006). As mentioned in the introductory paragraphs, a general role of callose-containing papillae in preventing or slowing fungal penetration can be questioned, since papillae are deposited at sites where pathogen attack is successful or unsuccessful (Aist, 1976). In this regard, the penetration success of the adapted powdery mildew *Golovinomyces cichoracearum* appears unaltered in *pmr4* mutants (C. A. Voigt and S. C. Somerville, unpublished results). Also, penetration success by the adapted powdery mildew *G. orontii* is similar in the partially resistant *mlo2* and *mlo2 pmr4* double mutants (C. Consonni and R. Panstruga, unpublished result). Thus, the absence of wound and pathogen-inducible callose deposition does not lead to enhanced penetration success, suggesting that callose does not act as a significant barrier to penetration by adapted powdery mildew species.

In addition to a proposed role as a barrier, callose-rich papillae were proposed to be part of a wound repair mechanism to preserve the integrity of the cell (Wheeler, 1974) and facilitate healing (Aist, 1976; Israel et al., 1980). It has also been suggested that papillae form an effective permeability barrier between fungal hyphae and plant host cells, impairing movement of toxic fungal materials to plant cell membrane (Aist, 1976). Studies on Arabidopsis *pmr4* mutants lacking the stress-induced callose synthase GSL5 suggest an unexpected role for callose or the GSL5 protein in dampening the salicylic acid signal transduction pathway (Nishimura et al. 2003). Double mutant and microarray analyses showed that the salicylic acid pathway was hyperactivated in the *pmr4* mutant relative to wild type (Nishimura et al., 2003). This dampening of salicylic acid signalling can potentially be explained by assuming that callose acts as a barrier shielding access of fungal MAMP to plant plasma membrane PRR. The lack of callose deposition in the *pmr4* mutant might then unmask plant PRR that recognise fungal wall polysaccharides or other elicitors. In support of this idea, certain branched (1,3;1,6)-β-oligoglucosides can be released from fungal wall polysaccharides (Bartnicki-Garcia, 1968) and a protein that binds glucan elicitors (GEBP = glucan elicitor binding protein) has been purified from the membrane fraction of soybean root cells. Its expression both in tobacco suspension cultured cells and in *Escherichia coli* facilitated binding to glucan elicitors, suggesting that this glucan elicitor receptor may signal plant defence responses (Umemoto et al., 1997). In addition to mixed linkage β-glucans, fungal cell walls contain chitin (*N*-acetylchitooligosaccharides), which is an elicitor of plant defence responses, including callose deposition, in both monocots and dicots (Fig. 1F) (Kauss, 1985; Köhle et al., 1985; Shibuya and Minami, 2001). Receptors for this MAMP have been identified in rice and Arabidopsis. A high-affinity binding protein for chitin, CEBiP (chitin oligosaccharide elicitor binding protein), was isolated from the plasma membrane of suspension-cultured rice cells (Kaku et al., 2006). In Arabidopsis, the receptor-like kinase CERK1 was identified to be essential for chitin elicitor signalling (Miya et al., 2007). Also, disease resistance against the incompatible fungus, *Alternaria brassicicola,* was partly impaired in *cerk1* mutants. Unlike chitin, linear (1,3)-β-oligoglucosides (i.e. callose fragments), which would also be released from fungal cell walls by plant (1,3)-β-glucanases, do not elicit plant defences (Côté and Hahn, 1994).

Plants synthesize and secrete into their cell walls and vacuoles chitinases and glucanases that degrade fungal cell wall components. These degradative enzymes are thought to weaken the fungal cell wall and slow growth, in addition to potentially generating biologically active oligosaccharides that elicit further plant defence responses. Plant (1,3)-β-glucanases form a highly diverse family of hydrolytic enzymes. Differences can be found in gene expression (Cabello et al., 1994; Garcia-Garcia et al., 1994; Romero et al., 1998; Xu et al., 1992), in their chemical structure, and in their subcellular localization (Simmons, 1994). Pathogen attack or environmental stress can induce (1,3)-β-glucanase transcript accumulation (Kaufmann et al., 1987; Simmons et al., 1992). Data from tobacco also indicate an involvement of pathogenesis-related proteins in regulation of (1,3)-β-glucanase activity (Riviere et al., 2008). A biological function in plant defence has been deduced based on their coordinated induction, together with chitinases (Cordero et al., 1994), and their inhibition of *in vitro* growth and sporulation of pathogenic fungi (Jach et al., 1995; Mauch et al., 1988; Vögeli et al., 1988). An example for this is the increase of (1,3)-β-glucanase activity in cucumber leaves, which was associated with induced resistance against the fungal pathogen *Colletotrichum lagenarium* (Ji and Kuc, 1995). A correlation between (1,3)-β-glucanase level and resistance was also shown for pearl millet during interaction with the downy mildew pathogen *Sclerospora graminicola* (Kini et al., 2000). To counteract this plant defence and limit the release of elicitor active oligosaccharides, fungi may have evolved inhibitors to block plant degradative enzymes (Rose et al., 2002). In *Pisum sativum*, (1,3)-β-glucanase expression is induced during both compatible and incompatible fungus–plant interactions. However, only in incompatible interactions do (1,3)-β-glucanase mRNA levels remain high; in compatible interactions, mRNA levels rapidly decrease following an initial burst of expression (Chang et al., 1992). The basis for the attenuated expression in compatible interactions could potentially depend on fungal effectors that block (1,3)-β-glucanase gene expression, analogous to pathogen-triggered susceptibility proposed in the Zig-Zag model.

Interestingly, a different type of glucanase, the endo-(1,4)-β-glucanases, seem to be involved in pathogen response, too. In tomato, silencing of two endo-(1,4)-β-glucanases enhances callose deposition via an unknown mechanism and reduces susceptibility to the fungal necrotroph *Botrytis cinerea*, but increases susceptibility to the bacterial pathogen *Pseudomonas syringae*. These data suggest that enzymes of the general cell wall metabolism can be involved in plant–microbe interactions (Flors et al., 2007).

A different mechanism of suppression of plant defence by pathogenic fungi depends on lipases and the release of free fatty acids (FFAs) at infection sites. The inhibition of callose synthesis by unsaturated, but not saturated, FFAs was first shown by Kauss and Jeblick (1986) using isolated membranes of *Glycine max*. Similar to their results, we recently demonstrated the same effect on callose synthesis by isolated membrane fractions from wheat and Arabidopsis (C.A. Voigt, W. Schäfer and S.C. Somerville, unpublished results). Kauss and colleagues suggested that callose synthase depends on specific boundary phospholipids for optimal activity; FFAs do not meet this need (Kauss, 1987; Kauss and Jeblick, 1986). *Fusarium graminearum*, a necrotrophic fungal pathogen of cereals, appears to use this mechanism to inhibit callose deposition at infection sites and promote disease development. A lipase-deficient *F. graminearum* mutant exhibited strongly reduced virulence on wheat, which was associated with the formation of a callose barrier that appears to limit pathogen spread (Voigt et al., 2005). Inhibition of wheat callose synthases can potentially be explained by the release of FFAs at infection sites by the fungal lipase (Sandermann, 1978). This inhibition of callose synthesis facilitates pathogen spread throughout the wheat head in compatible interactions.

The Arabidopsis *PEN1* gene and its barley homologue, *ROR2*, encode syntaxins that are required for plant defences that limit fungal penetration across plant cell walls (Collins et al., 2003). Syntaxins are members of the SNARE (synaptosome-associated protein receptor) superfamily of proteins that mediate membrane fusion events. PEN1 and ROR2 proteins are recruited to plasma membrane microdomains at penetration sites immediately beneath powdery mildew appressoria (Assaad et al., 2004; Bhat et al., 2005). There, they are thought to function in the tethering of exocytic vesicles and the discharge of vesicle cargo into the apoplast (Assaad et al., 2004; Bhat et al., 2005). In the *pen1-1* mutant, papilla formation is delayed (Assaad et al., 2004; Bhat et al., 2005), suggesting involvement of PEN1/ROR2 in timely and localized deposition of host cell wall materials in papillae. Specific seven-transmembrane MLO (mildew resistance locus O) family members negatively regulate PEN1/ROR2-mediated secretion at sites of attempted pathogen invasion (Assaad et al., 2004; Collins et al., 2003; Consonni et al., 2006). Recessive *mlo* mutants in both Arabidopsis (*mlo2 mlo6 mlo12*) and barley (*mlo*) are highly resistant to their respective co-evolved powdery mildew pathogens and this resistance is associated with more rapid deposition of callose-rich papillae (Consonni et al., 2006; Miklis et al., 2007; Wolter et al., 1993). The precise roles of the PEN1/ROR2 syntaxins and the MLO protein in modulating callose deposition at sites of fungal attack are unknown.

Additional types of proteins have been identified that impact callose regulation during fungal and oomycete infection. These include putative Toll interleukin-1 receptor nucleotide binding (TIR-NB) proteins in Arabidopsis (Staal et al., 2008), the pathogen-derived Nep1 (necrosis and ethylene-inducing peptide1)-like proteins (Qutob et al., 2006), downy mildew effector proteins ATR (*Arabidopsis thaliana* recognised) 1 and ATR13 (Sohn et al., 2007), and monogenic genes *Ol-1*, *ol-2* and *Ol-4* from the tomato powdery mildew *Oidium neolycopersici* encoding proteins of unknown function (Li et al., 2007). Even though it is not known how these proteins trigger callose synthesis, they exemplify the complexity of callose regulation.

In addition to its frequently observed occurrence in papillae, callose encases complete host cells and pathogen haustoria, especially in incompatible interactions (Fig. 1C–E) (Jacobs et al., 2003; Kuzuya et al., 2006; Zimmerli et al., 2004). Haustoria are feeding structures formed by some pathogens, notably obligate biotrophic fungi and oomycetes. They develop in the extracellular space of plant cells and are encased in specialized plant membranes that appear continuous with host plasma membranes. The expansion of the haustoria increases surface contact with the invaded cell, presumably to maximize nutrient transfer from the host to the pathogen (Adam et al., 1999; Mendgen and Hahn, 2002; Szabo and Bushnell, 2001; Voegele et al., 2001). Release of fungal effectors into the invaded plant cell from haustoria might further increase fungal virulence (Ellis et al., 2006; Jacobs et al., 2003; Szabo and Bushnell, 2001). The role of the callose surrounding pathogen haustoria is not known, but it is presumed to restrict further haustorial growth and to limit nutrient and signal molecule exchange with the cytoplasm of attacked host cells.

## Callose in Plant–Bacteria Interactions

In contrast to mammalian intracellular pathogenic bacteria, most plant pathogenic bacteria are extracellular pathogens, colonizing intercellular spaces in host tissues. Many pathogenic bacteria have evolved an efficient system to deliver effector proteins into host cells called the type III secretion system (TTSS), which is encoded by *HRP* (hypersensitive reaction and pathogenicity) genes (Alfano and Collmer, 1997; Buttner and Bonas, 2002; Cornelis and Van Gijsegem, 2000; He, 1998; Lindgren, 1997). As outlined in the Zig-Zag model above, these bacterial effectors can modulate plant defences, stimulating R-mediated resistance in some cases and suppressing PRR-mediated defences in other cases. Effectors are known that either suppress the generalized programmed cell death machinery (Abramovitch et al., 2003; Bretz et al., 2003; Espinosa et al., 2003; Jamir et al., 2004) or suppress specific defence signalling pathways (Kim et al., 2005a; Mackey et al., 2003; Reuber and Ausubel, 1996; Ritter and Dangl, 1996; Tsiamis et al., 2000). Other effectors have an impact on basal plant defences, including the suppression of plant cell wall modifications like the formation of callose-rich papillae (Aist and Williams, 1971; DebRoy et al., 2004; Hauck et al., 2003; Jakobek et al., 1993; Keshavarzi et al., 2004; Kim et al., 2005b; Mercer et al., 1974; Mims et al., 2000; Sargent et al., 1973; Sherwood and Vance, 1976). For example, only *Xanthomonas campestris* pv. *vesicatoria* strains with an operational TTSS prevent or fail to elicit papillae formation in pepper (Brown et al., 1995). In contrast, infections with *hrp* mutants of *X. campestris* pv. *vesicatoria* and *P. syringae* pv. *phaseolicola*, as well as a saprophytic bacterium, elicit abundant papillae formation (Fig. 1G) (Bestwick et al., 1995; Brown et al., 1995; Brown et al., 1998). These results suggest that pathogenic bacteria secret one or multiple suppressors via the TTSS that suppress callose formation in some manner. Two *P. syringae* effectors, AvrRpt2 and AvrRpm1, inhibit MAMP-induced signalling and compromise the plant basal defence system, including callose deposition (Kim et al., 2005a). Hauck et al. (2003) showed that one function of the effector AvrPto was to suppress the callose deposition normally elicited by *hrp* mutants. A similar effect was also shown for AvrPtoB (de Torres et al., 2006). AvrPto is one of at least 33, and perhaps as many as 50, TTSS effectors that are secreted and/or translocated by the bacterial pathogen strain *P. syringae* pv. tomato DC3000 alone, leaving open the possibility that additional bacterial effectors may be identified that act directly or indirectly on the activation, regulation, synthesis or localized deposition of callose at infection sites (Alfano and Collmer, 2004; Boch et al., 2002; Chang et al., 2005; Fouts et al., 2002; Greenberg and Vinatzer, 2003; Guttman et al., 2002; Lindeberg et al., 2006; Petnicki-Ocwieja et al., 2002; Salanoubat et al., 2002; Zwiesler-Vollick et al., 2002).

Plants perceive pathogen MAMP elicitors, of which flagellin is the best characterized to date, and activate defence responses (Gomez-Gomez and Boller, 2000; Gomez-Gomez and Boller, 2002). Flagellin and a synthetic 22-amino-acid peptide (flg22) from a conserved flagellin domain are sufficient to induce callose deposition along with other defence responses (Fig. 1H) (Zipfel et al., 2004). This callose deposition is dependent on the callose synthase gene *GSL5*, responsible for wound- and fungal-induced callose deposition (Jacobs et al., 2003; Nishimura et al., 2003). These flg22-elicited defences, including callose deposition, are dependent on the flg22 receptor, flagellin sensitive 2 (FLS2), which is an LRR transmembrane receptor kinase (Chinchilla et al., 2006; Suarez-Rodriguez et al., 2007).

In addition to a range of extracellular enzymes, such as proteases, pectinases and endoglucanases, the extracellular polysaccharide xanthan is essential for pathogenesis of the pathogenic bacterium *Xanthomonas campestris* pv. *campestris*, the causal agent of black rot disease of cruciferous plants (Dow and Daniels, 2000; Newman et al., 1994; Vojnov et al., 2001). In contrast to *X. campestris* pv. *campestris* wild-type strains, mutant strains that either do not produce xanthan (Vojnov et al., 1998) or produce a truncated xanthan (Heyraud et al., 1998; Vojnov et al., 2002) fail both to cause disease and to induce callose deposition in both *Nicotiana benthamiana* and Arabidopsis (Yun et al., 2006). Virulence of the mutant strains was restored if plants were treated with exogenous xanthan. Yun et al. (2006) speculate that xanthan fragments might bind extracellular $Ca^{2+}$, depleting local available stores and thereby preventing callose synthase activation (Tempio and Zatz, 1981). Numerous pathogenic bacteria, including *P. syringae*, *Ralstonia solanacerarum*, *Erwinia amilovora*, and *E. stewartii*, have the ability to produce negatively charged exopolysaccharides in plants, which have been correlated with virulence and may act in a similar manner by binding extracellular $Ca^{2+}$ (Aslam et al., 2008; Dolph et al., 1988; Geier and Geider, 1993; Kao et al., 1992; Saile et al., 1997; Yu et al., 1999).

## Callose in Plant–virus Interactions

After an initial infection, some plant viruses replicate and spread from cell to cell through plasmodesmata, plasma-membrane-lined channels that cross plant cell walls, providing a connection between adjacent cells to facilitate cell-to-cell transport of molecules and the organization of functional symplastic domains (Fig. 3) (Levy et al., 2007b). Viral particles can be observed in plasmodesmata of virus-infected plant tissue by electron microscopy (Allison and Shalla, 1974; Esau et al., 1967; Roberts and Lucas, 1990; Weintraub et al., 1976). As the infection progresses, virus particles reach and enter the vascular system, and then spread systemically with the flow of photoassimilates. In uninoculated organs distal from the site of inoculation, virions can exit the host vasculature to infect non-vascular tissues (Leisner and Howell, 1993; Oparka and Cruz, 2000; Rhee et al., 2000).

Plasmodesmata can transiently change from being closed to open to dilated. A putative mechanism that could produce these focused changes in the plasmodesmal channels is the alternating deposition and hydrolysis of callose in the cell wall surrounding plasmodesmata (Heinlein and Epel, 2004; Turner et al., 1994). Reversible callose accumulation at plasmodesmata is observed during developmental processes (Rinne and van der Schoot, 1998; Ruan et al., 2004) and also in response to stress (Radford et al., 1998; Sivaguru et al., 2000). The amount

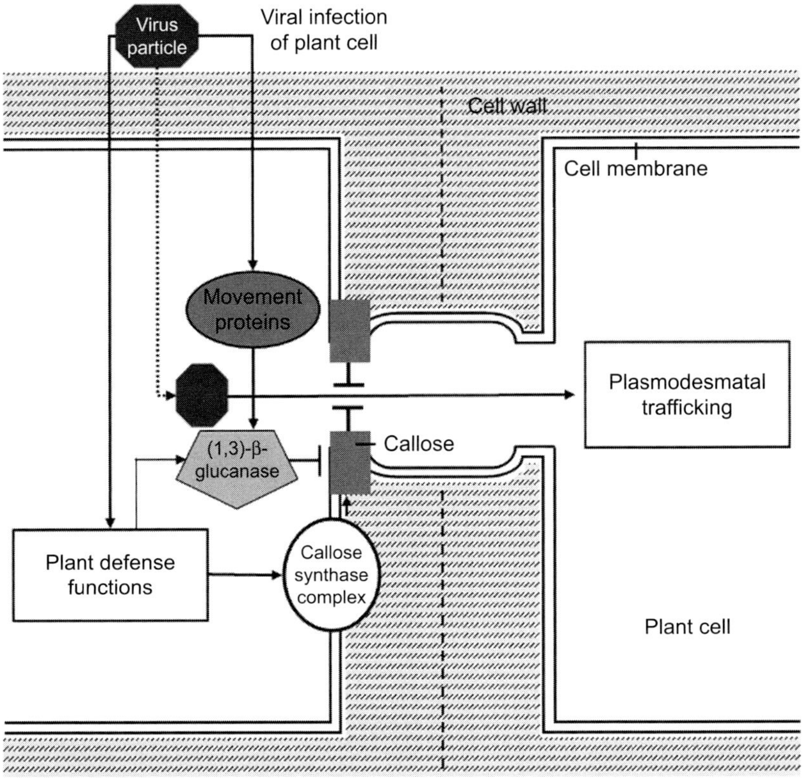


**Fig. 3: A model emphasizing the proposed roles of class I (1,3)-β-glucanase (GLU I) and viral movement protein in regulating plasmodesmatal trafficking and virus spread (Bucher et al., 2001; Levy et al. 2007b).**

of callose deposited at the plasmodesmata is negatively correlated with the plasmodesmal size exclusion limit (the size of the largest molecule that will move from cell to cell), as monitored by the movement of fluorescent dyes attached to molecules of varying sizes (Radford et al., 1998; Radford and White, 2001; Rinne et al., 2005; Sivaguru et al., 2000; Wolf et al., 1991). Support for the hypothesis that callose levels regulate plasmodesmal size exclusion limits was provided by Levy et al. (2007b). They showed that a specific (1,3)-β-glucanase (AtBG) mediated callose turnover at the plasmodesmata in Arabidopsis and further suggested that callose levels at plasmodesmata depend on the balance between synthesis and hydrolysis (Levy et al., 2007a). In AtBG T-DNA insertion mutants, both the number of cells to which free GFP diffused from a single cell and the coefficient of conductivity of plasmodesmata were lower than in wild-type plants. The identity of the callose synthase(s) responsible for callose deposition at plasmodesmata is currently unknown.

Transport through plasmodesmata is usually limited to small molecules less than 1 kDa in size (Lucas, 1995), and therefore these channels are too small to allow virions to move between cells. Accordingly, most plant viruses encode movement proteins that are thought to modify plasmodesmata or in some manner facilitate cell-to-cell movement of viral particles (Beachy and Heinlein, 2000; Deom et al., 1992; Lucas, 1995; Zambryski, 1995). Based on their findings of the plasmodesmal-associated (1,3)-β-glucanase AtBG in Arabidopsis, Levy et al. (2007b) suggest that the viruses recruit and/or activate a (1,3)-β-glucanase at the plasmodesmata to degrade callose, resulting in widening of plasmodesmal channels. In support of this suggestion, tobacco anti-sense mutants deficient in class I (1,3)-β-glucanases displayed a marked reduction in lesion size, lesion number and virus yield in the local lesion response to tobacco mosaic virus (TMV). A similar response to tobacco necrosis virus was observed in *Nicotiana sylvestris* (1,3)-β-glucanase mutants (Beffa et al., 1996). Since callose deposition in and surrounding TMV-induced lesions was increased in (1,3)-β-glucanase-deficient tobacco mutants, callose seemed to act as a physical barrier to virus spread. Subsequent experiments showed that the intercellular trafficking via plasmodesmata of TMV, recombinant potato virus X, cucumber mosaic virus, or the movement protein 3a of a cucumber mosaic virus was delayed in these tobacco mutants, showing that impaired movement was not limited to a particular virus (Iglesias and Meins, 2000). In support of these results, Bucher et al. (2001) inserted a (1,3)-β-glucanase gene into the TMV genome and found the size of local lesions was consistently increased, suggesting overexpression of the glucanase-facilitated viral spread. Enzymatically active (1,3)-β-glucanase was required for this effect on lesion size. In contrast, viruses expressing antisense (1,3)-β-glucanase constructs elicited smaller lesions.

The localization of class I (1,3)-β-glucanases in the lumen of the endoplasmic reticulum and the vacuole raises the question of how these proteins might interact with or be regulated by viral movement proteins in order to modify callose levels surrounding plasmodesmal channels through the plant cell wall. Tobacco ankyrin-repeat proteins of unknown biochemical function were identified as interacting with class I (1,3)-β-glucanases in a yeast two-hybrid screen by Levy et al. (2007b). In addition, three tobacco factors termed TBK 12K-interacting proteins bound to the potato virus X triple block protein TGB 12K, which is required for cell-to-cell movement and interacts with a class I (1,3)-β-glucanase (Fridborg et al., 2003). Similar proteins, which interacted with both (1,3)-β-glucanase and chitinase, were also found in *Nicotiana plumbaginifiolia* (Wirdnam et al., 2004). Even though the ankyrin-repeat-containing proteins might connect viral movement proteins and (1,3)-β-glucanases *in vitro*, their interaction *in vivo* is hindered by the occurrence of these proteins in different subcellular compartments (Fridborg et al., 2003; Rinne et al., 2001; Wirdnam et al., 2004). Since movement proteins and class I (1,3)-β-glucanases are localized in the endoplasmic reticulum membrane system (Beachy and Heinlein, 2000), Levy et al. (2007b) suggest that the binding between domains exposed on the cytoplasmic face of the endoplasmic reticulum and the cytoplasmically localized ankyrin-repeat-containing proteins may serve to target vesicles derived from the endoplasmic reticulum and containing class I (1,3)-β-glucanases to the cell wall where they can digest callose encasing plasmodesmata.

An additional barrier to viral spread can be found in the plant vasculature. Here, callose deposits are thought to stop systemic spread of viral particles. Tobacco plants exposed to low levels of cadmium ions showed a strong decrease in the systemic spread of turnip vein-clearing tobamovirus. This block of viral spread was associated with a gene encoding a glycine-rich protein, cdiGRP, which was specifically induced by low concentrations of cadmium (Ueki and Citovsky, 2002). This gene product occurs in the cell walls of plant vascular tissue and is associated with enhanced callose deposits in this tissue. Constitutive expression of cdiGRP, which was accompanied by increased callose accumulation in the vasculature, inhibited systemic transport of turnip vein-clearing tobamovirus similar to exposure to cadmium ions. Conversely, reductions in cdiGRP allowed the turnip vein-clearing tobamovirus to move readily in the vascular system and to infect distal parts of the plant tissue, even in the presence of cadmium. How cdiGRP might regulate viral systemic movement in plants and enhance callose deposition in the vasculature upon viral infection is unknown (Ueki and Citovsky, 2002).

## Conclusions

In recent decades, significant advances in our understanding of plant–pathogen interactions have been made. However, central questions about the constituents of the callose synthase complex and the role of callose in biotic stress are still unanswered. A majority of the evidence connecting microbial attack and signalling pathways with callose deposition is based on correlative data. The *GSL5* null mutants, the *pmr4* mutants of Arabidopsis, do not show enhanced penetration success or disease development by fungal pathogens as would be expected, thus calling into question many of the proposed roles for callose in fungus–plant interactions. However, additional tests of diverse pathogens on plants deficient in or overexpressing pathogen-induced callose synthase components are needed to determine whether these results with the *pmr4* mutants can be generalized.

The regulation of callose synthesis is largely unknown in plants under pathogen attack. There is some indication that stress-induced callose synthesis is activated by post-translational events, like $Ca^{2+}$ influx into cells and phosphorylation of callose synthase components. However, proof for these regulatory mechanisms is not available for plant callose synthases. Since callose synthase is thought to be an enzyme complex consisting of an unknown number of proteins in addition to the GSL proteins, the assembly of the complex and post-translational modifications of GSL companion proteins is likely to lead to multifaceted regulation of callose synthesis. In addition, the amount of callose accumulation at infection sites may also depend on the balance between synthesis and degradation as noted for the proposed regulation of plasmodesmatal channel size. Plants secrete (1,3)-β-glucanases to break down fungal cell walls, but these degradative enzymes may also act on plant-derived callose deposits, although this point has not been examined to date.

In addition, the subcellular localization or local enzymatic activation of the callose synthase complex at sites of pathogen attack represents an additional point of potential regulation of callose synthesis. For example, the deposition of callose in papillae would require either re-localization of the callose synthase complexes, selective activation of callose synthase complexes close to penetration sites, or the transportation of callose to these sites. An interesting feature of callose deposits at infection sites is that they also occur in plant cells adjacent to attacked cells, suggesting a diffusible signal is generated during the infection sequence that can cross cell boundaries to activate localized callose synthesis and deposition. Our lack of knowledge of callose synthase regulation is no doubt the basis for the ambiguity about the role of callose in plant–pathogen interactions that remains.

The *pmr4* mutants are interesting in a second context in that they suggest a link between the presence of a functional GSL5 and a dampening of salicylic acid signal transduction. The basis for the connection between callose deposition in the cell wall and salicylic acid signalling is unknown, but represents a new putative role for callose in plant–pathogen interactions. Due to the lack of comparable mutants in other plants, it is not clear whether this putative function of callose in signalling is specific to Arabidopsis or not. The availability of specific callose-deficient mutants from diverse plants might help to answer this question and could lead to a broader understanding of the role of callose in biotic stress.

## Acknowledgements

We thank Janine May, Stephan Wenkel, and Shundai Li for helpful discussions. C.A. Voigt is supported by a postdoctoral research fellowship from the Deutsche Forschungsgemeinschaft.

The work in the S.C. Somerville laboratory is supported in part by the Carnegie Institution of Science, the National Science Foundation and the U.S. Department of Energy.

## References

Abramovitch, R. B., Kim, Y. J., Chen, S., Dickman, M. B., & Martin, G. B. (2003). *Pseudomonas* type III effector AvrPtoB induces plant disease susceptibility by inhibition of host programmed cell death. *EMBO Journal*, *22*, 60–69.

Adam, L., Ellwood, S., Wilson, I., Saenz, G., Xiao, S., Oliver, R. P., Turner, J. G., & Somerville, S. (1999). Comparison of *Erysiphe cichoracearum* and *E. cruciferarum* and a survey of 360 *Arabidopsis thaliana* accessions for resistance to these two powdery mildew pathogens. *Molecular Plant-Microbe Interactions*, *12*, 1031–1043.

Aist, J. R. (1976). Papillae and related wound plugs of plant cells. *Annual Review of Plant Phytopathology*, *14*, 145–163.

Aist, J. R. (1983). Structural responses as resistance mechanisms. In J. A. Bailey & B. J. Deverall (Eds.), *The Dynamics of Host Defense* (pp. 33–70). New York: Academic Press.

Aist, J. R., & Williams, P. H. (1971). The cytology and kinetics of cabbage root hair penetration by *Plasmodiophora brassicae*. *Canadian Journal of Botany*, *49*, 2023–2034.

Alfano, J. R., & Collmer, A. (1997). The type III (Hrp) secretion pathway of plant pathogenic bacteria: trafficking harpins, Avr proteins, and death. *Journal of Bacteriology*, *179*, 5655–5662.

Alfano, J. R., & Collmer, A. (2004). Type III secretion system effector proteins: double agents in bacterial disease and plant defense. *Annual Review Phytopathology*, *42*, 385–414.

Allison, A. V., & Shalla, T. A. (1974). The ultrastructure of local lesions induced by potato virus X: A sequence of cytological events in the course of infection. *Phytopathology*, *64*, 784–793.

Amor, Y., Haigler, C. H., Johnson, S., Wainscott, M., & Delmer, D. P. (1995). A membrane-associated form of sucrose synthase and its potential role in synthesis of cellulose and callose in plants. *Proceedings of the National Academy of Sciences of the United States of America*, *92*, 9353–9357.

Aslam, S. N., Newman, M.-A., Erbs, G., Morrissey, K. L., Chinchilla, D., Boller, T., Jensen, T. T., De Castro, C., Ierano, T., Molinaro, A., Jackson, R. W., Knight, M. R., & Cooper, R. M. (2008). Bacterial polysaccharides suppress induced innate immunity by calcium chelation. *Current Biology*, *18*, 1078–1083.

Aspinall, G. O., & Kessler, G. (1957). *The Structure of Callose from the Grape Vine*. London: Chemistry and Industry p. 1296.

Assaad, F. F., Qiu, J. L., Youngs, H., Ehrhardt, D., Zimmerli, L., Kalde, M., Wanner, G., Peck, S. C., Edwards, H., Ramonell, K., Somerville, C. R., & Thordal-Christensen, H. (2004). The PEN1 syntaxin defines a novel cellular compartment upon fungal attack and is required for the timely assembly of papillae. *Molecular Biology of the Cell*, *15*, 5118–5129.

Ausubel, F. M. (2005). Are innate immune signaling pathways in plants and animals conserved? *Nature Immunology*, *6*, 973–979.

Ausubel, F. M., Katagiri, F., Mindrinos, M., & Glazebrook, J. (1995). Use of *Arabidopsis thaliana* defense-related mutants to dissect the plant response to pathogens. *Proceedings of the National Academy of Sciences of the United States of America*, *92*, 4189–4196.

Bailey, J. A. (1983). Biological perspectives of host-pathogen interactions. In J. A. Bailey & B. J. Deverall (Eds.), *The Dynamics of Host Defense* (pp. 1–32). New York: Academic Press.

Bartnicki-Garcia, S. (1968). Cell wall chemistry, morphogenesis, and taxonomy of fungi. *Annual Review of Microbiology*, *22*, 87–108.

Bayles, C. J., Ghemawat, M. S., & Aist, J. R. (1990). Inhibition by 2-deoxy-D-glucose of callose formation, papilla deposition, and resistance to powdery mildew in *ml-o* barley mutant. *Molecular Plant Pathology*, *36*, 63–72.

Beachy, R. N., & Heinlein, M. (2000). Role of P30 in replication and spread of TMV. *Traffic*, *1*, 540–544.

Beffa, R. S., Hofer, R. M., Thomas, M., & Meins, F., Jr. (1996) Decreased susceptibility to viral disease of β-1,3-glucanase-deficient plants generated by antisense transformation. *Plant Cell*, *8*, 1001–1011.

Benhamou, N. (1995). Immunocytochemistry of plant defense mechanisms induced upon microbial attack. *Microscopy Research and Technique*, *31*, 63–78.

Bestwick, C. S., Bennett, M. H., & Mansfield, J. W. (1995). Hrp mutant of *Pseudomonas syringae* pv *phaseolicola* induces cell wall alterations but not membrane damage leading to the hypersensitive reaction in lettuce. *Plant Physiology*, *108*, 503–516.

Bhat, R. A., Miklis, M., Schmelzer, E., Schulze-Lefert, P., & Panstruga, R. (2005). Recruitment and interaction dynamics of plant penetration resistance components in a plasma membrane microdomain. *Proceedings of the National Academy of Sciences of the United States of America*, *102*, 3135–3140.

Blume, B., Nurnberger, T., Nass, N., & Scheel, D. (2000). Receptor-mediated increase in cytoplasmic free calcium required for activation of pathogen defense in parsley. *Plant Cell*, *12*, 1425–1440.

Boch, J., Joardar, V., Gao, L., Robertson, T. L., Lim, M., & Kunkel, B. N. (2002). Identification of *Pseudomonas syringae* pv. tomato genes induced during infection of *Arabidopsis thaliana*. *Molecular Microbiology*, *44*, 73–88.

Bretz, J. R., Mock, N. M., Charity, J. C., Zeyad, S., Baker, C. J., & Hutcheson, S. W. (2003). A translocated protein tyrosine phosphatase of *Pseudomonas syringae* pv. tomato DC3000 modulates plant defense response to infection. *Molecular Microbiology*, *49*, 389–400.

Brown, I., Mansfield, J., & Bonas, U. (1995). *hrp* genes in *Xanthomonas campestris* pv. *vesicatoria* determine ability to suppress papilla deposition in pepper mesophyll cells. *Molecular Plant-Microbe Interactions*, *8*, 825–836.

Brown, I., Trethowan, J., Kerry, M., Mansfield, J., & Bolwell, G. P. (1998). Localization of components of the oxidative cross-linking of glycoproteins and of callose synthesis in papillae formed during the interaction between non-pathogenic strains of *Xanthomonas campestris* and French bean mesophyll cells. *Plant Journal*, *15*, 333–343.

Brownfield, L., Ford, K., Doblin, M. S., Newbigin, E., Read, S., & Bacic, A. (2007). Proteomic and biochemical evidence links the callose synthase in *Nicotiana alata* pollen tubes to the product of the *NaGSL1* gene. *Plant Journal*, *52*, 147–156.

Bucher, G. L., Tarina, C., Heinlein, M., Di Serio, F., Meins, F., Jr., & Iglesias, V. A. (2001) Local expression of enzymatically active class I β-1,3-glucanase enhances symptoms of TMV infection in tobacco. *Plant Journal*, *28*, 361–369.

Bulone, V., Fincher, G. B., & Stone, B. A. (1995). *In vitro* synthesis of a microfibrillar (1→3)-β-glucan by a ryegrass (*Lolium multiflorum*) endosperm (1→3)-β-glucan synthase enriched by product entrapment. *Plant Journal*, *8*, 213–225.

Buttner, D., & Bonas, U. (2002). Getting across–bacterial type III effector proteins on their way to the plant cell. *EMBO Journal*, *21*, 5313–5322.

Cabello, F., Jorrin, J. V., & Tena, M. (1994). Chitinase and β-1,3-glucanase activities in chickpea (*Cicer arietinum*). Induction of different isoenzymes in response to wounding and ethephon. *Physiologia Plantarum*, *92*, 654–660.

Cabib, E., Roh, D. H., Schmidt, M., Crotti, L. B., & Varma, A. (2001). The yeast cell wall and septum as paradigms of cell growth and morphogenesis. *Journal of Biological Chemistry*, *276*, 19679–19682.

Calonge, T. M., Arellano, M., Coll, P. M., & Perez, P. (2003). Rga5p is a specific Rho1p GTPase-activating protein that regulates cell integrity in *Schizosaccharomyces pombe*. *Molecular Microbiology*, *47*, 507–518.

Chang, J. H., Urbach, J. M., Law, T. F., Arnold, L. W., Hu, A., Gombar, S., Grant, S. R., Ausubel, F. M., & Dangl, J. L. (2005). A high-throughput, near-saturating screen for type III effector genes from *Pseudomonas syringae*. *Proceedings of the National Academy of Sciences of the United States of America*, *102*, 2549–2554.

Chang, M. M., Hadwiger, L. A., & Horovitz, D. (1992). Molecular characterization of a pea β-1,3-glucanase induced by *Fusarium solani* and chitosan challenge. *Plant Molecular Biology*, *20*, 609–618.

Chinchilla, D., Bauer, Z., Regenass, M., Boller, T., & Felix, G. (2006). The *Arabidopsis* receptor kinase FLS2 binds flg22 and determines the specificity of flagellin perception. *Plant Cell*, *18*, 465–476.

Chinchilla, D., Zipfel, C., Robatzek, S., Kemmerling, B., Nurnberger, T., Jones, J. D., Felix, G., & Boller, T. (2007). A flagellin-induced complex of the receptor FLS2 and BAK1 initiates plant defense. *Nature*, *448*, 497–500.

Collins, N. C., Thordal-Christensen, H., Lipka, V., Bau, S., Kombrink, E., Qiu, J. L., Huckelhoven, R., Stein, M., Freialdenhoven, A., Somerville, S. C., & Schulze-Lefert, P. (2003). SNARE-protein-mediated disease resistance at the plant cell wall. *Nature*, *425*, 973–977.

Consonni, C., Humphry, M. E., Hartmann, H. A., Livaja, M., Durner, J., Westphal, L., Vogel, J., Lipka, V., Kemmerling, B., Schulze-Lefert, P., Somerville, S. C., & Panstruga, R. (2006). Conserved requirement for a plant host cell protein in powdery mildew pathogenesis. *Nature Genetics*, *38*, 716–720.

Cordero, M. J., Raventos, D., & Sansegundo, B. (1994). Differential expression and induction of chitinases and beta-1,3-glucanases in response to fungal infection during germination of maize seeds. *Molecular Plant-Microbe Interactions*, *7*, 23–31.

Cornelis, G. R., & Van Gijsegem, F. (2000). Assembly and function of type III secretory systems. *Annual Review of Microbiology*, *54*, 735–774.

Côté, F., & Hahn, M. G. (1994). Oligosaccharins: structures and signal transduction. *Plant Molecular Biology*, *26*, 1379–1411.

Cui, X., Shin, H., Song, C., Laosinchai, W., Amano, Y., & Brown, R. M., Jr. (2001) A putative plant homolog of the yeast β-1,3-glucan synthase subunit FKS1 from cotton (*Gossypium hirsutum* L.) fibers. *Planta*, *213*, 223–230.

Dangl, J. L., & Jones, J. D. (2001). Plant pathogens and integrated defense responses to infection. *Nature*, *411*, 826–833.

de Torres, M., Mansfield, J. W., Grabov, N., Brown, I. R., Ammouneh, H., Tsiamis, G., Forsyth, A., Robatzek, S., Grant, M., & Boch, J. (2006). *Pseudomonas syringae* effector AvrPtoB suppresses basal defense in Arabidopsis. *Plant Journal*, *47*, 368–382.

DeBary, A. (1863). Recherches sur le développement de quelques champignons parasites. *Annales des Sciences Naturelles. Botanique et Biologie Végétale.*, *20*, 5–148.

DebRoy, S., Thilmony, R., Kwack, Y. B., Nomura, K., & He, S. Y. (2004). A family of conserved bacterial effectors inhibits salicylic acid-mediated basal immunity and promotes disease necrosis in plants. *Proceedings of the National Academy of Sciences of the United States of America*, *101*, 9927–9932.

Delmer, D. P., & Amor, Y. (1995). Cellulose biosynthesis. *Plant Cell*, *7*, 987–1000.

Delmer, D. P., Solomon, M., & Read, S. M. (1991). Direct photolabeling with [$^{32}$P]UDP-glucose for identification of a subunit of cotton fiber callose synthase. *Plant Physiology*, *95*, 556–563.

Deom, C. M., Lapidot, M., & Beachy, R. N. (1992). Plant virus movement proteins. *Cell*, *69*, 221–224.

Deverall, B. J. (1977). *Defense Mechanisms of Plants*. Cambridge: Cambridge University Press.

Dhugga, K. S., & Ray, P. M. (1994). Purification of 1,3-β-D-glucan synthase activity from pea tissue. Two polypeptides of 55 kDa and 70 kDa copurify with enzyme activity. *European Journal of Biochemistry*, *220*, 943–953.

Dijkgraaf, G. J., Abe, M., Ohya, Y., & Bussey, H. (2002). Mutations in Fks1p affect the cell wall content of β-1,3- and β-1,6-glucan in *Saccharomyces cerevisiae*. *Yeast*, *19*, 671–690.

Doblin, M. S., De Melis, L., Newbigin, E., Bacic, A., & Read, S. M. (2001). Pollen tubes of *Nicotiana alata* express two genes from different β-glucan synthase families. *Plant Physiology*, *125*, 2040–2052.

Dolph, P. J., Majerczak, D. R., & Coplin, D. L. (1988). Characterization of a gene cluster for exopolysaccharide biosynthesis and virulence in *Erwinia stewartii*. *Journal of Bacteriology*, *170*, 865–871.

Dong, X., Hong, Z., Sivaramakrishnan, M., Mahfouz, M., & Verma, D. P. (2005). Callose synthase (CalS5) is required for exine formation during microgametogenesis and for pollen viability in Arabidopsis. *Plant Journal*, *42*, 315–328.

Donofrio, N. M., & Delaney, T. P. (2001). Abnormal callose response phenotype and hypersusceptibility to *Peronospora parasitica* in defense-compromised Arabidopsis *nim1-1* and salicylate hydroxylase-expressing plants. *Molecular Plant-Microbe Interactions*, *14*, 439–450.

Douglas, C. M., Foor, F., Marrinan, J. A., Morin, N., Nielsen, J. B., Dahl, A. M., Mazur, P., Baginsky, W., Li, W., el-Sherbeini, M., et al. (1994). The *Saccharomyces cerevisiae FKS1* (*ETG1*) gene encodes an integral membrane protein which is a subunit of 1,3-β-D-glucan synthase. *Proceedings of the National Academy of Sciences of the United States of America*, *91*, 12907–12911.

Dow, J. M., & Daniels, M. J. (2000). *Xylella* genomics and bacterial pathogenicity to plants. *Yeast*, *17*, 263–271.

Ellis, J., Catanzariti, A. M., & Dodds, P. (2006). The problem of how fungal and oomycete avirulence proteins enter plant cells. *Trends in Plant Science*, *11*, 61–63.

Enns, L. C., Kanaoka, M. M., Torii, K. U., Comai, L., Okada, K., & Cleland, R. E. (2005). Two callose synthases, GSL1 and GSL5, play an essential and redundant role in plant and pollen development and in fertility. *Plant Molecular Biology*, *58*, 333–349.

Esau, K., Cronshaw, J., & Hoefert, L. L. (1967). Relation of beet yellows virus to the phloem and to movement in the sieve tube. *Journal of Cell Biology*, *32*, 71–87.

Espinosa, A., Guo, M., Tam, V. C., Fu, Z. Q., & Alfano, J. R. (2003). The *Pseudomonas syringae* type III-secreted protein HopPtoD2 possesses protein tyrosine phosphatase activity and suppresses programmed cell death in plants. *Molecular Microbiology*, *49*, 377–387.

Flor, H. H. (1947). Inheritance of reaction to rust in flax. *Journal of Agricultural and Resource*, *74*, 241–262.

Flor, H. H. (1971). Current status of the gene-for-gene concept. *Annual Review of Phytopathology*, *9*, 275–296.

Flors, V., Leyva Mde, L., Vicedo, B., Finiti, I., Real, M. D., Garcia-Agustin, P., Bennett, A. B., & Gonzalez-Bosch, C. (2007). Absence of the endo-β-1,4-glucanases Cel1 and Cel2 reduces susceptibility to *Botrytis cinerea* in tomato. *Plant Journal*, *52*, 1027–1040.

Fouts, D. E., Abramovitch, R. B., Alfano, J. R., Baldo, A. M., Buell, C. R., Cartinhour, S., Chatterjee, A. K., D'Ascenzo, M., Gwinn, M. L., Lazarowitz, S. G., Lin, N. C., Martin, G. B., Rehm, A. H., Schneider, D. J., van Dijk, K., Tang, X., & Collmer, A. (2002). Genomewide identification of *Pseudomonas syringae* pv. tomato DC3000 promoters controlled by the HrpL alternative sigma factor. *Proceedings of the National Academy of Sciences of the United States of America*, *99*, 2275–2280.

Fridborg, I., Grainger, J., Page, A., Coleman, M., Findlay, K., & Angell, S. (2003). TIP, a novel host factor linking callose degradation with the cell-to-cell movement of Potato virus X. *Molecular Plant-Microbe Interactions*, *16*, 132–140.

Fritz, J. H., Ferrero, R. L., Philpott, D. J., & Girardin, S. E. (2006). Nod-like proteins in immunity, inflammation and disease. *Nature Immunology*, *7*, 1250–1257.

Frost, D. J., Read, S. M., Drake, R. R., Haley, B. E., & Wasserman, B. P. (1990). Identification of the UDP-glucose-binding polypeptide of callose synthase from *Beta vulgaris* L. by photoaffinity labeling with 5-azido-UDP-glucose. *Journal of Biological Chemistry*, *265*, 2162–2167.

Garcia-Garcia, F., Schmelzer, E., Hahlbrock, K., & Roxby, R. (1994). Differential expression of chitinase and β-1,3-glucanase genes in various tissues of potato plants. *Zeitschrift Fur Naturforschung*, *49C*, 195–203.

Geier, G., & Geider, K. (1993). Characterization and influence on virulence of the levansucrase gene from the fireblight pathogen *Erwinia amylovora*. *Physiological and Molecular Plant Pathology*, *42*, 387–404.

Glazebrook, J. (2005). Contrasting mechanisms of defense against biotrophic and necrotrophic pathogens. *Annual Review of Phytopathology*, *43*, 205–227.

Gomez-Gomez, L., & Boller, T. (2000). FLS2: an LRR receptor-like kinase involved in the perception of the bacterial elicitor flagellin in Arabidopsis. *Molecular Cell*, *5*, 1003–1011.

Gomez-Gomez, L., & Boller, T. (2002). Flagellin perception: a paradigm for innate immunity. *Trends in Plant Science*, *7*, 251–256.

Grant, M., Brown, I., Adams, S., Knight, M., Ainslie, A., & Mansfield, J. (2000). The *RPM1* plant disease resistance gene facilitates a rapid and sustained increase in cytosolic calcium that is necessary for the oxidative burst and hypersensitive cell death. *Plant Journal*, *23*, 441–450.

Greenberg, J. T., & Vinatzer, B. A. (2003). Identifying type III effectors of plant pathogens and analyzing their interaction with plant cells. *Current Opinion in Microbiology*, *6*, 20–28.

Gus-Mayer, S., Naton, B., Hahlbrock, K., & Schmelzer, E. (1998). Local mechanical stimulation induces components of the pathogen defense response in parsley. *Proceedings of the National Academy of Sciences of the United States of America*, *95*, 8398–8403.

Guttman, D. S., Vinatzer, B. A., Sarkar, S. F., Ranall, M. V., Kettler, G., & Greenberg, J. T. (2002). A functional screen for the type III (Hrp) secretome of the plant pathogen *Pseudomonas syringae*. *Science*, *295*, 1722–1726.

Hammond-Kosack, K. E., & Jones, J. D. (1996). Resistance gene-dependent plant defense responses. *Plant Cell*, *8*, 1773–1791.

Hardham, A. R., Jones, D. A., & Takemoto, D. (2007). Cytoskeleton and cell wall function in penetration resistance. *Current Opinion in Plant Biology*, *10*, 342–348.

Hauck, P., Thilmony, R., & He, S. Y. (2003). A *Pseudomonas syringae* type III effector suppresses cell wall-based extracellular defense in susceptible Arabidopsis plants. *Proceedings of the National Academy of Sciences of the United States of America*, *100*, 8577–8582.

He, S. Y. (1998). Type III protein secretion systems in plant and animal pathogenic bacteria. *Annual Review of Phytopathology*, *36*, 363–392.

Heath, M. C. (2000). Nonhost resistance and nonspecific plant defenses. *Current Opinion in Plant Biology*, *3*, 315–319.

Heinlein, M., & Epel, B. L. (2004). Macromolecular transport and signaling through plasmodesmata. *International Review of Cytology*, *235*, 93–164.

Heyraud, A., Sayah, B., Vojnov, A., Colin-Morel, P., Gey, C., Geremia, R. A., & Dankert, M. (1998). Structure of an extracellular mannosylated cellulose produced by a mutant strain of *Xanthomonas campestris*. *Cellular and Molecular Biology*, *44*, 447–454.

Him, J. L., Pelosi, L., Chanzy, H., Putaux, J. L., & Bulone, V. (2001). Biosynthesis of (1→3)-β-D-glucan (callose) by detergent extracts of a microsomal fraction from *Arabidopsis thaliana*. *European Journal of Biochemistry*, *268*, 4628–4638.

Hong, Z., Delauney, A. J., & Verma, D. P. (2001). A cell plate-specific callose synthase and its interaction with phragmoplastin. *Plant Cell*, *13*, 755–768.

Iglesias, V. A., & Meins, F., Jr. (2000) Movement of plant viruses is delayed in a β-1,3-glucanase-deficient mutant showing a reduced plasmodesmatal size exclusion limit and enhanced callose deposition. *Plant Journal*, *21*, 157–166.

Israel, H. W., Wilson, R. G., Aist, J. R., & Kunoh, H. (1980). Cell wall appositions and plant disease resistance: acoustic microscopy of papillae that block fungal ingress. *Proceedings of the National Academy of Sciences of the United States of America*, *77*, 2046–2049.

Jach, G., Gornhardt, B., Mundy, J., Logemann, J., Pinsdorf, E., Leah, R., Schell, J., & Maas, C. (1995). Enhanced quantitative resistance against fungal disease by combinatorial expression of different barley antifungal proteins in transgenic tobacco. *Plant Journal*, *8*, 97–109.

Jacobs, A. K., Lipka, V., Burton, R. A., Panstruga, R., Strizhov, N., Schulze-Lefert, P., & Fincher, G. B. (2003). An Arabidopsis callose synthase, GSL5, is required for wound and papillary callose formation. *Plant Cell*, *15*, 2503–2513.

Jakobek, J. L., Smith, J. A., & Lindgren, P. B. (1993). Suppression of bean defense responses by *Pseudomonas syringae*. *Plant Cell*, *5*, 57–63.

Jamir, Y., Guo, M., Oh, H. S., Petnicki-Ocwieja, T., Chen, S., Tang, X., Dickman, M. B., Collmer, A., & Alfano, J. R. (2004). Identification of *Pseudomonas syringae* type III effectors that can suppress programmed cell death in plants and yeast. *Plant Journal*, *37*, 554–565.

Ji, C., & Kuc, J. (1995). Purification and characterization of an acidic β-1,3-glucanase from cucumber and its relationship to systemic disease resistance induced by *Colletotrichum lagenarium* and tobacco necrosis virus. *Molecular Plant-Microbe Interactions*, *8*, 899–905.

Jones, D. A., & Takemoto, D. (2004). Plant innate immunity-direct and indirect recognition of general and specific pathogen-associated molecules. *Current Opinion in Immunology*, *16*, 48–62.

Jones, J. D., & Dangl, J. L. (2006). The plant immune system. *Nature*, *444*, 323–329.

Kaku, H., Nishizawa, Y., Ishii-Minami, N., Akimoto-Tomiyama, C., Dohmae, N., Takio, K., Minami, E., & Shibuya, N. (2006). Plant cells recognize chitin fragments for defense signaling through a plasma membrane receptor. *Proceedings of the National Academy of Sciences of the United States of America*, *103*, 11086–11091.

Kao, C. C., Barlow, E., & Sequeira, L. (1992). Extracellular polysaccharide is required for wild-type virulence of *Pseudomonas solanacearum*. *Journal of Bacteriology*, *174*, 1068–1071.

Kaufmann, S., Legrand, M., Geoffrey, P., & Fritig, B. (1987). Biological function of 'pathogenesis-related' protein: four PR proteins from tobacco have 1,3-β-glucanase activity. *EMBO Journal*, *6*, 3209–3212.

Kauss, H. (1985). Callose biosynthesis as a $Ca^{2+}$-regulated process and possible relations to the induction of other metabolic changes. *Journal of Cell Science*, *2*(Suppl.), 89–103.

Kauss, H. (1987). Some aspects of calcium-dependent regulation in plant metabolism. *Annual Review of Plant Physiology*, *38*, 47–72.

Kauss, H. (1991). The role of the plasma membrane in host-pathogen interactions. In C. Larsson & I. M. Moller (Eds.), *The Plant Plasma Membrane* (pp. 320–350). Berlin: Springer-Verlag.

Kauss, H., & Jeblick, W. (1986). Influence of free fatty acids, lysophosphatidylcholine, platelet-activating factor, acylcarnitine, and Echinocandin B on 1,3-β-D-glucan synthase and callose synthesis. *Plant Physiology*, *80*, 7–13.

Kauss, H., Kohle, H., & Jeblick, W. (1983). Proteolytic activation and stimulation by $Ca^{2+}$ of glucan synthase from soybean cells. *FEBS Letters*, *158*, 84–88.

Keshavarzi, M., Soylu, S., Brown, I., Bonas, U., Nicole, M., Rossiter, J., & Mansfield, J. (2004). Basal defenses induced in pepper by lipopolysaccharides are suppressed by *Xanthomonas campestris* pv. *vesicatoria*. *Molecular Plant-Microbe Interactions*, *17*, 805–815.

Kim, H. S., Desveaux, D., Singer, A. U., Patel, P., Sondek, J., & Dangl, J. L. (2005a). The *Pseudomonas syringae* effector AvrRpt2 cleaves its C-terminally acylated target, RIN4, from Arabidopsis membranes to block RPM1 activation. *Proceedings of the National Academy of Sciences of the United States of America*, *102*, 6496–6501.

Kim, M. G., da Cunha, L., McFall, A. J., Belkhadir, Y., DebRoy, S., Dangl, J. L., & Mackey, D. (2005b). Two *Pseudomonas syringae* type III effectors inhibit RIN4-regulated basal defense in Arabidopsis. *Cell*, *121*, 749–759.

Kini, K. R., Vasanthi, N. S., & Shetty, H. S. (2000). Induction of β-1,3-glucanase in seedlings of pearl millet in response to infection by *Sclerospora graminicola. European Journal of Plant Pathology*, *106*, 267–274.

Kobayashi, I., Kobayashi, Y., Yamaoka, N., & Kunoh, H. (1992). Recognition of a pathogen and a non-pathogen by barley coleoptile cells. III. Responses of microtubules and actin filaments in barley coleoptile cells to penetration attempts. *Canadian Journal of Botany*, *70*, 1815–1823.

Koh, S., Andre, A., Edwards, H., Ehrhardt, D., & Somerville, S. (2005). *Arabidopsis thaliana* subcellular responses to compatible *Erysiphe cichoracearum* infections. *Plant Journal*, *44*, 516–529.

Köhle, H., Jeblick, W., Poten, F., Blaschek, W., & Kauss, H. (1985). Chitosan-elicited callose synthesis in soybean cells as a $Ca^{2+}$-dependent process. *Plant Physiology*, *77*, 544–551.

Kudlicka, K., & Brown, R. M., Jr (1997). Cellulose and callose biosynthesis in higher plants. I. Solubilization and separation of (1→3)- and (1→4)-β-glucan synthase activities from mung bean. *Plant Physiology*, *115*, 643–656.

Kudlicka, K., Brown, R. M., Jr., Li, L., Lee, J. H., Shin, H., & Kuga, S. (1995) β-glucan synthesis in the cotton fiber. IV. *In vitro* assembly of the cellulose I allomorph. *Plant Physiology*, *107*, 111–123.

Kuzuya, M., Yashiro, K., Tomita, K., & Ezura, H. (2006). Powdery mildew (*Podosphaera xanthii*) resistance in melon is categorized into two types based on inhibition of the infection processes. *Journal of Experimental Botany*, *57*, 2093–2100.

Lamb, C., & Dixon, R. A. (1997). The oxidative burst in plant disease resistance. *Annual Review of Plant Physiology and Plant Molecular Biology*, *48*, 251–275.

Leipe, D. D., Koonin, E. V., & Aravind, L. (2004). STAND, a class of P-loop NTPases including animal and plant regulators of programmed cell death: multiple, complex domain architectures, unusual phyletic patterns, and evolution by horizontal gene transfer. *Journal of Molecular Biology*, *343*, 1–28.

Leisner, S. M., & Howell, S. H. (1993). Long-distance movement of viruses in plants. *Trends in Microbiology*, *1*, 314–317.

Leon, J., Rojo, E., & Sanchez-Serrano, J. J. (2001). Wound signaling in plants. *Journal of Experimental Botany*, *52*, 1–9.

Leon, J., Rojo, E., Titarenko, E., & Sanchez-Serrano, J. J. (1998). Jasmonic acid-dependent and -independent wound signal transduction pathways are differentially regulated by $Ca^{2+}$/calmodulin in *Arabidopsis thaliana. Molecular and General Genetics*, *258*, 412–419.

Levy, A., Erlanger, M., Rosenthal, M., & Epel, B. L. (2007a). A plasmodesmata-associated β-1,3-glucanase in Arabidopsis. *Plant Journal*, *49*, 669–682.

Levy, A., Guenoune-Gelbart, D., & Epel, B. L. (2007b). β-1,3-glucanases: plasmodesmal gate keepers for intercellular communication. *Plant Signalling and Behavior*, *2*, 288–290.

Li, C., Bonnema, G., Che, D., Dong, L., Lindhout, P., Visser, R., & Bai, Y. (2007). Biochemical and molecular mechanisms involved in monogenic resistance responses to tomato powdery mildew. *Molecular Plant-Microbe Interactions*, *20*, 1161–1172.

Li, J., Burton, R. A., Harvey, A. J., Hrmova, M., Wardak, A. Z., Stone, B. A., & Fincher, G. B. (2003). Biochemical evidence linking a putative callose synthase gene with (1→3)-β-D-glucan biosynthesis in barley. Plant. *Plant Molecular Biology*, *53*, 213–225.

Li, L., & Brown, R. M., Jr. (1993). β-glucan synthesis in the cotton fiber. II. Regulation and kinetic properties of β-glucan synthases. *Plant Physiology*, *101*, 1143–1148.

Lindeberg, M., Cartinhour, S., Myers, C. R., Schechter, L. M., Schneider, D. J., & Collmer, A. (2006). Closing the circle on the discovery of genes encoding *Hrp* regulon members and type III secretion system effectors in the genomes of three model *Pseudomonas syringae* strains. *Molecular Plant-Microbe Interactions*, *19*, 1151–1158.

Lindgren, P. B. (1997). The role of *hrp* genes during plant-bacterial interactions. *Annual Review of Phytopathology*, *35*, 129–152.

Lipka, V., Dittgen, J., Bednarek, P., Bhat, R., Wiermer, M., Stein, M., Landtag, J., Brandt, W., Rosahl, S., Scheel, D., Llorente, F., Molina, A., Parker, J., Somerville, S., & Schulze-Lefert, P. (2005). Pre- and postinvasion defenses both contribute to nonhost resistance in Arabidopsis. *Science*, *310*, 1180–1183.

Lucas, W. J. (1995). Plasmodesmata: intercellular channels for macromolecular transport in plants. *Current Opinion in Cell Biology*, *7*, 673–680.

Mackey, D., Belkhadir, Y., Alonso, J. M., Ecker, J. R., & Dangl, J. L. (2003). Arabidopsis RIN4 is a target of the type III virulence effector AvrRpt2 and modulates RPS2-mediated resistance. *Cell*, *112*, 379–389.

Mangin, L. (1895). Recherches sur les Péronosporées. *Bulletin de la Société d'Histoire Naturelle d'Autun*, *8*, 55–108.

Martin, G. B., Bogdanove, A. J., & Sessa, G. (2003). Understanding the functions of plant disease resistance proteins. *Annual Review of Plant Physiology*, *54*, 23–61.

Mauch-Mani, B., & Mauch, F. (2005). The role of abscisic acid in plant-pathogen interactions. *Current Opinion in Plant Biology*, *8*, 409–414.

Mauch, F., Mauch-Mani, B., & Boller, T. (1988). Antifungal hydrolases in pea tissue: II. Inhibition of fungal growth by combinations of chitinase and β-1,3-glucanase. Plant. *Plant Physiology*, *88*, 936–942.

McCormack, B. A., Gregory, A. C., Kerry, M. E., Smith, C., & Bolwell, G. P. (1997). Purification of an elicitor-induced glucan synthase (callose synthase) from suspension cultures of French bean (*Phaseolus vulgaris* L.): purification and immunolocation of a probable Mr-65,000 subunit of the enzyme. *Planta*, *203*, 196–203.

Meikle, P. J., Bonig, I., Hoogengraad, N. J., Clarke, A. E., & Stone, B. A. (1991). The location of (1→3)-β-glucans in the walls of pollen tubes of *Nicotiana alata* using a (1→3)-β-glucan specific monoclonal antibody. *Planta*, *185*, 1–8.

Mendgen, K., & Hahn, M. (2002). Plant infection and the establishment of fungal biotrophy. *Trends in Plant Science*, *7*, 352–356.

Mercer, P. C., Wood, R. K. S., & Greenwood, A. D. (1974). Resistance to anthracnose of French bean. *Physiological and Molecular Plant Pathology*, *4*, 291–306.

Miklis, M., Consonni, C., Bhat, R. A., Lipka, V., Schulze-Lefert, P., & Panstruga, R. (2007). Barley MLO modulates actin-dependent and actin-independent antifungal defense pathways at the cell periphery. *Plant Physiology*, *144*, 1132–1143.

Mims, C. W., Sewall, T. C., & Richardson, E. A. (2000). Ultrastructure of the host-pathogen relationship in *Entomosporium* leaf spot disease of *Photinia*. *International Journal of Plant Sciences*, *161*, 291–295.

Mittler, R., & Lam, E. (1996). Sacrifice in the face of foes: pathogen-induced programmed cell death in plants. *Trends in Microbiology*, *4*, 10–15.

Miya, A., Albert, P., Shinya, T., Desaki, Y., Ichimura, K., Shirasu, K., Narusaka, Y., Kawakami, N., Kaku, H., & Shibuya, N. (2007). CERK1, a LysM receptor kinase, is essential for chitin elicitor signaling in Arabidopsis. *Proceedings of the National Academy of Sciences of the United States of America*, *104*, 19613–19618.

Newman, M. A., Conrads-Strauch, J., Scofield, G., Daniels, M. J., & Dow, J. M. (1994). Defense-related gene induction in *Brassica campestris* in response to defined mutants of *Xanthomonas campestris* with altered pathogenicity. *Molecular Plant-Microbe Interactions*, *7*, 553–563.

Nimchuk, Z., Eulgem, T., Holt, B. F., 3rd, & Dangl, J. L. (2003). Recognition and response in the plant immune system. *Annual Review of Genetics*, *37*, 579–609.

Nishimura, M. T., Stein, M., Hou, B. H., Vogel, J. P., Edwards, H., & Somerville, S. C. (2003). Loss of a callose synthase results in salicylic acid-dependent disease resistance. *Science*, *301*, 969–972.

Nürnberger, T., Brunner, F., Kemmerling, B., & Piater, L. (2004). Innate immunity in plants and animals: striking similarities and obvious differences. *Immunological Reviews*, *198*, 249–266.

Oparka, K. J., & Cruz, S. S. (2000). THE GREAT ESCAPE: Phloem transport and unloading of macromolecules. *Annual Review of Plant Physiology and Plant Molecular Biology*, *51*, 323–347.

Østergaard, L., Petersen, M., Mattsson, O., & Mundy, J. (2002). An Arabidopsis callose synthase. *Plant Molecular Biology*, *49*, 559–566.

Petnicki-Ocwieja, T., Schneider, D. J., Tam, V. C., Chancey, S. T., Shan, L., Jamir, Y., Schechter, L. M., Janes, M. D., Buell, C. R., Tang, X., Collmer, A., & Alfano, J. R. (2002). Genomewide identification of proteins secreted by the Hrp type III protein secretion system of *Pseudomonas syringae* pv. tomato DC3000. *Proceedings of the National Academy of Sciences of the United States of America*, *99*, 7652–7657.

Qadota, H., Python, C. P., Inoue, S. B., Arisawa, M., Anraku, Y., Zheng, Y., Watanabe, T., Levin, D. E., & Ohya, Y. (1996). Identification of yeast Rho1p GTPase as a regulatory subunit of 1,3-β-glucan synthase. *Science*, *272*, 279–281.

Qutob, D., Kemmerling, B., Brunner, F., Kufner, I., Engelhardt, S., Gust, A. A., Luberacki, B., Seitz, H. U., Stahl, D., Rauhut, T., Glawischnig, E., Schween, G., Lacombe, B., Watanabe, N., Lam, E., Schlichting, R., Scheel, D., Nau, K., Dodt, G., Hubert, D., Gijzen, M., & Nurnberger, T. (2006). Phytotoxicity and innate immune responses induced by Nep1-like proteins. *Plant Cell*, *18*, 3721–3744.

Radford, J. E., Vesk, M., & Overall, R. L. (1998). Callose deposition at plasmodesmata. *Protoplasma*, *201*, 30–37.

Radford, J. E., & White, R. G. (2001). Effects of tissue-preparation-induced callose synthesis on estimates of plasmodesma size exclusion limits. *Protoplasma*, *216*, 47–55.

Ramming, D. W., Hinrichs, H. A., & Richardson, P. E. (1973). Sequential staining of callose by aniline blue and lacmoid for fluorescence and regular microscopy on a durable preparation of the same specimen. *Stain Technology*, *48*, 133–134.

Read, S. M., & Delmer, D. P. (1987). Inhibition of mung bean UDP-glucose:(1→3)-β-glucan synthase by UDP-pyridoxal: evidence for an active-site amino group. *Plant Physiology*, *85*, 1008–1015.

Reuber, T. L., & Ausubel, F. M. (1996). Isolation of Arabidopsis genes that differentiate between resistance responses mediated by the *RPS2* and *RPM1* disease resistance genes. *Plant Cell*, *8*, 241–249.

Rhee, Y., Tzfira, T., Chen, M.-H., Waigmann, E., & Citovsky, V. (2000). Cell-to-cell movement of tobacco mosaic virus: enigmas and explanations. *Molecular Plant Pathology*, *1*, 33–39.

Richmond, T. A., & Somerville, C. R. (2000). The cellulose synthase superfamily. *Plant Physiology*, *124*, 495–498.

Rinne, P. L., Kaikuranta, P. M., & van der Schoot, C. (2001). The shoot apical meristem restores its symplasmic organization during chilling-induced release from dormancy. *Plant Journal*, *26*, 249–264.

Rinne, P. L., van den Boogaard, R., Mensink, M. G., Kopperud, C., Kormelink, R., Goldbach, R., & van der Schoot, C. (2005). Tobacco plants respond to the constitutive expression of the tospovirus movement protein NS(M) with a heat-reversible sealing of plasmodesmata that impairs development. *Plant Journal*, *43*, 688–707.

Rinne, P. L., & van der Schoot, C. (1998). Symplasmic fields in the tunica of the shoot apical meristem coordinate morphogenetic events. *Development*, *125*, 1477–1485.

Ritter, C., & Dangl, J. L. (1996). Interference between two specific pathogen recognition events mediated by distinct plant disease resistance genes. *Plant Cell*, *8*, 251–257.

Riviere, M. P., Marais, A., Ponchet, M., Willats, W., & Galiana, E. (2008). Silencing of acidic pathogenesis-related *PR-1* genes increases extracellular β-(1→3)-glucanase activity at the onset of tobacco defense reactions. *Journal of Experimental Botany*, *59*, 1225–1239.

Roberts, A. W., & Lucas, W. J. (1990). Plasmodesmata. *Annual Review of Plant Physiology and Plant Molecular Biology*, *41*, 369–419.

Romero, GAO., Simons, C., Llanecita, M., Doohan, M., Thomas, B. R., & Rodriguez, R. L. (1998). Characterization of rice endo-β-glucanase genes (Gns2-Gns14) defines a new subgroup within the gene family. *Gene*, *223*, 311–320.

Ron, M., & Avni, A. (2004). The receptor for the fungal elicitor ethylene-inducing xylanase is a member of a resistance-like gene family in tomato. *Plant Cell*, *16*, 1604–1615.

Rose, J. K., Ham, K. S., Darvill, A. G., & Albersheim, P. (2002). Molecular cloning and characterization of glucanase inhibitor proteins: coevolution of a counterdefense mechanism by plant pathogens. *Plant Cell*, *14*, 1329–1345.

Ruan, Y. L., Xu, S. M., White, R., & Furbank, R. T. (2004). Genotypic and developmental evidence for the role of plasmodesmatal regulation in cotton fiber elongation mediated by callose turnover. *Plant Physiology*, *136*, 4104–4113.

Russo, V. M., & Bushnell, W. R. (1989). Responses of barley cells to puncture by microneedles and to attempted penetration by *Erysiphe graminis* f. sp. *hordei*. *Canadian Journal of Botany*, *67*, 2912–2921.

Ryals, J. A., Neuenschwander, U. H., Willits, M. G., Molina, A., Steiner, H. Y., & Hunt, M. D. (1996). Systemic acquired resistance. *Plant Cell*, *8*, 1809–1819.

Saile, E., McGarvey, J. A., Schell, M. A., & Denny, T. P. (1997). Role of extracellular polysaccharide and endoglucanase in root invasion and colonization of tomato plants by *Ralstonia solanacearum*. *Phytopathology*, *87*, 1264–1271.

Salanoubat, M., Genin, S., Artiguenave, F., Gouzy, J., Mangenot, S., Arlat, M., Billault, A., Brottier, P., Camus, J. C., Cattolico, L., Chandler, M., Choisne, N., Claudel-Renard, C., Cunnac, S., Demange, N., Gaspin, C., Lavie, M., Moisan, A., Robert, C., Saurin, W., Schiex, T., Siguier, P., Thebault, P., Whalen, M., Wincker, P., Levy, M., Weissenbach, J., & Boucher, C. A. (2002). Genome sequence of the plant pathogen *Ralstonia solanacearum*. *Nature*, *415*, 497–502.

Sandermann, H., Jr. (1978) Regulation of membrane enzymes by lipids. *Biochimica et Biophysica Acta*, *515*, 209–237.

Sargent, J. A., Tommerup, I. C., & Ingram, D. S. (1973). The penetration of a susceptible lettuce variety by the downy mildew fungus *Bremia lactucae* Regel. *Physiological and Molecular Plant Pathology*, *3*, 231–239.

Schlupmann, H., Bacic, A., & Read, S. M. (1993). A novel callose synthase from pollen tubes of *Nicotiana*. *Planta*, *191*, 470–481.

Schulze-Lefert, P. (2004). Knocking on the heaven's wall: pathogenesis of and resistance to biotrophic fungi at the cell wall. *Current Opinion in Plant Biology*, *7*, 377–383.

Sequeira, L. (1979). The acquisition of systemic resistance by prior inoculation. In J. M. Daly & I. Uritani (Eds.), *Recognition and Specificity in Plant Host-Parasite Interactions* (pp. 231–251). Tokyo: Japan Scientific Societies Press.

Sherwood, R. T., & Vance, C. P. (1976). Histochemistry of papillae formed in reed canarygrass leaves in response to noninfecting pathogenic fungi. *Phytopathology*, *66*, 503–510.

Shibuya, N., & Minami, E. (2001). Oligosaccharide signaling for defense responses in plant. *Physiological and Molecular Plant Pathology*, *59*, 223–233.

Shimada, C., Lipka, V., O'Connell, R., Okuno, T., Schulze-Lefert, P., & Takano, Y. (2006). Nonhost resistance in *Arabidopsis-Colletotrichum* interactions acts at the cell periphery and requires actin filament function. *Molecular Plant-Microbe Interactions*, *19*, 270–279.

Simmons, C. R. (1994). The physiology and molecular biology of plant 1,3-β-D-glucanases and 1,3;1,4-β-D-glucanases. *Critical Reviews in Plant Sciences*, *13*, 325–387.

Simmons, C. R., Litts, J. C., Huang, N., & Rodriguez, R. L. (1992). Structure of a rice β-glucanase gene regulated by ethylene, cytokinin, wounding, salicylic acid and fungal elicitors. *Plant Molecular Biology*, *18*, 33–45.

Sivaguru, M., Fujiwara, T., Samaj, J., Baluska, F., Yang, Z., Osawa, H., Maeda, T., Mori, T., Volkmann, D., & Matsumoto, H. (2000). Aluminum-induced 1→3-β-D-glucan inhibits cell-to-cell trafficking of molecules through plasmodesmata. A new mechanism of aluminum toxicity in plants. *Plant Physiology*, *124*, 991–1006.

Skou, J. P., Jorgensen, J. H., & Lilholt, U. (1984). Comparative studies on callose formation in powdery mildew compatible and incompatible barley. *Phytopathologische Zeitschrift, 109*, 147–168.

Smith, H. (1978). Recognition and defence in plants. *Nature, 273*, 266–268.

Sohn, K. H., Lei, R., Nemri, A., & Jones, J. D. (2007). The downy mildew effector proteins ATR1 and ATR13 promote disease susceptibility in *Arabidopsis thaliana*. *Plant Cell, 19*, 4077–4090.

Staal, J., Kaliff, M., Dewaele, E., Persson, M., & Dixelius, C. (2008). *RLM3*, a TIR domain encoding gene involved in broad-range immunity of Arabidopsis to necrotrophic fungal pathogens. *Plant Journal, 55*, 188–200.

Stein, M., Dittgen, J., Sánchez-Rodriguez, C., Hou, B.-H., Molina, A., Schulze-Lefert, P., Lipka, V., & Somerville, S. (2006). *Arabidopsis* PEN3/PDR8, an ATP binding cassette transporter, contributes to nonhost resistance to inappropriate pathogens that enter by direct penetration. *Plant Cell, 18*, 731–746.

Stolzenburg, M. C., Aist, J. R., & Israel, H. W. (1984). The role of papillae in resistance to powdery mildew conditioned by the *ml-o* gene in barley. I. Correlative evidence. *Physiological and Molecular Plant Pathology, 35*, 337–346.

Stone, B. A., & Clarke, A. E. (1992). *Chemistry and Biology of (1→3)-β-Glucans*. Bundoora, Australia: La Trobe University Press.

Suarez-Rodriguez, M. C., Adams-Phillips, L., Liu, Y., Wang, H., Su, S. H., Jester, P. J., Zhang, S., Bent, A. F., & Krysan, P. J. (2007). MEKK1 is required for flg22-induced MPK4 activation in *Arabidopsis* plants. *Plant Physiology, 143*, 661–669.

Szabo, L. J., & Bushnell, W. R. (2001). Hidden robbers: the role of fungal haustoria in parasitism of plants. *Proceedings of the National Academy of Sciences of the United States of America, 98*, 7654–7655.

Takemoto, D., Jones, D. A., & Hardham, A. R. (2003). GFP-tagging of cell components reveals the dynamics of subcellular re-organization in response to infection of *Arabidopsis* by oomycete pathogens. *Plant Journal, 33*, 775–792.

Tempio, J. S., & Zatz, J. L. (1981). Interaction of xanthan gum with suspended solids. *Journal of Pharmaceutical Sciences, 70*, 554–558.

Thordal-Christensen, H. (2003). Fresh insights into processes of nonhost resistance. *Current Opinion in Plant Biology, 6*, 351–357.

Thordal-Christensen, H., Zhang, Z., Wei, Y., & Collinge, D. (1997). Subcellular localization of $H_2O_2$ in plants. $H_2O_2$ accumulation in papillae and hypersensitive response during the barley-powdery mildew interaction. *Plant Journal*, *11*, 1187–1194.

Ting, J. P., & Davis, B. K. (2005). CATERPILLER: a novel gene family important in immunity, cell death, and diseases. *Annual Review of Immunology*, *23*, 387–414.

Töller, A., Brownfield, L., Neu, C., Twell, D., & Schulze-Lefert, P. (2008). Dual function of Arabidopsis glucan synthase-like genes *GSL8* and *GSL10* in male gametophyte development and plant growth. *Plant Journal*, *54*, 911–923.

Tsiamis, G., Mansfield, J. W., Hockenhull, R., Jackson, R. W., Sesma, A., Athanassopoulos, E., Bennett, M. A., Stevens, C., Vivian, A., Taylor, J. D., & Murillo, J. (2000). Cultivar-specific avirulence and virulence functions assigned to avrPphF in *Pseudomonas syringae* pv. *phaseolicola*, the cause of bean halo-blight disease. *EMBO Journal*, *19*, 3204–3214.

Turner, A., Bacic, A., Harris, P. J., & Read, S. M. (1998). Membrane fractionation and enrichment of callose synthase from pollen tubes of *Nicotiana alata* Link et Otto. *Planta*, *205*, 380–388.

Turner, A., Wells, B., & Roberts, K. (1994). Plasmodesmata of maize root tips: structure and composition. *Journal of Cell Science*, *107*, 3351–3361.

Ueki, S., & Citovsky, V. (2002). The systemic movement of a tobamovirus is inhibited by a cadmium-ion-induced glycine-rich protein. *Nature Cell Biology*, *4*, 478–486.

Umemoto, N., Kakitani, M., Iwamatsu, A., Yoshikawa, M., Yamaoka, N., & Ishida, I. (1997). The structure and function of a soybean β-glucan-elicitor-binding protein. *Proceedings of the National Academy of Sciences of the United States of America*, *94*, 1029–1034.

van der Biezen, E. A., & Jones, J. D. G. (1998). Plant disease resistance proteins and the gene-for-gene concept. *Trends in Biochemical Sciences*, *23*, 454–456.

Vance, C. P., & Sherwood, R. T. (1976). Cycloheximide treatments implicate papilla formation in resistance of reed canarygrass to fungi. *Phytopathology*, *66*, 498–502.

Verma, D. P., & Hong, Z. (2001). Plant callose synthase complexes. *Plant Molecular Biology*, *47*, 693–701.

Voegele, R. T., Struck, C., Hahn, M., & Mendgen, K. (2001). The role of haustoria in sugar supply during infection of broad bean by the rust fungus *Uromyces fabae*. *Proceedings of the National Academy of Sciences of the United States of America*, *98*, 8133–8138.

Vögeli, U., Meins, J. F., & Boller, T. (1988). Co-ordinated regulation of chitinase and β-1,3-glucanase in bean leaves. *Planta*, *174*, 364–372.

Voigt, C. A., Schäfer, W., & Salomon, S. (2005). A secreted lipase of *Fusarium graminearum* is a virulence factor required for infection of cereals. *Plant Journal*, *42*, 364–375.

Vojnov, A. A., Bassi, D. E., Daniels, M. J., & Dankert, M. A. (2002). Biosynthesis of a substituted cellulose from a mutant strain of *Xanthomonas campestris*. *Carbohydrate Research*, *337*, 315–326.

Vojnov, A. A., Slater, H., Daniels, M. J., & Dow, J. M. (2001). Expression of the *gum* operon directing xanthan biosynthesis in *Xanthomonas campestris* and its regulation in planta. *Molecular Plant-Microbe Interactions*, *14*, 768–774.

Vojnov, A. A., Zorreguieta, A., Dow, J. M., Daniels, M. J., & Dankert, M. A. (1998). Evidence for a role for the *gumB* and *gumC* gene products in the formation of xanthan from its pentasaccharide repeating unit by *Xanthomonas campestris*. *Microbiology*, *144*, 1487–1493.

Weintraub, M., Ragetli, H. W., & Leung, E. (1976). Elongated virus particles in plasmodesmata. *Journal of Ultrastructure Research*, *56*, 351–364.

Wheeler, H. (1974). Cell wall and plasmalemma modifications in diseased and injured plant tissue. *Canadian Journal of Botany*, *52*, 1005–1009.

Wilmanski, J. M., Petnicki-Ocwieja, T., & Kobayashi, K. S. (2008). NLR proteins: integral members of innate immunity and mediators of inflammatory diseases. *Journal of Leukocyte Biology*, *83*, 13–30.

Wirdnam, C., Motoyama, A., Arn-Bouldoires, E., van Eeden, S., Iglesias, A., & Meins, F., Jr. (2004) Altered expression of an ankyrin-repeat protein results in leaf abnormalities, necrotic lesions, and the elaboration of a systemic signal. *Plant Molecular Biology*, *56*, 717–730.

Wolf, S., Deom, C. M., Beachy, R., & Lucas, W. J. (1991). Plasmodesmatal function is probed using transgenic tobacco plants that express a virus movement protein. *Plant Cell*, *3*, 593–604.

Wolter, M., Hollricher, K., Salamini, F., & Schulze-Lefert, P. (1993). The *mlo* resistance alleles to powdery mildew infection in barley trigger a developmentally controlled defense mimic phenotype. *Molecular and General Genetics*, *239*, 122–128.

Xu, P., Wang, J., & Fincher, G. B. (1992). Evolution and differential expression of the (1→3)-beta-glucan endohydrolase-encoding gene family in barley, *Hordeum vulgare*. *Gene*, *120*, 157–165.

Yu, J., Penaloza-Vazquez, A., Chakrabarty, A. M., & Bender, C. L. (1999). Involvement of the exopolysaccharide alginate in the virulence and epiphytic fitness of *Pseudomonas syringae* pv. *syringae*. *Molecular Microbiology*, *33*, 712–720.

Yun, M. H., Torres, P. S., El Oirdi, M., Rigano, L. A., Gonzalez-Lamothe, R., Marano, M. R., Castagnaro, A. P., Dankert, M. A., Bouarab, K., & Vojnov, A. A. (2006). Xanthan induces plant susceptibility by suppressing callose deposition. *Plant Physiology*, *141*, 178–187.

Zambryski, P. (1995). Plasmodesmata: plant channels for molecules on the move. *Science*, *270*, 1943–1944.

Zimmerli, L., Stein, M., Lipka, V., Schulze-Lefert, P., & Somerville, S. (2004). Host and non-host pathogens elicit different jasmonate/ethylene responses in *Arabidopsis*. *Plant Journal*, *40*, 633–646.

Zipfel, C., Kunze, G., Chinchilla, D., Caniard, A., Jones, J. D., Boller, T., & Felix, G. (2006). Perception of the bacterial PAMP EF-Tu by the receptor EFR restricts Agrobacterium-mediated transformation. *Cell*, *125*, 749–760.

Zipfel, C., Robatzek, S., Navarro, L., Oakeley, E. J., Jones, J. D., Felix, G., & Boller, T. (2004). Bacterial disease resistance in *Arabidopsis* through flagellin perception. *Nature*, *428*, 764–767.

Zwiesler-Vollick, J., Plovanich-Jones, A. E., Nomura, K., Bandyopadhyay, S., Joardar, V., Kunkel, B. N., & He, S. Y. (2002). Identification of novel *hrp*-regulated genes through functional genomic analysis of the *Pseudomonas syringae* pv. tomato DC3000 genome. *Molecular Microbiology*, *45*, 1207–1218.

CHAPTER 4.5.1

# Biological and Immunological Aspects of Innate Defence Mechanisms Activated by (1,3)-β-Glucans and Related Polysaccharides in Invertebrates

**Lage Cerenius[1], Shun-ichiro Kawabata[2] and Kenneth Söderhäll[1]**
[1]*Department of Comparative Physiology, Uppsala University, Uppsala, Sweden*
[2]*Department of Biology, Faculty of Sciences, Kyushu University, Fukuoka, Japan*

(1,3)-β-glucans are powerful stimulants of a wide variety of innate defence reactions in invertebrates. These polysaccharides exert a great influence on reactions such as induction of antimicrobial peptides, cellular defence such as encapsulation and phagocytosis, and on the melanization and coagulation cascades. In most cases, these reactions set up an effective defence against microorganisms containing (1,3)-β-glucans or (1,3;1,6)-β-glucans in their outer structures (i.e. mainly fungi and oomycetes).

## I.A. Introduction

(1,3)-β-Glucans trigger several of the arms of the innate immunity responses of invertebrates. They are able to initiate cellular immune reactions, e.g. encapsulation and phagocytosis (Jiravanichpaisal et al., 2006), induce synthesis of antimicrobial peptides (AMPs) (Lemaitre and Hoffmann, 2007), and trigger coagulation (Theopold et al., 2004) and melanization (Cerenius and Söderhäll, 2004) cascades. These responses are initiated by specific recognition proteins for these polysaccharides present in the plasma or on the blood cells (haemocytes). As a result the animal is able to defend itself successfully against most fungal and oomycete pathogens, so that they do not pose a serious threat for most invertebrates. However, certain fungi, notably the entomopathogens, have evolved means to circumvent these defence measures and may cause fatal infections. One mechanism used by some of these specialized fungi is to produce stages lacking a protoplastic cell wall (thus avoiding the exposure of (1,3)-β-glucans to the immune system) when present in the haemocoel of the insect. The processes of AMP synthesis, coagulation, melanization and cellular immunity are for clarity dealt with separately below, although it can not be emphasized enough that these processes are interconnected, and therefore (1,3)-β-glucans exert a multitude of effects on the invertebrate innate immune system.

## I.B.1. Induction of AMP Synthesis

The most detailed picture of how this occurs stems from studies on *Drosophila*. In this species, (1,3)-β-glucans activate the Toll pathway and induce the production of AMPs with antifungal activities, such as drosomycin and metchnikowin (for a detailed review see Lemaitre and Hoffmann, 2007). There are several pathways that lead to Toll activation, but of most relevance here is the binding of the glucans to the GNBP3 protein (the other two guanine nucleotide-binding proteins (GNBPs) in *Drosophila* are primarily involved in sensing the presence of bacterial peptidoglycan) which, after several intermediate steps in the plasma, will result in the proteolytic processing of the pro-form of the cytokine spätzle into an active cytokine capable of binding the membrane receptor protein Toll (Fig. 1). Thus, the fly senses at least partially a fungal infection, from the presence of (1,3)-β-glucans as detected by the specific binding protein, GNBP3 (Ferrandon et al., 2007). The engagement of the Toll receptor will trigger an intracellular signal cascade in the cytosol that finally will cause release and nuclear uptake of the transcription factors Dif and Dorsal. This will lead to the induction of synthesis of several AMPs, among them drosomycin and metchnikowin, by allowing their corresponding genes to be transcribed. Thus, insect Toll, in contrast to its vertebrate counterpart, does not bind microbial products directly. The other major pathway leading to AMP synthesis in the fly, i.e. imd (immune deficiency), is not affected by GNBP3, and therefore some specificity with respect to which antimicrobial compounds are made is achieved, since the IMD pathway at least partially controls other AMP-producing genes. The result is the production of AMPs particularly effective in combating fungi. *Drosophila* GNBP3 mutants (as well as other mutants of the Toll pathway) are highly susceptible to fungal infections, a fact that can at least be partly attributed to lack of induction of the Toll pathway by (1,3)-β-glucans in these animals. Fungi may, however, activate the Toll-dependent genes by other means, independently of the presence of (1,3)-β-glucans, e.g. by releasing extracellular proteinases, and possibly other virulence factors as well, that act directly or indirectly on the spätzle-processing enzyme without involvement of GNBP3 (Gottar et al., 2006). Although less investigated, it

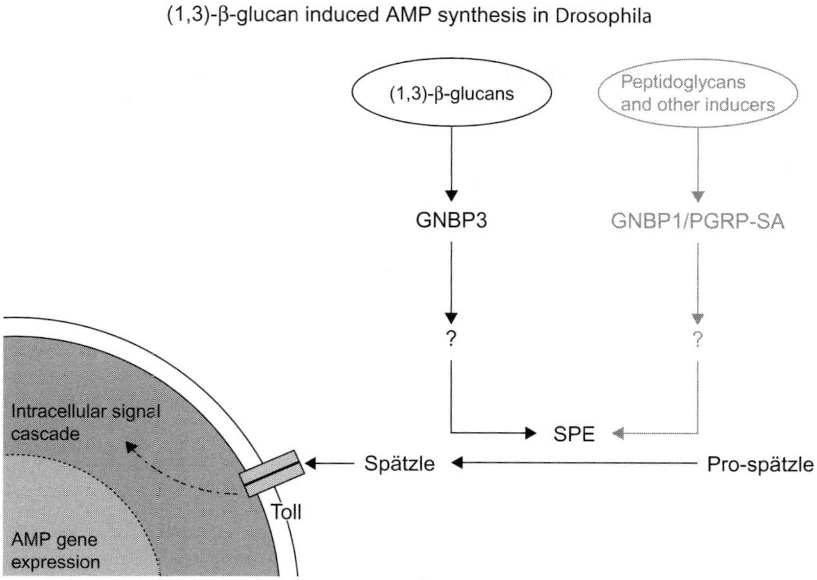


**Fig. 1: 1,3-β-glucan-induced antimicrobial peptide (AMP) synthesis in *Drosophila*. The presence of 1,3-β-glucans will be sensed by the so-called Gram-negative bacteria-binding protein (GNBP) 3, which triggers a proteolytic cascade (still not completely known) that cleaves the zymogenic proform of the spätzle-processing enzyme (SPE) to the active form. Active SPE cleaves pro-spätzle to its active form, spätzle, a protein with capacity to engage the Toll receptor at the cell membrane. SPE may be activated also via other proteolytic cascades triggered by e.g. peptidoglycans through GNBP1 and peptidoglycan recognition protein-SA (PGRP-SA) (indicated by grey text). The activated Toll will initiate an intracellular signalling cascade that ultimately will result in transcription of genes coding for AMPs.**

is likely that AMP synthesis is induced in a similar way in other arthropods as GNBP homologues capable of binding (1,3)-β-glucans are widespread, and their presence has been noted in other insects (Ochiai and Ashida, 2000) as well as other invertebrates (Lee et al., 2000). In most cases, however, such GNBP proteins have so far mainly been implicated in other defence reactions, notably in the melanization cascade (Section I.B.3.).

## I.B.2. Coagulation

Haemolymph clotting is an important innate immune reaction by which bleeding is controlled and micoorganisms entrapped and immobilized. The blood clotting mechanisms vary greatly between different groups of invertebrates. These mechanisms have been particularly well studied in horseshoe crabs and in crustaceans. In horseshoe crabs, e.g. *Tachypleus tridentatus*, clotting is achieved by a proteolytic cascade terminating with conversion of the soluble coagulogen protein into coagulin fibres consisting of coagulin (Iwanaga and Lee, 2005), whereas in crustaceans a transglutaminase modifies the clotting protein that subsequently polymerizes (Hall et al., 1999).

The major host defence system in the horseshoe crab *T. tridentatus* is carried in the haemolymph that contains one type of granular haemocyte comprising 99% of all haemocytes (Toh et al., 1991). The granular haemocyte is filled with two types of granules which selectively store coagulation factors and other immune proteins, such as lectins and antimicrobial peptides (Iwanaga et al., 1998; Shigenaga et al., 1993). The granular haemocyte is highly sensitive to lipopolysaccharides (LPS), and the stimulation by LPS leads to the immediate secretion of the defence molecules via LPS-induced haemocyte exocytosis (Ariki et al., 2004; Koshiba et al., 2007). The coagulation cascade of horseshoe crabs is composed of a clottable protein coagulogen and four serine protease zymogens, including factor C, factor G, factor B and the proclotting enzyme (Fig. 2). Factor C contains several short consensus repeats or sushi domains, typically associated with deuterostome complement, but in general the clotting factors are not structurally similar to complement proteins.

Factor G is one of the most important pattern-recognition proteins in the horseshoe crab's innate immune system and functions as a sensitive biosensor for (1,3)-β-glucans, which trigger the sequential activation of the serine protease zymogens leading to the conversion of coagulogen to insoluble coagulin homopolymers through head-to-tail interaction (Bergner et al., 1996; Kawasaki et al., 2000). Factor G is a heterodimeric serine protease zymogen composed of two non-covalently associated subunits, subunit α and subunit β (Seki et al., 1994). Subunit α contains only one domain of a serine protease zymogen, and subunit β functions as a pattern-recognition subunit, which is composed of three types of glycosidase-like domain: a (1,3)-β-glucanase A1-like domain composed of one unit, a xylanase A-like domain composed of three repeating units, and a xylanase Z-like domain composed of two repeating units with 87% sequence identity named Z1 and Z2. Among the three types of glycosidase-like domains in the subunit, each of the Z1 and Z2 units in the xylanase Z-like domain has been identified as an independent binding site for (1,3)-β-glucans, suggesting that the duplicated binding sites for (1,3)-β-glucans in the subunit may increase in avidity to allow the stable and specific recognition of pathogens (Takaki et al., 2002).

Moreover, LPS has been shown to induce the secretion of transglutaminase (TGase) from the haemocyte cytosol through an unknown mechanism (Osaki and Kawabata, 2004; Osaki et al.,

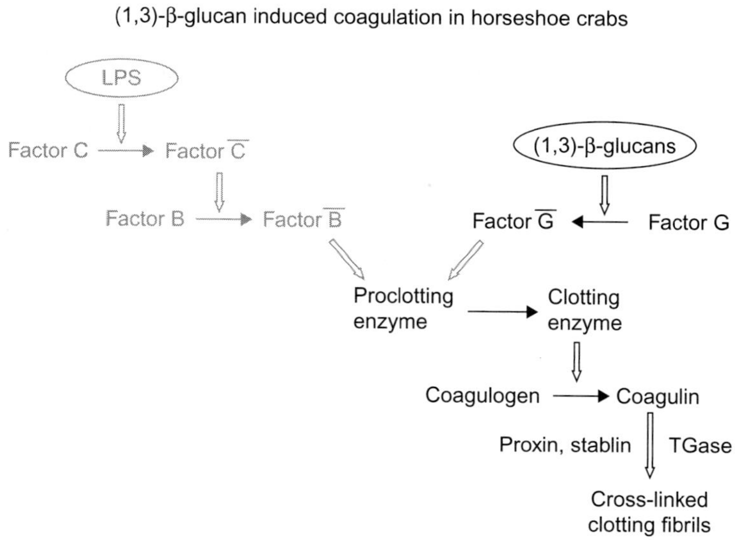


**Fig. 2: 1,3-β-glucan-induced coagulation in horseshoe crabs. Stimulation of haemocytes by microbial polysaccharides causes the release of components of the coagulation system. In plasma, proclotting enzyme may be processed into clotting enzyme by the 1,3-β-glucan-sensitive factor G or via another branch consisting of factor C and B (grey text). The latter pathway is triggered by lipopolysaccharide (LPS). The active clotting enzyme will bring about the conversion of coagulogen to insoluble coagulin homopolymers. The resulting coagulin polymers form stable clotting fibrils by transglutaminase-dependent cross-linking with haemocyte or haemolymph-derived proteins, such as proxin and stablin.**

2002). Horseshoe crab TGase shows a significant sequence similarity with mammalian TGase members (33~38% identity) (Tokunaga et al., 1993a; Tokunaga et al., 1993b). Although horseshoe crab TGase neither catalyses monodansylcadaverine incorporation into coagulin nor cross-links coagulin intermolecularly, TGase cross-links coagulin with a haemocyte-derived protein, named proxin (proline-rich protein for protein cross-linking) (Osaki et al., 2002). In addition, TGase cross-links proxin with another haemocyte-derived protein named stablin (clot-stabilizing protein), resulting in more stably clotting fibrils (Matsuda et al., 2007). Stablin is a component of the clotting fibrils and contributes to the proper formation of the clotting fibrils. Stablin also interacts with LPS and lipoteichoic acids and exhibits bacterial agglutinating activity, which possibly contributes to the bacterial immobilizing activity of the clotting fibrils. At least three additional haemolymph proteins, $\alpha_2$-macroglobulin, C-reactive proteins and haemocyanin, may be involved in the formation of the clotting fibrils (Isakova and Armstrong, 2003).

## I.B.3. Melanization Cascade

Melanin production is a common response to microbial intruders (Fig. 3) as well as mechanical injuries in most invertebrates (for reviews see e.g. Cerenius et al. 2008, Cerenius and Söderhäll, 2004; Söderhäll and Cerenius, 1998). The presence of (1,3)-β-glucans in extremely low concentrations trigger the melanization cascade or as it is commonly called the prophenoloxidase-activating system (proPO system). The specificity of the reaction is high; other glucans such as cellulose or dextran are without effect, and in crustaceans, where the conditions for activation have been thoroughly investigated, a linear (1,3)-β-oligoglucoside, laminaripentaose, suffices to induce activation of the proPO system (Söderhäll and Unestam, 1979). Larger (1,3)-β-glucans such as curdlan (containing solely 1,3-β-linkages) as well as laminarins (also containing some 1,6-β-linkages) are efficient triggers of the proPO system. Several intermediates in the pathways to melanin exert cytotoxic and microcidal activities, and will thus retard microbial growth (Zhao et al., 2007). If melanin production is prevented by for example specific gene knock-outs, the immune capacity of the animal is often significantly reduced (Beck and Strand, 2007; Eleftherianos et al., 2007; Liu et al., 2007). The melanin is produced by phenoloxidase (PO), an enzyme that is synthesized as a zymogenic precursor that is activated by a proteolytic cascade. The cascade is initiated by binding of (1,3)-β-glucans, peptidoglycans and other elicitors to their respective recognition/binding proteins, and this initial recognition event will trigger a chain of proteolytic enzymes that

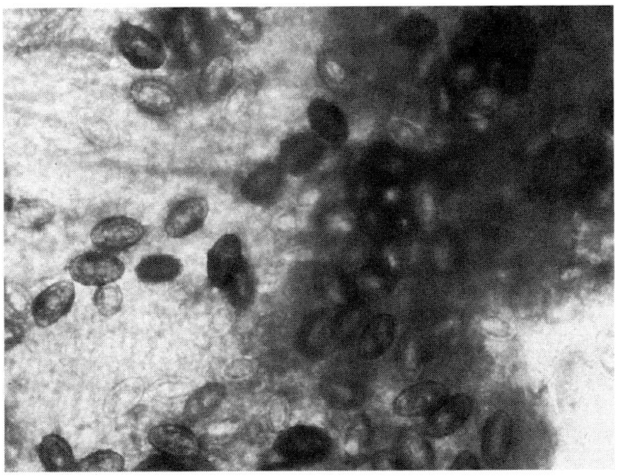

**Fig. 3: Melanization of a microbial pathogen. The parasite *Psorospermium haeckeli* becomes melanized (brown colouring) in haemolymph of the freshwater crayfish *Astacus astacus*. The colour specifications refer to colours in panels.**

terminates by the PO-activating proteinase(s) (Fig. 4). The latter produces catalytically active PO by a limited and very specific proteolysis of the proPO protein. In some insects the PO needs to bind to other factors such as certain serine proteinase homologues, in addition of undergoing proteolytic cleavage, in order to express PO activity (Gupta et al., 2005; Piao et al., 2005). Since toxic substances, potentially damaging tissues of the producing organism, appear during melanin production, careful temporal and spatial control of proPO activation and PO activity is needed. This is brought about by a complex system of controlling factors; for example, serine proteinase inhibitors that prevent excessive activation of the cascade

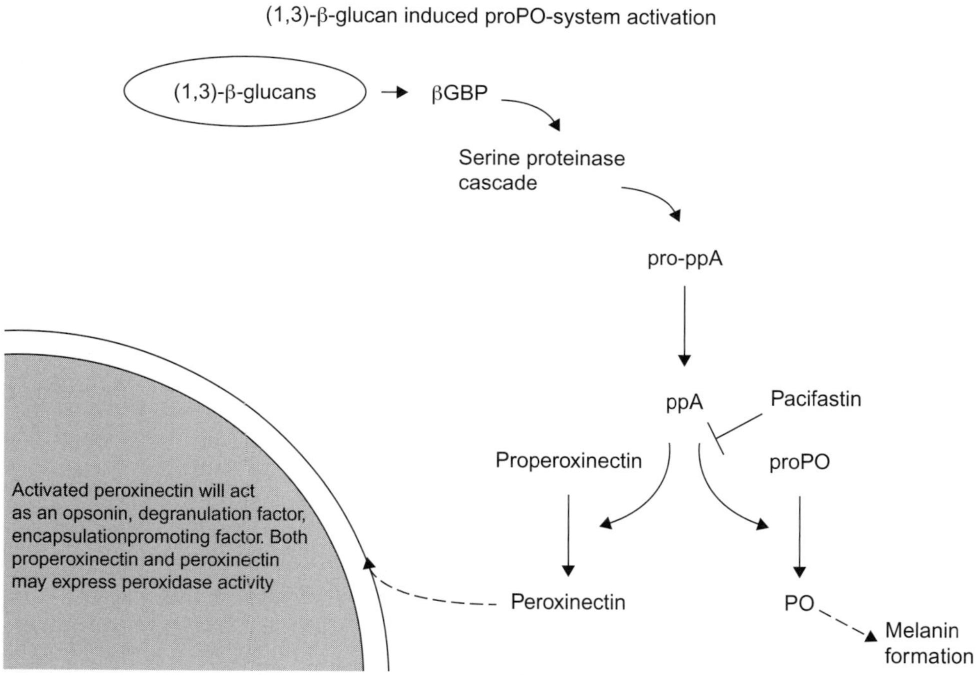


**Fig. 4: 1,3-β-Glucan-induced proPO activation in crustaceans. The presence of 1,3-β-glucans is sensed by glucan-binding proteins (βGBPs) in the plasma as well as by haemocyte membrane receptors (not shown) and will trigger a proteolytic cascade leading to the proteolytic processing of zymogenic pro-phenoloxidase-activating enzyme (pro-ppA). Active ppA proteolytically processes inactive zymogenic prophenoloxidase (proPO) into active PO, capable of producing toxic quinone intermediates as well as melanin. The active ppA will also cause the conversion of pro-peroxinectin into peroxinectin, a protein greatly stimulating several cellular defence reactions such as phagocytosis and encapsulation. Active ppA is negatively regulated by pacifastin, a large heterodimeric serine proteinase inhibitor.**

(Liang et al., 1997; Tong and Kanost, 2005) and other factors interfering with melanin production of activated PO (Zhao et al., 2005). Furthermore, active PO is linked to other proteins and solid surfaces, which also will contribute to restrict the spread of toxic compounds.

The (1,3)-β-glucan-induced activation of the proPO system is mediated by a variety of different proteins that specifically bind to these polysaccharides. The first glucan-binding proteins involved in proPO activation to be indentified were purified and biochemically characterized from the insects *Blaberus craniifer* (Söderhäll et al., 1988) and *Bombyx mori* (Ochiai and Ashida, 1988), and the first to be cloned was from the freshwater crayfish *Pacifastacus leniusculus* (Cerenius et al., 1994). In *P. leniusculus* , a (1,3)-β-glucan-binding protein present in plasma (Duvic and Söderhäll, 1990) was found, after having been reacted with (1,3)-β-glucans, to bind to a specific haemocyte membrane receptor and trigger exoctyosis of proPO components from the cell (Duvic and Söderhäll, 1992). Thus, this binding to the membrane receptor ensures that the initial immune response will be further augmented by the released blood cell components. The crayfish glucan-binding protein does not express any measurable (1,3)-β-glucanase activity (Cerenius et al., 1994), so it is possible that the bound complex is ultimately taken up by some blood cells to clear the plasma of the continuous presence of immunogenic (1,3)-β-glucans that may exhaust the immune system.

As indicated above many invertebrates and especially arthropods possess proteins homologous to *Drosophila* GNBP3 that mediate glucan-triggered activation of the melanization cascade. In arthropods, such proteins and their biological activity have been characterized in detail in silk worm, *B. mori*, (Ochiai and Ashida, 2000), freshwater crayfish, *P. leniusculus* (Lee et al., 2000), the tobacco hornworm *Manduca sexta* (Ma and Kanost, 2000), and the moth *Plodia interpunctella* (Fabrick et al., 2004). A GNBP homologue involved in proPO activation has also been cloned and characterized from the earthworm *Eisenia foetida* (Beschin et al., 1998). Interestingly, a similar protein has been found in the sponge, *Suberites domuncula*, that is capable of binding (1,3)-β-glucans, and whose expression is strongly stimulated by the presence of such glucans and is localized to tissues likely to come into contact with exogenous glucans (Perovic-Ottstadt et al., 2004). However, an immunological function for the sponge GNBP remains to be established.

(1,3)-β-Glucan-binding proteins will, in addition to promoting melanization, participate in other humoral immune reactions such as the agglutination of microorganisms. It should be stressed, however, that during proPO activation opsonic and encapsulation-promoting activities are produced that may be difficult to separate from possible lectin-like activities independent of the proPO system. Several lectin or lectin-like activities with (1,3)-β-glucan specificity have been described in different arthropods, such as the cockroach *Blaberus discoidalis* (Chen et al., 1999) and the silk worm *B. mori* (Miyake et al., 2005). However, their biological activities and their full significance remain to be established.

## I.B.4. Cellular Immunity

Invertebrate blood cells react to (1,3)-β-glucans by releasing factors augmenting the cellular defence, such as encapsulation promoting factors, opsonins, triggers of coagulation and, as discussed above, components of the melanization cascade. These reactions may be triggered by plasma recognition proteins, membrane receptor proteins for (1,3)-β-glucans (although any invertebrate homologues to the vertebrate dectins (Brown, 2006) have not yet been found), or binding of a complex of glucan and plasma-binding proteins to a membrane receptor as indicated above. This engagement of blood cell receptors will lead to the release of proPO-activating factors as described above, but also to the production of factors stimulating blood cells to phagocytose, encapsulate foreign objects, produce cytokines to stimulate further blood cell production and differentiation, etc. For a comprehensive review of these aspects the reader is referred to Jiravanichpaisal et al. (2006). Here, a few relatively well investigated cellular immune reactions stimulated by (1,3)-β-glucans in crustaceans will be reiterated. In crustacean plasma, a large serine proteinase homologue (masquerade-like protein) (Huang et al., 2000) is present. Upon binding to (1,3)-β-glucans or Gram-negative bacteria this protein is proteolytically processed into five subunits, one of which has cell adhesion activity and will mediate blood cell adhesion, and processed masquerade-protein will act as an opsonin aiding clearance of microorganisms from the blood. (1,3)-β-Glucans trigger the production of additional opsonins. For example, these glucans stimulate release of proPO-system components together with a myeloperoxidase homologue, properoxinectin, from the granular type of haemocytes (Johansson et al., 1995; Johansson and Söderhäll, 1988). The activation of the proPO cascade generates active proPO-activating enzyme, which besides cleaving proPO into catalytcally active PO also, from properoxinectin, generates peroxinectin with cell adhesion, opsonin and encapsulation promoting activities (Jiravanichpaisal et al., 2006; Kobayashi et al., 1990). The reaction is greatly stimulated by the presence of (1,3)-β-glucans in the plasma and, thus, these polysaccharides are involved both in triggering release of immune components from the haemocytes and in their subsequent activation in the extracellular milieu. The active peroxinectin promotes for example phagocytosis and the encapsulation of larger objects (Fig. 3). This protein is also an active peroxidase, so it is likely that is has a dual role in affecting blood cell behaviour and the enzymatic production of antimicrobial compounds.

Administration of (1,3)-β-glucans into whole animals has dramatic effects on haemocyte number and composition (Smith and Söderhäll, 1983) and may also affect gene expression (Cerenius et al., 2003). In crustaceans, haemocyte numbers may temporarily drop by some 90% upon injecting (1,3)-β-glucans. These haemocytes are partly replenished by blood cells in temporary storage but the immune stimulation will also lead to blood cell renewal by haematopoiesis (Söderhäll et al., 2003). Blood cell differentiation is stimulated by cytokines released from immunologically active haemocytes. Such a cytokine, named astakine since it contains a prokineticin sequence motif, was discovered in freshwater crayfish and has been studied in detail in several crustaceans (Söderhäll et al., 2005). It is released into plasma from circulating mature haemocytes and will stimulate the haematopoietic organ into producing new mature granular haemocytes. Mature haemocytes, in contrast to haematopoietic cells, will produce proPO, and the ability to express the proPO gene is a hallmark of the mature granular cell.

The effects of (1,3)-β-glucans on the arthropod immune system have led to the production of commercial products containing these glucans; for example, yeast cell wall preparations to be used during farming of shrimps and other shellfish. However, careful monitoring of several immune parameters by quantitative polymerase chain reaction, and other methods, of a number of relevant gene activities has failed to provide data indicating any significant long-term effects on immune capacity by exposure to such glucan preparations (Hauton et al., 2007; Hauton et al., 2005). On the contrary, the animal has to spend a considerable amount of energy in disarming these powerful immunostimulants to avoid exhausting its defence capacity and to be able to funnel resources into other reactions necessary for survival.

## References


Ariki, S., Koori, K., Osaki, T., Motoyama, K., Inamori, K. I., & Kawabata, S. (2004). A serine protease zymogen functions as a pattern-recognition receptor for lipopolysaccharides. *Proceedings of the National Academy of Sciences of the United States of America*, *101*, 953–958.

Beck, M. H., & Strand, M. R. (2007). A novel polydnavirus protein inhibits the insect prophenoloxidase activation pathway. *Proceedings of the National Academy of Sciences*, *104*, 19267–19272.

Bergner, A., Oganessyan, V., Muta, T., Iwanaga, S., Typke, D., Huber, R., & Bode, W. (1996). Crystal structure of coagulogen, the clotting protein from horseshoe crab: A structural homologue of nerve growth factor. *EMBO Journal*, *15*, 6789–6797.

Beschin, A., Bilej, M., Hanssens, F., Raymakers, J., Van Dyck, E., Revets, H., Brys, L., Gomez, J., De Baetselier, P., & Timmermans, M. (1998). Identification and cloning of a glucan- and lipopolysaccharide-binding protein from *Eisenia foetida* earthworm involved in the activation of prophenoloxidase cascade. *Journal of Biological Chemistry*, *273*, 24948–24954.

Brown, G. D. (2006). Dectin-1: a signalling non-TLR pattern-recognition receptor. *Nature Reviews Immunology*, *6*, 33–43.

Cerenius, L., Lee, B. L., & Söderhäll, K. (2008). The proPO-system, pros and cos for its role in invertebarte immunity. *Trends in Immunology*, *29*, 263–271.

Cerenius, L., Bangyeekhun, E., Keyser, P., Söderhäll, I., & Söderhäll, K. (2003). Host prophenoloxidase expression in freshwater crayfish is linked to increased resistance to the crayfish plague fungus, *Aphanomyces astaci*. *Cell Microbiol*, *5*, 353–357.

Cerenius, L., Liang, Z. C., Duvic, B., Keyser, P., Hellman, U., Palva, E. T., Iwanaga, S., & Söderhäll, K. (1994). Structure and biological activity of a 1,3-beta-D-glucan-binding protein in crustacean blood. *Journal of Biological Chemistry*, *269*, 29462–29467.

Cerenius, L., & Söderhäll, K. (2004). The prophenoloxidase-activating system in invertebrates. *Immunological Reviews*, *198*, 116–126.

Chen, C. L., Rowley, A. F., Newton, R. P., & Ratcliffe, N. A. (1999). Identification, purification and properties of a beta-1,3-glucan-specific lectin from the serum of the cockroach, *Blaberus discoidalis* which is implicated in immune defence reactions. *Comparative Biochemistry and Physiology B-Biochemistry & Molecular Biology*, *122*, 309–319.

Duvic, B., & Söderhäll, K. (1990). Purification and characterization of a beta-1,3-Glucan binding-protein from plasma of the crayfish *Pacifastacus leniusculus*. *Journal of Biological Chemistry*, *265*, 9327–9332.

Duvic, B., & Söderhäll, K. (1992). Purification and partial characterization of a beta-1,3-Glucan-binding-protein membrane-receptor from blood-cells of the crayfish *Pacifastacus leniusculus*. *European Journal of Biochemistry*, *207*, 223–228.

Eleftherianos, I., Boundy, S., Joyce, S. A., Aslam, S., Marshall, J. W., Cox, R. J., Simpson, T. J., Clarke, D. J., ffrench-Constant, R. H., & Reynolds, S. E. (2007). An antibiotic produced by an insect-pathogenic bacterium suppresses host defenses through phenoloxidase inhibition. *Proceedings of the National Academy of Sciences of the United States of America*, *104*, 2419–2424.

Fabrick, J. A., Baker, J. E., & Kanost, M. R. (2004). Innate immunity in a pyralid moth-Functional evaluation of domains from a beta-1,3-glucan recognition protein. *Journal of Biological Chemistry*, *279*, 26605–26611.

Ferrandon, D., Imler, J. L., Hetru, C., & Hoffmann, J. A. (2007). The *Drosophila* systemic immune response: sensing and signalling during bacterial and fungal infections. *Nature Reviews Immunology*, *7*, 862–874.

Gottar, M., Gobert, V., Matskevich, A. A., Reichhart, J. M., Wang, C. S., Butt, T. M., Belvin, M., Hoffmann, J. A., & Ferrandon, D. (2006). Dual detection of fungal infections in *Drosophila* via recognition of glucans and sensing of virulence factors. *Cell*, *127*, 1425–1437.

Gupta, S., Wang, Y., & Jiang, H. (2005). *Manduca sexta* prophenoloxidase (proPO) activation requires proPO-activating proteinase (PAP) and serine proteinase homologs (SPHs) simultaneously. *Insect Biochemistry and Molecular Biology*, *35*, 241–248.

Hall, M., Wang, R. G., van Antwerpen, R., Sottrup-Jensen, L., & Söderhäll, K. (1999). The crayfish plasma clotting protein: A vitellogenin-related protein responsible for clot formation in crustacean blood. *Proceedings of the National Academy of Sciences of the United States of America*, *96*, 1965–1970.

Hauton, C., Brockton, V., & Smith, V. J. (2007). Changes in immune gene expression and resistance to bacterial infection in lobster (*Homarus gammarus*) post-larval stage VI following acute or chronic exposure to immune stimulating compounds. *Molecular Immunology*, *44*, 443–450.

Hauton, C., Hammond, J. A., & Smith, V. J. (2005). Real-time PCR quantification of the in vitro effects of crustacean immunostimulants on gene expression in lobster (*Homarus gammarus*) granular haemocytes. *Developmental and Comparative Immunology*, *29*, 33–42.

Huang, T. S., Wang, H., Lee, S. Y., Johansson, M. W., Söderhäll, K., & Cerenius, L. (2000). A cell adhesion protein from the crayfish *Pacifastacus leniusculus*, a serine proteinase homologue similar to *Drosophila* masquerade. *Journal of Biological Chemistry*, *275*, 9996–10001.

Isakova, V., & Armstrong, P. B. (2003). Imprisonment in a death-row cell: The fates of microbes entrapped in the *Limulus* blood clot. *Biological Bulletin*, *205*, 203–204.

Iwanaga, S., Kawabata, S., & Muta, T. (1998). New types of clotting factors and defense molecules found in horseshoe crab hemolymph: Their structures and functions. *Journal of Biochemistry*, *123*, 1–15.

Iwanaga, S., & Lee, B. L. (2005). Recent advances in the innate immunity of invertebrate animals. *Journal of Biochemistry and Molecular Biology*, *38*, 128–150.

Jiravanichpaisal, P., Lee, B. L., & Söderhäll, K. (2006). Cell-mediated immunity in arthropods: Hematopoiesis, coagulation, melanization and opsonization. *Immunobiology*, *211*, 213–236.

Johansson, M. W., Lind, M. I., Holmblad, T., Thörnqvist, P. O., & Söderhäll, K. (1995). Peroxinectin, a novel cell-adhesion protein from crayfish blood. *Biochemical and Biophysical Research Communications*, *216*, 1079–1087.

Johansson, M. W., & Söderhäll, K. (1988). Isolation and purification of a cell adhesion factor from crayfish blood cells. *Journal of Cell Biology*, *106*, 1795–1803.

Kawasaki, H., Nose, T., Muta, T., Iwanaga, S., Shimohigashi, Y., & Kawabata, S. (2000). Head-to-tail polymerization of coagulin, a clottable protein of the horseshoe crab. *Journal of Biological Chemistry*, *275*, 35297–35301.

Kobayashi, M., Johansson, M. W., & Söderhäll, K. (1990). The 76 kD cell-adhesion factor from crayfish hemocytes promotes encapsulation in vitro. *Cell and Tissue Research*, *260*, 13–18.

Koshiba, T., Hashii, T., & Kawabata, S. (2007). A structural perspective on the interaction between lipopolysaccharide and Factor C, a receptor involved in recognition of Gram-negative bacteria. *Journal of Biological Chemistry*, *282*, 3962–3967.

Lee, S. Y., Wang, R. G., & Söderhäll, K. (2000). A lipopolysaccharide- and beta-1,3-glucan-binding protein from hemocytes of the freshwater crayfish *Pacifastacus leniusculus*-Purification, characterization, and cDNA cloning. *Journal of Biological Chemistry*, *275*, 1337–1343.

Lemaitre, B., & Hoffmann, J. (2007). The host defense of *Drosophila melanogaster*. *Annual Review of Immunology*, *25*, 697–743.

Liang, Z. C., SottrupJensen, L., Aspan, A., Hall, M., & Söderhäll, K. (1997). Pacifastin, a novel 155-kDa heterodimeric proteinase inhibitor containing a unique transferrin chain. *Proceedings of the National Academy of Sciences of the United States of America*, *94*, 6682–6687.

Liu, H., Jiravanichpaisal, P., Cerenius, L., Lee, B. L., Söderhäll, I., & Söderhäll, K. (2007). Phenoloxidase is an important component of the defense against *Aeromonas hydrophila* infection in a crustacean, *Pacifastacus leniusculus*. *Journal of Biological Chemistry*, *282*, 33593–33598.

Ma, C. C., & Kanost, M. R. (2000). A beta 1,3-glucan recognition protein from an insect, *Manduca sexta*, agglutinates microorganisms and activates the phenoloxidase cascade. *Journal of Biological Chemistry*, *275*, 7505–7514.

Matsuda, Y., Osaki, T., Hashii, T., Koshiba, T., & Kawabata, S. (2007). A cysteine-rich protein from an arthropod stabilizes clotting mesh and immobilizes bacteria at injury sites. *Journal of Biological Chemistry*, *282*, 33545–33552.

Miyake, S., Yamano, Y., & Morishima, I. (2005). Promoting protein, a silkworm hemolymph protein promoting in vitro replication of nucleopolyhedrovirus, binds to beta-glucans. *Bioscience Biotechnology and Biochemistry*, *69*, 2012–2014.

Ochiai, M., & Ashida, M. (1988). Purification of a beta-1,3-glucan recognition protein in the prophenoloxidase activating system from hemolymph of the silkworm, *Bombyx mori*. *Journal of Biological Chemistry*, *263*, 12056–12062.

Ochiai, M., & Ashida, M. (2000). A pattern-recognition protein for beta-1,3-glucan-The binding domain and the cDNA cloning of beta-1,3-glucan recognition protein from the silkworm, *Bombyx mori*. *Journal of Biological Chemistry*, *275*, 4995–5002.

Osaki, T., & Kawabata, S. (2004). Structure and function of coagulogen, a clottable protein in horseshoe crabs. *Cellular and Molecular Life Sciences*, *61*, 1257–1265.

Osaki, T., Okino, N., Tokunaga, F., Iwanaga, S., & Kawabata, S. (2002). Proline-rich cell surface antigens of horseshoe crab hemocytes are substrates for protein cross-linking with a clotting protein coagulin. *Journal of Biological Chemistry*, *277*, 40084–40090.

Perovic-Ottstadt, S., Adell, T., Proksch, P., Wiens, M., Korzhev, M., Gamulin, V., Muller, I. M., & Muller, W. E. G. (2004). A (1 -> 3)-beta-D-glucan recognition protein from the sponge *Suberites domuncula*-Mediated activation of fibrinogen-like protein and epidermal growth factor gene expression. *European Journal of Biochemistry*, *271*, 1924–1937.

Piao, S., Song, Y. L., Kim, J. H., Park, S. Y., Park, J. W., Lee, B. L., Oh, B. H., & Ha, N. C. (2005). Crystal structure of a clip-domain serine protease and functional roles of the clip domains. *EMBO Journal*, *24*, 4404–4414.

Seki, N., Muta, T., Oda, T., Iwaki, D., Kuma, K., Miyata, T., & Iwanaga, S. (1994). Horseshoe-crab (1,3)-beta-D-glucan-sensitive coagulation Factor-G-a serine-protease zymogen heterodimer with similarities to beta-glucan-binding proteins. *Journal of Biological Chemistry*, *269*, 1370–1374.

Shigenaga, T., Takayenoki, Y., Kawasaki, S., Seki, N., Muta, T., Toh, Y., Ito, A., & Iwanaga, S. (1993). Separation of large and small granules from horseshoe-crab (*Tachypleus tridentatus*) hemocytes and characterization of their components. *Journal of Biochemistry*, *114*, 307–316.

Smith, V. J., & Söderhäll, K. (1983). Beta-1,3 glucan activation of crustacean hemocytes *in vitro* and *in vivo*. *Biological Bulletin*, *164*, 299–314.

Söderhäll, I., Bangyeekhun, E., Mayo, S., & Söderhäll, K. (2003). Hemocyte production and maturation in an invertebrate animal; proliferation and gene expression in hematopoietic stem cells of *Pacifastacus leniusculus*. *Developmental and Comparative Immunology*, *27*, 661–672.

Söderhäll, I., Kim, Y. A., Jiravanichpaisal, P., Lee, S. Y., & Söderhäll, K. (2005). An ancient role for a prokineticin domain in invertebrate hematopoiesis. *Journal of Immunology*, *174*, 6153–6160.

Söderhäll, K., & Cerenius, L. (1998). Role of the prophenoloxidase-activating system in invertebrate immunity. *Current Opinion in Immunology*, *10*, 23–28.

Söderhäll, K., Rögener, W., Söderhäll, I., Newton, R. P., & Ratcliffe, N. A. (1988). The properties and purification of a *Blaberus craniifer* plasma-protein which enhances the activation of hemocyte prophenoloxidase by a beta-1,3-Glucan. *Insect Biochemistry*, *18*, 323–330.

Söderhäll, K., & Unestam, T. (1979). Activation of serum prophenoloxidase in arthropod immunity. The specificity of cell wall glucan activation and activation by purified fungal glycoproteins of crayfish phenoloxidase. *Canadian Journal of Microbiology*, *25*, 406–414.

Takaki, Y., Seki, N., Kawabata, S., Iwanaga, S., & Muta, T. (2002). Duplicated binding sites for (1 -> 3)-beta-D-glucan in the horseshoe crab coagulation factor G-Implications for a molecular basis of the pattern recognition in innate immunity. *Journal of Biological Chemistry*, *277*, 14281–14287.

Theopold, U., Schmidt, O., Söderhäll, K., & Dushay, M. S. (2004). Coagulation in arthropods: defence, wound closure and healing. *Trends in Immunology*, *25*, 289–294.

Toh, Y., Mizutani, A., Tokunaga, F., Muta, T., & Iwanaga, S. (1991). Morphology of the granular hemocytes of the Japanese horseshoe-crab *Tachypleus tridentatus* and immunocytochemical localization of clotting factors and antimicrobial substances. *Cell and Tissue Research*, *266*, 137–147.

Tokunaga, F., Muta, T., Iwanaga, S., Ichinose, A., Davie, E. W., Kuma, K., & Miyata, T. (1993a). Limulus hemocyte transglutaminase-cDNA cloning, amino-acid-sequence, and tissue localization. *Journal of Biological Chemistry*, *268*, 262–268.

Tokunaga, F., Yamada, M., Miyata, T., Ding, Y. L., Hiranagakawabata, M., Muta, T., Iwanaga, S., Ichinose, A., & Davie, E. W. (1993b). Limulus hemocyte transglutaminase – Its purification and characterization, and identification of the intracellular substrates. *Journal of Biological Chemistry*, *268*, 252–261.

Tong, Y., & Kanost, M. R. (2005). *Manduca sexta* serpin-4 and serpin-5 inhibit the prophenol oxidase activation pathway: cDNA cloning, protein expression, and characterization. *Journal of Biological Chemistry*, *280*, 14923–14931.

Zhao, M. Y., Söderhäll, I., Park, J. W., Ma, Y. G., Osaki, T., Ha, N. C., Wu, C. F., Söderhäll, K., & Lee, B. L. (2005). A novel 43-kDa protein as a negative regulatory component of phenoloxidase-induced melanin synthesis. *Journal of Biological Chemistry*, *280*, 24744–24751.

Zhao, P., Li, J., Wang, Y., & Jiang, H. (2007). Broad-spectrum antimicrobial activity of the reactive compounds generated in vitro by *Manduca sexta* phenoloxidase. *Insect Biochemistry and Molecular Biology*, *37*, 952–959.

CHAPTER 4.5.2

# (1,3)-β-Glucans in Innate Immunity: Mammalian Systems

**Gordon D. Brown[1] and David L. Williams[2]**

[1]*Institute of Infectious Disease and Molecular Medicine, Faculty of Health Sciences, University of Cape Town, Cape Town, South Africa*

[2]*Departments of Surgery and Pharmacology, James H. Quillen College of Medicine, East Tennessee State University, Johnson City, TN, USA*

The ability of (1,3)-β-glucans to stimulate innate immunity and modulate disease outcome has been known for more than four decades. However, the vast majority of these studies were phenomenological in nature. What was lacking was a mechanistic understanding of how (1,3)-β-glucans modulate innate immunity and alter disease outcome. Recent advances have dramatically increased our understanding of how glucans modulate innate immunity at the cellular and molecular level. The discovery of membrane-associated glucan-specific pattern recognition receptors, such as Dectin-1, has provided insights into how (1,3)-β-glucans are recognised by cells. New light has also been shed on the intracellular signal transduction pathways that are activated by glucans and how modulation of these signalling pathways ameliorates disease. This chapter focuses on recent advances in our understanding of the mechanisms by which (1,3)-β-glucans modulate innate immunity in mammals, and how modulation of innate immunity with glucan alters host response to disease.

## I.A. Introduction

There is an extensive literature demonstrating the ability of (1,3)-β-glucans to stimulate innate immunity and modulate disease outcome in mammalian systems (Bohn and BeMiller, 1995; Chen and Seviour, 2007; Williams et al., 1996; Williams, 1997; Williams et al., 2004b; Williams et al., 2004a). This literature has been the subject of numerous in-depth reviews (Bohn and BeMiller, 1995; Williams et al., 1996; Williams, 1997; Williams et al., 2004b; Williams et al., 2004a). There are also a number of clinical reports describing the prophylactic and/or therapeutic effects of glucans (Babineau et al., 1994a; Babineau et al., 1994b; Browder et al., 1990; Takahashi et al., 2006). However, most of these studies were phenomenological in nature; that is, they reported the effect of glucan administration in a given disease state. What was lacking was a mechanistic understanding of how (1,3)-β-glucans modulate innate immunity and subsequently alter the response to disease. Over the past decade, dramatic advances have been made in our understanding of the mechanisms of (1,3)-β-glucan immunobiology. This chapter focuses on recent advances in our understanding of the mechanisms by which (1,3)-β-glucans modulate innate immunity in mammals, and how modulation of innate immunity by (1,3)-β-glucans alters host response to disease.

## I.B. Mammalian receptors for (1,3)-β-glucans

In contrast to invertebrates, discussed in Chapter 4.1.2, (1,3)-β-glucan recognition in vertebrates occurs primarily through cell surface receptors. These receptors were first identified on human monocytes in the mid-1980s (Czop, 1986) and have subsequently been described on many other immune and non-immune cells, including neutrophils, macrophages, natural killer (NK) cells, eosinophils, alveolar epithelial cells, endothelial cells and fibroblasts (reviewed in Williams, 1997; Brown and Gordon, 2003). (1,3)-β-Glucan recognition by these cells is thought to involve multiple receptors (Battle et al., 1998; Kougias et al., 2001; Mueller et al., 2000), and at least four receptors have been identified, including lactosylceramide, scavenger receptors, complement receptor 3 (CR3) and Dectin-1. Of these molecules, Dectin-1 appears to be the major receptor for (1,3)-β-glucans on leukocytes and is capable of mediating many of the biological activities of these carbohydrates.

### *I.B.1. Lactosylceramide*

Lactosylceramide (LacCer, CDw17 or Galβ4Glcβ1Cer) is a glycosphingolipid consisting of a hydrophobic ceramide lipid and a hydrophilic sugar moiety and is found in microdomains on the plasma membranes of many cells. The ability of LacCer to recognise (1,3)-β-glucans was first demonstrated biochemically, using isolated human leukocyte membrane components (Zimmerman et al., 1998). Although relatively little is known about the role of LacCer in (1,3)-β-glucan mediated immunomodulation, the interaction of this glycosphingolipid with these polysaccharides has been shown to induce a variety of cellular responses, such as the production of cytokines including MIP-2 and TNF (Evans et al., 2005; Hahn et al., 2003), activation of NF-κB (Evans et al., 2005), and enhancement of the oxidative burst and anti-microbial functions of leukocytes (Wakshull et al., 1999). LacCer is thought to mediate the attachment of many microbes and may be an adhesion receptor for pathogens, such as *Candida albicans* (Jimenez-Lucho et al., 1990; Karlsson, 1989). Indeed, LacCer may play an important role in the innate response to these pathogens, especially on non-immune cells (Evans et al., 2005; Hahn et al., 2003; Jimenez-Lucho et al., 1990). How LacCer mediates intracellular signal transduction is unknown, but it may involve clustering and subsequent activation of the src kinase, Lyn (Iwabuchi and Nagaoka, 2002).

### I.B.2. Scavenger receptors

Scavenger receptors recognise modified low-density lipoproteins and consist of a structurally heterogenous family of glycoprotein receptors. Although a specific receptor has not been identified, scavenger receptors have been implicated in (1,3)-β-glucan recognition (Dushkin et al., 1996; Rice et al., 2002; Vereschagin et al., 1998). The ability of soluble (1,3)-β-glucans to inhibit the interaction of isolated monocyte membranes with classic scavenger receptor ligands is probably the best supporting evidence (Rice et al., 2002), although the ability of (1,3)-β-glucans to interact with these receptors may be related to their charge, as the affinity of these interactions was found to be affected by the charge of the molecules (Rice et al., 2002). It is notable that at least one invertebrate scavenger receptor (Drosophila SR-CI) recognising these carbohydrates has been identified (Pearson et al., 1995), and this receptor is discussed in Chapter 4.1.1.

### I.B.3. Complement receptor 3 (CR3)

Complement receptor 3 (CR3, Mac-1, $\alpha_m\beta_2$) is an integrin composed of two non-covalently linked chains, CD11b and CD18. CD11b is unique to CR3, whereas CD18 is found in all $\beta_2$ integrins. CR3 can be up-regulated following cellular activation and has been described on monocytes, macrophages, dendritic cells, neutrophils, eosinophils, NK cells, and on some T and B cells (Ross, 2000). Integrins are involved in immunity and cell–extracellular matrix interactions, in which they play a role in physical attachment as well as in intracellular signalling. The importance of integrins is exemplified by diseases such as leukocyte adhesion deficiency, where the loss of CD18 (and hence all $\beta_2$ integrins) causes life-threatening recurrent infections (Ehlers, 2000).

CR3 recognises a number of endogenous ligands (including intercellular adhesion molecules, extracellular matrix proteins, plasma proteins and proteases), but also acts as an opsonic receptor for the complement component, iC3b, and as a non-opsonic receptor for a variety of exogenous ligands. Importantly, CR3 possesses a distinct lectin domain, that maps to a site

C-terminal to the I-domain, which can recognise selected carbohydrates including (1,3)-β-glucan (Diamond et al., 1993; Ross et al., 1999; Thornton et al., 1996). Like other integrins, CR3 can mediate intracellular signalling following ligand binding, the so-called 'outside-in' signalling, but can also mediate 'inside-out' signalling, in which the affinity of the receptor for its ligand is influenced (Coppolino and Dedhar, 2000; Giancotti and Ruoslahti, 1999; Plow and Zhang, 1997). Signalling via CR3 can induce a number of cellular responses, including cytotoxicity, phagocytosis, adhesion and migration. However, some of these responses, such as cytotoxicity (Xia et al., 1999) and phagocytosis (Wright and Silverstein, 1982), require a second stimulus. Recently, it has been shown that CR3-mediated phagocytosis requires translocation of this receptor into LacCer-enriched lipid rafts, and signalling via Lyn kinase (Nakayama et al., 2007), whereas CR3-mediated cytotoxicity was shown to involve signalling via the Syk phosphatidylinositol 3-kinase (Li et al., 2006).

CR3 was first identified as a (1,3)-β-glucan receptor more than two decades ago and has been implicated in a number of cellular responses to these polysaccharides (Ross et al., 1985). The involvement of this receptor in the anti-tumourogenic properties of (1,3)-β-glucans is perhaps the best characterized, where the binding of these polysaccharides to the lectin domain primes leukocytes for CR3-dependent cytotoxicity of iC3b-coated target cells (Vetvicka et al., 1996; Xia et al., 1999). CR3 has also been implicated in the β-glucan-mediated enhancement of complement-mediated haematopoietic recovery, neutrophil chemotaxis, adhesion and transendothelial migration (Cramer et al., 2006; Harler et al., 1999; LeBlanc et al., 2006; Tsikitis et al., 2004; Xia et al., 2002; Yan et al., 1999). Interestingly, CR3 is now thought to recognise only low molecular weight (1,3)-β-glucans, generated from high molecular weight (1,3)-β-glucans through the actions of macrophages and other cells (Hong et al., 2004; Li et al., 2007). CR3 was originally proposed to be the principal (1,3)-β-glucan receptor on leukocytes (Ross, 2000), but the identification of Dectin-1 (see Section D) and the ability of CR3-deficient leukocytes to still recognise and respond to (1,3)-β-glucans suggest that CR3 may only play a minor role (Brown et al., 2002; Gantner et al., 2005; Li et al., 2007; Mueller et al., 1996).

### *I.B.4. Dectin-1*

Dectin-1, or dentritic cell-associated C-type lectin-1 (CLEC7A), was first identified in a dendritic cell line as a receptor for T-cells (Ariizumi et al., 2000), but was subsequently found to be a (1,3)-β-glucan receptor, following functional screening of a macrophage-derived cDNA expression library (Brown and Gordon, 2001). Dectin-1 is a member of the group V C-type lectin-like receptors (Zelensky and Gready, 2005) and is a type-II trans-membrane glycoprotein possessing a single non-classical C-type lectin-like domain, a stalk region, and a cytoplasmic tail which contains an immunoreceptor tyrosine-based activation-like motif (ITAM). Dectin-1 is alternatively spliced into two major and a number of minor isoforms. The major isoforms differ by the presence or absence of the extracellular stalk domain and appear to have slightly different functionalities, at least in transfected cell lines *in vitro* (Heinsbroek et al., 2006; Willment et al., 2001). One minor human isoform (hDectin-1E), which lacks the stalk and transmembrane domains, was recently shown to be located primarily in the cytoplasm and capable of interacting with a cytoplasmic scaffold protein, RanBPM (Ran Binding Protein in the Microtubule-organizing centre) (Murrin and Talbot, 2007; Xie et al., 2006). Dectin-1 is also N-glycosylated, a post-translation modification which contributes to the surface expression and function of this receptor (Kato et al., 2006).

Dectin-1 is expressed by a variety of leukocytes, with slight differences in expression between human and mouse. In mouse, the receptor is expressed in many tissues and largely by myeloid cells, including macrophages, monocytes, dendritic cells and neutrophils, although the receptor is also expressed on a subpopulation of T cells (Reid et al., 2004; Taylor et al., 2002). Alveolar macrophages and inflammatory cell populations show the highest levels of Dectin-1 expression (Reid et al., 2004; Taylor et al., 2002); however, Dectin-1 is absent on marginal zone macrophages, which suggests that this receptor is not involved in the trapping and clearance of antigen, nor is the receptor expressed in immune-privileged tissues, such as the eye and the brain. A variety of cytokines and other biological response modifiers, including (1,3)-β-glucans, are able to significantly influence the levels of Dectin-1 expression on leukocytes (Ozment-Skelton et al., 2006; Willment et al., 2003). The human receptor is similarly expressed, but is also found on B-cells, mast cells and eosinophils, and does not appear to be induced under inflammatory conditions (Olynych et al., 2006; Willment et al., 2005).

Despite its identification as a receptor for an undefined endogenous T-cell ligand (Ariizumi et al., 2000), Dectin-1 is now generally accepted to be a pattern recognition receptor for (1,3)-β-glucans (Brown and Gordon, 2001; Janeway, Jr., 1992). Dectin-1 can recognise both soluble and particulate (1,3)-β-glucans from a variety of sources, including fungi and bacteria (Brown and Gordon, 2001). As mentioned above, CR3 was originally proposed to be the major (1,3)-β-glucan receptor (Thornton et al., 1996), but a variety of approaches, including the use of blocking monoclonal antibodies and cells from receptor deficient mice, have clearly demonstrated that Dectin-1 is the primary receptor for (1,3)-β-glucans on leukocytes (Brown et al., 2002; Taylor et al., 2006; Underhill et al., 2005).

Dectin-1 specifically recognises (1,3)-linked β-glucans in a metal ion-independent fashion, and does not recognise monomers or polymers with other linkages (Adams et al., 2008; Brown and Gordon, 2001; Palma et al., 2005). Recent evidence indicates that the affinity of Dectin-1 for (1,3)-β-glucans is influenced by the polymer chain length and degree of branching, but that Dectin-1 can have extremely high affinity interactions for certain glucans, including those, such as (1,3)-β-glucan phosphate, which are used as biological response modifiers (Adams et al., 2008). Like the other non-classical C-type lectin-like receptors, Dectin-1 lacks the residues typically involved in calcium coordination and carbohydrate recognition and the mechanism by which this receptor recognises (1,3)-β-glucan ligand is unknown. However, mutational analysis has shown that two residues, $Trp^{221}$ and $His^{223}$, in the carbohydrate recognition domain are an essential requirement for (1,3)-β-glucan binding (Adachi et al., 2004). The structure of the carbohydrate recognition domain of Dectin-1 has been determined, and these two residues, which are highly conserved in all identified Dectin-1 homologs, flank a shallow groove on the surface of this receptor, which may be the ligand binding site (Brown et al., 2007). Structural analysis of Dectin-1 has also revealed that Dectin-1 can dimerize through a novel interface, creating another groove in which the (1,3)-β-linked trisaccharide (laminaritriose) was found to bind (Brown et al., 2007). However, given that the minimum oligosaccharide that is functionally recognised by Dectin-1 is an octasaccharide, the physiological significance of this potential binding site is unclear (Adams et al., 2008).

Dectin-1 is able to mediate a number of cellular responses following (1,3)-β-glucan binding, including ligand uptake through phagocytosis and endocytosis (Herre et al., 2004a), $PLA_2$ and COX activation (Suram et al., 2006), the respiratory burst (Gantner et al., 2005; Underhill et al., 2005) and the induction of a variety of cytokines and chemokines, including TNF-α, MIP-2, IL-23, IL-6, IL-2 and IL-10 (Brown et al., 2003; Gantner et al., 2003; LeibundGut-Landmann et al., 2007; Rogers et al., 2005; Steele et al., 2003). The signalling cascades leading to these various responses are complex and not yet fully understood, but require the cytoplasmic ITAM-like motif of Dectin-1, which becomes tyrosine phosphorylated upon ligand binding (Gantner et al., 2003) presumably by Src family kinases (Olsson and Sundler, 2007). The ITAM-like motif of this receptor is reminiscent of the ITAM motifs found in other activation receptors, such as DAP12 (Lanier et al., 1998) and the Fc receptors (van den Herik-Oudijk et al., 1995). However, unlike these other dual tyrosine-based ITAM-containing receptors, only the membrane proximal tyrosine residue in the cytoplasmic tail of Dectin-1 is required for signalling (Brown et al., 2003; Gantner et al., 2003; Herre et al., 2004a; Rogers et al., 2005; Underhill et al., 2005).

Unexpectedly, this single tyrosine-based ITAM-like motif was shown to recruit Syk kinase (Fuller et al., 2007; Rogers et al., 2005; Underhill et al., 2005), an activity which was previously thought to strictly require the dual tyrosine residues found in traditional ITAM motifs. Although signalling via Syk appears to mediate the majority of the cellular functions ascribed to Dectin-1, including the respiratory burst and cytokine production, some responses, such as phagocytosis in macrophages, do not involve signalling through this kinase (Herre et al., 2004a; LeibundGut-Landmann et al., 2007; Rogers et al., 2005; Underhill et al., 2005). The requirement for Syk may, however, be cell type specific, as phagocytosis in dendritic cells, for example, demonstrated a partial requirement for Syk (Rogers et al., 2005). The Syk-independent signalling pathways are unknown and probably novel. In myeloid cells, downstream signalling from Syk has been shown to involve ERK (Slack et al., 2007), the novel adaptor CARD9, which couples to Bcl10 and regulates Bcl10-Malt1-mediated NF-kappaB activation (Gross et al., 2006), and nuclear factor of activated T-cells (NFAT) (Goodridge et al., 2007).

While Dectin-1 signalling is sufficient for the induction of certain cytokines, such as IL-10 and IL-23, the induction of proinflammatory cytokines and chemokines requires collaborative signalling from the Toll-like receptors (TLR). Signals from TLR2 (Underhill et al., 1999) and TLR6 (Ozinsky et al., 2000) were shown to be required for the induction of cytokines, such as TNF-$\alpha$ and MIP-2 (Brown et al., 2003; Gantner et al., 2003), in response to zymosan, a yeast cell wall-derived particle which contains branch-on-branch (1,3;1,6)-β-glucans and TLR agonists (Gantner et al., 2003; Ikeda et al., 2008). Using highly purified, receptor-specific reagents, Dectin-1 was recently shown to synergistically enhance TLR-mediated proinflammatory cytokine and chemokine production (Dennehy et al., 2008). This collaborative signalling required both the Dectin-1/Syk and TLR2/MYD88 pathways, and resulted in enhanced translocation of NF*k*B to the nucleus (Dennehy et al., 2008). Furthermore, Dectin-1 could collaborate with multiple MyD88-coupled TLRs (Dennehy et al., 2008). It should be noted, however, that in certain cells, such as alveolar macrophages, Dectin-1 may be able to directly trigger pro-inflammatory cytokine production (Steele et al., 2005).

How signalling from Dectin-1 integrates with the TLR pathway is unclear. Dectin-1 and TLR2 co-localize upon binding of complex ligands such as zymosan (Herre et al., 2004a) and it is possible that these receptors form a signalling complex. Dectin-1 does associate with tetraspanins, including CD63 and CD37 (Meyer-Wentrup et al., 2007; Underhill et al., 2005), suggesting that it might be part of a tetraspanin-mediated supramolecular signalling complex, such as has been implicated in T- and B-cell receptor signalling (Levy and Shoham, 2005). The association of Dectin-1 with CD37, in particular, has been shown to be involved in the expression of Dectin-1 at the leukocyte cell surface and the regulation of Dectin-1-mediated IL-6 production (Meyer-Wentrup et al., 2007).

Dectin-1 probably plays a major role in (1,3)-β-glucan-mediated immunomodulation, although this has not yet been formally demonstrated *in vivo*. As we have seen, signalling via Dectin-1 can induce a variety of cellular responses, many of which have long been known to be induced by (1,3)-β-glucans. It is very likely that beneficial effects of these polysaccharides, particularly their anti-infective activities, stem, at least in part, from the ability of Dectin-1 to enhance TLR-mediated cytokine production (Dennehy et al., 2008). Furthermore, recent results using blocking monoclonal antibodies *in vivo* indicate that Dectin-1 may also be involved in the anti-cancer activity of (1,3)-β-glucans (Ikeda et al., 2007).

Through its ability to recognise (1,3)-β-glucans, Dectin-1 also plays a major role in the innate recognition of and response to fungal pathogens. Fungal cell walls can consist of more than 50% (1,3)-β-glucan, some of which, such as the (1,3;1,6) side-chain-branched β-glucans, are exposed on the cell surface, although possibly restricted to certain areas, such as the scar tissue remaining at the point of cell duplication (Gantner et al., 2005). Dectin-1 can recognise several fungal species, including *Candida* (Brown et al., 2003; Dennehy and Brown, 2007), *Pneumocystis* (Steele et al., 2003), *Saccharomyces* (Backer et al., 2008; Brown et al., 2003), *Coccidiodes* (Viriyakosol et al., 2005) and *Aspergillus* (Gersuk et al., 2006; Hohl et al., 2005; Steele et al., 2005). Recognition of these fungi by Dectin-1 induces many responses, including fungal uptake and killing, cytokine production and the induction of adaptive immunity (reviewed in Brown, 2006; Netea et al., 2008). The central role of Dectin-1 in immunity to certain fungal species has been demonstrated using Dectin-1 deficient mice (Nakamura et al., 2007; Saijo et al., 2006; Taylor et al., 2006), and the growing evidence that fungal pathogens may mask their (1,3)-β-glucan with mannan or α-glucan to avoid recognition through this receptor (Brown, 2006; Rappleye et al., 2007; Wheeler and Fink, 2006). In addition to fungi, there is also evidence that Dectin-1 may play a role in the immune response to mycobacteria, but as these organisms lack (1,3)-β-glucan, it is unclear how they are recognised by this receptor (Rothfuchs et al., 2007; Yadav and Schorey, 2006). Although Dectin-1 may play a protective and /or beneficial role, as described above, this receptor may also be involved in autoimmunity and disease. Fungal (1,3)-β-glucans, including curdlan, zymosan and intact organisms, were shown to induce autoimmune forms of rheumatoid arthritis in genetically susceptible mice, and inhibition of Dectin-1 function prevented the induction of the disease (Yoshitomi et al., 2005). This suggests that the immune responses triggered by Dectin-1 may induce autoimmune diseases in certain genetic backgrounds. Dectin-1 could also be involved in fungal-induced respiratory disorders, such as allergic bronchopulmonary aspergillosis, and although the mechanisms behind these disorders are not firmly established, this receptor is responsible for pulmonary inflammation following exposure to fungi, such as *Aspergillus fumigatus* (Steele et al., 2005).

## I.C. Structure/activity relationships between (1,3)-β-glucans and Dectin-1 in mammalian systems

As indicated above, the weight of experimental evidence points to Dectin-1 as the primary mammalian receptor for (1,3)-β-glucans (Brown et al., 2002; Brown et al., 2003). Previous reports indicate that the physicochemical properties of glucans (e.g. primary structure, polymer size, surface charge, solution conformation and side-chain branching) may be important for recognition and interaction with pattern recognition receptors in the innate immune system (Mueller et al., 1996; Mueller et al., 2000). Many of the early studies on the structure–activity relationships of glucans employed competitive binding experiments with transformed and/or immortalized cell lines, rather than pure receptor, and in some cases the polysaccharides employed were not critically characterized or they were not homogeneous glucans. However, recent studies have shed new light on how the structure of (1,3)-β-glucans influences recognition by membrane-associated innate immune pattern recognition receptors (Adams et al., 2008; Mueller et al., 1996; Mueller et al., 2000). Using a library of natural product and synthetic (1,3)-, (1,6)- and (1,3;1,6)-β glucans as well as non-glucan polymers, Adams et al. (2008) have demonstrated that recombinant murine Dectin-1 is highly specific for glucans that have a (1,3)-β-glucopyranosyl backbone. Dectin-1 did not recognise non-β-linked polysaccharides such as *Saccharomyces cerevisiae* mannan or the (1,4;1,6)-α-glucan, pullulan, nor did Dectin-1 interact with barley (1,3;1,4)-β-glucan, although there are reports using cell-based assays which suggest that (1,3;1,4) β glucans are recognised (Brown and Gordon, 2001). While Dectin-1 is highly specific for (1,3)-β-glucans, it does not recognise all (1,3)-β-glucans equally. Dectin-1 differentially interacted with (1,3)-β-glucans over a very wide range of binding affinities (2.6 mM to 2.2 pM) (Adams et al., 2008). Indeed, one of the most surprising observations that emerged from this study was the remarkably high affinity interaction of Dectin-1 with certain (1,3)-β-glucans (2.2 pM) (Adams et al., 2008). These authors also reported that Dectin-1 can recognise synthetic oligomeric? (1,3)- and (1,3;1,6)-β-glucan ligands, and that binding affinity increased when the (1,3)-β-oligoglucoside contained a single glucose side-chain branch (Adams et al., 2008).

Lowe and colleagues have reported that a linear (1,3)-β-hepta oligoglucoside is the minimum binding subunit for recognition by membrane-associated receptors (Lowe et al., 2001).

This study employed isolated membranes from a human U937 promonocytic cell line in a cell-based binding assay. Brown and Gordon (2001) reported that the minimum binding unit for Dectin-1 is more complex than a linear heptasaccharide. Palma et al. (2005) reported that the minimum (1,3)-β-glucan-binding subunit for Dectin-1 contains 10–11 glucose subunits. They employed neoglycolipid (1,3)-β-glucan ligands, but did not conduct competitive binding analyses or establish affinity. They speculated that the neoglycolipid nature of their ligands may have influenced the interaction such that the minimum (1,3)-β-glucan recognition motif was over-estimated. Adams et al. (2008) reported that Dectin-1 recognition of glucan ligands requires a backbone chain length of at least seven glucose subunits and at least one glucose side-chain branch. Thus, the data of Adams et al. (2008) support and extend the conclusion of Brown and Gordon (2001) that the minimum binding unit for Dectin-1 is more complex than a linear heptasaccharide. It is important to distinguish between the minimum (1,3)-β-glucan polymer size and structure that is required for Dectin-1 recognition and binding versus the minimum structure that is required for induction of biological activity. The reports discussed indicate that the minimum glucan recognition subunit for Dectin-1 is a polymer containing between eight and 11 (1,3)-β-linked glucose subunits with a single side-chain branch. However, (1,3)-β glucans of this size have not been shown to reliably stimulate intracellular signaling or exert biologic effects when administered parenterally. *In vivo* studies in mice suggest that (1,3;1,6)-β-glucans derived from *S. cerevisiae* and composed of >70 glucose subunits are required for induction of intracellular signalling and expression of biological activity (Williams et al., 1991). It has been speculated that, in addition to being recognised and bound by membrane-associated receptors, the (1,3)-β-glucan polymer must be of sufficient size to cross-link receptors on the cell surface as a prerequisite for induction of biological activity; however, this has not been proven unequivocally.

## I.D. *In vivo* pharmacokinetics, pharmacodynamics and bioavailability of glucans

There are several reports which focus on the pharmacokinetics of individual (1,3)-β-glucans (Miura et al., 1996; Yoshida et al., 1996; Yoshida et al., 1997). Some of these reports are clouded by a failure to fully characterize the carbohydrate used, but nevertheless they tend to support the conclusion that intravenously administered (1,3)-β-glucans have similar half-lives (Miura et al., 1996; Yoshida et al., 1996; Yoshida et al., 1997). It has also been reported that (1,3)-β-glucans are eventually deposited in the liver and spleen (Miura et al., 1996; Yoshida et al., 1996; Yoshida et al., 1997). Glucan derived from *Grifola frondosa* is reported to have a $t_{1/2}$ of 5.4 and 6.4 h following systemic administration in autoimmune-prone and normal mice, respectively, with about 70% of the (1,3)-β-glucan recovered in the liver and spleen (Miura et al., 1996). Following intravenous administration in rabbits, Yoshida et al. (1996, 1997) reported a distribution $t_{1/2}$ of less than 5 min for a 92-kDa radiolabelled (1,3)-β-glucan isolated from *Candida albicans*. Intravascular clearance studies suggested that the rapid distribution phase was followed by a prolonged elimination phase of several hours (Yoshida et al., 1996). Most of the radiolabelled (1,3)-β-glucan (>97%) was associated with cell-free plasma, while radioactivity associated with blood cells was initially found in platelets and later (2 h) distributed to polymorphonuclear leukocytes and red blood cells (Yoshida et al., 1996). More than 80% of the material was contained in the liver and 10% in the kidney after 24 h. There is also evidence that uptake into cells may differ not only for individual polysaccharides but also for various cells or tissues. Substantial differences were reported for accumulation of glucuronoxylmannan from *Cryptococcus neoformans* by macrophages and neutrophils (Monari et al., 2003). While neutrophils rapidly ingested polysaccharides, which were expelled or degraded after 1 h, macrophages continued to accumulate glucuronoxylmannan for up to 1 week (Monari et al., 2003). This supports the view that cells may respond differentially to (1,3)-β-glucans and related polysaccharides based on a number of parameters.

Rice et al. (2004) performed a comparative study of the pharmacokinetics of three well-characterized (1,3)-β-glucans following intravenous administration, i.e. glucan phosphate, laminarin and scleroglucan, which had different molecular sizes, branching frequencies, solution conformations and polymer charge (Rice et al., 2004). The total volume into which the (1,3)-β-glucans distributed in the body, i.e. ($V_D$ß) values and distribution compartment as determined $t_{1/2}$ values, were similar for all three (1,3)-β-glucans. Volume of distribution ($V_D$) values with an estimated plasma concentration at time=0 ($C_P0$) were lowest for glucan phosphate (100±19 mL/kg), compared with values for laminarin (252±99 mL/kg), and scleroglucan (261±130 mg/kg). Thus, all three (1,3)-β-glucans had similar pharmacokinetics, even though their physicochemical properties varied considerably. Rice et al. (2004) concluded that the *in vivo* clearance of glucans from the blood was largely independent of molecular size, branching frequency, polymer charge and solution conformation. This is consistent with observations from other laboratories. Interestingly, there is also evidence that the three (1,3)-β-glucans studied by Rice et al. (2004) differed in bioactivity. Laminarin does not stimulate innate immunity, while (1,3)-β-glucan phosphate and scleroglucan increase immune function (Williams et al., 1999b). However, all three (1,3)-β-glucans are bound and internalized by pattern recognition receptors on a variety of cell types (Mueller et al., 2000). Their data suggest that the bioactivity of (1,3)-β-glucans is not dramatically influenced by *in vivo* pharmacokinetics.

## I.E. *In vivo* effect of systemic glucan administration on Dectin-1 levels

As noted above, Dectin-1 is the primary receptor for (1,3)-β-glucans and mediates the internalization and inflammatory response to free and pathogen-associated (1,3)-β-glucans (Brown et al., 2002; Brown et al., 2003; Brown and Gordon, 2001). *In vitro* cell-based studies have shown that Dectin-1 will bind free (1,3)-β-glucans as well as whole *Candida albicans* and *Saccharomyces cerevisiae* cells in a (1,3)-β-glucan-dependent manner, and upon binding the Dectin-1/glucan complex is rapidly internalized (Ariizumi et al., 2000; Brown et al., 2002; Brown et al., 2003; Brown and Gordon, 2001; Herre et al., 2004a; Taylor et al., 2002). Ozment-Skelton et al., (2006) have reported a similar effect following systemic glucan administration in mice (Ozment-Skelton et al., 2006). Specifically, they examined the effect of intravenous glucan administration on circulating leukocyte surface Dectin-1 levels, and observed that a single intravenous administration of a water-soluble (1,3)-β-glucan phosphate resulted in a significant reduction in leukocyte Dectin-1 levels from 3 h to 7 days. This reduction was due to a loss of Dectin-1 from the surface of blood neutrophils and monocytes, which was likely due to internalization of the Dectin/(1,3)-β glucan complex. The intravenous administration of non-glucan polysaccharides did not decrease the percentage of Dectin-1-positive blood leukocytes, suggesting that this effect was specific for (1,3)-β-glucans. One of the most interesting observations was the prolonged effect of (1,3)-β-glucan administration on circulating leukocyte Dectin-1 levels. This is particularly significant in light of the relatively short half-life of (1,3)-β-glucans in the blood (Rice et al., 2004). Based on these data it is reasonable to speculate that even a single administration of a bioactive (1,3)-β-glucan can have effects that far exceed the plasma/serum clearance of the polysaccharide. This may relate to the rapid internalization of the Dectin-1/(1,3)-β-glucan complex and the activation of multiple intracellular signalling pathways (Wei et al., 2002b; Williams et al., 1999b; Williams et al., 2006).

## I.F. *In vivo* effects of (1,3)-β-glucans in animal models of disease

Over the past four decades literally hundreds of published works have appeared which describe the anti-infective, anti-tumour and immunobiological activities of many glucan preparations in a variety of disease states. This literature has been the subject of numerous reviews (Bohn and BeMiller, 1995; Brown and Gordon, 2003; Dennehy and Brown, 2007; Di Luzio et al., 1985; Herre et al., 2004b; Suzuki et al., 1996; Williams et al., 1996; Williams, 1997; Williams et al., 2003b; Williams et al., 2004b; Williams and Di Luzio, 1985a). However, the mechanisms associated with (1,3)-β-glucan-induced protection are only emerging. What follows is a brief review of our current knowledge of the cellular and molecular mechanisms of (1,3)-β-glucan-induced protection in animal models of disease.

### I.F.1. Bacterial infections

(1,3)-β-Glucans have been reported to increase resistance to a variety of bacterial pathogens, in a variety of animal models (Gualde et al., 1985; Kournikakis et al., 2003; Kupfahl et al., 2007; Liang et al., 1998; Persson et al., 2003; Rice et al., 2005; Williams et al., 1978; Williams et al., 1983; Williams et al., 1999b; Williams and Di Luzio, 1980; Williams and DiLuzio, 1980). The beneficial effect of (1,3)-β-glucan in sepsis have been attributed to a diverse array of effects, including stimulation of innate immunity, increased bacterial clearance, increased bactericidal activity, modulation of cytokine production and other non-specific effects (Liang et al., 1998; Williams et al., 1996). Wakshull and colleagues reported that the microbicidal activity of a PGG-(1,3)-β-glucan involves increased leukocyte numbers, enhanced oxidative burst and microbicidal activity as well as stimulation of neutrophil NFκB activation (Liang et al., 1998; Wakshull et al., 1999). However, the precise cellular and molecular mechanisms associated with (1,3)-β-glucan-induced protection against infectious diseases have not been fully elucidated. Furthermore, while the data discussed above suggest potential mechanisms of action, they also present an interesting paradox in light of recent results. By way of example, it has also been shown that (1,3)-β-glucans will decrease morbidity, maintain cardiovascular function and increase survival outcome in murine models of fulminating polymicrobial sepsis (Ha et al., 2006; Williams et al., 1999b). The pathophysiology of this experimental infection is known to involve the host overexpressing inflammatory mediators which result in a systemic inflammatory response that culminates in severe shock, multi-organ failure and death (Williams et al., 1999a; Williams et al., 1999b). Thus, a major question was how a pro-inflammatory agent, even a mild pro-inflammatory agent such as a (1,3)-β-glucan, might ameliorate a disease that has a significant inflammatory component. Indeed, agents such as (1,3)-β-glucan would seem to be contraindicated in diseases that have a significant inflammatory process. Recent data have shed new light on this paradox and in doing so have identified a novel mechanism of action for (1,3)-β-glucans in certain infectious and inflammatory diseases. In the normal host, (1,3)-β-glucan binding to its cognate receptors activates intracellular signalling pathways which are associated with stimulation of innate immunity (Battle et al., 1998; Williams et al., 1999b; Williams et al., 2000; Williams et al., 2003a). In striking contrast, administration of (1,3)-β-glucan ligands in the presence of an inflammatory and/or septic insult results in blunting of early increases in tissue transcription factor activity and cytokine transcription that are associated with the host inflammatory response to the injury (Ha et al., 2006; Williams et al., 1999b; Williams et al.,

2000). Preventing early activation of pro-inflammatory transcription factors, such as NFκB and NF-IL6, positively correlates with maintenance of cardiac function and improved survival outcome in septic shock (Ha et al., 2006; Williams et al., 1999b). It is important to note that (1,3)-β-glucans did not suppress NFκB or NF-IL6 levels, but rather prevented the dramatic increase in transcription factor activation that is observed in septic mice.

The mechanism by which (1,3)-β-glucans attenuate pro-inflammatory responses may relate to their ability to stimulate the phophoinositide-3-kinase/Akt (PI3k/Akt)-dependent signalling pathways (Li et al., 2003; Williams et al., 2004a; Williams et al., 2006). The phosphoinositide 3-kinases (PI3Ks) are a conserved family of signal transduction enzymes which are involved in regulating cellular proliferation and survival (Cantley, 2002; Fruman and Cantley, 2002). More specifically, the PI3Ks and the downstream serine/threonine kinase Akt (also known as protein kinase B (PKB) regulate cellular activation, inflammatory responses, chemotaxis and apoptosis (Cantley, 2002). Williams et al., (2004a) reported that (1,3)-β-glucan-phosphate-induced protection in murine polymicrobial sepsis is mediated through a PI3k/Akt-dependent mechanism. They treated mice with the PI3K inhibitor wortmannin or LY294002 in the presence and absence of (1,3)-β-glucan and sepsis. PI3K inhibition completely eliminated the protective effect of (1,3)-β-glucans, indicating that protection against septic mortality was mediated through PI3K (Williams et al., 2004a). Thus, manipulation of the endogenous PI3k/Akt signaling pathway by (1,3)-β-glucans may represent a new therapeutic approach to the management of important diseases.

### I.F.2. Fungal infections

The discovery of Dectin-1 as the primary (1,3)-β-glucan pattern recognition receptor in mammalian systems has dramatically advanced our knowledge of how the innate immune system recognizes and interacts with fungal (1,3)-β-glucans (Brown and Gordon, 2003; Brown and Gordon, 2005; Herre et al., 2004b). It has been reported that (1,3)-β-glucan administration will increase resistance to fungal infections (Rice et al., 2005; Williams et al., 1978). These findings have led many investigators to speculate that Dectin-1 plays a central role in the innate immune response to infection, specifically fungal infections (Dennehy and Brown, 2007). These data also imply that (1,3)-β-glucan mediates its anti-infective activity through a Dectin-1-dependent mechanism. However, the role of Dectin-1 in the *in vivo* response to fungal infections is controversial. Taylor et al. (2006) have reported that mice which are genetically deficient in Dectin-1 show increased mortality following *Candida albicans* infection. In striking contrast, Saijo et al. (2006) reported that survival during *C. albicans* infection is similar in Dectin-1 knock-out and wild-type mice. However, Saijo et al. (2006) also reported that Dectin-1 plays a role in the response to *Pneumocystis carinii* infection. Interestingly, the increase in susceptibility of Dectin-1 knock-out mice to *P. carinii* was observed only at early time periods, and the authors concluded that Dectin-1 was not required for 'chronic infection by *P. carinii*'. The precise reason for the discrepancies between the reports of (Taylor et al., 2006) and (Saijo et al., 2006) are not clear at this time. One possible explanation relates to the genetic background of the Dectin-1-deficient mice. Three different genetic backgrounds were employed to generate the Dectin-1-deficient mice (Saijo et al., 2006; Taylor et al., 2006). Taylor et al. (2006) employed a 129/Sv background for their studies whereas Saijo et al. (2006) employed Dectin-1-deficient mice on the Balb/c background for the *P. carinii* studies, and a C57Bl/6J background was used for the *C. albicans* studies. In addition, different strains and challenge doses of *C. albicans* were employed (Saijo et al., 2006; Taylor et al., 2006). Nakamura et al. (2007) have recently reported that Dectin-1 is not required for innate host defense against *Cryptococcus neoformans*. These investigators employed a Dectin-1-deficient mouse strain on the C57Bl/6J background.

### *I.F.3. Viral infections*

There are a limited number of reports on the anti-viral effect of various glucan preparations (Jung et al., 2004; Marchetti et al., 1996; Schaeffer and Krylov, 2000; Wang et al., 2007; Williams and Di Luzio, 1985b; Williams and DiLuzio, 1980). Most anti-viral (1,3)-β-glucan studies have focused on sulfated homopolysaccharides and heteropolysaccharides (Schaeffer and Krylov, 2000; Wang et al., 2007). Schaeffer and Krylov have reviewed the anti-viral effects of various natural products (Schaeffer and Krylov, 2000). They concluded that sulfated homopolysaccharides are more potent that sulfated heteropolysaccharides (Schaeffer and Krylov, 2000). With respect to (1,3)-β-glucans, most of the anti-viral research has focused on curdlan sulfate (Aoki et al., 1991; Aoki et al., 1992; Gordon et al., 1994; Gordon et al., 1997; Schaeffer and Krylov, 2000; Uryu, 1993; Wang et al., 2007). Curdlan sulfate has been reported to have anti-HIV activity (Schaeffer and Krylov, 2000; Wang et al., 2007). This has been attributed to binding of the virion by curdlan sulfate, interference with viral attachment to the cell, decreased viral penetration into the cell, as well as delaying the events that precede reverse transcription and/or beta-chemokine and cytokine production (Jagodzinski et al., 1994). However, the literature suggests that the curdlan is not that critical to the anti-viral activity. The presence of the sulfate is thought to be essential for anti-viral activity and the potency of the polysaccharide is thought to increase with the degree of sulfation (Jagodzinski et al., 1994). Indeed, non-β-linked glucan homopolysaccharides have also been reported to exhibit anti-viral activity. As an example, Qiu et al. have reported that sulfated

$\alpha$-glucans exert anti-dengue virus activity (Qiu et al., 2007), while Davis and colleagues have reported that orally administered oat (1,3;1,4)-$\beta$-glucan will decrease the morbidity and mortality associated with herpes simplex virus 1 (HSV-1) and exercise stress (Davis et al., 2005). There are a few reports indicating that non-sulfated (1,3)-$\beta$-glucans can exert anti-viral activity both *in vitro* and *in vivo* (Davis et al., 2005; Jung et al., 2004; Marchetti et al., 1996; Williams and Di Luzio, 1985b; Williams and DiLuzio, 1980). As an example, Marchetti et al. (1996) have reported that a neutral (1,3)-$\beta$-glucan, isolated from *Sclerotium glucanicum*, inhibits replication of HSV-1 in cultured cells. The relative dearth of information regarding the anti-viral effects of (1,3)-$\beta$-glucans suggest that this is an area ripe for investigation. The discovery of Dectin-1 as a specific (1,3)-$\beta$-glucan receptor may provide new avenues of investigation regarding potential anti-viral effects of non-sulfated glucans.

### I.F.4. Anti-tumor activity of glucans

The ability of glucans to inhibit tumour growth in a variety of experimental tumour models is well established (Adachi et al., 1989; Chihara, 1994; Di Luzio et al., 1979; Mansell and Di Luzio, 1976; Mayberry and Lane, 1993; Stone and Clarke, 1992; Zakany et al., 1980). In most cases, the (1,3)-$\beta$-glucan was administered prophylactically and the end-points were tumour growth, tumour volume, degree of metastases and/or survival (Chihara, 1994; Stone and Clarke, 1992; Zakany et al., 1980). The anti-tumour efficacy of (1,3)-$\beta$-glucans appears to relate to the type of tumour, the genetic background of the host animal, the dose of (1,3)-$\beta$-glucans, the route and timing of (1,3)-$\beta$-glucan administration, as well as the tumour load. This work has been the subject of several in-depth reviews (Chen and Seviour, 2007; Chihara, 1994; Stone and Clarke, 1992; Yan et al., 2005). However, interpretation of these data is complicated by the fact that many of the published works do not provide details about the nature or homogeneity of the (1,3)-$\beta$-glucans being evaluated. In addition, most of the studies were phenomenological in nature and did not address mechanisms. In some cases the anti-tumour effect of the (1,3)-$\beta$-glucans was attributed to modulation of macrophage, lymphocyte, neutrophil and/or NK cell activity (Cook et al., 1978; Di Renzo et al., 1991; Hamuro and Chihara, 1985; Hong et al., 2004). The most significant body of work on the anti-tumour mechanisms of glucans was done by the late Gordon Ross (Ross et al., 1985; Ross et al., 1999). Ross and colleagues identified the complement receptor 3 (CR3, Mac-1, $\alpha_m\beta_2$) as a receptor for (1,3)-$\beta$-glucans more than two decades ago (Ross et al., 1985; Ross et al., 1999). The role of CR3 as a glucan receptor was discussed earlier in this chapter. With respect to glucan-induced anti-tumour activity, Ross and colleagues identified $\beta$-glucans as 'specific' biological response modifiers that use endogenous antibodies to target tumours for cytotoxicity by neutrophils and complement (Yan et al., 1999). More specifically, these investigators proposed that (1,3)-β-glucans could "override the normal resistance of iC3b-opsonized tumors to the cytotoxic activations of phagocytes...", thus allowing effector mechanisms to mediate tumoricidal activity (Ross et al., 1999). Hong and colleagues have recently advanced the concept that orally administered glucans will exert an adjuvant effect when combined with exogenously administered anti-tumour antibodies that activate complement (Hong et al., 2003; Hong et al., 2004). (1,3)-β-Glucans are known to exert adjuvant activity (Mohagheghpour et al., 1995; Ormstad et al., 2000; Williams et al., 1989). In addition, recent evidence points to at least two mechanisms for the absorption of (1,3)-β-glucans from the gastrointestinal tract (Hong et al., 2004; Rice et al., 2005). Thus, this approach may have merit. However, as noted above, the ability of CR3-deficient cells to recognise (1,3)-β-glucans through a Dectin-1-dependent mechanism leaves open the possibility that Dectin-1 may play a role in the anti-tumour efficacy of glucans. In support of this concept, Ikeda et al. (2007) have recently reported that anti-Dectin-1 antibodies will inhibit the anti-tumour efficacy of schizophyllan (SPG) glucan in a murine sarcoma 180 tumor model. These investigators examined differences in tumour size over a 21-day period as their endpoint, but they did not provide any data regarding the effect of anti-Dectin-1 antibodies on survival outcome in SPG-treated tumor-bearing mice. Nevertheless, these data suggest that SPG interaction with Dectin-1 plays some role in the anti-tumor efficacy of this glucan.

### *I.F.5. Ischemia/reperfusion injury*

A growing body of evidence suggests that innate immune and inflammatory pathways participate in myocardial ischaemia/reperfusion (I/R) injury and congestive heart failure (Barry, 1994; Bozkurt et al., 1998; Bryant et al., 1998; Kapadia et al., 1998; Kiechl et al., 2001; Oral et al., 1995; Torre-Amione et al., 1996). Several recent papers have reported that (1,3)-β-glucan administration will ameliorate organ injury following I/R (Araujo-Filho et al., 2006; Bayrak et al., 2008; Bolcal et al., 2007; Li et al., 2001; Medeiros et al., 2006; Sandvik et al., 2007). Orally administered glucan has been reported to protect the kidney from I/R injury (Bayrak et al., 2008). In this study, (1,3)-β-glucan blunted the I/R-induced increase in serum urea and creatinine that was seen in the control I/R group (Bayrak et al., 2008). Medeiros et al. (2006), and Araujo-Filho et al. (2006) reported that parenteral (1,3)-β-glucan administration decreases bacterial translocation in small bowel I/R injury. They also reported that the muscosal damage associated with small bowel I/R was ameliorated in (1,3)-β-glucan-treated animals (Medeiros et al., 2006). Aarsaether et al. (2006) have translated these basic experimental observations to the clinical setting by pre-treating patients undergoing coronary artery bypass grafting with oral administration of a (1,3;1,6)-β-glucan. They observed that the compound was well tolerated and that it lowered creatinine kinase isoenzyme MB in the study group, which suggests decreased injury to the myocardium. Li et al. (2001) reported that (1,3:1,6)-β-glucan phosphate, a water-soluble derivative of *S. cerevisiae* (Li et al., 2003), will ameliorate cardiac I/R injury in either a prophylactic or therapeutic regimen. Pre-treatment of rats with (1,3;1,6)-β-glucan phosphate significantly reduced infarct size/area at risk (47%) when compared with the control group. Similarly, infarct size/area at risk was significantly reduced in (1,3;1,6)-β-glucan phosphate pre-treated animals compared with controls (Li et al., 2003). Of greater clinical importance was their finding that intravenous administration of (1,3;1,6)-β-glucan phosphate 5 min after the initiation of ischaemia significantly reduced infarct size/area at risk (Li et al., 2003). These authors also investigated the cellular mechanisms associated with glucan-induced cardioprotection in I/R injury and found that (1,3;1,6)-β-glucan phosphate administration blunts TLR4-mediated NFκB activation in the ischemic myocardium (Li et al., 2003). They also reported that the (1,3;1,6)-β-glucan phosphate induces a concomitant activation of PI3K/Akt signalling in rat hearts subjected to I/R (Li et al., 2003).

### I.F.6. Wound repair

A number of reports indicate that topical or systemic (1,3)-β-glucan administration enhances wound healing in skin and intestinal anastomoses (Chen, 1992; Compton et al., 1996; Leibovich and Danon, 1980; Portera et al., 1997; Wolk and Danon, 1985). Berdal et al. have reported that (1,3)-β-glucan improves wound healing in spontaneously diabetic mice (Berdal et al., 2007). Lyuksutova et al. (2005) reported that (1,3;1,6)-β-glucan phosphate treatment attenuates burn-induced inflammation and increases resistance to *P. aeruginosa* burn wound infection in an experimental model of burn injury. The mechanisms by which (1,3;1,6)-β-glucan phosphate enhances the reparative process have been attributed to increased macrophage activation and infiltration into the wound milieu, stimulating tissue granulation, collagen deposition, re-epithelialization and increased wound tensile strength (Chen, 1992; Compton et al., 1996; Leibovich and Danon, 1980; Portera et al., 1997; Wolk and Danon, 1985). The presence of (1,3)-β-glucan receptors on primary cultures of human fibroblasts has been reported (Kougias et al., 2001). (1,3)-β-Glucan stimulates primary human dermal fibroblast collagen biosynthesis through a nuclear factor-1 (NF-1)-dependent mechanism (Wei et al., 2002b). Additionally, (1,3)-β-glucan stimulates fibroblast AP-1 and Sp1 activation in a time-dependent manner, although the temporal kinetics varied between the two transcription factors; that is, AP-1 binding activity was increased at early time intervals (<12 h), while Sp1 nuclear binding activity was increased at later time intervals (Wei et al., 2002a). Further, (Wei et al., 2002a) reported that (1,3)-β-glucan stimulates human fibroblast expression of neurotrophin 3

(NT-3), platelet-derived growth factor A (PDGF-A), platelet-derived growth factor B (PDGF-B), acidic fibroblast growth factor (aFGF), basic fibroblast growth factor (bFGF), transforming growth factor alpha (TGFalpha), transforming growth factor beta (TGFbeta) and vascular endothelial growth factor (VEGF) mRNA. Lowe et al. (2002) found two (1,3)-β-glucan binding sites on primary human vascular endothelial cells. In endothelial cells, (1,3)-β-glucan stimulated NFκB nuclear binding activity and IL-8 expression, but not vascular endothelial growth factor, in a time-dependent manner (Lowe et al., 2002). These data indicate that (1,3)-β-glucans directly and/or indirectly modulate the activity of diverse cells and growth factors that are central to the reparative process. Delatte et al. (2001) have translated these experimental data to the clinical setting. They reported on the effectiveness of a (1,3)-β-glucan/collagen preparation in the treatment of partial thickness burns in a retrospective study of 225 paediatric burn patients, 43 of whom were treated with glucan/collagen. They concluded that paediatric burns can be effectively treated with (1,3)-β-glucan/collagen mixtures, and that this preparation markedly simplified wound care and significantly decreased post-injury pain (Delatte et al., 2001).

## I.G. Anti-inflammatory activity of glucans

The weight of evidence indicates that (1,3)-β-glucans will stimulate innate immunity and activate pro-inflammatory responses (Brown and Gordon, 2003; Williams et al., 2004b). However, there is a growing body of evidence which suggests that (1,3)-β-glucans can elicit an anti-inflammatory response in certain clinically relevant models of disease (Li et al., 2003; Williams et al., 1999b; Williams et al., 2004a). As was noted above, Williams et al. (1999b) and Li et al. (2003) have shown that (1,3)-β-glucan administration can attenuate early activation of pro-inflammatory transcription factors, such as NFκB and NF-IL6. The 'blunting' of pro-inflammatory transcription factor activation positively correlates with reduced morbidity and improved outcome in septic shock (Williams et al., 1999b) and I/R injury (Li et al., 2003). Initially, the anti-inflammatory effect of (1,3)-β-glucans was thought to be mediated via inhibition of the alternative pathway of complement. However, recent studies have identified activation of the PI3k/Akt signalling pathway as a potential mechanism of glucan-mediated anti-inflammatory responses (Li et al., 2003; Williams et al., 2004a). PI3K is an endogenous compensatory mechanism that suppresses pro-inflammatory and apoptotic processes in response to inflammatory injury (Fukao and Koyasu, 2003; Guha and Mackman, 2002; Li et al., 2003; Williams et al., 2004a). The PI3K/Akt pathway is also thought to play a pivotal role in the maintenance of homeostasis and integrity of the immune response during inflammatory injury (Fukao and Koyasu, 2003; Guha and Mackman, 2002; Li et al., 2003; Williams et al.,

2004a). The data suggest that (1,3)-β-glucan activation of the PI3K pathway may be an effective approach for the prevention and/or treatment of septic and inflammatory sequelae.

The anti-inflammatory effect of (1,3)-β-glucans has also been reported in environmental settings. Rylander and colleagues have studied the relationship of glucan exposure to indoor air related symptoms, allergy and asthma (Rylander et al., 1992; Rylander, 1996; Rylander and Lin, 2000; Thorn et al., 2001). Their results indicate that the response to inhaled (1,3)-β-glucans in humans is quite different from that observed following inhalation of bacterial endotoxin (Thorn et al., 2001). By way of example, endotoxin causes a predominately neutrophilic inflammatory response, while (1,3)-β-glucans does not induce neutrophil recruitment. Indeed, Rylander and Lin, (2000) speculate that (1,3)-β-glucans may prime airway cells for simultaneous or subsequent exposure to another agent. They reported that (1,3)-β-glucans potentiated ovalbumin-induced eosinophilia and IgE responses (Rylander and Lin, 2000). At the same time, some inflammatory responses were shown to be down-regulated (Rylander and Lin, 2000). Specifically, TNFα production is decreased under certain conditions (Rylander and Lin, 2000).

It should be noted that the term 'anti-inflammatory' may be a misnomer, when applied to (1,3)-β-glucans. The data support the concept of (1,3)-β-glucans eliciting an anti-inflammatory response, but it is not a classic anti-inflammatory response such as might be observed with corticosteroids. Indeed, the effect of (1,3)-β-glucans appears to be quite specific and can be linked to individual signalling pathways, i.e. PI3k/Akt (Li et al., 2003; Williams et al., 2004a). Furthermore, the response is not immunosuppressive, since normal or above normal levels of these important transcription factors are maintained (Williams et al., 1999b). In fact, immune competence is increased, as demonstrated by clinical observations that (1,3)-β-glucan administration stimulates conversion from anergy (immunosuppression) in trauma patients (Browder et al., 1990).

## I.H. Stimulation of innate immunity following oral glucan administration

Most *in vivo* studies with (1,3)-β-glucans have focused on parenteral administration via the intravenous, subcutaneous or intraperitoneal routes. However, as noted above there is a small, but growing literature which indicates that (1,3)-β-glucans may be orally effective as well. Hong et al. (2004) and Cheung et al. (2002) have reported that orally administered (1,3)-β-glucans function as potent anti-tumour adjuvants when combined with antibodies that recognise epitopes on tumours. Suzuki et al. (1989) reported that a branched water-soluble (1,3)-β-glucan increased splenocyte mitogenic response and NK cell activity following five daily oral administrations. In addition, oral supplementation with the (1,3)-β-glucan inhibited tumour growth of syngeneic Meth A fibrosarcoma, IMC carcinoma and Lewis lung carcinoma (Suzuki et al., 1989). In another study, Suzuki et al. (1990) reported that oral administration of glucan to $CDF_1$ mice increased peritoneal macrophage functional activity as denoted by increased acid phosphatase activity, increased phagocytic activity, increased killing of *Candida* and increased interleukin-1 (IL-1) production. These investigators also reported the oral efficacy of (1,3)-β-glucans in C3H/HeJ mice which are endotoxin hyporesponsive, indicating the effect is not due to increased uptake of endotoxin from the gut following oral (1,3)-β-glucan administration (Suzuki et al., 1989; Suzuki et al., 1990). Nicoletti et al. have reported that a branched glucan derived from *C. albicans* exerted immunoadjuvant activity following oral administration (Nicoletti et al., 1992). Orally administered (1,3)-β-glucans increased serum levels of IL-12 in mice (Rice et al., 2005). Interestingly, these authors reported that IL-6 was elevated at 8 hours following oral administration of water insoluble, microparticulate glucan, even though there was no evidence for oral absorption of the insoluble glucan (Rice et al., 2005). However, oral administration of a water soluble glucan, which is absorbed by the gastrointestinal tract, increased survival of mice challenged with *Staphylococcus aureus or Candida albicans* (Rice et al., 2005). Baran et al. (2007) have reported that orally administered whole yeast (1,3)-β-glucan particles converts non-protective Th2 responses to a more protective Th1 phenotype in a murine model of mammary carcinoma. Oral (1,3)-β-glucans are also approved by the FDA for lowering serum cholesterol. Oral (1,3)-β-glucans has also been reported to decrease post-prandial glucose surge in Type II diabetes (Braaten et al., 1994; Wuersch and Pi-Sunyer, 1997).

Not all reports have demonstrated biological activity following oral (1,3)-β-glucan administration. Chihara, (1985), Ohno et al., (1984), Wu et al., (1998),and Dritz et al., (1995) reported that dietary supplementation with (1,3)-β-glucans derived from various sources did not exert any significant immunomodulatory effects. Dritz et al. (1995). reported on the effect of dietary supplementation with (1,3)-β-glucan in weanling pigs. Part of the rationale for this study was to determine whether oral (1,3)-β-glucan administration could effectively modulate innate immunity in domestic animals, such that they are more resistant to opportunistic infections, thus reducing the need for exogenous antibiotics. However, this investigation revealed a "complex interaction" between oral (1,3)-β-glucan, growth performance and susceptibility to *Streptococcus suis.* While the pigs receiving (1,3)-β-glucan showed greater weight gain, they also showed an increased susceptibility to *Streptococcus suis* infection. Wu et al. (1998) reported that dietary supplementation with protein-bound (1,3)-β-glucan did not enhance immune function, suggesting that the complex of glucan and protein may not be biologically active.

An important caveat is that virtually every study discussed above employed a different oral (1,3)-β-glucan preparation. In some instances, details about the physical state, chemical nature and/or homogeneity of the (1,3)-β-glucans employed were provided, but this was not true in all cases. The variety of (1,3)-β-glucan preparations employed makes it difficult to determine why some preparations were bioactive after oral administration and others were not. Nevertheless, the weight of evidence suggests that some orally administered (1,3)-β-glucans can interact with intestinal cells and cross the gastrointestinal barrier, where they can exert systemic effects on innate immunity.

Based on the emerging interest in orally administered (1,3)-β-glucans, Rice et al.(2005) have examined the comparative pharmacokinetics of three water-soluble (1,3)-β-glucans (1,3)-β-glucans with varying physicochemical characteristics following oral administration. They reported that maximum plasma concentrations for the three (1,3)-β-glucans varied considerably with peak plasma levels occurring between 0.5 to 12 hrs. At 24 hrs after oral administration scleroglucan was completely eliminated from the serum, while approximately 25% of the glucan phosphate and laminarin remained in the serum at 24 hrs. It is interesting to note that the largest glucan studied, scleroglucan, had a Mw of $\sim 1 \times 10^6$ g/mol. Despite its relatively large size it was completely eliminated from the serum in 24 hrs, while (1,3)-β-glucans with smaller Mw remained (Rice et al., 2005). This suggests that absorption/transport of glucans across the gastrointestinal tract is complex. In all cases, bioavailability of the orally administered (1,3)-β-glucans was relatively low (0.5 to 4.9%). Interestingly, the liver did not significantly contribute to clearance of plasma (1,3)-β-glucans. Thus, these (1,3)-β-glucans do not appear to be removed from the plasma via a first pass effect in the liver. It was noted that orally administered (1,3)-β-glucans were bound and internalized by a sub-population of intestinal epithelial cells and gut-associated lymphoid tissue (GALT) cells. Internalization of (1,3)-β-glucans by intestinal epithelial cells was not Dectin-1 dependent (Rice et al., 2005), although gut associated lymphoid tissue (GALT) expression of Dectin-1 and TLR2, but not TLR4, increased following oral administration of glucan. These investigators also examined the fate of water insoluble, particulate glucan (1,3)-β-glucan after oral administration (Rice et al., 2005)and found that levels of (1,3)-β-glucan were undetectable suggesting that only soluble (1,3)-β-glucans are directly absorbed into the circulation. However, Hong et al.(2004). have reported that particulate (1,3)-β-glucans glucans in the gastrointestinal tract are internalized by macrophages, which transport the glucan to various sites throughout the body, and slowly degrade the particulate (1,3)-β-glucans, releasing a bioactive soluble (1,3)-β-glucan product.

When considered as a whole, these data suggest that (1,3)-β-glucans can be taken up from the gastrointestinal tract and that orally administered (1,3)-β-glucans can modulate innate immunity. The data also suggest that the physical state of the (1,3)-β-glucan may dictate the mechanism by which the (1,3)-β-glucan is transported across the gastrointestinal barrier into the systemic circulation.

## I.I. Conclusions

The ability of (1,3)-β-glucans to stimulate innate immunity has been known for more than four decades. However, recent advances in our knowledge in particular the discovery that Dectin is the primary (1,3)-β-glucan recognition protein have significantly increased our understanding of how (1,3)-β-glucans modulate innate immunity at the cellular and molecular level. These findings will be important in advancing our knowledge of the role that glucan plays as a pathogen associated molecular pattern and they may also be useful in the development of (1,3)-β-glucan based immunotherapeutics.

## Acknowledgements

We thank the Wellcome Trust, CANSA South Africa, National Research Foundation and Medical Research Council (SA) for funding. G.D.B. is a Wellcome Trust Senior Research Fellow in Biomedical Science in South Africa. This work was also supported, in part, by Public Health Service grant GM53522 from the National Institute of General Medical Sciences to D.L.W.

## References

Adachi, Y., Ishii, T., Ikeda, Y., Hoshino, A., Tamura, H., Aketagawa, J., Tanaka, S., & Ohno, N. (2004). Characterization of beta-glucan recognition site on C-Type lectin, Dectin 1. *Infection and Immunity*, *72*, 4159–4171.

Adachi, Y., Ohno, N., Ohsawa, M., Sato, K.d., Oikawa, S., & Toshiro, Y. (1989). Physicochemical properties and antitumor activities of chemically modified derivatives of antitumor glucan "Grifolan LE" from *Grifola frondosa*. *Chemical and Pharmaceutical Bulletin*, *37*, 1838–1843.

Adams, E. L., Rice, P. J., Graves, B., Ensley, H. E., Yu, H., Brown, G. D., Gordon, S., Monteiro, M. A., Papp-Szabo, E., Lowman, D. W., Power, T. D., Wempe, M. F., & Williams, D. L. (2008). Differential high affinity interaction of Dectin-1 with natural or synthetic glucans is dependent upon primary structure and is influenced by polymer chain length and side chain branching. *Journal of Pharmacolology and Experimental Therapeutics*. Epub ahead of print.

Aoki, T., Kaneko, Y., Nguyen, T., Stefanski, M. S., Ting, R. C., & Manak, M. M. (1992). Curdlan sulfate and HIV-1: II. In vitro long-term treatment of HIV-1 infection with curdlan sulfate. *AIDS Research and Human Retroviruses*, *8*, 605–612.

Aoki, T., Kaneko, Y., Stefanski, M. S., Nguyen, T., & Ting, R. C. (1991). Curdlan sulfate and HIV-1. I. In vitro inhibitory effects of curdlan sulfate on HIV-1 infection. *AIDS Research and Human Retroviruses*, *7*, 409–415.

Araujo-Filho, I., Rego, A. C., Pinheiro, L. A., Azevedo, I. M., Medeiros, V. B., Brandao-Neto, J., & Medeiros, A. C. (2006). Prevention of bacterial translocation using beta-(1-3)-D-glucan in small bowel ischemia and reperfusion in rats. *Acta cirúrgica brasileira*, *21*, 18–22.

Ariizumi, K., Shen, G.-L., Shikano, S., Xu, S., Ritter, R. I., Kumamoto, T., Edelbaum, D., Morita, A., Bergstresser, P. R., & Takashima, A. (2000). Identification of a novel, dendritic cell-associated molecule, Dectin-1, by subtractive cDNA cloning. *Journal of Biological Chemistry*, *275*, 20157–20167.

Babineau, T. J., Hackford, A., Kenler, A., Bistrian, B., Forse, R. A., Fairchild, P. G., Heard, S., Keroack, M., Causha, P., & Benotti, P. (1994a). A Phase II multicenter, double-blind, randomized, placebo-controlled study of three dosages of an immunomodulator (pgg-glucan) in high-risk surgical patients. *Archives of Surgery*, *129*, 1204–1210.

Babineau, T. J., Marcello, P., Swails, W., Kenler, A., Bistrian, B., & Forse, R. A. (1994b). Randomized phase I/II trial of a macrophage-specific immunomodulator (PGG-glucan) in high-risk surgical patients. *Annals of Surgery*, *220*, 601–609.

Backer, R., van Leeuwen, F., Kraal, G., & den Haan, J. M. (2008). CD8(-) dendritic cells preferentially cross-present *Saccharomyces cerevisiae* antigens. *European Journal of Immunology*. Epub ahead of print.

Baran, J., Allendorf, D. J., Hong, F., & Ross, G. D. (2007). Oral beta-glucan adjuvant therapy converts nonprotective Th2 response to protective Th1 cell-mediated immune response in mammary tumor-bearing mice. *Folia histochemica et cytobiologica*, *45*, 107–114.

Barry, W. H. (1994). Mechanisms of immune-mediated myocyte injury. *Circulation*, *89*, 2421–2432.

Battle, J., Ha, T., Li, C., Della Beffa, V., Rice, P., Kalbfleisch, J., Browder, W., & Williams, D. (1998). Ligand binding to the (1→3)-beta-D-glucan receptor stimulates NFκB activation, but not apoptosis in U937 cells. *Biochemical and Biophysical Research Communications*, *249*, 499–504.

Bayrak, O., Turgut, F., Karatas, O. F., Cimentepe, E., Bayrak, R., Catal, F., Atis, O., Akcay, A., & Unal, D. (2008). Oral beta-glucan protects kidney against ischemia/reperfusion injury in rats. *American Journal of Nephrology*, *28*, 190–196.

Berdal, M., Appelbom, H. I., Eikrem, J. H., Lund, A., Zykova, S., Busund, L. T., Seljelid, R., & Jenssen, T. (2007). Aminated beta-1,3-D-glucan improves wound healing in diabetic db/db mice. *Wound Repair and Regeneration*, *15*, 825–832.

Bohn, J. A., & BeMiller, J. N. (1995). (1-3)-Beta-d-glucans as biological response modifiers: a review of structure-functional activity relationships. *Carbohydrate Polymers*, *28*, 3–14.

Bolcal, C., Yildirim, V., Doganci, S., Sargin, M., Aydin, A., Kuralay, E., Ozal, E., Demirkilic, U., Oz, B. S., Sayal, A., & Tatar, H. (2007). Do N-acetylcystein, beta-glucan, and coenzyme A10 mollify myocardial ischemia-reperfusion injury? *The heart surgery forum*, *10*, E222–E227.

Bozkurt, B., Jribb, S. B., Clubb, F. J., Michael, L. H., Didenko, V. V., Hornsby, P. J., Yukihiro, S., Oral, H., Spinale, F. G., & Mann, D. L. (1998). Pathophysiologically relevant concentrations of tumor necrosis factor-α promote progressive left ventricular dysfunction and remodeling in rats. *Circulation*, *97*, 1382–1391.

Braaten, J. T., Scott, F. W., Wood, P. J., Riedel, K. D., Wolynetz, M. S., Brule, D., & Collins, M. W. (1994). High beta-glucan oat bran and oat gum reduce postprandial blood glucose and insulin subjects with and without Type 2 diabetes. *Diabetic Medicine*, *11*, 312–318.

Browder, W., Williams, D., Pretus, H., Olivero, G., Enrichens, F., Mao, P., & Franchello, A. (1990). Beneficial effect of enhanced macrophage function in the trauma patient. *Annals of Surgery*, *211*, 605–613.

Brown, G. D. (2006). Dectin-1: a signalling non-TLR pattern-recognition receptor. *Nature Reviews Immunology*, *6*, 33–43.

Brown, G. D., & Gordon, S. (2001). Immune Recognition: A new receptor for beta-glucans. *Nature*, *413*, 36–37.

Brown, G. D., & Gordon, S. (2003). Fungal β-glucans and mammalian immunity. *Immunity*, *19*, 311–315.

Brown, G. D., & Gordon, S. (2005). Immune recognition of fungal beta-glucans. *Cell Microbiology*, *7*, 471–479.

Brown, G. D., Herre, J., Williams, D. L., Willment, J. A., Marshall, A. S. J., & Gordon, S. (2003). Dectin-1 mediates the biological effects of β-glucans. *Journal of Experimental medicine*, *197*, 1119–1124.

Brown, G. D., Taylor, P. R., Reid, D. M., Willment, J. A., Williams, D. L., Martinez-Pomares, L., Wong, S. Y. C., & Gordon, S. (2002). Dectin-1 is a major β-glucan receptor on macrophages. *Journal of Experimental medicine*, *296*, 412.

Brown, J., O'callaghan, C., Marshall, A., Gilbert, R., Siebold, C., Gordon, S., Brown, G., & Jones, E. (2007). Structure of the fungal beta-glucan-binding immune receptor dectin-1: Implications for function. *Protein Science*, *16*, 1042–1052.

Bryant, D., Becker, L., Richardson, J., Shelton, J., Franco, F., Peshock, R., Thompson, M., & Giroir, B. (1998). Cardiac failure in transgenic mice with myocardial expression of tumor necrosis factor-alpha. *Circulation*, *97*, 1375–1381.

Cantley, L. C. (2002). The phosphoinositide 3-kinase pathway. *Science*, *296*, 1655–1657.

Chen, J., & Seviour, R. (2007). Medicinal importance of fungal beta-(1-3),(1-6)-glucans. *Mycology Research*, *111*, 635–652.

Chen, Y.L., 1992, Effects of collagen and glucan on the healing of burn wounds in mice, *Zhonghua Zheng Xing Shao Shang Wai Ke Za Shi*, *8*, 54–56.

Cheung, N. K., Modak, S., Vickers, A., & Knuckles, B. (2002). Orally administered beta-glucans enhance anti-tumor effects of monoclonal antibodies. *Cancer Immunology and Immunotherapy*, *51*, 557–564.

Chihara, G. (1985). Immunotherapeutic agents from plant origin. *Springer Seminars in Immunopathology*, *8*, 387–400.

Chihara, G. (1994). *Preclinical evaluations of lentinan in animal models.* National Cancer Research Institute of Tokyo 189–197.

Compton, R., Williams, D., & Browder, W. (1996). The beneficial effect of enhanced macrophage function on the healing of bowel anastomoses. *The American Surgeon*, *62*, 14–18.

Cook, J. A., Taylor, D., Cohen, C., Rodrigue, J., Malshet, V., & Di Luzio, N. R. (1978). Comparative evaluation of the role of macrophages and lymphocytes in mediating the antitumor action of glucan. In M. A. Chirigos (Ed.), *Immune Modulation and Control of Neoplasia by Adjuvant Therapy* (pp. 183–194). New York: Raven Press.

Coppolino, M. G., & Dedhar, S. (2000). Bi-directional signal transduction by integrin receptors. *International Journal of Biochemistry and Cell Biology*, *32*, 171–188.

Cramer, D. E., Allendorf, D. J., Baran, J. T., Hansen, R., Marroquin, J., Li, B., Ratajczak, J., Ratajczak, M. Z., & Yan, J. (2006). Beta-glucan enhances complement-mediated hematopoietic recovery after bone marrow injury. *Blood*, *107*, 835–840.

Czop, J. K. (1986). The role of β-glucan receptors in blood and tissue leukocytes in phagocytosis and metabolic activation. *Pathology and immunopathology research*, *5*, 286–296.

Davis, J. M., Murphy, E. A., Brown, A. S., Carmichael, M. D., Ghaffar, A., & Mayer, E. P. (2005). Effects of oat beta-glucan on innate immunity and infection after exercise stress. *Medicine and science in sports and exercise*, *36*, 1321–1327.

Delatte, S. J., Evans, J., Hebra, A., Adamson, W., Othersen, H. B., & Tagge, E. P. (2001). Effectiveness of beta-glucan collagen for treatment of partial-thickness burns in children. *Journal of Pediatric Surgery*, *36*, 113–118.

Dennehy, K. M., & Brown, G. D. (2007). The role of the beta-glucan receptor Dectin-1 in control of fungal infection. *Journal of Leukocyte Biology*, *82*, 253–258.

Dennehy, K. M., Ferwerda, G., Faro-Trindade, I., Pyz, E., Willment, J. A., Taylor, P. R., Kerrigan, A., Tsoni, S. V., Gordon, S., Meyer-Wentrup, F., Adema, G. J., Kullberg, B. J., Schweighoffer, E.,

Tybulewicz, V., Mora-Montes, H. M., Gow, N. A., Williams, D. L., Netea, M. G., & Brown, G. D. (2008). Syk kinase is required for collaborative cytokine production induced through Dectin-1 and Toll-like receptors. *European Journal of Immunology*, *38*, 500–506.

Di Luzio, N. R., Williams, D. L., McNamee, R. B., Edwards, B. F., & Kitahama, A. (1979). Comparative tumor-inhibitory and anti-bacterial activity of soluble and particulate glucan. *International Journal of Cancer*, *24*, 773–779.

Di Luzio, N. R., Williams, D. L., Sherwood, E. R., & Browder, I. W. (1985). Modification of diverse experimental immunosuppressive states by glucan. *Survey of Immunologic Research*, *4*, 160–167.

Di Renzo, L., Yefenof, E., & Klein, E. (1991). The function of human NK cells is enhanced by β-glucan, a ligand of CR3 (CD11b/CD18). *European Journal of Immunology*, *21*, 1755–1758.

Diamond, M. S., Garcia-Aguilar, J., Bickford, J. K., Corbi, A. L., & Springer, T. A. (1993). The I domain is a major recognition site on the leukocyte integrin Mac-1 (CD11b/CD18) for four distinct adhesion ligands. *Journal of Cell Biology*, *120*, 1031–1043.

Dritz, S. S., Shi, J., Kielian, T. L., Goodband, R. D., Nelssen, J. L., Tokach, M. D., Chengappa, M. M., Smith, J. E., & Belcha, F. (1995). Influence of dietary beta-glucan on growth performance, nonspecific immunity, and resistance to *Streptococcus suis* infection in weanling pigs. *Journal of Animal Science*, *73*, 3341–3350.

Dushkin, M. I., Safina, A. F., Vereschagin, E. I., & Schwartz, Y. S. (1996). Carboxymethylated beta-1,3-glucan inhibits the binding and degradation of acetylated low density lipoproteins in macrophages *in vitro* and modulates their plasma clearance *in vivo*. *Cell Biochemistry and Function*, *14*, 209–217.

Ehlers, M. R. (2000). CR3: a general purpose adhesion-recognition receptor essential for innate immunity. *Microbes and Infection*, *2*, 289–294.

Evans, S. E., Hahn, P. Y., McCann, F., Kottom, T. J., Pavlovic, Z. V., & Limper, A. H. (2005). Pneumocystis cell wall beta-glucans stimulate alveolar epithelial cell chemokine generation through nuclear factor-kappaB-dependent mechanisms. *American journal of respiratory cell and molecular biology*, *32*, 490–497.

Fruman, D. A., & Cantley, L. C. (2002). Phosphoinositide 3-kinase in immunological systems. *Seminars in Immunology*, *14*, 7–18.

Fukao, T., & Koyasu, S. (2003). PI3K and negative regulation of TLR signaling. *Trends in Immunology*, *24*, 358–363.

Fuller, G. L. J., Williams, J. A. E., Tomlinson, M. G., Eble, J. A., Hanna, S. L., Pöhlmann, S., Suzuki-Inoue, K., Ozaki, Y., Watson, S. P., & Pearce, A. C. (2007). The C-type lectin receptors CLEC-2 and Dectin-1, but not DC-SIGN, signal via a novel YXXL-dependent signalling cascade. *Journal of Biological Chemistry*, *282*, 12397–12409.

Gantner, B. N., Simmons, R. M., Canavera, S. J., Akira, S., & Underhill, D. M. (2003). Collaborative induction of inflammatory responses by Dectin-1 and toll-like receptor 2. *Journal of Experimental medicine*, *197*, 1107–1117.

Gantner, B. N., Simmons, R. M., & Underhill, D. M. (2005). Dectin-1 mediates macrophage recognition of Candida albicans yeast but not filaments. *The EMBO Journal*, *24*, 1277–1286.

Gersuk, G. M., Underhill, D. M., Zhu, L., & Marr, K. A. (2006). Dectin-1 and TLRs permit macrophages to distinguish between different *Aspergillus fumigatus* cellular states. *Journal of Immunology*, *176*, 3717–3724.

Giancotti, F. G., & Ruoslahti, E. (1999). Integrin signaling. *Science*, *285*, 1028–1032.

Goodridge, H. S., Simmons, R. M., & Underhill, D. M. (2007). Dectin-1 stimulation by *Candida albicans* yeast or zymosan triggers NFAT activation in macrophages and dendritic cells. *Journal of Immunology*, *178*, 3107–3115.

Gordon, M., Deeks, S., De Marzo, C., Goodgame, J., Guralnik, M., Lang, W., Mimura, T., Pearce, D., & Kaneko, Y. (1997). Curdlan sulfate (CRDS) in a 21-day intravenous tolerance study in human immunodeficiency virus (HIV) and cytomegalovirus (CMV) infected patients: indication of anti-CMV activity with low toxicity. *Journal of Medicine*, *28*, 108–128.

Gordon, M., Guralnik, M., Kaneko, Y., Mimura, T., Baker, M., & Lang, W. (1994). A phase I study of curdlan sulfate--an HIV inhibitor. Tolerance, pharmacokinetics and effects on coagulation and on CD4 lymphocytes. *Journal of Medicine*, *25*, 163–180.

Gross, O., Gewies, A., Finger, K., Schafer, M., Sparwasser, T., Peschel, C., Forster, I., & Ruland, J. (2006). Card9 controls a non-TLR signalling pathway for innate anti-fungal immunity. *Nature*, *442*, 651–656.

Gualde, N., Rigaud, M., & Goodwin, J. S. (1985). Induction of suppressor cells from peripheral blood T cells by 15-hydroperoxyeicosatetraenoic acid (15-HPETE). *Journal of Immunology*, *135*, 3424–3429.

Guha, M., & Mackman, N. (2002). The PI3K-Akt pathway limits LPS activation of signaling pathways and expression of inflammatory mediators in human monocytic cells. *Journal of Biological Chemistry*, *277*, 32124–32132.

Ha, T., Hua, F., Grant, D., Xia, Y., Ma, J., Gao, X., Kelley, J., Williams, D. L., Kablfleisch, J., Browder, I. W., & Li, C. (2006). Glucan phosphate attenuates cardiac dysfunction and inhibits cardiac MIF expression and apoptosis in septic mice. *American Journal of Physiology: Heart and Circulatory Physiology*, *291*, H1910–H1918.

Hahn, P. Y., Evans, S. E., Kottom, T. J., Standing, J. E., Pagano, R. E., & Limper, A. H. (2003). *Pneumocystis carinii* cell wall beta-glucan induces release of macrophage inflammatory protein-2 from alveolar epithelial cells via a lactosylceramide-mediated mechanism. *Journal of Biological Chemistry*, *278*, 2043–2050.

Hamuro, J., & Chihara, G. (1985). Lentinan, a T-cell-oriented immunopotentiator: Its experimental and clinical applications and possible mechanism of immune modulation. In R. L. Fenichel & M. A. Chirigos (Eds.), *Immune Modulation Agents and Their Mechanisms* (pp. 409–436). New York: Marcel Dekker, Inc.

Harler, M. B., Wakshull, E., Filardo, E. J., Albina, J. E., & Reichner, J. S. (1999). Promotion of neutrophil chemotaxis through differential regulation of beta 1 and beta 2 integrins. *Journal of Immunology*, *162*, 6792–6799.

Heinsbroek, S. E. M., Taylor, P. R., Rosas, M., Willment, J. A., Williams, D. L., Gordon, S., & Brown, G. D. (2006). Expression of functionally different Dectin-1 isoforms by murine macrophages. *Journal of Immunology*, *176*, 5513–5518.

Herre, J., Marshall, A. S. J., Caron, E., Edwards, A. D., Williams, D. L., Schweighoffer, E., Tybulewicz, V., Reis e Sousa, C., Gordon, S. E., & Brown, G. D. (2004a). Dectin-1 uses novel mechanisms for yeast phagocytosis in macrophages. *Blood*, *104*, 4038–4045.

Herre, J., Willment, J. A., Gordon, S., & Brown, G. D. (2004b). The role of Dectin-1 in antifungal immunity. *Critical Reviews in Immunolology*, *24*, 193–203.

Hohl, T. M., Van Epps, H. L., Rivera, A., Morgan, L. A., Chen, P. L., Feldmesser, M., & Pamer, E. G. (2005). *Aspergillus fumigatus* triggers inflammatory responses by stage-specific beta-glucan display. *PLOS Pathogens*, *1*, e30.

Hong, F., Hansen, R. D., Yan, J., Allendorf, D. J., Baran, J. T., Ostroff, G. R., & Ross, G. D. (2003). Beta-glucan functions as an adjuvant for monoclonal antibody immunotherapy by recruiting tumoricidal granulocyes as killer cells. *Cancer Research*, *63*, 9023–9031.

Hong, F., Yan, J., Baran, J. T., Allendorf, D. J., Hansen, R. D., Ostroff, G. R., Xing, P. X., Cheung, N.-K. V., & Ross, G. D. (2004). Mechanism by which orally administered β-1, 3-glucans enhance the tumoricidal activity of antitumor monoclonal antibodies in murine tumor models. *Journal of Immunology*, *173*, 797–806.

Ikeda, Y., Adachi, Y., Ishii, T., Miura, N., Tamura, H., & Ohno, N. (2008). Dissociation of Toll-like receptor 2-mediated innate immune response to Zymosan by organic solvent-treatment without loss of Dectin-1 reactivity. *Biological and Pharmaceutical Bulletin*, *31*, 13–18.

Ikeda, Y., Adachi, Y., Ishii, T., Tamura, H., Aketagawa, J., Tanaka, S., & Ohno, N. (2007). Blocking effect of anti-Dectin-1 antibodies on the anti-tumor activity of 1,3-beta-glucan and the binding of Dectin-1 to 1,3-beta-glucan. *Biological and Pharmaceutical Bulletin*, *30*, 1384–1389.

Iwabuchi, K., & Nagaoka, I. (2002). Lackosylceramide-enriched glycosphingolipid signaling domain mediates superoxide generation from human neutrophils. *Blood*, *100*, 1454–1464.

Jagodzinski, P. P., Wiaderkiewicz, R., Kurzawski, G., Kloczewiak, M., Nakashima, H., Hyjek, j. E., Yamamoto, N., Uryu, T., Kaneko, Y., Posner, M. R., et al. (1994). Mechanism of the inhibitory effect of curdlan sulfate on HIV-1. *Virology*, *202*, 735–745.

Janeway, C. A., Jr. (1992) The immune system evolved to discriminate infectious nonself from noninfectious self. *Immunology Today*, *13*, 11–16.

Jimenez-Lucho, V., Ginsburg, V., & Krivan, H. C. (1990). *Cryptococcus neoformans, Candida albicans,* and other fungi bind specifically to the Glycosphingolipic lactosylceramide (Galβ1-4G1cβ1-1Cer), a possible adhesion receptor for yeasts. *Infection and Immunity*, *58*, 2085–2090.

Jung, K., Ha, Y., Ha, S. K., Han, D. U., Kim, D. W., Moon, W. K., & Chae, C. (2004). Antiviral effect of *Saccharomyces cerevisiae* beta-glucan to swine influenza virus by increased production of interferon-gamma and nitric oxide. *Journal of veterinary medicine. B, Infectious diseases and veterinary public health*, *51*, 72–76.

Kapadia, S., Dibbs, Z., Kurrelmeyer, K., Kalra, D., Seta, Y., Wang, F., Buzkurt, B., Oral, H., Sivasubramanian, N., & Mann, D. L. (1998). The role of cytokines in the failing human heart. *Cardiology Clinics*, *16*, 645–656.

Karlsson, K. A. (1989). Animal glycosphingolipids as membrane attachment sites for bacteria. *Annual Review of Biochemistry*, *58*, 309–350.

Kato, Y., Adachi, Y., & Ohno, N. (2006). Contribution of N-linked oligosaccharides to the expression and functions of beta-glucan receptor, Dectin-1. *Biological and Pharmaceutical Bulletin*, *29*, 1580–1586.

Kiechl, S., Egger, G., Mayr, M., Wiedermann, C. J., Bonora, E., Oberhollenzer, F., Muggeo, M., Xu, Q., Wick, G., Poewe, W., & Willeit, J. (2001). Chronic infections and the risk of carotid atherosclerosis. *Circulation*, *103*, 1064.

Kougias, P., Wei, D., Rice, P. J., Ensley, H. E., Kalbfleisch, J., Williams, D. L., & Browder, I. W. (2001). Normal human fibroblasts express pattern recognition receptors for fungal (1→3)-β-D-glucans. *Infection and Immunity*, *69*, 3933–3938.

Kournikakis, B., Mandeville, R., Brousseau, P., & Ostroff, G. (2003). Anthrax-protective effects of yeast beta 1,3 glucans. *Medscape General Medicine*, *5*, 1–10.

Kupfahl, C., Geginat, G., & Hof, H. (2007). Lentinan has a stimulatory effect on innate and adaptive immunity against murine *Listeria monocytogenes* infection. *International Immunopharmacology*, *6*, 686–696.

Lanier, L. L., Corliss, B. C., Wu, J., Leong, C., & Phillips, J. H. (1998). Immunoreceptor DAP12 bearing a tyrosine-based activation motif is involved in activating NK cells. *Nature*, *391*, 703–707.

LeBlanc, B. W., Albina, J. E., & Reichner, J. S. (2006). The effect of PGG-{beta}-glucan on neutrophil chemotaxis in vivo. *Journal of Leukocyte Biology*, *79*, 667–675.

Leibovich, S. J., & Danon, D. (1980). Promotion of wound repair in mice by application of glucan. *Journal of the Reticuloendothelial Society*, *27*, 1–11.

LeibundGut-Landmann, S., Groß, O., Robinson, M. J., Osorio, F., Slack, E. C., Tsoni, S. V., Schweighoffer, E., Tybulewicz, V., Brown, G. D., Ruland, J., & e Sousa, C. R. (2007). Syk- and CARD9-dependent coupling of innate immunity to the induction of T helper cells that produce interleukin 17. *Nature Immunology*, *8*, 630–638.

Levy, S., & Shoham, T. (2005). The tetraspanin web modulates immune-signalling complexes. *Nature Reviews Immunology*, *5*, 136–148.

Li, B., Allendorf, D. J., Hansen, R., Marroquin, J., Ding, C., Cramer, D. E., & Yan, J. (2006). Yeast beta-glucan amplifies phagocyte killing of iC3b-opsonized tumor cells via complement receptor 3-Syk-phosphatidylinositol 3-kinase pathway. *Journal of Immunology*, *177*, 1661–1669.

Li, B., Cramer, D., Wagner, S., Hansen, R., King, C., Kakar, S., Ding, C., & Yan, J. (2007). Yeast glucan particles activate murine resident macrophages to secrete proinflammatory cytokines via MyD88- and Syk kinase-dependent pathways. *Clinical Immunology*, *124*, 170–181.

Li, C., Ha, T., Kelley, J., Gao, X., Qiu, Y., Kao, R. L., Browder, W., & Williams, D. L. (2003). Modulating Toll-like receptor mediated signaling by (1→3)-β-D-glucan rapidly induces cardioprotection. *Cardiovascular Research*, *61*, 538–547.

Li, C., Kao, R. L., Ha, T., Kelley, J., Browder, I. W., & Williams, D. L. (2001). Early activation of IKKβ during in vivo myocardial ischemia. *American Journal of Physiology: Heart and Circulatory Physiology*, *280*, H1264–H1271.

Liang, J., Melican, D., Cafro, L., Palace, G., Fisette, L., Armstrong, R., & Patchen, M. L. (1998). Enhanced clearance of a multiple antibiotic resistant *Staphylococcus aureus* in rats treated with PGG-glucan is associated with increased leukocyte counts and increased neutrophil oxidative burst activity. *International Journal of Immunopharmacology*, *20*, 595–614.

Lowe, E., Rice, P., Ha, T., Li, C., Kelley, J., Ensley, H., Lopez-Perez, J., Kalbfleisch, J., Lowman, D., Margl, P., Browder, W., & Williams, D. (2001). A (1-3)-β-D-linked heptasaccharide is the unit ligand for glucan pattern recognition receptors on human monocytes. *Microbes and Infection*, *3*, 789–797.

Lowe, E. P., Wei, D., Rice, P. J., Li, C., Kalbfleisch, J., Browder, I. W., & Williams, D. L. (2002). Human vascular endothelial cells express pattern recognition receptors for fungal glucans which stimulates nuclear factor κB activation and Interleukin 8 production. *The American Surgeon*, *68*, 508–517.

Lyuksutova, O. I., Murphey, E. D., Toliver-Kinsky, T. E., Lin, C. Y., Cui, W., Williams, D. L., & Sherwood, E. R. (2005). Glucan phosphate treatment attenuates burn-induced inflammation and improves resistance to *Pseudomonas Aeruginosa* burn wound infection. *Shock*, *23*, 224–232.

Mansell, P. W. A., & Di Luzio, N. R. (1976). The *in vivo* destruction of human tumor by glucan activated macrophages. In M. Fink (Ed.), *Role of Macrophage in Neoplasia* (pp. 227–243). New York: Academic Press.

Marchetti, M., Pisani, S., Pietropaolo, V., Saganti, L., Nicoletti, R., Degener, A., & Orsi, N. (1996). Antiviral effect of a polysaccharide from *Sclerotium glucanicum* towards herpes simplex virus type 1 infection. *Planta Medica*, *62*, 303–307.

Mayberry, W. R., & Lane, J. R. (1993). Sequential alkaline saponification/acid hydrolysis/esterification: A one-tube method with enhanced recovery of both cyclopropane and hydroxylated fatty acids. *Journal of Microbiological Methods*, *18*, 21–32.

Medeiros, A. C., Chacon, D. A., Sales, V. S., Egito, E. S., Brandao-Neto, J., Pinheiro, L. A., & Carvalho, M. R. (2006). Glucan and glutamine reduce bacterial translocation in rats subjected to intestinal ischemia-reperfusion. *Journal of Investigative Surgery*, *19*, 39–46.

Meyer-Wentrup, F., Figdor, C. G., Ansems, M., Brossart, P., Wright, M. D., Adema, G. J., & van Spriel, A. B. (2007). Dectin-1 interaction with tetraspanin CD37 inhibits IL-6 production. *Journal of Immunology*, *178*, 154–162.

Miura, N. N., Ohno, N., Aketagawa, J., Tamura, H., Tanaka, S., & Yadomae, T. (1996). Blood clearance of (1→3)-beta-D-glucan in MRL lpr/lpr mice. *FEMS Immunology and Medical Microbiology*, *13*, 51–57.

Mohagheghpour, N., Dawson, M., Hobbs, P., Judd, A., Winant, R., Dousman, L., Waldeck, L., Hokama, L., Tuse, D., Kos, F., et al. (1995). Glucans as immunological adjuvants. *Advances in Experimental Medicine and Biology*, *383*, 13–22.

Monari, C., Retini, C., Casadevall, A., Netski, D., Bistoni, F., Kozel, T. R., & Vecchiarelli, A. (2003). Differences in outcome of the interaction between *Cryptococcus neoformans* glucuronoxylomannan and human monocytes and neutrophils. *European Journal of Immunology*, *33*, 1041–1051.

Mueller, A., Raptis, J., Rice, P. J., Kalbfleisch, J. H., Stout, R. D., Ensley, H. E., Browder, W., & Williams, D. L. (2000). The influence of glucan polymer structure and solution conformation on binding to (1→3)-β-D-glucan receptors in a human monocyte-like cell line. *Glycobiology*, *10*, 339–346.

Mueller, A., Rice, P. J., Ensley, H. E., Coogan, P. S., Kalbfleisch, J. H., Kelley, J. L., Love, E. J., Portera, C. A., Ha, T., Browder, I. W., & Williams, D. L. (1996). Receptor binding and internalization of water-soluble (1→3)-beta-D-glucan biologic response modifier in two monocyte/macrophage cell lines. *Journal of Immunology*, *156*, 3418–3425.

Murrin, L. C., & Talbot, J. N. (2007). RanBPM, a scaffolding protein in the immune and nervous systems. *Journal of Neuroimmune Pharmacology*, *2*, 290–295.

Nakamura, K., Kinjo, T., Saijo, S., Miyazato, A., Adachi, Y., Ohno, N., Fujita, J., Kaku, M., Iwakura, Y., & Kawakami, K. (2007). Dectin-1 is not required for the host defense to *Cryptococcus neoformans*. *Microbiology and Immunology*, *51*, 1115–1119.

Nakayama, H., Yoshizaki, F., Prinetti, A., Sonnino, S., Mauri, L., Takamori, K., Ogawa, H., & Iwabuchi, K. (2007). Lyn-coupled LacCer-enriched lipid rafts are requird for CD11b/CD18-mediated neutrophil phagocytosis of nonopsonized microorganisms. *Journal of Leukocyte Biology*, eprint.

Netea, M. G., Brown, G. D., Kullberg, B. J., & Gow, N. A. (2008). An integrated model of the recognition of *Candida albicans* by the innate immune system. *Nature Reviews. Microbiology*, *6*, 67–78.

Nicoletti, A., Nicoletti, G., Ferraro, G., Palmieri, G., Mattaboni, P., & Germogli, R. (1992). Preliminary evaluation of immunoadjuvant activity of an orally administered glucan extracted from *Candida albicans*. *Arzneimittelforschung*, *42*, 1246–1250.

Ohno, N., Suzuki, I., Oikawa, S., Sato, K., Miyazaki, T., & Yadomae, T. (1984). Antitumor activity and structural characterization of glucans extracted from cultured fruit bodies of *Grifola frondosa*. *Chemical Pharmaceutical Bulletin*, *32*, 1142–1151.

Olsson, S., & Sundler, R. (2007). The macrophage beta-glucan receptor mediates arachidonate release induced by zymosan: Essential role for Src family kinases. *Molecular Immunology*, *44*, 1509–1515.

Olynych, T. J., Jakeman, D. L., & Marshall, J. S. (2006). Fungal zymosan induces leukotriene production by human mast cells through a dectin-1-dependent mechanism. *Journal of Allergy and Clinical Immunology*, *118*, 837–843.

Oral, H., Kapadia, S., Nakano, M., Torre-Amione, G., Lee, J., Lee-Jackson, D., Young, J. B., & Mann, D. L. (1995). Tumor necrosis factor-alpha and the failing human heart. *Clinical Cardiology*, *18*, IV20–IV27.

Ormstad, H., Groeng, E. C., Lovik, M., & Hetland, G. (2000). The fungal cell wall component β-1, 3-glucan has an adjuvant effect on the allergic response to ovalbumin in mice. *Journal of Toxicology and Environmental Health, Part A*, *61*, 55–67.

Ozinsky, A., Underhill, D. M., Fontenot, J. D., Hajjar, A. M., Smith, K. D., Wilson, C. B., Schroeder, L., & Aderem, A. (2000). The repertoire for pattern recognition of pathogens by the innate immune system is defined by cooperation between Toll-like receptors. *Proceedings of the National Academy of Sciences of the United States of America*, *97*, 13766–13771.

Ozment-Skelton, T. R., Goldman, M. P., Gordon, S., Brown, G. D., & Williams, D. L. (2006). Prolonged reduction of leukocyte membrane-associated Dectin-1 levels following β-glucan administration. *Journal of Pharmacolology and Experimental Therapeutics*, *318*, 540–546.

Palma, A. S., Feizi, T., Zhang, Y., Stoll, M. S., Lawson, A. M., Diaz-Rodrigues, E., Campanero-Rhodes, M. A., Costa, J., Gordon, S., Brown, G. D., & Chai, W. (2005). Ligands for the β-glucan receptor, Dectin-1, assigned using 'designer' microarrays of oligosaccharides (neoglycolipids) generated from glucan polysaccharides. *Journal of Biological Chemistry*, *281*, 5771–5779.

Pearson, A., Lux, A., & Kreieger, M. (1995). Expression cloning of dSR-CI, a class C macrophage-specific scavenger receptor from *Drosophila melanogaster*. *Proceedings of the National Academy of Sciences of the United States of America*, *92*, 4056–4060.

Persson, W. K., Gronlund, U., & Johannisson, A. (2003). Intramammary infusion of beta 1,3-glucan for prevention and treatment of *Staphylococcus aureaus* mastitis. *Journal of veterinary medicine. B, Infectious diseases and veterinary public health*, *50*, 121–127.

Plow, E. F., & Zhang, L. (1997). A MAC-1 attack: Integrin functions directly challenged in knockout mice. *Journal of Clinical Investigation*, *99*, 1146.

Portera, C. A., Love, E. J., Memore, L., Zhang, L., Mueller, A., Browder, W., & Williams, D. L. (1997). Effect of macrophage stimulation on collagen biosynthesis in the healing wound. *The American Surgeon*, *63*, 125–131.

Qiu, H., Tang, W., Tong, X., Ding, K., & Zuo, J. (2007). Structure elucidation and sulfated derivatives preparation of two alpha-D-glucans from *Gastrodia elata* Bl. and their anti-dengue virus bioactivities. *Carbohydrate Research*, *342*, 2230–2236.

Rappleye, C. A., Eissenberg, L. G., & Goldman, W. E. (2007). Histoplasma capsulatum alpha-(1,3)-glucan blocks innate immune recognition by the beta-glucan receptor. *Proceedings of the National Academy of Sciences of the United States of America*, *104*, 1366–1370.

Reid, D. M., Montoya, M., Taylor, P. R., Borrow, P., Gordon, S., Brown, G. D., & Wong, S. Y. (2004). Expression of the beta-glucan receptor, Dectin-1, on murine leukocytes in situ correlates with its function in pathogen recognition and reveals potential roles in leukocyte interactions. *Journal of Leukocyte Biology*, *76*, 86–94.

Rice, P. J., Adams, E. L., Ozment-Skelton, T., Gonzalez, A. J., Goldman, M. P., Lockhart, B. E., Barker, L. A., Breuel, K. F., DePonti, W. K., Kalbfleisch, J. H., Ensley, H. E., Brown, G. D., Gordon, S., & Williams, D. L. (2005). Oral delivery and gastrointestinal absorption of soluble glucans stimulate increased resistance to infectious challenge. *Journal of Pharmacology and Experimental Therapeutics*, *314*, 1079–1086.

Rice, P. J., Kelley, J. L., Kogan, G., Ensley, H. E., Kalbfleisch, J. H., Browder, I. W., & Williams, D. L. (2002). Human monocyte scavenger receptors are pattern recognition receptors for (1-3)-β-D-glucans. *Leukocyte Biology*, *72*, 140–146.

Rice, P. J., Lockhart, B. E., Barker, L. A., Adams, E. L., Ensley, H. E., & Williams, D. L. (2004). Pharmacokinetics of fungal (1-3)-β-D-glucans following intravenous administration in rats. *International Immunopharmacology*, *4*, 1209–1215.

Rogers, N. C., Slack, E. C., Edwards, A. D., Nolte, M. A., Schulz, O., Schweighoffer, E., Williams, D. L., Gordon, S., Tybulewicz, V. L., Brown, G. D., & Reis e Sousa, C. (2005). Syk-dependent cytokine induction by Dectin-1 reveals a novel pattern recognition pathway for C type lectins. *Immunity*, *22*, 507–517.

Ross, G. D. (2000). Regulation of the adhesion versus cytotoxic functions of the Mac-1/CR3/alphaM-beta2-integrin glycoprotein. *Critical Reviews in Immunology*, *20*, 197–222.

Ross, G. D., Cain, J. A., & Lachmann, P. J. (1985). Membrane complement receptor type three ($CR_3$) has lectin-like properties analogous to bovine conglutinin and functions as a receptor for zymosan and rabbit erythrocytes as well as a receptor for iC3b[1]. *The Journal of Immunology*, *134*, 3307–3315.

Ross, G. D., Vetvicka, V., Yan, J., Xia, Y., & Vetvickova, J. (1999). Therapeutic intervention with complement and beta-glucan in cancer. *Immunopharmacology*, *42*, 61–74.

Rothfuchs, A. G., Bafica, A., Feng, C. G., Egen, J. G., Williams, D. L., Brown, G. D., & Sher, A. (2007). Dectin-1 interaction with mycobacterium tuberculosis leads to enhanced IL-12p40 production by splenic dendritic cells. *Journal of Immunology*, *179*, 3463–3471.

Rylander, R. (1996). Airway responsiveness and chest symptoms after inhalation of endotoxin or (1→3)-beta-D-glucan. *Indoor Built Environment*, *5*, 106–111.

Rylander, R., & Lin, R. H. (2000). (1-3)-β-D-glucan – relationship to indoor air-related symptoms, allergy and asthma. *Toxicology*, *152*, 47–52.

Rylander, R., Persson, K., Goto, H., Yuasa, K., & Tanaka, S. (1992). Airborne Beta-1,3-glucan may be related to symptoms in sick buildings. *Indoor Environment*, *1*, 263–267.

Saijo, S., Fujikado, N., Furuta, T., Chung, S., Kotaki, H., Seki, K., Sudo, K., Akira, S., Adachi, Y., Ohno, N., Kinjo, T., Nakamura, K., Kawakami, K., & Iwakura, Y. (2006). Dectin-1 is required for host defense against *Pneumocystic carinii* but not against *Candida albicans*. *Nature Immunology*, *8*, 39–46.

Sandvik, A., Wang, Y. Y., Morton, H. C., Aasen, A. O., Wang, J. E., & Johansen, F. E. (2007). Oral and systemic administration of beta-glucan protects against lipopolysaccharide-induced shock and organ injury in rats. *Clinical and Experimental Immunology*, *148*, 168–177.

Schaeffer, D. J., & Krylov, V. S. (2000). Anti-HIV activity of extracts and compounds from algae and cyanobacteria. *Ecotoxicology and Environmental Safety*, *45*, 208–227.

Slack, E. C., Robinson, M. J., Hernanz-Falcon, P., Brown, G. D., Williams, D. L., Schweighoffer, E., Tybulewicz, V. L., & Reis e Sousa, C. (2007). Syk-dependent ERK activation regulates IL-2 and IL-10 production by DC stimulated with zymosan. *European Journal of Immunology*, *37*, 1600–1612.

Steele, C., Marrero, L., Swain, S., Harmsen, A. G., Zheng, M., Brown, G. D., Gordon, S., Shellito, J. D., & Kolls, J. K. (2003). Alveolar macrophage-mediated killing of *Pneumocystic carinii* f. sp. *muris* involves molecular recognition by the Dectin-1 β-glucan receptor. *Journal of Experimental medicine*, *198*, 1677–1688.

Steele, C., Rapaka, R. R., Metz, A., Pop, S. M., Williams, D. L., Gordon, S., Kolls, J. K., & Brown, G. D. (2005). The beta-glucan receptor Dectin-1 recognizes specific morphologies of *Aspergillus fumigatus*. *PLOS Pathogens*, *1*, e42.

Stone, B. A., & Clarke, A. E. (1992). *Chemistry and biology of (1-3)-β-D-glucan*. Melbourne: La Trobe University Press.

Suram, S., Brown, G. D., Ghosh, M., Gordon, S., Loper, R., Taylor, P. R., Akira, S., Uematsu, S., Williams, D. L., & Leslie, C. C. (2006). Regulation of cytosolic phospholipase A2 activation and cyclooxgenase 2 expression in macrophages by the beta-glucan receptor. *Journal of Biological Chemistry*, *281*, 5506–5514.

Suzuki, I., Hashimoto, K., Ohno, N., Tanaka, H., & Yadomae, T. (1989). Immunomodulation by orally administered beta-glucan in mice. *International Journal of Immunopharmacology*, *11*, 761–769.

Suzuki, I., Tanaka, H., Kinoshita, A., Oikawa, S., Osawa, M., & Yadomae, T. (1990). Effect of orally administered beta-glucan on macrophage function in mice. *International Journal of Immunopharmacology*, *12*, 675–684.

Suzuki, M., Takatsuki, F., Maeda, Y. Y., Hamuro, J., & Chihara, G. (1996). Antitumor and immunological activity of lentinan in comparison with LPS. *International Journal of Immunopharmacology*, *16*, 463–468.

Takahashi, T., Fujisaki, M., Hirahata, S., Maeda, D., Tokura, H., Ohyama, T., & Irino, T. (2006). A case of advanced gastric cancer with multiple liver metastases successfully treated with TS-1/low dose CDDP/lentinan combination chemotherapy. *Gan To Kagaku Ryoho*, *33*, 2061–2063.

Taylor, P. R., Brown, G. D., Reid, D. M., Willment, J. A., Martinez-Pomares, L., Gordon, S., & Wong, S. W. C. (2002). The β-glucan receptor, Dectin-1, is predominantly expressed on the surface of cells of the monocyte/macrophage and neutrophil lineages. *Journal of Immunology*, *169*, 3876–3882.

Taylor, P. R., Tsoni, S. V., Willment, J. A., Dennehy, K. M., Rosas, M., Findon, H., Haynes, K., Steele, C., Botto, M., Gordon, S., & Brown, G. D. (2006). Dectin-1 is required for β-glucan recognition and control of fungal infection. *Nature Immunology*, *8*, 31–38.

Thorn, J., Beijer, L., & Rylander, R. (2001). Effects after inhalation of (1-3)-β-D-glucan in healthy humans. *Mediators of Inflammation*, *10*, 173–178.

Thornton, B. P., Vetvicka, V., Pitman, M., Goldman, R. C., & Ross, G. D. (1996). Analysis of the sugar specificity and molecular location of the beta-glucan-binding lectin site of complement receptor type 3 (CD11b/CD18). *The Journal of Immunology*, *156*, 1235–1246.

Torre-Amione, G., Kapadia, S., Lee, J., Durand, J. B., Bies, R. D., Young, J. B., & Mann, D. L. (1996). Tumor necrosis factor-alpha and tumor necrosis factor receptors in the failing human heart. *Circulation*, *93*, 704–711.

Tsikitis, V. L., Morin, N. A., Harrington, E. O., Albina, J. E., & Reichner, J. S. (2004). The lectin-like domain of complement receptor 3 protects endothelial barrier function from activated neutrophils. *Journal of Immunology*, *173*, 1284–1291.

Underhill, D. M., Ozinsky, A., Hajjar, A. M., Stevens, A., Wilson, C. B., Bassetti, M., & Aderem, A. (1999). The toll-like receptor 2 is recruited to macrophage phagosomes and discriminates between pathogens. *Nature*, *401*, 811–815.

Underhill, D. M., Rossnagle, E., Lowell, C. A., & Simmons, R. M. (2005). Dectin-1 activates Syk tyrosine kinase in a dynamic subset of macrophages for reactive oxygen production. *Blood*, *106*, 2543–2550.

Uryu, T. (1993). Sulfated curdlan and sulfated alkyl oligosaccharide as anti-viral agent. *Nippon Rinsho*, *51*, 164–168.

van den Herik-Oudijk, I. E., Capel, P. J., van der Bruggen, T., & van de Winkel, J. G. (1995). Identification of signaling motifs within human Fc gamma RIIa and Fc gamma RIIb isoforms. *Blood*, *84*, 2202–2211.

Vereschagin, E. I., Van Lambalgen, A. A., Dushkin, M. I., Schwartz, Y. S., Polyakov, L., Heemskerk, A., Huisman, E., Thijs, L. G., & Van den Bos, G. C. (1998). Soluble glucan protects against endotoxin shock in the rat: The role of the scavenger receptor. *Shock*, *9*, 193–198.

Vetvicka, V., Thornton, B. P., & Ross, G. D. (1996). Soluble beta-glucan polysaccharide binding to the lectin site of neutrophil or natural killer cell complement receptor type 3 (CD11b/CD18) generates a primed state of the receptor capable of mediating cytotoxicity of iC3b-opsonized target cells. *Journal of Clinical Investigation*, *98*, 50–61.

Viriyakosol, S., Fierer, J., Brown, G. D., & Kirkland, T. N. (2005). Innate immunity to the pathogenic fungus *Coccidioides posadasii* is dependent on Toll-like receptor 2 and Dectin-1. *Infection and Immunity*, *73*, 1553–1560.

Wakshull, E., Brunke-Reese, D., Lindermuth, J., Fisette, L., Nathans, R. S., Crowley, J. J., Tufts, J. C., Zimmerman, J., Mackin, W., & Adams, D. S. (1999). PGG-Glucan, a soluble beta-(1,3)-glucan, enhances the oxidative burst response, microbicidal activity, and activates an NF-kappaB-like factor in human PMN: Evidence for a glycosphingolipid beta-(1,3)-glucan receptor. *Immunopharmacology*, *41*, 89–107.

Wang, S. C., Bligh, S. W., Shi, S. S., Wang, Z. T., Hu, Z. B., Crowder, J., Branford-White, C., & Vella, C. (2007). Structural features and anti-HIV-1 activity of novel polysaccharides from red algae *Grateloupia longifolia* and *Grateloupia filicina*. *International Journal of Biological Macromolecules*, *41*, 369–375.

Wei, D., Williams, D., & Browder, W. (2002a). Activation of AP-1 and SP1 correlates with wound growth factor gene expression in glucan-treated human fibroblasts. *International Immunopharmacology*, *2*, 1163–1172.

Wei, D., Zhang, L., Williams, D. L., & Browder, I. W. (2002b). Glucan stimulates human dermal fibroblast collagen biosynthesis through a nuclear factor-1 dependent mechanism. *Wound Repair and Regeneration*, *10*, 161–168.

Wheeler, R. T., & Fink, G. R. (2006). A drug-sensitive genetic network masks fungi from the immune system. *PLOS Pathogens*, *2*, e35.

Williams, D. L. (1997). Overview of (1→3)-β-D-glucan immunobiology. *Mediators of Inflammation*, *6*, 247–250.

Williams, D. L., Browder, I. W., & DiLuzio, N. R. (1983). Immunotherapeutic modification of *Escherichia coli*-induced experimental peritonitis and bacteremia by glucan. *Surgery*, *93*, 448–454.

Williams, D. L., Cook, J. A., Hoffmann, E. O., & DiLuzio, N. R. (1978). Protective effect of glucan in experimentally induced candidiasis. *Journal of the Reticuloendothelial Society*, *23*, 479–490.

Williams, D. L., & Di Luzio, N. R. (1980). Glucan induced modification of experimental *Staphylococcus aureus* infection in normal, leukemic and immunosuppressed mice. In M. R. Escobar & H. Friedman (Eds.), *Macrophages and Lymphocytes, Part A* (pp. 291–306). New York: Plenum Publishing Corporation.

Williams, D. L., & Di Luzio, N. R. (1985a). Immunopharmacologic modification of experimental viral diseases by glucan. *EOS - Rivista di immunologia ed immunofarmacologia*, *2*, 78–82.

Williams, D. L., & Di Luzio, N. R. (1985b). Immunopharmacologic modification of experimental viral diseases by glucan. *EOS Journal of Immunology and Immunopharmacology*, *5*, 78–82.

Williams, D. L., & DiLuzio, N. R. (1980). Glucan-induced modification of murine viral hepatitis. *Science*, *208*, 67–69.

Williams, D. L., Ha, T., Li, C., Kalbfleisch, J. H., & Ferguson, D. A., Jr. (1999a) Early activation of hepatic NFkB and NF-IL6 in polymicrobial sepsis correlates with bacteremia, cytokine expression and mortality. *Annals of Surgery*, *230*, 95–104.

Williams, D. L., Ha, T., Li, C., Kalbfleisch, J. H., Laffan, J. J., & Ferguson, D. A. (1999b). Inhibiting early activation of tissue nuclear factor-κB and nuclear factor interleukin 6 with (1→3)-β-D-glucan increases long-term survival in polymicrobial sepsis. *Surgery*, *126*, 54–65.

Williams, D. L., Ha, T., Li, C., Kalbfleisch, J. H., Schweitzer, J., Vogt, W., & Browder, I. W. (2003a). Modulation of tissue toll-like receptor 2 and 4 during the early phases of polymicrobial sepsis correlates with mortality. *Critical Care Medicine*, *31*, 1808–1818.

Williams, D. L., Ha, T., Li, C., Laffan, J., Kalbfleisch, J., & Browder, W. (2000). Inhibition of LPS induced NFκB activation by a glucan ligand involves down regulation of IKKβ kinase activity and altered phosphorylation and degradation of IκBα. *Shock*, *13*, 446–452.

Williams, D. L., Li, C., Ha, T., Ozment-Skelton, T., Kalbfleisch, J. H., Preiszner, J., Brooks, L., Breuel, K., & Schweitzer, J. B. (2004a). Modulation of the phosphoinositide 3-Kinase pathway alters innate resistance to polymicrobial sepsis. *Journal of Immunology*, *172*, 449–456.

Williams, D. L., Lowman, D. W., & Ensley, H. E. (2004b). Introduction to the chemistry and immunobiology of β-glucans. In S. H. Young & V. Castranova (Eds.), *Toxicology of 1→3-beta-glucans. Glucans as a marker for fungal exposure* (pp. 1–34). New York: Taylor & Francis.

Williams, D. L., Mueller, A., & Browder, W. (1996). Glucan-based macrophage stimulators: A review of their anti-infective potential. *Clinical Immunotherapy*, *5*, 392–399.

Williams, D. L., Ozment-Skelton, T., & Li, C. (2006). Modulation of the phosphoinositide-3-kinase signaling pathway alters host resistance to sepsis, inflammation, and ischemia/reperfusion injury. *Shock*, *25*, 432–439.

Williams, D. L., Pretus, H. A., McNamee, R. B., Jones, E. L., Ensley, H. E., Browder, I. W., & Di Luzio, N. R. (1991). Development, physicochemical characterization and preclinical efficacy evaluation of a water soluble glucan sulfate derived from *Saccharomyces cerevisiae*. *Immunopharmacology*, *22*, 139–156.

Williams, D. L., Rice, P. J., Herre, J., Willment, J. A., Taylor, P. R., Gordon, S., & Brown, G. D. (2003b). Recognition of fungal glucans by pattern recognition receptors. *Recent Developments in Carbohydrate Research*, *1*, 49–66.

Williams, D. L., Yaeger, R. G., Pretus, H. A., Browder, I. W., McNamee, R. B., & Jones, E. L. (1989). Immunization against *Trypanosoma cruzi:* Adjuvant effect of glucan. *International Journal of Immunopharmacology*, *11*, 403–410.

Willment, J. A., Gordon, S., & Brown, G. D. (2001). Characterization of the human β-glucan receptor and its alternatively spliced isoforms. *Journal of Biological Chemistry*, *276*, 43818–43823.

Willment, J. A., Lin, H.-H., Reid, D. M., Taylor, P. R., Williams, D. L., Wong, S. Y. C., Gordon, S., & Brown, G. D. (2003). Dectin-1 expression and function are enhanced on alternatively activated and GM-CSF-treated macrophages and are negatively regulated by IL-10, dexamethasone, and lipopolysaccharide. *Journal of Immunology*, *171*, 4569–4573.

Willment, J. A., Marshall, A. S. J., Reid, D. M., Williams, D. L., Wong, S. Y. C., Gordon, S., & Brown, G. D. (2005). The human β-glucan receptor is widely expressed and functionally equivalent to murine Dectin-1 on primary cells. *European Journal of Immunology*, *35*, 1539–1547.

Wolk, M., & Danon, D. (1985). Promotion of wound healing by yeast glucan evaluated on single animals. *Medical Biology*, *63*, 73–80.

Wright, S. D., & Silverstein, S. C. (1982). Tumor-promoting phorbol esters stimulate C3b and C3b' receptor-mediated phagocytosis in cultured human monocytes. *Journal of Experimental medicine*, *156*, 1149–1164.

Wu, D. Y., Han, S. N., Bronson, R. T., Smith, D. E., & Meydani, S. N. (1998). Dietary supplementation with mushroom-derived protein-bound glucan does not enhance immune function in young and old mice. *Journal of Nutrition*, *128*, 193–197.

Wuersch, P., & Pi-Sunyer, F. X. (1997). The role of viscous soluble fiber in the metabolic control of diabetes – a review with special emphasis on cereals rich in beta-glucan. *Diabetes Care*, *20*, 1774–1780.

Xia, Y., Borland, G., Huang, J., Mizukami, I. F., Petty, H. R., Todd, R. F., III, & Ross, G. D. (2002). Function of the lectin domain of Mac-1/complement receptor type 3 (CD11b/CD18) in regulating neutrophil adhesion. *Journal of Immunology* *169*, 6417–6426.

Xia, Y., Vetvicka, V., Yan, J., Hanikyrova, M., Mayadas, T., & Goss, G. D. (1999). The beta-glucan-binding lectin site of mouse CR3 (CD11b/CD18) and its function in generating a primed state of the receptor that mediates cytotoxic activation in response to iC3b-opsonized target cells. *Journal of Immunology*, *162*, 2281–2290.

Xie, J., Sun, M., Guo, L., Liu, W., Jiang, J., Chen, X., Zhou, L., & Gu, J. (2006). Human Dectin-1 isoform E is a cytoplasmic protein and interacts with Ran BPM. *Biochemical and Biophysical Research Communications*, *347*, 1067–1073.

Yadav, M., & Schorey, J. S. (2006). The β-glucan receptor Dectin-1 functions together with TLR2 to mediated macrophage activation by mycobacteria. *Blood*, *108*, 3168–3175.

Yan, J., Allendorf, D. J., & Brandley, B. (2005). Yeast whole glucan particle (WGP) beta-glucan in conjunction with antitumour monoclonal antibodies to treat cancer. *Expert Opinion on Biological Therapy*, *5*, 691–702.

Yan, J., Vetvicka, V., Xia, Y., Coxon, A., Carroll, M. C., Mayadas, T. N., & Ross, G. D. (1999). β-Glucan, a "Specific" biologic response modifier that uses antibodies to target tumors for cytotoxic recognition by leukocyte complement receptor Type 3 (CD11b/CD18). *Journal of Immunology*, *163*, 3045–3052.

Yoshida, M., Roth, R. I., Grunfeld, C., Feingold, K. R., & Levin, J. (1996). Soluble (1-3)-beta-D-glucan purified from *Candida albicans*: Biologic effects and distribution in blood and organs in rabbits. *Journal of Laboratory Clinical Medicine*, *128*, 103–114.

Yoshida, M., Roth, R. I., Grunfeld, C., Feingold, K. R., & Levin, J. (1997). Pharmacokinetics, biological effects, and distribution of (1→3)-β-D-glucan in blood and organs in rabbits. *Mediators of Inflammation*, *6*, 279–283.

Yoshitomi, H., Sakaguchi, N., Kobayashi, K., Brown, G. D., Tagami, T., Sakihama, T., Hirota, K., Tanaka, S., Nomura, T., Miki, I., Gordon, S., Akira, S., Nakamura, T., & Sakaguchi, S. (2005). A role for fungal {beta}-glucans and their receptor Dectin-1 in the induction of autoimmune arthritis in genetically susceptible mice. *Journal of Experimental medicine*, *201*, 949–960.

Zakany, J., Chihara, G., & Fachet, J. (1980). Effect of lentinan on tumor growth in murine allogeneic and syngeneic hosts. *International Journal of Cancer*, *25*, 371–376.

Zelensky, A. N., & Gready, J. E. (2005). The C-type lectin-like domain superfamily. *FEBS Journal*, *272*, 6179–6217.

Zimmerman, J. W., Lindermuth, J., Fish, P. A., Palace, G. P., Stevenson, T. T., & DeMong, D. E. (1998). A novel carbohydrate-glycosphingolipid interaction between a beta-(1-3)-glucan immunomodulator, PGG-glucan, and lactosylceramide of human leukocytes. *Journal of Biological Chemistry*, *273*, 22014–22020.

CHAPTER 4.6

# Distribution, Fine Structure and Function of (1,3;1,4)-β-Glucans in the Grasses and Other Taxa

**Philip J. Harris[1] and Geoffrey B. Fincher[2]**

[1]*School of Biological Sciences, The University of Auckland, Auckland, New Zealand*

[2]*Australian Centre for Plant Functional Genomics, School of Agriculture, Food and Wine, University of Adelaide, Waite Campus, Glen Osmond, SA, Australia*

(1,3;1,4)-β-Glucans occur mostly in the non-lignified cell walls of cereal grains and the vegetative organs of grasses (family Poaceae). The amounts of (1,3;1,4)-β-glucans in cereal grains, where they occur particularly in the walls of the aleurone and starchy endosperm, vary widely among and within different species. (1,3;1,4)-β-Glucans also occur in the walls of families related to the Poaceae ('core Poales'), the monilophyte genus *Equisetum*, some bryophytes, some green and red algae, some lichens, one fungus and a chromalveolate. (1,3;1,4)-β-Glucans are thought to form a gel-like matrix in cell walls, between the reinforcing cellulose microfibrils. Additionally, (1,3;1,4)-β-glucans may sometimes function as a source of metabolizable energy. Genes have recently been identified that encode enzymes involved in the biosynthesis of (1,3;1,4)-β-glucans, and it should soon be possible to understand how (1,3;1,4)-β-glucans with different fine structures are synthesized, and to manipulate the biosynthesis of (1,3;1,4)-β-glucans for specific commercial purposes.

## 4.6.1. Introduction

From *Deschampsia antarctica* on the Antarctic peninsula, through temperate zone and subtropical grassland species, to hardy tropical crops, grasses (family Poaceae) have become major components of, and in some case dominate, plant ecosystems around the world. Despite their relatively recent appearance in evolutionary terms (Paterson et al., 2004), grasses now comprise 10 550 species (Mabberley, 2008) and cover 20% or more of the Earth's land mass (Gaut, 2002), and their domestication has been inextricably linked with the development of human societies (Sweeney and McCouch, 2007). Grass species such as wheat, rice, maize, sorghum, sugar cane, millet, barley and oats provide us with much of our daily caloric requirements. Furthermore, humans rely heavily on forage and fodder grasses for livestock production and, quite logically, we are looking to grasses as sources of biomass for future biofuel production (Harris and Stone, 2008). The formidable evolutionary success of the grasses has been attributed, in part at least, to their dynamic and plastic genomes, particularly with respect to transitory and stable polyploidization events (Levy and Feldman, 2002; Paterson et al., 2004; Linder and Rudall, 2005; Bennetzen, 2007; Salse et al., 2008), and their associated ability to adapt to hostile or competitive environments.

One notable characteristic of grasses is the presence in their cell walls of (1,3;1,4)-β-glucans. Although grasses are the largest family that have walls containing (1,3;1,4)-β-glucans, these polysaccharides also occur in the walls of related families and in some other taxa. This review will focus on the (1,3;1,4)-β-glucans in grasses, but we will also outline the distribution of the polysaccharide in more general terms and, in particular, how the differences in fine structure of the polysaccharide might have influenced the function and distribution of (1,3;1,4)-β-glucans. The fine structures of the (1,3;1,4)-β-glucans would indicate that the grasses have evolved a relatively unusual chemical strategy to generate a non-cellulosic wall polysaccharide that possesses the physicochemical properties required of a component of the gel-like matrix phase of walls.

In addition to the presence of (1,3;1,4)-β-glucans, the primary walls of the vegetative tissues of grasses and the other families of the graminoid clade in the order Poales (see Section 4.6.5) have much lower amounts of pectins and xyloglucans, but higher amounts of heteroxylans, than walls of eudicotyledons and non-commelinid monocotyledons (Fincher and Stone, 2004; Harris, 2005). The walls of other commelinid monocotyledons have intermediate compositions, but the palms (Arecales) are exceptional in having wall compositions similar to those of eudicotyledons and non-commelinid monocotyledons. The non-lignified walls of cereal grains, where the concentrations of (1,3;1,4)-β-glucans are usually much higher than in equivalent walls of vegetative tissues, also have high concentrations of heteroxylans and low concentrations of pectins (Table 1). Xyloglucans are only minor components of grain walls.

Further to their role as key components of cell walls in the grasses, (1,3;1,4)-β-D-glucans are also important in various commercial processes that involve the utilization of cereal grains. Thus, (1,3;1,4)-β-D-glucans are considered to be undesirable in the malting and brewing industries, where they can be extracted from the grain in the brewery and, if they are not sufficiently depolymerized during the malting process, can form solutions of high viscosity.

**Table 1: Composition of walls of specific cell types in cereal grains (weight %)[a]**

| Species | Cell type | Heteroxylan | (1,3;1,4)-β-glucan | Cellulose | Glucomannan | Pectin |
|---|---|---|---|---|---|---|
| Wheat | Aleurone | 65 | 29 | 2 | 2 | – |
| (*Triticum aestivum*) | Starchy endosperm | 70 | 20 | 4 | 7 | – |
| Barley | Aleurone | 71 | 26 | 2 | 2 | – |
| (*Hordeum vulgare*) | Starchy endosperm[b] | 20 | 75 | 2 | 2 | – |
| Rice (*Oryza sativa*) | Starchy endosperm | 27 | 20 | 28 | 15 | 3 |

[a]Table modified from Fincher and Stone (2004) and Stone (2006).
[b]The composition of the starchy endosperm cell walls of oats is similar to that of barley (Miller and Fulcher, 1995; Miller et al., 1995).

Such solutions are difficult to filter and can significantly slow beer production, and can contribute to the formation of beer hazes in the final product. Similarly, (1,3;1,4)-β-D-glucans in stock feed formulations for monogastric animals have anti-nutritive properties because they increase the viscosity of gut contents, which in turn slows down the digestive process and results in reduced growth rates. More recently, the same effect of increased viscosity of gut contents has been recognised to bring beneficial effects in human health, specifically through lowering serum cholesterol, reducing glycaemic index and hence the risk of type II diabetes and obesity, and also possibly in lowering the risk of colorectal and certain other cancers (Morgan, 2000; Brennan and Cleary, 2005; Harris and Smith, 2006; Wood, 2007). These practical applications, together with emerging interest in wall polysaccharides of grasses in the context of biomass production for bioethanol industries, has led to renewed interest in enzymes and corresponding genes that direct the biosynthesis of the (1,3;1,4)-β-D-glucans. In the latter case, increasing the amount of (1,3;1,4)-β-D-glucan in vegetative tissues could enhance the conversion and fermentation phases of bioethanol production. Early biochemical studies revealed that the biosynthetic enzymes were membrane proteins, but the identification of genes that mediate in (1,3;1,4)-β-D-glucan biosynthesis has only recently been made possible through genome sequencing programs and comparative genomics, linked through powerful bioinformatics capabilities.

In the sections below, the fine structures of (1,3;1,4)-β-D-glucans from grasses are briefly reviewed, although these have been dealt with in more detail in Chapter 2.1. The biological roles of (1,3;1,4)-β-D-glucans will be discussed in relation to their fine structures, and the distribution of the polysaccharide in different taxa will be reviewed.

## 4.6.2. Fine Structure of (1,3;1,4)-β-D-Glucans

Although the chemical structures and physicochemical properties of (1,3;1,4)-β-glucans of the grasses are dealt with in detail in Chapters 2.1 and 2.2, it is important to briefly summarize these properties here, as a general introduction to the occurrence of (1,3;1,4)-β-glucans in walls of the grasses and other taxa, and to emphasize the importance of fine structure on (1,3;1,4)-β-glucan function in the walls.

The (1,3;1,4)-β-glucans of the grasses consist of unbranched and unsubstituted chains of β-glucopyranosyl monomers polymerized through (1,3)- and (1,4)-linkages. The ratio of (1,4)- to (1,3)-linkages is reasonably constant in the grasses and lies in the range 2.2–2.6:1 (Fincher and Stone, 2004). The well characterized (1,3;1,4)-β-glucans from the starchy endosperm of barley grains predominantly contain blocks of two or three adjacent (1,4)-linked β-glucosyl residues, which are separated by single (1,3)-linked β-glucosyl residues. About 10% by weight of this polysaccharide is made up of longer blocks of adjacent (1,4)-linked β-glucosyl residues, which consist of four to over 10 adjacent (1,4)-linked β-glucosyl residues (Woodward et al., 1983a). The water-soluble (1,3;1,4)-β-glucans from barley grain are relatively long molecules that can have degrees of polymerization (DP) of over 1000 glucosyl residues and adopt an extended, asymmetrical conformation in aqueous media. They have axial ratios of about 100, which indicate that the molecule is about 100 times as long as it is wide (Woodward et al., 1983b).

The (1,4)-linked and (1,3)-linked β-glucosyl residues in (1,3;1,4)-β-glucans from the grasses are not arranged randomly, but equally they are not arranged in regular, repeating sequences. The evidence to support this notion is afforded by the observation that (1,3;1,4)-β-D-glucans from barley and many other grasses consist mostly of two or three adjacent (1,4)-linked β-glucosyl residues separated by a single (1,3)-linked β-glucosyl residue, as noted above. If the residues were arranged randomly along the chain, then we would expect to find adjacent (1,3)-β-glucosyl residues at a relatively high frequency, given that these residues constitute about 30% of the molecule. Indeed, Buliga et al. (1986) showed through molecular modelling that the physicochemical properties of a barley water-soluble (1,3;1,4)-β-glucan could be accurately predicted if the (1,3)-β-glucosyl residues were present as single, individual components of the polysaccharide, and that the introduction of just one pair of adjacent (1,3)-β-glucosyl residues in any polysaccharide chain would cause major deviations of the physicochemical properties away from those that had been measured experimentally.

Thus, (1,3;1,4)-β-glucans from the grasses have been considered as co-polymers of cellotriosyl and cellotetraosyl residues linked by single (1,3)-β-linkages, with occasional longer

**Table 2: Molar ratios of cellotriosyl to cellotetraosyl units in the (1,3;1,4)-β-glucans of different species**

| Species | Ratio of cellotriosyl: cellotetraosyl units | Reference |
|---|---|---|
| *Equisetum arvense* | 0.1–0.05: 1 | Sørensen et al. (2008) |
| *Equisetum fluviatile* | 0.1: 1 | Fry et al. (2008) |
| Oats (*Avena sativa*)[a] | 1.5–2.3: 1 | Lazaridou and Biliaderis (2007) |
| Barley (*Hordeum vulgare*)[a] | 1.8–3.5: 1 | Lazaridou and Biliaderis (2007) |
| Rye (*Secale cereale*)[a] | 1.9–3.0: 1 | Lazaridou and Biliaderis (2007) |
| Wheat (*Triticum aestivum*)[a] | 3.0–4.5: 1 | Lazaridou and Biliaderis (2007) |
| Icelandic moss (*Cetraria islandica*) | 20.2–24.6: 1 | Wood et al. (1991, 1994) Lazaridou et al. (2004) |

[a] Grains.

cellodextrin units between the single (1,3)-β-linkages (Fincher and Stone, 2004). Staudte et al. (1983) used Markov chain analysis to show that the cellotriosyl and cellotetraosyl units are arranged at random along the polysaccharide chain. The ratio of tri- to tetrasaccharide units varies among species. In barley the ratio is 1.8–3.5:1 (Table 2), but (1,3;1,4)-β-glucan preparations with different ratios can be extracted, using different solvents and different temperatures (Fincher and Stone, 2004; Lazaridou and Biliaderis, 2007). Thus, (1,3;1,4)-β-glucan preparations are typically polydisperse with respect to molecular size and fine structure. Furthermore, barleys with waxy starch have (1,3;1,4)-β-glucans with significantly higher cellotriosyl:cellotetraosyl ratios than barleys with normal starch (Wood et al., 2003).

## 4.6.3. Biological Role of (1,3;1,4)-β-Glucans in Cell Walls

### *Formation of Gel-Like Matrices in the Wall*

The asymmetrical conformations of (1,3;1,4)-β-glucans, coupled with their high DP, enable them to form a gel-like matrix between the reinforcing cellulosic microfibrils in the wall. Gel-like matrix phase polysaccharides are believed to function in walls by providing porosity during active growth, when the exchange of water, nutrients and other small molecules such as phytohormones between adjacent cells and the vasculature is essential. In addition, a gel-like matrix will contribute additional strength to the wall, together with a degree of flexibility and elasticity. The structures of most non-cellulosic wall polysaccharides, including (1,3;1,4)-β-glucans, also allow limited, local intermolecular alignments that are potentially stabilized by hydrogen bonding between hydroxyl groups and/or the O atoms of the pyranosyl or furanosyl rings. Such limited intermolecular alignments would account for 'junction zone' formation that is believed to occur during gel formation and have also been implicated in non-covalent interactions between the non-cellulosic polysaccharides themselves and with cellulose microfibrils (Mohnen, 2008; York and O'Neill, 2008).

It appears that two chemical routes have been adopted to enable the non-cellulosic matrix phase polysaccharides of cell walls to meet these functional requirements. In many cases, non-cellulosic matrix phase polysaccharides consist of a molecular backbone that has the potential to aggregate and form microfibrils of the type formed by cellulose. However, this is prevented because the backbones of the polysaccharides are substituted with monosaccharide residues or with short oligosaccharide chains. These substituents preclude extensive intermolecular alignment of the molecules and hence would interfere with microfibril formation. Examples of this strategy are afforded by the xyloglucans, in which xylosyl and short oligosaccharide substituents will prevent extensive alignment of the cellulosic (1,4)-β-glucan backbone; the heteroxylans, in which arabinosyl, glucuronosyl or short oligosaccharide substituents will prevent aggregation of the regular (1,4)-β-xylan backbone of these polysaccharides (Andrewartha et al., 1979); and the galactoglucomannans, where protruding galactosyl substituents will prevent aggregation of the regular backbone of (1,4)-β-glucosyl and/or (1,4)-β-mannosyl residues (McCleary and Matheson, 1986).

In the xyloglucans, heteroxylans and galactoglucomannans, internal regions where the backbone is not substituted permit some local intermolecular alignment with similarly unsubstituted regions of other matrix phase polysaccharides, or with cellulose microfibrils. These 'junction zones' occur over limited sections but allow gel formation or non-covalent interaction between the non-cellulosic and cellulosic components of the wall. Plants appear to have adopted a similar overall strategy of main chain substituents to prevent extensive alignment of the backbone structure of pectic polysaccharides, which is based on a regularly shaped (1,4)-α-galacturonan chain, but which also includes regions of alternating 4-linked galacturonosyl and 2-linked rhamnosyl residues (Mohnen, 2008). In this case, a number of structurally complex oligosaccharide substituents are appended to the backbone, but unsubstituted homogalacturonan regions might allow limited alignment of polysaccharide chains, depending on the degree of methyl esterification and acetylation of the homogalacturonan chain.

In the (1,3;1,4)-β-glucans of the grasses, a different molecular strategy appears to have been adopted for the production of an asymmetrical non-cellulosic polysaccharide that will not aggregate into microfibrils. In this case, while the backbone of the polysaccharide again appears to be based on a cellulosic (1,4)-β-glucan chain, the limitation on alignment

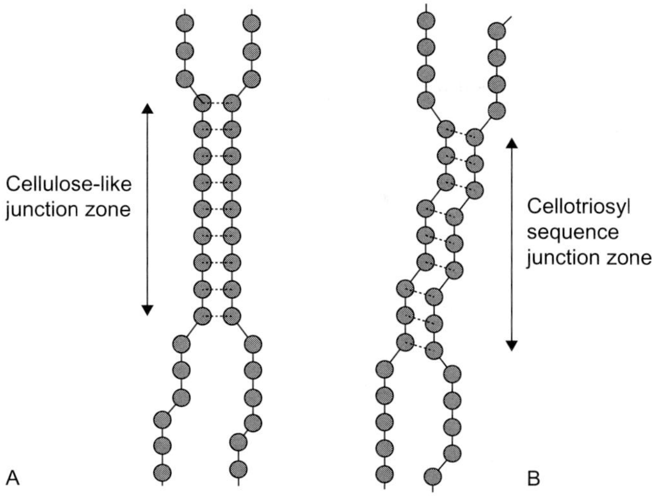


**Fig. 1: Diagram showing the interactions (A) between the cellulose-like regions and (B) between the sequential cellotriosyl units of (1,3;1,4)-β-glucan molecules. Reproduced from Tosh et al. (2004) with permission.**

and aggregation of the molecules is provided not through the addition of substituents, but rather through the irregular insertion of the single (1,3)-β-glucosyl residues into the 'cellulosic' backbone, which concurrently introduce irregularly spaced molecular 'kinks' into the polysaccharide chain. The irregular overall conformation of the polysaccharide backbone prevents extensive intermolecular interactions and microfibril formation, but the relatively long blocks of up to 12 or more adjacent (1,4)-linked β-glucosyl residues offer the potential for junction zone formation between individual (1,3;1,4)-β-glucan chains (Fig. 1), between (1,3;1,4)-β-glucans and other non-cellulosic wall polysaccharides such as heteroxylans, and between (1,3;1,4)-β-glucans and cellulose.

### *The effects of fine structure on biological function*

The potential success of the conformationally irregular backbone strategy for preventing the aggregation of polysaccharides into microfibrils will depend to a large extent on the relative proportions of cellotriosyl, cellotetraosyl, and higher oligodextrins in the polysaccharide. The ratios of cellotriosyl to cellotetraosyl units are known to vary widely among species (Table 2). If the polysaccharide consists mostly of cellotriosyl units, it will have a high cellotriosyl: cellotetraosyl ratio and as this ratio becomes higher the molecule will become more regular.

It follows that a more regular molecule will more readily align over more extended regions (Fig. 1). The (1,3;1,4)-β-glucan from *Cetraria islandica* (Icelandic moss), which is also known as lichenin or lichenan, has a very high cellotriosyl:cellotetraosyl ratio (20.2–24.6:1) and therefore a more regular structure than many of the (1,3;1,4)-β-glucans from grasses (Wood et al., 1991, 1994; Lazaridou et al., 2004) (Table 2). Consequently, solutions of lichenin form gels very readily (Lazaridou et al., 2004; Tosh et al., 2004). At the other end of the scale, if the polysaccharide consists mostly of cellotetraosyl residues, its cellotriosyl: cellotetraosyl ratio will be low and again it will tend to a more regular shape as the ratio decreases. Thus, the (1,3;1,4)-β-glucan from the monilophyte (fern *sensu lato*) *Equisetum* consists mostly of cellotetraosyl units rather than cellotriosyl units, with a cellotriosyl: cellotetraosyl ratio of 0.1 or less (Table 2), does not have the longer blocks of adjacent (1,4)-β-glucosyl residues (Sørenson et al., 2008), and is not readily soluble in aqueous media (Fry et al., 2008).

Even within the grasses cellotriosyl:cellotetraosyl ratios vary considerably and are often correlated both with the amount of (1,3;1,4)-β-glucan in the wall and with its solubility. Thus, low ratios appear to be associated with higher abundance in the walls and with increased water solubility. To illustrate this point it is noteworthy that in wheat grain walls the ratios range from 3.0:1 to 4.5:1, in barley 1.8 to 3.5:1 and in oats 1.5 to 2.3:1 (Lazaridou and Biliaderis, 2007) (Table 2). In wheat grain much of the (1,3;1,4)-β-glucan is insoluble in warm water, consistent with its relatively high cellotriosyl:cellotetraosyl ratio (Li et al., 2006; Lazaridou and Biliaderis, 2007), whereas the polysaccharides from barley and oats, which have lower cellotriosyl:cellotetraosyl ratios, are considerably more soluble (Table 3) (see Section 4.6.8). Furthermore, the concentrations of (1,3;1,4)-β-glucans in wheat grain are lower than those in barley and oats (see Section 4.6.8 and Table 3), which are relatively rich sources of (1,3;1,4)-β-glucans that are more soluble than the polysaccharide from wheat (Fincher and Stone, 2004).

These data and observations suggest that backbone irregularity is a key determinant of the physicochemical properties of (1,3;1,4)-β-glucans (Lazaridou et al., 2004) and that the ability of species to synthesize high molecular weight (1,3;1,4)-β-glucans with irregularity in their structures that do not readily align or aggregate might be a prerequisite for the use of this polysaccharide as a major component of the matrix phase of walls. In other words, the evolutionary adoption of (1,3;1,4)-β-glucans as a matrix phase component of cell walls might be dependent upon a capacity to fine-tune the structure of the polysaccharide and might account for the differences between (1,3;1,4)-β-glucans in different taxa.

**Table 3: The concentrations of (1,3;1,4)-β-glucans (g per 100 g dry weight) in whole cereal grains**

| Cereal species | Total (1,3;1,4)-β-glucan | Water-soluble (1,3;1,4)-β-glucan (% of total) | Reference |
|---|---|---|---|
| Barley (*Hordeum vulgare*) | 2–20 | ~65 (Average) | See text |
| Oats (*Avena sativa*) | 3–8 | ~82 (Average) | See text |
| Sorghum (*Sorghum bicolor*) | 1.1-6.2 | - | Ogbonna and Egunwu (1994) |
| Rye (*Secale cereale*) | 1.3–2.7 | <1 | See text |
| Maize (*Zea mays*) | 0.8–1.7 | - | Fincher and Stone (2004) |
| Triticale (x *Triticosecale*) | 0.3–1.2 | 0 | Stone and Clarke (1992) |
| Wheat (*Triticum aestivum*) | 0.5–1.0 | <3 | Beresford and Stone (1983) |
| Durum wheat (*Triticum durum*) | 0.5–0.6 | 0 | Beresford and Stone (1983) |
| Rice (*Oryza sativa*) | 0.13 | ~20 | Anderson et al. (1978) |

### *(1,3;1,4)-β-Glucans as a Source of Metabolizable Energy*

There are indications that (1,3;1,4)-β-glucans might also function in the grasses as a short to medium term source of glucose and hence metabolizable energy in the wall. In many cases (1,3;1,4)-β-glucans appear transiently in vegetative tissues (McCann et al., 2007) and it has been shown that (1,3;1,4)-β-glucans in walls of barley seedlings are re-mobilized when the leaves are placed in the dark. The glucose released by hydrolysis of wall (1,3;1,4)-β-glucans might provide an alternative source of energy under conditions of sugar depletion (Roulin and Feller, 2001; Roulin et al., 2002). Glucose recovered from the high concentrations of (1,3;1,4)-β-glucans in walls of the starchy endosperm of some grasses certainly makes a major contribution to available metabolizable energy in the grain following germination (Morrall and Briggs, 1978). Large changes in concentrations of (1,3;1,4)-β-glucans can occur during normal growth and development. In barley coleoptiles (1,3;1,4)-β-glucans increase to 10 mol% of walls during the elongation phase of growth but rapidly decrease to 1 mol% following the cessation of growth (Gibeaut et al., 2005). Similar patterns of (1,3;1,4)-β-glucan deposition and removal are observed in maize coleoptiles (see Section 4.6.5) (Carpita and Gibeaut, 1993; Inouhe and Nevins, 1998; Kim et al., 2000).

## 4.6.4. Distribution of (1,3;1,4)-β-Glucans in Plants and Other Taxa

It has often been assumed that (1,3;1,4)-β-glucans exist only in the grasses, but there is considerable evidence that (1,3;1,4)-β-glucans are found in the walls of certain other taxa.

It appears that the adoption of (1,3;1,4)-β-glucans as a major wall polysaccharide in these other species is not as widespread as it is in species of the grasses, and this might be related to differences in fine structure of the polysaccharide and associated differences in the functional suitability of the particular (1,3;1,4)-β-glucan in the wall. The distribution of (1,3;1,4)-β-glucans in different taxa is summarized in the sections below, where special emphasis is placed on the fine structural features of the polysaccharide if these have been defined. The tree of life based on Cavalier-Smith's (2004) concept of six kingdoms, and those plant lineages containing (1,3;1,4)-β-glucans are shown in Figs 1 and 2, respectively, of Chapter 4.7.

## Plants Other Than Angiosperms

In the plant kingdom (Cavalier-Smith, 2004), (1,3;1,4)-β-glucans have been found in a number of taxa outside the angiosperms. In the red algae, a sulfated (1,3;1,4)-β-glucan has been reported in the cell walls of *Kappaphycus alvarezii* (Gigartinales) (Lechat et al., 2000). In the green algae, there have been two reports of (1,3;1,4)-β-glucans. The chlorophyte alga *Ulva lactuca* contains a polysaccharide that is hydrolysed by (1,3;1,4)-β-glucan endo-hydrolase, but which also contains xylose (Popper and Fry, 2003). The charophyte alga *Micrasterias* (a desmid) has secondary walls and pores that are labelled with a monoclonal antibody specific for (1,3;1,4)-β-glucans (Eder et al., 2008).

In the bryophytes, only the leafy liverwort *Lophocolea bidentata* has been reported to contain a polysaccharide hydrolysed by (1,3;1,4)-β-glucan endo-hydrolase, but this polysaccharide is not always detectable and also contains arabinose (Popper and Fry, 2003). In the monilophytes (ferns *sensu lato*), (1,3;1,4)-β-glucans have recently been found in the cell walls of *Equisetum* spp. (Fry et al., 2008; Sørensen et al., 2008). Unlike the (1,3;1,4)-β-glucans of grasses, they consist mostly of cellotetraosyl units rather than cellotriosyl units (Table 2), and cellodextrin units with a DP greater than 7 are absent. Using a monoclonal antibody in conjunction with immunofluorescence and immunogold microscopy, it was shown that the (1,3;1,4)-β-glucans occur in the walls of all cell types in *E. arvense* stems with the exception of cell walls in the vascular tissue (Sørensen et al., 2008).

## Lichens and Fungi

The occurrence of (1,3;1,4)-β-glucans has been reported in many lichens, which are a symbiotic association between a mycobiont (fungus), usually an ascomycete, and a photobiont, usually a green alga or a cyanobacterium. The most detailed studies have been performed on the (1,3;1,4)-β-glucan (lichenin or lichenan) produced by *Cetraria islandica* (Icelandic moss).

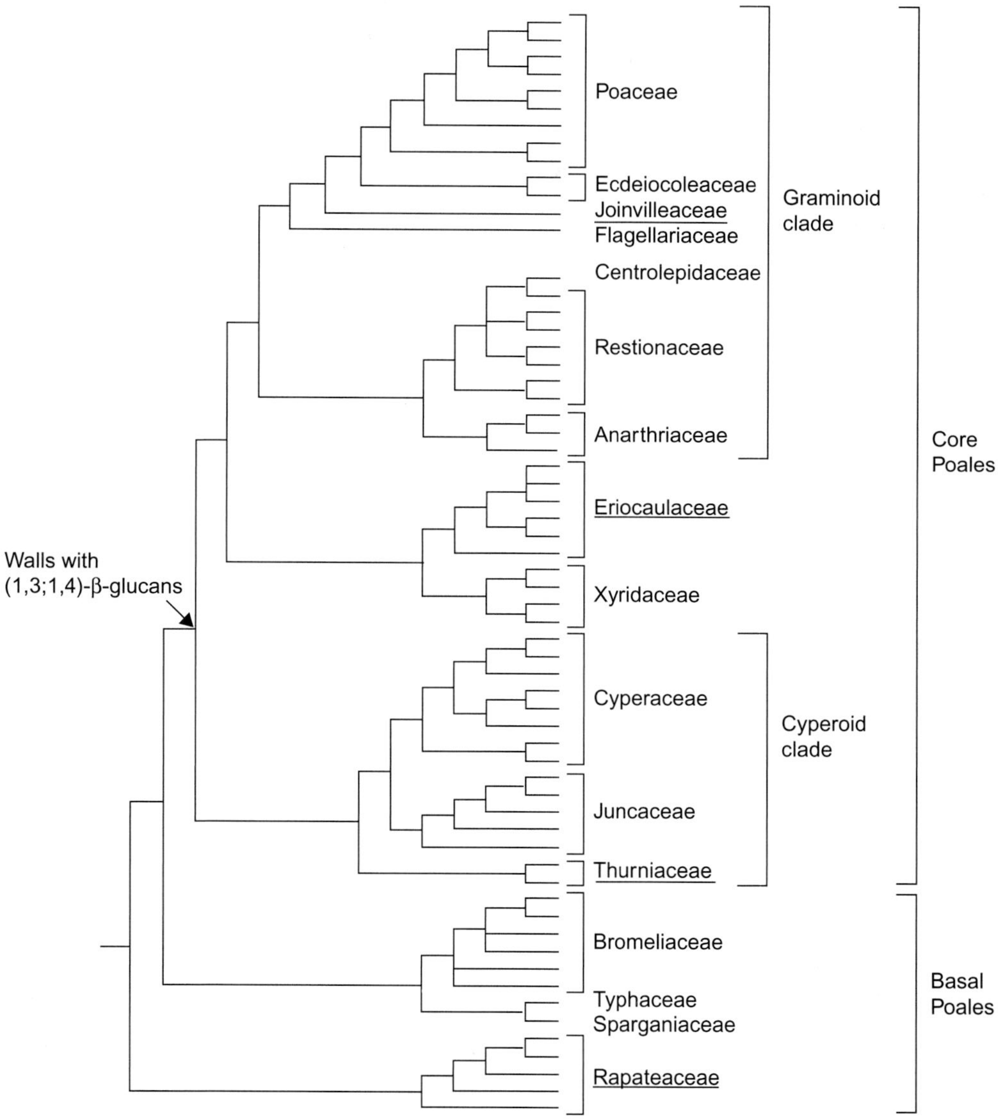


**Fig. 2: The distribution of (1,3;1,4)-β-glucans in the cell walls of vegetative organs (leaves or stems) of the enlarged order Poales, as defined by the Angiosperm Phylogeny Group (APG, 2003). The phylogenetic tree is based on the nucleotide sequences of the genes *rbcL* and *atpB* (Bremer, 2002). (1,3;1,4)-β-Glucans were detected by a specific enzymatic assay (Smith and Harris, 1999) and by immunogold labelling with a specific monoclonal antibody (Trethwey and Harris, 2002; Trethewey et al., 2005). Families that are underlined have not been investigated.**

This (1,3;1,4)-β-glucan differs from those characterized from grasses in having a much higher proportion of cellotriosyl to cellotetraosyl units (20.2–24.6:1, Table 2), with these two types of units accounting for 75% compared with more than 90% of the molecule. Many other species of lichens contain similar, lichenin-like (1,3;1,4)-β-glucans and their distribution is of taxonomic importance (Olafsdottir and Ingolfsdottir, 2001). (1,3;1,4)-β-Glucans have been found in many species in the Parmeliaceae, including other species of *Cetraria*, as well as *Evernia*, *Newropogon*, *Parmelia*, *Parmotrema*, *Rimelia* and *Usnea* (Olafsdottir and Ingolfsdottir, 2001; Carbonero et al., 2005). The ratio of (1,4)- to (1,3)-linkages varies from 2.3:1, found in the glucans from *Cetraria islandica*, to 0.3:1, found in the (1,3;1,4)-β-glucans from *Evernia prunastri*. (1,3;1,4)-β-Glucans have also been found in species from the families Alectoriaceae (*Alectoria*) and Rocellaceae (*Rocella*) (Olafsdottir and Ingolfsdottir, 2001). Immunogold labelling using a monoclonal antibody raised against barley (1,3;1,4)-β-glucan showed that the (1,3;1,4)-β-glucan in *Cetraria islandica* is located in the thick outer wall of the mycobiont, whereas none is associated with the cell wall or protoplast of the photobiont, the chlorophyte alga *Trebouxia* ( Honegger and Haisch, 2001).

Despite their widespread occurrence in lichens, (1,3;1,4)-β-glucans have been reported in the cell walls of only one non-lichen fungus, the ascomycete *Aspergillus fumigatus*, where they occur as part of a very complex, alkali-insoluble heteropolysaccharide (Fontaine et al., 2000).

### Chromalveolates

In the chromalveolates (see Chapter 4.7), a (1,3;1,4)-β-glucan occurs in the cell walls of the xanthophyte alga (a chromist) *Monodus subterraneus* (Ford and Percival, 1969), and a putative (1,3;1,4)-β-glucan has been reported from the dinoflagellate *Peridinium westii* (an alveolate) (Nevo and Sharon, 1969).

### Prokaryotes

Although (1,3;1,4)-β-glucans have so far not been found in prokaryotes, the isolation of oligosaccharides of β-Glc*p* containing (1,3)- and (1,4)-links has been reported from the Gram-positive bacterium *Sarcina ventriculi* (Lee and Hollingsworth, 1997).

## 4.6.5. Distribution of (1,3;1,4)-β-Glucans in Angiosperms

Because of their economic importance, much is known about the (1,3;1,4)-β-glucans that occur in cereal grains. However, (1,3;1,4)-β-glucans are also found in the cell walls of vegetative organs of grasses as well as in related families. (1,3;1,4)-β-Glucans have been reported in the cell walls of coleoptiles, leaves, stems, roots and suspension cultures (Wilkie, 1979; Stone and Clarke, 1992; Carpita, 1996; Gibeaut et al., 2005). Only low concentrations of (1,3;1,4)-β-glucans are present in cell walls of tissues that are mostly meristematic, but are synthesized during cell expansion. Increases in the concentrations of (1,3;1,4)-β-glucans during expansion have been studied, particularly in elongating coleoptiles. For example, in maize (*Zea mays*) coleoptiles, the (1,3;1,4)-β-glucan concentration in the cell walls increased from about 2.5% at day one to 20% at day four, when the coleoptiles were almost fully elongated, but decreased to 6% at day six (Kim et al., 2000). A similar pattern of change in (1,3;1,4)-β-glucan concentration during coleoptile development was found by Inouhe and Nevins (1998) in the same species and by Gibeaut et al. (2005) in barley (*Hordeum vulgare*) (see Section 4.6.3). In deepwater rice (*Oryza sativa*), higher concentrations of (1,3;1,4)-β-glucans were found in the walls of the zone of cell elongation than in the walls of the meristematic zone below and the zone of cell differentiation above (Sauter and Kende, 1992). Nevertheless, significant concentrations of (1,3;1,4)-β-glucans have been reported in preparations composed of mostly non-lignified primary cell walls from mature tissues. For example, such a preparation from the pith of mature stem internodes of perennial ryegrass (*Lolium perenne*) contained 8.8% (1,3;1,4)-β-glucan (Smith and Harris, 1999).

An initial study in angiosperms indicated that (1,3;1,4)-β-glucans occurred in the non-lignified cell walls of only the Poaceae (Stinard and Nevins, 1980). However, this study did not include families closely related to the Poaceae. A subsequent examination was undertaken in non-lignified cell walls isolated from species in families, in addition to the Poaceae, that comprised the Poales as defined by Dahlgren et al. (1985) (now often referred to as the graminoid clade) (Fig. 2), using a specific enzymatic assay (McCleary and Codd, 1991). The presence of (1,3;1,4)-β-glucans was demonstrated in the families Anarthriaceae, Centrolepidaceae, Ecdeiocoleaceae, Flagellariaceae and Restionaceae; plant material from the Joinvilleaceae was not available (Smith and Harris, 1999). The highest concentrations of (1,3;1,4)-β-glucans were found in the cell walls from *Flagellaria indica* (Flagellariaceae) (3.2%) and *Ecdeiocolea monostachya* (Ecdeiocoleaceae) (1.7%) and the lowest concentrations in the cell walls of the Restionaceae (0–0.1%). (1,3;1,4)-β-Glucans have also been identified in the non-lignified walls of *Flagellaria guineensis* (Popper and Fry, 2004).

In an extension of the study of Smith and Harris (1999), Trethewey et al. (2005) used immunogold labelling with a monoclonal antibody to (1,3;1,4)-β-glucans (Meikle et al., 1994) to investigate the presence of these polysaccharides in the cell walls of vegetative organs (leaves or stems) of families in the enlarged order Poales, as defined by the Angiosperm Phylogeny Group (APG, 2003) (Fig. 2). In addition to the Poaceae, species were examined from the

Flagellariaceae, Restionaceae, Xyridaceae, Cyperaceae, Juncaceae, Bromeliaceae, Typhaceae and Sparganiaceae. This study showed that immunogold labelling is a very sensitive and specific method for detecting (1,3;1,4)-β-glucans in cell walls. The density of labelling of non-lignified walls varied among the families. Labelling was heavy in the Poaceae and Flagellariaceae (Fig. 3A), light and very light in the Restionaceae and Xyridaceae (Fig. 3D), very light in the Cyperaceae and Juncaceae, and no labelling was detected in the Bromeliaceae, Typhaceae or Sparganiacae. Those families that showed labelling of the non-lignified cell walls also showed labelling of the lignified cell walls of the xylem tracheary elements and sclerenchyma fibres. The families that have (1,3;1,4)-β-glucans in their cell walls are all members of a major clade, referred to as the 'core Poales', resolved in phylogenetic trees of the enlarged order Poales, which was constructed using nucleotide sequences of genes (Bremer, 2002; Linder and Rudall, 2005) (Fig. 2). The families in which no (1,3;1,4)-β-glucans were detected are all members of the 'basal Poales', the other group in this order recognised. Thus, although further families remain to be examined, the presence of (1,3;1,4)-β-glucans in cell walls may be synapomorphic (i.e. a shared, derived character) for the 'core Poales'. If this is so, *CslF* genes, which have been identified as encoding glycosyltransferases involved in the biosynthesis of (1,3;1,4)-β-glucans in the Poaceae (Burton et al., 2006), may have evolved in extinct taxa that gave rise to the modern 'core Poales' clade (see also Chapter 4.7).

## 4.6.6. Location of (1,3;1,4)-β-Glucans in Vegetative Organs of Poaceae

Immunofluorescence microscopy using polyclonal antibodies against (1,3;1,4)-β-glucans showed that these polysaccharides occur widely in the cell walls of the coleoptiles of barley, oats, maize and rice (Hoson and Nevins, 1989). Immunogold light microscopy, using a monoclonal antibody against (1,3;1,4)-β-glucans (Meikle et al., 1994), also showed the widespread occurrence of these polysaccharides in the leaves of maize (Sinha and Lynch, 1998).

More recently, the same monoclonal antibody against (1,3;1,4)-β-glucans was used in conjunction with immunogold electron microscopy to determine the locations of (1,3;1,4)-β-glucans in the walls of the coleoptile, first leaf and root tip of barley (*Hordeum vulgare*) seedlings (Trethewey and Harris, 2002). Except for those of the outer root cap cells, all the walls were labelled. The labelling density on the non-lignified walls was heavy in the coleoptile and leaf (Fig. 3A), but only light in the root tip (Fig. 3B). Two types of distribution of the label occurred on non-lignified walls. In the coleoptile (except in the epidermis and in the two layers of parenchyma thereunder) and in the leaf, labelling was observed throughout the wall (Fig. 3A), but in

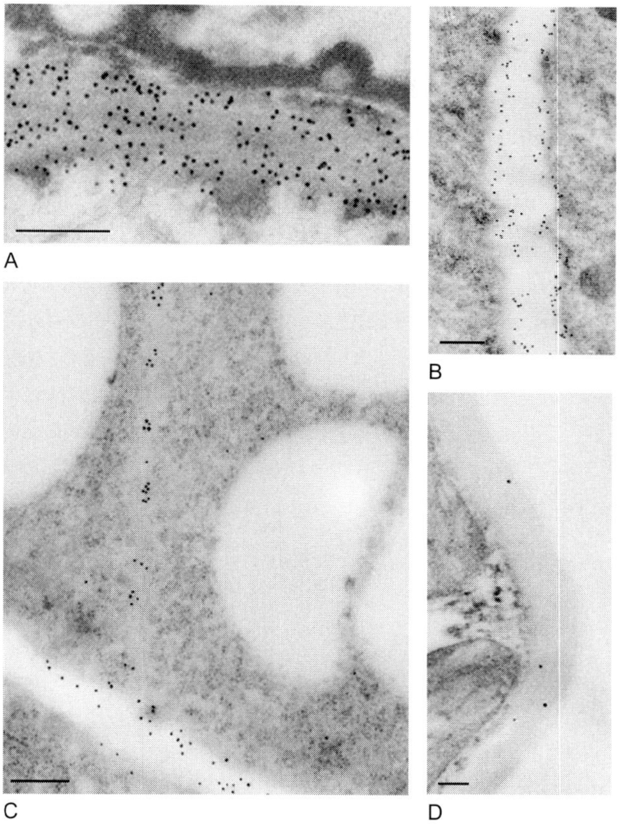

Fig. 3: Transmission electron micrographs of sections of vegetative organs of Poales species immunogold labelled with a (1,3;1,4)-β-glucan monoclonal antibody. (A) Walls of contiguous parenchyma cells (in the second and third layers from the epidermis) from a 6-day-old coleoptile of a barley seedling. The walls show heavy labelling with colloidal gold particles throughout. (B) Walls of contiguous root cap initial cells from a 4-day-old barley seedling. The walls show moderate labelling with colloidal gold particles only adjacent to the plasma membranes. (C) A cell plate formed after the division of a cortical cell in the root of a 4-day-old barley seedling. Clusters of colloidal gold particles are present over the cell plate. (D) Walls of contiguous subepidermal chlorenchyma cells from an immature stem of *Leptocarpus similis* (Restionaceae). The walls show very light labelling with colloidal gold particles only adjacent to the plasma membranes. Bars: (A, B, C) 250 nm, (D) 200 nm. Modified from Trethewey and Harris (2002) and Trethewey et al. (2005).

the root tips labelling was only found adjacent to the plasma membrane (Fig. 3B). In the epidermis of the coleoptile, labelling of the outer periclinal walls was confined to a band adjacent to the plasma membrane that constituted approximately one quarter of the total width of the wall. In the two layers of parenchyma under this, label was present only adjacent to the plasma membrane. In the root meristem, young, newly formed cell walls were lightly labelled throughout, but in cell plates the labelling occurred in small clusters (Fig. 3C) in a pattern similar to that found using a monoclonal antibody to (1,3)-β-glucan. In addition to non-lignified cell walls, the lignified walls of the few sclerenchyma fibres and xylem tracheary elements present were labelled mostly over the compound middle lamella (middle lamella and primary wall), but small amounts were also detected over the secondary wall in the xylem tracheary elements. This suggested that most of the (1,3;1,4)-β-glucans in the walls of these two cell types are laid down at an early developmental stage. Although (1,3;1,4)-β-glucans are not usually considered components of lignified cell walls, indications of the presence of small amounts of (1,3;1,4)-β-glucans have been obtained from linkage analyses of the wall polysaccharides (Chesson et al., 1985; Smith and Harris, 1999).

As part of a study on the cell walls of maize coleoptiles, the same monoclonal antibody to (1,3;1,4)-β-glucans was used in conjunction with immunogold electron microscopy to determine the locations of (1,3;1,4)-β-glucans in the walls (Carpita et al., 2001). As in barley coleoptiles, the non-lignified cell walls were heavily labelled, with the parenchyma walls more heavily labelled than those of the epidermal cells. However, there were some differences in the distribution of the label that could be species related or because the coleoptiles of the two species were harvested at different stages. In addition to using immunogold labelling, Carpita et al. (2001) isolated epidermal and mesophyll cell walls from the coleoptiles and estimated the (1,3;1,4)-β-glucan concentration in these by linkage analysis to be about 6% and 20%, respectively.

In the immunogold labelling study to determine the distribution of (1,3;1,4)-β-glucans in the cell walls of different families of the Poales (Trethewey et al., 2005), the heavy labelling of the non-lignified cell walls of the leaves of *Lolium multiflorum* (Poaceae) and *Flagellaria indica* (Flagellariaceae) was mostly throughout the walls as in barley seedlings (Fig. 3A). In species from the other Poales families that showed wall labelling, the distribution of labelling on the epidermal walls varied, but on all parenchyma walls was adjacent to the plasma membrane (Fig. 3D). As in barley seedlings (Trethewey and Harris, 2002), the lignified walls of sclerenchyma fibres and xylem tracheary elements of the other species all showed some labelling. The compound lamella was lightly labelled in *L. multiflorum* and *F. indica* in both cell types, but in other species was either unlabelled or only an occasional gold particle was present. The secondary wall was only very lightly labelled in all species except those in the Restionaceae and Xyridaceae, where it was lightly or sometimes moderately labelled.

## 4.6.7. Role of (1,3;1,4)-β-Glucans in the Cell-Wall Architecture of the Poaceae

Transmission electron microscopy of the cell walls of maize roots after preparation using fast-freeze, deep-etch and rotary shadowing revealed what appeared to be cross-links between the cellulose microfibrils (Satiat-Jeunemaitre et al., 1992). The identity of these cross-links remains uncertain, as does the exact role of (1,3;1,4)-β-glucans in the cell wall architecture of the Poaceae and other families with walls containing (1,3;1,4)-β-glucans. Models of the architecture of these walls have been proposed in which there are two co-extensive, but independent, polymer networks (Carpita and Gibeaut, 1993; Carpita, 1996; Carpita et al., 2001). Initially, (1,3;1,4)-β-glucans and glucuronoarabinoxylans were considered to form cross-links (Carpita and Gibeaut, 1993; Carpita, 1996), but as the result of a study of the cell walls of maize coleoptile epidermal walls, Carpita et al. (2001) no longer consider that (1,3;1,4)-β-glucans form cross-links. In this study, the walls were examined by field-emission scanning electron microscopy and analysed chemically after sequential chemical extraction of matrix wall polymers and extraction of (1,3;1,4)-β-glucans using a specific (1,3;1,4)-β-glucan hydrolase. About 70% of the (1,3;1,4)-β-glucan remained associated with the wall after sequential extraction of over half of the non-cellulosic material with sodium chlorite and sodium hydroxide. From this, the authors concluded that much of the (1,3;1,4)-β-glucans are tightly associated with microfibrils, rather than forming a gel-like matrix (see Section 4.6.3). Furthermore, the techniques they used did not show cross-links as demonstrated by Satiat-Jeunemaitre et al. (1992), but their results do not exclude (1,3;1,4)-β-glucans as cross-linking polymers between cellulosic microfibrils.

## 4.6.8. Concentrations of (1,3;1,4)-β-Glucans in Whole Cereal Grains

The concentrations of (1,3;1,4)-β-glucans in whole grains vary among different cereal species (Table 3). Within a particular species, the concentration varies with genotype and is influenced by environmental conditions. In oats (*Avena sativa*) and barley (*Hordeum vulgare*), which, of the commonly grown cereal species, contain the highest concentrations of (1,3;1,4)-β-glucans, the ranking of genotypes is usually consistent in different environments (MacGregor and Fincher, 1993; Cervantes-Martinez et al., 2001). The total concentration of (1,3;1,4)-β-glucans in barley is most frequently within the range 4–7% w/w (Stone and Clarke, 1992; MacGregor and Fincher, 1993), which is similar to the 3–6% range for oats

(Welch and Lloyd, 1989; Miller et al., 1993). Averaged across genotypes, Lee et al. (1997) found that the total (1,3;1,4)-β-glucan concentrations in oats and barley were similar, but that oats had a higher proportion of water-soluble glucan, with an average of 82% for oats compared with 65% for barley.

Plant breeders aim to reduce the concentrations of (1,3;1,4)-β-glucans when developing cultivars of barley and oats for animal feed and cultivars of barley for malting. However, their aim when developing cultivars of barley and oats for human food is to increase the concentrations of (1,3;1,4)-β-glucans. Large-scale preparations rich in (1,3;1,4)-β-glucans have been obtained from barley and oats as additives to bread, crackers and other foods (Morgan, 2000; Harris and Smith, 2006). The total concentration of (1,3;1,4)-β-glucan in grain is a quantitative trait controlled by the additive effects of between three and five genetic factors (Han et al., 1995; Igartua et al., 2002; Kim et al., 2004; Molina-Cano et al., 2007). One of these quantitative trait loci (QTL) on chromosome 2 H was shown to contain *HvCslF* genes that encode enzymes responsible for (1,3;1,4)-β-glucan synthesis (Burton et al., 2006) and other *HvCslF* genes also mapped to positions that coincided with QTLs for grain (1,3;1,4)-β-glucan (Burton et al., 2008).

Currently, the international standard method for the quantification of (1,3;1,4)-β-glucans is the method developed by McCleary and Glennie-Holmes (1985) and streamlined by McCleary and Codd (1991), which uses a specific (1,3;1,4)-β-glucanase. However, the method is relatively slow, and in developing both low and high (1,3;1,4)-β-glucan cultivars, it is important to have high-throughput methods so that hundreds or even thousands of lines can be assayed per day. Such rapid methods include enzyme-linked immunosorbent assays (ELISAs) with monoclonal antibodies specific to (1,3;1,4)-β-glucans (Meikle et al., 1994; Rampitsch et al., 2003, 2006), near-infrared spectroscopy (Czuchajowska et al., 1992; Szczodrak et al., 1992; Blakeney and Flinn, 2005; de Sá and Palmer, 2006), and Calcofluor binding used in conjunction with flow-injection analysis (Jørgensen, 1988).

In barley, the concentrations of (1,3;1,4)-β-glucans vary depending on the type of starch, with genotypes containing starch with higher or lower proportions of amylose than normal having higher concentrations of (1,3;1,4)-β-glucans. For example, in a comparison of barley genotypes, those with high amylose starch (37.3–41.8% amylose), waxy starch (3.8–6.0% amylose), zero amylose waxy starch (no amylose) and normal starch (23.8–27.1% amylose) contained 6.9–8.2%, 5.5–8.1%, 5.8–7.0% and 3.3–6.3% (1,3;1,4)-β-glucan, respectively (Izydorczyk et al., 2000). The proportion of total (1,3;1,4)-β-glucan soluble in water was less for the high-amylose barleys.

Barley genotypes with particularly high and low concentrations of (1,3;1,4)-β-glucans have been developed. For example, the Montana Agricultural Experimental Station developed cv. Pronashonupana, a hull-less waxy barley, with a (1,3;1,4)-β-glucan concentration of 14.9% (Andersson et al., 1999), and five high-lysine barley mutants have been identified with (1,3;1,4)-β-glucan concentrations of 13.3–19.8% (Munck et al., 2004). At the opposite extreme, barley mutants have been described with (1,3;1,4)-β-glucan concentrations as low as 2% (Aastrup, 1983).

So far, there has been less success in developing oat genotypes with very high concentrations of (1,3;1,4)-β-glucans. For example, the oat genotype N979 developed at Iowa State University and containing 8.1% (1,3;1,4)-β-glucan is described as having a high concentration (Sayar et al., 2005). However, considerable interspecific and intraspecific variation in (1,3;1,4)-β-glucan concentrations (2.2–11.3%) was found in a study of nine other *Avena* species, indicating the potential usefulness of such species in oat breeding (Welch et al., 1991). The highest (1,3;1,4)-β-glucan concentrations (5.7–11.3%) were found in the nine *A. atlantica* genotypes studied.

Cereals other than oats and barley have lower (1,3;1,4)-β-glucan concentrations (Table 3). The range for rye (*Secale cereale*) is 1.3–2.7% (Henry, 1987; Hansen et al., 2003; Salmenkallio-Marttila and Hovinen, 2005) and, in contrast to oats and barley, the (1,3;1,4)-β-glucans are mostly water insoluble (Anderson et al., 1978; Brummer et al., 2008). Even lower concentrations of (1,3;1,4)-β-glucans, almost all insoluble in water at 65°C, occur in wheat (*Triticum aestivum*) grains (0.5–1.0%; 29 cvs) (Beresford and Stone, 1983). Durum wheat (*Triticum durum*) had a (1,3;1,4)-β-glucan concentration range of 0.5–0.6%, with similar water solubility (Beresford and Stone, 1983). Few data are available for the concentrations of (1,3;1,4)-β-glucans in rice grains. Anderson et al. (1978) reported a concentration of 0.13% for cv. Baru, with a similar solubility to wheat (1,3;1,4)-β-glucans.

## 4.6.9. Effects of the Environment on (1,3;1,4)-β-Glucan Concentrations in Whole Grains

There have been many field studies comparing the concentrations of grain (1,3;1,4)-β-glucans in the same cultivars of barley and oats grown at the same or different sites in different years and subjected to different fertilizer treatments. Nevertheless, because of the complexity of field environments, the environmental factors that influence (1,3;1,4)-β-glucan concentrations are poorly understood (Stone and Clarke, 1992; MacGregor and Fincher, 1993). Many of these studies on both oats and barley have suggested that water availability is an important factor, with drought conditions causing an increase in (1,3;1,4)-β-glucan concentrations and high rainfall causing a decrease (Peterson, 1991; Welch et al., 1991; MacGregor and Fincher, 1993; Brunner and Freed, 1994; Peterson et al., 1995; Pérez-Vendrell et al., 1996; Zhang et al., 2001). It has also been suggested that similar effects operate on (1,3;1,4)-β-glucan concentrations in rye (Hansen et al., 2003; Salmenkallio-Marttila and Hovenen, 2005). On the other hand, a study on oats in North Dakota indicated that precipitation in July and August was positively correlated with (1,3;1,4)-β-glucan concentrations (Doehlert et al., 2001).

Application of nitrogen fertilizers to the soil in field trials has also been suggested by some to increase (1,3;1,4)-β-glucan concentrations in oats (Welch et al., 1991; Baur and Geisler, 1996), but Humphreys et al. (1994) and Weightman et al. (2004) found no significant effects. In contrast, application of nitrogen to oats in the form of foliar urea resulted in significantly increased (1,3;1,4)-β-glucan concentrations (Weightman et al., 2004).

Few experiments have been done in which the effects of individual environmental factors have been examined using controlled environments. However, one series of such experiments on barley showed that water stress applied at mid grain-fill significantly decreased the (1,3;1,4)-β-glucan content of mature grains, but when applied near the end of grain-fill had no effect (MacNicol et al., 1993). Heat stress, another environmental factor implicated in increasing the concentrations of (1,3;1,4)-β-glucans (Pérez-Vendrell et al., 1996), applied at either of these stages had no effect. There is an urgent need for further carefully conducted experiments in controlled environments in which environmental treatments are applied at specific times during grain fill.

## 4.6.10. Within Plant Variation in (1,3;1,4)-β-Glucan Concentrations in Whole Grains

In addition to effects of the environment, a study of three barley cultivars showed within-plant variation in (1,3;1,4)-β-glucan concentrations of grains (Zhang et al., 2002). Plants with only one spike had grains with significantly higher concentrations than plants with two or three spikes, and within a spike, grains in the middle had significantly higher concentrations than those at the top or bottom of the spike.

## 4.6.11. Distribution of (1,3;1,4)-β-Glucans in the Cell Walls of Mature Grains

Several different approaches have been used to determine the distribution of (1,3;1,4)-β-glucans in the walls of the different tissues and cell types in mature grains (Fig. 4): the

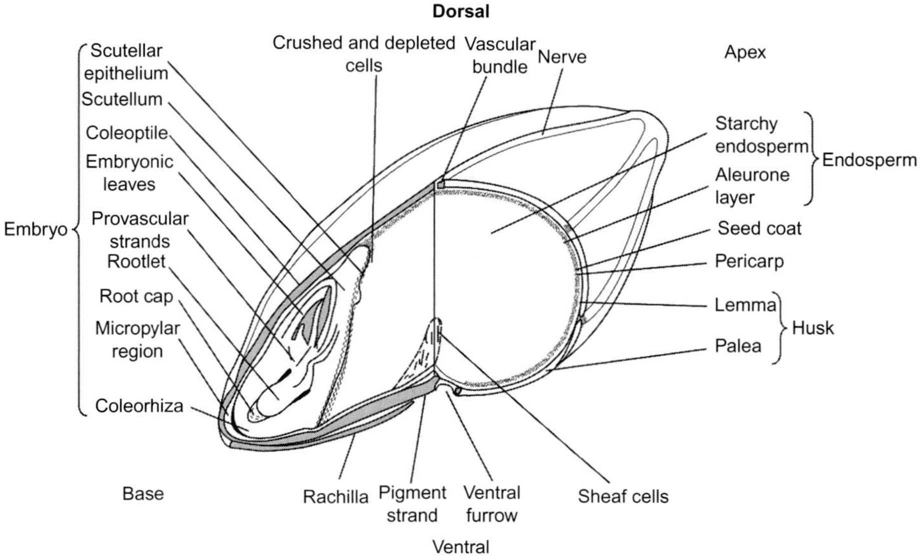


**Fig. 4: Diagram of the barley grain, showing the different organs, tissues and cell types. Reproduced from Briggs (1978) with permission.**

walls of different cell types have been isolated and analysed; Calcofluor has been used as a fluorescent histochemical probe; antibodies have been used in immunocytochemical studies; and fractions have been successively pearled from grains and analysed.

## *Isolation and Analysis of the Walls of Specific Cell Types*

The walls of starchy endosperm cells have been isolated, and the concentrations of (1,3;1,4)-β-glucans determined, from the grains of barley (Fincher, 1975; Ballance and Manners, 1978), oats (Miller and Fulcher, 1995; Miller et al., 1995), wheat (Bacic and Stone, 1980) and rice (Shibuya et al., 1985; Stone, 2006) (Table 1). The concentrations of (1,3;1,4)-β-glucans in the cell walls of oats, 75–78%, was slightly greater than in the cell walls of barley, 70–75%. The starchy endosperm cell walls of oats were isolated from two cultivars, Marion and OA516-2. The grains of cv. Marion and cv. OA516-2 had (1,3;1,4)-β-glucan concentrations of 6.4 and 3.8%, respectively, and yielded walls with 78% and 75% (1,3;1,4)-β-glucans, respectively, although these concentrations may be slightly underestimated (Miller and Fulcher, 1995; Miller et al., 1995). In contrast to oats and barley, isolated starchy endosperm cell walls of both wheat and rice contained only 20% (1,3;1,4)-β-glucan.

The walls of aleurone cells have been isolated from barley and wheat grains and contained 26% and 29% (1,3;1,4)-β-glucans, respectively (Bacic and Stone, 1981) (Table 1). Analyses of a lignified cell wall preparation isolated from wheat beeswing bran, comprising the outer layers of the pericarp (epidermis and hypodermis), showed no evidence of (1,3;1,4)-β-glucans (Ring and Selvendran, 1980; DuPont and Selvendran, 1987).

### *Histochemical Location Using Calcofluor*

Initial studies showed that the fluorescent brightening agent Calcofluor (Calcofluor White M2R New) can be used to locate (1,3;1,4)-β-glucans histochemically in cell walls of the grains of oats, barley and wheat (Wood et al., 1983). Miller and Fulcher (1994) used scanning microspectrofluorometry to determine the relative fluorescence of wall-bound Calcofluor and hence the relative wall concentrations of (1,3;1,4)-β-glucans in five cultivars of oats, with grain (1,3;1,4)-β-glucan concentrations of 3.7–6.4%, and four cultivars of barley, with grain (1,3;1,4)-β-glucan concentrations of 2.8–11%. In all the cultivars of oats, there was a high concentration of bound Calcofluor in the walls of a region of the starchy endosperm adjacent to the embryo, known as the depleted layer. Furthermore, except in the cultivar Marion, which had the highest grain (1,3;1,4)-β-glucan concentration (6.4%), the walls of the starchy endosperm cells just below the aleurone layer (the sub-aleurone layer) were thicker than in the interior of the starchy endosperm and had a higher concentration of bound Calcofluor. In cv. Marion, the sub-aleurone walls were thicker, but did not have higher concentrations of bound Calcofluor. In contrast to oats, the distribution of (1,3;1,4)-β-glucans in the cell walls of the starchy endosperm of barley grains was uniform with no higher concentration of bound Calcofluor in the cell walls of the sub-aleurone layer. Bhatty et al. (1991) also examined sections of three barley grains of three cultivars stained with Calcofluor and found that the thickness of the walls of the starchy endosperm cells was greatest for cultivars that had higher concentrations of (1,3;1,4)-β-glucans in the grains. However, unlike Miller and Fulcher (1994), they found that higher concentrations of bound Calcofluor occurred in the cell walls of the sub-aleurone layer, particularly in the two cultivars examined with the lower glucan concentrations (4.0% and 4.9% compared with 5.4%). These results are consistent with those obtained using pearling (see below).

### *Location Using Immunocytochemistry*

The distribution of (1,3;1,4)-β-glucans in the cell walls of mature wheat grains has been investigated through immunofluorescence microscopy (Guillon et al., 2004) and immunogold electron microscopy (Meikle et al., 1994; Guillon et al., 2004), using a monoclonal antibody specific for (1,3;1,4)-β-glucans (Meikle et al., 1994). The walls of the aleurone and the sub-aleurone starchy endosperm cells were strongly labelled. The walls of the other starchy endosperm cells were also labelled but less strongly, and those in the central region of the grain were labelled only adjacent to the plasma membrane. The colloidal-gold labelling of the aleurone cell walls differed between the two studies. Meikle et al. (1994) showed even labelling, whereas Guillon et al. (2004) showed a multilamellar labelling pattern. In both studies, the middle lamella region between adjacent aleurone cells at the seed coat face of the aleurone layer showed little labelling. The stronger labelling of the aleurone cell walls than the starchy endosperm cell walls is consistent with the higher concentration of (1,3;1,4)-β-glucans found in isolated walls of wheat aleurone cells, compared with starchy endosperm cells (Bacic and Stone, 1980, 1981).

### Pearling and Chemical Analysis of Fractions

Another approach to examining the distribution of (1,3;1,4)-β-glucans in grains is to mechanically abrade (pearl) away successive fractions, analyse the concentrations of (1,3;1,4)-β-glucans in fractions and relate the fractions to grain anatomy. For example, Zheng et al. (2000) pearled the grains of nine cultivars of hull-less barley, producing successive fractions that were each 10% of the original grain weight. In two cultivars with low concentrations of (1,3;1,4)-β-glucans (4.3% and 6.5%), they showed that the (1,3;1,4)-β-glucan concentration was greatest in the sub-aleurone region of the starchy endosperm and declined slightly towards the inner layers. In cultivars with higher concentrations of (1,3;1,4)-β-glucans, the polysaccharides were distributed throughout the starchy endosperm, although the central core tended to contain less than the mid-layers. In all cultivars, the lowest concentration of (1,3;1,4)-β-glucan occurred in the fraction containing the pericarp-testa; the aleurone-containing fraction had a (1,3;1,4)-β-glucan concentration intermediate between that of the pericarp-testa fraction and the first starchy endosperm-containing fraction.

A similar approach, using a PeriTec debranning system followed by a pilot flour mill, was used to produce an aleurone-rich and a pericarp-rich fraction from wheat grain (Harris et al., 2005). The (1,3;1,4)-β-glucans were found mostly in the aleurone-rich fraction.

## 4.6.12. Future Studies

We are now in a strong position to examine further the distribution, fine structure and function of (1,3;1,4)-β-glucans in the cell walls of grasses and other taxa. In particular, functional genomics technologies have recently been applied in the identification of genes that mediate (1,3;1,4)-β-glucan synthesis in cereals, namely the *CslF* cellulose synthase-like genes (Burton et al., 2006) and provide a platform from which we can now identify additional genes that will almost certainly be involved in the process. Thus, transformation of Arabidopsis lines with rice *CslF* genes resulted in the appearance of (1,3;1,4)-β-glucans in walls of the transgenic plants. Because Arabidopsis normally has no *CslF* genes or (1,3;1,4)-β-glucans in its walls, this represented the key evidence for the function of these genes in (1,3;1,4)-β-glucan synthesis in plants (Burton et al., 2006). Do the so-called 'grass-specific' *CslH* and *CslJ* genes (Fincher, 2009a) also mediate in (1,3;1,4)-β-glucan synthesis? Will similar genes be discovered in the other taxa that also have (1,3;1,4)-β-glucans in their walls? The identification of the (1,3;1,4)-β-glucan synthase genes has placed us in a position to investigate genetic and environmental factors that affect concentrations of (1,3;1,4)-β-glucans both in vegetative organs and in grain, and, in particular, to determine how enzymes encoded by these genes combine to synthesize polysaccharides with the structural features of (1,3;1,4)-β-glucans. For example, it should be possible to determine if the different CslF enzymes, and possibly CslH and CslJ enzymes, synthesize (1,3;1,4)-β-glucans with different ratios of (1,3)-β-glucosyl to (1,4)-β-glucosyl residues or with different proportions of cellotriosyl and cellotetraosyl units (Fincher, 2009b). Investigating the existence and expression levels of these genes in the various taxa discussed in this review, particularly with respect to the amounts and fine structures of (1,3;1,4)-β-glucans in those taxa, should cast further light on the regulation and biological significance of (1,3;1,4)-β-glucan synthesis more generally. Although the determination of the three-dimensional structures of membrane proteins remains very difficult, attempts to obtain structural information on these enzymes are underway and, if successful, are likely to yield a good deal of information about the molecular mechanisms of (1,3;1,4)-β-glucan assembly. As mentioned earlier, as our knowledge of the synthetic processes of wall polysaccharides becomes more precise and detailed, so our ability to manipulate wall biosynthesis to improve the performance of plant crops and pasture species for applications in human and animal nutrition, biofuels production and other end-uses of grasses will be enhanced.

## Acknowledgements

Financial support from the Australian Research Council, the Grains Research and Development Corporation and the CSIRO Food Futures Collaboration Fund, through grants to GBF, and from the University of Auckland and the New Zealand Flour Millers Association, through grants to PJH, is gratefully acknowledged.

## References

Aastrup, S. (1983). Selection and characterization of low β-glucan mutants in barley. *Carlsberg Research Communications*, *48*, 307–316.

Anderson, M. A., Cook, J. A., & Stone, B. A. (1978). Enzymatic determination of 1,3:1,4-β-glucans in barley grain and other cereals. *Journal of the Institute of Brewing*, *84*, 233–239.

Andersson, A. A. M., Andersson, R., Autio, K., & Åman, P. (1999). Chemical composition and micro-structure of two naked waxy barleys. *Journal of Cereal Science*, *30*, 183–191.

Andrewartha, K. A., Phillips, D. R., & Stone, B. A. (1979). Solution properties of wheat-flour arab-inoxylans and enzymatically modified arabinoxylans. *Carbohydrate Research*, *77*, 191–204.

APG. (2003). An update of the Angiosperm Phylogeny Group classification for the orders and families of flowering plants: APG II. *Botanical Journal of the Linnean Society*, *141*, 399–436.

Bacic, A., & Stone, B. (1980). A (1→3)- and (1→4)-linked β-D-glucan in the endosperm cell-walls of wheat. *Carbohydrate Research*, *82*, 372–377.

Bacic, A., & Stone, B. A. (1981). Chemistry and organisation of aleurone cell wall components from wheat and barley. *Australian Journal of Plant Physiology*, *8*, 475–495.

Ballance, G. M., & Manners, D. J. (1978). Structural analysis and enzymic solubilization of barley endosperm cell-walls. *Carbohydrate Research*, *61*, 107–118.

Baur, S. K., & Geisler, G. (1996). β-Glucan content in caryopses of oat varieties with regard to cultiva-tion year and nitrogen level. *Journal of Agronomy and Crop Science*, *176*, 5–14.

Bennetzen, J. L. (2007). Patterns in grass genome evolution. *Current Opinion in Plant Biology*, *10*, 176–181.

Beresford, G., & Stone, B. A. (1983). (1→3),(1→4)-β-D-glucan content of *Triticum* grains. *Journal of Cereal Science*, *1*, 111–114.

Bhatty, R. S., MacGregor, A. W., & Rossnagel, B. G. (1991). Total and acid-soluble β-glucan content of hulless barley and its relationship to acid-extract viscosity. *Cereal Chemistry*, *68*, 221–227.

Blakeney, A. B., & Flinn, P. C. (2005). Determination of non-starch polysaccharides in cereal grains with near-infrared reflectance spectroscopy. *Molecular Nutrition and Food Research*, *49*, 546–550.

Bremer, K. (2002). Gondwanan evolution of the grass alliance of families (Poales). *Evolution*, *56*, 1374–1387.

Brennan, C. S., & Cleary, L. J. (2005). The potential use of (1→3,1→4)-β-D-glucans as functional food ingredients. *Journal of Cereal Science*, *42*, 1–13.

Briggs, D. E. (1978). *Barley*. London: Chapman and Hall.

Brummer, Y., Jones, S., Tosh, S. M., & Wood, P. J. (2008). Extraction and physicochemical characterization of rye β-glucan and effects of barium on polysaccharide molecular weight. *Cereal Chemistry*, *85*, 174–181.

Brunner, B. R., & Freed, R. D. (1994). Oat grain β-glucan content as affected by nitrogen level, location and year. *Crop Science*, *34*, 473–476.

Buliga, G. S., Brandt, D. A., & Fincher, G. B. (1986). The sequence statistics and solution conformation of a barley (1→3,1→4)-β-D-glucan. *Carbohydrate Research*, *157*, 139–156.

Burton, R. A., Wilson, S. M., Hrmova, M., Harvey, A. J., Shirley, N. J., Medhurst, A., Stone, B. A., Newbigin, E. J., Bacic, A., & Fincher, G. B. (2006). Cellulose synthase-like *CslF* genes mediate the synthesis of cell wall (1,3;1,4)-β-D-glucans. *Science*, *311*, 1940–1942.

Burton, R. A., Jobling, S. A., Harvey, A. J., Shirley, N. J., Mather, D. E., Bacic, A., & Fincher, G. B. (2008). The genetics and transcriptional profiles of the cellulose synthase-like *HvCslF* gene family in barley (*Hordeum vulgare* L.). *Plant Physiology*, *146*, 1821–1833.

Carbonero, E. R., Montai, A. V., Mellinger, C. G., Eliasaro, S., Sassaki, G. L., Gorin, P. A. J., & Iacomini, M. (2005). Glucans of lichenized fungi: Significance for taxonomy of the genera *Parmotrema and Rimelia*. *Phytochemistry*, *66*, 929–934.

Carpita, N. C. (1996). Structure and biogenesis of the cell walls of grasses. *Annual Review of Plant Physiology and Plant Molecular Biology*, *47*, 445–476.

Carpita, N. C., Defernez, M., Findlay, K., Wells, B., Shoue, D. A., Catchpole, G., Wilson, R. H., & McCann, M. C. (2001). Cell wall architecture of the elongating maize coleoptile. *Plant Physiology*, *127*, 551–565.

Carpita, N. C., & Gibeaut, D. M. (1993). Structural models of primary cell walls in flowering plants: Consistency of molecular structure with the physical properties of the walls during growth. *The Plant Journal*, *3*, 1–30.

Cavalier-Smith, T. (2004). Only six kingdoms of life. *Proceedings of the Royal Society London B*, *271*, 1251–1262.

Cervantes-Martinez, C. T., Frey, K. J., White, P. J., Wesenberg, D. M., & Holland, J. B. (2001). Selection for greater β-glucan content in oat grain. *Crop Science*, *41*, 1085–1091.

Chesson, A., Gordon, A. H., & Lomax, J. A. (1985). Methylation analysis of mesophyll, epidermis, and fibre cell-walls isolated from the leaves of perennial and Italian ryegrass. *Carbohydrate Research*, *141*, 137–147.

Czuchajowska, Z., Szczodrak, J., & Pomeranz, Y. (1992). Characterization and estimation of barley polysaccharides by near-infrared spectroscopy. I. Barleys, starches, and β-D-glucans. *Cereal Chemistry*, *69*, 413–418.

Dahlgren, R. M. T., Clifford, H. T., & Yeo, P. F. (1985). *The families of the monocotyledons*. New York: Springer-Verlag.

de Sá, R. M., & Palmer, G. H. (2006). Analysis of β-glucan in single grains of barley and malt using NIR-spectroscopy. *Journal of the Institute of Brewing*, *112*, 9–16.

Doehlert, D. C., McMullen, M. S., & Hammond, J. J. (2001). Genotypic and environmental effects on grain yield and quality of oat grown in North Dakota. *Crop Science*, *41*, 1066–1072.

DuPont, S. M., & Selvendran, R. R. (1987). Hemicellulosic polymers from the cell walls of beeswing wheat bran: Part 1, polymers solubilized by alkali at 2°. *Carbohydrate Research*, *163*, 99–113.

Eder, M., Tenhaken, R., Driouich, A., & Lütz-Meindl, U. (2008). Occurrence and characterization of arabinogalactan-like proteins and hemicelluloses in *Micrasterias* (Streptophyta). *Journal of Phycology*, *44*, 1221–1234.

Fincher, G. B. (1975). Morphology and chemical composition of barley endosperm walls. *Journal of the Institute of Brewing*, *81*, 347–349.

Fincher, G. B. (2009a). Revolutionary times in our understanding of cell wall biosynthesis and re-modelling in the grasses. *Plant Physiology*, *149*, 27–37.

Fincher, G. B. (2009b). Exploring the evolution of (1,3;1,4)-β-D-glucans in plant cell walls: Comparative genomics can help! *Current Opinion in Plant Biology, 12*, 140–147.

Fincher, G. B., & Stone, B. A. (2004). Chemistry of nonstarch polysaccharides. In C. Wrigley, H. Corke, & C. E. Walker (Eds.), *Encyclopedia of grain science* (pp. 206–223). Oxford: Elsevier.

Fontaine, T., Simene, C., Dubreucq, G., Adam, O., Delepierre, M., Lemoinei, J., Vorgias, C. E., Diaquin, M., & Latgé, J.-P. (2000). Molecular organization of the alkali-insoluble fraction of *Aspergillus fumigatus* cell wall. *Journal of Biological Chemistry*, *275*, 27594–27607.

Ford, C. W., & Percival, E. (1969). Polysaccharides synthesised by *Monodus subterraneus*. Part II. The cell-wall glucan. *Journal of the Chemical Society*, 3014–3016.

Fry, S. C., Nesselrode, B. H. W. A., Miller, J. G., & Mewburn, B. R. (2008). Mixed-linkage (1→3,1→4)-β-D-glucan is a major hemicellulose of *Equisetum* (horsetail) cell walls. *New Phytologist*, *179*, 104–115.

Gaut, B. S. (2002). Evolutionary dynamics of grass genomes. *New Phytologist*, *154*, 15–28.

Gibeaut, D. M., Pauly, M., Bacic, A., & Fincher, G. B. (2005). Changes in cell wall polysaccharides in developing barley (*Hordeum vulgare*) coleoptiles. *Planta*, *221*, 729–738.

Guillon, F., Tranquet, O., Quillien, L., Utille, J.-P., Ortiz, J. J. O., & Saulnier, L. (2004). Generation of polyclonal and monoclonal antibodies against arabinoxylans and their use for immunocytochemical location of arabinoxylans in cell walls of endosperm of wheat. *Journal of Cereal Science*, *40*, 167–182.

Han, F., Ullrich, S. E., Chirat, S., Menteur, S., Jestin, L., Sarrafi, A., Hayes, P. M., Jones, B. L., Blake, T. K., Wesenberg, D. M., Kleinhofs, A., & Kilianm, A. (1995). Mapping of β-glucan content and β-glucanase activity loci in barley grain and malt. *Theoretical and Applied Genetics*, *91*, 921–927.

Hansen, H. B., Rasmussen, C. V., Knudsen, K. E. B., & Hansen, A. (2003). Effects of genotype and harvest year on content and composition of dietary fibre in rye (*Secale cereale* L.) grain. *Journal of the Science of Food and Agriculture*, *83*, 76–85.

Harris, P. J. (2005). Diversity in plant cell walls. In R. J. Henry (Ed.), *Plant diversity and evolution: Genotypic and phenotypic variation in higher plants* (pp. 201–227). Wallingford, Oxon, UK: CAB International Publishing.

Harris, P. J., & Smith, B. G. (2006). Plant cell walls and cell-wall polysaccharides: Structures, properties and uses in food products. *International Journal of Food Science and Technology*, *41*, 129–143.

Harris, P. J., & Stone, B. A. (2008). Chemistry and molecular organization of plant cell walls. In M. E. Himmel (Ed.), *Biomass recalcitrance: Deconstructing the plant cell wall for bioenergy* (pp. 61–93). Oxford: Blackwell Publishing.

Harris, P. J., Chavan, R. R., & Ferguson, L. R. (2005). Production and characterisation of two wheat-bran fractions: An aleurone-rich and a pericarp-rich fraction. *Molecular Nutrition and Food Research*, *49*, 536–545.

Henry, R. J. (1987). Pentosan and (1→3),(1→4)-β-glucan concentrations in endosperm and wholegrain of wheat, barley, oats and rye. *Journal of Cereal Science*, *6*, 253–258.

Honegger, R., & Haisch, A. (2001). Immunocytochemical location of the (1→3)(1→4)-β-glucan lichenin in the lichen-forming ascomycete *Cetraria islandica* (Icelandic moss). *New Phytologist*, *150*, 739–746.

Hoson, T., & Nevins, D. J. (1989). Antibodies as probes for the study of location and metabolism of (1→3),(1→4)-β-D-glucans. *Physiologia Plantarum*, *75*, 452–457.

Humphreys, D. G., Smith, D. L., & Mather, D. E. (1994). Nitrogen fertilizer and seeding date induced changes in protein, oil and β-glucan contents of four oat cultivars. *Journal of Cereal Science*, *20*, 283–290.

Igartua, E., Hayes, P. M., Thomas, W. T. B, Meyer, R., & Mather, D. E. (2002). Genetic control of quantitative grain and malt quality traits in barley. *Journal of Crop Production*, *5*, 131–164.

Inouhe, M., & Nevins, D. J. (1998). Changes in the activities and polypeptide levels of exo- and endo-glucanases in cell walls during developmental growth of *Zea mays* coleoptiles. *Plant and Cell Physiology*, *39*, 762–768.

Izydorczyk, M. S., Storsley, J., Labossiere, D., MacGregor, A. W., & Rossnagel, B. G. (2000). Variation in total and soluble β-glucan content in hulless barley: Effects of thermal, physical, and enzymic treatments. *Journal of Agricultural and Food Chemistry*, *48*, 982–989.

Jørgensen, K. G. (1988). Quantification of high molecular weight (1→3),(1→4)-β-D-glucan using Calcofluor complex formation and flow injection analysis: I. Analytical principle and its standardization. *Carlsberg Research Communications*, *53*, 277–285.

Kim, J.-B., Olek, A. T., & Carpita, N. C. (2000). Cell wall and membrane-associated exo-β-D-glucanases from developing maize seedlings. *Plant Physiology*, *123*, 471–485.

Kim, H. S., Park, K. G., Baek, S. B., Son, Y. G., Lee, C. W., Kim, J. C., Kim, J. G., & Nam, J. H. (2004). Inheritance of (1→3)(1→4)-β-D-glucan content in barley (*Hordeum vulgare* L.). *Proceeding of the 9th international barley genetics symposium* (pp. 543–548). Brno, Czech Republic.

Lazaridou, A., & Biliaderis, C. G. (2007). Molecular aspects of cereal β-glucan functionality: Physical properties, technological applications and physiological effects. *Journal of Cereal Science*, *46*, 101–118.

Lazaridou, A., Biliaderis, C. G., Micha-Screttas, M., & Steele, B. R. (2004). A comparative study on structure-function relations of mixed-linkage (1→3),(1→4) linear β-D-glucans. *Food Hydrocolloids*, *18*, 837–855.

Lechat, H., Amat, M., Mazoyer, J., Buléon, A., & Lahaye, M. (2000). Structure and distribution of glucomannan and sulfated glucan in the cell walls of the red alga *Kappaphycus alvarezii* (Gigartinales, Rhodophyta). *Journal of Phycology*, *36*, 891–902.

Lee, C. J., Horsley, R. D., Manthey, F. A., & Schwarz, P. B. (1997). Comparisons of β-glucan content of barley and oat. *Cereal Chemistry*, *74*, 571–575.

Lee, J., & Hollingsworth, R. I. (1997). Oligosaccharide β-glucans with unusual linkages from *Sarcina venriculi*. *Carbohydrate Research*, *304*, 133–141.

Levy, A. A., & Feldman, M. (2002). The impact of polyploidy on grass genome evolution. *Plant Physiology*, *130*, 1587–1593.

Li, W., Cui, S. W., & Kakuda, Y. (2006). Extraction, fractionation, structural and physical characterization of wheat β-D-glucans. *Carbohydrate Polymers*, *63*, 408–416.

Linder, H. P., & Rudall, P. J. (2005). Evolutionary history of Poales. *Annual Review of Ecology Evolution and Systematics*, *36*, 107–124.

Mabberley, D. J. (2008). *Mabberley's plant book: A portable dictionary of plants, their classification and uses*. Cambridge: Cambridge University Press.

MacGregor, A. W., & Fincher, G. B. (1993). Carbohydrates of the barley grain. In A. W. MacGregor & R. S. Bhatty (Eds.), *Barley: Chemistry and technology* (pp. 73–130). St Paul, MN, USA: American Association of Cereal Chemists Inc.

MacNicol, P. K., Jacobsen, J. V., Keys, M. M., & Stuart, I. M. (1993). Effects of heat and water stress on malt quality and grain parameters of Schooner barley grown in cabinets. *Journal of Cereal Science*, *18*, 61–68.

McCann, M. C., Defernez, M., Urbanowicz, B. R., Tewari, J. C., Langewisch, T., Olek, A., Wells, B., Wilson, R. H., & Carpita, N. C. (2007). Neural network analyses of infrared spectra for classifying cell wall architectures. *Plant Physiology*, *143*, 1314–1326.

McCleary, B. V., & Codd, R. (1991). Measurement of (1→3, 1→4)-β-D-glucan in barley and oats: A streamlined enzymic procedure. *Journal of the Science of Food and Agriculture*, *55*, 303–312.

McCleary, B. V., & Glennie-Holmes, M. (1985). Enzymic quantification of (1→3, 1→4)-β-D-glucan in barley and malt. *Journal of the Institute of Brewing*, *91*, 285–295.

McCleary, B. V., & Matheson, N. K. (1986). Enzymic analysis of polysaccharide structure. *Advances in Carbohydrate Chemistry and Biochemistry*, *44*, 147–276.

Meikle, P. J., Hoogenraad, N. J., Bonig, I., Clarke, A. E., & Stone, B. A. (1994). A (1→3, 1→4)-β-glucan-specific monoclonal antibody and its use in the quantification and immunocytochemical location of (1→3, 1→4)-β-glucans. *The Plant Journal*, *5*, 1–9.

Miller, S. S., & Fulcher, R. G. (1994). Distribution of (1→3, 1→4)-β-D-glucan in kernels of oats and barley using microspectrofluorometry. *Cereal Chemistry*, *71*, 64–68.

Miller, S. S., & Fulcher, R. G. (1995). Oat endosperm cell walls: II. Hot-water solubilization and enzymatic digestion of the wall. *Cereal Chemistry*, *72*, 428–432.

Miller, S. S., Fulcher, R. G., Sen, A., & Arnason, J. T. (1995). Oat endosperm cell walls: 1. Isolation, composition, and comparison with other tissues. *Cereal Chemistry*, *72*, 421–427.

Miller, S. S., Wood, P. J., Pietrzak, L. N., & Fulcher, R. G. (1993). Mixed linkage β-glucan, protein content, and kernel weight in *Avena* species. *Cereal Chemistry*, *70*, 231–233.

Mohnen, D. (2008). Pectin structure and biosynthesis. *Current Opinion in Plant Biology*, *11*, 266–277.

Molina-Cano, J.-L., Moralejo, M., Elía, M., Muñoz, P., Russell, J. R., Pérez-Vendrell, A. M., Ciudad, F., & Swanston, J. S. (2007). QTL analysis of a cross between European and North American malting barleys reveals a putative candidate gene for β-glucan content on chromosome 1H. *Molecular Breeding*, *19*, 275–284.

Morgan, K. (2000). Cereal β-glucans. In G. O. Phillips & P. A. Williams (Eds.), *Handbook of hydrocolloids* (pp. 287–307). Cambridge: Woodhead Publishing.

Morrall, P., & Briggs, D. E. (1978). Changes in cell-wall polysaccharides of germinating grains. *Phytochemistry*, *17*, 1495–1502.

Munck, L., Møller, B., Jacobsen, S., & Søndergaard, I. (2004). Near infrared spectra indicate specific mutant endosperm genes and reveal a new mechanism for substituting starch with (1→3, 1→4)-β-glucan in barley. *Journal of Cereal Science*, *40*, 213–222.

Nevo, Z., & Sharon, N. (1969). The cell wall of *Peridinium westii*, a non-cellulosic glucan. *Biochimica et Biophysica Acta*, *173*, 161–175.

Ogbonna, A. C., & Egunwu, A. L. (1994). β-Glucan degradation in malting sorghum. *World Journal of Microbiology and Biotechnology*, *10*, 595–596.

Olafsdottir, E. S., & Ingolfsdottir, K. (2001). Polysaccharides from lichens: Structural characteristics and biological activity. *Planta Medica*, *67*, 199–208.

Paterson, A. H., Bowers, J. E., & Chapman, B. A. (2004). Ancient polyploidization predating divergence of the cereals, and its consequences for comparative genomics. *Proceedings of the National Academy of Sciences USA*, *101*, 9903–9908.

Pérez-Vendrell, A. M., Brufau, J., Molina-Cano, J. L., Francesch, M., & Guasch, J. (1996). Effects of cultivar and environment on β-(1,3)-(1,4)-D-glucan content and acid extract viscosity of Spanish barleys. *Journal of Cereal Science*, *23*, 285–292.

Peterson, D. M. (1991). Genotype and environmental effects on oat β-glucan concentration. *Crop Science*, *31*, 1517–1520.

Peterson, D. M., Wesenberg, D. M., & Burrup, D. E. (1995). β-Glucan content and its relationship to agronomic characteristics in elite oat germplasm. *Crop Science*, *35*, 965–970.

Popper, Z. A., & Fry, S. C. (2003). Primary cell wall composition of bryophytes and charophytes. *Annals of Botany*, *91*, 1–12.

Popper, Z. A., & Fry, S. C. (2004). Primary cell wall composition of pteridophytes and spermatophytes. *New Phytologist*, *164*, 165–174.

Rampitsch, C., Ames, N., Storsley, J., & Marien, L. (2003). Development of a monoclonal antibody-based enzyme-linked immunosorbent assay to quantify soluble β-glucans in oats and barley. *Journal of Agricultural and Food Chemistry*, *51*, 5882–5887.

Rampitsch, C., Bizimungu, B., Ames, N., & Rothwell, L. (2006). Early generation β-glucan selection in oat using a monoclonal antibody-based enzyme-linked immunosorbent assay. *Cereal Chemistry*, *83*, 510–512.

Ring, S. G., & Selvendran, R. R. (1980). Isolation and analysis of cell wall material from beeswing wheat bran (*Triticum aestivum*). *Phytochemistry*, *19*, 1723–1730.

Roulin, S., & Feller, U. (2001). Reversible accumulation of (1→3),(1→4)-β-glucan endohydrolase in wheat leaves under sugar depletion. *Journal of Experimental Botany*, *52*, 2323–2332.

Roulin, S., Buchala, A. J., & Fincher, G. B. (2002). Induction of (1→3),(1→4)-β-D-glucan hydrolases in leaves of dark-incubated barley seedlings. *Planta*, *215*, 51–59.

Salmenkallio-Marttila, M., & Hovinen, S. (2005). Enzyme activities, dietary fibre components and rheological properties of wholemeal flours from rye cultivars grown in Finland. *Journal of the Science of Food and Agriculture*, *85*, 1350–1356.

Salse, J., Bolot, S., Throude, M., Jouffe, V., Piegu, B., Quraishi, U. M., Calcagno, T., Cooke, R., Delseny, M., & Feuillet, C. (2008). Identification and characterization of shared duplications between rice and wheat provide new insight into grass genome evolution. *Plant Cell*, *20*, 11–24.

Satiat-Jeunemaitre, B., Martin, B., & Hawes, C. (1992). Plant cell wall architecture is revealed by rapid-freezing and deep-etching. *Protoplasma*, *167*, 33–42.

Sauter, M., & Kende, H. (1992). Levels of β-glucan and lignin in elongating internodes of deepwater rice. *Plant and Cell Physiology*, *33*, 1089–1097.

Sayar, S., Jannink, J.-L., & White, P. J. (2005). *In vitro* bile acid binding of flours from oat lines varying in percentage and molecular weight distribution of β-glucan. *Journal of Agricultural and Food Chemistry*, *53*, 8797–8803.

Shibuya, N., Nakane, R., Yasui, A., Tanaka, K., & Iwasaki, T. (1985). Comparative studies on cell wall preparations from rice bran, germ, and endosperm. *Cereal Chemistry*, *62*, 252–258.

Sinha, N., & Lynch, M. (1998). Fused organs in the *adherent1* mutation in maize show altered epidermal walls with no perturbations in tissue identities. *Planta*, *206*, 184–195.

Smith, B. G., & Harris, P. J. (1999). The polysaccharide composition of Poales cell walls: Poaceae cell walls are not unique. *Biochemical Systematics and Ecology*, *27*, 33–53.

Sørensen, I., Pettolino, F. A., Wilson, S. M., Doblin, M. S., Johansen, B., Bacic, A., & Willats, W. G. T. (2008). Mixed linkage (1→3),(1→4)-β-D-glucan is not unique to the Poales and is an abundant component of cell walls *Equisetum arvense*. *Plant Journal*, *54*, 510–521.

Staudte, R. G., Woodward, J. R., Fincher, G. B., & Stone, B. A. (1983). Water-soluble (1→3),(1→4)-β-D-glucans from barley (*Hordeum vulgare*) endosperm. III. Distribution of cellotriosyl and cellotetraosyl residues. *Carbohydrate Polymers*, *3*, 299–312.

Stinard, P. S., & Nevins, D. J. (1980). Distribution of noncellulosic β-D-glucans in grasses and other monocots. *Phytochemistry*, *19*, 1467–1468.

Stone, B. A. (2006). Cell walls of cereal grains. *Cereal Foods World*, *51*, 62–65.

Stone, B. A., & Clarke, A. E. (1992). *Chemistry and biology of (1→3)-β-glucans*. Victoria, Australia: La Trobe University Press, La Trobe University.

Sweeney, M., & McCouch, S. (2007). The complex history of the domestication of rice. *Annals of Botany*, *100*, 951–957.

Szczodrak, J., Czuchajowska, Z., & Pomeranz, Y. (1992). Characterization and estimation of barley polysaccharides by near-infrared spectroscopy. II. Estimation of total β-D-glucans. *Cereal Chemistry*, *69*, 419–423.

Tosh, S. M., Brummer, Y., Wood, P. J., Wang, Q., & Weisz, J. (2004). Evaluation of structure in the formation of gels by structurally diverse (1→3)(1→4)-β-D-glucans from four cereal and one lichen species. *Carbohydrate Polymers*, *57*, 249–259.

Trethewey, J. A. K., Campbell, L. M., & Harris, P. J. (2005). (1→3),(1→4)-β-D-Glucans in the cell walls of the Poales (*sensu lato*): An immunogold labelling study using a monoclonal antibody. *American Journal of Botany*, *92*, 1669–1683.

Trethewey, J. A. K., & Harris, P. J. (2002). Location of (1→3)- and (1→3), (1→4)-β-D-glucans in vegetative cell walls of barley (*Hordeum vulgare*) using immunolabelling. *New Phytologist*, *154*, 347–358.

Weightman, R. M., Heywood, C., Wade, A., & South, J. B. (2004). Relationship between grain (1→3,1→4)-β-D-glucan concentration and the response of winter-sown oats to contrasting forms of applied nitrogen. *Journal of Cereal Science*, *40*, 81–86.

Welch, R. W., Leggett, J. M., & Lloyd, J. D. (1991). Variation in the kernel (1→3)(1→4)-β-D-glucan content of oat cultivars and wild *Avena* species and its relationship to other characteristics. *Journal of Cereal Science*, *13*, 173–178.

Welch, R. W., & Lloyd, J. D. (1989). Kernel (1→3)(1→4)-β-D-glucan content of oat genotypes. *Journal of Cereal Science*, *9*, 35–40.

Wilkie, K. C. B. (1979). The hemicelluloses of grasses and cereals. *Advances in Carbohydrate Chemistry and Biochemistry*, *36*, 215–264.

Wood, P. J. (2007). Cereal β-glucans in diet and health. *Journal of Cereal Science*, *46*, 230–238.

Wood, P. J., Fulcher, R. G., & Stone, B. A. (1983). Studies on the specificity of interaction of cereal cell wall components with Congo Red and Calcofluor. Specific detection and histochemistry of (1→3),(1→4)-β-D-glucans. *Journal of Cereal Science*, *1*, 95–100.

Wood, P. J., Weisz, J., & Blackwell, B. A. (1991). Molecular characterization of cereal β-D-glucans. Structural analysis of oat β-D-glucan and rapid structural evaluation of β-D-glucans from different sources by high-performance liquid chromatography of oligosaccharides released by lichenase. *Cereal Chemistry*, *68*, 31–39.

Wood, P. J., Weisz, J., & Blackwell, B. A. (1994). Structural studies of (1→3),(1→4)-β-D-glucans by $^{13}C$-nuclear magnetic resonance spectroscopy and by rapid analysis of cellulose-like regions using high-performance anion-exchange chromatography of oligosaccharides released by lichenase. *Cereal Chemistry*, *71*, 301–307.

Wood, P. J., Weisz, J., Beer, M. U., Newman, C. W., & Newman, R. K. (2003). Structure of (1→3)(1→4)-β-D-glucan in waxy and nonwaxy barley. *Cereal Chemistry*, *80*, 329–332.

Woodward, J. R., Phillips, D. R., & Fincher, G. B. (1983a). Water-soluble (1→3),(1→4)-β-D-glucans from barley (*Hordeum vulgare*) endosperm. I. Physicochemical properties. *Carbohydrate Polymers*, *3*, 143–156.

Woodward, J. R., Fincher, G. B., & Stone, B. A. (1983b). Water-soluble (1→3), (1→4)-β-D-glucans from barley (*Hordeum vulgare*) endosperm. II. Fine structure. *Carbohydrate Polymers*, *3*, 207–225.

York, W. S., & O'Neill, M. A. (2008). Biochemical control of xylan biosynthesis – Which end is up? *Current Opinion in Plant Biology*, *11*, 258–265.

Zhang, G., Chen, J., & Wang, J. (2002). Variation in barley endosperm β-glucan content in three barley cultivars as a function of spike number and within-spike position. *Journal of Cereal Science*, *35*, 99–101.

Zhang, G., Chen, J., Wang, J., & Ding, S. (2001). Cultivar and environmental effects on (1→3,1→4)-β-D-glucan and protein content in malting barley. *Journal of Cereal Science*, *34*, 295–301.

Zheng, G. H., Rossnagel, B. G., Tyler, R. T., & Bhatty, R. S. (2000). Distribution of β-glucan in the grain of hull-less barley. *Cereal Chemistry*, *77*, 140–144.

CHAPTER 4.7

# Evolutionary Aspects of (1,3)-β-Glucans and Related Polysaccharides

**Philip J. Harris[1] and Bruce A. Stone[2,*]**

[1]*School of Biological Sciences, The University of Auckland, Auckland, New Zealand*

[2]*Department of Biochemistry, La Trobe University, Bundoora, Victoria, Australia*

(1,3)-β-Glucans and related polysaccharides are widely, but selectively, distributed among the different lineages in the tree of life. They are found throughout plants (red algae, green algae and embryophytes) and, in the chromalveolates, they occur in some alveolates, including dinoflagellates, and in many chromistans (chrysophytes, diatoms, brown algae, oomycetes and some haptophytes). In cabozoans, they occur in the chloroplast-containing euglenozoans and chorarachneans. They also occur in many fungi. In the different lineages, their structure has evolved to accommodate their various functions. Genes encoding (1,3)-β-glucan synthases fall into two distinct glycosyltransferase (GT) families: the GT2 family, which is found only in Gram-negative eubacteria and belongs to an ancient gene family that occurs throughout the eubacteria, and the GT48 family, which is found in angiosperms and in fungi, including yeasts. The synthesis of (1,3;1,4)-β-glucans in the Poaceae involves GT2 proteins, and cognate genes, belonging to the cellulose synthase-like (*CSL*) family. Two of these proteins, CSLF and CSLH, occur in the Poaceae, but not in eudicotyledons.

## 4.7.1. Introduction

The (1,3)-β-glucans and related polysaccharides are widely, but selectively, distributed among the different lineages in the tree of life (Fig. 1). They are found in both Gram-negative and Gram-positive eubacteria but not in the archaebacteria. In the fungi, they occur in all ascomycetes, basidiomycetes and some zygomycetes; however, except for a single,

*Passed away 28 June 2008.

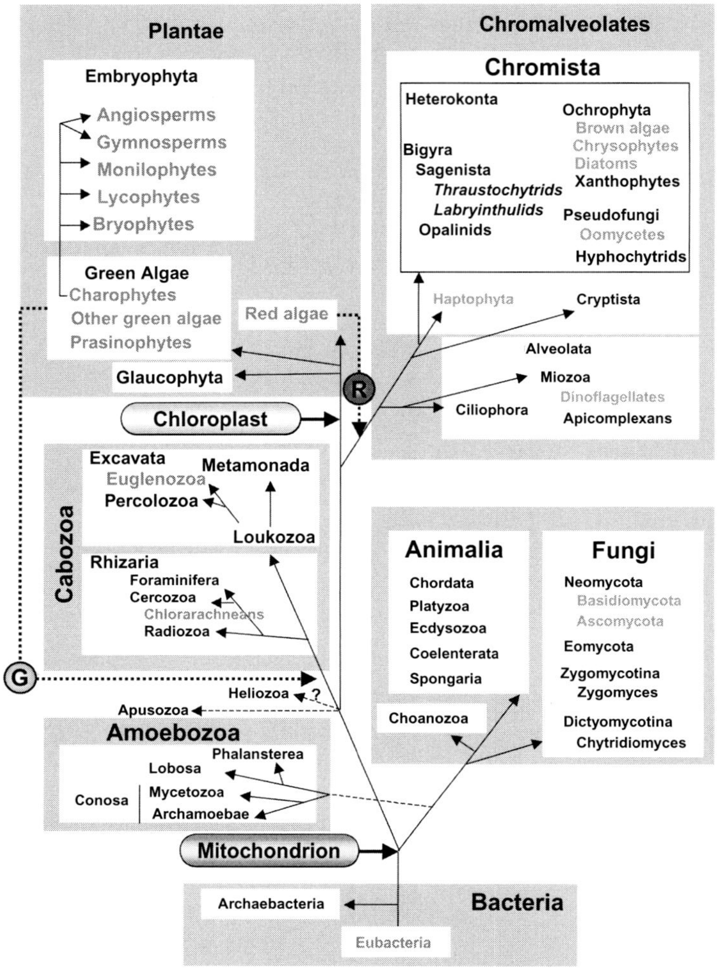


**Fig. 1: Occurrence of (1,3)-β-glucans (grey lettering) in the tree of life based on Cavalier-Smith's (2004) concept of six kingdoms of life. The tree of life shows the appearance of mitochondria and chloroplasts and the secondary 'enslavement' of a green alga by an amoeboid cell to give the cabozoan kingdom, and a similar 'enslavement' of a red alga by an amoeboid cell to give the chromistan kingdom. (After Cavalier-Smith, 2004.)**

unconfirmed report of a phosphate ester of a (1,3)-β-glucan (onuphic acid) (Pautard and Zola, 1967), related polysaccharides have not been found in the animal kingdom or among the amoebozoans. Within the plant kingdom (Plantae), (1,3)-β-glucans are found throughout the red algae, green algae and embryophytes. Within the chromalveolates, they occur in some alveolates (dinoflagellates) and in many chromistans, including the chrysophytes, diatoms, brown algae, oomycetes and some haptophytes. They have also been reported in the cabozoans, where they occur in the chloroplast-containing euglenozoans and chlorarachneans (chlorarachniophytes) (McFadden, 2001) (see Chapter 2.1 for details).

The forms and functions of the (1,3)-β-glucans differ amongst the various lineages. Thus, in the eubacteria, an unbranched (1,3)-β-glucan and the side-chain-branched (1,3;1,2)-β-glucan form protective capsules around cells, and the cyclic (1,3;1,6)-β-glucans are involved in osmoregulation (see Chapter 4.1). In basidiomycetes and ascomycetes, two forms are found: a branch-on-branch (1,3;1,6)-β-glucan that is covalently linked to other polysaccharides and to protein and forms part of the cell wall, and a side-chain-branched (1,3;1,6)-β-glucan that occurs on the surfaces of hyphae (see Chapter 4.3). An unbranched (1,3)-β-glucan (pachyman) is found as a carbohydrate reserve in the sclerotia of certain basidomycete fungi. Unbranched (1,3)-β-glucans are also found as cell wall components in members of the entomophthorean group of zygomycete fungi (see Chapter 4.3). In euglenozoans, the unbranched (1,3)-β-glucan paramylon serves as a carbohydrate reserve. In plants (Plantae), the unbranched (1,3)-β-glucan callose is ubiquitous and is deposited as a result of both abiotic (see Chapter 4.4.4) and biotic (see Chapter 4.4.5) stress; it is also a transient component of the cell plate in cytokinesis and is a component of the specialized cell wall of pollen tubes and other structures involved in microsporogenesis, macrosporogenesis, plamodesmatal transport and in the sieve plates of phloem sieve tube elements (see Chapters 4.4.1, 4.4.2 and 4.4.3). Amongst the chromistans, two forms of (1,3)-β-glucans are encountered. The form in oomycete cell walls is probably a branch-on-branch (1,3;1,6)-β-glucan, whereas side-chain-branched (1,3;1,6)-β-glucans in the brown algae (laminarin), oomycetes (mycolaminarin), chrysophytes and diatoms (chrysolaminarin) have a carbohydrate reserve function (see Chapter 4.2).

Thus, the basic linear water-insoluble (1,3)-β-glucan structure has, in many situations, evolved to give more complex side-chain-branched, branch-on-branch and cyclic-(1,3;1,6)-β-glucans. The architecture of these molecules is aligned to structural, storage and other functions.

## 4.7.2. Genes Encoding (1,3)-β-Glucan Synthases

The genes known so far that encode (1,3)-β-glucan synthases fall into two distinct families. The first is the GT2 family found only in Gram-negative eubacteria where synthases for the linear (1,3)-β-glucan curdlan, the cyclic (1,3)-β-glucan and the side-chain-branched (1,3;1,2)-β-glucan have been described (see Chapter 3.3.1). These are members of a family of enzymes that include cellulose and hyaluronan synthases and belong to an ancient gene family found throughout the eubacteria. The eubacterial (1,3)-β-glucan synthase genes were presumably derived from an ancestral GT2 protein-encoding gene, for example for a cellulose synthase with which it has strong homology (see Chapter 3.3.1). Nevertheless, outside the eubacteria these genes have not been shown to participate in the synthesis of (1,3)-β-glucans.

The second is the GT48 family found in angiosperms (Plantae) and in fungi, including yeasts. Although the GT48 proteins are required for (1,3)-β-glucan synthesis in angiosperms and fungi, their functional significance as the catalytic sub-unit in the process is not well established. However, the GT48 proteins (FKS proteins) from plants and from fungi, including yeasts, show strong homology, 30% identity and 50% similarity (see Chapter 3.3.4). In yeasts, the FKS gene product acts in association with branching enzymes to give the branch-on-branch core of the wall heteropolysaccharide (see Chapters 2.1 and 4.3). Although evidence is lacking, the same enzyme is presumably responsible for the synthesis of the (1,3)-β-glucan backbone in side-chain-branched (1,3:1,6)-β-glucans found on the surfaces of fungal hyphae and in fruiting bodies and in chromistan taxa (see Chapter 4.3).

Although the genes encoding family GT48 proteins occur in fungi, including yeasts, and plants (angiosperms), no genes for GT48 proteins have been detected in the basal prokaryote (archaebacterial or eubacterial) groups although, given their strong relatedness, presumably they must have been present prior to the divergence of the animal/fungal lineages from those that eventually led to the plant-chromalveolate groups. One possibility is that prior to the appearance of these major groups, genes that encode protein domain structures evolved independently, but subsequently fused to give the GT48 gene.

We are missing information about the identity of the enzymes involved in the synthesis of the chromistan (1,3;1,6)-β-glucans and the euglenid and chlorarachnean (1,3)-β-glucans. However, it may be relevant that the chromistan haptophyte *Pavlova* also produces a linear storage (1,3)-β-glucan, so the same type of synthase may be involved in the synthesis of both types of glucans.

## 4.7.3. Occurrence of (1,3;1,4)-β-Glucans

In addition to polysaccharides with (1,3)-β-glucan backbones, (1,3;1,4)-β-glucans have been found in the walls of several taxa (Fig. 2) (Chapters 2.1 and 4.6). They are best known in the walls of the Poaceae (grass and cereal family) and, as lichenin, in the mycobiont walls of the lichen *Cetraria islandica* (Iceland moss). (1,3;1,4)-β-Glucans are found in other lichens and in one non-lichenized fungus, the ascomycete *Aspergillus fumigatus,* as part of the cell wall (1.3;1,6)-branch-on-branch complex. (see Chapter 4.3). Within the plants (Plantae), they also

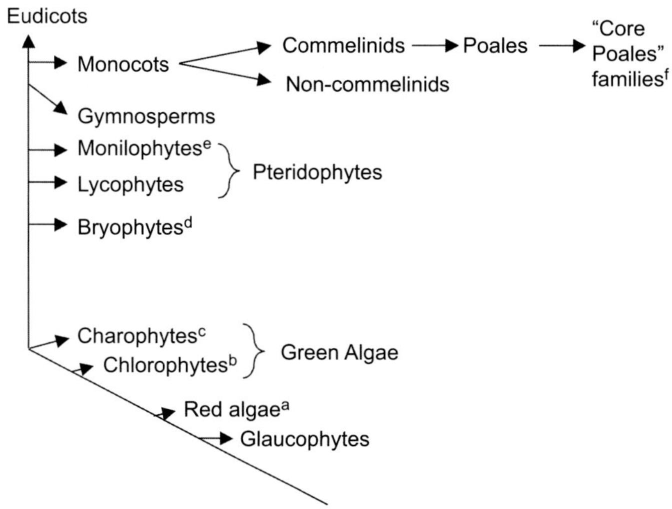


**Fig. 2: Occurrence of (1,3;1,4)-β-glucans in lineages from the glaucophytes, red algae and green algae to the commelinid Poales: a. the red alga *Kappapheus* contains the sulfated form; b. the glucan in the chlorophyte alga *Ulva* contains xylose; c. in the desmid *Micrasterias* (a charophyte alga); d. the glucan in the liverwort *Locopholea* (a bryophyte) contains arabinose; e. in the monilophyte genus *Equisetum* (horsetails); f. in the Poaceae and related 'core Poales' families in the Poales *sensu lato*.**

occur in varying amounts in the walls of many of the families related to the Poaceae ('core Poales') in the order Poales (*sensu lato*) (Trethewey et al., 2005), in the walls of the monilophyte genus *Equisetum* (horsetail) (Fry et al., 2008b; Sørensen et al., 2008), and in the charophyte alga *Micrasterias* (a desmid) (Eder et al., 2008). A sulfated (1,3;1,4)-β-glucan has been found in the cell wall matrix of the red alga *Kappapheus*. The proportion of (1,3)- to (1,4)-linkages and the length of uninterrupted (1,4)-linkages in the (1,3;1,4)-β-glucans varies among the different taxa. (1,3;1,4)-β-Glucans containing xylose and arabinose residues occur in the chlorophyte alga *Ulva* and the liverwort (bryophyte) *Lophocolea,* respectively. A putative (1,3;1,4)-β-glucan has been reported from an alveolate (dinoflagellate) and in the cell walls of a chromistan (xanthophyte) (see Chapter 2.1). There are no reports of (1,3;1,4)-β-glucan in cabozoans, the animal kingdom, amoebozoans or prokaryotes.

## 4.7.4. Genes Encoding (1,3;1,4)-β-Glucan Synthases

The synthesis of (1,3;1,4)-β-glucans in the Poaceae, and probably in related families, involves GT2 cellulose synthase-like (CSL) proteins belonging to the cellulose synthase super family.

Two of these, CSLF and CSLH, are found in species of the Poaceae but not in eudicotyledons (Burton et al., 2006, 2008; Doblin et al., 2009). Their exact role in the synthesis of these unique linear (1,3;1,4)-β-glucans is yet to be clarified (see Chapter 4.6). Nothing is known about the genes encoding proteins involved in the synthesis of (1,3;1,4)-β-glucans in other taxa.

## 4.7.5. Evolutionary Significance of (1,3)- and (1,3;1,4)-β-Glucans

In contrast to gymnosperms, angiosperms, including basal angiosperms, have callose in the walls of their pollen tubes, which also contain callosic plugs (see Chapters 3.3.4 and 4.4.3). Williams (2008) has associated this callose with faster and longer-distance growth of the pollen tubes, enabling the evolution of highly diversified carpels (flowers and fruits) in phylogenetically diversified angiosperms. This contrasts with the slow growth of gymnosperm pollen tubes. Callose also occurs in the cell plate, although transiently, whenever it is formed during cell division, suggesting that it is necessary for the function of the cell plate. This inference can be made from a survey of the occurrence of callose at cell division in green algae and a range of embryophytes (bryophytes, monilophytes and seed plants) (Scherp et al., 2001). With only two exceptions, cell plates containing callose were found in all taxa except for single-celled green algae. Similarly, callose occurs in the walls of cells involved in carbohydrate transport: phloem sieve plates and sieve cells throughout the vascular plants, as well as in the leptoids of mosses and sieve-tube-like elements (trumpet hyphae) in brown algae (Stone and Clarke, 1992). This association indicates a functional association between callose and the transport function of these cells, and a possible association between callose and the initial and subsequent evolution of these cell types. The occurrence, often transient, of callose in the walls of other cell types, especially those concerned with reproduction, may also have evolutionary consequences.

The gene or genes involved in the biosynthesis of (1,3; 1,4)-β-glucans in the cell walls of many of the families of the Poales (*sensu lato*) ('core Poales') may well have evolved separately from those involved in similar polysaccharides in the phylogenetically quite separate *Equisetum.* Nevertheless, Fry et al. (2008b) suggested that the polysaccharides in both groups of plants may be involved, for example, in the rapid basal growth of the internodes. Furthermore, Fry et al. (2008b) noted that walls of both *Equisetum* and the Poaceae (as well as other 'core Poales') are highly silicified and suggested a possible structural association between the occurrence of (1,3; 1,4)-β-glucans and silica. Fry et al. (2008a) have shown that in *Equisetum* a MLG:xyloglucan endotransglucosylase (MXE) is able to transfer (1,3; 1,4)-β-glucans to xyloglucans to remodel the wall and/or strengthen it during growth. MXE was also found in charophyte algae (especially *Coleochaete scutata*), the closest related extant algal group to embryophtes. Hrmova et al. (2007) demonstrated that barley endotransglycosylase/hydrolase can also catalyse the formation of covalent linkages between xyloglucans and cellulose, and between xyloglucans and (1,3;1,4)-β-D-glucans.

## 4.7.6. Conclusion

Although the distribution of (1,3)- and (1,3;1,4)-β-glucans in the tree of life is reasonably clear, many taxa have so far not been examined for the possible presence of these polymers. Furthermore, much remains to be learnt about the evolution of the genes for their biosynthesis, in particular the origin of the genes for the GT48 proteins and the identity of the genes for (1,3;1,4)-β-glucan synthesis in the various taxa in which they are found.

## References

Burton, R. A., Wilson, S. M., Hrmova, M., Harvey, A. J., Shirley, N. J., Medhurst, A., Stone, B. A., Newbigin, E. J., Bacic, A., & Fincher, G. B. (2006). Cellulose synthase-like *CslF* genes mediate the synthesis of cell wall (1,3;1,4)-β-D-glucans. *Science*, *311*, 1940–1942.

Burton, R. A., Jobling, S. A., Harvey, A. J., Shirley, N. J., Mather, D. E., Bacic, A., & Fincher, G. B. (2008). The genetics and transcriptional profiles of the cellulose synthase-like *HvCslF* gene family in barley. *Plant Physiology*, *146*, 1821–1833.

Cavalier-Smith, T. (2004). Only six kingdoms of life. *Proceedings of the Royal Society of London B*, *271*, 1251–1262.

Doblin, M. S., Pettolino, F., Wilson, S. M., Campbell, R., Burton, R. A., Fincher, G. B., Newbigin, E. & Bacic, A. (2009). A barley *cellulose synthase-like CSLH* gene mediates (1,3;1,4)-β-D-glucan synthesis in transgenic *Arabidopsis*. *Proceedings of the National Academy of Sciences of the United States of America*. *106*, 5996–6601.

Eder, M., Tenhaken, R., Driouich, A., & Lütz-Meindl, U. (2008). Occurrence and characterization of arabinogalactan-like proteins and hemicelluloses in *Micrasterias* (Streptophyta). *Journal of Phycology*, *44*, 1221–1234.

Fry, S. C., Mohler, K. E., Nesselrode, B. H. W. A., & Franková, L. (2008a). Mixed-linkage β-glucan: Xyloglucan endotransglucosylase, a novel wall-remodelling enzyme from *Equisetum* (horsetails) and charophytic algae. *Plant Journal*, *55*, 240–252.

Fry, S. C., Nesselrode, B. H. W. A., Miller, J. G., & Mewburn, B. R. (2008b). Mixed-linkage (1→3,1→4)-β-D-glucan is a major hemicellulose of *Equisetum* (horsetail) cell walls. *New Phytologist*, *179*, 104–115.

Hrmova, M., Farkas, V., Lahnstein, J., & Fincher, G. B. (2007). A barley xyloglucan xyloglucosyl transferase covalently links xyloglucan, cellulosic substrates, and (1,3;1,4)-β-D-glucans. *Journal of Biological Chemistry*, *282*, 12951–12961.

McFadden, G. I. (2001). Chlorarachniophytes. In: *Encyclopedia of life sciences*. John Wiley & Sons, Ltd. pp. 1–3. www.els.net

Pautard, F. G. E., & Zola, H. (1967). The location of onuphic acid in *Hyalinoecia tubicola*. *Journal of Histochemistry and Cytochemistry*, *15*, 737–744.

Scherp, P., Grotha, R., & Kutschera, U. (2001). Occurrence and phylogenetic significance of cytokinesis-related callose in green algae, bryophytes, ferns and seed plants. *Plant Cell Reports*, *20*, 143–149.

Sørensen, I., Pettolino, F. A., Wilson, S. M., Doblin, M. S., Johansen, B., Bacic, A., & Willats, W. G. T. (2008). Mixed linkage (1→3),(1→4)-β-D-glucan is not unique to the Poales and is an abundant component of cell walls *Equisetum arvense*. *Plant Journal*, *54*, 510–521.

Stone, B. A., & Clarke, A. E. (1992). *Chemistry and biology of (1→3)-β-glucans*. Victoria, Australia: La Trobe University Press, La Trobe University.

Trethewey, J. A. K., Campbell, L. M., & Harris, P. J. (2005). (1→3),(1→4)-β-D-Glucans in the cell walls of the Poales *(sensu lato)*: An immunogold labelling study using a monoclonal antibody. *American Journal of Botany*, *92*, 1669–1683.

Williams, J. H. (2008). Novelties of the flowering plant pollen tube underlie diversification of a key life history stage. *Proceedings of the National Academy of Sciences of the United States of America*, *105*, 11259–11263.

# Index

Page numbers follwed by *f* indicate figures.


## A

Abiotical stresses induce Pd callose, 447–448
Abiotic stresses, callose in, 446–447
*Acer pseudoplatanus*, 449
*Acetobacter xylinum*, 331
*Achlya*
  cytoplasmic (1,3)-β-glucan of, 366–367
*Achlya bisexualis*, 241, 367
*Achnanthes brevipes*, 362
*Achyla bisexualis*, 15
Acidic fibroblast growth factor (aFGF), 597
Acidic terpenoids, 271
Acremonium, 73
*Acremonium persicinum*, 139
Actin filaments, 428
*Actinidia deliciosa*, 441
AFM. *See* Atomic force microscopy (AFM)
*Agaricus bisporus*, 394
*Agrobacterium*, 328, 329
  10C3K strain of, 328
  growth of, 331*f*
  LTU50 strain of, 333, 333*f*
  scanning electron microscopy on, 331
*Agrobacterium* curdlan synthase, 288
*Agrobacterium* LTU50, 202
*Agrobacterium tumefaciens*, 202, 329
*Alaria angusta*, 370
Alkaline peeling, 11
*Allium cepa*, 441, 503
*Alternaria brassicicola*, 535
Aluminium
  induced callose formation, 505–507
  induced inhibition of root elongation, 506
  reducing $K^+$ efflux, 500
  toxicity, 504
  treated plants, 447
Alveolation
  in syncytial systems, 429–432
Alveoli, 429, 430*f*
1-Aminocyclopropane-1-carboxylic acid, 507
*Amphibolis griffithi*, 474
*Amphiobolus*, 476
Amphotericin B, 270, 271, 300, 305
AMP-producing genes, 564
AMP synthesis, 563
  induction of, 564–565
Anarthriaceae, 633
Angiosperms, 465
Aniline Blue, 11, 265, 267, 295, 309, 397
  fluorochrome, 283
  fluorophore in, 61
Anoxia, 510–511
Anticlinal walls (AW), 429, 430*f*, 431
  callose and, 431–432
Antimicrobial peptides (AMPs), 563
*Aphanomyces astaci*, 240–241, 243, 246
*Apodachlya*, 408
Apomictic embryo sacs, callose in, 487
  events of embryo sac formation in, 488
*Arabidopsis*, 286
  pollen grains, 471–472
*Arabidopsis thaliana*, 441, 501, 526
Archaebacteria, 655
Aromatic cradle, 179
*Arum alpinum*, 476
Ascomycete side-chain-branched (1,3;1,6)-β-glucans, 16
Ash
  seasonal variation of, in *Laminaria hyperborea*, 372*f*
*Aspergillus fumigatus*, 261–262, 388, 392, 394, 587, 632, 658
  cell wall glucan, 18
  wall composition of, 393*f*
*Aspergillus nidulans*, 261–262, 394
*Aspergillus niger*, 141
*Astasia ocellata*, 355

*AtBG_pap* gene, 453
AtBG T-DNA insertion mutants, 540
*AtCalS1*gene, 501
AtGSL6, 302
AtGSL8 and AtGSL10 proteins, 481
*Atgsl10* mutants, 482
Atomic force microscopy (AFM), 60
*Avena sativa*, 91
AW. *See* Anticlinal walls (AW)
*Azorhizobium caulinodans*, 19
*Azospirillum brasilense*, 19, 336
Azuki bean, 303

## B

Bacillariophyceae, 236. *See also* Diatoms
*Bacillus*, 329
*Bacillus amyloliquefaciens*, 12
*Bacillus circulans*, 188
*Bacillus halodurans*, 182
*Bacillus subtilis*, 136
Bacitracin, 224
Barley endotransglycosylase/ hydrolase, 661
Barley grain, 641
Basic fibroblast growth factor (bFGF), 596
Basidiomycete side-chain-branched (1,3;1,6)-β-glucans, 16
*Bathycoccus sp.* BAN7, 311
Bcl10-Malt1-mediated NF-kappaB activation, 585
(1,3)-β-D-glucan endohydrolases
  GH17 family, 122–128
  GH26 family, 132–133
(1,6)-β-D-glucan endohydrolases, 139
β-D-glucan glucohydrolases
  in auxin-mediated cell, 145
  in pollen development, 146
(1,3)-β-D-glucan phosphorylases, 153–154
(1,3;1,4)-β-D-glucans, 140, 141, 622
  in commercial processes, 622
  fine structure of, 624–625
    cellotriosyl and cellotetraosyl residues, 624
    molar ratios of, cellotriosyl to cellotetraosyl units, 625
  from grasses, 623
β-D-glucosidases of family GH1
  action pattern, 142–143
  catalytic mechanism, 143
  classification of, 139–140
  3D structures, 143–145
  properties of, 140
  substrate specificity, 140–142
β-D-glycoside exo-hydrolases of family GH3
  action pattern, 147–148
  catalytic mechanism, 148–150
  classification of, 145
  3D structures, 150–151
  properties of, 145–146
  substrate specificity, 146–147
    structural determinants of, 151–153
*Betula pubescens*, 441
(1,3)-β-glucanase
  to break down fungal cell walls, 543
  degradation of transient callose wall, 471
  in endoplasmic reticulum membrane system, 542
  expressed by tapetum, 470
  gene into TMV genome, 541
  in regulation of cell-to-cell transport, 502
  roles in plasmodesmatal trafficking and, 540
  transgenes expression in tapetum, 475
β-glucan biosynthesis in protozoans and chromistans, molecular biology of, 248–249
(1,3)-β-glucan endo-hydrolases
  activity in early spring, 449
  catalysing callose degradation, 443
  genes encoding, 452
  importance of, 452
  limited/absent expression of gene, 452
  localization, 444, 451
  plasmodesmal associated, 453
  *Ulva lactuc,* polysaccharide hydrolysed by, 630
  vesicles transportation to, 453
  virus recruiting and/or activating, 452
(1,3)-β-glucan in embryophytes, synthesis of, 285
  biochemical identification of GSL proteins, 296–298
  callose synthase complexes, 298–303
  functional analysis of GSL genes, 294–296
  GSL gene family, 288, 292–294
  GSL proteins, 286–288, 287*f*
β-glucan recognition proteins (BGRP), 189
(1,3)-β-glucans, 563
  anti-inflammatory activity, 597–598
  anti-tumor activity of, 594–595
  on arthropod immune system, 572
  bacterial infections, 591–592
  biosensor for, 568
  chemical characteristics of, 5–10
  chemical structure variants of, 62–67
  conformations of, 48–52, 65–67
  derivatives of, 24–27
  detection of, 11
  dietary supplementation with, 599
  dissolution of, 11–12
  distribution, 655–656

in embryophytes, 284–285
evolutionary significance of, 660–661
fluorescence spectroscopy, 61–62
forms and functions of, 657
fungal infections, 592–593
in fungi. *See* Fungi
gelation behaviour of, 93–103
induced AMP synthesis, in *Drosophila*, 565*f*
induced coagulation, in horseshoe crabs, 567*f*
induced proPO activation, in crustaceans, 569*f*
ischemia/reperfusion injury, 595–596
isolation of, from cell walls, 389–391
labelling of, using Dectin1-Fc chimeric protein, 398*f*
in lichens, 406–407
linear, 399, 407
localization of, 399*f*
location of, in fungi, 409*f*
mammalian receptors for, 580
  complement receptor 3, 581–582
  dectin-1, 582–587
  lactosylceramide, 580–581
  scavenger receptors, 581
microscopy, 60–61
occurrence of, 658
in oomycetes
  oomycete cell wall, 407–408
  storage (1,3)-β-glucans, 408–409
SAXS, 61
sedimentation analysis, 61
solid state $^{13}C$ NMR of, 58–60
solid state conformations in, 52–67
to stimulate innate immunity and, 579
and stimulation, of innate immunity, 598–600
structural variations in, 396–405
structure/activity relationships with, 587–588. *See also* Dectin-1
structure and molecular organization of, in yeast and module cell walls, 391–396
structures of, 12
synthases, 241–242
triggering release of immune components from, 571
and viral infections, 593–594
*in vivo*
  *effect, administration on*, 590
  effects, in animal models of disease, 590–591
  pharmacokinetics, pharmacodynamics and bioavailability, 588–589
and wound repair, 596–597
x-ray diffraction of, 54–58
yeast cell wall structure, 260–261, 260*f*
(1,3;1,2)-β-glucans
  side-chain-branched, 20
(1,3;1,4)-β-glucans, 178, 621, 622, 629. *See also* (1,3; 1,4)-β-D-glucans
  in aleurone-rich fraction, 643
  among different taxa, 659
  biological role of, 625
    concentrations of, 629
    effects of fine structure on, 627–628
    formation of gel-like matrices in, 625–627
    as source of metabolizable energy, 629
  in cell walls of mature wheat grains, 642
  cereal, 21–22
  from *Cetraria islandica*, 628
  chlorophyte, 23–24
  chromalveolate, 23–24
  chromistan, 23–24
  concentrations and locations, in cereal grains, 637–639
    in barley, 638–639
    cereals other than oats and barley, 639
    developing oat genotypes with, 639
    quantification of, 638
    reducing concentrations of, 638
    in whole grains, 637–638
  in cultivars of oats, 642
  distribution of, 629, 631
    in angiosperms, 632–634
    in cell walls, of mature grains, 640–643
    chromalveolates, 632
    lichens and fungi, 630–632
    plants other than angiosperms, 630
    prokaryotes, 632
  effects of environment on, 639–640
  environmental factor implicated in, 640
  *Equisetum arvens*e, 22
  evolutionary significance of, 660–661
  fungal cell wall, 23
  genes, in biosynthesis of, 660
  in grasses, 21–22, 622
  interactions, 627
  lichen, 23
  liverwort, 23
  location vegetative organs, of Poaceae, 634–636
    transmission electron micrographs, 635
  occurrence of, 658–659
  within plant variation in, 640
  rhodophyte sulfated, 24
  role in cell-wall architecture, of Poaceae, 637
  stock feed formulations for, 623

(1,3;1,4)-β-glucans (*Continued*)
in walls of *E. arvense*, 630
in walls of starchy endosperm, 629
in wheat grain, 628
(1,3;1,6)-β-glucans
branch-on-branch, 16–19, 392
chemical structure of, 357, 358*f*
cyclic, 19–20
side-chain-branched, 14–16
(1,6)-β-glucans
localization of, 399*f*
β-glucans biosynthesis, hypothetical model for, 234*f*
β-glucans biosynthesis in Protozoans and Chromistans, biochemistry of
assay conditions and kinetic parameters, 239–240
biosynthesis of side-chain-branched (1,3;1,6)-β-glucans, 248
model organisms, 235–236
modulation of (1,3)-β-glucan synthase activities in vitro, 240–244
product characterization, 246–248
purification of (1,3)-β-glucan synthases, 244–246
subcellular localization of β-glucan synthases and preparation of cell-free extracts, 236–239
(1,3)-β-glucan synthases
movement on cell wall surface, 267*f*
fungal *FKS* genes, role in, 442–443
genes encoding, 452, 657–658
and Pd conductivity, 445
(1,3;1,4)-β-glucan synthases, 659–660
(1,3)-β-glucosidic linkages, 176
(1,4)-β-glycosidic linkages, 393
BGRP. *See* β-glucan recognition proteins (BGRP)
*Bh*CBM6, 182
(1,3)-β-homopolymers, 175, 176
Bioassays, 341
Biofilm matrix, 388
Biosynthesis of linear β-glucans, hypothetical model for, 234
Biosynthetic enzymes for (1-3)-β-glucans
catalytic subunit, 261–263
cell wall integrity checkpoint in yeast, 269–270
cross talk in the regulation of other cell wall components, 270–271
glucan synthase inhibitors, 271–272
glucan synthesis in spore formation, 267–269
other proteins associated with glucan synthase activity, 264
regulation of (1,3)-β-glucan synthesis in yeast
factors involved in the regulation of glucan synthase activity via Rho1p, 264–265
location of (1,3)-β-glucan synthase, 266
movement of Fks1p in the plasma membrane, 266–267
post-translational modification of Rho1p, 265
transcriptional regulation of the FKS genes, 265–266
regulatory subunit, 263–264
Biosynthetic enzymes for (1-6)-β-glucans, 272–273
Biotic stresses, callose in, 525–531
herbivorous attack, 451–452
viral infection, 452–453
hydrolysis of callose and gating of Pd, 454
*Blaberus craniifer*, 570
*Blaberus discoidalis*, 571
*Blastomyces dermatitidis*, 400
(1,3)-β-linked gluco-oligosaccharides, 243
*Blumeria graminis*, 293, 532
Boltzmann constant, 70
*Botrychium virginianum*, 448
*Botryosphaeria rhodina*, 16
*Botrytis cinerea*, 388, 536
*Bradyrhizobium japonicum*, 5, 19, 336
nodulation and, 338–339
strains, 19
*Brevicoryne brassicae*, 143
Bromeliaceae, 634
Brown algae. *See* Phaeophyceae
*Bryophyllum daigremontiana*, 202
Bryophytes, 23
*Bsp*CBM28, 185


## C


$Ca^{2+}$ influx, 543
Calcofluor white, 271
Callose, 11, 14, 283, 284, 364
accumulation, 294*f*
AW and, 431–432
deposition, 305–306, 432–434
after microbial attack, 528
function of, in cytokinesis, 434–435
periclinal wall and, 431–432
plant cytokinesis and, 425
in pollen tubes, 473
and role in, flowering plant, 465
Callose during development, 448
cotton fibre elongation, 451
definitive and dormancy callose, 449–451
sieve plate pore development, 448
Callose formation
and anoxia, 510–511

and gravitropic response, 513
kinetics Al-induced and digitonin-induced, 503
metal toxicity and, 502–508
aluminium, 504–507
manganese (Mn), 507–508
in response to abiotic stress, 499–502
and temperature stress, 511–513
TTSS suppressing, 538
wounding, osmotic stress and, 508–510
Callose localization
in Pd and sieve plates, 441–442
localization of callose, 442
raised collar structures, 442
Callose SCW, during microgamete development, 476
Callose synthase (CalS), 300, 432
cellulose and, 512
components of, 530
gene *GSL5*, 538
genes coding for, 514
inhibition of, 536
in plants, 442
preparations, 299
molecular weight of major polypeptides present in, 297–298
purify and characterization, 529
regulatory mechanisms, 300*f*, 543
role in regulating of, 531
Callose synthesis, regulation of, 303–304, 306–309
synthesis of (1,3)-β-glucans in chlorophytes, 309–311
synthesis of (1,3)-β-glucans in rhodophytes, 311–312
Calmodulin, 302
Calnexin, 301
CalS. *See* Callose synthase (CalS)
*Candida albicans*, 153, 261, 262, 271, 272, 388, 400, 599
cell wall glucan, 18
*Candida* enzyme, 153
*Candida glabrata*, 396
Capsular polysaccharide (CPS), 342
Carbohydrate-Active enZymes (CAZy) database, 121, 139
Carbohydrate-binding modules (CBM), 172–174
biochemical analysis of
CBM13, 188–189
CBM39, 189–191
classification, 173–174
ligand binding profiles for, 172
structural comparison of, 191–192
structure-function of
CBM4, 176–179
CBM6, 179–182
CBM11, 182–183
CBM17, 183–185
CBM22, 186
CBM28, 185
CBM43, 186–187
three-dimensional structures for, 177
Carboxymethylation, 72
κ-carrageenan, 134
Caspofungin, 270, 271
Catalytic acid, 123
Catalytic base, 123
CAZy. *See* Carbohydrate-Active enZymes (CAZy) database
CBM4, 176–179, 191
aromatic cradle, 179
*Cf* CBM4-1, 179
*Tm*CBM4-2, 177–178, 179
CBM6, 179–182, 191
CBM11, 182–183
CBM13, 188–189
CBM17, 183–185
CBM22, 186
CBM28, 185
CBM43, 186–187
Cellotriosyl, 63
Cell plate, formation of, 428–429
Cell plate assembly matrix (CPAM), 428
Cellular immunity, 571–572
Cellular metabolism, global and curdlan, 335
Cellulins, 408
*Cellulomonas*, 329, 335
*Cellulomonas falvigena*, 13
*Cellulomonas fimi*, 149, 177
*Cellulomonas flavigena*
floc formation and, 334
KU, 332
Cellulose colocalization, with callose, 472
Cellulose synthase-like (CSL) proteins, 287–288
*Cellulosimicrobium cellulans*, 188
*Cellvibrio mixtus* endoglucanse 5 A (*Cm*CBM6-2), 180
Cell wall glucan
*Aspergillus fumigatus*, 18
*Candida albicans*, 18
*Pythium aphanidermatum*, 18–19
*Saccharomyces cerevisiae*, 16–18
Cell wall polysaccharides, 367–368
Cell walls, fungal
(1,3)-β-glucans, 388
isolation of (1,3)-β-glucans from cell walls, 389–391
structural variations, 396–405
structure and molecular organization of, in yeast and module cell walls, 391–396
chemical composition of yeasts and fungi, 389*t*
fractionation. *See* Fractionation, of cell walls
protocol for extraction of components, 390, 391*f*
Cell walls and β-glucans in yeast, 260–261, 260*f*
Centrolepidaceae, 633

Cereal β-glucans
behaviour of concentrated solutions, 91–93
gelation of, 100–103
physical properties of, 22
solution properties of, 81–82
*Cerk1* mutants, 535
*Cetraria islandica*, 407, 630, 658
*Cf*CBM4-1, 179
*Chaetoceros brevis*, 362
*Chaetoceros mulleri*, 16
Chain association, 49–52
Chain shape, 49
Chain stiffness, 48
CHAPS. *See* 3-[(3-cholamidopropyl)-dimethylammonio]-1-propane sulfonate (CHAPS)
CHAPSO. *See* (3-[(3-cholamidopropyl)-dimethylammonio]-2-hydroxy-1-propane sulfonate) (CHAPSO)
Charophytes, 23
Chitin, 11, 261
Chitinases, 535
Chitin hydrolases, 527
Chitin oligosaccharide elicitor binding protein, 534
*Chlamydomonas reinhardtii*, 309, 310
Chlorarachniophyceae. *See* Chlorarachniophytes
Chlorarachniophyta. *See* Chlorarachniophytes
Chlorarachniophytes, 371–372
(3-[(3-cholamidopropyl)-dimethylammonio]-2-hydroxy-1-propane sulfonate) (CHAPSO), 238
3-[(3-cholamidopropyl)-dimethylammonio]-1-propane sulfonate (CHAPS), 238, 245, 296
detergent extraction, 297
*Chondrus crispus*, 312
*Chorda filum*, 370
Chromatin immunoprecipitation (ChIP), 265
Chromistans
types and structures of polysaccharides in, 354*t*
*Chrysochromulina*, 357
Chrysolaminarin and spores, 362
Chrysolaminarins, 15, 236, 357, 359, 364
diel changes in, 361*f*, 362
formation of resting cells and, 362
formation of spores and, 362
Chrysophyceae. *See* Chrysophytes
Chrysophytes, 364–365
*Clostridium cellulovorans* cellulase (CcCBM17), 183–185
*Clostridium stercorarium* xylanase (*Cs*CBM6-3), 180
*Clostridium thermocellum*, 132
*Clostridium thermocellum* xylanase (*Ct*CBM6), 180
Coat protein, 452
*Coccidioides immitis*, 400
*Coleochaete scutata*, 660
*Collema leptosporum*, 407
*Colletotrichum lagenarium*, 535
*Colletotrichum lindemutianum*, 154
Complement receptor 3, 581–582
Conduritol B epoxide, 149
Congo Red, 244, 271
Conidia
comparison of wall composition between, and Hyphae, 399
antifungal compounds, 401–402
cell wall mutants, 402
dimorphic forms, 400–401
environmental conditions, 401
Coomassie Blue, 243, 245
*Coprinus cinereus*, 394
*Coscinodiscus concinnus*, 362
*Coscinodiscus nobilis*, 359
Covalently bound wall proteins (CWP)
GPI-anchored, 396
schematic representation of, in yeast, 395*f*
COX activation, 584
Cox-Merz rule, 83
CP. *See* Cross polarization (CP)
CPAM. *See* Cell plate assembly matrix (CPAM)
CPS. *See* Capsular polysaccharide (CPS)
CR3. *See* Complement receptor 3
*Craspedostauros australis*, 359
CrdASC gene, 203–205
CR3-deficient leukocytes, 582
CrdR gene, 205–206
CrdS, 288
C-reactive proteins, 567
Critical gelation point, 93
CR3-mediated phagocytosis, 582
Cross polarization (CP), 53
Cryogelation
of water-soluble polymers, 103
*Cryptococcus neoformans*, 388, 400, 589, 593
*Cs*CBM6-3, 191
*CslF* cellulose synthase-like genes, 644
*Ct*CBM11, 183
*Ct*CBM22-2, 186
C-type lectin-like receptors, 584
Curdlan, 13
behaviour of concentrated solutions, 84–85
biosynthesis locus, 202
chain, 214–216
crdASC gene, 203–205
crdR gene, 205–206
enzymology, 211
formation and functional roles, 331–336
gelation of, 93–96
global cellular metabolism and, 335

membrane typology, 211–212
ntrBC, 206
ntrC, 207–208
ntrYX gene, 208–209
occurrence, 328–330, 330*t*
pSSA gene, 205
relA gene, 209–211
Curdlan sulfate, 593
CWP. *See* Covalently bound wall proteins (CWP)
*Cyanidioschyzon merolae*, 311
Cyclic (1,3)-β-glucans
formation and functional roles
hypo-osmotic conditions, adaptation to, 337–338
nodulation and $N_2$-fixation, 338–340
suppression of plant defense response, 340–342
occurrence, 336–337
Cyclic (1,3;1,6)-β-glucans, 19–20, 216–222
bioassays and, 341
formation and functional roles
hypo-osmotic conditions, adaptation to, 337–338
nodulation and $N_2$-fixation, 338–340
suppression of plant defense response, 340–342
molecular genetics of, 216–219
ndvB genes, 219–221
ndvC genes, 221–222
ndvD genes, 221–222
ndv genes, 216–219
occurrence, 336–337
structure of, 19
*Cyclotella cryptica*, 236, 246, 362
*Cylindrotheca closterium*, 362–363
*Cylindrotheca fusiformis*, 359
Cyperaceae, 634
*Cystophora scalaris*, 370
*Cystoseira barbata*, 371
*Cystoseira crinita*, 371
*CYT1* gene, 499
Cytokines, 571
Cytokinesis, 398, 470
function of callose in, 434–435
plant, callose and, 425
Cytoplasmic (1,3)-β-glucan of *Achlya*, 366–367

## D

Daylily. *See Hemerocallis fulva*
Dectin-1, 581–587, 593, 595. *See also* (1,3)-β-glucans
with CD37, 585
to enhance TLR-mediated cytokine, 586
expressing in leukocytes, 583
inducing autoimmune diseases, 586
in mammalian systems, 587–588
mediating cellular responses, 584
mediating IL-6 production, 586
receptor for (1,3)-β-glucans, 590
recognising synthetic oligomeric, 587
role in, innate immune response, 592
signalling, 585
Dectin-1, 190–191
Dectin1-Fc chimeric protein
labelling of (1,3)-β-glucans using, 398*f*
Degree of polymerization (DP), 11, 355
Denaturation – renaturation (DR) process, 88
Dentritic cell-associated C-type lectin-1 (CLEC7A). *See* Dectin-1
2-Deoxy-D-glucose (DDG), 446
*Deschampsia antarctica*, 621
*DEX1* gene, 475
*Dianthus caryophyllus*, 441
Diatoms, 358
callose, 364
chrysolaminarin, 359–364
transmission electron micrographs of, 360*f*
*Dictyonema glabratum*, 407
*Dictyostelium discoideum*, 246–247
Digitonin, 296
Dilute solution
definitions of molecular parameters, 69–80
Dimethyl sulfoxide (DMSO), 13, 47
Dinoflagellates, 371
Dinophyceae. *See* Dinoflagellates
Dinophyta. *See* Dinoflagellates
Disaccharides, 342
Dithiothreitol, 244, 270
Dityrosine, 267
DMSO. *See* Dimethyl sulfoxide (DMSO)
DP. *See* Degree of polymerization (DP)
2D-PAGE analysis, 245
DR. *See* Denaturation – renaturation (DR) process
*Drosophila* GNBP3, 570
*Dunaliella salina*, 311
Dye
phenoxazone, 11
triphenylmethane, 11
Dynactin, 269

## E

Ecdeiocoleaceae, 633
*Ecdeiocolea monostachya*, 633
Echinocandin drugs, 271
Echinocandins, 271
LY303366, 264
*Ecklonia radiata*, 370
EGTA, 302
*Eisenia bicyclis*, 12, 139, 370
*Eisenia foetida*, 570
*Emiliania*, 357
*Emiliania huxleyi*, 357
Endoplasmic reticulum (ER), 439
*Equisetum arvense*
(1,3;1,4)-β-glucans, 22

*Equisetum* spp, 630, 660
*Erwinia amilovora*, 539
*Escherichia coli*, 303, 390
EST databases, 249
Eubacterial (1,3)-β-glucan synthase genes, 658
*Euglena gracilis*, 153, 235–236, 238, 241–242, 244, 245, 246, 354
  biosynthesis of paramylon in, 356
Euglenids, 354–356
Euglenophyceae. *See* Euglenids
Euglenophyta. *See* Euglenids
*Evernia*, 632
ExoI, 151

## F

FAB. *See* Fast atom bombardment (FAB)
Factor C, 566
Factor G, 566
Fast atom bombardment (FAB), 370
FK506, 266
*FKS* gene, 443, 530, 658
Fks1p-GFP, 267
FKS proteins, 658
Flagellariaceae, 633, 634
*Flagellaria guineensis*, 633
*Flagellaria indica*, 633
Flagellin sensitive 2 (FLS2), 538
*Flavobacterium meningosepticum*, 149
Flg22 receptor, 538
Flory universal viscosity constant, 71
Fluorescence resonance energy transfer (FRET), 62
Fluorescence spectroscopy, 61–62
Fractionation, of cell walls, 390
  protocols for, 390, 391*f*
*Fraxinus excelsior*, 449
Free fatty acids (FFAs), 536
FRET. *See* Fluorescence resonance energy transfer (FRET)
Fungi
  (1,3)-β-glucans in
    cell walls. *See* Cell walls, fungal
    fungal and yeast, extracellular polysaccharides, 387–388
    location of, 409*f*
  biofilms, 388
*Fusarium graminearum*, 536

## G

Galactomannan, 18
*Galdieria sulphuraria*, 311–312
Gelation behaviour
  of (1,3)-β-glucans, 93–103
*Gemini pollen 1* ( *gem1* ) mutants, 482
Gentiobiose – enzyme models, 152
GH16, 189–190
GH17 (1,3)-β-D-glucan endohydrolases, 122–128
  catalytic mechanism, 124–128
  classification of, 122
  properties of, 122
  structure of, 123–124
  substrate specificities of, 122–123
  transglycosylation reactions, 128
GH17 (1,3;1,4)-β-D-glucan endohydrolases, 128–132
  catalytic mechanism, 131
  classification of, 128–129
  3D structure of, 131–132
  properties of, 129–130
GH26 (1,3;1,4)-β-D-glucan endohydrolases, 132–133
GH17 enzymes, 121
GH16 family
  evolution of, 138–139
*GhGSL1* in cotton fibres, 443
Gigartinales, 630
GIP-1 protein, 155
*GLS* genes, 479
Glucanase, 453
Glucans, 259
Glucan synthase-like*(GSL)* gene, 530
*Gluconacetobacter xylinus*, 242
Glucuronoxylmannan, 589
Glyceollins, soybean, 341
*Glycine max*, 302, 512
Glycine-rich protein, cdiGRP, 542
Glycogens, 366–367, 408–409
Glycoprotein, 243
Glycosidic oxygen
  protonation of, 125
Glycosylation, 309
Glycosylphosphatidylinositol, 17
Glycosylphosphatidylinositol-anchored protein, 444
Glycosyl-phosphatidylinositol (GPI), 261
Glycosylphosphatidylinositol (GPI)-anchor, 395–396
GNBP3 protein, 564
GNBP proteins, 565
Golden algae. *See* Chrysophytes
*Golovinomyces cichoracearum*, 532, 534
*Gossypium barbadense*, 441
*Gossypium hirsutum*, 285, 292, 441
GPI-anchor. *See* Glycosyl phosphatidylinositol (GPI)-anchor
Gram-negative bacteria, 332
Gram-positive *Cellulomonas*, 120
Gravitropism, 513
Green fluorescent protein (GFP), 443
*Grifola frondosa*, 588
*GSL* gene family members, in Arabidopsis, 531
*Gsl1/gls1gsl5/* + plants, 479
*GSL* (Glucan Synthase-Like) genes, 442, 443, 531
*Gsl1* mutation, 479
*GSL5* null mutants, 542
GSL proteins, 442, 543

GT2 cellulose synthase-like (CSL) proteins, 659
GT2 enzyme, 224
GT48 enzymes, 309
GT48 family, in angiosperms, 658
GT2 protein-encoding gene, 658
Guanine nucleotide-binding proteins (GNBPs), in *Drosophila*, 564
Gut associated lymphoid tissue (GALT) expression, 600
Gymnosperm pollen tubes, slow growth, 660
Gyration
  mean square radius of, 70

## H

Haemolymph clotting, 565–567
Haptophyceae, 357–358
Haptophyta. *See* Haptophyceae
*Haramonas dimorpha*, 15, 372
*Hemerocallis fulva*
  meristematic cells of
    cell plate in, 426*f*
    cytokinetic apparatus in, 426*f*
Heterokontophyta. *See* Diatoms
*Histoplasma capsulatum*, 400
*Hordeum vulgare*, 292, 298, 441, 633
Horseshoe crab TGase, 567
*Hrp* mutants, 538
Huggins constant, 89
Human isoform (hDectin-1E), 583
Hydrodynamic volume
  of polysaccharide, 48
*p*-hydroxymercuribenzoate, 244
Hyphae
  comparison of wall composition between, and Conidia, 399
    antifungal compounds, 401–402
    cell wall mutants, 402
    dimorphic forms, 400–401
    environmental conditions, 401
Hyphal, cell walls
  chemical composition of, 368*t*

## I

I.C. streptococcal type 37 (1,3;1,2)-β-glucan, 342
IgE responses, 598
IMC carcinoma, 598
IMD pathway, 564
Immunoreceptor tyrosine-based activation-like motif (ITAM), 583
Inositol 1,4,5-trisphosphate (IP3), 513
Ionomycin, 501
*Ishige okamurai*, 370

## J

Joinvilleaceae, 633
Juncaceae, 634

## K

*Kappapheus,* containing sulfated form glucan, 659
*Kappaphycus alvarezii*, 630
Keratin sulfate, 134
Kinesin, 479
K1 killer toxin, 271
KRE6, 248
KRE genes
  mutations in, 402
Kre5p, 272
Kuhn's length, 69

## L

Lactosylceramide, 580–581
*Lamellibrachia satsuma*, 248
*Laminaria cichorioides*, 370
*Laminaria digitata*, 122, 370
*Laminaria hyperborea*, 369, 370
  seasonal variation of laminarin and ash in, 372*f*
*Laminaria longicrusis*, 371
*Laminaria saccharina*, 176
Laminaribiose phosphorylase, 153
Laminarins, 14–15, 236, 370, 371, 589
  seasonal variation of, in *Laminaria hyperborea*, 372*f*
*Larix laricina*, 14
Lectin, 570
*Lemna gibba*, 508
Lentinan, 78
  gelation mechanism proposed for, 97
*Lentinus edodes*, 78
*Leptomitus lacteus*, 367
Leucine-rich repeats (LRRs), 526
Leucosin, 16, 359
Lewis lung carcinoma, 598
Lichens
  (1,3)-β-glucans in, 406–407
Light-induced paramylon
  kinetics of, 355, 356*f*
Lignin, 140
*Limpet pollen (lip)* mutants, 482
Linear (1,3)-β-glucans, 12–14, 234*f*, 399, 407. *See also* Curdlan
  dilute solution properties of, 70–73
  in walls, gametophytes and gametes, 466
Linear (1,3;1,4)-β-glucans, 20–24
Lipopolysaccharides (LPS), 331, 342, 566
Lipoteichoic acids, 567
*Lolium multiflorum*, 292
*Lolium perenne*, 633
*Lophocolea bidentata*, 23
LPS-induced haemocyte exocytosis, 566. *See also* Lipopolysaccharides (LPS)
LRR transmembrane receptor kinase, 538
*Lycopersicon esculentum*, 441

## M

*Macrocystis*, 368
Magic angle spinning (MAS), 18
*Magnolia kobus*, 449
MALDI. *See* Matrix-assisted laser desorption/ionization (MALDI)
MALS. *See* Multi-angle laser-light scattering (MALS)
*Malus domestica*, 441
MAMP-induced signalling, 538
*Manduca sexta*, 570
Mannan, 587
Mannoproteins, 16, 261
Marine diatom. *See* Diatoms
Mark-Houwink-Sakurada ( MHS), 81
  of cereal, 82
MAS. *See* Magic angle spinning (MAS)
Matrix-assisted laser desorption/ionization (MALDI), 370
Megagamete development, in *Arabidopsis*, 483–487
  different stages of, 485
  ovule development, 484
Melanin production, 568
Melanization, of microbial pathogen, 568
Melanization cascade, 568–571
Meristematic cells, 425
  cell plate formation in, 428
  cell plate in, of daylily, 426*f*
  cytokinetic apparatus in, of daylily, 426*f*
Metal toxicity, 502
Methanol, 54
α-methyl-D-glucoside, 242
MHS. *See* Mark-Houwink-Sakurada ( MHS)
*Micrasterias*, 630, 659
Microgamete development, 466
  appearance of callose during, 468
  in *Arabidopsis*, 466, 469
    meiosis to cytokinesis, 469–470
    premeiosis I and II, 469
    released microspore I, 470
    released microspore II, 470–471
    second mitotic division to, 471
  cell wall compositions and changes, 467
  functions for callose during, 473
    during cytokinesis, 476–480
    megagamete development, 482–483
    in post-meiotic male gamete development, 480–482
    in special cell wall, 474–478
  pollen development, 467
*Micromonas pusilla*, 311
Microsporogenesis, 466
Microtubules, 428
MLG:xyloglucan endotransglucosylase (MXE), 660
*Mlo* mutants, 536
MLO protein, 537
Mn toxicity, 507–508
*Monodus subterraneus*, 23, 372, 632
Moth bean *(Vigna aconitifolia)*, 308
Movement proteins (MPs), 452
  interactions with microtubule and microfilaments, 453
MS/MS proteomic analysis, 443
Multi-angle laser-light scattering (MALS), 11
Multinucleate cytoplasm, 429
Mung beans (Vigna radiata), 301, 307
Mutants
  cell wall, 402
Mycolaminarins, 15, 236, 237, 365, 366–367, 408
MyD88-coupled TLRs, 585

## N

*N*-acetyl-β-glucosaminidase, 146, 149, 151, 388
*N*-acetylchitooligosaccharides, 534
*N*-acetylglucosamine (GlcNAc), 261
*N*-acetylhexosamine residues, 15
*NaGSL1* gene, 443, 530
NaGSL1 protein, 295–296, 297, 308
NBS-LRR proteins, 526
NCD. *See* Nuclear-cytoplasmic domain (NCD)
NdvB genes, 219–221
NdvC genes, 221–222
NdvD genes, 221–222
Ndv genes, 216–219
Nebenlinien, 364
Neo β-glucans, 28
*Nereocystis*, 368
*Neurospora crassa*, 262, 263, 264, 288, 394
Neurotrophin 3 (NT-3), 596
*Newropogon*, 632
Newtonian plateau, 82
Newtonian viscosity, 90
*Nicotiana*, 284
*Nicotiana alata*, 14, 287, 297, 304, 308
*Nicotiana benthamiana*, 441, 539
*Nicotiana plumbaginifiolia*, 541
*Nicotiana sylvestris*, 541
*Nicotiana tabacum*, 441, 452
  pollen tube, 473
*Nilaparvata lugens*, 452
NMR. *See* Nuclear magnetic resonance (NMR)
*N*-myristoylation, 309
NO, in primary signals of anoxia, 511
Nodulation, 338–340
Non-Newtonian fluids, 82
NtrBC gene, 206
NtrB gene, 206–207
NtrC gene, 207–208
NtrYX gene, 208–209

Nuclear – cytoplasmic domain (NCD), 426
Nuclear factor of activated T-cells (NFAT), 585
Nuclear magnetic resonance (NMR), 47, 390
 in solid state, 53–54
Nystatin, 271

## O

*Ochromonas malhamensis*, 365
*Ochromonas tuberculatus*, 365
*Oe*CBM43A, 186–187
*Oe*CBM43B, 187
*Olea europaea*, 186
(1,3;1,4)-oligoglucosides, 24
Oligomers, 65
Oligosaccharide, 5
Oomycetes
 (1,3)-β-glucans in
  oomycete cell wall, 407–408
  storage (1,3)-β-glucans, 408–409
Oomycota, 365
 cellulin, 367
 cell wall polysaccharides, 367–368
 chemical structure of, 366*f*
 mycolaminarin, 366–367
Opsonins, 571
Optical rotatory dispersion (ORD)
 in curdlan, 72
ORD. *See* Optical rotatory dispersion (ORD)
*Oryza sativa*, 292
 cross-sections of, 430*f*
 wall development in endosperm of, 431*f*
*Ostreococcus tauri*, 310
Oxidation
 of scleroglucan, 80

## P

Pachyman, 13
*Paenibacillus*, 390
Papulacandin B, 244
Papulacandin B-resistant mutant, 262
Papulacandins, 271
*Paracoccidioides brasiliensis*, 396, 400
Paramylon, 13, 235, 242
Paramylon granules, 354, 355
 biosynthesis of, in *E. gracilis*, 356
 freeze-etch electron micrographs of, 355*f*
 light-induced, kinetics of, 355, 356*f*
Paramylon synthase, 245
*Parmelia*, 632
Parmeliaceae, 632
*Parmotrema*, 632
Pathogenesis-related (PR) proteins, 443
*Pavlova mesolychnon*, 13, 357
*Pavlova prymnesiida*, 357
Pd. *See* Plasmodesmata
Pd callose synthase-associated protein, 443
Pd regulation, by callose turnover, 444–446
Pellicle, 331
*Pen1-1* mutant, 536
PEN1/ROR2 syntaxins, 537
*Peranema trichophorum*, 13, 355
*Pergularia daemia*, 476
Periclinal wall
 callose and, 431–432
*Peridinium westii*, 23, 371, 632
Periodate-Schiff reaction, 11
*Peronospora parasitica*, 293, 532
PGG-(1,3)-β-glucan, 591
*Phaeocystis*, 357, 358
*Phaeocystis globosa*, 357
*Phaeodactylum tricornutum*, 248, 249, 359
Phaeophyceae, 368–371
*Phaseolus vulgaris*, 441, 443, 512
Phenoloxidase (PO), 568
Phenoxazone dye, 11
Phragmoplastin, 295, 302
Phragmoplasts, 425
 development in syncytial systems, 426–428, 427*f*
*Phythium debaryanum*, 408
Phytoalexins, 527
*Phytophthora*, 235
*Phytophthora cinnamomi*, 235, 243, 246, 247
*Phytophthora* genus, 248
*Phytophthora infestans*, 249, 368
*Phytophthora palmivora*, 237, 241, 256, 366
 spores, 408
*Phytophthora parasitica*, 15, 408
*Phytophthora ramorum*, 249
*Phytophthora* species, 248
*Phytopthora sojae*, 155, 238, 240, 246, 249, 340
*Pichia pastoris*, 390
PI3K/Akt signalling, 592, 596
PI3K inhibitor, 592
PI3K pathway, 597
*Pinnularia*, 359, 364
PIR. *See* Putative proteins with internal repeats (PIR)
Pir proteins, 17
*Pisum sativum*, 441, 512, 535
Plant–bacteria interactions
 callose in, 537–539
  extracellular enzymes for, 539
  MAMP elicitors, 538
  producing negatively charged exopolysaccharides in, 539
Plant defense response, suppression of, 340–342
Plant–fungi interactions
 and callose-containing papilla formation in, 533
 callose in, 532–537
Plant–microbe interactions
 enzymes of general cell wall, 536
 function of callose in, 525

Plant–pathogen interactions, 543
papillae, as defence mechanism, 529
role of callose in, 543
Plant–virus interactions
callose in, 539–542
AtBG T-DNA insertion mutants, 540
barrier to viral spread, 542
callose deposition, at plasmodesmata, 540
cdiGRP, role of, 542
expression antisense (1,3)-β-glucanase, 541
localization of, 541
movement proteins, role of, 541
Plasmalemma, 443
Plasmodesmata, 439. *See also* (1,3)-glucanase
Al treatment induced callose deposition around, 506
callose deposition at, 531
callose levels, 540
callose turnover at, 540
cell-to-cell transport processes through, 502
coefficient of conductivity of, 540
conductivity, 440–441
electron micrographs of, 440
flow resistance due to, 512
intercellular communication through, 507
intercellular trafficking via, 541
plant viruses replication through, 539
structural changes in, 441
structure of, 439–440
synthesis and hydrolysis of callose, 441
of virus-infected plant tissue, 539
Platelet-derived growth factor A (PDGF-A), 597
Platelet-derived growth factor B (PDGF-B), 597
*Pleurochrysis haptonemofera*, 357
*Pleurotus tuber-regium*, 80
*Plodia interpunctella*, 189, 570
*Pmr4* mutants, 534, 542, 543
Pneumococcal type 37 (1,3;1,2)-β-glucan
enzymology, 224
membrane topology, 223–224
molecular genetics, 222–223
*Pneumocystis*, 586
Poaceae, 658, 660
*Polygonum*-type development, 486
*Polygonum*-type embryo sacs, 485
*Polypodium vulgare*, 448
Polysaccharide chain
network formation, 52
Polysaccharide conformation
principles of, 48–52
Polysaccharides, 660
cell walls, of oomycetes, 367–368
chrysolaminarin and, 359–364
degradation of, 365
extracellular, fungal and yeast (1,3)-β-glucan, 387–388
storage, in euglenids, 354
types and structures of, in protozoans and chromistans, 354*t*
x-ray fibre diffraction of, 53
*Populus tremula*, 137
*Populus trichocarpa*, 288
*Poria cocos*, 13, 72
*Porphyra purpurea*, 312
*Poterioochromonas malhamensis*, 365
PPB. *See* Predictive preprophase band (PPB)
Predictive preprophase band (PPB), 425
Prenylation, 309
Product entrapment, 245, 261, 262, 263, 297
ProPO-activating enzyme, 571
ProPO system, 568
(1,3)-β-glucan-induced activation of, 570
Proteinaceous inhibitors
of plant, 154–155
Protein disulfide isomerase (PDI), 301
Protein kinase B (PKB), 592
Proteins
biochemical analysis of, 188–192
CBM. *See* Carbohydrate-binding modules (CBM)
GPI-anchored, 395–396
overview, 171–172
PIR. *See* Putative proteins with internal repeats (PIR)
structural properties of, 172–174
Protozoans
types and structures of polysaccharides in, 354*t*
Proxin, 567
Prymnesiophyceae. *See* Haptophyceae
*Pseudoalteromonas carragenovora*, 135
*Pseudomonas aeruginosa*, 596
*Pseudomonas syringae*, 536
*Psilotum nudum*, 448
PSSA gene, 205
Pullulan, 587
Putative proteins with internal repeats (PIR), 394
*Saccharomyces cerevisiae* and, 395
*Pythium aphanidermatum*, 408
cell wall glucan, 18–19

# Q

Quantazyme, 390
*Quartet* (*qrt*) mutations ( *qrt1-3* ), 480
*Quercus rubra*, 505

# R

*Ralstonia solanacerarum*, 539
*Ramalina celastri*, 407
*Ramalina usnea*, 407

RelA gene, 209–211
Resorcin blue, 11
Resorcinol blue, 16
Resting cells
chrysolaminarin and, 362
Restionaceae, 633, 634
*Reviews of Pure and Applied Chemistry*, 1
*Rhizobium*, 99, 329
*Rhizobium loti*, 19
*Rhizopus arrhizus*, 135
Rhodophyte sulfated (1,3;1,4)-β-glucans, 24
*Rhodothermus marinus* (*Rm*CBM4-2), 176
Rho1 GTPase, 531
Rho1p, 263
Rice. *See Oryza sativa*
*Rimelia*, 632
R/K-X-G-G, 263
ROP1, 302
R protein, 527

## S

*Saccharomyces cerevisiae*, 12, 16, 121, 251, 261, 262, 271, 272, 285, 389, 391–392, 394, 587
cell wall glucan, 16–18
PIR proteins and, 395
wall composition of, 393*f*
Salicylic acid signal transduction, 543
Saponins, 300
*Saprolegnia*, 235
*Saprolegnia monoica*, 237–238, 241, 242, 243, 244, 245–246, 247, 248–249
*Sarcina ventriculi*, 24, 632
SAXS. *See* Small angle X-ray scattering (SAXS)
Scanning electron microscopy, on *agrobacterium*, 331
Scavenger receptors, 581
*S*-cellobioside – enzyme complexes, 152
*Schizophyllan commune*, 66, 73
Schizophyllan (SPG) glucan, 595
*Schizophyllum commune*, 394
*Schizosaccharomyces pombe*, 263, 389, 392
localization of (1,3)-β-glucans in septum of, 399*f*
localization of (1,6)-β-glucans in septum of, 399*f*
Scleroglucan, 75, 589
Scleroglucan – borax complexes
gelation of, 96
*Sclerospora graminicola*, 535
*Sclerotium glucanicum*, 75
SDS-PAGE analysis, 245
SDS-PAGE chromatography, 154
SDS-PAGE gels, 243, 297
SEC-MALS/RI. *See* Size-exclusion chromatography with multi-angle light scattering/ refractive index technique (SEC-MALS/RI)
Sedimentation analysis, 61
*Selaginella kraussiana*, 448
*Sensu lato*, 660
*Serratia marcescens*, 390
*Sesbania rostrata*, 336
Shear thinning, 85
*Sidecar* mutant, 481
Side-chain-branched (1,3;1,6)-β-glucans
behavior of concentrated solutions, 85–88
concentrated solution properties of, 88–90
dilute solution properties of, 73–80
gelation of, 96–99
solution properties of charge, 80–81
Sieve plates. *See also* Plasmodesmata
callose deposited in, 446, 512
chemical stresses inducing callose in, 447
cytokinin promoting callose, 449
localization of callose at Pd and, 441–442
plamodesmatal transport and in, 657
pores, 439
sensitive response of phloem, 513
*Sinapsis alba*, 143
*Sinorhizobium*, 338
*Sinorhizobium meliloti*, 20, 339
Size-exclusion chromatography, 154
Size-exclusion chromatography with multi-angle light scattering/refractive index technique (SEC-MALS/RI), 81
*Skeletonema costatum*, 359
*S*-laminaribioside – enzyme complexes, 152
Small angle X-ray scattering (SAXS), 61
*Solanum lycopersicum*, 512
Solid state $^{13}$C NMR
of (1,3)-β-glucans, 58–60
Sophorose – enzyme models, 152
Sorbitol, 98
*S*-palmitoylation, 309
Sparganiaceae, 634
Specific cell types in cereal grains, composition, 623*t*
Spectroscopy, 54
fluorescence, 58, 61–62, 183–184
mobility-resolved, 54
near-infrared, 638
peptide mass fingerprinting and tandem mass, 288
Saturation Transfer Difference NMR, 142
*Sphaerotrichia divaricata*, 370
*Spinacea oleracea*, 298
Spindle pole body (SPB), 269
*Sporobolis africanus*, 441
*Sporothrix schenckii*, 401
*Sporotrichum dimorphosporum*, 12

STAND ATPases, 526
*Staphylococcus aureus*, 599
*Stauroneis amphioxys*, 359
Steady shear viscosity
  of curdlan, 72
  of scleroglucan, 87
*Stereocaulon ramulosum*, 407
*Streptococcus pneumoniae*, 5, 20, 222, 342. *See also* Pneumococcal type 37 (1,3;1,2)-β-glucanTts gene
*Streptococcus pyogenes*, 215, 342
*Streptococcus suis*, 599
*Strongylocentrotus purpuratus*, 135
*Suberites domuncula*, 572
Sucrose synthase (SuSy), 301–302, 433, 500–501
Sulfated heteropolysaccharides, 593
SuSy. *See* Sucrose synthase (SuSy)
SuSy activity in submerged potato roots, 511
Swi4-Swi6 cell cycle box-binding factor (SBF), 265
Syk independent signalling pathways, 585
Synaptosome-associated protein receptor, 536
Syncytial systems
  alveolation in, 429–432
  phragmoplast development in, 426–428, 427*f*
Synthetic β-oligoglucosides, 24

## T

*Tachypleus tridentatus*, 565
*Thalassiosira*, 364
*Thalassiosira pseudonana*, 248, 249, 359
*Thalassiosira weissflogii*, 16, 248
*Thermoplasma acidophilum*, 505
*Thermotoga neapolitana*, 390
TLR2/MYD88 pathways, 585
*Tm*CBM4-2, 177–178, 179, 191
TMV-induced lesions, 541
TMV infection, 452
TNF α production, 598
Toll interleukin-1 receptor nucleotide binding (TIR-NB) proteins, 537
Toll-like receptors (TLR), 585
Transforming growth factor alpha (TGFalpha), 597
Transforming growth factor beta (TGFbeta), 597
Transglutaminase (TGase), 566
Trichloroacetic acid, 243
*Trichoderma harzianum*, 390
Triphenylmethane dye, 11
*Triticum aestivum*, 441, 639
*Triticum durum*, 639
*Tropaeolum majus*, 137
Trypsin, 308
Tts gene, 222
  enzymology, 224
  membrane topology of, 223–224
TTSS effectors, 538
Tubular – vesicular network (TVN), 428
TVN. *See* Tubular-vesicular network (TVN)
*Two-in-one pollen* ( *tio* ) mutants, 482
Type III secretion system (TTSS), 537–538
Typhaceae, 634

## U

UDP-α-D-[$^{14}$ C]glucose, 246
UDP-α-D-glucose, 239–240
UDP-α-D-[$^{3}$ H]glucose, 246
UDP degrading enzyme, 241
UDP-Glc, 297
α –[$^{32}$ P]UDPglucse, 245
UDPglucose: glycoprotein glucosyltransferase (UGGT), 272
UDP-glucose-binding proteins, 245
UDP-glucose pyrophosphorylases, 362–363
UDP-glucose transferase (UGT1), 435
UGT1. *See* UDP-glucose transferase (UGT1)
*Ulva lactuca*, 23, 630
*Usnea*, 632

## V

Vascular endothelial growth factor (VEGF), 597
*Vibrio cholerae*, 151
*Vicia faba*, 510
*Vigna radiata*, 301
*Vigna sinensis*, 512
*Vigna unguiculata*, 503
Viscometers, 74
  Zimm-Crothers type, 85
*Vitis vinifera*, 14
*Volvox carteri f. nagariensis*, 311

## W

Winter-Chambon criterion
  for gelation, 97, 98
Wortmannin, 592

## X

*Xanthomonas campestris*, 538, 539
XEH. *See* Xyloglucan endohydrolases (XEH)
X-ray diffraction
  (1,3)-β-glucans, 54–58
  in polysaccharide chain conformation, 52–53
XTH. *See* Xyloglucan endotransglycosylases/ hydrolase (XTH) group
Xylanase Z-like domain, 566
Xyloglucan endohydrolases (XEH), 137
Xyloglucan endotransglycosylases/ hydrolase (XTH) group, 133, 137–138
Xyloglucans, 137
Xyridaceae, 634

## Y

*Yarrowia lipolytica*, 396
Yeasts

(1,3)-β-glucans extracellular polysaccharides, 387–388
chemical composition of cell wall of, 389*t*
KRE genes. *See* KRE genes
protocol for extraction of, 390, 391*f*
schematic representation of CWP in, 395*f*
structure and molecular organization of (1,3)-β-glucans, 391–396
Yeasts cell wall
localization of (1,3)-β-glucan in, 261
structure and (1,3)-β-glucan, 260
Young's modulus, 103

## Z

*Zea mays*, 302, 441, 512, 633
Zero shear viscosity, 82
Zimm plot analysis, 82

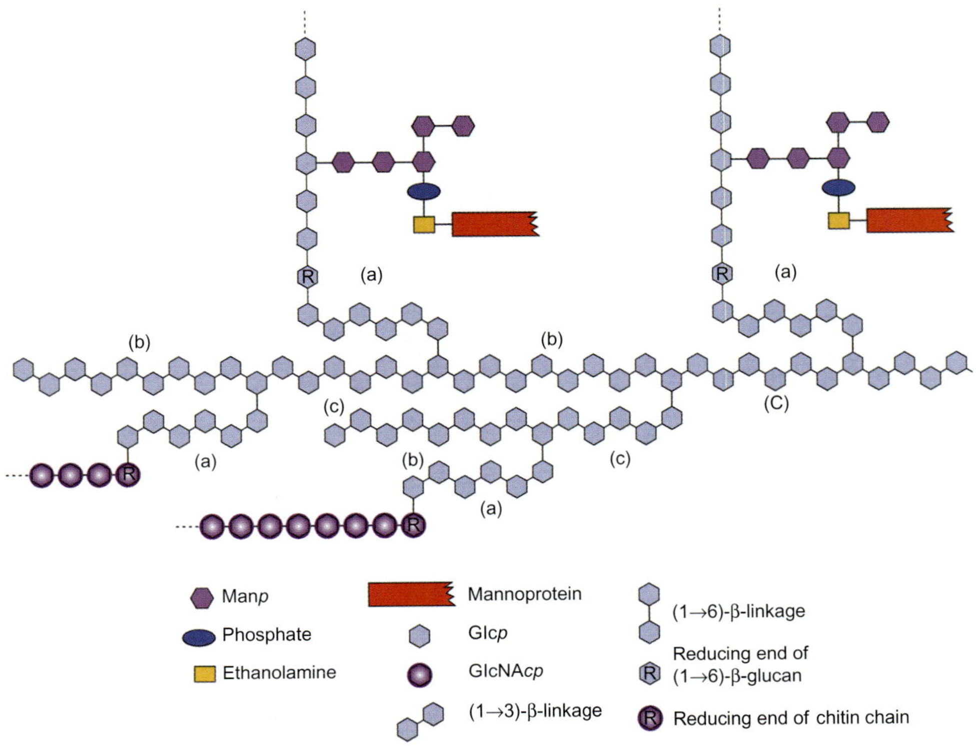


**Chapter 2.1, Fig 1C: Model of the cell wall of the yeast *Saccharomyces cerevisiae*. The branch-on-branch (1→3,1→6)-β-D-glucan forms the central component of the wall. The proportions of (a), (b) and (c) chains are about equal but their exact lengths are not known. The non-reducing ends of (a) and (b) chains are attachment sites for chitin chains at the plasma membrane surface of the wall and for the reducing ends of (1→6)-β-D-glucan chains at the outer surface of the wall. The (1→6)-β-D-glucan chains are in turn substituted by mannoproteins through the C-terminal amino acid of an amino acid-ethanolamine-phosphodiester-$(Man)_5$ remnant of a GPI anchor. Pir cell wall proteins (not shown) are linked directly to (1→3;1→6)-β-D-glucan. Chitin chains may also be attached to the (1→6)-β-D-glucan chains (not shown). After Manners et al. (1973); de Nobel et al. (2001); Kollar et al. (1997).**

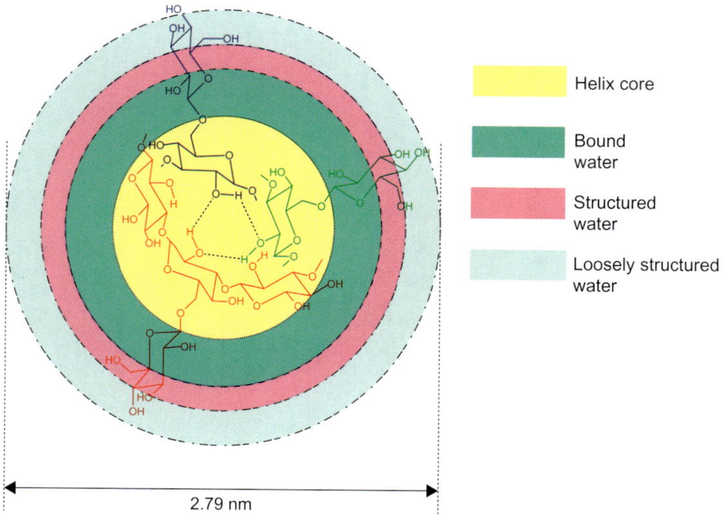


Chapter 2.2, Fig 15: Cross-sectional view of the schizophyllan triple helix I in water. The repeat unit of schizophyllan consists of three glucose residues and one side-chain glucose residue (only the red chain residues are shown in full). The circles represent the proposed boundaries of four water structure layers, with approximate diameters and critical weight fractions in parentheses: helix core–bound water (1.68 nm, 1.0), bound water–structured water (2.05 nm, 0.769), structured water–loosely structured water (2.31 nm, 0.644), and loosely structured water–free water (2.79 nm, 0.478). Each cylindrical shell is marked in a specific colour. From Yoshiba et al. (2003).

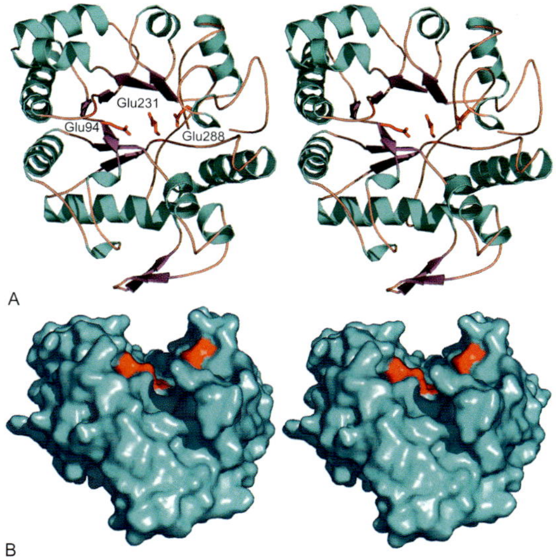


Chapter 3.1, Fig 1: Stereo representation of the 3D structure of a barley family GH17 (1,3)-β-D-glucanase isoenzyme GII (Protein Data Bank accession code 1GHS). (A) Ribbon diagram of the enzyme with three catalytic amino acids Glu94, Glu231 and Glu288 shown in sticks and coloured in red. (B) Molecular surface drawing of the enzyme with three catalytic amino acids Glu94, Glu231 and Glu288 (in red) positioned at the bottom of substrate-binding cleft. Adapted with permission from the National Academy of Sciences of USA (Varghese et al., 1994).

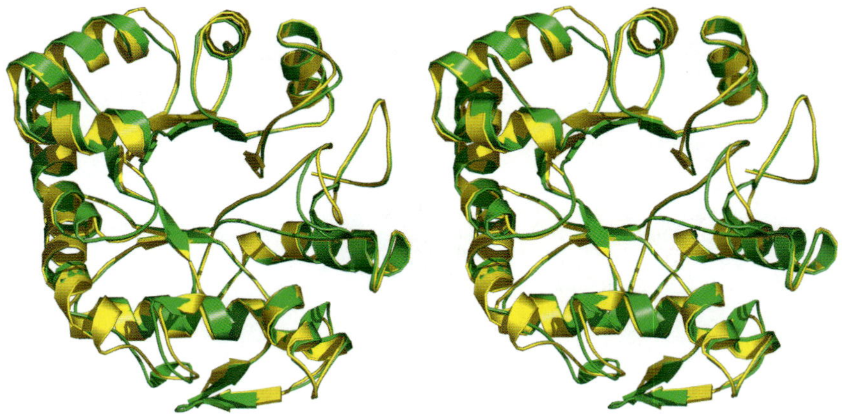

Chapter 3.1, Fig 4: Stereo representation of 3D structures of a barley (1,3)-β-D-glucanase isoenzyme GII (green) superposed on (1,3;1,4)-β-D-glucanase isoenzyme EII (yellow). The rmsd deviation in Cα positions between the two enzymes is 0.65 A for 278 residues. Adapted with permission from the National Academy of Sciences of USA (Varghese et al., 1994).

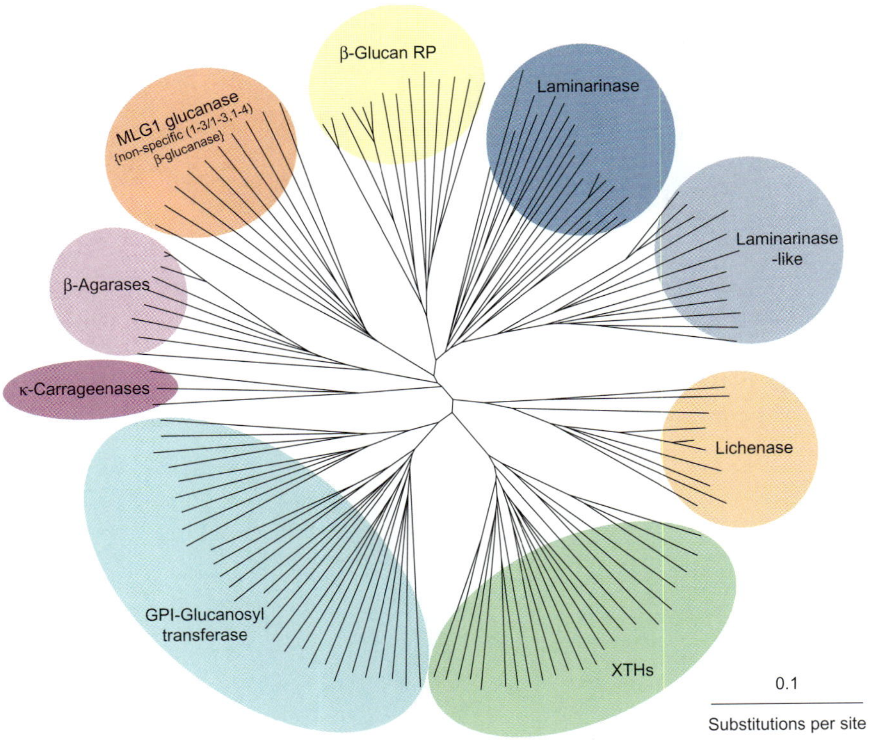


Chapter 3.1, Fig 5: Unrooted radial phylogenetic tree of selected family GH16 members. Amino acid sequences were aligned with ClustalX and branch lengths are drawn to scale. Nine subgroups are clustered into non-catalytic and catalytic proteins, the latter according to their substrate specificity.

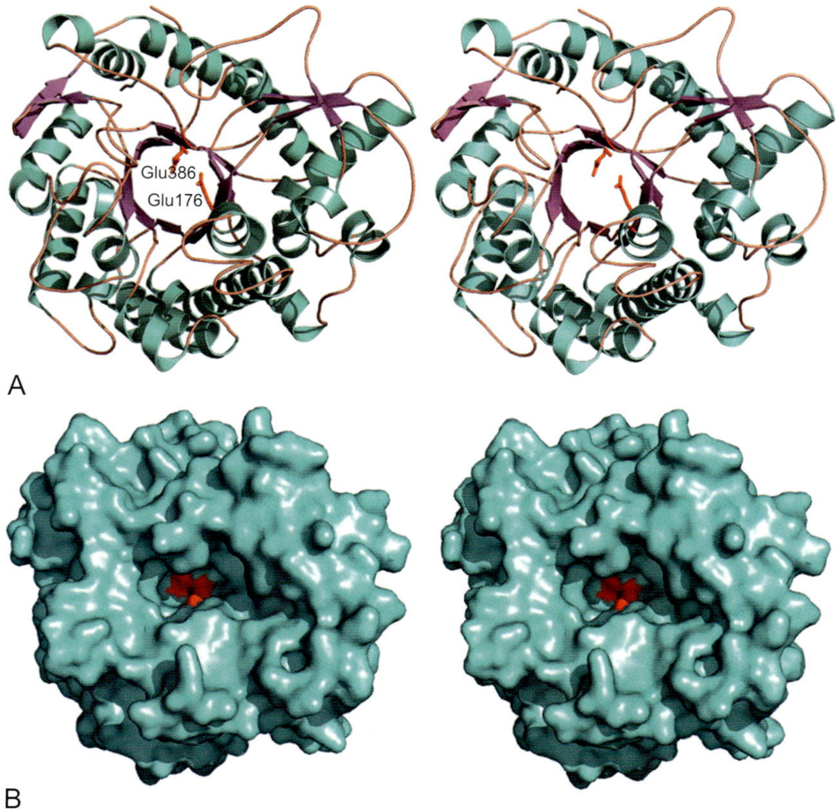


Chapter 3.1, Fig 7: Stereo representation of 3D structure of a rice β-D-glucosidase (Protein Data Bank accession code 2RGL). (A) Ribbon diagram of the enzyme with two catalytic amino acids Glu176 and Glu386 shown in sticks and coloured in red. (B) Molecular surface drawing of the enzyme with two catalytic amino acids Glu176 and Glu386 coloured in red. Adapted with permission from Elsevier Science (Chuenchor et al., 2008).

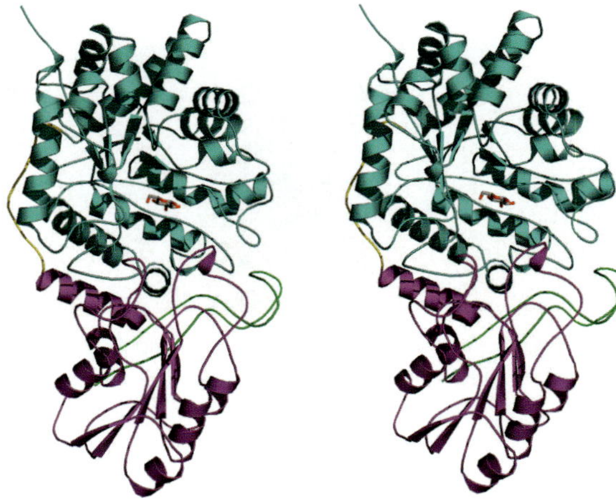

A

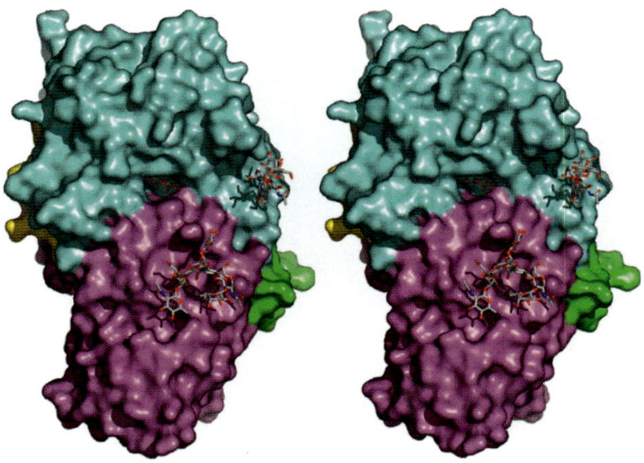

B

Chapter 3.1, Fig 9: Stereo representation of 3D structure of a barley β-D-glucan glucohydrolase (Protein Data Bank accession code 1EX1 and 1IEQ). (A) Ribbon diagram of the enzyme with bound glucose (in cpk colours), where domain 1, linker, domain 2, and the COOH-terminal antiparallel loop are in cyan, yellow, magenta and green, respectively. (B) Molecular surface drawing of the enzyme (colours as specified in panel A) with two occupied Asn221 and Asn498-linked glycosylation sites (cpk colours). Adapted with permission from Elsevier Science (Varghese et al., 1999; Hrmova et al., 2001).

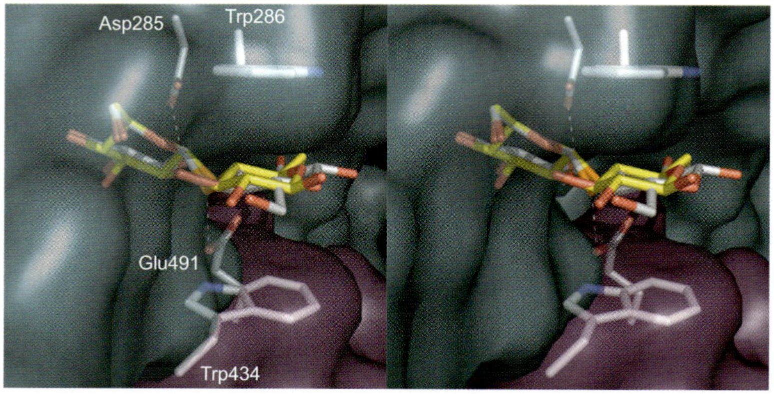


Chapter 3.1, Fig 10: Stereo representation of the active site of barley β-D-glucan glucohydrolase with bound *S*-cellobioside and *S*-laminaribioside moieties. The *S*-cellobioside and *S*-laminaribioside moieties are presented as sticks and atoms are coloured grey (carbons; yellow for *S*-laminaribioside), orange (sulfur), and red (oxygens). Transparent cyan and magenta represent the molecular surface of domain 1 and 2, respectively. Selected active site amino acid residues Asp285, Trp286, Trp434 and Glu491 are shown (grey, red and blue represent carbons, nitrogens and oxygens). The entrance to the active site is located towards the lower right hand corner. Adapted with permission from the American Society of Plant Biologists (Hrmova et al., 2002).

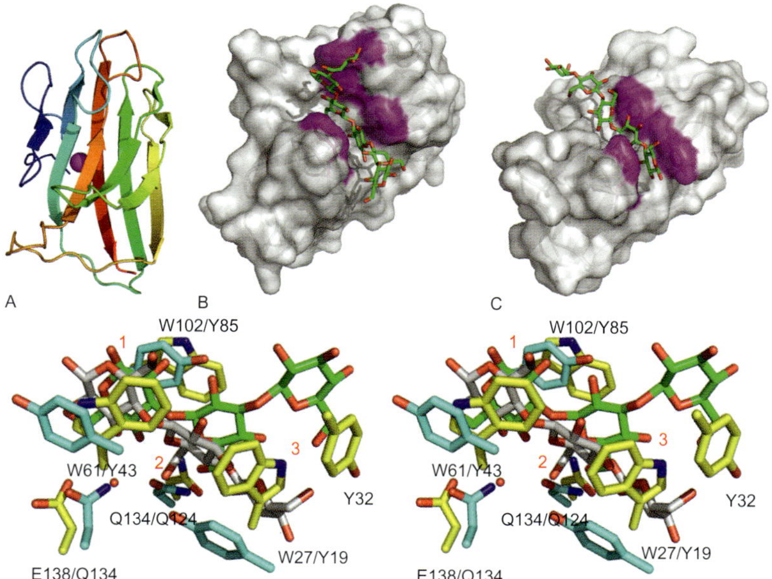


Chapter 3.2, Fig 3: Differential β-glucan recognition by family 4 CBMs. (A) *Tm*CBM4-2 (PDB ID: 1GUI) displayed in cartoon format. The structural calcium is shown as a sphere. (B) Solvent-accessible surface representation of *Tm*CBM4-2, a (1,3)-β-glucan binding CBM in complex with laminarihexaose (PDB ID: 1GUI). (C) *Cf*CBM4-1, a (1,4)-β-glucan binding CBM in complex with cellopentaose (PDB ID: 1GU3). Aromatic residues involved in ligand binding in (B) and (C) are shown. (D) Wall-eyed stereo view of superimposed complexes from (B) and (C). Amino acids involved in ligand binding from *Tm*CBM4-2 (Trp27, Trp61, Trp102, Gln134 and Glu138) and *Cf*CBM4-1 are shown (Tyr19, Tyr43, Tyr85, Gln124 and Gln134). Ligands are trimmed down to the three residues that occupy subsites 1 to 3 for clarity. Laminarihexaose and cellopentaose is rendered as sticks with the subsites numbered.

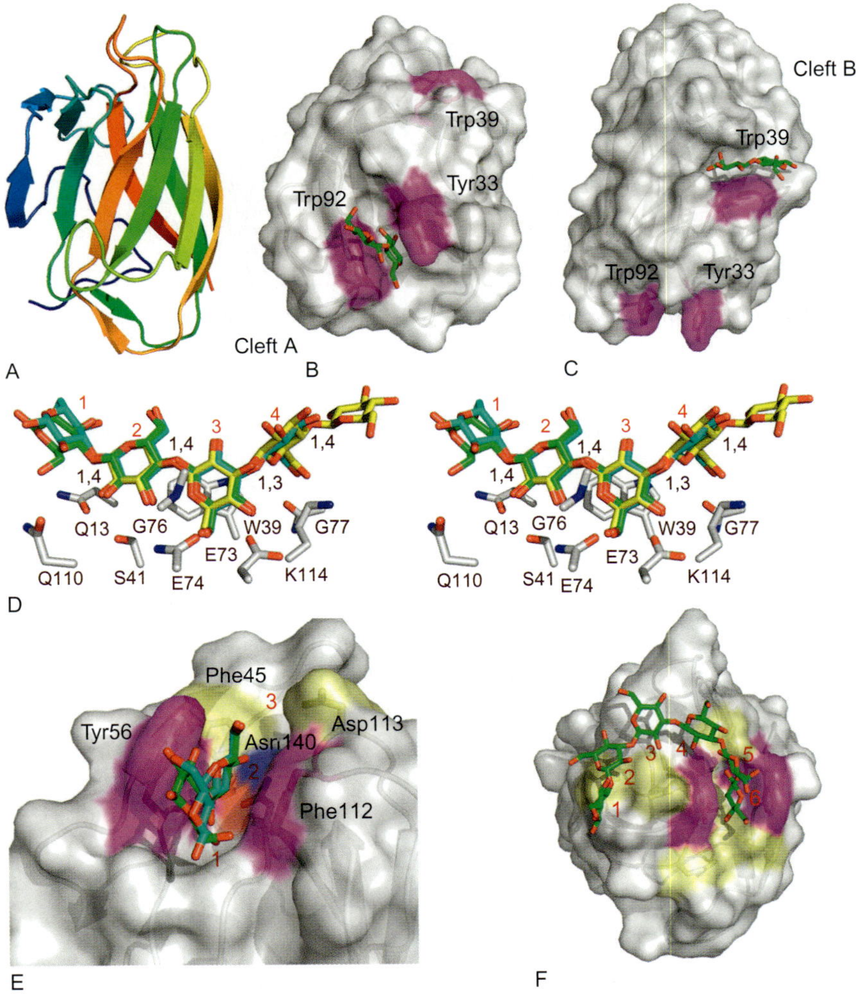


Chapter 3.2, Fig 4: (A) Secondary structure representation of unliganded *Cm*CBM6-2 in cartoon format (PDB ID: 1UXZ). (B) Cleft 'A' in *Cm*CBM6-2 in complex with cellobiose from chain B (PDB ID: 1UYX). The three surface aromatic residues Tyr33, Trp39 and Trp92 are shown. (C) Cleft 'B' is displayed in complex with cellobiose from chain A (PDB ID: 1UYX). (D) Superimposition of ligands bound in *Cm*CBM6-2: cellotriose (PDB ID: 1UYY), G3G4G3 (PDB ID: 1UY0), and G4G3G4 (PDB ID: 1UZ0). Only the amino acids from the cellotriose complex that define the subsite architecture (numbered 1 to 4) are shown. The linkages are labelled. (E) Solvent surface representation of *Cs*CBM6-3 of superimposed complexes with cellobiose and laminaribiose. The aromatic residues Tyr56 and Phe112 that define subsite 1 are shown. The oxygen and nitrogen of the Asn140 functional group which interacts with O3 and O4 of both ligands are displayed. Phe45 and Asp113, which form subsite 3 in the xylo-oligosaccharide complexes, are indicated. (F) Solvent-accessible surface model of BHCBM6 in complex with laminarihexaose.

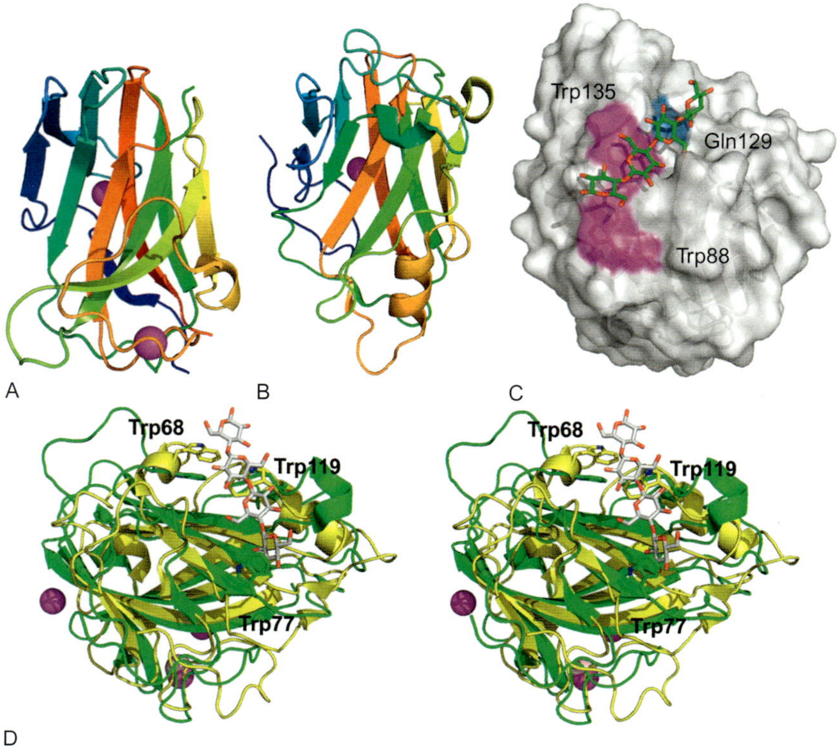


Chapter 3.2, Fig 5: (A) Cartoon representation of *Ct*CBM11 in complex with two calcium atoms shown as spheres (PDB ID: 1V0A). (B) Cartoon representation of *Cc*CBM17 in complex with a calcium atom shown as a sphere (PDB ID: 1J83). (C) Solvent accessible surface model of *Cc*CBM17 in complex with cellotetraose (PDB ID: 1J84). The shallow binding cleft is lined with two surface aromatics residues, Trp88 and Trp135, and Gln129. (D) Wall-eyed stereo view of *Bsp*CBM28 (PDB ID: 1UWW) overlaid with *Ct*CBM17 (PDB ID: 1V0A). The aromatic residues implicated in binding for *Bsp*CBM28 (Trp68, Trp77, and Trp119) are displayed. The cellotetraose ligand from *Ct*CBM17 is shown.

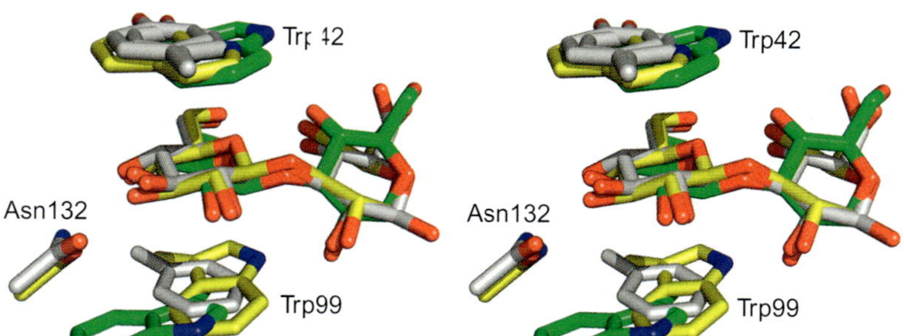


Chapter 3.2, Fig 9: Wall-eyed stereo view of structural alignment of the binding sites from (1,3)-β-glucanbinding CBMs. The ligands from *Bh*CBM6 (PDB ID: 1W9W), *Cs*CBM6-3 (PDB ID: 1OD3) and *Tm*CBM4-2 (PDB ID: 1GUI) have been superimposed and displayed along with key amino acids within the binding sites. Only the residues from *Bh*CBM6 are labelled. The two asparagines from the family 6 CBMs, which make critical contacts with the non-reducing O3 and O4, are shown.

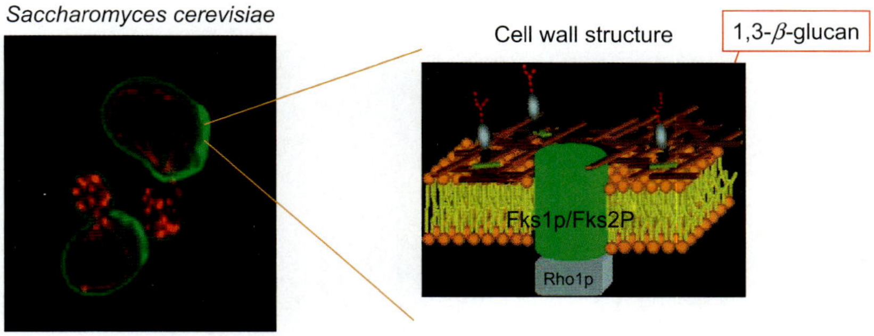


**Chapter 3.3.3, Fig 1: Yeast cell wall structure and (1,3)-β-glucan. In the left panel, yeast cells are stained with FITC-conjugated concanavalin A and Rhodamin phalloidin to visualize cell wall and actin filament. In the right panel, glucans and chitin are depicted in brown and yellow fibers, respectively, and mannoproteins are depicted white balls with a red tail.**

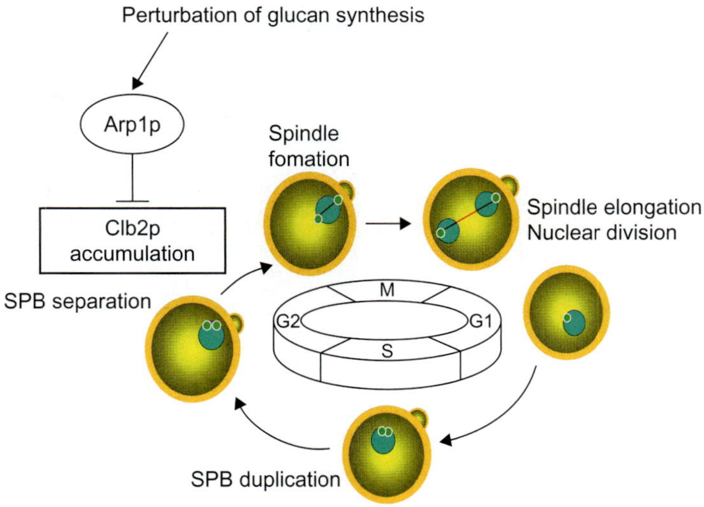


**Chapter 3.3.3, Fig 4: Cell wall integrity checkpoint as a novel checkpoint control. If cell wall synthesis is perturbed, Arp1p represses CLB2 mRNA and inhibits accumulation of Clb2p. This results in cell cycle arrested at G2/M boundary, with a tiny bud (because of defect in bud growth) and duplicated SPBs.**

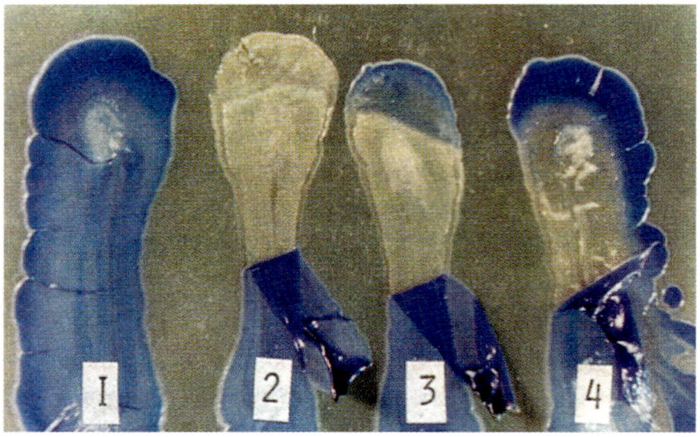


**Chapter 4.1, Fig 1: Growth of *Agrobacterium* NTK-u on the surface of agar medium containing the (1,3)-β-glucan-specific Aniline Blue dye after 5 days incubation (1) and thereafter for 1 h (2), 3 days (3) and 5 days (4) after stripping off of the curdlan pellicle. From Nakanishi et al. (1976).**

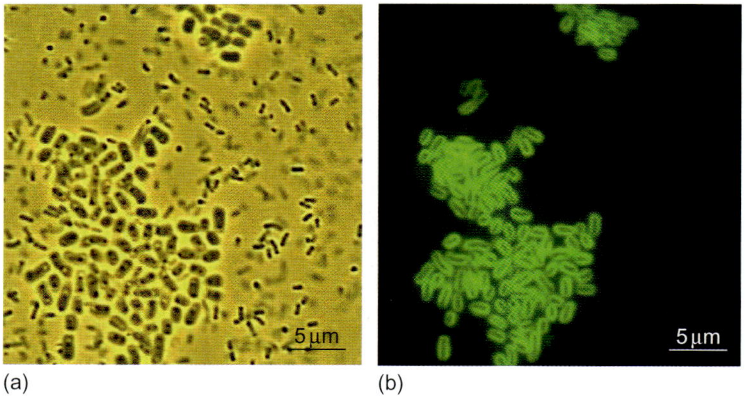


**Chapter 4.1, Fig 2: Phase contrast image (a) and UV fluorescence image (b) of cells of *Agrobacterium* LTU50 in the same field stained with the (1,3)-β-glucan specific Aniline Blue fluorochrome (Evans et al., 1984), showing small flocs of curdlan-encapsulated cells and curdlan-free, planktonic, cells. From McIntosh (2004).**

Chapter 4.2, Fig 8: *Laminaria hyperborea* kelp forest from Reine, Lofoten, Norway.

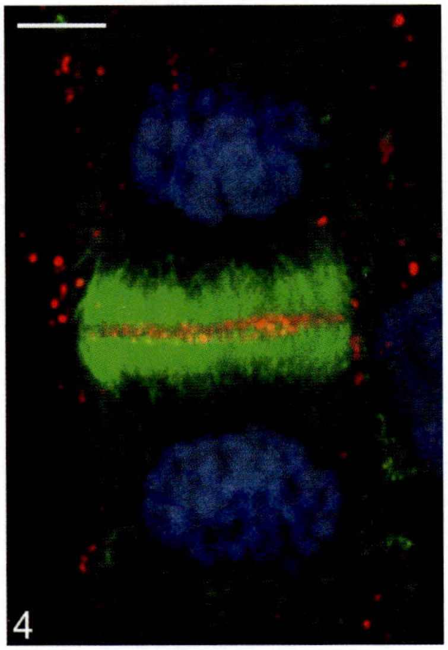


Chapter 4.4.1, Fig 4: The cell plate in a meristematic cell of daylily, *Hemerocallis fulva*, triple stained for microtubules, nuclei and callose. The cell plate grows as an enlarging disc-shaped structure in association with the phragmoplast that expands as a ring. Callose produced at the luminal membrane surfaces of the cell plate was detected with a (1,3)-ß-glucan antibody. Bar = 3.5 μm.

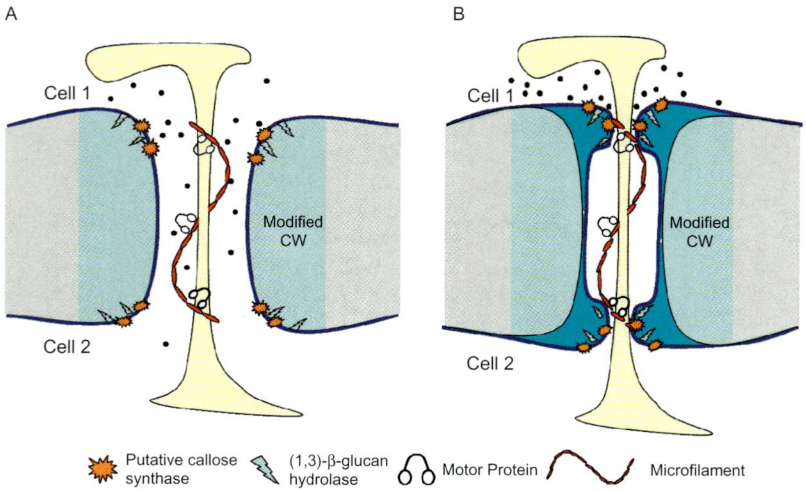


**Chapter 4.4.2, Fig 3: A schematic model for non-selective regulation of plasmodesmal permeability by callose. In the open state, the cell wall (CW) sheath immediately surrounding the Pd is devoid of cellulose and is composed in part of non-esterified pectin, callose, other non-cellulosic polyglucans and as yet uncharacterized proteins. Embedded in the Pd plasma membrane are callose synthase and (1,3)-β-glucan hydrolase. Cytoplasmic sleeve is open and non-targeted macromolecules (black spheres) diffuse according to electrochemical gradient and Stokes radius. (B) Deposition of callose in the apoplast (dark wall area) around the cytoplasmic sleeve, and especially at the cytoplasmic annulus, results in the formation of a sphincter that reduces the size of the annulus, creating a bottle neck for diffusion.**

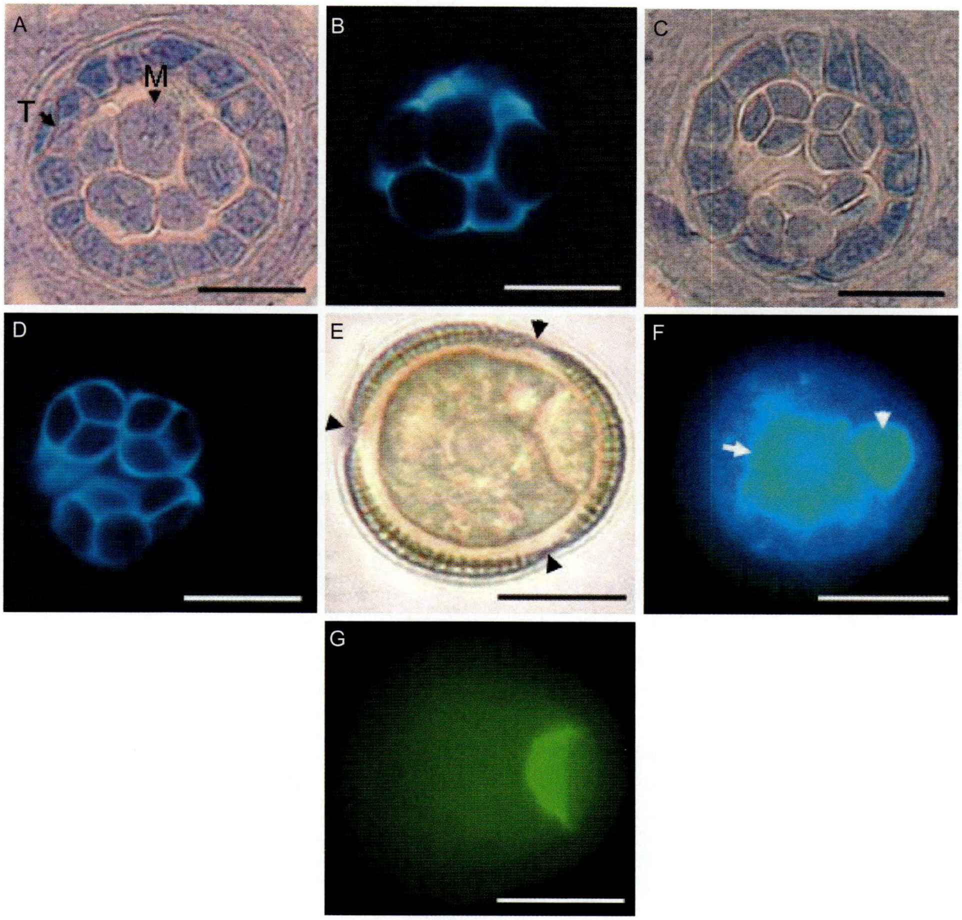


Chapter 4.4.3, Fig 2: The appearance of callose during microgamete development in a wild-type *Arabidopsis* plant. (A, C) Toluidine blue-stained cross-section. (E) Bright-field image. (B, D, F and G) Corresponding images viewed with UV illumination after Aniline Blue fluorochrome-staining (B, D and G) or 4′,6-diamidino-2-phenylindole (DAPI) staining (F). (A, B) Microsporocytes (M) undergoing meiosis (stage 3). The surrounding tapetal cell layer (T) is also evident. Callose in the common SCW surrounding each microsporocyte fluoresces when stained with Aniline Blue fluorochrome (B). (C, D) Microspores still in tetrads (stage 5) and surrounded by their individual SCWs. (E–G) *Arabidopsis* microspore after mitosis I. (E) Arrowheads indicate areas where the wall is thin and colpi will later form. (F) The arrowhead and arrow indicate the centrally placed vegetative nucleus and peripheral generative nucleus, respectively. (G) The callosic wall separating the generative and vegetative cells is clearly seen. Scale bars: (A–D) 10 μm; (E–G) 5 μm. (A–D reproduced from Figure 8 of Enns et al. (2005), with kind permission from Springer Science+Business media; E–G kindly provided by Professor David Twell, University of Leicester UK.)

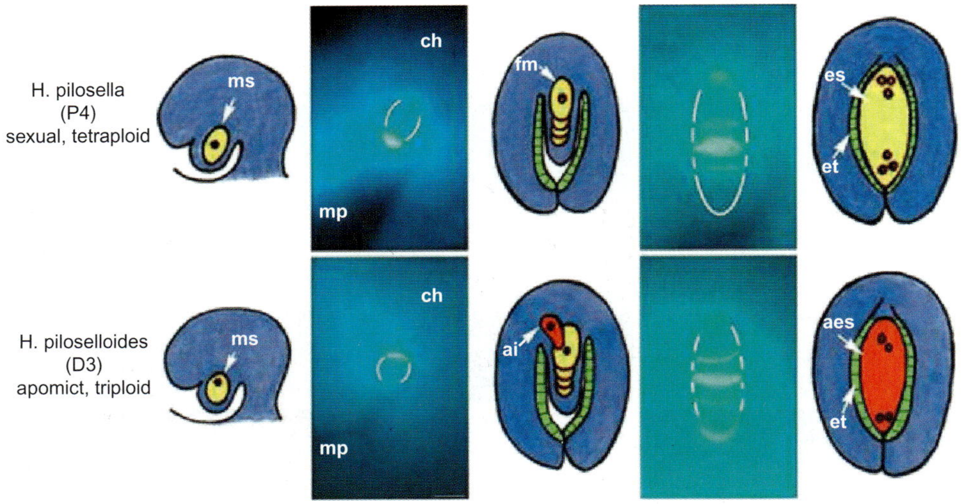


Chapter 4.4.3, Fig 7: Early events of embryo sac formation in ovules of a sexual and an apomictic variety of *Hieracium*. The early events of reduced embryo sac formation in a sexual plant and an apomict are coloured yellow, and aposporous embryo sac formation is coloured red. The presence of callose in the walls of the megasporocyte (ms) in ovules prior to meiosis and in ovules after meiosis is shown by Aniline Blue fluorochrome-staining viewed with UV light. aes, aposporous embryo sac; ai, aposporous initial; es, embryo sac; ch, chalazal end; et, endothelium; fm, functional megaspore; mp, micropylar end; ms, megasporocyte. (Reproduced with permission of American Society of Plant Biologists from Bicknell and Koltunow (2004); permission conveyed through Copyright Clearance Centre, Inc.)

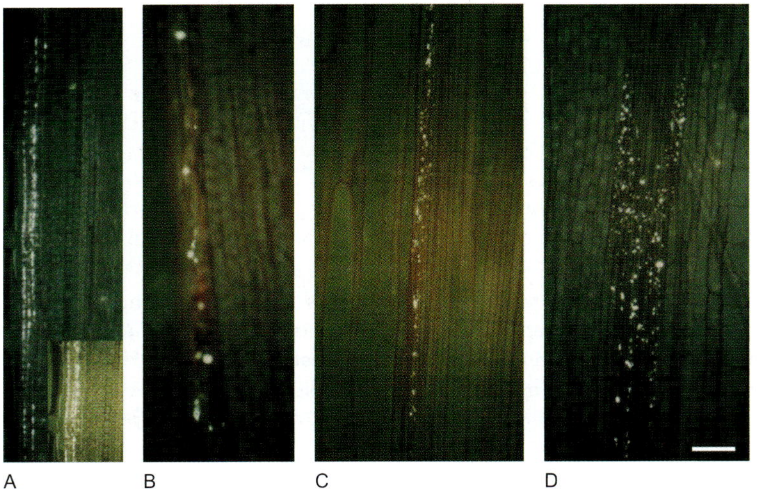


Chapter 4.4.4, Fig 3: Localization of callose by staining with sirofluor in longitudinal sections of root tips of soybean (*Glycine max*). Roots were treated with (A) 25 μM Al for 4 h, (B) 18 μM Pb for 8 h, (C) 1 μM Hg for 4 h, (D) 10 μM Cd for 8 h. Insert in (A) shows a magnification of the rhizodermis and outer cortex. Scale bar = 100 μm. (A. Diening, A. H. Wissemeier, W. J. Horst, unpublished results.)

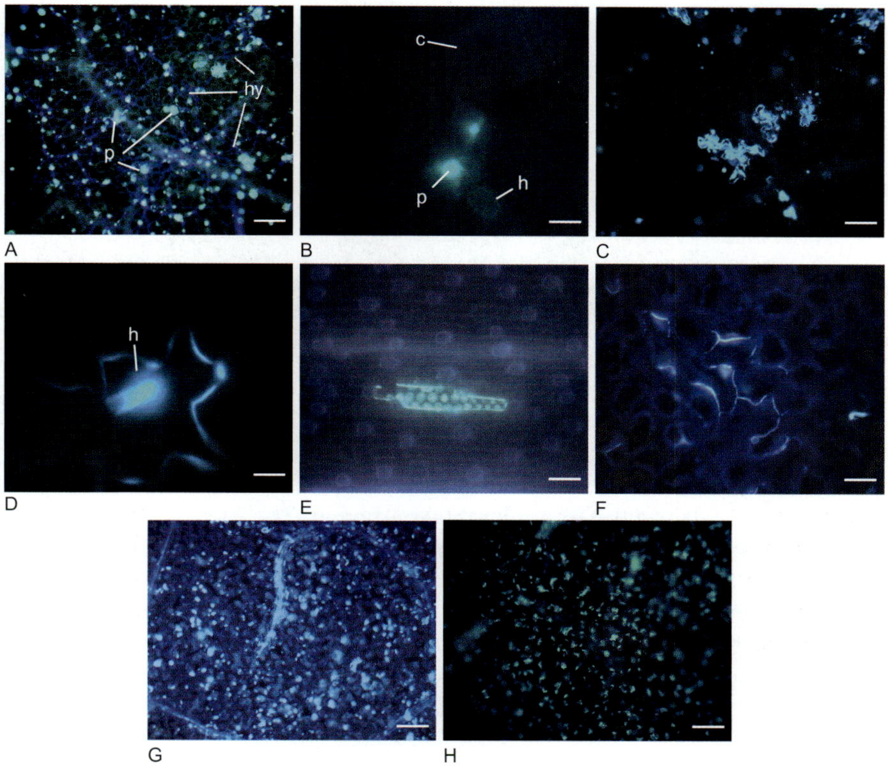


Chapter 4.4.5, Fig 1: Callose deposition after microbial attack. Fluorescence microscopy of treated plant leaf tissue was used to visualize callose deposits after staining with Aniline Blue (Ramming et al., 1973). (A) Callose-rich papillae (p) in Col-0 wild-type Arabidopsis leaves are deposited in response to attack with the virulent Arabidopsis powdery mildew, *Golovinomyces cichoracearum*. Picture was taken 3 days post inoculation. Fungal hyphae (hy) stained dark blue. Scale bar, 50 μm. (B) Magnification of a penetration site of a single conidium (c) illustrating callose deposition in papillae (p). Haustorium (h) formation has already started 1 day post inoculation of Col-0 wild-type Arabidopsis leaves with *G. cichoracearum*. Scale bar, 1 μm. (C) Callose deposits outlining attacked cells and papillae in Col-0 wild-type Arabidopsis leaves inoculated with the inappropriate barley powdery mildew, *Blumeria graminis* f. sp. *hordei*. Image was collected 3 days post inoculation. Scale bar, 50 μm. (D) Magnification of a single epidermal leaf cell ringed with callose 3 days post inoculation with *B. graminis* f. sp. *hordei*. The fungal haustorium (h) is also encased with callose. Scale bar, 1 μm. (E) Callose deposition in a leaf epidermal cell of the grass *Brachypodium distachyon* as a defence response in an incompatible interaction. Picture was taken 1 day post inoculation with *G. cichoracearum*. Scale bar, 2.5 μm. (F) Callose deposition in leaf epidermal cells of Arabidopsis wild type in response to chitin treatment. Chitin is a fungal 'microbe-associated molecular pattern' (MAMP). Picture was taken 1 day post spraying with a chitin solution. Scale bar, 2.5 μm. (G) Callose deposition in an Arabidopsis leaf 24 h after syringe infiltration of an attenuated bacterial *Pseudomonas syringae* strain into the apoplast. Callose deposits occur in mesophyll cells. Scale bar, 50 μm. (H) Callose deposition in Arabidopsis mesophyll cells 24 h after syringe infiltration of a solution containing a 22-amino-acid peptide (flg22) from a conserved bacterial flagellin. Flagellin is an important bacterial MAMP. Scale bar, 50 μm. (G and H courtesy of Bill Underwood, Carnegie Institution of Science, Stanford, CA, USA.)

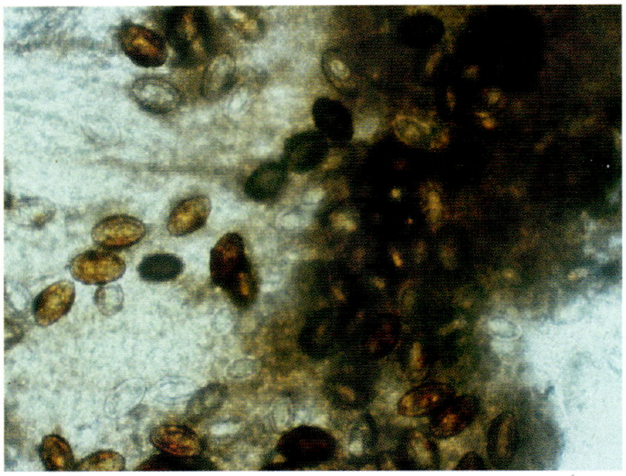

Chapter 4.5.1, Fig 3: Melanization of a microbial pathogen. The parasite *Psorospermium haeckeli* becomes melanized (brown colouring) in haemolymph of the freshwater crayfish *Astacus astacus*.